P9-DDO-623

Handbook of Phytoalexin Metabolism and Action

BOOKS IN SOILS, PLANTS, AND THE ENVIRONMENT

Soil Biochemistry, Volume 1, edited by A. D. McLaren and G. H. Peterson
Soil Biochemistry, Volume 2, edited by A. D. McLaren and J. Skujiņš
Soil Biochemistry, Volume 3, edited by E. A. Paul and A. D. McLaren
Soil Biochemistry, Volume 4, edited by E. A. Paul and A. D. McLaren
Soil Biochemistry, Volume 5, edited by E. A. Paul and J. N. Ladd
Soil Biochemistry, Volume 6, edited by Jean-Marc Bollag and G. Stotzky
Soil Biochemistry, Volume 7, edited by G. Stotzky and Jean-Marc Bollag
Soil Biochemistry, Volume 8, edited by Jean-Marc Bollag and G. Stotzky

Organic Chemicals in the Soil Environment, Volumes 1 and 2, edited by C. A. I. Goring and J. W. Hamaker
Humic Substances in the Environment, M. Schnitzer and S. U. Khan
Microbial Life in the Soil: An Introduction, T. Hattori
Principles of Soil Chemistry, Kim H. Tan
Soil Analysis: Instrumental Techniques and Related Procedures, edited by Keith A. Smith
Soil Reclamation Processes: Microbiological Analyses and Applications, edited by Robert L. Tate III and Donald A. Klein
Symbiotic Nitrogen Fixation Technology, edited by Gerald H. Elkan
Soil-Water Interactions: Mechanisms and Applications, edited by Shingo Iwata, Toshio Tabuchi, and Benno P. Warkentin
Soil Analysis: Modern Instrumental Techniques, Second Edition, edited by Keith A. Smith
Soil Analysis: Physical Methods, edited by Keith A. Smith and Chris E. Mullins
Growth and Mineral Nutrition of Field Crops, N. K. Fageria, V. C. Baligar, and Charles Allan Jones
Semiarid Lands and Deserts: Soil Resource and Reclamation, edited by J. Skujiņš
Plant Roots: The Hidden Half, edited by Yoav Waisel, Amram Eshel, and Uzi Kafkafi
Plant Biochemical Regulators, edited by Harold W. Gausman
Maximizing Crop Yields, N. K. Fageria
Transgenic Plants: Fundamentals and Applications, edited by Andrew Hiatt
Soil Microbial Ecology: Applications in Agricultural and Environmental Management, edited by F. Blaine Metting, Jr.
Principles of Soil Chemistry: Second Edition, Kim H. Tan

Water Flow in Soils, edited by Tsuyoshi Miyazaki
Handbook of Plant and Crop Stress, edited by Mohammad Pessarakli
Genetic Improvement of Field Crops, edited by Gustavo A. Slafer
Agricultural Field Experiments: Design and Analysis, Roger G. Petersen
Environmental Soil Science, Kim H. Tan
Mechanisms of Plant Growth and Improved Productivity: Modern Approaches, edited by Amarjit S. Basra
Selenium in the Environment, edited by W. T. Frankenberger, Jr., and Sally Benson
Plant–Environment Interactions, edited by Robert E. Wilkinson
Handbook of Plant and Crop Physiology, edited by Mohammad Pessarakli
Handbook of Phytoalexin Metabolism and Action, edited by M. Daniel and R. P. Purkayastha
Soil–Water Interactions: Mechanisms and Applications, Second Edition, Revised and Expanded, Shingo Iwata, Toshio Tabuchi, and Benno P. Warkentin
Stored-Grain Ecosystems, edited by Digvir S. Jayas, Noel D. G. White, and William E. Muir
Agrochemicals from Natural Products, edited by C. R. A. Godfrey

Additional Volumes in Preparation

Seed Development and Germination, edited by Jaime Kigel and Gad Galili

Nitrogen Fertilization in the Environment, edited by Peter Edward Bacon

Phytohormones in Soils: Microbial Production and Function, W. T. Frankenberger, Jr., and Muhammad Arshad

Handbook of Phytoalexin Metabolism and Action

edited by

M. Daniel
The Maharaja Sayajirao University of Baroda
Vadodara, Gujarat, India

R. P. Purkayastha
University of Calcutta
Calcutta, West Bengal, India

Marcel Dekker, Inc. New York • Basel • Hong Kong

Library of Congress Cataloging-in-Publication Data

Handbook of phytoalexin metabolism and action / edited by M. Daniel,
R. P. Purkayastha.
p. cm. — (Books in soils, plants, and the environment)
Includes bibliographical references and index.
ISBN 0-8247-9269-6
1. Phytoalexins—Metabolism. 2. Phytoalexins—Physiological effect. 3. Plants—Disease and pest resistance. 4. Phytoalexins--Metabolism—Research. 5. Phytoalexins—Physiological effect--Research. 6. Plants—Disease and pest resistance—Research.
I. Daniel, M., Dr. II. Purkayastha, R. P. III. Series.
QK898.P66H35 1994
581.2'9—dc20 94-22878
CIP

The publisher offers discounts on this book when ordered in bulk quantities. For more information, write to Special Sales/Professional Marketing at the address below.

This book is printed on acid-free paper.

MARCEL DEKKER, INC.
270 Madison Avenue, New York, New York 10016

Current printing (last digit):
10 9 8 7 6 5 4 3 2 1

PRINTED IN THE UNITED STATES OF AMERICA

Preface

Ever since Müller and Borger proposed the concept of phytoalexins in 1940, these compounds have fascinated plant pathologists engaged in unraveling the mysteries of disesase resistance in plants. More than 40 years later, Bailey and Mansfield, in their excellent treatise *Phytoalexins*, brought together a number of students of phytoalexins to share their experiences and views on these compounds. That book gave a fillip to research on these defensive chemicals, and much has happened since then, especially in the area of molecular biology of phytoalexin synthesis and gene regulation. With the great relevance of these studies to agriculture, microbiology, biochemistry, biotechnology, forestry, and horticulture, more and more research laboratories are laying greater emphasis on investigating phytoalexins. The relevant research literature has increased tremendously during the past two decades, and an overwhelmingly large wealth of information exists scattered in journals on various subjects. It is almost impossible to review all of them and, therefore, an attempt is made here to summarize the latest developments and to present the state of the art of phytoalexin research worldwide.

The approach here is that of a comprehensive handbook, but of the vast number of higher plants around us, we have focussed on specific, economically important species with respect to their phytoalexins, thereby avoiding broad generalizations. For a specific crop, the leading investigators were asked to summarize all the available data on that plant species, giving special emphasis to the work done in the authors' laboratory. New analytical techniques that have been developed as well as the latest unpublished results were also included in each paper. Thus each chapter contains much new information. Described are several important crops, such as barley, pea, soybean, cotton, groundnut (peanut), mustard, *Citrus*, grape,

Phaseolus, *Vigna*, *Dioscorea*, tobacco, alfalfa, red clover, potato, sweet potato, tea, pepper, sesame; medicinal plants, such as *Adhatoda*, *Trianthema*, *Withania*, and *Tylophora*; and various tree species, such as teak, *Cassia, Morinda, Eucalyptus, Syzygium, Mangifera, Anogeissus, Madhuca, Heterophragma, Spathodea, Zizyphus,* and *Carvia*. The topics covered are: the role of phytoalexins in disease resistance, artificial induction of phytoalexins, nature of elicitors, biosynthesis, detoxification, enzymatic conversion, mode of action, and molecular approaches such as cloning of cDNA-encoding elicitors, receptor sites in membranes, receptor substances, cellular responses and their regulation, signal molecules and systems, signal transduction and location, and regulation of disease resistance genes. A number of new phytoalexins are reported together with their structures and properties.

We hope that the researchers in plant pathology, biochemistry, microbiology, molecular biology, biotechnology, natural products chemistry, chemotaxonomy, and plant breeding, will find this book greatly useful. The methodology provided for each crop should encourage readers to devise their own experiments for research or teaching. This book will serve as an excellent reference for teachers and students of plant pathology and biotechnology. It is also hoped that this book will expose the need and promises of phytoalexin research to the scientific community and initiate many new researchers into the exciting field of plant–microbe interactions.

We are grateful to our contributors, who complied with our demands valiantly and courteously. We wish to thank our publisher, Marcel Dekker, Inc., especially Christine Dunn, Production Editor, and Russell Dekker, Editor-in-Chief, for their encouragement and expert guidance, which led to the prompt publication of this book.

M. Daniel
R. P. Purkayastha

Contents

Preface — *iii*
Contributors — *ix*

1. Progress in Phytoalexin Research During the Past 50 Years — 1
   R. P. Purkayastha

2. Phytoalexins and Host Specificity in Plant Diseases — 41
   Hachiro Oku and Tomonori Shiraishi

3. Elicitor and the Molecular Bases of Phytoalexin Elicitation — 61
   Masaaki Yoshikawa

4. Soybean Phytoalexins: Nature, Elicitation, Mode of Action, and Role — 69
   Jack Paxton

5. Cellular Biochemistry of Phenylpropanoid Responses of Soybean to Infection by *Phytophthora sojae* — 85
   Terrence Lee Graham

6. Culture Darkening, Cell Aggregate Size, and Phytoalexin Accumulation in Soybean Cell Suspensions Challenged with Biotic Agents — 117
   Robert M. Zacharius, William F. Fett, and Prakash G. Kadkade

7. Phytoalexins as a Factor in the Wilt Resistance of Cotton — 129
   M. H. Avazkhodjaev, S. S. Zeltser, H. V. Nuritdinova, and Raviprakash G. Dani

8. Cotton (*Gossypium hirsutum*) Strategies of Defense Expression: Accumulation of Unique *Aspergillus flavus* Control Substances 161
H. J. Zeringe, Jr.

9. Sesquiterpenoid Phytoalexins Synthesized in Cotton Leaves and Cotyledons During the Hypersensitive Response to *Xanthomonas campestris* pv. *malvacearum* 183
Margaret Essenberg and Margaret L. Pierce

10. Chemistry, Biology, and Role of Groundnut Phytoalexins in Resistance to Fungal Attack 199
P. V. Subba Rao and Richard N. Strange

11. Phytoalexins from the Crucifers 229
Thierry Rouxel, Albert Kollman, and Marie-Hélène Balesdent

12. Scoparone (6,7-Dimethoxycoumarin), a *Citrus* Phytoalexin Involved in Resistance to Pathogens 263
Uzi Afek and Abraham Sztejnberg

13. Stilbene Phytoalexins and Disease Resistance in *Vitis* 287
Wilhelm Dercks, L. L. Creasy, and C. J. Luczka-Bayles

14. Mode of Toxic Action of Vitaceae Stilbenes on Fungal Cells 317
Roger Pezet and Vincent Pont

15. Inducible Compounds in *Phaseolus, Vigna,* and *Dioscorea* Species 333
S. A. Adesanya and M. F. Roberts

16. Involvement of Phytoalexins in the Response of Phosphonate-Treated Plants to Infection by *Phytophthora* Species 375
P. Saindrenan and David I. Guest

17. Phytoalexins in Forage Legumes: Studies on Detoxification by Pathogens and the Role of Glycosidic Precursors in Roots 391
V. J. Higgins, Dallas K. Bates, and J. Hollands

18. Induction of Phytoalexin Synthesis in *Medicago sativa* (Lucerne)–*Verticillium albo-atrum* Interaction 405
Christopher J. Smith, J. Michael Milton, and J. Michael Williams

19. Stereoselective Synthesis of Spirovetivane-Type Phytoalexins 445
Chuzo Iwata and Yoshiji Takemoto

20. Enzymic Conversion of Furanosesquiterpene in *Ceratocystis fimbriata*-Infected Sweet Potato Root Tissue 467
Masayuki Fujita, Hiromasa Inoue, K. Oba, and Ikuzo Uritani

21. Defense Strategies of Tea (*Camellia sinensis*) Against Fungal Pathogens 485
B. N. Chakraborty, Usha Chakraborty, and A. Saha

22. Effects of Age-Related Resistance and Metalaxyl on Capsidiol Production in Pepper Plants Infected with *Phytophthora capsici* 503
Byung Kook Hwang

23. Induced Chemical Resistance in *Sesamum indicum* Against *Alternaria sesami* 525
R. K. S. Chauhan and B. M. Kulshrestha

24. Phytoalexins and Other Postinfectional Compounds of Some Economically Important Plants of India 533
M. Daniel

25. Possible Role of Phytoalexin Inducer Chemicals in Plant Disease Control 555
Asoke Kumar Sinha

Index *593*

Contributors

S. A. Adesanya, Ph.D. Department of Pharmacognosy, Faculty of Pharmacy, Obafemi Awolowo University, Ile-ife, Nigeria

Uzi Afek, Ph.D. Research Scientist, Department of Postharvest Science of Fresh Produce, The Volcani Center, Bet Dagan, Israel

M. H. Avazkhodjaev, D.Sc. Professor, Department of Plant Immunology, Institute of Experimental Biology of Plants, Academy of Sciences of Uzbekistan, Tashkent, Uzbekistan

Marie-Hélène Balesdent, Ph.D. Pathologie Végétale, Institut National de la Recherche Agronomique, Versailles, France

Dallas K. Bates, Ph.D. Associate Professor, Department of Chemistry, Michigan Technological University, Houghton, Michigan

B. N. Chakraborty, Ph.D. Botany Department, University of North Bengal, Darjeeling, West Bengal, India

Usha Chakraborty, Ph.D., F.P.S.I. Centre for Life Sciences, University of North Bengal, Darjeeling, West Bengal, India

R. K. S. Chauhan, Ph.D. School of Studies in Botany, Jiwaji University, Gwalior, Madhya Pradesh, India

L. L. Creasy, Ph.D. Professor of Pomology, Department of Fruit and Vegetable Science, Cornell University, Ithaca, New York

Raviprakash G. Dani, Ph.D. Scientist, Plant Breeding, Division of Crop Improvement, Central Institute for Cotton Research, Nagpur, Maharashtra, India

M. Daniel, Ph.D. Reader, Department of Botany, Faculty of Science, The Maharaja Sayajirao University of Baroda, Vadodara, Gujarat, India

Wilhelm Dercks, Ph.D. Professor, Fachbereich, Gartenbau, Fachhochschule Erfurt, Erfurt, Germany

Margaret Essenberg, Ph.D. Regents Professor, Department of Biochemistry and Molecular Biology, Oklahoma Agricultural Experiment Station, Oklahoma State University, Stillwater, Oklahoma

William F. Fett, Ph.D. Research Plant Pathologist, Department of Plant Science and Technology, Eastern Regional Research Center, Agricultural Research Service, U.S. Department of Agriculture, Philadelphia, Pennsylvania

Masayuki Fujita, Ph.D. Associate Professor, Department of Bioresource Science, Faculty of Agriculture, Kagawa University, Miki-cho, Kita-gun, Kagawa, Japan

Terrence Lee Graham, Ph.D. Associate Professor, Department of Plant Pathology, The Ohio State University, Columbus, Ohio

David I. Guest, Ph.D. Senior Lecturer, School of Botany, University of Melbourne, Parkville, Victoria, Australia

V. J. Higgins, Ph.D. Department of Botany, University of Toronto, Toronto, Ontario, Canada

J. Hollands, B.Sc. Department of Botany, University of Toronto, Toronto, Ontario, Canada

Byung Kook Hwang, Ph.D. Professor, Department of Agricultural Biology, Korea University, Seoul, Republic of Korea

Hiromasa Inoue, Ph.D. Assistant Professor, Laboratory of Oral Bacteriology, Kyushu Dental College, Kitakyushu, Japan

Chuzo Iwata, Ph.D. Faculty of Pharmaceutical Sciences, Osaka University, Osaka, Japan

Prakash G. Kadkade, M.Sc., Ph.D. Vice President and Chief Scientist, Phyton Inc., Ithaca, New York

Albert Kollman Pathologie Végétale, Institut Nationale de la Recherche Agronomique, Versailles, France

B. M. Kulshrestha, Ph.D. Department of Botany, K. R. G. College, Gwalior, Madhya Pradesh, India

C. J. Luczka-Bayles, M.S. Assistant Director, Flow Cytometry and Imaging Facility, Cornell Center for Advanced Technology in Biotechnology, Cornell University, Ithaca, New York

J. Michael Milton, Ph.D. Senior Lecturer in Botany, Department of Biological Sciences, University of Wales, Swansea, Wales

H. V. Nuritdinova, Ph.D. Institute of Experimental Biology of Plants, Academy of Sciences of Uzbekistan, Tashkent, Uzbekistan

K. Oba, Ph.D. Faculty of Domestic Science, Nagoya Women's University, Nagoya, Japan

Hachiro Oku, Ph.D. Professor Emeritus, Laboratory of Plant Pathology and Genetic Engineering, College of Agriculture, Okayama University, Okayama, Japan

Jack Paxton, Ph.D. Associate Professor, Department of Plant Pathology, University of Illinois, Urbana, Illinois

Roger Pezet, Ph.D. Staff Researcher, Phytopathology/Mycology Department, Swiss Federal Agricultural Research Station of Changins, Nyon, Switzerland

Margaret L. Pierce, Ph.D. Assistant Researcher, Department of Biochemistry and Molecular Biology, Oklahoma State University, Stillwater, Oklahoma

Vincent Pont Staff Researcher, Laboratory of Organic Chemistry, Swiss Federal Agricultural Research Station of Changins, Nyon, Switzerland

R. P. Purkayastha, Ph.D., D.I.C. Professor, Department of Botany, University of Calcutta, Calcutta, West Bengal, India

M. F. Roberts, Ph.D., D.Sc. Reader, Department of Pharmacognosy, School of Pharmacy, London University, London, England

Thierry Rouxel, Ph.D. Pathologie Végétale, Institut National de la Recherche Agronomique, Versailles, France

A. Saha, Ph.D. Botany Department, University of North Bengal, Darjeeling, West Bengal, India

P. Saindrenan, Ph.D. Scientific Researcher, Department of Plant Pathology, Institut de Biolgie Moléculaire des Plantes, C.N.R.S., Strasbourg, France

Tomonori Shiraishi, Ph.D. Professor, Laboratory of Plant Pathology and Genetic Engineering, College of Agriculture, Okayama University, Okayama, Japan

Asoke Kumar Sinha, Ph.D. Professor, Department of Plant Pathology, Bidhan Chandra Krishi Viswavidyalaya, Kalyani, West Bengal, India

Christopher J. Smith, Ph.D. Biochemistry Research Group, School of Biological Sciences, University of Wales, Swansea, Wales

Richard N. Strange, Ph.D. Department of Biology, University College London, London, England

P. V. Subba Rao, Ph.D.* Centre de Coopération Internationale en Recherche Agronomique pour le Développement (CIRAD), Montpellier, France

Abraham Sztejnberg, Ph.D. Associate Professor of Plant Pathology and Head, Department of Plant Pathology and Microbiology, Faculty of Agriculture, The Hebrew University of Jerusalem, Rehovot, Israel

Yoshiji Takcmoto, Ph.D. Faculty of Pharmaceutical Sciences, Osaka University, Osaka, Japan

Ikuzo Uritani, Ph.D.† Faculty of Domestic Science, Nagoya Women's University, Nagoya, Japan

J. Michael Williams, Ph.D. Department of Chemistry, University of Wales, Swansea, Wales

Masaaki Yoshikawa, Ph.D. Department of Biology, Faculty of Science, Hokkaido University, Sapporo, Japan

Robert M. Zacharius, Ph.D. Principal Consultant, R. M. Zacharius and Associates, Science Consultants, Highland, Maryland

S. S. Zeltser, Ph.D. Institute of Experimental Biology of Plants, Academy of Sciences of Uzbekistan, Tashkent, Uzbekistan

H. J. Zeringue, Jr. Chemist, Commodity Safety Research Unit, Southern Regional Research Center, Agricultural Research Service, U.S. Department of Agriculture, New Orleans, Louisiana

---

**Current affiliation:* Department of Biology, University College London, London, England
†*Current affiliation:* Chairman of Board of Trustees, Aichi Konan Gakuen School Corporation, Konan City, Japan

1

Progress in Phytoalexin Research During the Past 50 Years

R. P. Purkayastha
University of Calcutta, Calcutta, West Bengal, India

I. INTRODUCTION

Several landmarks have been established in the domain of plant sciences during the past half century but the discovery of phytoalexin is undoubtedly an outstanding one since it has opened up a new vista in plant pathology. Müller and Börger (1940) laid a firm foundation for the phytoalexin concept, although in the beginning it did not receive much researchers' attention. However, it has been studied with an increasing awareness for the last two decades but the controversy as to whether phytoalexin alone confers resistance to all plant diseases or this is one of the several mechanisms of disease resistance of plants remains.

A. Some Relevant Research Prior to Phytoalexin Theory Proposed

The response of plants to fungal infection was first recognized by Ward (1905) in England and subsequently by Bernard (1911) in France (Cruickshank, 1978). The restricted and unrestricted growth of the pathogen on resistant and susceptible hosts, respectively, clearly indicates the differential response of host cultivars to a pathogen. This fundamental discovery of responsive mechanisms of plants against pathogens probably forms the basis for the concept that an immune system exists in plants. Wingard (1928) first observed symptoms of recovery in tobacco ring spot virus–infected plants and presumed the infected plants to have acquired immunity against disease. In 1933, Chester also considered the possibility that plants react to bacterial or fungal infection in the same way as animals react to microorganisms or viral infections by producing specific antibodies. But he later realized that since plants have no circulatory system, the possibility of antibody formation in plants is highly improbable. Wallace (1940) also pointed out that symptoms of recovery in tobacco plants infected with sugarbeet curly top virus was due to the formation of protective substances in plants. It was also demonstrated that if the protective substances are transferred to other plants they may develop passive acquired immunity against the disease. Perhaps these reports prompted Müller and Börger to visualize that a functionally similar mechanism comparable to the antigen–antibody reaction in animals could also be operative in plants.

B. Experiments Lead to Phytoalexin Theory and Modern Phytoalexin Concept

Müller and Börger (1940) tested several varieties of potato tubers against virulent and avirulent strains of *Phytophthora infestans*. They observed that when the cut surface of a potato tuber was inoculated with an avirulent strain of *P. infestans* and reinoculated after 24 hr with a virulent strain, it failed to induce any symptom. It strongly suggested that the postinfectional production of an antifungal substance caused inhibition of fungal growth in host tissues. They have drawn a number of conclusions from their experiments that form the basic postulates of the phytoalexin theory. It would be worthwhile to include these conclusions as summarized by Cruickshank (1963): "1.) Phytoalexin inhibits the growth of the fungus in the hypersensitive tissue and this is formed or activated only when the parasite comes in contact with host cells; 2.) the defensive reaction takes place in living cells only; 3.) the inhibitory substance may be regarded as the product of 'necrobiosis' of the host cells; 4.) phytoalexin is nonspecific in its toxicity towards fungi; 5.) basic response of resistant and susceptible plants is similar but the speed of phytoalexin formation differs; 6.) the defense reaction is restricted to the tissue colonized by the fungus and its immediate neighborhood; 7.) the resistant state is 'acquired' after attempted infection and is not 'inherited'; 8.) the speed of host reaction is determined by the sensitivity of the host cell. This is specific and genotypically determined." According to the original definition of Müller and Börger (1940), phytoalexin is a chemical compound produced by living host cells only when these are invaded by a parasite and consequently necrobiosis occurs.

Müller (1956) redefined phytoalexins as "antibiotics" that are the result of an interaction of two different metabolic systems, host and parasite, and that inhibit the growth of microorganisms pathogenic to plants. The term *phytoalexin* (PA) means "warding off compound in plants" (*phyton*, Gr. for "plant"; *alexin*, Greek for "warding off compound"). Various definitions of phytoalexin have been proposed earlier by different researchers. But considering the present state of knowledge it has become imperative to modify its original definition. Phytoalexin may now be defined as an antimicrobial, low molecular weight, secondary metabolite formed de novo as a result of physical, chemical, or biological stress which resists or suppresses the activity of invaders, and its rate of production/accumulation depends either on host genotypes or both host and pathogen genotypes. It should be borne in mind that the same pathogen produces more PA in resistant cultivars than in susceptible ones. Again, two pathogens produce different amounts of PA in the same host cultivar and sometimes the host alone can also produce PA due to physical or chemical stress without any contact with the pathogen or its metabolites. A pathogen may be more or less sensitive to PA depending on its detoxifying ability. Phytoalexins are regarded as a class of compounds that are of diverse chemical nature. The occurrence of similar phytoalexin in different hosts and different phytoalexins in the same host have also been reported. Whatever may be the chemical nature of phytoalexins, the major questions are 1.) whether PA production is the only mechanism of resistance or immune response of a host to disease or is one of the multicomponent and coordinated mechanisms of disease resistance, 2.) whether PA is really involved in disease resistance of all plants, and 3.) whether PA could be successfully exploited for crop protection by induction or direct application or by genetic manipulation. If so, which process would be most effective, simple, durable, non-phytotoxic, and inexpensive remains to be worked out.

C. Early Work on Phytoalexins

Müller and Börger (1940) first detected an antifungal substance known as phytoalexin in a potato tuber which was produced in response to a fungal infection. Subsequently they carried out a series of experiments (1940–1961) to establish that phytoalexin was a defensive substance formed in plants only when they were attacked by an organism. From 1960 onward Cruickshank and his associates at the Division of Plant Industry, CSIRO, Australia continued research on phytoalexins and identified two new phytoalexins—pisatin and phaseollin—which they isolated from pea and bean, respectively. A number of papers on phytoalexins were also published by Uehara (1958a–c) from the Hiroshima Agricultural College, Japan. In 1963, Klarman and Gerdemann at the University of Maryland reported that resistance of soybean to three *Phytophthora* species was associated with the production of phytoalexin. In England, Purkayastha and Deverall initiated work on phytoalexin at the Imperial College of Science and Technology, London. They (1964, 1965) first demonstrated that the steady growth of *Botrytis fabae* and inhibition of growth of *B. cinerea* on leaves of *Vicia faba* (broad bean) were due to the production of phytoalexin. Earlier, Gaumann and Kern (1959a, 1959b) at the Eidg Technical Hochschule, Zurich isolated orchinol, an antifungal substance, from the infected (by *Rhizoctonia repens*) tuber of *Orchis militaris*. Similarly, another antifungal substance was also detected in infected (by *Ceratocystis fimbriata*) roots of sweet potato by Hiura (1943), which was identified by Kubota and Matsuura (1953) as ipomeamarone. Both ipomeamarone and orchinol were later considered as phytoalexins. Although antifungal substances were detected in several infected plants much before the phytoalexin concept came into existence, these were not regarded as induced defensive substances.

D. Merits and Demerits of Phytoalexin Research

It is an undeniable fact that formation of new antimicrobial substances (may be phytoalexins) in plants as a result of any stress is an indication of defense response. Hence activation of defense by any biotic or abiotic agent could be considered as a nonconventional method of disease control. Both resistant and susceptible hosts respond to their parasites (biotic agents) but the speed of response is always much higher in resistant cultivars than susceptible ones. It suggests that host genotypes react differently to a pathogen. A higher rate of phytoalexin production in plants may be deemed as an additional biochemical parameter for screening disease-resistant germplasm. Some phytoalexin-inducing chemicals (nonphytotoxic) are now being used to treat seeds or plants (as foliar spray) for inducing resistance. Besides, phytoalexin is a vast source of new natural products which could be exploited for some useful purposes. Some of them are known chemotherapeutants. High phytoalexin-yielding cultivar may be selected for breeding purpose as a source of resistance cultivar. Biogenetic relationships in plants are now determined on the basis of the chemical nature of phytoalexin, produced by the plant. For instance, Leguminosae (Fabaceae), Solanaceae, Compositae, and Convolvulaceae produce isoflavonoids, terpenoids, polyacetylenes, and furanosesquiterpenoids, respectively.

Several reports of the phytotoxic nature of PA, nondetection of PA in some infected host plants, rapid inactivation/degradation of PA by fungal enzymes or field conditions, the problem of identification of PA due to nonavailability of adequate amounts of material, and very little use of phytoalexins as chemotherapeutants have discouraged many phytoalexin workers in recent years. Besides, it is also doubtful as to whether phytoalexin alone controls plant resistance.

II. PROGRESS IN PHYTOALEXIN RESEARCH: AN OUTLINE

A survey of literature reveals the gradual progress in phytoalexin research during the past 50 years. At the onset, a brief outline is given below:

1940–1959: 1.) Detection, isolation, characterization; identification of new antifungal compounds (phytoalexins) and bioassay; 2.) physiological factors affecting PA production; 3.) testing sensitivity of different parasites/nonparasites to PA; 4.) induction of PA; 5.) mode of action; 6.) relation with disease resistance.

1960–1969: 1–6 continued. 7.) Mechanism of PA induction; 8.) host specificity and PA; 9.) biosynthesis of PA; 10.) degradation of PA; 11.) structure determination of PA in same host by different fungi; 12.) changes in host metabolism due to PA production; 13.) association of gene with PA production.

1970–1979: 1–13 continued. 14.) PAL and PA; 15.) PA production by bacteria and viruses; 16.) metabolism of PA; 17.) disease control by PA; 18.) elicitation of PA; 19.) isolation and characterization of elicitors; 20.) structure of elicitor; 21.) PA as a taxonomic character; 22.) PA and immunity; 23.) suppression of PA induction; 24.) mode of action of PA elicitors; 25.) tissue specific PA; 26.) elicitors of PA from bacteria; 27.) enzymes and detoxification of PA; 28.) PA–plant antigens—disease resistance.

1980–1989: 1–28 continued. 29.) Race-specific PA elicitors; 30.) mechanisms for the phytotoxicity of PA; 31.) factors affecting elicitation of PA; 32.) radioimmunoassay for the PA; 33.) PA chemistry and mode of action.

1990: 1–33 continued. 34.) Sensitive and rapid assay for antibacterial activity of PA; 35.) PA production in progeny of an interspecific cross; 36.) PA in nodules; 37.) antisera raised against resveratrol (a phytoalexin); 38.) gene-encoding enzymes of PA biosynthesis (i.e., phenylalanine ammonialyase, 4-coumarate:CoA ligase, chalcone synthase, chalcone isomerase, stilbene synthase, 3-hydroxymethyl-3-glutaryl coenzyme A reductase); 39.) isolation of stilbene synthase genes from grapevine and transfer to tobacco for increasing disease resistance in transgenic plants.

Extensive reviews on phytoalexins have been published by several workers (Cruickshank, 1963, 1978, 1980; Kuć, 1972, 1976; Ingham, 1972, 1973, 1982; Deverall, 1972, 1976; Purkayastha, 1973, 1985, 1986; Van Etten and Pueppke, 1976; Keen and Bruegger, 1977; Harborne and Ingham, 1978; Keen, 1981; Bailey and Mansfield, 1982; Sequiera, 1983; Yoshikawa, 1983; Kuć and Rash, 1985; Ebel, 1987; Wood, 1986; Mahadevan, 1991; Dixon, 1992). This chapter is intended to highlight the important developments in phytoalexin research during the past 50 years.

It appears from previous reviews that during the first 20 years phytoalexin research was restricted mainly to detection, isolation, identification, and characterization of a very few phytoalexins, factors affecting PA production, sensitivity of microorganisms to PA, and induction of PA for disease resistance in plants. Since 1970 scientists of several disciplines have realized the importance of phytoalexins and have developed novel approaches to study these natural products. Papers published up to 1980 could be classified into six broad categories: 1.) isolation of PA from a number of plant families (more than 20); 2.) biosynthesis of PA; 3.) toxicity of PA; 4.) detoxification of PA; 5.) role of PA in disease resistance; 6.) elicitors and elicitation of phytoalexins by fungi, bacteria or their metabolites, viruses, animals, chemical and physical agents. During the last 10 or 12 years a few more aspects have been added to PA research as mentioned earlier.

III. RECENT RESEARCH ON ELICITATION OF SOME COMMON PHYTOALEXINS

Elicitation of PA is usually caused by an elicitor, which may be either biotic or abiotic. A pathogen metabolite that stimulates PA production in a host may be defined as an elicitor according to Cruickshank (1980). But the term *specific elicitor* has been used for metabolites that stimulate "differential phytoalexin production on various host cultivars similar to the race of the fungus that produces them" (Keen, 1975). The presence of PA elicitor in cell-free extracts of fungi was first demonstrated by Uehara (1959), although the term elicitor was not mentioned. Subsequently, many workers have reported elicitation of PAs by various elicitors. It is interesting to notc that production of similar PA could be induced in a host by both biotic and abiotic agents. How is it possible? The mechanism is not yet clearly understood. According to Albersheim et al. (1986), plants can recognize oligosaccharide fragments of fungal cell walls that are released by enzymes present constitutively in the cell walls of plants. Apart from this, fungi and bacteria can also secrete enzymes capable of releasing oligosaccharide elicitors from the cell walls of plants. Sometimes injury of plants by microbes can cause the release of plant enzyme. This enzyme solubilizes fragments of plant cell walls that elicit PA production/accumulation. Albersheim et al. (1986) also suggested that abiotic elicitors such as heavy metals, UV light, organic solvents, and freezing are likely to work by activating plant enzymes that release oligogalacturonide elicitors from the walls of plant cells. There are several mechanisms by which PA accumulation can be activated. It is also not unlikely that a similar change may occur in the repetitive DNA (Kuć, 1987) due to treatment by more than one substance prior to the production of new messenger RNA. However, the sequence of events involved in the biosynthesis of PA could be studied at a molecular level. Several working models have been proposed (Albersheim and Anderson-Prouty, 1975; Yoshikawa, 1983; Ellingboe, 1982; Purkayastha, 1986; Gabriel et al., 1988) but all of them remain to be confirmed.

The rate of production of PAs is generally higher in case of incompatible host–parasite interaction (Akazawa and Wada, 1961; Cruickshank and Perrin, 1968; Partridge and Keen, 1976; Purkayastha et al., 1983). It is not unreasonable to speculate that elicitor molecules of avirulent pathogens are probably more stimulatory to PA biosynthesis. The degree of stimulation may depend on several factors such as quantity of elicitors released by the organism; speed of release; chemical nature of the elicitor; presence or absence of receptors in the host cell membrane, if present; strong or weak response of receptor; duration of treatment; and environmental conditions. Several comprehensive reviews pertaining to elicitors of PAs have been published (Albersheim and Anderson-Prouty, 1975; Callow, 1977; Darvill and Albersheim, 1984; Keen and Brueggar, 1977; Purkayastha, 1986; Yoshikawa, 1983; also see Smith et al., Yoshikawa and Paxton in this volume). Therefore, a brief account of the elicitation of some common phytoalexins (Fig. 1) is given below.

A. Pisatin

The isolates of *Fusarium solani* which differed in their pathogenicity also showed differential pisatin-eliciting potential. It was confirmed when their culture filtrates were tested on pea (Daniels and Hadwiger, 1976). There was a difference in the concentration of elicitor in the culture filtrates of isolates. The elicitor was fairly heat-stable and also stable in freezing, but eliciting activity was reduced significantly by pronase digestion. This strongly suggests that some of the activities were due to proteinaceous components. An interesting observation was made by Shiraishi et al. (1978) who detected both elicitor and suppressor of pisatin in the pycnospore germination fluid of *Mycosphaerella pinoides* (pea pathogen) (Crute et al., 1985). The elicitor and suppressor were high and low molecular weight substances, respectively. The suppressor usually counteracts the activity of elicitor. The components of suppressor were identified as low molecular weight peptides which inhibited pisatin accumulation in pea leaves when inoculated separately with nonpathogens like *Erysiphe graminis hordei* and *Stemphylium sarcinaeforme*. As a result pea leaves became susceptible to *S. sarcinaeforme*. Yamoto et al. (1986) demonstrated that pisatin could be induced in pea leaves by elicitors from *Mycosphaerella pinoides*, *M. melonis*, and *M. ligulicola*. Accumulation of pisatin increased after removal of epidermis and application of elicitors (high molecular weight compound $>10{,}000$ Da) from germination fluid of the fungus. The authors also studied the effect

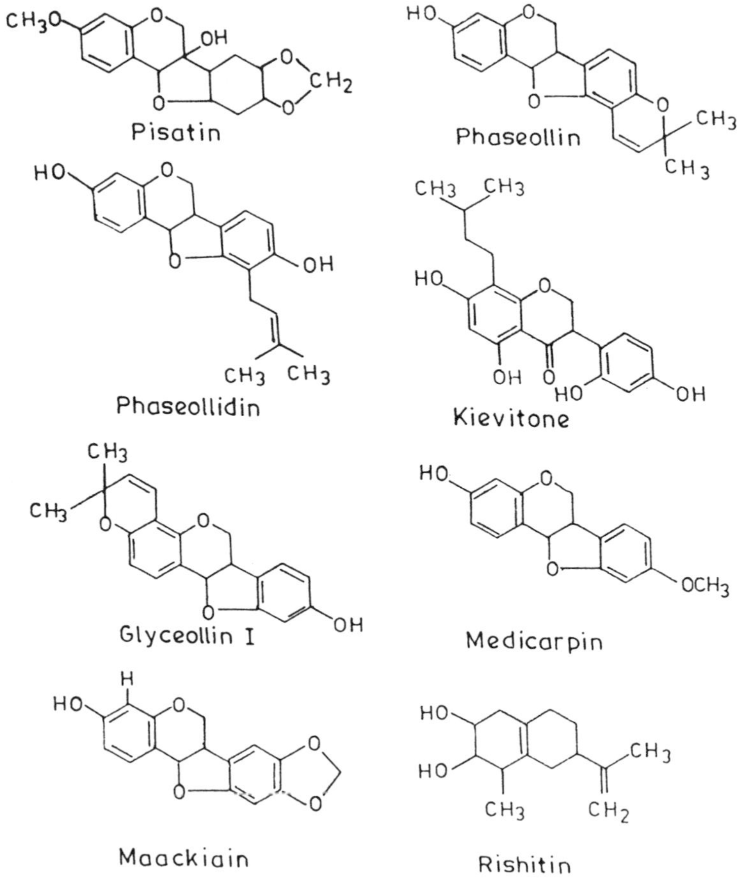

Figure 1 Structures of some common phytoalexins.

of a suppressor (low molecular weight compound <10,000 Da) of pisatin obtained from the germination fluid of *M. pinoides* which counteracted the activity of pisatin elicitors.

Elicitation of pisatin by various abiotic elicitors such as metallic salts (Uehara, 1963); mercuric chloride, sodium iodoacetate, sodium selenate (Perrin and Cruickshank, 1965); actinomycin D (2×10^{-5} M) (Schwochau and Hadwiger, 1968); cupric chloride and ethylene (Chalutz and Stahmann, 1969); and synthetic peptides (Hadwiger et al., 1971) has been reported. Hadwiger et al. (1976) raised an induced mutant of pea by treatment with sodium azide which was able to produce more pisatin without extraneous supply of sodium azide. Elicitation of pisatin by UV light was also noted

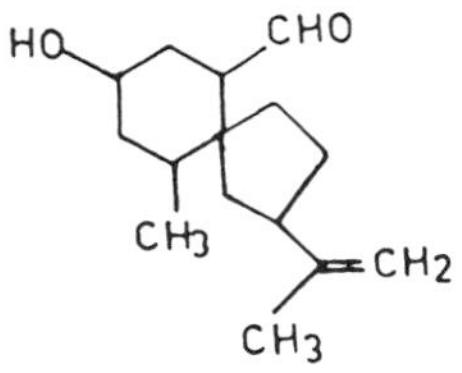

Lubimin

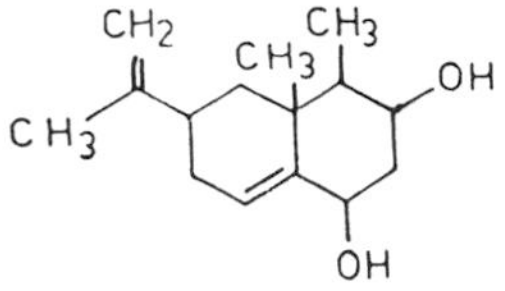

Capsidiol

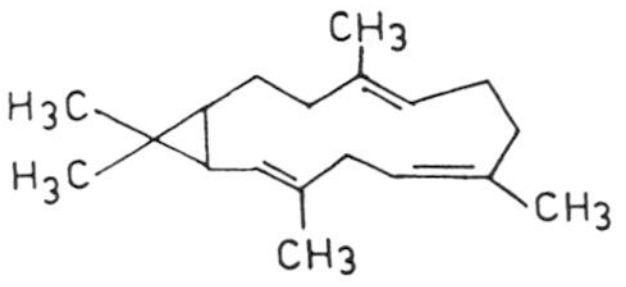

Casbene

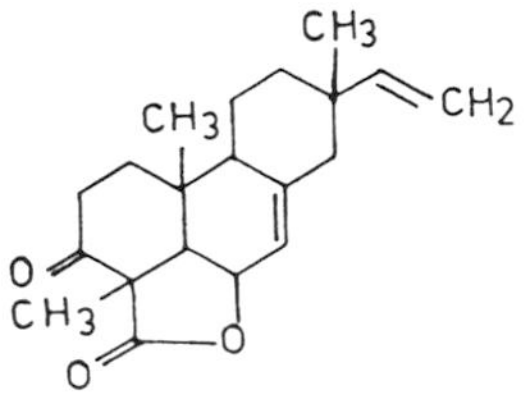

Momilactone A

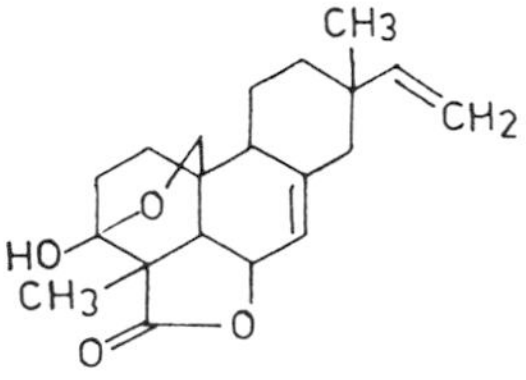

Momilactone B

by Hadwiger and Schwochau (1971). It appears from the above statements that pisatin production could be elicited in plants by physical, chemical, and biological agencies.

B. Phaseollin

An elicitor of phaseollin was isolated from the mycelial walls and culture filtrates of *Colletotrichum lindemuthianum*, which was identified as a polysaccharide. The molecular weight varied between 1 million and 5 million DA, and consisted predominantly of 3- and 4-linked glucosyl residues (Anderson-Prouty and Albersheim, 1975). An amount equivalent to 100 ng of glucose elicited a similar response in the bean tissue.

The regulation system of phaseollin synthesis in cell suspension cultures of dwarf french bean (*Phaseolus vulgaris*) was studied by Dixon and Christopher (1979). Considerable amount of phaseollin accumulated when french bean was treated with an elicitor from the cell wall of *C. lindemuthianum*.

But the elicitors isolated from the cell walls of *P. megasperma* var. *sojae* and *Botrytis cinerea* were less effective.

A carbohydrate-rich extracellular component from a race of *C. lindemuthianum* showed a high level of PA activity on a resistant cultivar "Dark Red" of kidney bean but not on the susceptible cultivar "Great Northern." Other extracellular components were also recognized as elicitors by both cultivars. It is noteworthy that the two cultivars of *Phaseolus vulgaris* displayed a differential response to extracellular components. These observations support the hypothesis that both general and specific mechanisms exist in race–cultivar interaction (Tepper and Anderson, 1986).

A glucan was isolated from the cell wall extracts of *Fusarium oxysporum* f. sp. *lycopersici* (Anderson, 1980a) and a polypeptide (monilicolin A) from mycelia of *Monilinia fructicola* (Cruickshank and Perrin, 1968). Both compounds elicited phaseollin production. Apart from biotic elicitors, several abiotic agents such as mercuric chloride (Fraile et al., 1980), abscisic acid, benzylaminopurine, silver nitrate (Stoessel and Magnalato, 1983), oxadiazone, herbicides (Rubin et al., 1983; Ricci and Rousse, 1983), ozone and SO_2 also elicited phaseollin in *P. vulgaris*.

C. Glyceollin

An elicitor of glyceollin was isolated from the mycelial cell wall of *Phytophthora megasperma* var. *sojae* by Ebel et al. (1976). This elicitor stimulated the activity of phenylalanine ammonialyase and also induced glyceollin production in soybean cell cultures. They concluded that the action of elicitors is not species or variety specific but is a part of the general defense response of plants. Ayers et al. (1976a) also isolated a branched β-glucan elicitor from both cell walls and culture filtrates of *P. megasperma*. The elicitor was considered to be predominant by a 3-3, 6-linked glucosyl residue. The hyphal cell walls of *P. megasperma* f. sp. *glycinea* (Pmg) also showed considerable eliciting activity which was due to three classes of carbohydrates. The activities of other glucans or glucomannans and mannans obtained from other sources were much lower as reported by Keen et al. (1983). Chitin and Chitosan also acted as elicitors of PAs in other plants but their activities were very low in the case of soybean cotyledons.

Purkayastha and Ghosh (1983) reported elicitor activity of fresh mycelial wall extract of *Myrothecium roridum*. Spores suspended in mycelial wall extract, drops placed on leaf surfaces of soybean, and incubated for 48 hr. The results of bioassay test revealed that the spores suspended in mycelial wall extract were more inhibitory than the spores suspended in sterile distilled water and incubated on leaf surfaces for a similar period. Mycelial wall extract induced greater production of glyceollin in soybean leaves.

Soybean leaves treated with only mycelial wall extract and mycelial wall extract containing conidia of *Colletotrichum dematium* var. *truncata* produced 73.22 and 217 μg/g (fresh wt) glyceollin, respectively, after 48 hr of incubation (Purkayastha and Banerjee, unpublished).

There is evidence that oligogalacturonides derived from the pectic polysaccharides of plant cell walls can serve as regulatory molecules that induce glyceollin accumulation in soybean (Davis et al., 1986). This is inconsistent with the hypothesis that oligogalacturonides play a major role in plant disease resistance.

An elicitor was also extracted from wounded, frozen cotyledons of soybean. When extract was heated at 95 °C for 10 min its eliciting activity was lost. It suggests that the elicitor is thermolabile. But addition of calcium chloride with elicitor enhanced eliciting activity. Lyon and Albersheim (1982) observed maximum accumulation of glyceollin when elicitor was applied immediately after the cutting of soybean cotyledons.

Elicitors extracted from the cell walls of *Saccharomyces cerevisiae* were identified as structural glucans. These are able to stimulate glyceollin accumulation in soybean (Albersheim et al., 1978). Specific elicitors of glyceollin were also detected in the cellular envelops of incompatible races of *Pseudomonas glycinea*. However, elicitor activity could not be detected in lipopolysaccharide preparations, exopolysaccharide fractions, or the culture fluids of various races of *P. glycinea*. Elicitors were solubilized with sodium dodecyl sulfate and then preparations from five bacterial races excepting one had similar specificity for elicitation of glyceollin in cotyledons of two soybean cultivars (Brueggar and Keen, 1979). These observations suggest that elicitors are not always race-specific.

Cahill and Ward (1989) compared the release of glyceollin elicitors into culture fluids of a metalaxyl-sensitive and a tolerant (50– > 500 μg/ml) isolate of *Phytophthora megasperma* f. sp. *glycinea* following addition of metalaxyl to the culture medium. The elicitor activity was increased markedly by metalaxyl treatment in culture fluids of the sensitive isolate but not in that of tolerant isolate. Most elicitor activity was detected in fractions (obtained by gel filtration) corresponding to the second of the two carbohydrate peaks. These findings are very interesting because stimulation of elicitor activity may cause more accumulation of PA which in turn may stimulate host defense responses.

Among the abiotic elicitors of glyceollin, gibberellic acid, sodium azide (Purkayastha and Chakraborty, 1985), cloxacilin (Purkayastha and Banerjee, 1990), and benzyl penicillin (Purkayastha and Banerjee, unpublished) are already known. Benzyl penicillin (100 ppm) and cloxacillin (100 ppm)-treated soybean leaves produced 17.67 μg/g (fresh wt of leaves) and 35.25 μg/g glyceollin, respectively. Ghosh and Purkayastha (1990) demonstrated that ajmalicine (100 ppm), an alkaloid, could induce glyceollin (84.80 μg/g fresh wt) in soybean leaves. They stated that cadmium chloride (10^{-4} M) is also an elicitor of glyceollin (67.52 μg/g) (Ghosh and Purkayastha, 1992).

D. Rishitin and Lubimin

Elicitation of rishitin formation in potato by a glucan and a lipoglycoprotein isolated from the cell wall and cytoplasm of *Phytophthora infestans* was noted earlier by Chalova et al. (1976, 1977). The elicitors of terpenoid PAs such as rishitin and lubimin, were also obtained from autoclaved sonicates of *P. infestans*, *P. parasitica*, *Pythium aphanidermatum*, *Achyla flagellata*, and *Aphanomyces euteiches* and were able to elicit PA production in potato tuber slices. Potatoes treated with glucan extracted from *P. megasperma* var. *sojae* produced a greater amount of rishitin (29 μg/g fresh wt) while glucan produced a lower amount (19.5 μg/g fresh wt) (Cline et al., 1978).

Elicitors of rishitin could play an important role in regulating defense reaction in potatoes (Terekhova et al., 1980). Two potato cultivars [Temp. (R.I.) and *Belorusskii rannii* (r)] having two different resistant genes were inoculated with two races of *P. infestans*. The tubers were extracted with acetone and the acetone extract contained more rishitin-inducing substances than alcohol extract of the same. It was suggested that the intensity of release of rishitin inducers depended on the races of *P. infestans*.

When spores of *P. infestans* were killed by freezing, followed by thawing, only sporangia and cystospores elicited terpene formation in the slices of potato. Even dead cystospores were also able to elicit rishitin and lubimin accumulation in potatoes (Henfling et al., 1980). Bostock et al. (1982) detected two fatty acids (eicosapentaenoic and arachidonic acids) in the mycelia of *P. infestans* which elicited sesquiterpenoid PA accumulation in potatoes.

An interesting observation was made by Metlitskii et al. (1984). They detected two types of substances in the cell walls of *P. infestans*, namely, elicitors and suppressors of defense reaction in potatoes. Elicitors were extracted from six races differing in genes for virulence in potatoes. Eventually these elicitors were tested on tuber discs carrying the resistant gene R_1. The extracts from incompatible races acted as elicitors of rishitin while those from the compatible races acted as suppressors of rishitin. Eliciting activity was found to be 50% higher than the control while the suppressing activity reached up to 200%. A pronase-sensitive, heat-labile elicitor of rishitin was obtained from the germination fluid of *P. infestans*. Rishitin accumulation in potato tuber slices was observed when treated with this elicitor. Gel permeation chromatography of germination fluid revealed the presence of several substances. However, active substances were precipitated with ammonium sulfate (Osman and Moreau, 1985).

An important role of lipid and nonlipid components of mycelial extract of *P. infestans* in the elicitation of PA in potato tuber tissue was pointed out by Bryan et al. (1985). There were two fractions of mycelial extract: lipid-containing and lipid-free fractions. The most active fraction was composed of carbohydrate, proteins, and lipid. A heat-released preparation of mycelia containing very low levels of eicosapentaenoic, arachidonic, and dihomo-γ-linolenic acids were found to be more active than a lipid extract of the mycelium. In 1986 Woodward and Pegg demonstrated that mycelial extract and culture filtrates of *Verticillium alboatrum* were able to elicit rishitin production in both resistant and susceptible isolines of tomato. The fungus grown in Czapek-dox medium for 33 days produced high molecular weight substance which elicited larger quantities of rishitin in susceptible than in resistant plants. But low molecular weight glucan obtained from 14-day-old culture filtrate of the fungus produced higher levels of rishitin in both resistant and susceptible lines. The nonspecificity of low molecular weight glucans is thought to preclude their involvement in single-gene resistance mechanisms.

Mycelia and spores of incompatible races of *P. infestans* and *Helminthosporium carbonum* induced rishitin and lubimin production in potato. The eliciting activity of spores and mycelia of *H. carbonum* was lost when treated with heat, ethanol, or liquid N_2. But *P. infestans* retained its eliciting activity even after heat or ethanol treatment. Gas chromatography mass spectrophotometric analysis of mycelial extract of *H. carbonum* failed to detect the presence of elicitors (arachidonic acid and eicosapentaenoic acid). The most significant observation was that inoculation of potato with a compatible race of *P. infestans* suppressed rishitin and lubimin accumulation in response to subsequent inoculation with an incompatible race of the same fungus. According to Zook and Kuć (1987), it is unlikely that the suppression by *P. infestans* is due to inhibition of the pathway for the synthesis of rishitin and lubimin in potato tissue.

In 1987, Rohwer et al. detected accumulation of rishitin and several structurally related sesquiterpene derivative in potato tuber when treated with culture filtrate of *P. megasperma* f. sp. *glycinea* but the rate of accumulation was rapid in non–host-incompatible interaction, less rapid in host-incompatible interaction, and slow in compatible interaction. The cause of differential response of host and nonhost cultivars to an elicitor remains yet to be elucidated.

Among the abiotic agents mercuric acetate (Cheema and Haard, 1978) was found to be an elicitor of potato PAs—rishitin and lubimin. Calcium and strontium (Sr^{2+}) ions also enhanced rishitin accumulation but not lubimin in potato tubers when treated with arachidonic acid (Zook et al., 1987). The same cations, however, in the presence of poly-L-lysine (PL) increased lubimin accumulation even greater than rishitin. On the contrary, Mg^{2+}, which did not affect arachidonic acid, elicited both rishitin and lubimin. But it inhibited the accumulation of PAs after treatment with PL. Zook et al. (1987) suggested that the mobilization of calcium may play a central regulatory role in the expression of PA accumulation in potato tissues after elicitation.

E. Momilactone

Momilactone (A and B) and oryzalexin (A, B, C, D) are two known PAs of rice. Induction of momilactone A in rice coleoptiles and leaf sheaths by gibberellic acid (GA_3) was recorded by Ghosal and Purkayastha (1984). Since GA is a degraded diterpene (Birch et al., 1958) it may act as a precursor. Besides, gibberellin-mediated enzyme production may also account for the elicitation of momilactone biosynthesis in plants. A fungicide known as 2,2,-dichloro-3,3-dimethylcyclopropanecarboxylic acid (WL 28323) has been found to activate the natural resistance of rice plants against blast disease caused by *Pyricularia oryzae* (Cartwright et al., 1977). The activity of this fungicide is unique because it does not itself stimulate momilactone production but rather increases the capacity of rice plants to synthesize more momilactones in response to fungal infection. Sodium azide and x-ray also elicited momilactone in rice (Purkayastha and Ghosal, unpublished).

F. Other Phytoalexins

A high molecular weight elicitor of casbene (a castor PA) was isolated from 3-day-old culture of *Rhizopus stolonifer*. The purified fraction showed eliciting activity which contained both protein and carbohydrates. Heat treatment (at 60°C or higher) for 15 min inactivated the elicitor. Higher concentration (2×10^{-8} M), however, increased (about 14-fold) casbene synthetase activity in extracts of treated split seedlings (Stekoll and West, 1978). Victorin, a host-specific toxin obtained from *Helminthosporium victoriae*, was tested for its ability to elicit the oat PA, avenalumin. This toxin elicited PA in *Avena sativa* as reported by Mayama et al. (1986).

An elicitor detected in the culture filtrate of *Chaetomium globosum* was found to be heat-labile. It is interesting to note that the elicitor activity was correlated with the activity of pectinolytic enzymes in the culture filtrate of the said fungus. The culture filtrates induced the release of heat-stable elicitors from carrot cell homogenates. The eliciting activities of culture filtrates of *Botrytis cinerea*, *Fusarium moniliforme*, and *Helminthosporium oryzae* were also recorded earlier (Amin et al., 1986). Elicitation of capsidiol formation in fruits of *Capsicum annuum* L. by copper sulfate (0.1 M), sodium nitrate (0.1 M), and chloramphenicol was demonstrated by Watson and Brooks (1984).

In 1986, Kessmann and Barz found that polymeric compounds obtained from *Ascochyta rabiei* could elicit PA production in cotyledons of *Cicer arietinum*. Even wounding could also induce PA. Accumulation of isoflavones biochanin A, formonetin, and their glucoside conjugates was detected in infected cotyledons. Wounding caused additional accumulation of the pterocarpan PAs, medicarpin and maackiain.

IV. STRUCTURE OF ELICITOR

Although a large number of biotic and abiotic elicitors of PAs have been reported by previous workers, very little work has been done so far on the structure of elicitors. Albersheim et al. (1986) presented the structure of elicitor-active heptaglucosides and identified the structure of active elicitor of PA. The structure is given in Fig. 2.

```
G→4G→4G→4G→4G→4G→4G→

G→6G→6G→6G→6G→
3  3
↑  ↑
G  G

G→6G→6G→6G→6G→
3          3
↑          ↑
G          G

      G→6G→6G→
      3      3
      ↑      ↑
G→6G→6G      G
```

```
┌─────────────────────────┐
│ G→6G→6G→6G→6G→          │
│   3        3            │
│   ↑        ↑            │
│   G        G            │
└─────────────────────────┘

G→6G→6G→6G→6G→
3         3
↑         ↑
G         G

G→6G→6G→6G→6G→
    3   3
    ↑   ↑
    G   G

G→6G→6G→6G→6G→
        3   3
        ↑   ↑
        G   G
```

Figure 2 Structures of eight hepta-β-glucoside alditols. Structure shown in box (right top) was active as an elicitor of PA accumulation. The seven other hepta-β-glucoside alditols were inactive as elicitors, even at the concentration 25 times more than that of the active elicitor (Albersheim et al., 1986).

V. DETOXIFICATION/DEGRADATION OF PHYTOALEXIN AND ROLE OF ENZYMES

Generally, PAs have been considered as a possible means by which plants attempt to defend themselves against parasitic attack. But some pathogens or strains of a pathogen are moderately or highly tolerant to these toxic compounds (phytoalexins) while others are sensitive to them. It has been conclusively demonstrated in a few diseases that virulent parasites can metabolize these toxic compounds and hence it is presumed that these parasites could have their own detoxification mechanisms. The study of such mechanisms of parasites is important because 1.) pathogenicity of an organism may depend on its ability to detoxify the host's toxic compounds, 2.) control of detoxification mechanism may help to enhance disease resistance of host plants, and 3.) genes for PA detoxification might be used in the pathogens for biocontrol of weeds in the crop field as suggested by Van Etten et al., 1989; (see also Higgins et al. in this volume).

Several pathogens of pea could metabolize PAs such as pisatin, maackiain, and medicarpin. For instance, *Nectria hematococca* can detoxify PAs as follows:

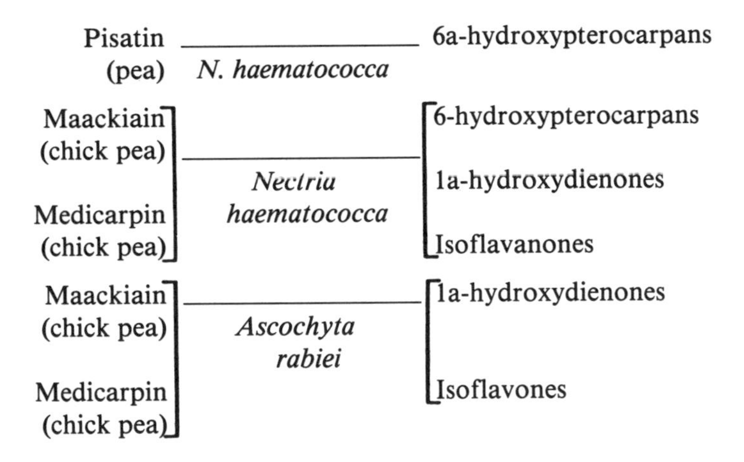

Pisatin demethylase, an enzyme of *N. haematococca* (an ascomycetous fungus), is a microsomal cytochrome P450 monooxygenase (Matthews and Van Etten, 1983). This enzyme catalyzes pisatin detoxification. Usually Pda genes of *N. haematococca* code for cytochrome P450 isozymes. All ascospore progeny of *N. haematococca* that were highly or moderately virulent on pea were Pda^+ and therefore tolerant to pisatin (McIntosh et al., 1989). A phenotype which was referred to as Pda^- showed lack of ability to demethylate pisatin, such isolate of *N. haematococca* was sensitive to pisatin and also nonpathogenic on pea (Van Etten et al., 1980). Evidence suggests that *N. haematococca* DNA fragment that confers the ability to demethylate pisatin contains the structural gene for cytochrome P450. Apart from *N. haematococca*, *Ascochyta pisi*, and *Fusarium oxysporum* f. sp. *pisi* can also degrade pisatin by demethylation (Fuchs et al., 1980; Sanz Platero and Fuchs, 1978). However, *P. pisi* and *Rhizoctonia solani* differ significantly in their rate of demethylation and in virulence. These findings indicate that each parasite requires a specific detoxification system for its pathogenicity. There are reports that some pathogens cannot metabolize the PAs of their own hosts. Therefore, PA metabolism cannot be strictly correlated with host specificity (Van Etten et al., 1989).

Fusarium solani f. sp. *phaseoli* can detoxify all four PAs of bean (*Phaseolus vulgaris*), namely kievitone, phaseollin, phaseollidin, and phaseollin isoflavan. An extracellular enzymatic system is involved in conversion of kievitone to kievitone hydrate (Fig. 3). Generally, kievitone hydrate is apparently mediated by a single enzyme designated as kievitone hydratase. Detoxification of kievitone occurs by hydration of the isopentenyl side chains (Van Etten et al., 1989).

A few pathogens (*F. solani* f. sp. *phaseoli*, *Colletotrichum lindemuthianum*) and nonpathogens (*Saptoria nodorum*, *Stemphylium botryosum*) have been found to detoxify phaseollin. The initial metabolites of phaseollin produced by the aforesaid organisms are shown below.

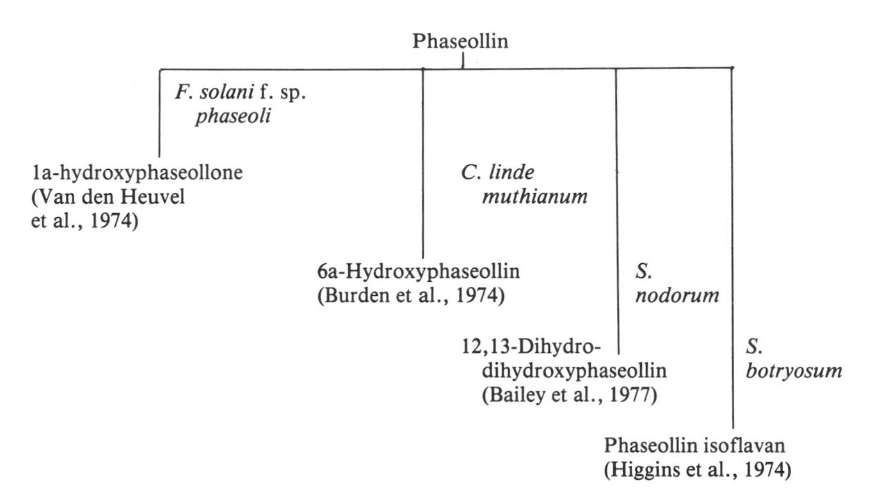

Detoxification of potato phytoalexin lubimin by *Gibberella pulicaris* (anamorph: *Fusarium sambucinum*) has also been reported by Desjardins et al. (1989). The pathway of lubimin metabolism has been proposed, the beginning of which is shown as follows:

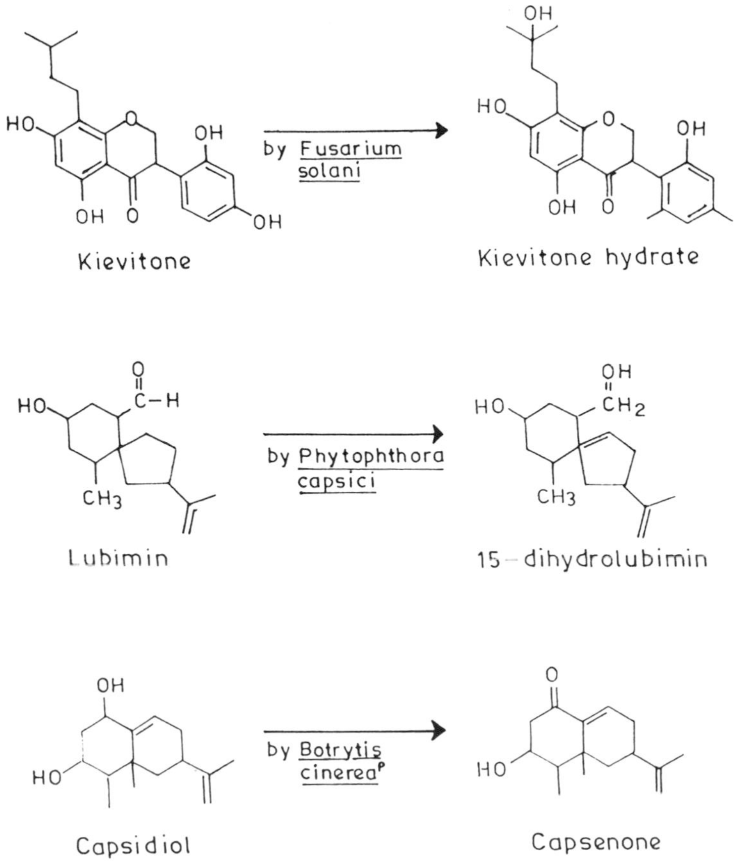

Figure 3 Detoxification of phytoalexins by different fungi.

Lubimin — 15-dihydrolubimin / 2-dehydrolubimin — isolubimin —— cyclodehydroisolubimin

Lubimin-sensitive isolate of *G. pulicaris* metabolized lubimin relatively slowly and produced only 15-dihydrolubimin and isolubimin. Both the products were toxic to the isolate but not cyclodehydroisolubimin. The reduction of the sesquiterpenoid PA lubimin has been demonstrated using

Phytophthora capsici, *Glomerella cingulata*, and *F. sulphureum* by Ward and Stoessl (1977). The transformation of PAs by reductive reaction has been studied in the metabolism of wyerone derivatives by *B. cinerea* and *B. fabae*. Both these fungi are able to reduce the ketones wyerone and wyerone epoxide to the respective alcohols wyerol and wyerol epoxide:

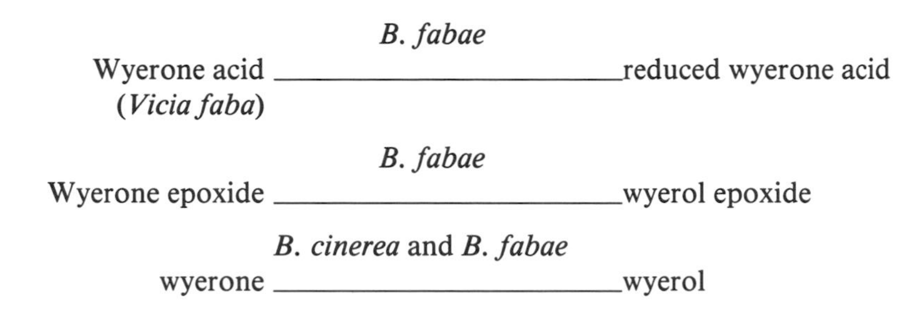

Ward and Stoessl (1977) reported that capsidiol (pepper PA) was degraded to capsenone in shake cultures of *B. cinerea* and *F. oxysporum* f. sp. *vasinfectum* but degradation was not detected in the cultures of *Phytophthora capsici* or *Monilinia fructicola*.

B. cinerea

Capsidiol ——————— capsenone
(*Capsicum frutescens*)

Ipomeamarone, a sweet potato PA, was degraded by *Botryodiplodia theobromae* and also by *B. cinerea*. It was detected when this phytoalexin was added to the mycelial suspension of these fungi and incubated for 24 hr at 25 °C. The degraded product was not identified.

Phytoalexin degradation/detoxification and fungal pathogenicity have been discussed in detail by several workers (Van Etten et al., 1982, 1989; Purkayastha, 1985; Boominathan et al., 1986; Mahadevan, 1991). Although an apparent correlation exists between the higher rate of PA degradation and greater pathogenicity of a species, it is not applicable in all cases. Any significant change in any factor(s) involved in the degradation process may alter this relationship. The enzymes associated with the degradation process may be intracellular or extracellular, constitutive or inducible, single or multiple. Usually more than one step may be involved in complete degradation of a toxic PA to a nontoxic product. These aspects are yet to be studied in detail in cases of newly reported PAs. Degraded products of several known PAs have not been identified so far.

VI. PHYTOALEXIN AND PLANT ANTIGENS RELATED TO DISEASE RESISTANCE

Purkayastha (1973) first attempted to show a relationship among PAs, plant antigens, and disease resistance on the basis of available information. The objectives are to determine whether both PAs and plant antigens are associated with disease resistance and, if so, whether both could be considered as reliable biochemical parameters for screening disease resistance germplasm in vitro. Since production of both PAs and plant antigens are genetically controlled (based on evidence), it was thought judicious to select these two stable parameters. After about 20 years of research it has become more or less certain that a relationship exists among the three. In 1978, Cruickshank stated that the net rate of PA accumulation was greater in resistant cultivars than in susceptible ones. He also suggested on the basis of the consistency of the cultivar data that PA concentrations for individual plants within a progeny are genetically inherited and that selections could be made on this basis. There is substantial evidence from several gene-for-gene systems that PAs accumulate in high quantities after inoculation of plants with avirulent but not with virulent parasite races (Keen, 1982). It suggests that both host and pathogen genes are involved in both susceptible and resistant disease reactions. Recently it was demonstrated conclusively that isolation of stilbene synthase genes (responsible for synthesizing the stilbene-type PA resveratrol when attacked by pathogens) from grapevine and transfer of these genes to tobacco could increase its resistance to *Botrytis cinerea* infection. This increased disease resistance in transgenic plants is due to an additional foreign phytoalexin (Hain et al., 1993). It suggests that foreign PA in a plant may also confer resistance to disease.

The involvement of antigens in disease reaction of plants was also reported by several workers. They found antigenic similarity in susceptible/compatible host–parasite interaction while antigenic disparity in resistant/incompatible interactions. The main reason for considering PA and plant antigens together is that both are believed to be involved in the disease reaction and probably production of PA depends on the interaction of host–parasite proteins or specific antigens. The dimer model proposed by Ellingboe (1982) also presented the similar view. The dimers are viewed as being formed by the primary protein products of the R genes and avr genes, directly binding one another in a protein–protein interaction. Alternatively, one of them might bind the gene or mRNA transcript of the other (Gabriel and Rolfe, 1990). The dimer formation was attributed to incompatible interaction (Ellingboe, 1982). The formation of dimer was also supported by Gabriel et al. (1988) who proposed a separate model known as the ion channel defense model. Whatever may be the interpretation of various models, the two basic factors common to all cases are that 1.) both host and pathogen genes or their products are involved in disease reaction and 2.) more or less PA accumulates in host tissues after infection. The involvement of antigen (protein) in disease reaction was also suspected when one of the host antigens was found to be missing after chemical induction of resistance in susceptible host cultivars (Chakraborty and Purkayastha, 1983; Purkayastha and Banerjee, 1990).

A number of previous workers presented conclusive evidence that resistant cultivars of different host species produced more PAs in response to fungal infection than the susceptible ones. Resistant cultivar of soybean Harosoy 63 produced seven times more PA than susceptible cultivar Harosoy when plants were inoculated with *Phytophthora megasperma* var. *sojae* (Klarman, 1968). Keen et al. (1971) also reported that PA (6a-hydroxyphaseollin) accumulated more rapidly (20–50 times faster) in the hypocotyls of Harosoy 63 than in Harosoy when inoculated with that pathogen. The resistant tomato plants also produced more rishitin (phytoalexin) than the susceptible plants (Tijamos and Smith, 1974) after inoculation with *Verticillium alboatrum*. The leaves of highly resistant sugarbeet showed greater amounts of β-vulgarin than susceptible ones when infected with *Cercospora beticola* (Johnson et al., 1976). Besides, monogenically resistant cowpeas accumulated 10-fold or more kievitone than the near-isogenic susceptible cultivar (to *Phytophthora vignae*) (Partridge and Keen, 1976). In 1981, Vaziri et al. noted that the rate of medicarpin production was higher in resistant cultivar of alfalfa following inoculation with *P. megasperma* f. sp. *medicaginis*. A number of resistant rice cultivars, such as Mahsuri, Rupsail, and Bad Kalamkati, produced greater amount of PA momilactone A than the susceptible cultivars (Jaya, IR-8, CR-126-42-1) when inoculated with *Acrocylindrium* (= *Sarocladium*) *oryzae* (Purkayastha et al., 1983). In a separate investigation, Purkayastha and Chakraborty (1983) demonstrated that UPSM-19, a resistant (to *Macrophomina phaseolina*) cultivar of soybean, was able to produce significantly higher amounts of glyceollin than a susceptible cultivar Soymax. The rapid accumulation of PA in resistant cultivar ILC 3279 of *Cicer arietinum* after inoculation with *Ascochyta rabiei* was also noted by Weigard et al. (1986). The average production of medicarpin in the resistant cultivar was found to be 20 μmol/g fresh wt, while it was only 5 μmol/g in susceptible cultivars. Kessmann and Barz (1987) prepared cell suspension cultures of *C. arietinum* from both resistant cultivar ILC 3279 and susceptible cultivar ILC 1429 and inoculated with *A. rabiei*. These two cultures differed in their accumulation of medicarpin and maackiain. In view of the above facts it can be presumed that PA is associated with disease resistance of many plants. Hahn et al. (1985), however, stated that phytoalexins may not be solely responsible for the initial inhibition of *Phytophthora megasperma* f. sp. *glycinea* in the resistant cultivar of soybean.

Involvement of PAs in plant disease resistance has been demonstrated by many workers but whether the absence of common antigens between host and parasite is also an important factor in disease resistance has not been ascertained. In this chapter, however, an attempt has been made to summarize the work done so far on host–pathogen antigens pertaining to disease resistance/susceptibility of plants with a view to evaluate how far the validity of "common antigens" or "matching proteins" concept is applicable in determining basic compatibility and the absence of same in basic incompatibility. PAs and plant antigens have always been considered separately by the researchers. No attempt was made to work on both aspects together and hence it was not possible to show any relationship between the two. Since the gene-for-gene concept of Flor (1956) is now well recognized, it is not inconceivable that host genes responsible for the production of matching proteins (with parasite) may also contain a PA suppressor gene(s). If a suppressor gene is eliminated or inactivated by any physical or chemical treatment, the susceptible plant may become temporarily resistant and produces/accumulates more PA. Since both resistant and susceptible cultivars contain PA-producing genes, the rate of production in susceptible cultivar is very low may be due to the presence of PA suppressor gene. Again, the same cultivar is able to produce more when treated with a suitable chemical and inoculated with the same pathogen. Recently, it was observed in a number of cases that chemical induction of resistance in susceptible host cultivars causes two changes: 1.) production of more phytoalexin and 2.) missing of a specific host antigen. The gene which was responsible for the missing antigen (protein) could be a PA suppressor gene. It would be interesting to trace the missing antigen and its relationship, if any, with the rate of PA production.

In 1948, Fedotova recognized the significance of antigenic relationship between host and parasite in disease susceptibility of plants. Subsequently, Dineen (1963) reported that susceptibility of host increases with antigenic similarity while resistance of the host is characterized by disparity of antigenic determinants. This concept is applicable in vertebrate animals but not plants according to Charudattan and DeVay (1972). However, experimental evidence suggests that the common antigen relationship is not an unnatural phenomenon and cannot be entirely ruled out on the basis of a few exceptions. An interesting observation was made by Doubly et al. (1960), who found that a specific antigen in each of the four races of *Melampsora lini* was commonly shared by only those lines of flax (*Linum usitatissimum*) that were susceptible to a particular race. This was later confirmed by

Peterman (1967). During the last three decades the common antigen relationship has been detected in a number of host–parasite systems (Table 1).

In view of the findings summarized in the table it is reasonable to speculate that common antigenic substances underlie compatibility of host and parasite albeit there are exceptions. For instance, *Phytophthora infestans* shared antigens in common with its hosts potato and tomato and also with its nonhost tobacco but not with nonsolanaceous species (Palmerly and Callow, 1978).

More than 100 host pathogen/non-pathogen combinations including several cultivars of soybean, rice, groundnut, pigeon pea, jute, and bean and their respective pathogens and nonpathogens were examined. The results of immunodiffusion and immunoelectrophoretic tests reveal that there is no common antigenic relationship between hosts and nonpathogens (45 combinations tested) while more than 50% combinations exhibited cross reactive antigens between known hosts and pathogens (67 combinations tested) irrespective of virulent and avirulent strains. No cross-reactive antigen was detected in immunodiffusion test using resistant hosts (10 cultivars) and their parasites but it was detected at a very low titer by enzyme-linked immunoassay in case of a resistant soybean cultivar UPSM-19. All known susceptible cultivars, however, showed common antigenic relationship (28 combinations tested). Absence and presence of common antigens in immunodiffusion test could be regarded as an indication of resistant (incompatible) and susceptible (compatible) reactions, respectively. The absence or failure to detect common antigens may be due to several reasons. The method of extraction of antigens as well as the age of plant tissues and culture of microbes have a marked influence on the yield of antigenic substance and may account for the failure to detect (DeVay and Adler, 1976). Besides, avirulence of pathogen, resistance of host, or low titer of antigen/antiserum of host/pathogen or due to any other technical error. The degree of susceptibility or resistance of a cultivar or virulence or avirulence of a pathogen cannot always be accurately determined on the basis of the results of immunodiffusion or immunoelectrophoretic tests because in a number of cases no correlation could be established between the two. The common antigen concept could be related with parasitism to some extent but not with pathogenicity. This view was also expressed earlier by a group of workers. It is necessary to mention that a threshold titer of both host and pathogen antigens is essential for compatible interaction. The examples cited above clearly indicate that host–pathogen antigens have a role in disease reaction. On the contrary, PA accumulation/production is also involved in disease reaction of plants. Therefore, it is not unnatural to speculate that a relationship exists among them. Both factors should be

Table 1 Presence/Absence of Common Antigenic Determinants Between Compatible Hosts and Parasites

Host	Parasite	Presence/ absence of common antigens	Ref.
1	2	3	4
Arachis hypogea (Groundnut)	*Macrophomina phaseolina*	+	Purkayastha and Ghosal (1987)
Avena sativa	*Ophiobolus graminis*	+	Abbot (1973)
Coffea arabica (Coffee)	*Hemileia vastatrix*	+	Alba et al. (1983)
Citrullus vulgaris (Watermelon)	*Fusarium semitectum*	+	Abd-El-Rehim et al. (1971)
Corchorus capsularis (Jute)	*Colletotrichum corchori*	+	Bhattacharyya and Purkayastha (1985)
Glycine max (Soybean)	*Macrophomina phaseolina*	+	Chakraborty and Purkayastha (1983)
	Colletotrichum dematium var. *truncata*	+	Purkayastha and Banerjee (1990)
	Myrothecium roridum	+	Ghosh and Purkayastha (1990)
Gossypium hirsutum (Cotton)	*Verticillium dahliae*	+	Charudattan and De Vay (1972)
	Fusarium oxysporum f. sp. *vasinfectum*	+	Venkataraman et al. (1973) Abd-El-Rehim et al. (1988)
	Fusarium solani	+	-do-
	Xanthomonas malvacearum	+	DeVay et al. (1967)
Helianthus annuus (Sunflower)	*Agrobacterium tumefaciens*	+	DeVay et al. (1970)
Ipomoea batatus (Sweet potato)	*Ceratocystis fimbriata*	+	DeVay et al. (1967)
Linum usitatissimum (Flax)	*Melampsora lini*	+	Doubly et al. (1960)

Table 1 Continued

Host	Parasite	Presence/ absence of common antigens	Ref.
1	2	3	4
Medicago sativa (Alfalfa)	*Corynebacterium insidiosum*	−	Carroll et al. (1972)
Oryza sativa (Rice)	*Acrocylindium oryzae* (=*Sarocladium oryzae*)	+	Purkayastha and Ghosal (1985)
Triticum aestivum (Wheat)	*Puccinia graminis*	−	Johnson (1962)
Zea mays (Maize)	*Ustilago maydis*	+	Wimalajeewa and DeVay (1971)
Solanum tuberosum (Potato)	*Phytophthora infestans*	+	Alba and DeVay (1985)
Hordeum aestivum (Barley)	*Erysiphe graminis* f. sp. *hordei*	+	Heide and Smedegaard Petersen (1985)
Lycopersicum esculentum (Tomato)	*Phytophthora infestans*	+	Palmerly and Callow (1978)
Trifolium repens (Clover)	*Rhizobium trifollii*	+	Dazzo and Hubbek (1975)
Cajanus cajan (Pigeon pea)	*Fusarium udum*	+	Purkayastha et al. (1991)
Gossypium arboreum (Cotton)	*Fusarium vasinfectum*	+	Kalyanasundaram et al. (1975)
Cotton roots	*Thielaviopsis basicola*	+	Guseva et al. (1979)
Abelmoschus esculentus (a member of cotton family)	*F. vasinfectum*	+	Balasubramanian and Kalyanasundaram (1978)

considered together for screening disease resistant/susceptible germplasm (Table 2).

There is evidence that alteration/regulation of specific host antigen(s) by chemicals may induce resistance in plants. For example, gibberellic acid (100 μg/ml) and sodium azide (100 μg/ml) altered antigenic pattern of a susceptible (to *Sarocladium oryzae*) rice cultivar Jaya (Ghoshal and Purkayastha, 1987); sodium azide (100 μg/ml) also changed the antigenic

Table 2 Rate of Phytoalexin Production/Accumulation and Presence/Absence of Cross-Reactive Antigens in Different Host–Parasite Systems

Parasite	Host cultivars	Reaction	Phyto-alexin μg/g fresh wt	Presence/absence of CRA[a]	Ref.
1	2	3	4	5	6
Acrocylindrium oryzae (= *Sarocladiun* oryzae)	*Rice* (leaf sheath)		*Momilactone* A		
	Cv. Mahsuri	Resistance	19.68	–	Purkayastha et al. (1983)
	Cv. Jaya	Susceptible	8.64	+	
Macrophomina phaseolina	*Soybean* (roots)		*Glyceollin*		
	Cv. UPSM-19	Resistance	421.00	–	Purkayastha and Chakraborty (1983)
	Cv. Soymax	Susceptible	265.00	+	
	Cv. R-184	Susceptible	279.00	+	
Colletotrichum dematium var. *truncata*	*Soybean* (leaves)		*Glyceollin*		
	Cv. UPSM-19	Resistance	176.76	–	Chattopadhyay (Bandyopadhyay) (1989)
	Cv. DS-178	Susceptible	71.10	+	
	Cv. PK-327	Susceptible	106.66	+	
Myrothecium roridum	*Soybean* (leaves)		*Glyceollin*		
	Cv. UPSM-19	Resistance	441.68	–	Ghosh (1990)
	Cv. DS-74-24-2	Susceptible	135.76	+	
M. phaseolina	*Groundnut* (root)		*Cis-3,5-dimethoxy stilbene*		
	Cv. RSHY-1	Resistance	32.66	–	Purkayastha (and Ghoshal (unpublished)
	Cv. TMV-2	Susceptible	10.56	+	

*CRA = Cross reactive antigens; + = CRA present; – = CRA not detected in immunodiffusion test.

pattern of a susceptible (to *M. phaseolina*) soybean cultivar Soymax (Chakraborty and Purkayastha, 1987). Recently it was reported that cloxacillin, an antibiotic, reduced anthracnose disease of soybean (cultivar Soymax) caused by *Colletotrichum dematium* var. *truncata* and altered the antigenic pattern of the susceptible host cultivar (Purkayastha and Banerjee, 1990). In another investigation, Purkayastha and Ghosh (1991) showed that cadmium chloride treatment caused disappearance of a CRA in soybean cultivar.

Higher production of PA and changes in antigenic pattern after chemical treatment of a susceptible host suggest that a relationship may exist among PA, host antigen, and disease resistance. The disappearance of particular host antigen(s) or proteins in treated plant could be due to the inactivation of a particular gene(s) which is accounted for the biosynthesis of missing proteins and that gene(s) may be a PA suppressor gene. Since several genes are responsible for PA production, inactivation of genes may cause lower production of PA. However, this observation is an important one and provides a very valuable basis for further investigations of a similar type of activity involving other host–parasite systems.

VII. EPILOG

In the past half century phytoalexin has been fully established as one of the major factors in plant disease resistance. Its application in plant protection, plant breeding, germplasm screening, and the study of biogenetic relationships among plants is now more or less known. But whether higher rate of production/accumulation of PA or inability of a parasite to degrade/detoxify PA is more important in plant disease resistance is not clear. Since PA detoxifying mechanisms of parasites differ in many respects, it is not unusual to find differential tolerance of pathogens to a PA. A specific gene which controls detoxification could be considered as a diagnostic character of that strain of a pathogen. Besides, pathogen's host range could also be determined using this diagnostic character (Van Etten et al., 1989).

Another important aspect of PA research is the elicitation of PA by biotic and abiotic agents along with identification and determination of the structure of elicitor. It is also not clearly understood as to how both biotic and abiotic agents elicit similar PA in a host plant. It is a matter of speculation that both may cause similar change in DNA sequence and consequently more or new mRNA or both could activate enzymes involved in PA biosynthesis. What are the factors affecting elicitor activity of a pathogen? How long does an elicitor remain active? Is an elicitor capable of eliciting similar amounts of PA in different plant tissues? If not, is it due to any difference in cell membrane receptors? These questions remain unanswered. A potent elicitor can activate host's resistance under certain conditions. Therefore, effective, stable, and nonphytotoxic elicitors should be identified for use in disease control. Apart from PA elicitors, genetic control of race–cultivar specificity and PA synthesis have been discussed earlier by several workers (Paxton, 1980; Dixon et al., 1983, 1992; Crute, 1985; Bailey, 1987). Gene-for-gene relationships have been demonstrated or suggested in at least 43 different plant–pathogen interactions (Gabriel and Rolfe, 1990). Both compatible and incompatible host–pathogen interactions depend on specific combination of host–pathogen genes. These genes may control both production of PA and host–pathogen antigens (proteins/enzymes). Because there is evidence that chemical induction of resistance in a susceptible cultivar sometimes alter antigenic pattern of host cultivar and also stimulates PA production. Hence it is expedient to identify the specific antigen(s) which disappears due to induction of resistance in susceptible host cultivar and to determine whether it has any relation with PA suppressor gene, if any, present in the susceptible host. Induction of disease resistance by alteration/regulation of specific host antigen(s) could be a novel approach to crop protection program. Recent evidence suggests that foreign PA in a transgenic plant also confers resistance to disease. It is not improbable that isolation of new PAs from many other plants in future may be more potent and nonphytotoxic like several other chemotherapeutants. However, PA research will continue until molecular mechanism of phytoimmunity is clearly known to the plant scientists.

REFERENCES

Abbott, L. K. (1973). Taxonomy and host specificity of *Ophiobolus graminis* Sacc., an application of electrophoretic and serological techniques. PhD thesis, Monash University, Clayton, Victoria, Australia, p. 200.

Abd-El-Rehim, M. A., Ibrahim, T. A., Michail, S. H., and Fadel, F. M. (1971). Serological and immunoelectrophoretical studies on resistant and susceptible watermelon varieties to *Fusarium semitectum* Berk and Rev. *Phytopathology Z. 71*:43–55.

Abd-El-Rehim, M. A., Abou-Taleb, E. M., and Johamy, A. (1988). Common antigen(s) in cotton to *Fusarium oxysporum* f. sp. *vasinfectum. Journal of Phytopathology 121*:217–223.

Akazawa, T., and Wada, K. (1961). Analytical study of ipomeamarone and chlorogenic acid alterations in sweet potato roots infected by *Ceratocystis fimbriata. Plant. Physiology 36*:139–141.

Alba, A. P. C., and DeVay, J. E. (1985). Detection of cross-reactive antigens between *Phytophthora infestans* (Mont.) de Bary and *Solanum* species by indirect enzyme linked immunosorbent assay. *Phytopathology Z. 112*:97–104.

Alba, A. P. C., Guzzo, S. D., Mahlow, M. F. P., and Moraes, W. B. C. (1983).

Common antigens in extracts of *Hemileia vastatrix* uredinospores and of *Coffea arabica* leaves and roots. *Fitopathologia Brasileira 8*:473–483.

Albersheim, P., and Anderson-Prouty, A. J. (1975). Carbohydrates, proteins, cell surfaces and the biochemistry of pathogenesis. *Annu. Rev. Plant Physiol. 26*: 31–52.

Albersheim, P., Valent, B. S., Hahn, M., Wade, M., and Cline, K. (1978). Elicitation and sites of formation of phytoalexins and induced resistance. 3rd Int. Congr. Plant Pathol., Munchen, Germany (Abstr.).

Albersheim, P., Darvill, A. G., Sharp, J. K., Davis, K. R., and Doares, S. H. (1986). Studies on the role of carbohydrates in host-microbe interactions. *Recognition in Microbe-Plant Symbiotic and Pathogenic Interactions* (Ben Lugtenberg, ed.), NATO AS1 Series, Vol. H4, Springer-Verlag, Heidelberg, pp. 1–13.

Amin, M., Kurosaki, F., and Nishi, A. (1986). Extracellular pectinolytic enzymes of fungi elicit phytoalexin accumulation in carrot suspension culture. *J. Gen. Microbiol. 132*:771–777.

Anderson, A. J. (1980a). Differences in biochemical compositions and elicitor activity of extra-cellular components produced by three races of fungal plant pathogen *Colletotrichum lindemuthianum. Can. J. Microbiol. 26*:1473–1479.

Anderson, A. J. (1980b). Studies on the structure and elicitor activity of fungal glucans. *Can. J. Bot. 58*:2343–2348.

Anderson-Prouty, A. J., and Albersheim, P. (1975). Host pathogen interactions. VIII. Isolation of a pathogen synthesized fraction rich in glucan that elicits a defense response in the pathogen's host. *Plant Physiol. 56*(2):286–291.

Ayers, A. R., Ebel, J., Finelli, F., Berger, N., and Albersheim, P. (1976a). Host–pathogen interactions. IX. Quantitative assays of elicitor activity and characterization of the elicitor present in the extracellular medium of cultures of *Phytophthora megasperma* var. *sojae. Plant Physiol. 57*:751–759.

Ayers, A. R., Ebel, J., Valent, B., and Albersheim, P. (1976b). Host–pathogen interactions. X. Fractionation and biological activity of an elicitor isolated from the mycelial walls of *Phytophthora megasperma* var. *sojae. Plant Physiol. 57*:760–765.

Ayers, A. R., Valent, B., Ebel, J., and Albersheim, P. (1976c). Host–pathogen interactions XI. Composition and structure of wall-released elicitor fractions. *Plant Physiol. 57*:766–774.

Bailey, J. A. (1987). Phytoalexins: a genetic view of their significance. In *Genetics and Plant Pathogenesis* (P. R. Day and G. J. Jellies eds.), pp. 233–244. Blackwell Scientific Publications.

Bailey, J. A., and Mansfield, J. W. (1982). *Phytoalexins*, Blackie, London.

Bailey, J. A., Burden, R. S., Mynett, A., and Brown, C. (1977). Metabolism of phaseollin by *Septoria nodorum* and other non-pathogens of *Phaseolus vulgaris. Phytochemistry 16*:1541–1544.

Balasubramanian, R., and Kalyanasundarm, R. (1978). Electrophoretic and antigenic studies of soluble proteins in *Fusarium* wilt of cotton. *Proc. Indian Acad. Sci. Sect. B. 87*(8):223–230.

Bernard, N. (1911). Sur la function fungicide des bulbs d'ophrydes. *Ann. Sci. Nat.* (*Bot.*) *14*:221–234.

Bhattacharya, B., and Purkayastha, R. P. (1985). Occurrence of common antigens in jute and *Collectotrichum corchori. Curr. Sci. 54*:251–252.

Birch, A. J., Rickards, R. W., and Smith, H. (1958). The biosynthesis of gibberellic acid. *Proc. Chem. Soc.*, p. 192.

Boominathan, K., Gurujeyalakshmi, G., and Mahadevan, A. (1986). Detoxification mechanisms of plant pathogens. *Vistas in Plant Pathology* (A. Varma and J. P. Varma, eds.), Malhotra, New Delhi, pp. 45–69.

Bostock, R. M., Lane, R. A., and Kuć, J. A. (1982). Factors affecting the elicitation of sesquiterpenoid phytoalexin accumulation by eicosapentaenoic acid and arachidonic acid in potato. *Plant Physiol. 70*:1417–1424.

Brueggar, B. B., and Keen, N. T. (1979). Specific elicitors of glyceollin accumulation in the *Pseudomonas glycinea*-soybean host parasite system. *Physiol. Plant Pathol. 15*:43–51.

Bryan, I. B., Rathmell, W. G., and Friend, J. (1985). The role of lipid and non-lipid components of *Phytophthora infestans* in the elicitation of the hypersensitive response in potato tuber tissue. *Physiol. Plant Pathol. 26*:331.

Burden, R. S., Bailey, J. A., and Vincent, G. G. (1974). Metabolism of phaseollin by *Colletotrichum lindemuthianum. Phytochemistry 13*:1789–1791.

Cahill, D. M., and Ward, E. W. B. (1989). Effects of metalaxyl on elicitor activity, stimulation of glyceollin production and growth of sensitive and tolerant isolates of *Phytophthora megasperma* f. sp. *glycinea. Physiol. Mol. Plant. Pathol. 35*(2):97–112.

Callow, J. A. (1977). Recognition, resistance and the role of plant lectins in host-parasite interactions. *Adv. Bot. Res. 4*:1–49.

Carroll, R. B., Lukezic, F. L., and Roslyn, G. L. (1972). Absence of a common antigen relationship between *Corynebacterium insidiosum* and *Medicago sativa* as a factor in disease development. *Phytopathology 62*:1351–1360.

Cartwright, D., Langcake, P., Pryce, R. J., Leworthy, D. P., and Ride, J. P., (1977). Chemical activation of host defence mechanism as a basis of crop protection. *Nature 267*:511–513.

Chakraborty, B. N., and Purkayastha, R. P. (1983). Serological relationship between *Macrophomina phaseolina* and soybean cultivars. *Physiol. Plant Pathol. 23*:197–205.

Chakraborty, B. N., and Purkayastha, R. P. (1987). Alteration in glyceollin synthesis and antigenic patterns after chemical induction of resistance in soybean to *Macrophomina phaseolina. Can. J. Microbiol. 33*:835–840.

Chalova, L. T., Baramidze, V. G., Yurgonova, L. A., D'yakov, Yu, T., Ozeretskovskya, O. L., and Metlitskii, L. V. (1977). Isolation and characterization of the inducer of protective responses of the potato from the cytoplasmic contents of the *Phytophthora* pathogen. *Transl. Dokl. Akad. Nauk. SSSR 235*:1215.

Chalova, L. T., Ozeretskouskya, O. L., Yurganova, L. A., Baramidze, V. G., Protosenko, M. A., D'Yakov, Yu, T., and Metlitskii, L. V. (1976). Metabolites of phytopathogenic fungi as elicitors of defense reactions in plants. *Trousb. Dokl. Akad. SSSR 230*:722–725.

Chalutz, E., and Stahmann, M. A. (1969). Induction of pisatin by ethylene. *Phytopathology 59*:1972–1973.

Charudattan, R., and DeVay, J. E. (1972). Common antigens among varieties of *Gossypium hirsutum* and isolates of *Fusarium* and *Verticillium* sp. *Phytopathology 62*:230–234.

Chattopadhyay, Ratna (Mrs. Bandyopadhyay) (1989). Studies on phytoalexin and plant antigens in relation to resistance of soybean (*Glycine max* (L.) Merrill) cultivars to Anthracnose disease. PhD thesis, University of Calcutta, pp. 1–194.

Cheema, A. S., and Haard, N. F. (1978). Induction of rishitin and lubimin in potato tuber discs by non-specific elicitors and the influence of storage conditions. *Physiol. Plant Pathol. 13*:233–240.

Chester, K. S. (1933). The problem of acquired physiological immunity in plants. *Quart. Rev. Biol. 8*:129–154.

Cline, K., Wade, M., and Albersheim, P. (1978). Host pathogen interactions. XV. Fungal glucans which elicit phytoalexin accumulation in soybean also elicit the accumulation of phytoalexins in other plants. *Plant Physiol. 62*(6):918–921.

Cruickshank, I. A. M. (1963). Phytoalexins. *Annu. Rev. Phytopathol. 1*:351–374.

Cruickshank, I. A. M. (1978). A review of the role of phytoalexins in disease resistance mechanisms. *Pontificae Academiae Scientiarum Scripta Varia* (21) *IV*(2):1–61.

Cruickshank, I. A. M. (1980). Defenses triggered by the invader: chemical defenses. *Plant Disease: An Advanced Treatise*, Vol. 5 (J. G. Horsfall and E. B. Cowling, eds.), Academic Press, London, pp. 247–267.

Cruickshank, I. A. M., and Perrin, D. R. (1968). The isolation and partial characterization of monilicolin A, a polypeptide with phaseollin-inducing activity from *Monilinia fructicola. Life Sci. 7*:449–458.

Crute, I. R. (1985). The genetic bases of relationships between Microbial Parasites and their hosts. *Mechanisms of Resistance to Plant Diseases* (R. S. S. Fraser, ed.), Martinus Nijhoff, Dodrecht, pp. 80–142.

Crute, R., De Wit, P. J. G. M., and Wade, M. (1985). Mechanisms by which genetically controlled resistance and virulence influence host colonization by fungal and bacterial parasites. *Mechanisms of Resistance to Plant Diseases* (R. S. S. Fraser, ed.), Martinus Nijhoff, Boston, pp. 197–284.

Daniels, D. L., and Hadwiger, L. A. (1976). Pisatin-inducing components in filtrates of a virulent and avirulent *Fusarium solani* cultures. *Physiol. Plant Pathol. 8*:9–19.

Darvill, A. G., and Albersheim, P. (1984). Phytoalexins and their elicitors: a defense against microbial infection in plants. *Annu. Rev. Plant Physiol. 35*:243–275.

Davis, K. R., Darvill, A. C., Albersheim, P., and Dek, A. (1986). Host-pathogen interactions. XXIX. Oligo-galacturonides released from sodium polypectate by endogalacturonic acid lyase are elicitors of Phytoalexins in soybean. *Plant Physiol. 80*(2):568–577.

Dazzo, F. B., and D. Hubbek (1975). Cross-reactive antigens and lectins as determinants of symbiotic specificity in the *Rhizobium* clover association. *Appl. Microbiol. 30*:1017–1033.

Desjardins, A. E., Gardner, H. W., and Plattner, R. D. (1989). Detoxification of potato phytoalexin lubimin by *Gibberella pulicaris*. *Phytochemistry 28*:431–437.

DeVay, J. E., and Adler, H. F. (1976). Antigens common to hosts and parasites. *Annu. Rev. Microbiol. 30*:147–168.

DeVay, J. E., Schnathorst, W. C., and Foda, M. S. (1967). The dynamic role of molecular constituents in plant–parasite interactions (C. J. Mirocha and I. Uritani, eds.), Bruce, Minneapolis, pp. 313–328.

DeVay, J. E., Romani, R. J., Monadjem, A. M., and Etzler, M. (1970). Induction of Phytoprecipitins in sunflower gall tissue in response to infection by *Agrobacterium tumefaciens*. *Phytopathology 60*:1289 (Abstr.).

Deverall, B. J. (1972). Phytoalexins and disease resistance. *Proc. R. Soc. Lond. B. 181*:233–246.

Deverall, B. J. (1976). Current perspectives in research on phytoalexins. *Biochemical Aspects of Plant-Parasite Relationship* (J. Friend and D. R. Threlfall, eds.), Academic Press, New York, pp. 207–223.

Dineen, J. K. (1963). Antigenic relationships between host and parasite. *Nature 197*: 268–269.

Dixon, R. A. (1992). Characterised defence response genes. *Molecular Plant Pathology: A Practical Approach*, Vol. 1 (S. J. Gurr, M. J. McPherson, and D. J. Bowles, eds.), Oxford University Press, Oxford, p. 216.

Dixon, R. A., and J. L. Christopher (1979). Stimulation of de novo synthesis of L-phenylalanine ammonia lyase in relation to phytoalexin accumulation in *Colletotrichum lindemuthianum* elicitor-treated cell suspension cultures of French bean (*Phaseolus vulgaris*). *Biochem. Biophys. Acta 586*(3):453–463.

Dixon, R. A., Dey, P. M., and Lamb, C. J. (1983). Phytoalexins: enzymology and molecular biology. *Advances in Enzymology and Related Areas of Molecular Biology* (A. Meister, ed.), Wiley, New York, pp. 1–135.

Doubley, J. A., Flor, H. H., and Glagett, C. O. (1960). Relation of antigens of *Melampsora lini* and *Linum* to resistance and susceptibility. *Science 131*:229.

Ebel, J. (1986). Phytoalexin synthesis: the biochemical analysis of the induction process. *Annu. Rev. Phytopathol. 24*:235–264.

Ebel, J., Ayers, A. R., and Albersheim, P. (1976). Host pathogen interactions. XII. Response of suspension cultured soybean cells to the elicitor isolated from *Phytophthora megasperma* var. *sojae*, a fungal pathogen of soybean. *Plant Physiol. 57*:775–779.

Ellingboe, A. H. (1982). Genetical aspects of active defence. *Active Defense Mechanisms in Plants* (R. K. S. Wood, ed.), Plenum Press, New York, pp. 179–192.

Fedotova, T. I. (1948). Significance of individual proteins of seed in the manifestation of the resistance of plants to diseases. *Trudy Leninnger Inst., Zasheb. Rast. Sborn. 1*:61–71.

Flor, H. H. (1956). The complementary genic systems in flax and flax rust. *Adv. Genet. 8*:29–54.

Fraile, A., Garcia, A. F., and Sagasta, E. M. (1980). Phytoalexin accumulation in bean (*Phaseolus vulgaris*) after infection with *Botrytis cinerea* and treated with mercuric chloride. *Physiol. Plant Pathol. 16*:9–18.

Fuchs, A., De Vries, F. W., and Sanz, M. P. (1980). The mechanism of pisatin degradation by *Fusarium oxysporum* f. sp. *pisi*. *Physiol. Plant Pathol. 16*:114–133.

Gabriel, D. W., and Rolfe, B. G. (1990). Working models of specific recognition in Plant-microbe interactions. *Annu. Rev. Phytopathol. 28*:365–391.

Gabriel, D. W., Loschke, D. C., and Rolfe, B. G. (1988). Gene-for-gene recognition: the ion channel defense model. *Molecular Genetics of Plant-Microbe Interaction* (R. Palacios, D. P. S. Verma, and M. N. St. Paul, eds.), APS Press, p. 314.

Gaumann, E., and Kern, H. (1959a). Uber die isolierung and den chemischen Nadiweis des orchinobs. *Phytopathol. Z. 35*:347–356.

Gaumann, E., and Kern, H. (1959b). Uber chiemische Abwehrreaktionenbei orchideen. *Phytopath. Z. 36*:1–26.

Ghosh, Sumita (1990). Studies on host–parasite interactions with special reference to *Myrothecium* leaf spot disease of soybean. PhD thesis, University of Calcutta, p. 222.

Ghosh, S., and Purkayastha, R. P. (1988). Immunoserological studies on soybean–*Myrothecium* interactions, 75th Session of the Indian Science Congress, Pune (Abstr.).

Ghosh, Sumita, and Purkayastha, R. P. (1990). Analysis of host-parasite cross reactive antigens in relation to *Myrothecium*-infection of soybean. *Indian J. Exp. Biol. 28*:1–5.

Ghosh, Sumita, and Purkayastha, R. P. (1992). Elicitation of glyceollin by natural products and metallic salts. *Int. J. Trop. Plant Dis. 10*:99–108.

Ghoshal, A., and Purkayastha, R. P. (1984). Elicitation of momilactone by gibberellin in rice. *Curr. Sci. 53*:506–507.

Ghoshal, A., and Purkayastha, R. P. (1987). Biochemical responses of rice (*Oryza sativa* L.) leaves to some abiotic elicitors of phytoalexin. *Indian J. Exp. Biol. 25*:395–399.

Guseva, N. N., Gromova, B. B., and Lanlas, E. S. (1979). Cross-reacting antigens of pathogens and their host plants. *Acta Phytopathol. Acad. Scient. Hung. 14*: 449.

Hadwiger, L. A., Jafri, A., Von Broembseu, S., and Eddy, R., Jr. (1971). Mode of pisatin induction, increased template activity and dye-binding capacity of chromatin isolated from polypeptide treated pea pods. *Plant Physiol. 53*:52–63.

Hadwiger, L. A., and Schwochau, M. E. (1971). Ultra-violet light-induced formation of pisatin and phenylalanine ammonia lyase. *Plant Physiol. 47*:346–351.

Hadwiger, L. A., Sander, C., Eddyvean, J., and Ralston, J. (1976). Sodium-azide-induced mutants of peas that accumulate pisatin. *Phytopathology 66*:629–630.

Hahn, M. G., Bonhoff, A., and Grisebach, H. (1985). Quantitative localisation of the phytoalexin glyceollin in relation to fungal hyphae in soybean roots infected with *Phytophthora megasperma* f. sp. *glycinea*. *Plant Physiol. 77*:591–601.

Harborne, J. B., and Ingham, J. L. (1978). *Biochemical Aspects of Plant and Animal Coevolution*. Academic Press, London, pp. 343–405.

Henfling, J. W. D. M., Bostock, R. M., and Kuć, J. (1980). Cell walls of *Phytophthora infestans* contain an elicitor of terpene accumulation in potato tubers. *Phytopathology 70*(8):772–776.

Heide, M., and Smedegaard-Petersen, V. (1985). Common antigens between barley powdery mildew and their relation to resistance and susceptibility. *Can. J. Plant Pathol. 7*(4):341–346.

Hein, R., Reif, H. J., Krause, E., Langebartels, R., Kindl, H., Vornam, B., Wiese, W. Schmetzer, E., Schreier, P. H., Stocker, R. H., and Stenzel, K. (1993). Disease resistance results from foreign phytoalexin expression in a novel plant. *Nature 361*:153–156.

Higgins, V. J., Stoessl, A. and Heath, M. C. (1974). Conversion of phaseollin to phaseollin-isoflavone by *Stemphyllium botryosum*. *Phytopathology 64*:105–107.

Hiura, M. (1943). Studies in storage and rot of sweet potato. *Sci. Rep. Gifu Agric. Coll. Jpn. 50*:1–5.

Ingham, J. L. (1972). Phytoalexins and other natural products as factors in plant disease resistance. *Bot. Rev. 38*:343–424.

Ingham, J. L. (1973). Disease resistance in higher plants. The concept of pre-infectional and post-infectional resistance. *Phytopathol. Z. 78*:314–335.

Ingham, J. L. (1982). Phytoalexins from the Leguminosae. *Phytoalexins* (John A. Bailey and John W. Mansfield, eds.), Blackie, London, pp. 21–80.

Johnson, R. (1962). Serological studies on wheat and wheat stem rust. MS thesis, University of Saskatchewan, Saskatoonsask, Canada, p. 57.

Johnson, G., Maag, D. D., Johnson, D. K., and Thomas, R. D. (1976). The possible role of phytoalexins in the resistance of sugarbeet (*Beta vulgaris*) to *Cercospora beticola*. *Physiol. Plant Pathol. 8*:225–230.

Kalyanasundaram, R., Lakshminarashimam, C., and Venkataraman, S. (1975). Common antigens in host–parasite relationship. *Curr. Sci. 44*(2):55–56.

Keen, N. T. (1975). Specific elicitors of plant phytoalexin production: Determinants of race specificity in pathogens. *Science 187*:74–75.

Keen, N. T. (1981). Evaluation of the role of phytoalexins. *Plant Disease Control* (R. C. Staples and Gary H. Joenniesson, eds.), Wiley, New York, pp. 155–157.

Keen, N. T. (1982). Specific recognition in gene-for-gene host–parasite systems. *Advances in Plant Pathology*, Vol. 1 (P. H. Williams and D. Ingram, eds.), Academic Press, New York, pp. 35–81.

Keen, N. T., and Brueggar, B. (1977). Phytoalexins and chemicals that elicit their production in plants. *ASC Symp. Ser. 62*:1–26.

Keen, N. T., Sims, J. J., Erwin, D. C., Rice, E., and J. E. Partridge (1971). 6-α-Hydroxyphaseollin, an antifungal chemical induced in soybean hypocotyls by *Phytophthora megasperma* var. *sojae*. *Phytopathology 61*:1084–1089.

Keen, N. T., Yoshikawa, M., and Wang, M. C. (1983). Phytoalexin elicitor activity of carbohydrate from *Phytophthora megasperma* f. sp. *glycinea* and other sources. *Plant Physiol. 71*(3):466–471.

Kessmann, H., and Barz, W. (1986). Elicitation and suppression of phytoalexin and isoflavone accumulation in cotyledons of *Cicer arietinum* L. as caused by wounding and by polymeric compounds from the fungus *Ascochyta rabiei*. *J. Phytopathol.* *117*(4):321.

Kessmann, H., and Barz, W. (1987). Accumulation of isoflavones and pterocarpan phytoalexins in cell suspension cultures of different cultivars of chickpea (*Cicer arietinum*). *Plant Cell Rep.* *6*(1):55.

Klarman, W. L. (1968). The importance of phytoalexins in determining resistance of soybeans to three isolates of *Phytophthora*. *Neth. J. Plant Pathol. Suppl. I*: 171–175.

Klarman, W. L., and Gerdemann, J. W. (1963). Resistance of soybeans to three *Phytophthora* species due to the production of a phytoalexin. *Phytopathology* *53*:1317–1320.

Kubota, T., and Matsuura, T. (1953). Chemical studies on the black rot disease of sweet potato. *J. Chem. Soc. Jpn.* (Pure Chem Sect) *74*:248–251.

Kuć, J. (1972). Phytoalexins. *Annu. Rev. Phytopathol.* *10*:207–232.

Kuć, J. (1976). Phytoalexins. *Encyclopedia of Plant Physiology, Vol. 4, Physiological Plant Pathology* (R. Heitefuss and P. H. Williams, eds.), Springer-Verlag, Berlin, pp. 637–652.

Kuć, J. (1987). Plant immunization and its applicability for disease control. *Innovative Approaches to Plant Disease Control* (Ilan Chet, ed.), John Wiley and Sons, New York, pp. 255–274.

Kuć, J., and Rash, S. J. (1985). Phytoalexins. *Arch. Biochem. Biophys.* *232*:455–472.

Lyon, G. D., and Albersheim, P. (1982). Host pathogen interactions. 21. Extraction of a host-labile elicitor of phytoalexin accumulation from frozen soybean (*Glycine max*, cultivar Wayne) stems. *Plant Physiol.* *70*(2):406–409.

Mahadevan, A. (1991). *Post-infectional Defence Mechanisms*. Today and Tomorrow's Printers and Publishers, New Delhi.

Matthews, D. E., and Van Etten, H. D. (1983). Detoxification of the phytoalexin pisatin by a fungal cytochrome. *Arch. Biochem. Biophys.* *224*(2):494–505.

McIntosh, S. F., Matthews, D. E., and Van Etten, H. D. (1989). Two additional genes for pisatin demethylation and their relationship to the pathogenicity of *Nectria haematococca* on pea. *Annu. Rev. Phytopathol.* *27*:143–164.

Mayama, S., Tani, T., Ueno, T., Midland, S. L., Sims, J. J., and Keen, N. T. (1986). The purification of victorin and its phytoalexin elicitor activity in oat leaves. *Physiol. Mol. Plant Pathol.* *29*(1):1–18.

Metlitskii, L. V., Ozeretskovskya, O. L., Vasyukova, N. I., Sabirdina, M. S., and Chalenko, G. I. (1984). Substances in the cell walls of *Phytophthora infestans* as inducers and suppressor of defence reactions in potato. *Dokl. Akad. Nauk. SSSR* *274*(4):1020.

Müller, K. O., and Börger, H. (1940). Experimentelle Untersuchungen Uber die *Phytophthora* Resistenz der Kartoffel. *Arb. Biol. Reichsanst.* Land-Forstwirtsch, Berlin-Dahlem *23*:189–231.

Müller, K. O. (1956). Einige einfache Versuchezum nachweis vow Phytoalexinen. *Phytopath. Z.* *27*:237–254.

Osman, S., and Moreau, R. (1985). Potato phytoalexin elicitors in *Phytophthora infestans* spore germination fluids. *Plant Sci.* *41*(3):205–209.

Palmerly, R. A., and Callow, J. A. (1978). Common antigens in extracts of *Phytophthora infestans* and potatoes. *Physiol. Plant Pathol. 12*:241–248.

Partridge, J. E., and Keen, N. T. (1976). Association of the phytoalexin kievitone with single resistance of cowpeas to *Phytophthora vignae. Phytopathology 66*: 426–429.

Paxton, J. (1980). A new working definition of the term "Phytoalexin." *Plant Dis. 64*:734.

Perrin, D. R., and Cruickshank, I. A. M. (1965). Studies on phytoalexins. VII. Chemical stimulation of pisatin formation in *Pisum sativum* L. *Aust. J. Biol. Sci. 18*:803–816.

Peterman, M. A. (1967). Relation of antigens in selected host-parasite systems of *Linum usitatissimum* and *Melampsora lini*. MS thesis, North Dakota State University, Fargo, p. 62.

Purkayastha, R. P. (1973). Phytoalexins: plant antigens and disease resistance. *Sci. Culture 39*:528–535.

Purkayastha, R. P. (1985). Phytoalexins. *Frontiers in Applied Microbiology*, Vol. 1. (K. G. Mukherji, N. C. Pathak, Ved Pal Singh, eds.), Lucknow Print House, pp. 363–407.

Purkayastha, R. P. (1986). Elicitors and elicitation of phytoalexins. *Vistas in Plant Pathology* (A. Varma and J. P. Verma, eds.), Malhotra, New Delhi, pp. 25–44.

Purkayastha, R. P., and Banerjee, R. (1990). Immunoserological studies on cloxacillin-induced resistance of soybean against anthracnose. Zeitschrift fur Pflanzenkrankheiten und Pflanzenschutz. *J. Plant Dis. Prot. 97*(4):349–359.

Purkayastha, R. P., and Chakraborty, B. N. (1983). Immunoelectrophoretic analysis of plant antigens in relation to biosynthesis of phytoalexin and disease resistance of soybean. *Trop. Plant Sci. Res. 1*(1):89–96.

Purkayastha, R. P., and Chakraborty, B. N. (1985). Induction of disease resistance and associated changes in *Macrophomina*-infected soybean. National Seminar on Advances in Mycology and Plant Pathology, Punjab University, Chandigarh (Abstr).

Purkayastha, R. P., and Deverall, B. J. (1964). A phytoalexin type of reaction in the *Botrytis*-infected leaves of bean (*Vicia faba* L.). *Nature 201*:938–939.

Purkayastha, R. P., and Deverall, B. J. (1965). The growth of *Botrytis fabae* and *B. cinerea* in leaves of bean (*Vicia faba* L.). *Ann. Appl. Biol. 56*:139–147.

Purkayastha, R. P., and Ghosh, S. (1983). Elicitation and inhibition of phytoalexin biosynthesis in *Myrothecium*-infected soybean leaves. *Indian J. Exp. Biol. 21*: 216–218.

Purkayastha, R. P., and Ghoshal, A. (1985). Analysis of cross-reactive antigens of *Acrocylindrium oryzae* and rice in relation to sheath rot disease. *Physiol. Plant Pathol. 27*:245–252.

Purkayastha, R. P., and Ghosal, A. (1987). Immunoserological studies on root rot of groundnut (*Arachis hypogea* L.). *Can. J. Microbiol. 33*:647–651.

Purkayastha, R. P., and Ghosh, Sumita (1992). Heavy metal salt inducing disease resistance and altering specific antigen of soybean leaves. *Int. J. Trop. Plant Dis. 10*:131–142.

Purkayastha, R. P., Ghoshal, A., and Biswas, S. (1983). Production of momilactone associated with resistance of rice cultivars to sheath rot disease. *Curr. Sci. 52*(3):131–132.

Purkayastha, R. P., Ghosal, A., Garai, M., and Ghosh, S. (1991). Cross-reactive antigens as determinants of susceptibility of pigeon pea to Fusarial wilt. In *Botanical Researches in India* (N. C. Aery and B. L. Chaudhury, eds.), Himanshu, Udaipur, pp. 508–513.

Ricci, P., and Rousse, G. (1983). Absence of induction of phytoalexins and resistance in carnation by the selective herbicide oxadiazone. *Acta Hort. 141*:109–113.

Rohwer, F., Fritzmeier, K. H., Scheel, D., and Hahlbrock, K. (1987). Biochemical reaction of different tissues of potato (*Solanum tuberosum*) to zoospores or elicitors from *Phytophthora infestans*. Accumulation of sesquiterpenoid phytoalexin. *Planta 170*(4):556–561.

Rubin, B., Penner, D., and Saettler, A. W. (1983). Induction of isoflavonoid production in *Phaseolus vulgaris* leaves by ozone, sulfur dioxide and herbicide stress. *Environ. Toxicol. Chem. 2*(3):295–306.

Sanz Platero, de M., and Fuchs, A. (1978). Degradation of pisatin, an antimicrobial compound produced by *Pisum sativum. Phytopathol. Mediterr. 17*:14–17.

Schwochau, M. E., and Hadwiger, L. A. (1968). Stimulation of pisatin production in *Pisum sativum* by Actinomycin D and other compounds. *Arch. Biochem. Biophys. 126*:731–733.

Sequiera, L. (1983). Mechanisms of induced resistance in plants. *Annu. Rev. Microbiol. 37*:51–79.

Shirashi, T., Oku, H., Yamashita, M., and Ouchi, S. (1978). Elicitor and suppressor of pisatin induction in spore germination fluid of pea pathogen *Mycosphaerella pinoides. Ann. Phytopathol. Soc. Jpn. 44*:659–665.

Stekoll, M., and West, C. A. (1978). Purification and properties of an elicitor of castor bean phytoalexin from culture filtrates of the fungus *Rhizopus stolonifer. Plant Physiol.* (Bethesda) *61*:38–45.

Stoessel, P., and Magnalato, D. (1983). Phytoalexins in *Phaseolus vulgaris* and *Glycine max* induced by chemical treatment, microbial contamination and fungal infection. *Experientia 39*:153–154.

Tepper, C. S., and Anderson, A. J. (1986). Two cultivars of bean display a differential response to extracellular components from *Colletotrichum lindemuthianum. Physiol. Mol. Plant Pathol. 29*(3):411–420.

Terekhova, V. A., Mustafa, M., and D'Yakov, Yu. T. (1980). Effect of metabolites of potato tubers on the induction of the phytoalexin, rishitin by the fungus *Phytophthora infestans. Biol. nauké* (*Mos.*) *2*:83–87.

Tjamos, E. C., and Smith, I. M. (1974). The role of phytoalexins in the resistance of tomato to *Verticillium* wilt. *Physiol. Plant Pathol. 4*:249–259.

Uehara, K. (1958a). On the production of phytoalexin by the host plant as a result of interaction between the rice plant and the blast fungus (*Pyricularia oryzae* Cav.). *Ann. Phytopathol. Soc. Jpn. 23*:127–130.

Uehara, K. (1958b). On some properties of phytoalexin produced as result of the interaction between *Pisum sativum* L. and *Ascochyta pisi* Lib. I. on the activity affected by ultra-violet irradiation and on some physico-chemical properties of Phytoalexin. *Ann. Phytopathol. Soc. Jpn. 23*:230–234.

Uehara, K. (1958c). On the phytoalexin of the soybean pool in reaction to *Fusarium* sp., the causal fungus of pod blight. I. Some experiments on the phytoalexin production as affected by host plant conditions and on the nature of the phytoalexin produced. *Ann. Phytopathol. Soc. Jpn. 23*:225–229.

Uehara, K. (1959). On the phytoalexin production of the soybean pod in reaction to *Fusarium* sp. the causal fungus of pod blight. *Ann. Phytopathol. Soc. Jpn. 24*: 224–228.

Uehara, K. (1963). On the production of phytoalexin by metallic salts. *Bull. Hiroshima Agr. Coll. 2*:41–44.

Van Etten, H. D., and Pueppke, S. G. (1976). Isoflavonoid phytoalexins. *Biochemical Aspects of Plant Parasite Relationships* (J. Friend and D. R. Threlfall, eds.), Academic Press, New York, pp. 230–289.

Van Etten, H. D., Matthews, P. S., Tegtmier, K. J., Dietert, M. F., and Stein, J. I. (1980). The association of pisatin tolerance and demethylation with virulence on pea in *Nectria haematococca. Physiol. Plant Pathol. 16*:257–268.

Van Etten, H. D., Matthews, D. E., and Smith, D. A. (1982). Metabolism of Phytoalexins. In *Phytoalexins* (J. A. Bailey and J. W. Mansfield, eds.), Blackie, Glasgow, pp. 181–217.

Van Etten, H. D., Matthews, D. E., and Matthews, P. S. (1989a). Phytoalexin detoxification: importance for pathogenicity and practical implications. *Annu. Rev. Phytopathol. 27*:143–164.

Van Etten, H., Matthews, D., Matthews, P., Miao, V., Moloney, A., and Straney, D. (1989b). A family of genes for phytoalexin detoxification in the plant pathogen *Nectria haematococca. Molecules in Plants and Plant-Microbe Interactions* (B. Lugtenberg, ed.), Springer-Verlag, Berlin, pp. 219–228.

Van den Heuvel, Van Etten, H. D., Serum, J. W., Coffen, D. L., and Williams, T. H. (1974). Identification of α-hydroxyphaseollone, a phaseollin metabolite produced by *Fusarium solani. Phytochemistry 13*:1129–1131.

Vaziri, A., Keen, N. T., and Erwin, D. C. (1981). Correlation of medicarpin production with resistant to *Phytophthora megasperma* f. sp. medicoginin alfalfa (*Medicago sativa*) seedlings. *Phytopathology 11*(2):1235–1238.

Venkataraman, S., Lakshimnarasimhan, C., and Kalyansundaram, R. (1973). Antigenic determinants of host pathogen specificity, 2nd Int. Congr. Plant Pathol. 958 (Abstr.).

Wallace, J. M. (1940). Evidence of passive immunization of tobacco, *Nicotiana tabacum* from the virus of curly top. *Phytopathology 30*:673–679.

Ward, E. W. B., and Stoessl, A. (1972). Post-infectional inhibitors from plants. III. Detoxification of capsidiol, an antifungal compound from peppers. *Phytopathology 62*:1186–1187.

Ward, E. W. B., and Stoessl, A. (1977). Phytoalexins from potatoes: evidence for the conversion of lubimin to 15-dihydrolubimin by fungi. *Phytopathology 67*: 468–471.

Ward, H. M. (1905). Recent researches on the parasitism of fungi. *Ann. Bot. 19*:1–54.

Watson, D. G., and Brooks, C. J. W. (1984). Formation of capsidiol in *Capsicum annuum* fruits in response to non-specific elicitors. *Physiol. Plant Pathol. 24*: 331–337.

Weigard, F., Koster, J., Wettziers, H. C., and Barz, W. (1986). Accumulation of phytoalexins and isoflavone glucosides in a resistant and a susceptible cultivar of *Cicer arietinum* during infection with *Ascochyta rabiei. J. Phytopathol. 115*(3):214–221.

Wimalajeewa, D. L. S., and De Vay, J. E. (1971). The occurrence and characterization of a common antigen relationship between *Ustilago maydis* and *Zea mays. Physiol. Plant Pathol. 1*:523–535.

Wingard, S. A. (1928). Hosts and symptoms of ring spot: a virus disease of plants. *J. Agric. Res. 37*(3):127–153.

Wood, R. K. S. (1986). Elicitors—What of? *Vistas in Plant Pathology* (A. Varma and J. P. Verma, eds.), Malhotra, New Delhi, pp. 13–24.

Woodward, S., and Pegg, G. F. (1986). Rishitin accumulation elicited in resistant and susceptible isolines of tomato by mycelial extracts and filtrates from cultures of *Verticillum albo-atrum. Physiol. Mol. Plant Pathol. 29*(3):337–347.

Yamoto, Y., Oku, H., Shirashi, T., Ouchi, S., and Koshizawa, K. (1986). Nonspecific induction of pisatin and local resistance in Pea leaves by elicitors from *Mycosphaerella pinoides, M. melonis* and *M. lignilicola* and the effect of suppressor from *M. pinoids* Jr. *Phytopathology 117*(2):136.

Yoshikawa, M. (1983). Macromolecules, recognition and the triggering of resistance. *Biochemical Plant Pathology* (J. A. Callow, ed.), John Wiley & Sons, New York, pp. 267–298.

Zook, M. N., and Kuć, J. A. (1987). Differences in phytoalexin elicitation by *Phytophthora infestans* and *Helminthosporium carbonum* in potato. *Phytopathology 77*(8):1217.

Zook, M. N., Rush, J. S., and Kuć, J. A. (1987). A role for Ca^{2+} in elicitation of rishitin and lubimin accumulation in potato tuber tissue. *Plant Physiol. 84*(2): 520–525.

2
Phytoalexins and Host Specificity in Plant Diseases

Hachiro Oku and Tomonori Shiraishi
College of Agriculture, Okayama University, Okayama, Japan

I. INTRODUCTION

One of the main interests in host–parasite interaction has been the role of phytoalexins in plant diseases for about five decades. Reviews by Ingham (1970), Cruickshank et al. (1971), and Kuć (1972) show that the role of phytoalexins in disease resistance is rather diverse depending on host–parasite combinations. However, the roles were found to be more important than generally believed when we conducted precise and careful experiments. Here we describe our results and discuss on the role of phytoalexins in host–parasite specificity in plant diseases.

In addition, taking the rate of phytoalexin accumulation as an indication of a defense reaction, we found that some pathogenic fungi secrete factors which suppress defense reactions of hosts. The factors, which are called *suppressors*, are demonstrated to be responsible for the host-specific pathogenicity of these fungal plant pathogens.

The mechanisms of elicitation and suppression of phytoalexin biosynthesis and associated defense reactions are also described in this chapter.

II. PHYTOALEXINS IN OBLIGATE PARASITISM

It had been widely believed that phytoalexins may not be involved in the host–parasite relation of obligate parasitic diseases. This concept comes from the idea that obligate parasitism is a kind of symbiosis and substances harmful to parasites, such as phytoalexins, should not be produced by host plants.

However, the possibility of phytoalexin formation in downy mildew of tobacco was provided by Cruickshank and Mandryk (1960) and Shepherd and Mandryk (1962, 1963). Leath and Rowell (1970) suggested that a phytoalexin might be responsible for the inability of *Puccinia graminis* to infect corn leaves, although they did not show any direct evidence for its production. Bailey and Ingham (1971) reported that phaseollin was detected in bean leaves inoculated with *Uromyces appendiculatus* only when the leaves responded with cellular browning, but they did not describe the role of phaseollin in the resistance of bean against rust fungus. Klarman and Hammerschlag (1972) extracted hydroxyphaseollin from soybean leaves which had been infected with tobacco necrosis virus.

Oku et al. (1975c) first presented evidence that phytoalexin may participate in resistance of barley to powdery mildew disease and then proved the role of pisatin in resistance of pea against powdery mildew fungus (Oku et al., 1975a). Later on, many reports appeared indicating that phytoalexin may play some roles in resistance of obligate parasitic diseases. We will describe our results on barley and pea powdery mildew diseases.

A. Phytoalexin Activity in Barley Powdery Mildew

Previously, Ouchi et al. (1974a,b) found that the preliminary inoculation with an incompatible race of *Erysiphe graminis* induccd resistance to a primarily compatible race and that the preliminary inoculation with the compatible race rendered barley leaves accessible to primarily incompatible races. On the basis of these results we proposed the hypothesis that the primary recognition of invading microbes as compatible or incompatible may be controlled by an all-or-nothing type of gene action (Ouchi et al., 1974b). During the analysis of these race–cultivar interactions, we found that the growth of the secondary hyphae of a compatible race is markedly inhibited on the resistance-induced barley leaves. This fact strongly suggested the participation of phytoalexin-like substance in the resistance to powdery mildew disease. Therefore, we conducted experiments and elucidated that an antifungal activity was induced in barley leaves by infection with powdery mildew fungi (Oku et al., 1975b,c).

The primary leaves of 8- to 10-day-old seedlings of barley cultivars having different compatibilities with race 1, Kobinkatagi, HES4, Nos. 21 and 241 were densely inoculated with fresh conidia of *E. graminis* f. sp. *hordei* race 1 with a soft hair brush. The inoculum density was adjusted to give more than five conidia per epidermal cell. The inoculated plants were incubated at 20°C for an appropriate time and the leaf surface was rubbed with a wet cotton ball to remove the fungal material as much as possible in order to negate a possible spore density effect. The adaxial surface of the inoculated leaves was injured slightly by pressing gently with the tip of a glass capillary. Then the droplets of deionized water were placed on the injured parts of the leaves and incubated 20°C for 4 hr to extract the phytoalexin. The droplets were then collected by a glass capillary. The droplets similarly collected from noninoculated leaves served as control. These diffusates (0.025 ml) were placed in small agarose blocks (25 mm^2 × 2 mm) on glass slides and allowed to diffuse into agarose for 3 hr at room temperature. The antifungal activity was tested by placing the conidia of *E. graminis hordei* race 1 on diffusate-impregnated agarose blocks. After incubation at 20°C for 16 hr, the percentage of germination was determined on the basis of conidia germinated with a characteristic germ tube. Percent inhibition was calculated as a ratio of germination in the diffusate from inoculated leaves to that in the exudates from noninoculated control.

The results of the time course study for the production of phytoalexin activity are indicated in Fig. 1.

Thus, the phytoalexin activity was found in barley powdery mildew disease. The isolation and characterization of phytoalexin from powdery mildew barley has not yet been accomplished, but the active factor responsible for this antifungal activity should be termed as phytoalexin in Müller's definition (Müller, 1956) because it could not be detected in noninfected healthy leaves.

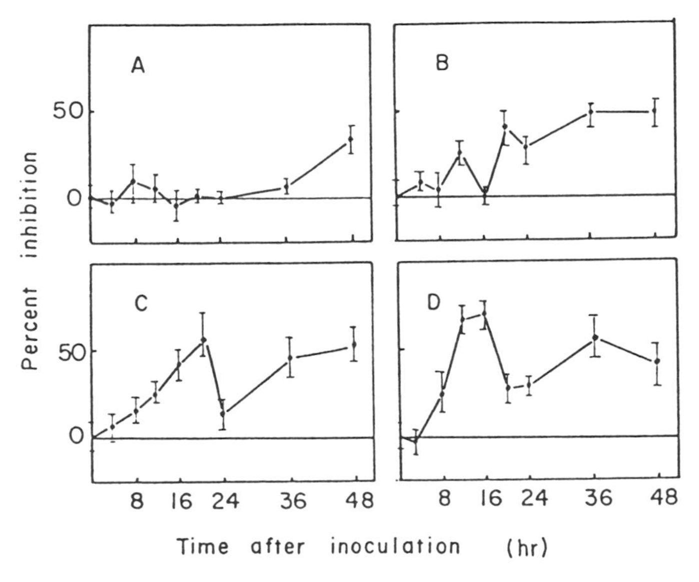

Figure 1 Comparison of phytoalexin induction in some cultivar–race interactions of barley powdery mildew disease. A = Kobinkatagi-race 1 (reaction type 4); B = No. 21-race 1 (reaction type 3); C = No. 241-race 1 (reaction type 2); D = H. E. S. 4-race 1 (reaction type 0). Percent inhibition was calculated as follows:

$$100 - \left(\frac{\%\text{ germination in exudates from inoculated leaves}}{\%\text{ germination in exudates from noninoculated leaves}} \times 100\right)$$

B. Pisatin in Powdery Mildewed Pea

Further evidence for the production of phytoalexin in obligate parasitic disease was provided by Oku et al. (1975a). They densely inoculated young pea seedlings grown in a growth chamber with conidia of *Erysiphe pisi* and extracted an antifungal principle from diseased seedlings. The active substance was purified in crystalline form and identified as pisatin by physicochemical analysis. Oku et al. obtained 6 mg of purified pisatin from 40 g fresh wt of powdery mildewed pea seedlings.

III. PRODUCTIVITY OF TWO PHASES OF PHYTOALEXINS AND THEIR RESPECTIVE ROLES IN DISEASE RESISTANCE

As evident from Fig. 1, phytoalexin activity was detectable 8–20 hr after inoculation in incompatible combinations of barley powdery mildew disease. The activity was proportional to the degree of incompatibility between host cultivar and parasite. We considered the activity at this stage to be first-phase phytoalexin. Phytoalexin activity first became detectable 8 hr after inoculation, coinciding with the time taken to induce resistance and also with the time of fungal penetration into host epidermal cell. Phytoalexin activity was also detected in the later stages of the pathogenesis of barley powdery mildew around the fungal colonies formed on the leaves of the compatible host (Table 1). Phytoalexin activity at this stage was called second-phase phytoalexin.

The first-phase phytoalexin seems to play a role in determining host–parasite specificity because the activity was produced in parallel with the degree of incompatibility. This fact also indicates that first-phase phytoalexin might be produced as a result of recognition by the host cell of a

Table 1 Accumulation of Phytoalexin in Tissue Surrounding Fungal Colonies of *Erysiphe gaminis* Formed on Compatible Cultivar[a]

Age of colonies (days)	Percent inhibition[b]
0	0[c]
15	17
25	41

[a]Tested with Kobinkatagi–Race 1 combination.
[b]Estimated from germ tube growth of *Cochliobolus miyabeanus*.
[c]Significantly different at 95% level of confidence.

parasite to be rejected. If so, the production of first-phase phytoalexin should be suppressed in accessibility-induced leaves and increased resistance–induced barley leaves.

A barley cultivar, Kobinkatagi, was first inoculated with a compatible (*E. graminis hordei* race 1) or incompatible race (*E. graminis tritici* race t2), incubated for 48 hr, and the inoculum removed and then inoculated with a second race. The effect of the inoculation on the phytoalexin production in response to the second fungus was compared (Fig. 2) (Oku et al., 1975b).

The production of first-phase phytoalexin in incompatible interaction was suppressed when the barley leaves were primarily inoculated with a compatible race. Conversely, the preliminary inoculation with an incompatible race conditioned the leaves to produce the same level of phytoalexin on subsequent inoculation regardless of the compatibility of the challenger race.

Similarly, two phases of phytoalexin production were found in pisatin accumulation when inoculated with compatible or incompatible pathogens on pea leaves (Oku et al., 1975a; Shiraishi et al., 1977). That is, pisatin was detectable 12–15 hr after inoculation with the incompatible powdery mildew fungus *E. graminis*. The inoculation of resistant pea cultivar to powdery mildew disease (Resistant Stratagem, with *E. pisi* or *E. graminis*) also induced first-phase pisatin.

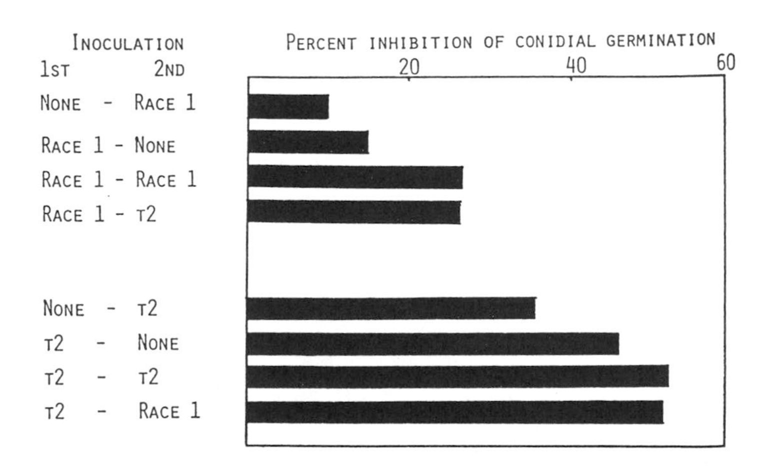

Figure 2 Effect of accessibility and resistance induction on phytoalexin production in barley leaves. Accessibility (or resistance) was induced in barley leaves (cv. Kobinkatagi) by inoculation with race 1 (or t2), and phytoalexin production was estimated 12 hr after second inoculation.

In contrast, the inoculation of pea leaves with compatible pathogens, *E. pisi* or *Mycosphaerella pinodes*, did not elicit first-phase pisatin but induced second-phase pisatin (21 hr after inoculation).

Here the absence of first-phase pisatin in pea leaves seems to baffle our understanding, if we consider that pisatin is only an antifungal substance, because both pathogens *E. pisi* and *M. pinodes* are highly tolerant to pisatin (Oku et al., 1975a; Shiraishi et al., 1978a). In other words, these fungi are unnecessary to suppress the pisatin production because their growth is not inhibited by pisatin.

These results lead us to an idea that pisatin has another function besides its antifungal activity, and we examined the effect of artificial administration of pisatin at early stages of infection by these fungi on the establishment of infection.

As indicated in Fig. 3, the infection of pea leaves by *E. pisi* was strongly inhibited when pisatin was administered within 15 hr after inoculation even at concentrations which do not exert any effect on conidial germination, i.e., 30 or 100 ppm. Administration of pisatin at 16 or 19 hr postinoculation however, did not give any significant effect on infection establishment by *E. pisi* (Oku et al., 1976).

Similarly, *M. pinodes* is highly tolerant to pisatin as assessed by the inhibitory activity to spore germination or germ tube elongation (ED_{50} = 500 ppm). This fungus, however, could not establish infection on pea if a very low concentration (less than 50 ppm) of pisatin was administered artificially into the spore inoculum. The perforation of cellophane sheet by germ tube of *M. pinodes* was also inhibited by the same concentration of pisatin that inhibited infection on the host, showing that infection-inhibiting activity of pisatin is rather a direct action to the pathogen than the host-mediated action (Shiraishi et al., 1978a).

These results obtained by experiments with *E. pisi* and *M. pinodes* suggest that if we direct our attention to the inhibitory effect of phytoalexins on the infection establishment of a pathogen, their role in determining host–parasite specificity should be much more important than has been considered (Christenson and Hadwiger, 1973; Király et al., 1972; Pueppke and VanEtten, 1974). The results also suggest that some substances which were synthesized during the infection process but have no antibiotic activity may play a significant part in determining host–parasite specificity. In this connection, Berard et al. (1972) reported that a diffusate from the incompatible interaction of bean anthracnose protected the bean plant from the compatible pathogen though the diffusate had no antibiotic activity. In fact, we detected a substance which has no antifungal activity but has infection-inhibiting activity in pea leaves 1 hr after treatment with elicitor from *M. pinodes*. This substance, an infection inhibitor in our terms, is

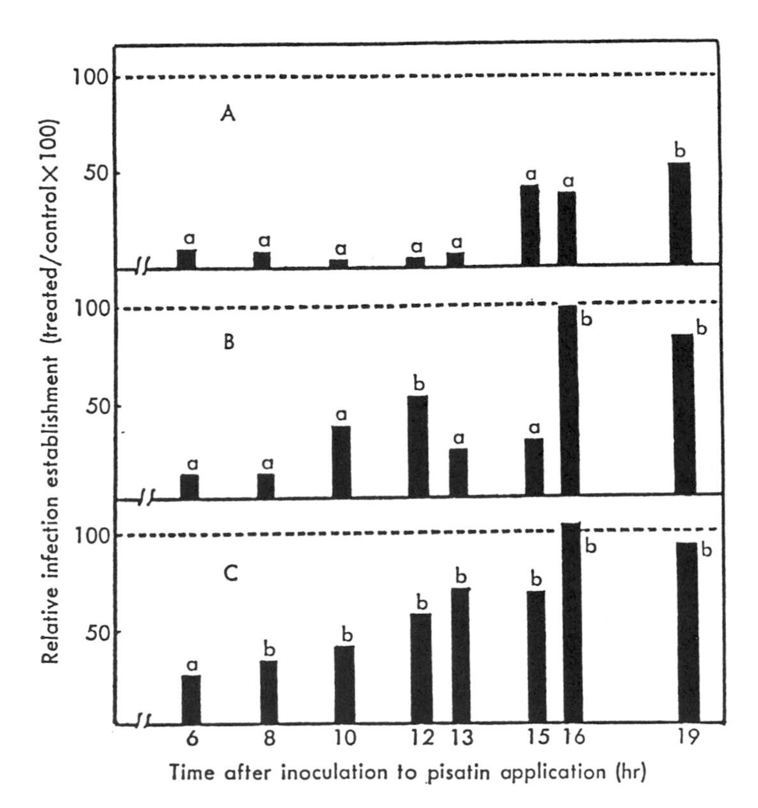

Figure 3 Effect of pisatin on infection establishment of *E. pisi* on pea leaves. After the lower epidermis of half leaves was removed, the opposite surface was inoculated with *E. pisi*, and the stripped mesophyll tissue was brought into direct contact with (A) 100 ppm, (B) 30 ppm, or (C) 10 ppm pisatin solution at the above time intervals. Infection frequency was estimated 48 hr after inoculation. (a) represents a significant difference and (b) a nonsignificant difference from control respectively at $p = 0.05$.

not yet be characterized but the molecular weight is 329 as determined by mass spectrography (Yamamoto et al., 1986).

The inoculation of the pathogenic fungi *E. pisi* and *M. pinodes* on pea leaves did not induce pisatin biosynthesis until the infection was completed, unlike the nonpathogens *E. graminis* and *Stemphyllium sarcinaeforme*, which induced pisatin to a detectable level by 12 hr after inoculation. These facts strongly suggest that the pathogenic fungi have an ability to suppress the first step of the defense reaction, hence pisatin production, in order to avoid the inhibitory action of pisatin on penetration.

Table 2 Antifungal Activity of Second-Phase Phytoalexin Against Several Races of *Erysiphe graminis*

Race	Percent inhibition[a]
Race 1	98.6[b]
Hr 74	94.5
Hr 4	97.0
E. graminis triciti t2	97.0

[a]Refer to Fig. 1.
[b]No significant difference at 95% level of confidence.

The role of second-phase phytoalexin may be diverse according to the sensitivity of pathogenic fungus against phytoalexin produced by the host plant.

Table 2 shows the sensitivity of several races of *E. graminis* to the phytoalexin solution obtained from the surrounding barley leaf tissue colonized by powdery mildew fungus. Thus, all races of *E. graminis* are similarly sensitive to barley phytoalexin. In these cases second-phase phytoalexin might possibly be responsible for the resistance against colony development. As a matter of fact, the colony of barley powdery mildew fungus usually grows very rapidly in the early stage of pathogenesis and gradually levels off as the incubation time is prolonged, finally stopping the growth at the later stage. The phytoalexin accumulation was most prominent at the tissues surrounding the colony that ceased to grow. In this connection, Pierre (1971) suggested that phaseollin and another unidentified phytoalexin induced in bean plants may play a role in restricting the size of the lesions produced in bean leaves. The same is true of medicarpin in alfalfa (Higgins, 1972) and also of rishitin in potato tubers (Sato et al., 1971).

In contrast, second-phase phytoalexins have nothing to do with the development of lesions formed by the pathogens which are tolerant to the host phytoalexin.

As described above, pea powdery mildew fungus is highly tolerant to pisatin. Therefore, once infection is established, the fungus colonizes freely on pea plant tissues, even though high levels of pisatin accumulate.

Heavy inoculation with *E. pisi* causes wilting of inoculated pea leaves. In inoculated leaves, approximately 300 μg/g fresh weight of pisatin accumulates 4 days after inoculation (Oku et al., 1975a), when wilting begins. This fact suggests that pisatin is the main cause of wilting because such a powerful toxin with a wilting effect cannot be produced by the pathogenic fungus which is a typical obligate parasite. In fact, 300 μg/ml of pisatin causes complete burst of isolated protoplast from mesophyll layers of pea leaves (Shiraishi et al., 1975). Thus, the main cause of wilting in pea leaves heavily infected with powdery mildew fungus may not be the fungus itself, but pisatin produced by the host leaves affecting the plasma membrane of the same host cells. In other words, the wilting by this disease may be said to be a self-destruction effect resulting from a defense mechanism.

IV. MECHANISM OF SUPPRESSION OF DEFENSE REACTION INCLUDING THE ACCUMULATION OF FIRST-PHASE PHYTOALEXIN

As described, the pathogenic fungus seemed to have some mechanisms to suppress defense reactions including the accumulation of first-phase phytoalexins. Therefore, we searched for substances which suppress the defense reaction of host using a pea–*M. pinodes* system.

Since first-phase phytoalexins accumulate at an early stage of pathogenesis, our search has been focused on the pycnospore germination fluid of the pathogenic fungus *M. pinodes*. As the result, we obtained both elicitor and suppressor of pisatin biosynthesis from spore germination fluid (Shiraishi et al., 1978b). That is, dialyzation or ultrafiltration of germination fluid gave an elicitor for pisatin synthesis in high molecular weight fraction and suppressor in low molecular weight fraction. The elicitor was found to be a polysaccharide (M_W ~70,000). At least two kinds of suppressors are contained in spore germination fluid and they are glycopeptides.

As shown in Table 3, suppressor counteracts the activity of elicitor

Table 3 Effect of Fractions Prepared from Spore Germination Fluid of *Mycosphaerella pinodes* on Pisatin Induction in Pea Leaves

Treatment with	Pisatin accumulated[a] (μg/g of fresh leaves)
Distilled water (control)	0
Concentrated germination fluid	0
High molecular weight fraction	27.2
Low molecular weight fraction	0
Mixture of low and high molecular weight fraction (1 : 1)	0

[a]Pisatin was determined after 24 hr incubation at 20°C.

Table 4 Correlation Between the Leguminous Host Range of *Mycosphaerella pinodes* and Induction of Susceptibility in Those Plants to *Alternaria alternata* by the Suppressor (F5) from *M. pinodes*

Legume species	Degree of colonization by		
	M. pinodes	*A. alternaria* (15B)	*A. alternaria* (15B) + F5
Arachis hypogaea	0	0	0
Glycine max	0–1	0	0
Lespedeza buergeri	2	0	2
Lotus corniculatus	0	0	0
L. bicolor	0	0	0
Medicago sativa	1	0	1
Millettia japonica	2	0	1
Pisum sativum	4	0	4
Trifolium pratense	1	0	1
T. repens	0	0	0
Vicia faba	0	0	0
Vigna sinensis	0	0	0

Number indicates the amount of infection hyphae; from nothing (0) to abundant formation (4).

(Shiraishi et al., 1978b). In the presence of suppressor, eight species of pea nonpathogens established infection on pea leaves. Among these, *Alternaria alternata* 15B, an avirulent isolate of pear pathogen, grew and colonized on pea and formed conidia on suppressor-treated pea leaves 4–7 days later. Further, as shown in Table 4, *Alternaria* 15B could infect five species of leguminous plants to which *M. pinodes* was pathogenic in the presence of the suppressor, but not on other plant species. In other words, the specificity of the biological activity of the suppressor coincided with the host range of the producer fungus, *M. pinodes* (Oku et al., 1980).

Thus, it was found that suppressor acts not only as suppressor of pisatin biosynthesis but also as the determinant of pathogenicity of *M. pinodes* by suppressing the other defense reactions.

The same type of suppressors were found in the spore germination fluid of *M. melonis* and *M. ligulicola* (Oku et al., 1987).

Since the suppressor causes no visible injury to pea leaf tissue and isolated protoplast, it can hardly be a host-specific toxin.

A mannan glycoprotein isolated from culture filtrate of *Phytophtora megasperma* f. sp. *glycinea* was reported to suppress the glyceollin accumulation in soybean in a race-specific manner (Ziegler and Pontzen, 1982).

As the mode of action, we recently found that the suppressor inhibits the plasma membrane ATPase (Yoshioka et al., 1990). The inhibitory activity of the suppressor from *M. pinodes* is nonspecific in vitro, i.e., active to plasma membrane ATPases prepared from pea, bean, cowpea, soybean, and barley, contradicting the result shown in Fig. 4. However, the activity was found to be specific at the tissue level (Shiraishi et al., 1991) by electron microscopy according to the method of Hall et al. (1980). The principle of this method is that when the ATPase in the tissue is active, the phosphate released from added ATP is precipitated as lead base by addition of lead nitrate, but not when ATPase is inactive. As shown in Fig. 5, vanadate

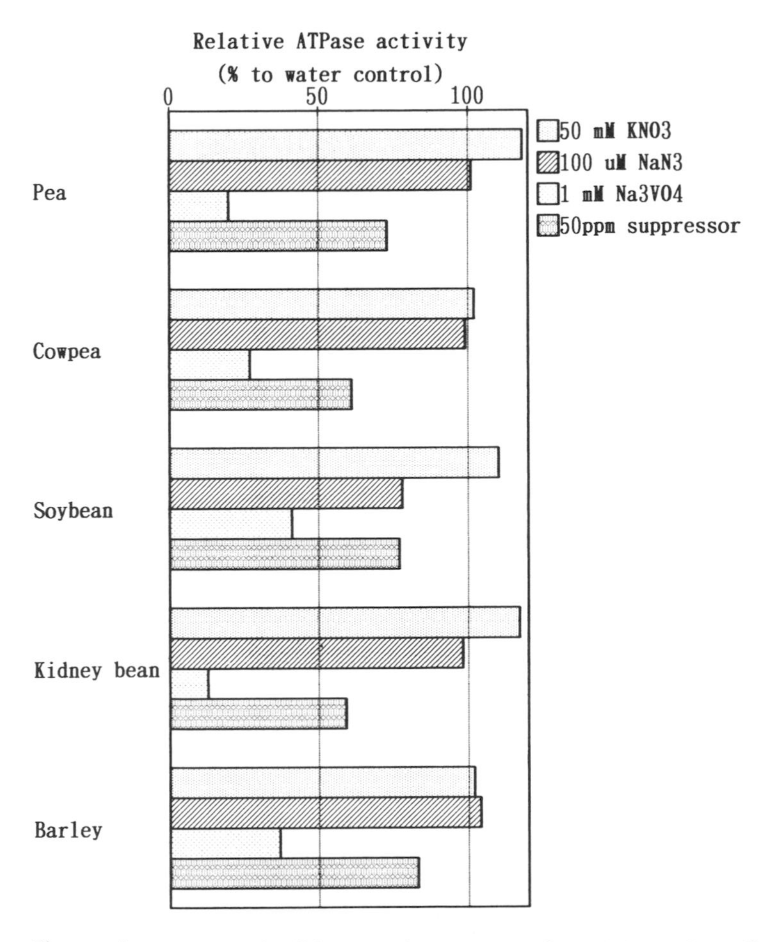

Figure 4 Effect of inhibitors and suppressor from *Mycosphaerella pinodes* on ATPases prepared from plasma membrane fractions of various plants.

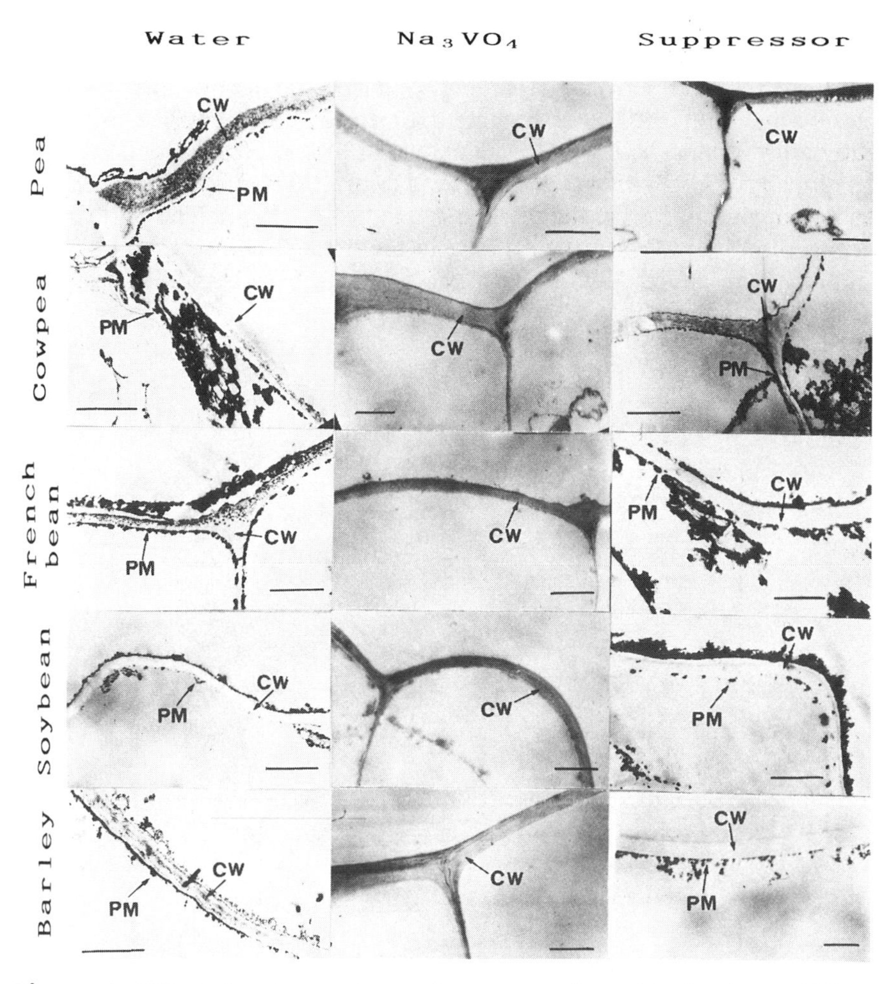

Figure 5 Effect of orthovanadate and suppressor from *M. pinodes* on ATPase activity of various plants in vivo detected by lead precipitation procedures (see text).

inhibits ATPase nonspecifically but the suppressor inhibits ATPase only of pea.

Further, this specific inhibition of ATPase was also found at the interface of host and parasite (Shiraishi et al., 1991). That is, as shown in Fig. 6, *M. pinodes* inhibits ATPase of pea until 6 hr after inoculation but *M. ligulicola*, a chrysanthemum pathogen, does not.

Since the suppressor inhibits the decline of pH of droplets of elicitor solution or water placed on pea leaves, the suppressor may inhibit the

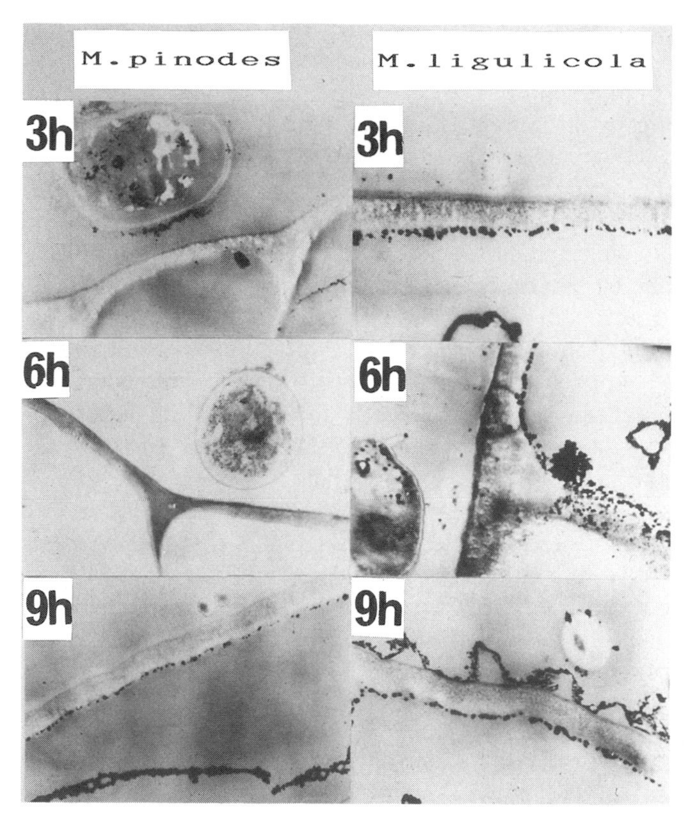

Figure 6 Evidence that pea plasma membrane ATPase is suppressed by the pea pathogen *M. pinodes* but not by the nonpathogen *M. ligulicola* at host-parasite interfaces.

proton pump ATPase of pea plasma membrane, hence the activity of pea cells to defend. The inhibitory activity of suppressor on pea plasma membrane ATPase in vivo was found to be temporary.

V. PARTICIPATION OF PROTEIN KINASE IN ELICITATION OF THE DEFENSE REACTION

Signal transduction in the elicitation of a defense reaction is a matter of worldwide interest for physiological plant pathologists.

Since almost all elicitors secreted by pathogens are high molecular weight substances, it is hardly considered that elicitors directly activate the gene for resistance. Therefore, many scientists believe there to be receptors in plant cell surface for elicitors and several scientists demonstrate the binding of elicitors to receptors (Ebel and Grisebach, 1988; Yoshikawa et al., 1983). As the result of interaction between elicitor and receptor, some signal substances which directly or indirectly activate the gene for resistance may be produced and so activate; hence defense reactions.

We have recently found that protein kinase may play an important role in the process of elicitation of defense reaction in pea plant (Shiraishi et al., 1990). Verapamil, a Ca^{2+} channel blocker, and K-252a, a strong inhibitor of protein kinase, inhibit pisatin accumulation in pea epicotyl which had been treated with elicitor from *M. pinodes*. Since $LaCl_3$ and EGTA did not inhibit pisatin accumulation and, further, since verapamil inhibits pisatin accumulation even if applied 6 hr after the elicitor treatment, when the pisatin biosynthetic pathway has already been activated, it can hardly be considered that Ca^{2+} plays a role as second messenger for signaling of pisatin biosynthesis in pea tissue induced by elicitor. On the other hand, K-252a inhibits pisatin accumulation when applied to pea epicotyl only before the elicitor treatment. This fact suggests that signal transduction occurs very rapidly after the elicitor treatment and, further, that protein kinase may play a key role in the signal transduction for pisatin biosynthesis because K-252a inhibits markedly the in vitro phosphorylation of pea plasma membrane proteins.

Rapid and transient phosphorylation of specific proteins by treatment with fungal elicitors was reported in parsley cell suspension cultures by Dietrich et al. (1990), and they also suggest that protein phosphorylation is involved in the signal transduction process following elicitor treatment.

As for the substrate of protein kinase, Harrison et al. (1991) suggested that the phosphorylation of regulatory proteins binding to silencer region may regulate the expression of CHS gene in bean because the DNA binding activity of the proteins is lost by treatment with alkaline phosphatase. Datta and Cashmore (1989) reported that the phosphorylation of nuclear protein binding to promoters of certain photoregulated genes decreases the binding activity. Kobayashi (1991) found, using his system of barley and *Erysiphe pisi*, a non–host resistance combination, that the rearrangement of cytoplasmic strand, mainly composed of actin filaments, plays a very important role in the expression of defense reaction. That is, several cytoplasmic strands composed of actin filaments are rearranged and appeared under the appressoria when inoculated. However, we found that K-252a completely inhibits the appearance of actin filaments. This is not direct evidence, but it is possible to assume that the phosphorylation of actin by protein kinase may be one of the processes for the expression of defense reaction not only in pea but also in barley.

VI. REGULATION OF EXPRESSION OF GENES FOR PHYTOALEXIN BIOSYNTHESIS

Much information is available to indicate that the infection of plants by plant pathogens induces various kinds of isozymes which are not detected in healthy plant tissues, and that the newly synthesized isozymes are the result of net protein synthesis. These facts suggest that the genes encoding new proteins are activated by infection.

Recent progress of genetic engineering enables these phenomena to be understood at the gene level. Suspension-cultured plant cells are generally used for this type of experiment because the signal molecules distribute simultaneously to each plant cell. However, in some plants or organs the resistance is reported to be expressed at the tissue or organ level but not at the cellular level. For example, Tomiyama et al. (1967) clarified that potato tuber tissue needs more than 10 cell layers in order to keep resistance against late blight disease.

Anyway, the feature of activation of genes encoding phytoalexin biosynthesis was examined using cultured bean cells by several scientists. Isolated bean cells cultured in the dark were treated with elicitor prepared from *Colletotrichum lindemuthianum*, and the amount of mature mRNA accumulated in the cells was determined by northern blot hybridization analysis with cDNA of phenylalanine ammonia-lyase (PAL), chalcone synthase (CHS), and chalcone isomerase (CHI), which are key enzymes for isoflavonoid phytoalexin biosynthesis. Results show that all mRNA (PAL, CHS, and CHI) accumulate very rapidly, reaching a maximum level 3 hr after treatment with elicitor, and then decrease to the original level (Edwards et al., 1985; Hahlbrock and Scheel, 1981, Lamb et al., 1989; Mehdy and Lamb, 1987; Ryder et al., 1984).

We studied the effect of elicitor and suppressor isolated from the pea pathogen *M. pinodes* on the defense reaction of pea with respect to the pisatin biosynthesis at enzyme (PAL and CHS, which are key enzymes for pisatin biosynthesis) and their gene activation, using pea epicotyl tissues (Yamada et al., 1989). Pea epicotyl grown in the dark has a minimal amount of PAL and CHS. However, treatment of etiolated pea epicotyl tissue with elicitor activates the accumulation of PAL and CHS mRNAs within an hr, followed by an increase in enzyme activity, and then pisatin biosynthesis. Concomitant presence of suppressor with elicitor results in the delay of the gene transcriptions for 3 hr and increase of PAL enzyme activity for 6 hr. As the result, pisatin accumulation is delayed for 6–9 hr in pea epicotyl. Thus it was demonstrated that the pisatin biosynthesis is activated in pea epicotyl tissues by treatment with elicitor at the gene level, and the suppressor of the pathogenic fungus suppresses the expression of genes responsible for the defense reaction.

It should be noted that our results are in close agreement with the in situ hybridization experiment of the accumulation of potato PAL mRNA reported by Cuypers et al., (1988). They found a remarkable difference in the timing of PAL mRNA accumulation between compatible and incompatible interactions of late blight disease. A marked increase in accumulation of PAL mRNA was observed 3 hr after inoculation with an incompatible race at the infection site, but it was 6 hr after inoculation with a compatible race. The coincidence of a 3-hr delay in the accumulation of PAL mRNA in compatible interaction with our result of suppressor-treated pea epicotyl suggests the idea that some factors such as suppressors might get involved in the compatibility of potato late blight disease.

In our experimental results, the expression and suppression of PAL and CHS genes are coordinately regulated by treatment with elicitor and suppressor. Coordinate activation of transcription of PAL and CHS genes is also reported in bean suspension cultured cells as well as bean hypocotyl by treatment with elicitor (Edwards et al., 1985; Hahlbrock and Scheel, 1989; Lamb et al., 1989; Mehdy and Lamb, 1987; Ryder et al., 1984).

VII. EPILOG

In this chapter, the importance of phytoalexins in resistance mechanisms of plant was described from biological and biochemical perspectives.

There is a large amount of information on the mechanism of resistance of plants against invading microorganisms but little is known about the pathogenicity of pathogens, especially how pathogens escape the defense reaction of host plant to infect and colonize these plants. Host-specific toxins are certainly pathogenicity factors, but they have been found in limited genera of pathogenic fungi and are hardly considered to be the determinant of pathogenicity in obligate parasites.

The most important description in this chapter may be the finding of suppression of defense reaction and the mode of action. Though the known data on suppressors are limited at the present time, the biological evidence shows that many more suppressors will be found in the future to be determinants of pathogenicity.

Another major interest on our part is to clarify the series of molecular events in the process of defense reaction elicitation.

Progress in these basic fields of plant pathology may shed light for the development of new control measure of plant diseases without polluting the environment.

REFERENCES

Baily, J. A., and Ingham, J. L. (1971). Phaseollin accumulation in bean (*Phaseolus vulgaris*) in response to infection by tobacco necrosis virus and rust *Uromyces appendiculatus*. *Physiol. Plant Pathol. 1*:451–456.

Berard, D. F., Kuć, J., and Williams, E. B. (1972). A cultivar specific protection factor from incompatible interactions of green bean with *Colletotrichum lindemuthianum*. *Physiol. Plant Pathol. 2*:123–128.

Christenson, J. A., and Hadwiger, L. A. (1973). Induction of pisatin formation in pea foot region by pathogenic and nonpathogenic clones of *Fusarium solani*. *Phytopathology 63*:784–790.

Cruickshank, I. A. M., and Mandryk, M. (1960). The effect of stem infection of tobacco with *Peronospora tabacina* Adam on foliage reaction to blue mold. *J. Aust. Inst. Agric. Sci. 26*:369–372.

Cruickshank, I. A. M., Briggs, D., and Perrin, R. (1971). Phytoalexins as determinants of disease reaction in plants. *J. Indian Bot. Soc. Golden Jubilee 50A*: 1–11.

Cuypers, B., Schmelzer, E., and Hahlbrock, K. (1988). In situ localization of rapidly accumulated phenylalanine ammonia-lyase mRNA around penetration sites of *Phytophthora infestans* in potato leaves. *Mol. Plant–Microbe Interact. 1*:157–160.

Datta, N., and Cashmore, A. R. (1989). Binding of pea nuclear protein to promoters of certain phosphorylated genes is modulated by phosphorylation. *Plant Cell 1*:1069–1077.

Dietrich, A., Mayers, J. E., and Hahlblock, K. (1990). Fungal elicitor triggers rapid, transient, and specific protein phosphorylation in parsley cell suspension cultures. *J. Biol. Chem. 265*:6360–6368.

Ebel, J., and Grisebach, H. (1988). Defense strategies of soybean against the fungus *Phytophthora megasperma* f. sp. *glycinea*: a molecular analysis. *Trends Biochem. Sci. 13*:23–27.

Edwards, K., Cramer, C. L., Bolwell, G. P., Dixon, R. A., Schuch, W., and Lamb, C. J. (1985). Rapid and transient induction of phenylalanine ammonia-lyase mRNA in elicitor treated bean cells. *Proc. Natl. Acad. Sci. USA 82*:6731–6735.

Hahlbrock, K., and Scheel, D. (1989). Physiology and molecular biology of phenylpropanoid metabolism. *Ann. Rev. Plant Physiol. 40*:347–369.

Hall, J. L., Browning, A. J., and Harvey, D. M. R. (1980). The validity of the lead precipitation technique for the localization of ATPase activity in plant cells. *Protoplasma 104*:193–200.

Harrison, M. J., Lawton, M. A., Lamb, C. J., and Dixon, R. A. (1991). Characterization of a nuclear protein that binds to three elements with the silencer region of a bean chalcone synthase. *Proc. Natl. Acad. Sci. USA, 88*:2515–2519.

Higgins, V. J. (1972). Role of phytoalexin medicarpin in three leaf spot diseases of alfalfa. *Physiol. Plant Pathol. 2*:289–300.

Ingham, J. L. (1970). Phytoalexins and other natural products as in plant disease resistance. *Bot. Rev. 38*:343–423.

Király, Z., Berna, N., and Ersek, T. (1972). Hypersensitivity as a consequence, not the cause, of plant resistance to infection. *Nature 239*:456–458.

Kobayashi, I. (1991). Recognition of *Erysiphe pisi* by barley coleolptile cells and the induction of inaccessibility, PhD thesis, Okayama University, pp. 26–42.

Klarman, V. L., and Hammerschlag, F. (1972). Production of the phytoalexin, hydroxyphaseollin, in soybean leaves inoculated with tobacco necrosis virus. *Phytopathology 62*:719–721.

Kuć, J. (1972). Phytoalexins. *Ann. Rev. Phytopathol. 10*:207–232.

Lamb, C. J., Lawson, M. A., Dron, M., and Dixon, R. A. (1989). Signal transduction mechanism for activation of plant defenses against microbial attack. *Cell 56*:215–224.

Leath, K. T., and Rowell, J. B. (1970). Nutritional and inhibitory factors in the resistance of *Zea mays* to *Puccinia graminis. Phytopathology 60*:1097–1100.

Mehdy, M. C., and Lamb, C. J. (1987). Chalcone isomerase cDNA cloning and mRNA induction by fungal elicitor, wounding and infection. *EMBO J. 6*:1527–1533.

Müller, K. O. (1956). Einige Einfache Versuche zum Nachweis von Phytoalexinen. *Phytopathol. Z. 27*:237–259.

Oku, H., Ouchi, S., Shiraishi, T., and Baba, T. (1975a). Pisatin production in powdery mildewed pea seedling. *Phytopathology 65*:1263–1267.

Oku, H., Ouchi, S., Shiraishi, T., Baba, T., and Miyagawa, H. (1975b). Phytoalexin production in barley powdery mildew as affected by thermal and biological predispositions. *Proc. Jpn. Acad. 51*:198–201.

Oku, H., Ouchi, S., Shiraishi, T., Komoto, Y., and Oki, K. (1975c). Phytoalexin activity in barley powdery mildew. *Ann. Phytopathol. Soc. Jpn. 41*:185–191.

Oku, H., Shiraishi, T., and Ouchi, S. (1976). Effect of preliminary administration of pisatin to pea leaf tissues on subsequent infection by *Erysiphe pisi* DC. *Ann. Phytopathol. Soc. Jpn. 42*:597–600.

Oku, H., Shiraishi, T., and Ouchi, S. (1987). Role of specific suppressors in pathogenesis of *Mycosphaerella pinodes. Molecular Determinants of Plant Diseases* (S. Nishimura, C. D. Vance, and N. Doke, eds.), Japan Scientific Press, Tokyo/Springer-Verlag, Berlin, pp. 145–156.

Oku, H., Shiraishi, T., Ouchi, S. Ishiura, M., and Matuseda, R. (1980). A new determinant of pathogenicity in plant disease. *Naturwissenschaften 67*:310–311.

Ouchi, S., Oku, H., Hibino, C., and Akiyama, I. (1974a). Induction of accessibility and resistance in leaves of barley by some races of *Erysiphe graminis. Phytopathol. Z. 79*:24–34.

Ouchi, S., Oku, H., Hibino, C., and Akiyama, I. (1974b). Induction of accessibility to a nonpathogen by preliminary inoculation with a pathogen. *Phytopathol. Z. 79*:142–154.

Pierre, R. (1971). Phytoalexin induction in beans resistant or susceptible to *Fusarium* and *Thielaviopsis. Phytopathology 61*:332–337.

Pueppke, S. G., and VanEtten, H. D. (1974). Pisatin accumulation and lesion development in peas infected with *Aphanomyces euteiches*, *Fusarium solani* f. sp. *pisi*, or *Rhizoctonia solani. Phytopathology 64*:1433–1440.

Ryder, T. B., Cramer, C. J., Bell, J. N., Robbins, M. P., Dixon, R. A., and Lamb, C. J. (1984). Elicitor rapidly induces chalcone synthase mRNA in *Phaseolus vulgaris* cells at the onset of the phytoalexin defense response. *Proc. Natl. Acad. Sci. USA, 81*:5724–5728.

Sato, N., Kitazawa, K., and Tomiyama, K. (1971). The role of rishitin in localizing the invading hyphae of *Phytophthora infestans* in infection sites at the cut surfaces of potato tuber. *Physiol. Plant Pathol. 1*:289–295.

Shephard, C. J., and Mandryk, M. (1962). Auto inhibitors of germination and sporulation in *Peronospora tabacina* Adam. *Trans. Br. Mycol. Soc. 45*:233–244.

Shephard, C. J., and Mandryk, M. (1963). Germination of conidia of *Peronospora tabacina* Adam. II. Germination in vivo. *Aust. J. Biol. Sci. 16*:77–87.

Shiraishi, T., Araki, M., Ysohioka, H., Kobayashi, I., Yamada, T., Ichinose, Y., Kunoh, H., and Oku, H. (1991). Inhibition of ATPase activity in pea plasma membranes in situ by a suppressor from a pea pathogen, *Mycosphaerella pinodes. Plant Cell Physiol. 32:*1067–1075.

Shiraishi, T., Hori, N., Yamada, T., and Oku, H. (1990). Suppression of pisatin accumulation by an inhibitor of protein kinase. *Ann. Phytopathol. Soc. Jpn. 56*:261–264.

Shiraishi, T., Oku, H., Isono, M., and Ouchi, S. (1975). The injurious effect of pisatin on the plasma membrane of pea. *Plant Cell Physiol. 16*:939–942.

Shiraishi, T., Oku, H., Ouchi, S., and Tsuji, Y. (1977). Local accumulation of pisastin in tissues of pea seedlings infected by powdery mildew fungi. *Phytopathol. Z. 88*:131–135.

Shiraishi, T., Oku, H., Tsuji, Y., and Ouchi, S. (1978a). Inhibitory effect of pisatin on infection process of *Mycosphaerella pinodes* on pea. *Ann. Phytopathol. Soc. Jpn. 44*:641–645.

Shiraishi, T., Oku, H., Yamashita, M., and Ouchi, S. (1978b). Elicitor and suppressor of pisatin induction in spore germination fluid of pea pathogen, *Mycosphaerella pinodes. Ann. Phytopathol. Soc. Jpn. 44*:659–665.

Tomiyama, K., Sakai, R., Sakuma, T., and Ishizaka, N. (1967). The role of polyphenols in the defense reaction in plants induced by infection. *The Dynamic Role of Molecular Constituents in Plant-Parasite Interaction* (C. J. Mirocha and I. Uritani, eds.), Bruce, St. Paul, MN, pp. 165–182.

Yamada, T., Hashimoto, H., Shiraishi, T., and Oku, H. (1989). Suppression of pisastin, phenylalanine ammonia-lyase mRNA and chalchone synthase mRNA accumulation by a putative pathogenicity factor from the fungus *Mycosphaerella pinodes. Mol. Plant-Microbe Interact. 2*:256–261.

Yamamoto, Y., Oku, H., Shiraishi, T., Ouchi, S., and Koshizawa, K. (1986). Non-specific induction of pisatin and local resistance in pea leaves by elicitors from *Mycosphaerella pinodes, M. melonis* and *M. ligulicola* and the effect of suppressor from *M. pinodes. J. Phytopathol. 117*:136–143.

Yoshikawa, M., Keen, N. T., and Wang, M. C. (1983). A receptor on soybean membranes for a fungal elicitor of phytoalexin accumulation. *Plant Physiol. 73*:497–506.

Yoshioka, H., Shiraishi, T., Yamada, T., Ichinose, Y., and Oku, H. (1990). Sup-

pression of pisatin production and ATPase activity in pea plasma membrane by orthovanadate, verapamil and a suppressor from *Mycosphaerella pinodes. Plant Cell Physiol. 38*:1139–1146.
Ziegler, E., and Pontzen, R. (1982). Specific inhibition of glucan-elicited glyceollin accumulation in soybeans by an extracellular mannan-glycoprotein of *Phytophthora megasperma* f. sp. *glycinea. Physiol. Plant Pathol. 20*:321–332.

3

Elicitor and the Molecular Bases of Phytoalexin Elicitation

Masaaki Yoshikawa
Hokkaido University, Sapporo, Japan

I. INTRODUCTION

Disease resistance in many plant–fungal pathogen interactions has been suggested to be due to inducible production of antibiotic low molecular weight compounds, i.e., phytoalexins (Keen, 1981). The induction of phytoalexins in infected plants is presumed to be mediated by an initial recognition process between plants and pathogens which involves detection of certain unique molecules of pathogen origin, termed elicitors, by recognitional receptor-like molecules in plants, thereby setting off a cascade of biochemical events leading ultimately to phytoalexin accumulation (Yoshikawa, 1983; Yoshikawa and Masago, 1982). However, detailed mechanisms involved in each biochemical process leading to the phytoalexin production are poorly understood. The main subject of this chapter is the molecular basis of the elicitation of glyceollin, a phytoalexin (Yoshikawa et al., 1978a) produced by the expression of monogenic resistance in soybean (*Glycine max*) to incompatible races of *Phytophthora megasperma* f. sp. *glycinea* (Pmg).

II. ELICITORS OF PATHOGEN ORIGIN THAT FUNCTION IN GLYCEOLLIN PRODUCTION IN VIVO

Isolated mycelial walls of many fungi possess potent elicitor activity to induce phytoalexin accumulation in plants. There are, however, several unresolved questions regarding the in vivo involvement of the wall-associated elicitors. A major detracting argument arises from the observation that active elicitor moieties can only be extracted from fungal walls by severe treatments such as autoclaving or exposure to acids or alkalines, which are unlikely to exist in biological environment. This raises the question of how normally insoluble elicitor molecules on or in the fungal walls may come in to contact with the corresponding recognitional molecules in plant cells during natural infection processes. Although unnaturally extracted elicitors are frequently used in many types of studies including biological and structural analysis, the possibility exists that such elicitors may not be those functioning in phytoalexin elicitation in fungus-infected plant tissues (Yoshikawa, 1983).

Our studies (Yoshikawa et al., 1981) with the soybean–Pmg system indicated that highly active carbohydrate elicitors of glyceollin production were released into a soluble form from insoluble mycelial walls of the fungus by a factor contained in soybean tissues. These studies provided a new insight for the in situ production of soluble elicitors which could be more efficiently recognized by plant cells. The elicitor release occurred as rapidly as 2 min after incubation of mycelial walls or actively growing hyphae with soybean tissues, suggesting that the process may be important as the earliest plant–pathogen interaction leading to the induction of glyceollin production.

The factor capable of releasing elicitors was purified from soybean cotyledons to apparent homogeneity and shown to possess β-1,3-endoglucanase activity (Keen and Yoshikawa, 1983). Identity of the factor to the glucanase was further supported by the facts that β-1,3-endoglucanase purified from the bacterial culture of *Arthrobacter luteus* possessed similar elicitor-releasing activity and that antibodies raised against the highly purified soybean glucanase inhibited both the elicitor-releasing and glucanase activity to similar extents. Furthermore, pretreatment of soybean tissues with the antibodies partially inhibited glyceollin accumulation otherwise induced by mycelial walls, suggesting that the elicitor release indeed occurs in natural infection processes and the released elicitors are responsible for glyceollin elicitation in the infected tissues.

III. ACTIVITY AND STRUCTURE OF THE GLYCANASE-RELEASED ELICITORS

The glucanase-released Pmg elicitors active in glyceollin elicitation were heterogeneous in size, ranging from 1,000 to more than 100,000 Da when evaluated by gel filtration. Activity of different size fractions of the released elicitors to induce glyceollin accumulation in soybean tissues were at least 10–100 times higher, based on weight concentrations, than the previously reported elicitors extracted by autoclaving as well as acid and alkaline treatments, and the released elicitors accounted for more than 90% of the total elicitor activity of the native mycelial walls.

Tentative structures of the released elicitors were deduced by use of sugar, ^{13}C-NMR, and enzymatic analysis (Fig. 1). Hepta-β-D-glucopyranoside (G7), the smallest elicitor-active molecule obtained by acid hydrolysis

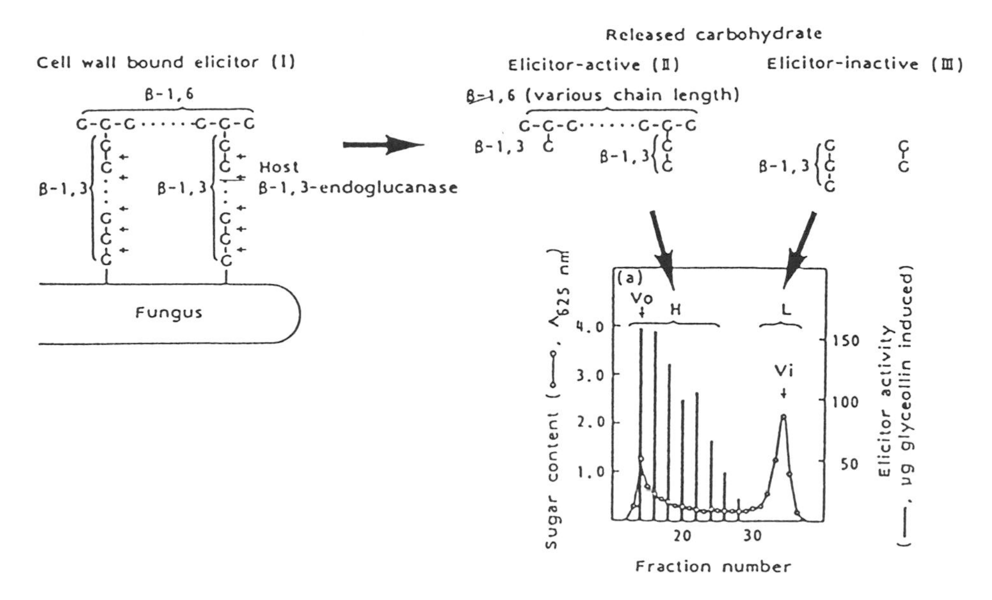

Figure 1 Tentatively proposed structures of elicitors bound to cell walls of *Phytophthora megasperma* f. sp. *glycinea* and its released forms due to attack by soybean β-1,3-endoglucanase. Inserted figure (A) is an elution profile of total carbohydrates released by soybean glucanase on Sephadex G-100. β-1,6-Glucans of various chain length are bound to cell walls through β-1,3 side chains (I). Upon infection, β-1,3 side chains are attacked by endo-type soybean β-1,3-glucanase, resulting in the release of elicitor-active β-1,6 chains of various chain length with side chains of β-1,3-linked one- or two-glucose moieties [II, correspond to fraction H in (A)] and di- or trimer of β-1,3-glucans [III, correspond to fraction L in (A)] derived from the endoglucanase attack of β-1,3 side chains of cell wall-bound elicitors. Small arrows indicate the sites for glucanase attack.

of Pmg cell walls (Sharp et al., 1984), was chemically synthesized and was also used for structural comparison. ^{13}C-NMR and sugar analysis indicated the presence of β-1,6- and β-1,3-glucose linkages for various size fractions of the released elicitors. The released elicitors, as well as G7, were not decomposed by β-1,3-endoglucanase, suggesting the main chain to be β-1,6-linked. In contrast, β-1,3-exoglucanase partially degraded both the elicitors resulting in complete loss of elicitor activity. These results indicate that the released elicitors are composed of β-1,6-linked main chains of different length and originally bound to fungal cell walls by β-1,3-linked side chains. Elicitors are thus released, upon infection, due to the attack of the side chains by the host β-1,3-endoglucanase, leaving one or two of β-1,3-linked glucose moieties to each side chain of the released elicitors.

IV. MOLECULAR CLONING OF cDNA ENCODING THE ELICITOR-RELEASING FACTOR, β-1,3-ENDOGLUCANASE IN SOYBEAN

Soybean β-1,3-endoglucanase thus appears to be a key host component involved in the earliest soybean–Pmg interaction leading to the induction of a plant defense reaction, by releasing elicitor-active carbohydrates from mycelial walls. We therefore cloned and characterized cDNA encoding soybean β-1,3-endoglucanase to further elucidate the role of this enzyme in the expression of disease resistance (Takeuchi et al., 1990).

Several cDNA clones for the glucanase gene were obtained by antibody screening of a λ gt11 expression library prepared from soybean cotyledons. Hybrid-selected translation experiments indicated that the cloned cDNA encoded a 36-kDa precursor protein product that was specifically immunoprecipitated with β-1,3-endoglucanase antiserum. Nucleotide sequence of three independent clones revealed a single uninterrupted open reading frame of 1041 nucleotides, corresponding to a polypeptide of 347 residue long. The primary amino acid sequence of β-1,3-endoglucanase as deduced from the nucleotide sequence was confirmed by direct amino acid sequencing of trypsin digests of the glucanase, indicating that the cloned cDNAs were indeed those for the soybean glucanase. The soybean β-1,3-endoglucanase exhibited 53% amino acid homology to a β-1,3-glucanase cloned from cultured tobacco cells and 48% homology to a β-(1,3-1,4)-glucanase from barley.

E. coli cells expressing the cloned full-length cDNA (pEG488) synthesized a protein positive to the glucanase antiserum which, upon a solubilization and reconstitution, possessed both the β-1,3-endoglucanase and elicitor-releasing activities, firmly establishing that both activities were due to β-1,3-endoglucanase. Furthermore, tobacco plants transformed by *Agrobacterium tumefaciens* carrying pEG488 also synthesized a glucanase antiserum-positive protein. Our preliminary experiments indicated that the transgenic tobacco plants with higher levels of β-1,3-endoglucanase activity enhanced disease resistance to several fungal tobacco pathogen.

V. A SPECIFIC RECEPTOR ON SOYBEAN MEMBRANES FOR THE GLUCANASE-RELEASED ELICITORS

We previously demonstrated that soybean membranes contained a specific binding site for intercellular β-1,3-glucan of Pmg, mycolaminaran, which possessed weak elicitor activity (Yoshikawa et al., 1983). Further study was made to examine whether soybean membranes contain receptors specific for the glucanase-released elicitors which appear to play a crucial role in the elicitation of glyceollin accumulation in the fungus-infected soybean tissues.

A direct binding assay between the ^{14}C-labeled released elicitors and the isolated soybean membranes was used. Total binding of the ^{14}C-labeled released elicitors in a concentration-dependent manner, suggesting the existence of specific binding sites. A Scatchard plot of the binding data disclosed the presence of a single class of binding sites having a K_d value of 4 $\times$ 10^{-7} M and approximately 20 binding sites per cotyledon cell. Preincubation of soybean membranes above 50°C or with proteases abolished the specific binding. Higher binding activity was observed with a membrane fraction rich in plasma membrane–associated ATPase obtained by aqueous two-polymer-phase system. These results indicate the existence of heat-labile proteinaceous binding sites specific for released elicitors on soybean membranes, presumably plasma membranes. Furthermore, the binding activity of several carbohydrates obtained from Pmg and other sources or from chemical modification of the glucanase-released elicitors correlated with their elicitor activity to induce glyceollin accumulation in soybean cotyledons. Experiments are now being conducted to solubilize and isolate the binding site.

VI. TRANSCRIPTIONAL ACTIVATION OF THE GENES FOR GLYCEOLLIN BIOSYNTHETIC ENZYMES

Our previous study (Yoshikawa et al., 1978b) using transcriptional and translational inhibitors indicated the glyceollin production was mediated by de novo mRNA and protein synthesis. Further study directly measuring mRNA levels with the use of P-labeled cDNAs for phenylalanine ammonia-lyase (PAL), chalcone synthase (CHS), and chalcone isomerase (CHI) demonstrated that the glucanase-released elicitors as well as the native fungal cell walls induced the gene transcription of each enzyme involved in glyceollin biosynthesis within 1–2 hr after elicitor application. Concomitant to the induction of glyceollin biosynthesis, glyceollin-degrading activity present in soybean tissues was reduced after elicitor treatment. It may therefore be deduced that these two biochemical actions of the elicitors, the induction of glyceollin synthesis and the reduction of glyceollin turnover activity, synergistically lead to the massive glyceollin accumulation in the elicitor-treated soybean tissues.

As summarized in this chapter, research in the last several years has disclosed the outline of a sequence of biochemical events that appear to participate in invocation of disease defense reactions (Fig. 2). The processes probably involve pathogen-associated molecule (elicitor)–plant receptor interaction, formation of signal transmitting substances, and gene activation,

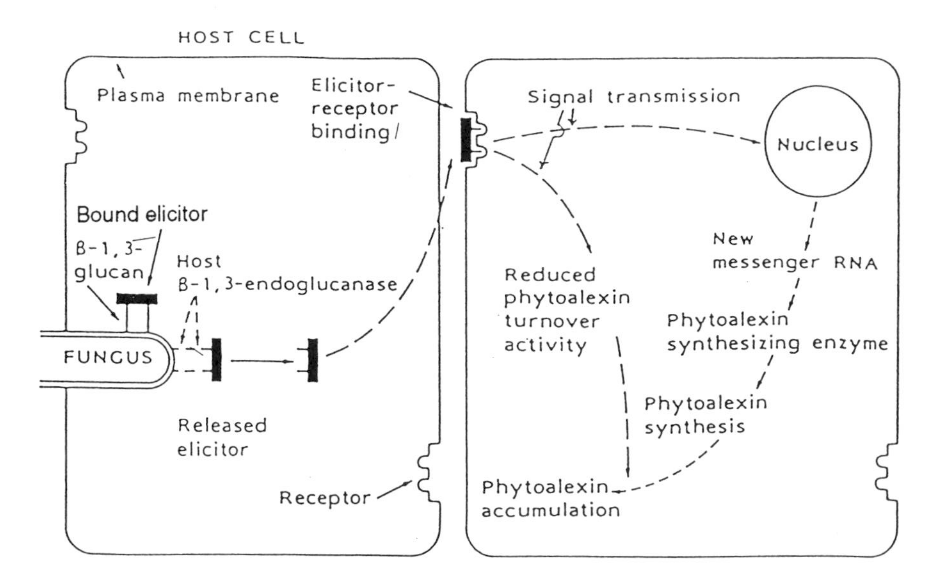

Figure 2 Schematic representation of possible sequence of biochemical events leading to plant defense reactions. The scheme exemplifies phytoalexin accumulation mechanism in the *Phytophthora megasperma* f. sp. *glycinea*-soybean interaction, but it may be applicable to other plant-pathogen systems. Contact of incompatible fungal races with host cells results in rapid release of phytoalexin elicitors from fungal cell wall surface, due to attack by β-1,3-endoglucanase constitutively present in host cells. The released elicitors then interact with the complementary receptors on plant plasma membrane. This interaction generates second messengers which transmit the signal to the nucleus where de novo transcription is invoked. The resulting new messenger RNA leads to the synthesis of enzymes involved in phytoalexin biosynthesis and the phytoalexin is formed. Levels of phytoalexin accumulation are accelerated by simultaneous inhibition of the phytoalexin-degrading system, which may also result from the elicitor-receptor interaction. Compatible fungal races may either possess elicitors that, upon release, cannot interact with the host receptors, or produce "suppressors" that interfere with the elicitor-receptor interaction or inhibit one of the subsequent host metabolic processes leading to the phytoalexin accumulation.

resulting in de novo biosynthesis of enzymes responsible for production of phytoalexins or other defense substances. At present the scheme is still hypothetical in may respects, however, and requires further evaluation of each process before molecular details in the expression of disease resistance can be fully understood.

REFERENCES

Keen, N. T. (1981). Evaluation of the role of phytoalexins. *Plant Disease Control* (R. C. Staples, ed.), John Wiley and Sons, New York, pp. 155–177.

Keen, N. T., and Yoshikawa, M. (1983). β-1,3-Endoglucanase from soybean releases elicitor-active carbohydrates from fungus cell walls. *Plant Physiol. 71*: 460–465.

Sharp, J. K., McNeil, M., and Albersheim, P. (1984). The primary structure of one elicitor-active and seven elicitor-inactive hexa(β-D-glucopyranosyl)-D-glucitols isolated from the mycelial walls of *Phytophthora megasperma* f. sp. *glycinea. J. Biol. Chem. 259*:11321–11336.

Takeuchi, Y., Yoshikawa, M., Takeba, G., Tanaka, K., Shibata, D., and Horino, O. (1990). Molecular cloning and ethylene induction of mRNA encoding a phytoalexin elicitor-releasing factor, β-1,3-endoglucanase, in soybean. *Plant Physiol. 93*:673–682.

Yoshikawa, M. (1978). Diverse modes of action of biotic and abiotic phytoalexin elicitors. *Nature 275*:546–547.

Yoshikawa, M. (1983). Macromolecules, recognition, and the triggering of resistance. *Biochemical Plant Pathology* (J. A. Callow, ed.), John Wiley and Sons, New York, pp. 267–298.

Yoshikawa, M., Keen, N. T., and Wang, M. C. (1983). A receptor on soybean membranes for a fungal elicitor of phytoalexin accumulation. *Plant Physiol. 73*:497–506.

Yoshikawa, M., and Masago, H. (1982). Biochemical mechanism of glyceollin accumulation in soybean. *Plant Infection: The Physiological and Biochemical Basis* (Y. Asada et al., eds.), Japan Sci. Soc. Press, Tokyo/Springer-Verlag, Berlin, pp. 265–280.

Yoshikawa, M., Matama, M., and Masago, H. (1981). Release of a soluble phytoalexin elicitor from mycelial walls of *Phytophthora megasperma* var. *sojae* by soybean tissues. *Plant Physiol. 67*:1032–1035.

Yoshikawa, M., Yamauchi, K., and Masago, H. (1978a). Glyceollin: its role in restricting fungal growth in resistant soybean hypocotyls infected with *Phytophthora megasperma* var. *sojae. Physiol. Plant Pathol. 12*:73–82.

Yoshikawa, M., Yamauchi, K., and Masago, H. (1978b). De novo messenger RNA and protein synthesis are required for phytoalexin-mediated disease resistance in soybean hypocotyls. *Plant Physiol. 61*:314–317.

Yoshikawa, M., Yamauchi, K., and Masago, H. (1979). Biosynthesis and biodegradation of glyceollin by soybean hypocotyls infected with *Phytophthora megasperma* var. *sojae. Physiol. Plant Pathol. 14*:157–169.

4

Soybean Phytoalexins: Elicitation, Nature, Mode of Action, and Role

Jack Paxton
University of Illinois, Urbana, Illinois

I. INTRODUCTION

I was pleased to be asked to write a chapter on my research and to discuss where phytoalexin research should go in the future. Some of my thoughts along these lines were discussed previously (Paxton, 1988b).

The objectives of my research are to understand how a plant "recognizes" a pathogen or pest, and what it does to defend itself after it recognizes the pathogen or pest. Armed with this information, we can improve disease control in soybeans and other crops. Through better understanding of plant disease, gained by study of a model system, scientists should be able to increase crop yields by preventing crop losses in an environmentally sound and sustainable way.

Soybeans, *Glycine max* [L.] Merr., are a major world crop. The value of U.S. production alone in 1991 was over $10 billion. *Phytophthora* root rot is a serious disease of soybeans and *Phytophthora sojae* Kauf. and Gerd. (Hansen and Maxwell, 1991) (previously known as *Phytophthora megasperma* Dreshs. f. sp. *glycinea* Kuan and Erwin) causes losses estimated at 3% of this crop on average each year (Plant Pathology Extension, University of Illinois, 1990). Some farmers' soybean crops can be decimated by this disease while others escape unscathed, as seen by above-ground symptoms. Evidence is accumulating, however, that a large toll is taken on soybean roots. This root loss is translated into yield losses which go unreported because they are considered normal. The plant replaces roots that have been rotted off at the expense of seed production.

Research in the summer of 1991 by Morris Huck and me (unpublished), using a minirhizotron, confirmed that root loss during the growing season in soybeans is substantial. But generally this root loss goes unobserved because of problems in observing roots (Paxton, 1974). Kittle and Gray (1982), established that soybean yields could be reduced by pathogens by at least 26%, based on fumigation of normal soils and use of foliar sprays to control a number of fungal diseases.

Phytophthora root rot of soybeans was chosen as a model system to study because of the value of the crop and the seriousness of the disease (Paxton, 1983). These factors led to the development by plant breeders of several cultivars and near-isolines of soybeans that differ from one another in one dominant gene, *Rps,* for resistance to races of the pathogen.

When I started my research on this disease a new cultivar of soybeans, Harosoy 63, had just been released by Professor R. L. Bernard. He created this cultivar by crossing Harosoy with Blackhawk (which was resistant to *Phytophthora* root rot) and then back-crossing the progeny with Harosoy for six generations to give a cultivar which was agronomically identical to Harosoy but now contained the *Rps1* gene, which conferred resistance to *Phytophthora* root rot. Little did we know at the time that the single dominant gene strategy of disease resistance would lead to the discovery of numerous races of the pathogen, several of which could attack this and other single, dominant genes that were subsequently found and used in soybean breeding for disease resistance. With this in mind, let's explore what we have learned about this disease.

II. ELICITORS

Previous work on elicitors of phytoalexins was summarized recently (Paxton, 1989). These compounds often arise from pathogen cell walls and seem to be specifically recognized by the plant cell membrane.

An important insight from my laboratory was Frank's discovery that fungi release elicitors, then called inducers, of phytoalexin production in soybeans (Frank and Paxton, 1971). We were the first to recognize that *P. sojae* cultures release compounds capable of inducing phytoalexin production in soybean plants. This elicitor appeared to be a glycoprotein with a molecular weight of about 10,000. We found elicitor activity in extracts of *Venturia inaequalis*, *Rhizoctonia solani*, *Diplodia zeae*, and *Helminthosporium turcicum*, suggesting that elicitors of phytoalexin production are common in fungi. We also found that plants may have the ability to stimulate the production of these elicitors. For this research we developed a useful cut-cotyledon bioassay that has been widely used in subsequent elicitor studies.

Ayers et al. (1976a,b) explored the composition of elicitors released from purified *P. sojae* cell walls after autoclaving. They found at least four fractions, with varying composition and ability to elicit defense responses on soybean cotyledons. They suggested that the elicitor activity of each fraction resides in the 3 and 3,6 highly branched glucan component of the fraction. They also suggested that the mannosyl residues, which represent about 1% of the undegraded glucan, participate in the activity of this molecule.

Keen, in one of the first studies looking for specific elicitors, which would explain race specificity in fungal pathogens, presented evidence that culture filtrates of race 1 of *P. sojae* contained elicitors that differed in some respects from race 3 of this fungus. Although he was not able to purify and identify this specific elicitor, he speculated that absence of this elicitor may be involved in the difference between race 3 and race 1 reactions of soybean cultivars (Keen, 1975). Keen and LeGrand (1980) later reported the isolation of specific elicitors extracted from the cell walls of *P. sojae* with an alkali treatment. Surface glycoproteins on *P. sojae* were suggested to play the role of race-specific phytoalexin elicitors. Isolated cell walls of *P. sojae* were extracted with 0.1 N NaOH at 0°C. The high molecular weight glycoproteins in this extract elicited glyceollin accumulation in a race-specific fashion. The glycoprotein contained only glucose and mannose as sugars. And while its activity was diminished by boiling or pronase treatment, the elicitor activity was destroyed by periodate treatment. This suggested that the carbohydrate portions are important for elicitor activity. As they point out, this observation is clouded by considerable variation in replicates, a high concentration requirement for activity, and a lack of total specificity.

The specificity in this case was not very clear-cut since the differences between the levels of glyceollin induced by elicitor preparations from virulent and avirulent races were small. This work remains to be confirmed but provides an interesting lead toward finding a very useful type of compound. We suggested that this type of compound could be very useful in plant breeding efforts (Paxton and Jacobsen, 1974).

Race-specific molecules that protect soybeans from *P. sojae* were found by Wade and Albersheim (1979). These appeared to be glycoproteins found in the incompatible races of the pathogen but not the compatible races. Introduction of this glycoprotein fraction into wounds 90 min before subsequent inoculation was sufficient to protect soybean plants. Subsequent work by Desjardins et al. (1982) indicated that glycoprotein fractions from both compatible and incompatible races of *P. sojae* could protect soybean plants from subsequent infection by this fungus. Variability in the bioassay unfortunately prevented further purification and identification of these components.

Bonhoff and Grisebach (1988) found that laminarin and polytran N also are effective elicitors of glyceollin, comparable to the glucan elicitor isolated from *P. sojae* by Sharp and coworkers (1984). Interestingly, on cut soybean cotyledons, the reverse was true in that laminarin was less effective in eliciting glyceollin accumulation. Furthermore, when they pretreated soybean roots with laminarin, they increased resistance of seedlings against *P. sojae*. Digitonin, when added with the laminarin, further increased this elicitation of glyceollin accumulation in intact roots. Soybean roots deposit callose when treated with digitonin but not when treated with various glucans.

An interesting observation was made by Yoshikawa et al. (1981) that soybean tissues contain an enzyme which released elicitor from *P. sojae* cell walls. This suggests a role for such enzymes in "recognition" of potential pathogens when they come in contact with soybean tissue, by specifically releasing fragments that are free to diffuse to binding sites on the cell membranes. At a recent Congress of Plant Pathology in Kyoto, Yoshikawa reported that he has succeeded in releasing very active elicitor fragments from the cell walls of *P. sojae* with a β-1,3-endoglucanase. These fragments appear to be slightly larger than the acid-hydrolyzed fragments studied by Sharp et al. (1984), and they were 10–100 times more active in elicitation of glyceollin in soybeans.

Transformation of this elicitor signal into gene activation also has received attention. Chappell et al. (1984) saw a rapid induction of ethylene biosynthesis in cultured parsley cells that had been treated with *P. sojae* elicitor. Ethylene production is a common response of plant cells to stresses such as pathogen attack. Parsley cell cultures respond by production of ethylene within 1 hr of treatment with this elicitor. Hauffe et al. (1985) showed that cultured parsley cells respond to this elicitor by producing *S*-adenosyl-L-methionine:bergaptol and *S*-adenosyl-L-methionine:xanthotoxol *O*-methytransferases. Kuhn et al. (1984) demonstrated that treatment of parsley cells with the *P. sojae* elicitor induced phenylalanine ammonia-lyase and 4-coumarate:CoA ligase mRNAs. Farmer (1985) demonstrated that 5 μg/ml of *P. sojae* elicitor suppressed lignin accumulation in treated soybean suspension cultures. Phenolics, normally polymerized into lignin, were probably diverted into glyceollin and other compounds. This also suggests that elicitor recognition is a common phenomenon in triggering various defense responses in plants. The importance of phenylalanine ammonia-lyase to this process in soybeans is suggested by the work of Simcox and Paxton (1984).

It should be noted, however, that ethylene may not have a direct role in the elicitation of glyceollin accumulation in soybean plants (Paradies et al., 1980).

Ziegler and Pontzen (1982) found an extracellular invertase in *P. sojae* which was capable of inhibiting glucan-elicited glyceollin accumulation in soybeans. This glycoprotein suppressor was active at 0.3–10 μg/ml and was race-specific. The invertase would only inhibit elicitation of glyceollin accumulation on soybean cultivars that could be attacked by the race which was the source of the invertase. This work has not been confirmed and remains an area worth exploring.

III. PHYTOALEXINS

Phytoalexins were first defined by Müller and Borger in 1941 and our understanding of them as progressed to a more recent definition (Paxton, 1981). They are compounds produced de novo in plants after microorganism attack or other stress and they play a role in plant disease and insect resistance. The precise role of these compounds is hard to determine as is true for finding and assaying many antifungal compounds (Paxton, 1991c; Vaillancourt and Paxton, 1986).

Glyceollin is the phytoalexin identified from soybeans and occurs as a series of isomers. Five isomers have been identified but their activity has not been studied. Some species of *Glycine* appear to produce almost exclusively one isomer and various tissues of soybean produce predominantly different isomers (Keen et al., 1986). Why this is the case is not understood but certainly is worth investigating.

A phytoalexin, PA_k, was identified in soybeans (Keen and Paxton, 1975) but has yet to be characterized. It is a yellow compound with a strong absorbance at 492 nm at pH 7, which shifts to 430 nm at pH 2. It may represent an oxidation product of glyceollin, but that has yet to be determined.

My laboratory also discovered phytoalexin accumulation in corn (Lim et al., 1970). Despite efforts by Simcox and other graduate students, this elusive compound or compounds has not been identified.

Several metabolic processes are activated when a plant is attacked by a pathogen (Paxton, 1991a). Phenylalanine ammonia-lyase is one of the first enzymes to be activated. This enzyme has been studied extensively, partly because the assay for it is relatively simple.

It is my opinion that crucial parts of plants, such as seeds, accumulate compounds that can serve as primary deterrents to pathogen and insect attack. Failing in this role, these compounds can then be converted to much more toxic compounds that serve as secondary defense. Phytoalexins, then, would be a secondary defense, and they are toxic to plant cells as well as pathogens and insects. Among the compounds naturally accumulated in soybeans, precursors of glyceollin are common, especially isoflavanoids in seeds. Isoflavanoids serve as precursors for glyceollins in soybean tissues and studies show these isoflavanoids to be held as complex conjugates in different concentrations in different tissues in soybeans (Graham, 1991; Graham and Graham, 1991).

IV. ROLE IN DISEASE AND INSECT RESISTANCE

The role of phytoalexins in plant disease and pest resistance has been very hard to determine with scientific precision. Regardless of this difficulty, considerable evidence has accumulated that phytoalexins play an important role in soybean disease resistance. Some of the first evidence in soybeans was that the removal of phytoalexins can cause a resistant plant to become susceptible (Klarman and Gerdemann, 1963). Later we found that addition of a phytoalexin back into the plant can make a susceptible plant become resistant (Chamberlain and Paxton, 1968). The phytoalexins accumulate in the plant after inoculation with a timing appropriate to inhibition of pathogen growth (Frank and Paxton, 1970). Another approach was our discovery that prior inoculation with a non-pathogen will make the susceptible plant become resistant to subsequent inoculation with what would have been a pathogen (Paxton and Chamberlain, 1967; Svoboda and Paxton, 1972). More evidence for the role of phytoalexins and an explanation of the impact of environment on plant disease was the finding that environmental conditions could affect phytoalexin accumulation (Murch and Paxton, 1979). We found that anaerobic conditions, such as would occur around soybean roots during flooding, drastically depressed the soybean plant's ability to accumulate glyceollin. This could explain why flooding is so important for disease development in the field.

Other environmental effects that influence glyceollin accumulation in soybeans are salinity and temperature (Murch and Paxton, 1980a, 1980b). All of these environmental effects on glyceollin accumulation in soybeans relate nicely to field conditions of wet, cold soils which are ideal for *Phytophthora* root rot development (Murch and Paxton, 1980c). Calcium ion seems to play an important role in elicitation and therefore disease resistance as well (Stäb and Ebel, 1987).

One of the best proofs of the role of phytoalexins in plant disease resistance comes from the work of Schafer et al. (1989). They transferred a gene for pisatin demethylase (pisatin is a phytoalexin in peas) into a corn pathogen and converted it to a pea pathogen as well. This clearly suggests that pathogens have to have the ability to deal with phytoalexins of their host in order to be pathogens. Our work would suggest that this is true for soybean *Phytophthora* root rot as well (Paxton, 1982).

It also is probable that a pathogen must have an extensive set of enzymes to be a pathogen. This could help explain the common occurrence of loss of virulence in pathogens held in culture. Some enzymes necessary for pathogenicity probably are not necessary for survival as saprophytes on artificial media.

V. MODE OF ACTION

The mode of action of biologically active compounds is generally of interest to scientists. It is surprising that more has not been done on the mode of action of phytoalexins. Our work on glyceollin suggests that glyceollin has at least two important modes of action, at physiological levels in plants. We discovered that glyceollin can inhibit electron transport at site 1 in soybean and corn mitochondria (Boydston et al., 1983). Later we also showed that glyceollin can inhibit ATPase in plasma membranes of both plant and pathogen (Giannini et al., 1988a,b). Tonoplast vesicles were almost twice as sensitive to glyceollin as plasma membrane vesicles. Cell membranes are most likely a site for both elicitation and toxic action of phytoalexins (Paxton, 1979).

Glyceollin is toxic to plant cells as well as pathogens and insects (Bhandal et al., 1987; Hart et al., 1983; Fischer et al., 1990b). This suggests that these compounds have evolved as a general stress response and may serve to protect the plant by creating very toxic environments at the cellular level.

VI. DISCUSSION

My initial studies on the nature of disease resistance in walnuts introduced me to compounds which can accumulate in plants and be stored in relatively nontoxic forms such as glycosides (Paxton and Wilson, 1965). The plant cell is armed with glycosides which are hydrolyzed rapidly to the aglycone, which is often more toxic. This happens when the cell is damaged and protects the plant from attack, much like our white blood cells produce free radicals to help destroy foreign organisms in our bodies.

Phytoalexins were discovered in soybeans by Dr. Gerdemann, shortly after the first phytoalexin was characterized in peas by Perrin and Bottomley (1961) in Australia.

Gerdemann's story of serendipity deserves to be retold here. Dr. Gerdemann was a mycologist and as such realized that *Phytophthora* species require thiamine to grow. He reasoned that one explanation for resistance of soybean varieties was that the resistant cultivars had tissues deficient in thiamine and the fungus therefore couldn't grow. As a typical mycologist without a large research budget, he set up a simple experiment in which he introduced a thiamine solution into a soybean stem via a string wick that had been threaded through the stem with a lacing needle (Fig. 1). When he then inoculated the plant in the wick wound, he found that the plant was susceptible. Being a good scientist, he also ran a control experiment in which just plain water was introduced into the wound via the wick, and the

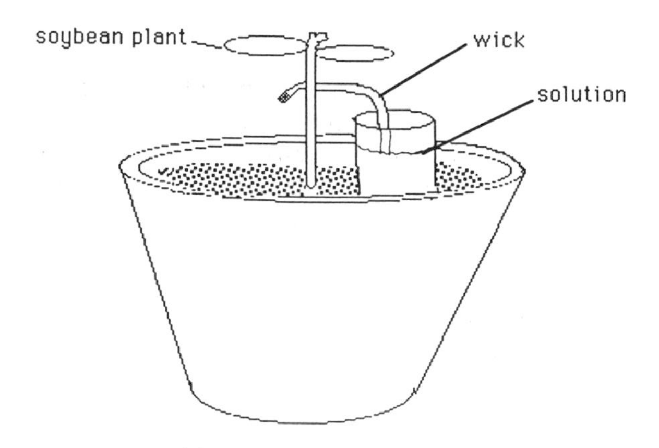

Figure 1 Wick experiment.

result was that the inoculated plant was again susceptible. It appeared that it was not something he was putting into the plant that made it susceptible, but rather something he was taking out of the plant. When he tested the exposed tip of the wick for materials that would inhibit the growth of *Phytophthora*, he found that nothing inhibitory to the growth of the pathogen was on the tip from the uninoculated plant. But the tip from the resistant inoculated plant contained compounds inhibitory to the growth of the pathogen, when added to media with the fungus.

During my first sabbatical, spent with Dr. Cruickshank in Australia, we explored the accumulation of phytoalexins in plant cells and found that the process could occur without profound changes in the cell or cell death (Paxton et al., 1974). The plant cells, however, were often subsequently killed by phytoalexins, which are toxic to the plant cell as well as the fungal or bacterial pathogen (Bhandal et al., 1987). There still is a need for study of cell death and how it relates to phytoalexin accumulation. Pectic substances released from cell walls on cell death can elicit phytoalexin accumulation and may be a signal to the plant that repair processes are needed.

Work with Dr. Kogan found that phytoalexins could deter insect attack on soybeans, just as they play a role in protecting soybeans from plant pathogens. Physiologically significant levels of glyceollin deter insect feeding by casual feeders but not true pests of soybeans, as would be anticipated (Hart et al., 1983; Fischer et al., 1990b). If true soybean pests were deterred by glyceollin it would suggest that the compound did not occur at feeding sites. The true soybean pest must have a method of blocking accumulation at the feeding site or excluding or detoxifying the compound. This point is confirmed in the work of Schafer et al. (1989).

VII. EPILOG

The study of disease resistance of soybeans by the researchers mentioned above has opened many doors to better understanding plant disease in general and therefore to improved control of plant diseases. These studies might even lead to a better understanding of how pollenation is controlled in plants (Fett et al., 1976).

One exciting discovery we made is the presence of elicitors of phytoalexin accumulation in cultures of the pathogen. These compounds have been partially identified and appear to be small carbohydrates. It appears that the plant produces enzymes, β-1,3-glucanases, which cleave the cell wall of the pathogen to release fragments that can be bound to receptors in the plant cell wall (Schmidt and Ebel, 1987). One concern is that the fragments used are released from fungal cell walls by autoclaving them under acid conditions and therefore may be artifacts.

Since they are carbohydrates, elicitors could be stable compounds without environmental detriment. Yet these compounds may hold the key to an interesting and valuable method of controlling plant diseases and pest attack. Since the plant responds to these compounds in the same way it would to a pathogen, the plant can be "vaccinated" against subsequent attack or protected from attack by having its surface covered with the elicitor (Paxton, 1973). When the pathogen gains ingress through a wound or direct penetration, the plant could be "notified" of the invasion and mount its inducible defense mechanisms. An example of the possibility of this approach is our work with Polytran L (Bhandal and Paxton, 1991) and the demonstration that elicitors can act on many different plants to cause accumulation of phytoalexins (Cline et al., 1978).

Another aspect worth exploring is the use of these elicitors to protect plants against insect attack (Kogan and Paxton, 1983; Paxton, 1991b). It appears that the same compounds interfere with insect feeding, and this feeding would logically introduce the elicitors into plant cells.

Much remains to be understood as to exactly what the true elicitor is, how it binds to a presumed receptor, what secondary signal is then sent to the plant cell nucleus to activate genes required for the accumulation of phytoalexins, and how these phytoalexins are then accumulated in the infection site to inhibit growth of the pathogen. This provides interesting leads to novel controls for plant diseases by using stable, natural elicitors to evoke the plant's normal defense mechanisms (Paxton, 1988a).

These mechanisms undoubtedly include the accumulation of phytoalexins that also have been described above. It is my opinion that several compounds that could have very important roles in plant disease resistance remain to be discovered. Some of these are compounds that have been variously described as nonexistent since they are so chemically reactive as to be destroyed when the plant cells are crushed to extract them from plant tissue. These compounds are highly reactive and generally would not survive the tituration of plant tissues. More studied examples are superoxide ion (Afanas'ev, 1989) and nitric oxide (Snyder and Bredt, 1992).

Another possibility is that compounds that can be formed in the process of extracting plant tissues. These compounds are not there until the researcher damages the tissue and releases enzymes or compartmentalized compounds that normally would not mix. My thesis research (Paxton and Wilson, 1965) depended on this phenomenon. The British chemist Daglish discovered that walnuts contained large quantities of hydrojuglone 4-glucoside but not juglone, as had been previously thought. This reduced quinone was kept reduced and made more water-soluble by the glucose attached in the 4 position. When cells are ruptured, as occurs in harvest or pathogen or insect attack, β-glucosidases are released which rapidly hydrolyze the compound. The aglycone is then readily oxidized to the very reactive naphthoquinone juglone. It is juglone which is very toxic to fungi and bacteria, and stains the hands brown when walnuts are dehulled.

I feel that there are several compounds that can be called phytoalexins in soybeans which have yet to be identified. These compounds are chemically reactive enough that they are not seen in crude extracts of soybean plants and more sophisticated chemical methods will be required to study them. One such method is electron paramagnetic resonance, and unpublished studies with Linn Belford and Robert Clarkson in the Chemistry Department at the University of Illinois suggest that free radicals (that would fill the above criterion of active compounds very difficult to chemically isolate) do indeed accumulate in inoculated soybean stems.

Synergism of these compounds in planta is yet another area that remains to be investigated. Phytoalexins are generally tested for activity on thin-layer chromatography plates or petri plates in relatively inert growth media. In the plant, with enzymes and active metabolism, these compounds undoubtedly interact with other compounds in many different and unanticipated ways. Some interesting work we have done suggests this is the case with several soybean compounds that can deter insect feeding (Fischer et al., 1990a).

Although phytoalexins show some tendencies to be chemically related in taxonomically related plants, this is not always the case (Kumar et al., 1984). Phytoalexins are not always closely related to disease resistance in soybeans (Paxton and Chamberlain, 1969).

Biological control is another area that needs much more attention. We have relied too much in the past to kill the pathogen with the most powerful chemical we can without endangering ourselves. It turns out that in the process we have exposed ourselves to hazards that were probably unnecessary (Duncan and Paxton, 1981). Biological control as pointed out by recent research can involve establishing native organisms on the plant that preclude the attack of pathogens simply by occupying a niche that is needed by the pathogen or producing something that makes the pathogen less fit for survival or less fit as a pathogen.

REFERENCES

Afanas'ev, I. B. (1989). *Superoxide Ion*, Vols. 1 and 2. CRC Press, Boca Raton, FL.

Ayers, A. R., Ebel, J., Valent, B., and Albersheim, P. (1976a). Host-pathogen interactions. X. Fractionation and biological activity of an elicitor isolated from the mycelium walls of *Phytophthora megasperma* var. *sojae. Plant Physiol. 57*:760-765.

Ayers, A. R., Valent, B., Ebel, J., and Albersheim, P. (1976b). Host-pathogen interactions XI. Composition and structure of wall-released elicitor fractions. *Plant Physiol. 57*:766-774.

Bhandal, I. S., and Paxton, J. (1991). Phytoalexin biosynthesis induced by the fungal glucan, Polytran L, in soybean, pea and sweet pepper tissues. *Journal of Agricultural and Food Chem. 39*:2156-2157.

Bhandal, I. S., Paxton, J. D., and Widholm, J. M. (1987). Effect of *Phytophthora megasperma* f. sp. *glycinea* culture filtrate and cell wall on glyceollin production and cell viability in cell suspension cultures of soybean (*Glycine max* [Merr.]). *Phytochemistry 26*:2691-2694.

Bonhoff, A., and Grisebach, H. (1988). Elicitor-induced accumulation of glyceollin and callose in soybean roots and localized resistance against *Phytophthora megasperma* f. sp. *glycinea*, *Plant Sci. 54*:203-209.

Boydston, R., Paxton, J. D., and Koeppe, D. E. (1983). Glyceollin: a site specific inhibitor of electron transport in isolated soybean mitochondria. *Plant Physiol. 72*:151-155.

Chamberlain, D. W., and Paxton, J. D. (1968). Protection of soybean plants by phytoalexin. *Phytopathology 58*:1349-1350.

Chappell, J., Hahlbrock, K., and Boller, T. (1984). Rapid induction of ethylene biosynthesis in cultured parsley cells by fungal elicitor and its relationship to the induction of phenylalanine ammonia-lyase, *Planta 161*:475-480.

Cline, K., Wade, M., and Albersheim, P. (1978). Host-pathogen interactions. XV. Fungal glucans which elicit phytoalexin accumulation in soybean also elicit the accumulation of phytoalexins in other plants. *Plant Physiol. 62*:918-921.

Desjardins, A., Ross, L. M., Spellman, M. W., Darvill, A. G., and Albersheim, P. (1982). Host-pathogen interactions. XX. Biological variation in the protection of soybeans from infection by *Phytophthora megasperma* f. sp. *glycinea. Plant Physiol. 69*:1046-1050.

Duncan, D. R., and Paxton, J. D. (1981). Trifluralin enhancement of *Phytophthora* root rot of soybean. *Plant Dis. 65*:435-436.

Farmer, E. (1985). Effects of fungal elicitor on lignin biosynthesis in cell suspension cultures of soybean. *Plant Physiol. 78*:338–342.

Fett, W. F., Paxton, J. D., and Dickinson, D. B. (1976). Studies on the self-incompatibility response of *Lilium longiflorum. Am. J. Bot. 63*:1104–1108.

Fischer, D., Kogan, M., and Paxton, J. (1990a). Deterrency of Mexican bean beetle (Coleoptera: Coccinellidae) feeding by free phenolic acids. *Journal of Entomological. Sci. 25*:230–238.

Fischer, D. C., Kogan, M., and Paxton, J. (1990b). Effect of glyceollin, a soybean phytoalexin, on feeding by three phytophagous beetles (Coleoptera: Chrysomelidae, Coleoptera: Coccinellidae): dose vs. response. *Environmental Entomol. 19*:1278–1282.

Frank, J. A., and Paxton, J. D. (1970). Time sequence for phytoalexin production in Harosoy and Harosoy 63 soybeans. *Phytopathology 60*:315–318.

Frank, J. A., and Paxton, J. D. (1971). An inducer of soybean phytoalexin and its role in the resistance of soybeans to *Phytophthora* rot. *Phytopathology 61*:954–958.

Giannini, J. L., Briskin, D. P., Holt, J. S., and Paxton, J. D. (1988a). Inhibition of plasma membrane and tonoplast H+ transporting ATPases by glyceollin. *Phytopathology 78*:1000–1003.

Giannini, J. L., Holt, J. S., Paxton, J., and Briskin, D. P. (1988b). Glyceollin effects on *Phytophthora megasperma* f. sp. *glycinea* plasma membrane H^+ and Ca^+/ conductance (Abstr). *Phytopathology 78*:1503.

Graham, T. L. (1991). Flavonoid and isoflavonoid distribution in developing soybean seedling tissues and in seed and root exudates. *Plant Physiol. 95*:594–603.

Graham, T. L., and Graham, M. Y. (1991). Glyceollin elicitors induce major but distinctly different shifts in isoflavonoid metabolism in proximal and distal soybean cell populations. *Mol. Plant Microbe Interact. 4*:60–68.

Hansen, E. M., and Maxwell, D. P. (1991). Species of the *Phytophthora megasperma* complex. *Mycologia 83*:376–381.

Hart, S. V., Kogan, M., and Paxton, J. D. (1983). Effect of soybean phytoalexins on the herbivorous insects Mexican bean beetle and soybean looper. *J. Chem. Ecol. 9*:657–672.

Hauffe, K. D., Hahlbrock, K., and Scheel, D. (1986). Elicitor-stimulated furanocoumarin biosynthesis in cultured parsley cells:S-adenosyl-L-methionine:bergaptol and S-adenosyl-L-methionine:xanthotoxol O-methyltransferases, *Z. Naturforsch. 41c*:228–239.

Keen, N. T. (1975). Specific elicitors of plant phytoalexin production: Determinants of race specificity in pathogens?, *Science 187*:74–75.

Keen, N. T., and LeGrand, M. (1980). Surface glycoproteins: evidence that they may function as the race specific phytoalexin elicitors of *Phytophthora megasperma* f. sp. *glycinea*, *Physiol. Plant Pathol. 17*:175–192.

Keen, N. T., and Paxton, J. D. (1975). Coordinate production of hydroxyphaseollin and the yellow-fluorescent compound PAk in soybeans resistant to *Phytophthora megasperma* var. *sojae. Phytopathology 65*:635–637.

Keen, N. T., Lyme, R. L., and Hymowitz, T. (1986). Phytoalexin production as a chemosystematic parameter within the genus Glycine. *Biochem. System. Ecol. 14*:481-486.

Kittle, D. R., and Gray, L. E. (1982). Response of soybeans and soybean pathogens to soil fumigation and foliar fungicide sprays. *Plant Dis. 68*:213-215.

Klarman, W. L., and Gerdemann, J. W. (1963). Resistance of soybeans to three *Phytophthora* species due to the production of a phytoalexin. *Phytopathology 53*:1317-1320.

Kogan, M., and Paxton, J. D. (1983). Natural inducers of plant resistance to insects. *Plant Resistance to Insects* (P. A. Hedin, ed.), American Chemical Society, New York, pp. 153-171.

Kuhn, D., Chappell, J., Boudet, A., and Hahlbrock, K. (1984). Induction of phenylalanine ammonia-lyase and 4-coumarate:CoA ligase mRNAs in cultured plant cells by UV light or fungal elicitor. *Proc. Nat. Acad. Sci. USA 81*:1102-1106.

Kumar, S., Shukla, R. S., Singh, K. P., Paxton, J. D., and Husain, A. (1984). Glyceollin: a phytoalexin in leaf-blight disease of *Costus speciosus. Phytopathology 74*:1349-1352.

Lim, S. M., Hooker, A. L., and Paxton, J. D. (1970). Isolation of phytoalexins from corn with monogenic resistance to *Helminthosporium turcicum. Phytopathology 60*:1071-1075.

Murch, R. S., and Paxton, J. D. (1979). Rhizosphere anaerobiosis and glyceollin accumulation in soybean. *Phytopathol. Z. 96*:91-94.

Murch, R. S., and Paxton, J. D. (1980a). Rhizosphere salinity and phytoalexin accumulation in soybean. *Plant and Soil 54*:163-167.

Murch, R. S., and Paxton, J. D. (1980b). Temperature and glyceollin accumulation in *Phytophthora*-resistant soybean. *Phytopathol. Z. 97*:282-285.

Murch, R. S., and Paxton, J. D. (1980c). Environmental stress and phytoalexin accumulation in soybean. *Bull. Soc. Bot. France Actual. Bot. 127*:151-155.

Paradies, I., Konze, J. R., Elstner, E. F., and Paxton, J. D. (1980). Ethylene: indicator but not inducer of phytoalexin synthesis in soybean. *Plant Physiol. 66*:1106-1109.

Paxton, J. D. (1973). *Plants "self" recognition may aid disease control.* Illinois Research, Univ. Ill. Agric. Exp. Sta. *15*(4):13.

Paxton, J. D. (1974). Phytoalexins, phenolics and other antibiotics in roots resistant to soil-borne fungi. *Biology and Control of Soil-borne Plant Pathogens* (C. W. Brudhl, ed.), American Phytopathological Society, St. Paul, MN, pp. 185-192.

Paxton, J. D. (1979). Elicitors of incompatible host responses: the role of host cell membranes. *Recognition and Specificity in Plant Host-Parasite Interactions* (J. M. Daly and I. Uritani, eds.), pp. 153-163.

Paxton, J. D. (1981). Phytoalexins—a working redefinition. *Phytopathol. Z. 101*: 106-109.

Paxton, J. D. (1982). Degradation of glyceollin by *Phytophthora megasperma* var. *sojae. Active Defense Mechanisms in Plants* (R. K. S. Wood, ed.), Plenum Press, New York, pp. 343-344.

Paxton, J. D. (1983). *Phytophthora* root and stem rot of soybean. *Biochemical*

Plant Pathology (J. A. Callow, ed.), John Wiley and Sons, London, pp. 19–27.

Paxton, J. D. (1988a). Fungal elicitors of phytoalexins and their potential use in agriculture. *Biologically Active Natural Products: Potential Use in Agriculture.* American Chemical Society Symp. Ser. No. 380, pp. 109–119.

Paxton, J. D. (1988b). Phytoalexins in plant-parasite interactions. *Experimental and Conceptual Plant Pathology.* Oxford and IBH, New Delhi, pp. 537–549.

Paxton, J. D. (1989). Fungal elicitors of plant phytoalexins. *Handbook of Natural Toxins* Vol. 6, Marcel Dekker, New York, pp. 439–457.

Paxton, J. D. (1991a). Biosynthesis and accumulation of legume phytoalexins. *Mycotoxins and Phytoalexins* (R. P. Sharma, ed.), Telford Press, pp. 485–499.

Paxton, J. D. (1991b). Phytoalexins and their potential role in the control of insect pests. American Chemical Society Symposium Series No. *449*, pp. 198–207.

Paxton, J. D. (1991c). Assays for antifungal activity. *Meth. Plant Biochem. 6*:33–46.

Paxton, J. D., and Chamberlain, D. W. (1967). Acquired local resistance of soybean plants to *Phytophthora* spp. *Phytopathology 57*:352–353.

Paxton, J. D., and Chamberlain, D. W. (1969). Phytoalexin production and disease resistance in soybeans as affected by age. *Phytopathology 59*:775–777.

Paxton, J. D., Cruickshank, I. A. M., and Goodchild, D. (1974). Phaseollin production by live bean endocarp. *Physiol. Plant Pathol. 4*:167–171.

Paxton, J. D., and Jacobsen, B. J. (1974). A new method for screening soybeans for *Phytophthora* root rot resistance. *Proc. Am. Phytopathol. Soc. 1*:67–68.

Paxton, J. D., and Wilson, E. E. (1965). Anatomical and physiological aspects of branch wilt disease of Persian walnut. *Phytopathology 55*:21–26.

Perrin, D. R., and Bottomley, W. (1961). Pisatin: an antifungal substance from *Pisum sativum*. *Nature 191*:76–77.

Puckett, J. L., and Paxton, J. D. (1986). Auxin starvation enhances glyceollin production in soybean cell suspensions. *Agron. Abstr. 6* (Abstr).

Schafer, W., Straney, D., and Ciuffetti, L. (1989). One enzyme makes a fungal pathogen but not a saprophyte, virulent on a new host plant. *Science 246*:247–249.

Schmidt, W., and Ebel, J. (1987). Specific binding of a fungal glucan phytoalexin elicitor to membrane fraction from soybean (*Glycine max*). *Proc. Natl. Acad. Sci. USA 84*:4117–4121.

Sharp, J. K., Albersheim, P., Ossowski, P., Pilotti, A., Garegg, P., and Lindberg, B. (1984). Comparison of the structures and elicitor activities of a synthetic and mycelial-wall-derived hexa (β-D-glucopyranosyl)-D-glucitol, *J. Biol. Chem. 259*: 11341–11345.

Simcox, K., and Paxton, J. D. (1984). Induced pathogenicity in glyphosate-treated soybeans. *Phytopathology 74*:1271.

Snyder, S. H., and Bredt, D. S. (1992). Biological roles of nitric oxide. *Sci. Am. 266*:68–77.

Stäb, M., and Ebel, J. (1987). Effects of Ca on phytoalexin induction by fungal elicitor in soybean cells, *Arch. Biochem. Biophys. 257*:416–423.

Svoboda, W. E., and Paxton, J. D. (1972). Phytoalexin production in locally cross-protected Harosoy and Harosoy-63 soybeans. *Phytopathology 62*:1457–1460.

Vaillancourt, L. J., and Paxton, J. D. (1986). Evaluation of biological assays for the elicitation of glyceollin in soybeans. *Phytopathology 76*:1112 (Abstr).

Wade, M., and Albersheim, P. (1979). Race-specific molecules that protect soybeans from *Phytophthora megasperma* var. *sojae. Proc. Natl. Acad. Sci. USA 76*:4433–4437.

Yoshikawa, M., Matama, M., and Masago, H. (1981). Release of a soluble phytoalexin elicitor from mycelial walls of *Phytophthora megasperma* var. *sojae* by soybean tissues. *Plant Physiol. 67*:1032–1035.

Ziegler, E., and Pontzen, R. (1982). Specific inhibition of glucan-elicited glyceollin accumulation in soybeans by an extracellular mannan-glycoprotein of *Phytophthora megasperma* f. sp. *glycinea. Physiol. Plant Pathol. 20*:321–331.

5

Cellular Biochemistry of Phenylpropanoid Responses of Soybean to Infection by *Phytophthora sojae*

Terrence Lee Graham
The Ohio State University, Columbus, Ohio

I. INTRODUCTION

A. General Background

The fungal pathogen *Phytophthora megasperma* Drechs. f. sp. *glycinea* (Hildeb.) Kuan and Erwin (recently redesignated as *P. sojae*) can infect all vegetative soybean organs (Sinclair, 1982). It causes symptoms ranging from rapid pre- and postemergence seedling damping off to slowly spreading lesions on older plant tissues. The soybean–*P. sojae* association provides an excellent system for the study of molecular aspects of host–pathogen interactions. Resistance to the pathogen in soybean is determined by major dominant *Rps* genes occurring at seven loci, with several allelic forms at two of these loci. There are at least 25 known races of *P. sojae* which are characterized by their different specific interactions with these *Rps* genes (Schmitthenner, 1985). Where a given *Rps* gene provides effective resistance to a given race of *P. sojae*, the interaction between host and pathogen is termed incompatible. Where a given *Rps* gene provides no resistance to a given race of *P. sojae*, the interaction is termed compatible.

Antibiotic pterocarpan phytoalexins have generally been accepted to play a role in the *Rps* gene-mediated restriction of the spread of *P. sojae* in incompatible interactions (Keen and Yoshikawa, 1982). Research to date has led to the characterization of four isomeric pterocarpan antibiotics now referred to as glyceollins I–IV (Burden and Bailey, 1975; Lyne and Mulheirn, 1978; Lyne et al., 1976; Partridge and Keen, 1977; Sims et al., 1972). Both the timing and magnitude of the accumulation of the glyceollins differ markedly in compatible and incompatible infections and are consistent with the proposed role of the glyceollins in race-specific resistance (Darvill and Albersheim, 1984; Ebel, 1986; Keen and Yoshikawa, 1982).

The biotic elicitors, which may be responsible for glyceollin elicitation in infected tissues, include $\beta 1 \rightarrow 3$, $\beta 1 \rightarrow 6$-linked cell wall glucans from *P. sojae* (Ayers et al., 1976; Sharp et al., 1984) and α-1,4-D-galacturonides from the plant cell wall (Nothnagel et al., 1983). A synergistic interaction between the biotic elicitors of host and pathogen origin has been reported (Davis et al., 1986).

Investigations on the regulation of glyceollin biosynthesis have focused primarily on three specific enzymes of early phenylpropanoid and flavonoid metabolism [phenylalanine ammonia lyase (PAL), chalcone synthase (CHS) and chalcone isomerase (CHI)]. An early research project suggested that increased activity of these enzymes may not be required for the earliest glyceollin accumulation in infected tissues (Partridge and Keen, 1977). However, later studies, employing direct assays of enzymatic activity (Bonhoff et al., 1986a,b; Borner and Grisebach, 1982), radiolabel incorporation from early precursors (Moesta and Grisebach, 1981; Yoshikawa et al., 1978; Zahringer et al., 1978), and messenger RNA measurements (Esnault et al., 1987; Habereder et al., 1989; Schmelzer et al., 1984) suggested that glyceollin biosynthesis is accompanied by increased transcription and activities of these enzymes and by transient synthesis of the isoflavone precursor of glyceollin, daidzein. Recently, race-specific induction of several of the later enzymes in glyceollin biosynthesis was also reported in infected roots (Bonhoff et al., 1986a,b).

Although *P. sojae* infects all soybean seedling organs, race-specific resistance, characterized by rapid and large accumulations of glyceollin, is expressed somewhat differently in each individual organ. Moreover, a series of important studies demonstrated that the age and/or developmental state of the specific organ infected by *P. sojae* as well as environmental conditions, particularly light, strongly influence race-specific accumulation of the glyceollins (Bhattacharyya and Ward, 1986a,b; Lazarovits et al., 1981; Paxton and Chamberlain, 1969; Ward and Buzzell, 1983; Ward and Lazarovits, 1982). Furthermore, specific cell populations within a given organ may be involved in glyceollin accumulation (Graham and Graham, 1991b; Hahn et al., 1985; Yoshikawa et al., 1978).

A few laboratories have begun to examine discrete cellular aspects of the responses of various plant tissues to infection or elicitor treatment (for a recent review, see Graham and Graham, 1991c). For instance, several reports have elegantly elucidated localized changes in gene expression or defense product accumulation at the site of attempted penetration by pathogens (Cuypers et al., 1988; Pierce and Essenberg, 1987; Schmelzer et al., 1989; Snyder and Nicholson, 1990). In several more recent studies, changes in the expression of specific genes (Stermer et al., 1990) or distinctly different accumulations of defense products (Graham and Graham, 1991a,b) have been documented in cell populations proximal and distal to the sites of infection or elicitor treatment. In our own work, using highly sensitive high-performance liquid chromatography (HPLC) metabolite-profiling procedures (Graham, 1991a), we have demonstrated that distinctly different shifts in the expression of the phenylpropanoid pathways occur in proximal and distal soybean cell populations in response to the *P. sojae* wall glucan elicitor (Graham and Graham, 1991a,b). As discussed in more detail below, similar responses occur in tissues infected by incompatible races of *P. sojae*. These multiple and complementary defense responses are carefully coordinated by the host both spatially and temporally to provide effective resistance.

In this chapter are described some of the progress being made in understanding the multiplicity of phenylpropanoid defense responses in soybean and the cellular biochemistry of their regulation. Cellular biochemistry refers to those aspects of the phytoalexin and other phenylpropanoid responses related to spatial and temporal coordination of cellular response, cellular specialization in response, cell-to-cell communication, and signal perception and transduction.

B. Special Constraints in Cellular Research and Approaches to Overcome Them

Since the pioneering studies of Yoshikawa and coworkers (1978) on cellular aspects of the soybean–*P. sojae* interaction, it has become increasingly apparent that a clear correlation of phytoalexin accumulations to the incompatible (hypersensitive) response requires that one quantitate phytoalexin levels in the very discrete cell populations that are immediately relevant to pathogen containment. Generally these are the cells at or just in advance of the infection front.

Since a natural infection front is often composed of just a few cells surrounding the point of attempted penetration by the pathogen, a key problem has been the development of protocols to allow the measurement of molecular responses at a cellular level. Methods such as in situ hybridization, immunolocalization, and HPLC have been particularly useful in measuring responses at the mRNA, protein, and product levels, respectively. A recent review summarizes the development and use of some of these methods, some of their limitations, and the insights they are providing us (Graham and Graham, 1991c).

Even though we have made considerable progress in measuring molecular responses in discrete cell populations, very few methods are sensitive enough to measure responses at a truly cellular level. Those that can be employed at a single-cell level (e.g., in situ hybridization and immunolocalization) are qualitative or at best semi-quantitative. This has led researchers to employ protocols which increase the size of the cell populations undergoing hypersensitivity. For example, Yoshikawa et al. (1978) circumvented the problem of sensitivity of detection by creating an artificially large infection front by uniformly inoculating the surface of a longitudinally wounded hypocotyl. The use of cell suspension cultures has also been employed to rapidly subject a large population of small cell clusters to a pathogen or elicitor. Another approach has been to take advantage of the fact that under certain conditions some hypersensitive responses progress through tissues as a very slowly spreading necrotic lesion. For instance, soybean cotyledons show very sharp and localized necrotic lesions to incompatible isolates of *Phytophthora sojae* under high light intensity, slowly spreading necrotic lesions under low light intensities, and virtually no race-specific resistance in the dark (Graham et al., 1990). One can gain a great deal of information from such a system by studying the responses under varying conditions and either analyzing cells at a given distance from the point of inoculation over time or sampling cell populations at various distances from the point of inoculation at a given time, or preferably both.

Another constraint to effective cellular research is that there are usually many tissues in a given organ and these tissues are sometimes themselves only a single cell layer thick. As pointed out by Barz and Hoesel (1979), distributions of flavonoids are often tissue-specific within an organ. Thus, work with hypocotyls at a cellular level is complicated by the fact that a pathogen may spread longitudinally within a tissue or radially through a series of different tissues. Research on roots shares these problems with the added complication that the longitudinal zone which is subject to infection is often near the root tip and undergoing rapid growth and developmental changes as well. Tissue culture again has the potential advantage of greater cellular homogeneity but suffers from the fact that the cells are often dedifferentiated, they are not in a natural environment, and it is very difficult to measure cell-to-cell communication in a meaningful way. In our laboratory, we began our studies with cell cultures (Pierson and Graham, 1987) but quickly shifted to the use of cotyledons when we realized the comparative architectural simplicity and cellular uniformity of this organ. Soybean cotyledons are largely composed of tightly aligned columns of mesophyll parenchyma cells of very uniform dimension. As we will describe in more detail below, these attributes and the tight packing of these cells make them nearly ideal in measuring cellular response and cell-to-cell communication.

A final major constraint to effective cellular research is the extraordinary complexity of the interactions taking place in infected tissues. The fact that one is trying to observe discrete molecular events in the interaction between two living systems leads to great difficulties in reproducibility. Moreover, there are almost certainly multiple signaling processes taking place, leading to overlapping or possibly conflicting effects on the process under measurement. These problems are greatly compounded by the fact that in most systems the genetics of either the host or pathogen (or both) is ambiguous, a fact that is often overlooked in the literature. For instance, although near-isogenic isolines of soybean are available carrying the various *Rps* resistance genes, nearly all work on the soybean–*P. sojae* interaction to date has used genetically undefined field or laboratory isolates of the fungus. Recent efforts at obtaining pure-breeding races of *P. sojae* have shown that although such isolates may display a particular race phenotype, they often harbor the genes which are determinant for other race characteristics as well (Bhat et al., 1993). Even though the interaction of such an isolate may show the appropriate gross symptomatic phenotype, it is completely unknown as to how such genetic heterogeneity may affect more subtle interactions measured at a molecular level. Also, different researchers use different isolates of a given race or favor the use of different races in their studies. In addition to heterogeneity in isolates as regards race characteristics, different isolates of a given race can differ greatly in their aggressiveness. The aggressiveness of a particular race can dramatically effect the spatial and temporal aspects of interaction with the host at a cellular level (Graham and Graham, unpublished).

Electrophoretic karyotyping of a large number of field and laboratory isolates of *P. sojae* (our results, unpublished) has confirmed the remarkable genetic heterogeneity of different isolates of *P. sojae*. As many as 10 different chromosomes are found in various *P. sojae* isolates, but no isolate carries all of these and some isolates share as few as one or two chromosomes in common with others.

In the soybean system, two complementary approaches are being undertaken to circumvent some of these problems associated with examination of intact infected tissues. First of all, pure-breeding races are being developed from defined field parentage (Bhat et al., 1993). These races and subsequent variants generated by mutation or transformation will greatly facilitate the interpretation of infection studies. Second, as described in detail below, we have determined that *P. sojae* wall glucan elicitor preparations not only induce all of the multiple phenylpropanoid responses seen in incompatible infections, but induce them with similar spatial and temporal coordination. This crucially important finding allows us to examine host responses to purified elicitor molecules and thus isolate and study very specific cellular responses of the host.

II. DEVELOPMENTAL REGULATION AND GENETICS OF THE DISTRIBUTION OF PHENYLPROPANOID DERIVED METABOLITES

Before one addresses the role of alterations in phenylpropanoid metabolites in the defense responses of various organs, it is critical to thoroughly understand the constitutive distribution of these metabolites within these organs and the developmental and genetic parameters which influence their turnover. Only in this manner can experiments on *induced* changes associated with infection or elicitor treatment be properly designed and interpreted.

A. Distribution and Developmental Regulation

Various soybean organs and root and seed exudates display characteristic and complex aromatic metabolite profiles (Graham, 1991b). Over a hundred individual UV-absorbing metabolites can be separated by gradient HPLC and only a handful of those in any given organ have been identified. However, the predominant constitutive metabolites characteristic of each of these organs have been isolated and characterized. These include the isoflavones (daidzein, genistein, glyceteine, formononetin, and biochanin A) and their conjugates, and the flavonols (kaempferol, quercetin, and isorhamnetin) and their conjugates. Generally, the isoflavones are the predominant metabolites in all young soybean seedling organs and in older cotyledons, stem and root tissues, and the flavonols are the major metabolites in older leaf tissues (Graham, 1991b).

Although present at detectable levels in soybean tissues, formononetin and biochanin A are comparatively minor metabolites. However, large constitutive pools of conjugates of daidzein, genistein (Graham, 1991b), and, occasionally, glyceteine (Morris et al., 1991) are present in various organs. The presence of high levels of conjugates of daidzein is of particular interest since daidzein is the first committed metabolite in the synthesis of glyceollin. Genistein, although not considered a precursor of glyceollin, possesses antibiotic activity against *P. sojae* (Rivera-Vargas et al., 1993). There is little information on the biological activities of glyceteine.

Soybean tissues generally contain low amounts of the free isoflavones (Graham, 1991b). The only exception is daidzein, which is sometimes present in substantial amounts in root tissues. The malonylglucosyl conjugates of the isoflavones are the predominant forms found constitutively in all soybean organs. The glucosyl conjugate of daidzein is present in lower amounts in all tissues, while the glucosyl conjugate of genistein is found at substantial levels only in lower hypocotyl sections and the crown. The malonylated conjugate of daidzein is the predominant metabolite in roots and hypocotyls, particularly in the root tip, where it reaches the highest levels in the seedling. Lower amounts of the malonylated conjugate of genistein are also present in the root and hypocotyl, and both daidzein and genistein conjugates are secreted by soybean roots. The detection of free daidzein and genistein in soybean root exudates may be the result of enzymatic hydrolysis during collection or processing of samples (Graham, 1991b). Very much lower amounts of conjugated glyceteine (peak 19.5 of Graham, 1991b) have since been identified by Morris et al. (1991), present predominantly in the lower hypocotyl.

In cotyledon tissues, malonylglucosyl conjugates of daidzein and genistein are at about equal and relatively high levels. Although both are present in very young leaves, the malonylated conjugate of genistein is the only major isoflavone in older leaves (Graham, 1991b).

The seed contains vast stores of the isoflavones (Graham, 1991b). The relative amounts and distribution in the various parts of the embryo are very similar to those seen in the counterparts of young germinated seedlings, suggesting that there may be very little net synthesis or turnover of these compounds in the first few days postgermination (Graham, unpublished). The deposition of daidzein and genistein and their conjugates has also been investigated during seed development and maturation (Graham and Olah, unpublished). As in seedling tissues, the free isoflavones do not accumulate to appreciable amounts during seed development. However, accumulation of the glucosyl conjugate of daidzein occurs first and then falls off as the malonylglucosyl conjugate of daidzein accumulates later in embryo development. The malonylglucosyl conjugate of genistein accumulates later than the malonylglucosyl conjugate of diadzein without prior accumulation of the corresponding glucosyl conjugate. The timing of these various accumulations is interesting and may reflect developmentally programmed induction of the corresponding enzymes.

During seed imbibition, the conjugates of both daidzein and genistein are released in a continuous but saturable manner (Graham, 1991b). If the imbibition medium is replaced, the exudate quickly reestablishes saturating amounts of the isoflavones. This suggests that the very large stores of daidzein and genistein present in the seed (and, as noted above, in the root) may be available in the rhizosphere, where they may play a role as chemoattractants for *P. sojae* (Morris and Ward, 1992).

Light has a particularly strong effect on the amounts of and distribution of the isoflavones in soybean tissues (Graham, 1991b). In the dark, isoflavone levels in the root tips are greatly reduced, while those in the cotyledon are higher. It is possible that at least part of the root isoflavone pools may be transported to the roots from the larger stores present in the cotyledons (Graham, 1991b).

Older soybean leaves are predominated by glycosides of the flavonols kaempferol, quercetin (Buttery and Buzzell, 1975), and isorhamnetin (LeVan and Graham, unpublished). The genetics underlying the formation of kaempferol and quercetin and their glycosides has been elegantly described by Buttery and Buzzel (1975). Parallel studies on isorhamnetin remain to be done. These various flavonols are not present at readily detectable levels except in the leaf (Graham, 1991b).

B. Genetics

Although the genetics of flavonol and flavonol glycoside formation in soybean has been described, there is little information available on the genetics of isoflavone and isoflavone conjugate formation. However, evidence to date suggests much less genetic diversity in the nature of the isoflavone aglycones and their conjugates. We have examined over 100 soybean lines of various genetic backgrounds and none of these vary markedly in the amounts or distributions of the isoflavones. Among the lines examined are: 1.) the Williams isoline series (Williams, Williams 79, Williams 82) carrying no *Rps* resistance genes and the *Rps* 1c and 1k genes respectively, 2.) the Harosoy differentials, carrying a wider range of *Rps* genes integrated back into the Harosoy background, 3.) soybeans from all of the various flavonol genetic groups described by Buttery and Buzzel (1975), 4.) the *Bradyrhizobium japonicum* response mutants of the cultivar Bragg (Abbasi et al., unpublished), including the nonnodulating mutants nod49 and nod139 (Carroll et al., 1986), and the supernodulating mutants nts382 and nts1007 (Carroll et al., 1985), and 5.) *Glycine sojae*. Thus the presence and distribution of the isoflavones is not influenced by the *Rps* genes, the genes controlling flavonol metabolism, or the genes controlling *Bradyrhizobium* nodulation. The fact that *Glycine sojae* contains the same compounds with similar organ-specific distribution (Abbasi et al., unpublished) underscores the lack of genetic diversity in these compounds within the genus.

III. MULTIPLICITY OF INDUCED PHENYLPROPANOID RESPONSES IN INFECTED AND ELICITOR TREATED TISSUES AND SPATIAL AND TEMPORAL ASPECTS OF THEIR REGULATION

The isoflavone daidzein is the first committed metabolite in the biosynthetic pathway to the soybean phytoalexin glyceollin. As noted above, daidzein and the related isoflavone genistein were previously thought to be transiently synthesized only after pathogen attack or upon treatment with elicitors of the isoflavonoid phytoalexins (Ebel, 1986). The discovery that very large pools of these isoflavones were present in soybean seedling tissues as preformed conjugates (Graham et al., 1990) thus raised obvious questions as to the potential role(s) of the conjugates in disease resistance and/or in glyceollin accumulation. In addition, the recent discovery that large depositions of cell wall phenolics occur in response to infection or elicitor treatment (Graham and Graham, 1991a) suggested a possible complementary role of these phenylpropanoids in defense as well.

As described below, studies on the relative roles of these various phenylpropanoids in infected and elicitor-treated tissues confirm that soybeans may possess a multitude of discrete yet complementary mechanisms of defense based on the differential control of various alternative phenylpropanoid pathways (Fig. 1). Intriguingly, different populations of cells appear to play distinct roles in the differential deployment of the responses. The host appears to coordinate these defense strategies in a highly sophisticated manner, not only through the regulation of complex metabolic pathways with complementary functions but through the careful orchestration of

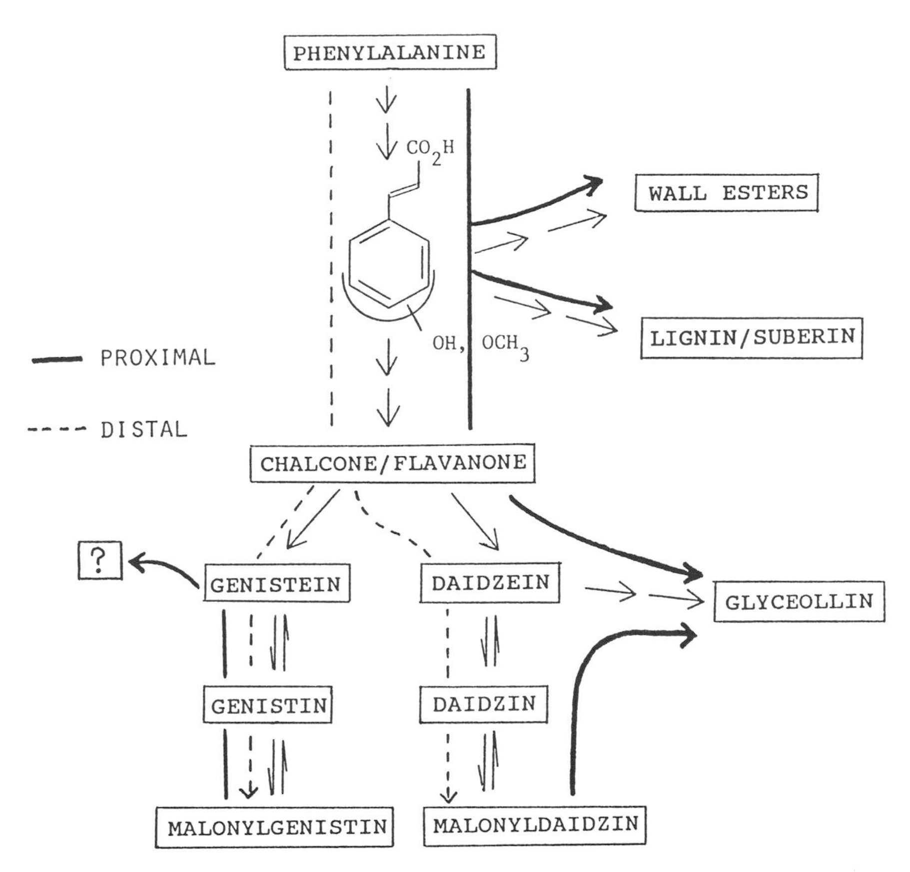

Figure 1 Phenylpropanoid defense responses of proximal and distal soybean cells.

cellular communication and response. In addition, organ- or tissue-specific, spatial, developmental, and environmental effects on glyceollin accumulation and/or resistance have been well documented in the soybean–*P. sojae* interaction. Our preliminary results suggest that some of these organ- or tissue-specific effects may also be due to specific differences in the regulation of the various alternative pathways outlined in Fig. 1.

A. Responses of Infected Tissues

Although all soybean seedling organs are infected by *P. sojae* and all of these tissues show expression of race-specific resistance conditioned by the *Rps* genes, we began our infection and elicitor studies with cotyledon tissues for several reasons (Graham et al., 1990). By far the most valuable aspect of the cotyledon system, however, is the fact that cotyledons are made up largely of tightly aligned columns of remarkably uniform mesophyll parenchyma cells. This attribute greatly simplifies the examination of spatial and temporal events relating to the infection front (Graham et al., 1990) and has provided us with a very powerful tool to examine cell-to-cell communication following elicitor treatment (Graham and Graham, 1991b).

Our studies with infected cotyledon tissues (Graham et al., 1990) demonstrated that there is a rapid and nearly complete net hydrolysis of both diadzein and genistein conjugates at the infection front in incompatible infections between 12 and 24 hr, followed by the accumulation of glyceollin during the period 24–36 hr. Both the stoichiometry and timing of the responses were consistent with a role of the hydrolysis of the daidzein conjugates in the accumulation of glyceollin. Free genistein transiently accumulated to levels as high as 800 nmol/g tissue between 24 and 48 hr. Subsequent research showed that genistein is toxic to *P. sojae* (Rivera-Vargas and Graham, 1993) at levels well below those released at the infection front. In compatible infections, the release of both isoflavones from their conjugates was much later, well after the infection front had progressed through the tissue, and very low levels of glyceollin accumulated only after 72 hr (Graham et al., 1990). Thus, in cotyledon tissues carrying the *Rps* 1c gene for resistance, infection results were consistent with two race-specific events contributing to the antibiotic containment of *P. sojae*. First is the rapid release of the antibiotic genistein and glyceollin precursor diadzein from preformed conjugates, and second is the somewhat later formation of the more elaborate antibiotic glyceollin from daidzein. Although limited net de novo synthesis of 5-deoxyisoflavones (total daidzein and glyceollin) was also seen in this study, suggesting that newly synthesized daidzein could also be contributing to the glyceollin accumulations (Graham et al., 1990), the results suggested a potentially more important role of release of daidzein from preformed conjugates in the early incompatible response. In contrast, net accumulations of daidzein, as its malonylated conjugate, occurred in tissues ahead of the infection front (Graham, unpublished results), suggesting that de novo synthesis and conjugation was favored over conjugate hydrolysis in these outlying tissues. These infection results were our first indications that distinctly different metabolic switches were being thrown in different cell populations in the overall incompatible response. The accumulation of isoflavone conjugates in cells ahead of the infection front is intriguing and suggests that the signal(s) which elicit isoflavone conjugate synthesis might spatially precede those for conjugate hydrolysis and glyceollin accumulation, thus "setting up" the cells in advance of the infection front to respond with a larger free isoflavone release (and thus genistein and glyceollin response) if the infection front progressed into this tissue. In essence, this distal cell response could raise the defense potential of these uninfected tissues.

Subsequent studies in our laboratory have examined the potential roles of the diadzein and genistein conjugates in infected hypocotyl, root, and leaf tissues as well. As expected, we have found that the design and interpretation of experiments with these organs (especially roots) is much more complicated due to the greater difficulty in sampling discrete tissues within these organs. However, to date these studies generally support the cotyledon results, including the net hydrolysis of conjugates at the infection front, the subsequent accumulation of glyceollin, and the net accumulation of the isoflavone conjugates ahead of the infection front. The relative contribution of the various responses outlined above for cotyledon tissues may be somewhat different in these organs due to differences in the amounts of daidzein and genistein in these tissues. On the other hand, based on very preliminary studies with hypocotyl and leaf tissues, Morris et al. (1991) suggest that the conjugates play only a "secondary" role in the glyceollin response to infection, even though the timing of hydrolysis of the conjugates in their studies is actually consistent with a primary role in glyceollin accumulation. Unfortunately, these studies (Morris et al., 1991) were very limited in their examination of specific temporal and spatial events, and those events which were followed were not correlated to the infection front per se, which was never localized.

B. Responses to the Glucan Elicitor from *P. sojae*

Results with infected tissues prompted us to reexamine the activity of the *P. sojae* wall glucan elicitor which previously was characterized for its elicitation of glyceollin (Darvill and Albersheim, 1984). We wished to determine if some of the isoflavone turnover events seen in incompatible infections could be triggered by these glucan preparations. Our initial finding, that the *P. sojae* wall glucan also induced large accumulations of the isoflavone conjugates in soybean tissues, was very intriguing (Graham and Graham, 1991b). Even more relevant to the results from infected tissues was the fact that, while glyceollin was induced in cells proximal to the site of elicitor treatment, the isoflavone conjugates accumulated in massive amounts in cells as many as 25–30 cells from the elicitor treatment. Thus it seemed that the glucan induced the two discrete (proximal and distal) *biosynthetic* events seen in infected tissues, but not the net hydrolysis of conjugates observed at the infection front. In these studies, however, we further reported that the isoflavones also showed net accumulation in proximal cells, the amounts relative to glyceollin depending on the amount of elicitor, the age of the cotyledons, and the assay conditions.

This differential control of the *relative* amounts of glyceollin and isoflavone synthesis in proximal cells suggested that we might still be monitoring a mixed cell response and prompted us to evaluate cellular events in proximal cells in more detail. As noted in our initial work (Graham and Graham, 1991b), the proximal cell layer (four cells thick) was enriched for "proximal" cells, but might contain some cells showing the "distal" response as well. This possibility led us to examine even thinner cell layers at various times and under varying conditions. As we selectively sampled cells more and more immediately proximal to the elicitor treatment, we uncovered a rapid net hydrolysis of the conjugates of both daidzein and genistein within the first 4 hr (Graham and Graham, unpublished). The levels of the malonylated conjugates of daidzein and genistein fall precipitously within this short period to levels often only a third of their original amount. Following this net hydrolysis, however, is a later period of renewed synthesis of daidzein conjugates. Depending on the conditions of the assay and the relative flux into glyceollin, total daidzein levels may partially recover or even undergo a net accumulation. Genistein conjugates, on the other hand, show little reaccumulation in this cell layer. In distal cells, conjugates of both isoflavones show net accumulations with no early hydrolysis.

Thus even the net hydrolysis of the isoflavone conjugates which accompanies the incompatible response at the infection front is seen in proximal cells in response to elicitor, although, for reasons we do not yet understand, the hydrolysis is more prolonged in infected tissues. Net isoflavone conjugate hydrolysis also occurs upon infection of alfalfa tissues by *Ascochyta imperfecta* (Olah and Sherwood, 1973). In these studies, the authors found that glycosidases of both host and pathogen origin could be detected in infected tissues but that isozymes corresponding to the pathogen predominated. A similar contribution of both host and pathogen enzymes may be operating in soybean. Although our results in elicitor-treated tissues suggest that host enzymes may contribute to this response, we have demonstrated the presence in *P. sojae* of enzymes capable of rapidly hydrolyzing the daidzein and genistein conjugates (Rivera-Vargas and Graham, 1993). Intriguingly, these enzyme activities are associated with the growing hyphal tip in *P. sojae*, which places them in a key position to have an impact on events at the infection front. The *P. sojae* wall glucan may activate only the host enzymes, giving rise to a more transient net effect. Since conjugate hydrolysis occurs much later in the compatible response (well after the infection front passes through the tissue), either the host or the pathogen enzymes or both may be temporarily suppressed in compatible infections. It is intriguing to speculate that control of conjugate hydrolysis at the infection front by the host and pathogen may play a very central role in defining the race-specific outcome of the interaction between *P. sojae* and soybean by contributing to, or failing to contribute to, the net and timely release of genistein and daidzein. To decipher the relative roles of host and pathogen in conjugate hydrolysis in planta will require that the enzymes of both host and pathogen origin be characterized and their regulation examined in situ during infection.

Although the HPLC profiling protocols gave us a very complete picture of the turnover of various soluble aromatic metabolites in discrete cell populations, they did not allow us to monitor the accumulation of wall-bound phenolics. Elicitor-induced accumulations of glyceollin in the cut cotyledon assay are nearly always accompanied by a browning of the cut cotyledon surface, an increase in its hydrophobicity and a remarkable toughening of the uppermost cell layer. This prompted us to examine the effects of the *P. sojae* wall glucan elicitor on peroxidase isozymes and phenolic polymer deposition. Although wounding alone caused the gradual induction of a specific group of anionic peroxidases and the subsequent deposition of both lignin and suberin like polymers, the *P. sojae* wall glucan was found to greatly increase the rapidity of this response (Graham and Graham, 1991a). Within just 4 hr of wall glucan treatment, phenolic polymer levels were over 10 times those in wounded controls. This more rapid induction of the phenolic polymers was accompanied by correspondingly earlier induction of the anionic peroxidases. Although the peroxidases were also induced substantially in distal cells, there was very little phenolic polymer deposition, possibly due to a lack of appropriate substrates or sufficient oxygen for this highly oxidative process.

Of additional interest was the fact that large quantities of monomeric hydroxycinnamic acids (coumaric and ferulic) as well as several unidentified phenolics were incorporated into the cell walls in ester linkages (Graham and Graham, 1991a). The accumulation of such wall-esterified phenolic acids occurs in other species as well and has previously been suggested to play a potential role in disease resistance (Hahlbrock and Scheel, 1989). Interestingly, the deposition of these acids would most likely be mediated by a transferase and not a peroxidase. Thus their deposition suggests yet another response that may be induced by *P. sojae* wall glucan. Their presence in the wall is intriguing and opens the possibility that these phenolics could serve as a store for future, more rapid wall polymer accumulations. This possibility would be even more interesting if the esterified hydroxycinnamic acids were induced in distal cells. We have yet to examine distal cells for wall-bound phenolic esters.

Events similar to those seen in response to the glucan elicitor are also induced at the infection front in incompatible infections (Graham and Graham, unpublished). Thus a very complex series of phenolic depositions occurs in the walls of cells proximal to the infection front or site of elicitor treatment. Of greatest interest, however, is the very early and massive nature of these responses when compared to isoflavone hydrolysis or glyceollin accumulation. The wall phenolic responses, taken together, represent not only a much earlier but a 17-fold greater commitment of phenylpropane skeletons than the glyceollin response (Graham and Graham, 1991a). It is possible that de novo synthesis of phenylpropanoids in proximal cells is predominantly directed into the wall defenses. This would explain the low level of net de novo accumulation of soluble phenolics, including daidzein in these tissues and the apparent use of the preformed daidzein conjugates for glyceollin synthesis. In distal cells, on the other hand, de novo synthesis is directed toward isoflavone synthesis and conjugation rather than the phenolic polymers.

The various events described at length above are summarized in Fig. 1. It is clear that *P. sojae* infection or elicitor treatment turn on a remarkably complex array of phenylpropanoid-related responses. These responses, in turn, are coordinated both spatially and temporally in a highly sophisticated manner by the host. We have seen that this coordination involves a very precise "gating" of metabolites into and out of the various phenylpropanoid pools (isoflavone conjugate, wall phenolic, and glyceollin) in several distinct cellular zones. Although *P. sojae* wall glucan preparations are apparently sufficient to trigger all of these cellular events, it is obvious that additional regulation of cellular response is taking place.

IV. SIGNAL MOLECULES FOR THE REGULATION OF CELLULAR RESPONSE

The multiplicity of soybean phenylpropanoid responses to infection or elicitor treatment, the sophisticated differential deployment of these responses in different cell populations, and the fact that *P. sojae* wall glucan preparations induce all of the various coordinated responses seen in incompatible infected tissues make this system a particularly rich one for studies on signal perception, signal transduction, and cell-to-cell communications. The ability to follow the various responses so clearly in soybean cotyledons has proven to be a particularly powerful tool for these studies for reasons alluded to above. A major emphasis of our lab is thus to begin to decipher the various molecular and cellular components involved in perception, signal generation, and signal transduction in response to the *P. sojae* wall glucan elicitor.

An obvious first step in such an effort, however, is to further characterize the signals that initiate or condition the various responses. This is particularly important due to the inherent heterogeneity of polysaccharide elicitor preparations. Questions we are currently addressing include: Are the various wall glucan responses described above due to a single elicitor or are there multiple elicitors present in the glucan preparations? Are there host factors, such as the pectic oligomers, that act as synergists or coelicitors? If so, how do these various pathogen- and host-derived primary signals work together to initiate or propagate the proximal and distal cell responses to the glucan elicitor? Is there actually a separate signal for the distal response? If so, is this generated in the plant as a result of the proximal response or is it released or processed from one of the primary proximal elicitor signals? Since age and organ developmental state influence the various responses, what are the roles of cellular growth regulators in the regulation of the various cell responses? Light and nutritional status also have profound effects on the responses. Are these later effects simply influencing the general physiology of the cell or do they have more specific and selective influences on signal perception and transduction?

To simplify discussions, we have defined two overall types of regulatory processes: 1.) primary elicitor or signal processes which serve to trigger the phenylpropanoid defense responses, and 2.) secondary molecular or cellular processes which may serve to condition or modify specific cellular response(s) to the primary signals. Although these various processes are highly intertwined, we have begun to gain some insights into their nature.

Being an external signal, *P. sojae* wall glucan per se is clearly a primary elicitor. However, as we have demonstrated, this elicitor differentially induces a variety of phenylpropanoid responses in different soybean cell populations, including isoflavone conjugate hydrolysis, isoflavone conjugate accumulation, glyceollin accumulation, wall-esterified hydroxycinnamic acid accumulation, and phenolic polymer deposition. These various activities could be due to 1.) the presence of more than one primary elicitor activity in the *P. sojae* wall glucan preparation, 2.) the generation, during infection or elicitor treatment, of additional primary or secondary elicitor signals in situ (e.g., host cell wall oligogalacturonides) which act independently or as coelicitors with *P. sojae* wall glucan, 3.) inherent differences in the responses of specific cell populations to the same elicitor, or 4.) differential *conditioning* of cellular responses to the elicitor by cellular growth regulators. As detailed below, preliminary evidence from our laboratory and others suggests that each of these mechanisms may operate to achieve differential control of these pathways in specific situations. What is not clear is how these various mechanisms of regulation work together to achieve the sophisticated and precise control of the various pathways suggested by infection and elicitor studies with various tissues.

A. Multiple Primary Elicitors

As noted above, we have demonstrated that the classical unfractionated *P. sojae* wall glucan preparation of Ayers et al. (1976) triggers all of the various phenylpropanoid responses that occur in incompatible infections. Moreover, the responses induced by the wall glucan and incompatible infections arc nearly identical in regard to the precise composition of the products and their timing and spatial coordination. This finding underscores the potential importance of the wall glucan as a primary determinant of defense gene stimulation in soybean and provides a point of focus for future studies.

The possible existence of multiple primary signals in the classical *P. sojae* wall glucan preparation has been documented in several cases. First of all, by its very nature, the branched β1,3/β1,6 glucan itself is heterogeneous in both size and position of branch points. Thus, depending on how they are generated, elicitor fragments of different sizes and activities can be released. Albersheim and coworkers have elegantly demonstrated that the highest per unit activity in acid hydrolysates is found in a seven-residue branched β1,3/β1,6 glucan (Sharp et al., 1984). However, using enzymatic hydrolyses with a soybean β-1,3-endoglucanase, Yoshikawa (1988) concluded that the highest glyceollin elicitor activity is present in the largest wall glucan fragments. Thus, it is still not totally clear as to which subfractions of the *P. sojae* glucan possess the most potent glyceollin elicitor activity in planta. Nonetheless, the enzymatic release of the glucan elicitor from the pathogen cell wall has been attributed to a now cloned β-1,3-endoglucanase (Takeuchi et al., 1990a,b) and elegant research on the receptor for highly defined glucan elicitors is well underway (Cheong and Hahn, 1991; Cheong et al., 1991; Cosio et al., 1992).

There is also evidence for elicitors with distinctly different activities in the *P. sojae* wall glucan preparations. In addition to the glucan elicitor active in soybean, Hahlbrock and coworkers identified a minor proteinaceous component of the *P. sojae* wall glucan preparation which serves as a specific elicitor of the furanocoumarin phytoalexins in parsley (Parker et al., 1988). In preliminary experiments, we have discovered that if an autoclaved *P. sojae* glucan preparation is filtered through 0.2-μm filters and thus separated into soluble and insoluble fractions, the insoluble wall glucan preparation is a very potent elicitor of isoflavone conjugates, glyceollin, and phenolic polymers. The soluble portion is a relatively poor elicitor of glyceollin or the phenolic polymers, but is an excellent elicitor of the isoflavone conjugates. Since this soluble elicitor also induces the distal buildup of the isoflavone conjugates, it may be at least partially responsible for the isoflavone response of soybean tissues to infection or unfractionated *P. sojae* wall glucan preparations. Whether this elicitor is a minor, nonglucan, component of the wall preparation or a smaller, defined fragment of the β1,3/β1,6 glucan is unknown.

Related to the *P. sojae* wall glucan is the much smaller cytoplasmic β1,3/β1,6 glucan mycolaminaran. Mycolaminaran is also a heterogeneous glucan containing 30–36 glucose residues. It exists in both phosphorylated and nonphosphorylated forms which vary in their relative levels during the life cycle of *Phytophthora*. At certain stages in the life cycle, the neutral form of mycolaminaran can reach levels as high as 30% of the dry weight of the mycelia (Wang and Bartnicki-Garcia, 1973, 1974). Mycolaminaran has been shown to be a relatively weak elicitor of glyceollin (Keen et al., 1983). Its effects on the other pathways outlined in Fig. 1 have not been examined. Structural analogs of molecules often play a role as competitive inhibitors or alternative ligands. Despite its structural similarity to the *P. sojae* wall glucan, the possible role of mycolaminaran in modifying or complementing the activity of the *P. sojae* wall glucan has also not been examined.

Host oligogalacturonides have also been proposed as primary elicitors or elicitor synergists (Darvill and Albersheim, 1984; Davis et al., 1986; Ebel, 1986). Although they could be classified as secondary, in the sense that they are released only after infection, since they are generated in the apoplast we will treat them in our discussions as primary cell signals. Like *P. sojae* wall glucan, pectic oligomers are by nature heterogeneous. Distinctly different activities are associated with oligomers of different sizes. While larger fragments, with an optimal size of 12 residues, are most active as glyceollin elicitors in soybean (Nothnagel et al., 1983), fragments of 2–10 residues are optimal as elicitors of protease inhibitors in potato (Ryan, 1981). Recently, the smaller oligogalacturonides (optimal activity of 7 residues) have also been shown to induce the deposition of lignin in castor bean cultures (Bruce and West, 1989).

Not surprisingly, pectate degrading enzymes have also been demonstrated to possess elicitor activity (Ebel, 1986). The activity of the pectate lyase from *Erwinia carotovora* has been particularly well characterized. This enzyme has been reported as a glyceollin elicitor and as a synergist to glyceollin elicitation by *P. sojae* wall glucan (Davis et al., 1986). Recently, the discovery of proteinaceous inhibitors of endopolygalacturonidase (Cervone et al., 1989) led to the hypothesis that the expression of these inhibitors in host tissues could lead to limited digestion of host cell wall pectin fractions and thus the accumulation of the oligogalacturonide elicitor synergists or to factors which lead to host cell death (Doares et al., 1989).

Preliminary experiments in our laboratory, using a purified pectate lyase preparation from *Aspergillus japonicus*, suggests one possible explanation for the synergism exerted by these preparations. Although we observed no changes in the various phenylpropanoids in response to pectate lyase alone, treatment of soybean cotyledon tissues with low levels of *P. sojae* wall glucan in the presence of pectate lyase led to an increase in the relative flux of metabolites into glyceollin and a marked decrease in the accumulation of isoflavone conjugates. Pectate lyase thus has the effect of shifting somewhat the balance of response to *P. sojae* wall glucan specifically away from the isoflavones and in favor of glyceollin. Since host cell wall degradation might be expected to occur at the infection front, this could be an additional factor contributing to the local accumulation of glyceollin vs. distal accumulation of isoflavones. Although a direct elicitor synergistic activity of the released pectic oligomers has been proposed, it seems possible that dissolution of the middle lamella of cells in the vicinity of the infection front may also simply expose more cells to the wall glucan elicitor, thus facilitating its activity.

B. Secondary Regulation and Conditioning of Cellular Responses

Wound-Associated Competency Factors for Cellular Response to the P. sojae *Wall Glucan Elicitor*

Classically, elicitors of the glyceollin response of soybeans have been assayed using a cut cotyledon assay in which the cut abaxial surface of excised cotyledons is exposed to a solution of elicitor (Frank and Paxton, 1971). Over a period of 24–48 hr the glyceollins are produced and diffuse into the elicitor droplet, which is then collected for UV or HPLC analysis. As described above, for many of our experiments, we have used a modification of this assay (Graham and Graham, 1991b) in which the elicitor droplet is allowed to dry onto the cut surface and a column of cells is then harvested from the cotyledon. The cell column is then thin-sliced into disks and the disks are extracted separately to allow examination of proximal and distal cell responses. Several years ago we also initiated a minimal-wounding cotyledon infiltration assay in which cotyledons on intact plants are infiltrated with elicitor by subepidermal injection and cotyledons are later harvested and extracted for analysis (Lundry et al., 1981). This assay, although not providing information on proximal and distal cells, allows us to examine the effects of elicitor under minimal wound conditions.

One of our earliest observations with the cotyledon infiltration assay was that elicitation of the glyceollin response required much higher levels of elicitor (cf. Figs. 2a and 2b). This was true with three different β-1,3/β-1,6–linked glucan preparations: *P. sojae* wall glucan, mycolaminaran, and yeast wall glucan. In addition to the marked increase in the amount of

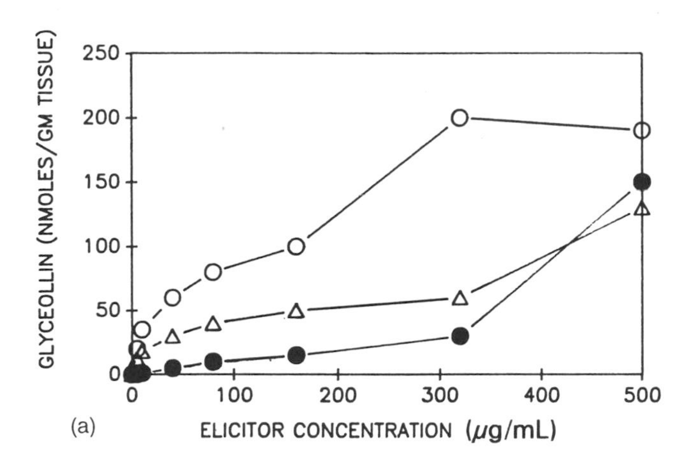

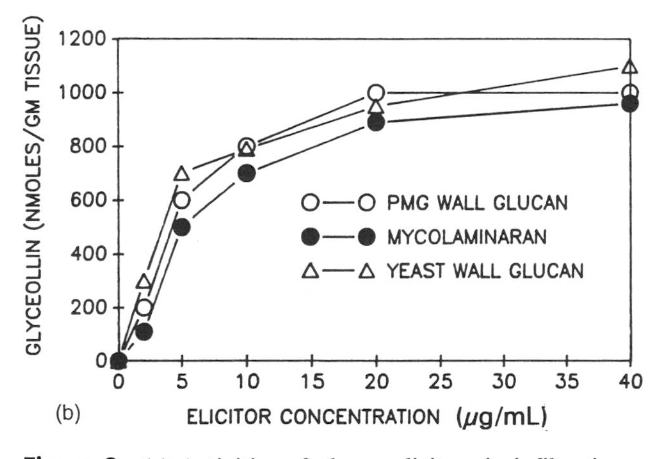

Figure 2 (a) Activities of glucan elicitors in infiltration asssay. (b) Activities of glucan elicitors in cut cot assay.

elicitor required for half-maximal elicitation, maximal glyceollin elicitation was significantly lowered and the assay more clearly differentiated between the activities of the various elicitors. Further analysis of cells within the cotyledon infiltration assay demonstrated that the responses shown in Fig. 2a were contributed solely by the tiny fraction of wounded cells at the point where the infiltration needle was inserted.

This result suggested to us that wounding not only greatly stimulates the glyceollin response but may be a prerequisite for elicitor activity. To test this hypothesis, we returned to the cut cotyledon assay. When the wounded surface of the cut cotyledon assay was immediately washed prior to elicitor application, the glyceollin response was greatly diminished and in some experiments completely abolished (Fig. 3). If washing was delayed for 1–3 hr, the response to elicitor increased nearly to the level of unwashed cotyledons. However, if washing were delayed even further (more than 4 hr), the cells were no longer responsive to elicitor regardless of washing (Fig. 3). These results suggest that a washable wound factor(s) may transiently induce a state of elicitor competency in the exposed cells.

To provide further evidence for such a factor(s), we attempted to restore elicitor competency to washed cotyledons by adding various amounts of the wound washings back to washed cotyledons. As shown in Fig. 4, elicitor competency is restored in a dose-responsive and saturable manner.

An obvious question is whether the wound factor affects only the glyceollin response or also affects isoflavone and phenolic polymer accumulations. Our preliminary results suggest that at least two wound factors may be present, one of which greatly increases the sensitivity of cells to elicita-

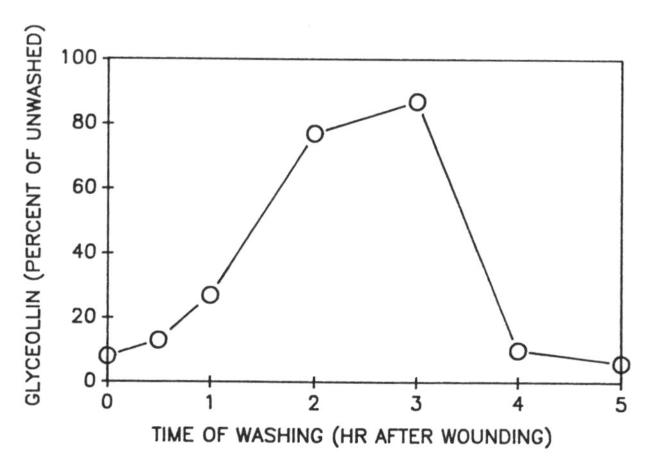

Figure 3 Time course of establishment of competent state.

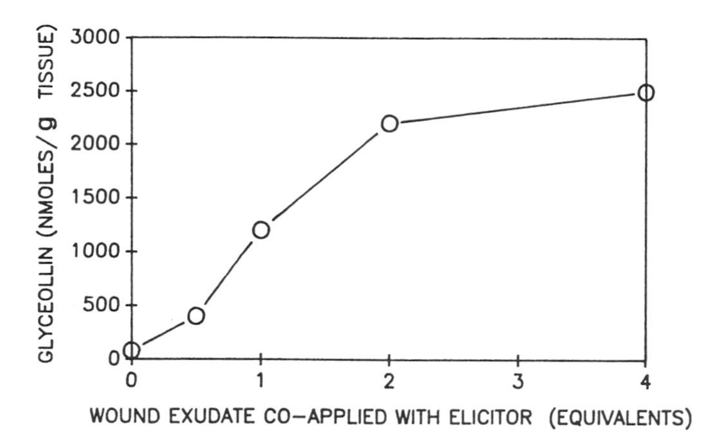

Figure 4 Restoration of elicitor competency.

tion of all the phenylpropanoid responses (i.e., dramatically lowers the amount of elicitor needed for half-maximal response). The other factor is required for specifically "gating" released or induced daidzein toward the formation of glyceollin.

The ability to restore elicitor competency to washed cells suggested that we could take two complementary approaches to the identification of the wound factor(s). Using restoration of elicitor competency as an assay, we could purify the wound factor(s) directly or we could test known wound-associated factors for their activity in restoration.

Since we found that the amount of the wound factor present and the effectiveness of washing in removing it was dependent on a number of physiological parameters of the soybean plants used as a source for the cotyledons (particularly light, age, and nutritional status), these conditions must be controlled if the assay is to be used effectively.

Many different wound-associated molecules and possible elicitor synergists have been reported. Some of these are listed in Table 1 along with their effects on restoration of competency to elicitor-incompetent (washed) cells. Although some of these factors had interesting effects on phenylpropanoid metabolism (alone or in combination with the wall glucan), none of them fully restored either aspect of elicitor competency to elicitor-incompetent cells. Perhaps of most interest were the lack of restoration activity by the pectate lyase [known as an elicitor synergist (Davis et al., 1986)] and the strongly inhibitory activity of jasmonic acid [a signal known to induce the protease inhibitors in solanaceous plants (Farmer and Ryan, 1992)] on glyceollin accumulation.

Purification of the competency factor(s) has just begun. At least one of

Table 1 Effects of Wound Factors or Potential Coelicitors on Glyceollin Response in Elicitor-Incompetent Cells[a]

Factor	Effect on soybean phenylpropanoid responses
Wound exudate	Strong stimulation of glyceollin and phenolic polymer accumulations Inhibition of daidzein and genistein conjugate accumulations
Abscisic acid	Inhibition of genistein conjugate accumulation (slight inhibition of glyceollin accumulation)
ACC/ethylene	No significant effect
Glutathione (reduced)	Inhibition of daidzein and genistein conjugate formation, stimulation of phenolic polymer accumulation (slight stimulation of glyceollin accumulation)
Jasmonic acid	Strong inhibition of glyceollin accumulation
Oxalic acid	(Slight inhibition of daidzein and genistein conjugate and glyceollin accumulations)
Pectate lyase	(Slight stimulation of daidzein and genistein conjugate accumulation, slight inhibition of glyceollin accumulation)
Salicylic acid	Stimulation of daidzein and genistein conjugate accumulation
Traumatic acid	No significant effect

[a]Reported here, for simplicity, are those effects only on washed cotyledon cells in the presence of the *P. sojae* wall glucan. Full details of other effects of these factors will be reported elsewhere. Effects in parentheses were minor.

the activities is stable to relatively long-term storage at −80°C and can be readily fractionated. Further characterization is underway.

We have thus demonstrated that wound-associated factors are required for the competency of soybean cells to respond to the *P. sojae* wall glucan elicitor. Both sensitivity of the cells to elicitor and specific “gating” of the general phenylpropanoid response to glyceollin may be affected. None of the known wound-associated factors tested fully restored competency to elicitor-incompetent cells.

We hypothesize that the wound factor(s) may be released from dead or dying cells in the hypersensitive lesion and that they may then condition neighboring cells at the infection front to become competent for elicitor responsiveness. Once identified these factors may shed important information on the regulation of resistance vs. susceptibility by both host and pathogen, possibly at a race-specific level.

Role of Cellular Growth Regulators in Mediating or Modulating Cellular Response.

A number of investigations have suggested that cellular growth regulators may alter the expression of defense-related genes. We review here only those experiments which deal directly with the regulation of phenylpropanoid metabolism or with the responses of discrete cell populations to infection or elicitor treatment.

Ethylene. Perhaps most convincing, though still somewhat ambiguous, are the possible roles that ethylene may play in defense responses. Ethylene and its immediate precursor, 1-aminocyclopropane carboxylic acid (ACC) have been implicated in the induction of such diverse responses as phenylalanine ammonium lyase (Chappell et al., 1984), peroxidases (Abeles et al., 1989), hydroxyproline-rich glycoproteins (Roby et al., 1986; Rumeau et al., 1988), protease inhibitors (Ryan, 1981), and hydrolytic enzymes such as β-glucanases and chitinases (Mauch and Staehelin, 1989) and proteases (Vera and Conejero, 1989). Thus, like *P. sojae* wall glucan, ethylene is associated with a large number of defense responses. Indeed, early increases in ACC and ethylene have been associated with *P. sojae* wall glucan treatment in parsley (Chappell et al., 1984) and soybean (Lambert and Graham, 1987). As with *P. sojae* wall glucan, however, in very few cases is the association of this growth regulator with a given response clearly understood.

As an example, although Chappell et al. (1984) demonstrated that the induction of ACC synthase is one of the earliest responses of parsley tissues to *P. sojae* wall glucan (<60 min), they conclude that the accumulation of ACC alone is not sufficient to account for the subsequent induction of phenylalanine ammonia lyase. A similar lack of a direct cause-and-effect relationship of ethylene to defense gene expression or phytoalexin accumulation, respectively, was found by Mauch et al. (1984) in pea and by Paradies et al. (1980) in soybean.

One of the clearest demonstrations of a role for ethylene in host defense gene expression is the recent work of Ecker and Davis (1987). These researchers were able to demonstrate ethylene induction of several defense gene mRNAs only when the uppermost layers of treated tissues were examined. Thus, as we have demonstrated in soybean, it is vital in research on host responses to signal molecules to examine the responses of highly discrete cell populations. Otherwise one may be examining a composite or average of many separate responses. Research is underway in our laboratory which we hope will help better define the role of ACC and ethylene in regulation of the various pathways outlined in Fig. 1. Although ACC and ethylene are early, wound-associated factors, as noted above, our preliminary studies do not implicate them as cellular competency factors.

The role of ethylene in the signal process thus remains somewhat unclear. However, two recent papers may have particular relevance to their role in soybean. In an examination of infection and elicitation in soybean roots, Reinhardt et al. (1991) demonstrated that even though the glucan elicitor induced glyceollin accumulation in this organ, the rapid burst in ethylene biosynthesis characteristic of incompatible infection was not seen with elicitor treatment. This may suggest that ethylene is not involved in responses to the glucan per se but may play another role in defense gene regulation. That role may in fact be the enzymatic release of elicitor from the pathogen in planta, since Yoshikawa et al. (1990) have shown that ethylene induces the activity of the host β-1,3-endoglucanases involved in glucan elicitor release.

Polyamines. A second growth regulator which has been investigated for its effects specifically on phytoalexin elicitation are the polyamines. Preliminary work in pea (Hadwiger et al., 1974) and in soybean (Graham, unpublished) has demonstrated that high levels (>500 μM) of the polyamines spermine and spermidine, but not the diamine putrescence, elicit the isoflavonoid phytoalexins pisatin and glyceollin, respectively. Although the concentrations required for elicitation are high, the fact that the polyamines are natural plant metabolites prompted a preliminary investigation of their possible role as internal messengers in pea (Teasdale and Hadwiger, 1977). It was concluded that changes in the concentrations of the polyamines in response to infection by *Fusarium* were too small to account directly for phytoalexin elicitation. However, the possibility that the polyamines might play a conditioning rather than a messenger role was not investigated.

In preliminary studies (Lambert and Graham, unpublished), we demonstrated that wounding of soybean tissues results in a transient burst of putrescine (to four times that in control tissues within 12 hr and back to control levels in 24 hr; Fig. 5a). Accompanying this burst in putrescine are transient and stoichiometric accumulations of both spermine and spermidine, which peak sharply at 24 hr (Fig. 5a). Treatment of wounded tissues with *P. sojae* wall glucan completely suppresses the putrescine and spermine accumulations (Fig. 5b), but markedly enhances and lengthens spermidine accumulation over the period 24–48 hr. Thus *P. sojae* wall glucan treatment affects wound-associated changes in polyamine levels. Although it is possible that the polyamines affect cellular competency, the late nature of these changes and our preliminary experiments on competency restoration with them (data not shown) do not implicate the polyamines in this function. Changes in polyamines, however, could condition or regulate later aspects of cellular response.

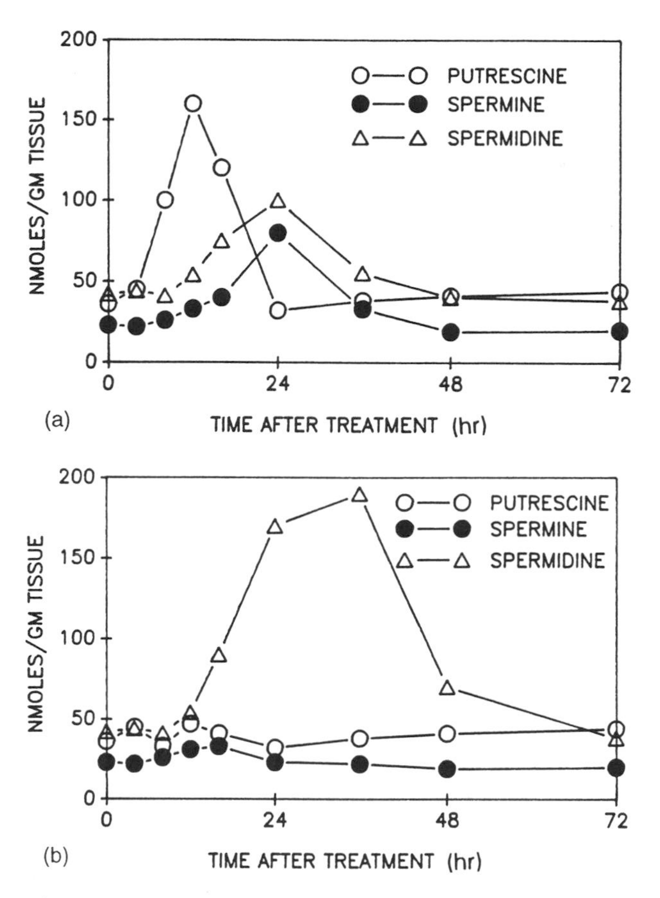

Figure 5 (a) Induced changes in polyamines: Williams cotyledons, wounded control. (b) Induced changes in polyamines: Williams cotyledons, wounded and PMG-glucan–treated.

Cytokinins. Although there has been considerable research on the roles of the cytokinins in green island formation in response to biotrophic fungi and in the growth abnormalities associated with fasciation and gall-forming diseases, there has been comparatively little research on the involvement of cytokinins in disease resistance.

However, a particularly clear demonstration of an effect of the cytokinins on phenylpropanoid metabolism comes from the work of Miller (1969). He demonstrated that 1 nM zeatin or 5 μM kinetin in the presence of an auxin stimulated the production of large quantities of two unidentified conjugates of daidzein, called compounds I and II, in soybean callus or suspension cultures. Neither cytokinins nor auxins alone led to the accumulation of these conjugates. We have repeated the observation of Miller in callus cultures; it is almost certain that Miller's compounds correspond to the glucosylated and malonylated conjugates of daidzein. Thus cytokinins may be involved in regulation of isoflavone and isoflavone conjugate turnover. How this specifically relates to proximal and distal cell events in elicitor-treated or infected tissues needs to be examined.

V. EPILOG

As noted above, our current research directions center on further characterization of the primary signals triggering the various phenylpropanoid responses and the secondary signals that condition cellular response. Our immediate objectives are to fractionate and characterize the various possible primary elicitor activities associated with the *P. sojae* wall glucan, mycolaminaran, and host oligogalacturonide preparations and the cellular conditioning activities associated with the wound competency factors. Complementary to this are our efforts to more fully characterize the role(s) that cellular growth regulators may play in mediating or conditioning the responses of different cell populations. Taken together, these research efforts should provide us with important clues as to the signal transduction processes per se. Cotyledon tissues have proven to be particularly valuable in such studies due to their relative cellular uniformity and to the much greater ease of treatment, analysis, and interpretation of results. Since all of the responses we have elucidated are expressed in cotyledons, we will continue to employ them for the initial characterizations of elicitor and growth regulator responses.

REFERENCES

Abeles, F. B., Biles, C. L., and Dunn, L. J. (1989). Hormonal regulation and distribution of peroxidase isozymes in the Cucurbitaceae. *Plant Physiol. 91*: 1609–1612.

Ayers, A. R., Ebel, J., Valent, B. S. and Albersheim, P. (1976). Host- pathogen interactions. X. Fractionation and biological activity of an elicitor isolated from the mycelial walls of *Phytophthora megasperma* var. *sojae*. *Plant Physiol. 57*: 760–765.

Barz, W., and Hoesel, W. (1979). Metabolism and degradation of phenolic compounds in plants. *The Biochemistry of Plant Phenolics* (Swain, T. et al., eds.), Academic Press, New York, pp. 339–369.

Bhat, R. G., McBlain, B. A., and Schmitthenner, A. F. (1993). Development of pure lines of *Phytophthora sojae* races. *Phytopathology 83*:473–477.

Bhattacharyya, M. K., and Ward, E. W. B. (1986a). Expression of gene-specific and age-related resistance and the accumulation of glyceollin in soybean leaves infected with *Phytophthora megasperma* f. sp. *glycinea. Physiol. Mol. Plant Pathol. 29*:105–111.

Bhattacharyya, M. K., and Ward, E. W. B. (1986b). Resistance, susceptibility and accumulation of glyceollins I-III in soybean organs inoculated with *Phytophthora megasperma* f. sp. *glycinea. Physiol. Mol. Plant Pathol. 29*:227–237.

Bonhoff, A., Loyal, R., Ebel, J., and Grisebach, H. (1986a). Race:cultivar specific induction of enzymes related to phytoalexin biosynthesis in soybean roots following infection with *Phytophthora megasperma* f. sp. *glycinea. Arch. Biochem. Biophys. 246*:149–154.

Bonhoff, A., Loyal, R., Feller, K., Ebel, J., and Grisebach, H. (1986b). Further investigations of race:cultivar–specific induction of enzymes related to phytoalexin biosynthesis in soybean roots following infection with *Phytophthora megasperma* f. sp. *glycinea. Hoppe-Seyler's Z. Biol. Chem. 367*:797–802.

Borner, H., and Grisebach, H. (1982). Enzyme induction in soybean infected by *Phytophthora megasperma* f. sp. *glycinea. Arch. Biochem. Biophys. 217*:65–71.

Bruce, R. J., and West, C. A. (1989). Elication of lignin biosynthesis and isoperoxidase activity by pectic fragments in suspension cultures of castor bean. *Plant Physiol. 91*:889–897.

Burden, R. S., and Bailey, J. A. (1975). Structure of the phytoalexin from soybean. *Phytochemistry 14*:1389–1390.

Buttery, B. R., and Buzzell, R. I. (1975). Soybean flavonol glycosides: identification and biochemical genetics. *Can. J. Bot. 53*:219–224.

Carroll, B. J., McNeil, D. L., and Gresshoff, P. M. (1985). Isolation and properties of soybean *Glycine max* (L.) Merr. mutants that nodulate in the presence of high nitrate concentrations. *Proc. Natl. Acad. Sci. USA 82*:4162–4166.

Carroll, B. J., McNeil, D. L., and Gresshoff, P. M. (1986). Mutagenesis of soybean (*Glycine max* (L.) Merr.) and the isolation of non-nodulating mutants. *Plant Sci. 47*:109–114.

Cervone, F., Hahn, M. G., De Lorenzo, G., Darvill, A., and Albersheim, P. (1989). Host–pathogen interactions. XXXIII. A plant protein converts a fungal pathogenesis factor into an elicitor of plant defense responses. *Plant Physiol. 90*: 542–548.

Chappell, J., Hahlbrock, K., and Boller, T. (1984). Rapid induction of ethylene biosynthesis in cultured parsley cells by fungal elicitor and its relationship to the induction of phenylalanine ammonia-lyase. *Planta 161*:475–480.

Cheong, J. J., and Hahn, M. G. (1991). A specific high-affinity binding site for the hepta-β-glucoside elicitor exists in soybean membranes. *Plant Cell 3*:137–148.

Cheong, J. J., Birberg, W., Fugedi, P., Pilotti, A., Garegg, P. J., Hong, N., Ogawa, T., and Hahn, M. G. (1991). Structure–activity relationships of oligo-β-glucoside elicitors of phytoalexin accumulation in soybean. *Plant Cell 3*:127–136.

Cosio, E. G., Frey, T., and Ebel, J. (1992). Identification of a high-affinity binding protein for a hepta-β-glucoside phytoalexin elicitor in soybean. *Eur. J. Biochem. 204*:1115–1123.

Cuypers, B., Schmelzer, E., and Hahlbrock, K. (1988). *In situ* localization of rapidly accumulated phenylalanine ammonia-lyase MRNA around penetration sites of *Phytophthora infestans* in potato leaves. *Mol. Plant Microbe Interact. 1*:157–160.

Darvill, A. G., and Albersheim, P. (1984). Phytoalexins and their elicitors—a defense against microbial infection in plants. *Ann. Rev. Plant Physiol. 35*:243–275.

Davis, K. R., Darvill, A. G., and Albersheim, P. (1986). Several biotic and abiotic elicitors act synergistically in the induction of phytoalexin accumulation in soybean. *Plant Mol. Biol. 6*:23–32.

Doares, S. H., Bucheli, P., Albersheim, P., and Darvill, A. G. (1989). Host-pathogen interactions. XXXIV. A heat labile activity secreted by a fungal phytopathogen releases fragments of plant cell walls that kill plant cells. *Mol. Plant Microbe Interact. 2*:346–353.

Ebel, J. (1986). Phytoalexins synthesis: the biochemical analysis of the induction process. *Ann. Rev. Phytopathol. 24*:235–264.

Ecker, J. R., and Davis, R. W. (1987). Plant defense genes are regulated by ethylene. *Proc. Natl. Acad. Sci. 84*:5202–5206.

Esnault, R., Chibbar, R. N., Lee, D., Van Huystee, R. B., and Ward, E. W. B. (1987). Early differences in production of mRNA's for phenylalanine ammonia-lyase and chalcone synthetase in resistant and susceptible cultivars of soybean inoculated with *Phytophthora megasperma* f. sp. *glycinea. Physiol. Mol. Plant Pathol. 30*:293–297.

Farmer, E. E., and Ryan, C. A. (1992). Octadecanoid precursors of jasmonic acid activate the synthesis of wound-inducible proteinase inhibitors. *Plant Cell 4*: 129–134.

Frank, J. A., and Paxton, J. D. (1971). An inducer of soybean phytoalexin and its role in the resistance of soybean to Phytophthora rot. *Phytopathology 61*:954–958.

Graham, T. L. (1991a). A rapid, high resolution HPLC profiling procedure for plant and microbial aromatic secondary metabolites. *Plant Physiol. 95*:584–593.

Graham, T. L. (1991b). Flavonoid and isoflavonoid distribution in developing soybean seedling tissues and in seed and root exudates. *Plant Physiol. 95*:594–603.

Graham, M. Y., and Graham, T. L. (1991a). Rapid accumulation of anionic peroxidases and phenolic polymers in soybean cotyledon tissues following treatment with *Phytophthora megasperma* f. sp. *glycinea* wall glucan. *Plant Physiol. 97*: 1445–1455.

Graham, T. L., and Graham, M. Y. (1991b). Glyceollin elicitors induce major but distinctly different shifts in isoflavonoid metabolism in proximal and distal soybean cell populations. *Mol. Plant Microbe Interact. 4*:60–68.

Graham, T. L., and Graham, M. Y. (1991c). Cellular coordination of molecular responses in plant defense. *Mol. Plant Microbe Interact. 4*:415–421.

Graham, T. L., Kim, J. E., and Graham, M. Y. (1990). Role of constitutive isoflavone conjugates in the accumulation of glyceollin in soybean infected with *Phytophthora megasperma. Mol. Plant Microbe Interact. 3*:157–166.

Habereder, H., Schröder, G., and Ebel, J. (1989). Rapid induction of phenylalanine ammonia-lyase and chalcone synthase mRNAs during fungus infection of soybean (*Glycine max* L.) roots or elicitor treatment of soybean cell cultures at the onset of phytoalexin synthesis. *Planta 177*:58–65.

Hadwiger, L. A., Jafri, A., von Broembsen, S., and Eddy, R. (1974). *Plant Physiol. 53*:52–55.

Hahlbrock, K., and Scheel, D. (1989). Physiology and Molecular Biology of Phenylpropanoid Metabolism. *Ann. Rev. Plant Physiol. 40*:347–369.

Hahn, M. G., Bonhoff, A., and Grisebach, H. (1985). Quantitative localization of the phytoalexin glyceollin I in relation to fungal hyphae in soybean roots infected with *Phytophthora megasperma* f. sp. *glycinea. Plant Physiol. 77*:591–601.

Ingham, J. L., Keen, N. T., Mulheirn, L. J., and Lyne, R. L. (1981). Inducibly-formed isoflavonoids from leaves of soybean. *Phytochemistry 20*:795–798.

Keen, N. T., and Yoshikawa, M. (1982). Physiology of disease and the nature of resistance to *Phytophthora. Phytophthora*: *Its Biology, Taxonomy, Ecology and Pathology* (D. C. Erwin, S. Bartnicki-Garcia, and P. H. Tsao, eds.), American Phytopathological Society, St. Paul, MN, pp. 279–298.

Keen, N. T., and Yoshikawa, M. (1983). β-1,3-Endoglucanase from soybean releases elicitor active fragments from fungus cell walls. *Plant Physiol. 71*:460–465.

Keen, N. T., Yoshikawa, M., and Wang, M. C. (1983). Phytoalexin elicitor activity of carbohydrates from *Phytophthora megasperma* f. sp. *glycinea* and other sources. *Plant Physiol. 71*:466–471.

Lambert, M. R., and Graham, T. L. (1987). Alternative induced resistance pathways in soybean and their regulation. *Phytopathology 77*:1739.

Lazarovits, G., Stoessel, R., and Ward, E. W. B. (1981). Age-related changes in specificity and glyceollin production in the hypocotyl reactions of soybean to *Phytophthora megasperma* var. *sojae. Phytopathology 71*:94–97.

Leube, J., and Grisebach, H. (1983). Further studies on induction of enzymes of phytoalexin synthesis in soybean and cultured soybean cells. *Z. Naturforsch. 38c*:730–735.

Lundry, D. R., Bass, J., Castanho, B., and Graham, T. L. (1981). Protection of soybean plants against disease by phytoalexin elicitors. *Plant Physiol. 67*:75.

Lyne, R. L., and Mulheirn, L. J. (1978). Minor pterocarpinoids of soybean. *Tetrahedron Lett. 34*:3127–3128.

Lyne, R. L., Mulheirn, L. J., and Leworthy, D. P. (1976). New pterocarpinoid phytoalexins of soybean. *J. Chem. Soc. Chem. Commun.*:497–498.

Mauch, F., and Staehelin, L. A. (1989). Functional implications of the subcellular localization of ethylene-induced chitinase and beta-1,3-glucanase in bean leaves. *Plant Cell. 1*:447–457.

Mauch, F., Hadwiger, L. A., and Boller, T. (1984). Ethylene: symptom, not signal for the induction of chitinase and beta-1,3-glucanase in pea pods by pathogens and elicitors. *Plant Physiol. 76*:607–611.

Miller, C. O. (1969). Control of deoxyisoflavone synthesis in soybean tissue. *Planta 87*:26–35.

Moesta, P., and Grisebach, H. (1981). Investigation of the mechanism of glyceollin accumulation in soybean infected by *Phytophthora megasperma* f. sp. *glycinea. Arch. Biochem. Biophys. 212*:462–467.

Morris, P. F., and Ward, E. W. B. (1992). Chemoattraction of zoospores of the soybean pathogen *Phytophthora sojae*, by isoflavones. *Physiol. Mol. Plant Pathol. 40*:17–22.

Morris, P. F., Savard, M. E., and Ward, E. W. B. (1991). Identification and accumulation of isoflavonoids and isoflavone glucosides in soybean leaves and hypocotyls in resistance responses to *Phytophthora megasperma* f. sp. *glycinea. Physiol. Mol. Plant Pathol. 39*:229–244.

Nothnagel, E. A., McNeil, M., Albersheim, P., and Dell, A. (1983). Host-pathogen interactions. XXII. A galacturonic acid oligosaccharide from plant cell walls elicits phytoalexins. *Plant Physiol. 71*:916–926.

Olah, A. F., and Sherwood, R. T. (1973). Glycosidase activity and flavonoid accumulation in alfalfa infected by *Ascochyta imperfecta. Phytopathology 63*:739–742.

Paradies, I., Konze, J. R., Elstner, E. F., and Paxton, J. (1980). *Plant Physiol. 66*: 1106.

Parker, J. E., Hahlbrock, K., and Scheel, D. (1988). Different cell wall components from *Phytophthora megasperma* f. sp. *glycinea* elicit phytoalexin production in soybean and parsley. *Planta 176*:75–82.

Partridge, J. E., and Keen, N. T. (1977). Soybean phytoalexins: rates of synthesis are not regulated by activation of initial enzymes in flavonoid biosynthesis. *Phytopathology 67*:50–55.

Paxton, J. D., and Chamberlain, D. W. (1969). Phytoalexin production and disease resistance in soybeans as affected by age. *Phytopathology 59*:775–777.

Pierce, M., and Essenberg, M. (1987). Localization of phytoalexins in fluorescent mesophyll cells isolated from bacterial blight-infected cotton cotyledons and separated from other cells by fluorescence-activated cell sorting. *Physiol. Mol. Plant Pathol. 31*:273–290.

Pierson, P. E., and Graham, T. L. (1987). A cellular system for biochemical studies on soybean-*Phytophthora* interactions. *Phytopathology 77*:1756.

Reinhardt, D., Wiemken, A., and Boller, T. (1991). Induction of ethylene biosynthesis in compatible and incompatible interactions of soybean roots with *Phytophthora megasperma* f. sp. *glycinea* and its relation to phytoalexin accumulation. *J. Plant Physiol. 138*:394–399.

Rivera-Vargas, L. I., Schmitthenner, A. F., and Graham, T. L. (1993). Flavonoid effects on and metabolism by *Phytophthora sojae. Phytochemistry 32*:851–857.

Roby, D., Toppan, A., and Esquerre-Tugaye, M. T. (1986). Cell surfaces in plant-micro-organism interactions. *Plant Physiol. 81*:228.

Rowlan, A. R., Hall, J. A., Barfield-Schneider, T., and Essenberg, M. (1991). Protection of cotton leaf palisade cells from light-activated toxicity of a phytoalexin by red epidermal cells. *Phytopathology 81*:1139.

Rumeau, D., Mazau, D., Panabieres, F., Delseny, M., and Esquerre-Tugaye, M. T. (1988). Accumulation of hydroxyproline-rich glycoprotein mRNAs in infected or ethylene treated melon plants. *Physiol. Mol. Plant Pathol. 33*:419–428.

Ryan, C. A. (1981). Proteinase inhibitors. *The Biochemistry of Plants*, Vol. 6 (P. K. Stumpf and E. E. Conn, eds.), Academic Press, New York, pp. 351–370.

Schmelzer, E., Borner, H., Grisebach, H., Ebel, J., and Hahlbrock, K. (1984). Phytoalexin synthesis in soybean (*Glycine max*). Similar time courses of mRNA induction in hypocotyls infected with a fungal pathogen and in cell cultures treated with fungal elicitor. *FEBS Let. 172*:59–63.

Schmelzer, E., Kruger-Lebus, S., and Hahlbrock, K. (1989). Temporal and spatial patterns of gene expression around sites of attempted fungal infection in parsley leaves. *Plant Cell 1*:993–1001.

Schmitthenner, A. F. (1985). Problems and progress in control of *Phytophthora* root rot of soybean. *Plant Dis. 69*:362–368.

Sharp, J. K., McNeil, M., and Albersheim, P. (1984). The primary structures of one elicitor-active and seven elicitor-inactive hexa (β-D-glucopyranosyl)-D-glucitols isolated from the mycelial walls of *Phytophthora megasperma* f. sp. *glycinea. J. Biol. Chem. 259*:11321–11336.

Sims, J. J., Keen, N. T., and Honwad, V. K. (1972). Hydroxyphaseolin, and induced antifungal compound from soybeans. *Phytochemistry 11*:827–828.

Sinclair, J. B. (1982). *Compendium of Soybean Diseases*, American Phytopathological Society, St. Paul, MN.

Snyder, B. A., and Nicholson, R. L. (1990). Synthesis of phytoalexins in sorghum as a site-specific response to fungal ingress. *Science 248*:1637–1639.

Stermer, B. A., Schmid, J., Lamb, C. J., and Dixon, R. A. (1990). Infection and stress activation of bean chalcone synthase promoters in transgenic tobacco. *Mol. Plant Microbe Interact. 3*:381–388.

Takeuchi, Y., Yoshikawa, M., and Horino, O. (1990a). Immunological evidence that β-1,3-endoglucanase is the major elicitor-releasing factor in soybean. *Ann. Phytopathol. Soc. Jpn. 56*:523–531.

Takeuchi, Y., Yoshikawa, M., Takeba, G., Tanaka, K., Shibata, D., and Horino, O. (1990b). Molecular cloning and ethylene induction of mRNA encoding a phytoalexin elicitor-releasing factor, β-1,3-endoglucanase, in soybean. *Plant Physiol. 93*:673–682.

Teasdale, J. R., and Hadwiger, L. A. (1977). Effect of pisatin-inducing fungi on pea polyamines. *Phytochemistry 16*:681–683.

Vera, P., and Conejero, V. (1989). The induction and accumulation of the pathogenesis-related P69 proteinase in tomato during citrus exocortis viroid infection and in response to chemical treatments. *Physiol. Mol. Plant Pathol. 34*:323–334.

Wang, M. C., and Bartnicki-Garcia, S. (1973). Novel phosphoglucans from the cytoplasm of *Phytophthora palmivora* and their selective occurrence in certain life cycle stages. *J. Biol. Chem. 248*:4112–4118.

Wang, M. C., and Bartnicki-Garcia, S. (1974). Mycolaminarans: storage β-1,3-glucans from the cytoplasm of the fungus *Phytophthora palmivora. Carbohydrate Res. 37*:331–338.

Ward, E. W. B., and Buzzell, R. I. (1983). Influence of light, temperature and wounding on the expression of soybean genes for resistance to *Phytophthora megasperma* f. sp. *glycinea*. *Physiol. Mol. Plant Pathol. 23*:401–409.

Ward, E. W. B., and Lazarovits, G. (1982). Temperature induced changes in specificity in the interaction of soybeans with *Phytophthora megasperma* f. sp. *glycinea*. *Phytopathology 72*:826–830.

Yoshikawa, M. (1978). Diverse modes of action of biotic and abiotic phytoalexin elicitors. *Nature 275*:546–547.

Yoshikawa, M. (1988). Molecular mechanisms for induction of host defenses in fungal diseases. *Molecular Strategies for Pathogenicity and Host Defense in Viral, Bacterial and Fungal Diseases* (R. Heitefuss and S. Ouchi, eds.), Satellite Meeting of the 5th International Congress for Plant Protection, August 22–23, Kyoto, pp. 3–7.

Yoshikawa, M., Yamamuchi, K., and Masago, H. (1978). Glyceollin: its role in restricting fungal growth in resistant soybean hypocotyls infected with *Phytophthora megasperma* var. *sojae*. *Physiol. Plant Pathol. 12*:73–82.

Zahringer, U., Ebel, J., and Grisebach, H. (1978). Induction of phytoalexin synthesis in soybean: elicitor-induced increase in enzyme activities of flavonoid biosynthesis and incorporation of mevalonate into glyceollin. *Arch. Biochem. Biophys. 188*:450–455.

6

Culture Darkening, Cell Aggregate Size, and Phytoalexin Accumulation in Soybean Cell Suspensions Challenged with Biotic Agents

Robert M. Zacharius
R. M. Zacharius and Associates, Science Consultants, Highland, Maryland

William F. Fett
Eastern Regional Research Center, Agricultural Research Service, U.S. Department of Agriculture, Philadelphia, Pennsylvania

Prakash G. Kadkade
Phyton Inc., Ithaca, New York

I. INTRODUCTION

Fett and Zacharius (1982, 1983) demonstrated that bacteria as well as fungal cell wall elicitors can induce the accumulation of the phytoalexin glyceollin in soybean (*Glycine max* L. Merr.) cell suspension cultures. Moreover, the typical hypersensitive response (HR) as defined by rapid host cell death displayed by the intact plant was not a prerequisite for phytoalexin induction in soybean cell cultures. The concentration of glyceollin produced declined with successive culture transfers. Some interesting responses by separate cell lines of the same cultivar to challenge with a biotic agent were observed, suggesting further study.

A subsequent study by Zacharius and Kalan (1990) revealed that soybean cell suspension cultures producing glyceollin when challenged with *Pseudomonas syringae* pv. *glycinea* (Psg) or fungal cell wall elicitor did not undergo an HR but rather darkened with a gradual decline in culture viability. A cell line, Sb-1 (cv. Mandarin), which failed to darken on challenge with Psg, also did not accumulate glyceollin within the cells or media. This Sb-1 culture was found to have rather low levels of constituent isoflavonoid which exhibited a small decline on exposure to biotic agents and produced only trace amounts of glyceollin. The other cell lines of cv. Mandarin having high levels of constituent isoflavonoids, exhibited a dramatic decline in their isoflavonoids along with accumulation of glyceollin following exposure to biotic agents.

High levels of constitutive isoflavonoids, particularly daidzein, seemed indicative of culture potential for glyceollin production concomitant with a decrease in daidzein, genistein, and coumestrol when either fungal wall elicitors or live bacteria was the stressing agent. However, cell cultures containing very high levels of these isoflavonoids did not accumulate higher levels of the phytoalexin than those with lesser levels. Augmentation of the weakly responsive Sb-1 culture with exogenously supplied isoflavonoids followed by fungal elicitor challenge had little measurable effect on glyceollin production, but resulted in a metabolic breakdown of the exogenously supplied daidzein and genistein.

Thus, on the basis of this study, there appeared to be a link between endogenous levels of cell isoflavonoids, ability for culture darkening, and glyceollin production. From other observations with cv. Mandarin cell suspensions, we noted the larger cell aggregates tended to darken more readily with exposure to fungal wall elicitor than the smaller aggregates. Therefore, our earlier studies were extended to include both smaller and larger cell suspension aggregations of soybean cv. Mandarin and cv. Clark challenged by live bacteria and fungal elicitor. We speculated that large cell aggregates may have a closer relationship to in vivo tissues of an organized structure than single cells or small cell aggregates.

II. METHODOLOGY

A. Suspension Cell Cultures

Calluses of soybean cv. Mandarin cell line Sb-4a were initiated from epicotyl tissue of 7- to 10-day old plants by the procedure described by Fett and Zacharius (1982). Calluses of soybean cv. Clark were also initiated from epicotyl explants on B5 media (Gamborg, 1975) containing 10 mg/liter 2,4-D and 0.21 mg/liter kinetin. A suspension cell culture was developed from friable calluses of each cultivar and grown in 1B5 media on a reciprocating shaker at 150 rpm at 26–27°C under 17.5 ± 4.5 μmol/m^2/sec photosynthetic photon flux (PPF).

B. Bacterial Cultures

Xanthomonas campestris pv. *glycines* strains XP175 and S-9-8 were grown overnight on nutrient agar at 28°C, suspended in sterile distilled water, washed three times, and resuspended in sterile distilled water to 1.0 OD at 600 nm. Strain XP175 is virulent and causes bacterial pustule disease on cv. Mandarin and cv. Clark while strain S-9-8 is avirulent (Fett, 1984).

C. Fungal Elicitor

A cell-free mycelial elicitor was prepared from *Phytophthora infestans* race 0 by the procedure of Alves et al. (1979) with an added final filtration step through a 0.45-μm Millipore filter before autoclaving. The elicitor preparation contained 6.4 mg dry wt per milliliter distilled water.

D. Interaction of Suspension Cells with Bacteria or Fungal Elicitors

Suspension cultures of each cultivar of soybean were grown and maintained in 60-ml volumes of 1B5 medium in 250-ml Delong flasks and were used 5 days after the last transfer. The contents of several flasks of Sb-4A or Sb-Clark were each aseptically sieved on 100-mesh wire screen to separate large, >150-μm aggregates and small, <150-μm aggregates. Aggregates of one size and culture were pooled and 10 ml of loosely packed aggregates was redistributed into flasks with 40 ml of fresh 1B5 media. Fungal elicitor was applied to each flask at 0.75 ml/50 ml of aggregated cell culture and each bacterial strain was added to give an initial concentration of approximately 1×10^7 CFU/ml. Each aggregate size/cultivar/biotic–elicitor interaction and controls were carried out with three replicates. All flasks were shaken at 150 rpm at 26–27°C under 17.5 $\pm$ 4.5 μmol/m^2/sec PPF.

Cell viability was followed by a dye exclusion test with 0.4% trypan blue (Phillips, 1973).

E. Isoflavonoid Extraction, Identification, and Quantitation

Following 68 hr of interaction, cells plus media were extracted by vortex mixing four times, each with an equal volume of chloroform (Zacharius and Kalan, 1990). The combined extracts were dried at ambient temperature under a stream of nitrogen and taken up in 1.0 ml methanol 0.5 g^{-1} dry weight. Components of the extract were separated qualitatively by thin-layer chromatography (TLC) on Analtech silica gel G plates (250 μm) irrigated with cyclohexane–ethyl acetate (1:1) (Zacharius and Kalan, 1984). Quantitation was performed by high-performance liquid chromatography (HPLC) using a 250-mm C_{18} reverse phase column (Whatman Partisil 10 ODS-325) attached to a Waters Model 6000A solvent delivery system with a Reodyne (70-10) loop injector valve. The column effluent was monitored with a Perkin-Elmer Model LC-55B spectrophotometric detector at 262 nm (genistein and daidzein), 343 nm (coumestrol), and 290 nm (glyceollins). Integrations were made with the Hewlett-Packard 3390A integrator and compared with those of reference compounds (Zacharius and Kalan, 1990).

The column eluant was 40% aqueous acetonitrile containing trifluoroacetic acid (0.4 ml/liter) at a flow rate of 1.5 ml/min. Retention times (min) were as follows: daidzein, 2.2; genistein, 3.3; formononetin, 4.5; coumestrol, 6.0; glyceollin isomers, 7.8, 8.0; biochanin A, 9.5. The presence of glyceollin, daidzein, genistein, and coumestrol in selected samples was further confirmed by comparison with authentic compounds by mass spectrometry using a Finnegan MAT 311A mass spectrometer.

F. Peroxidase Assay

The peroxidase assay was carried out according to the method of Worthington (1972).

III. RESULTS

The large cell aggregates of both Sb-Clark and Sb-4A dramatically darkened when challenged with either virulent strain XP175 or avirulent strain S-9-8 or the fungal elicitor. On the other hand, these challenges with the small aggregate cultures of either soybean cultivar produced no visual change in color or color intensity (Fig. 1).

Figure 1 Culture darkening 68 hr after inoculation of small or large cell aggregates of Sb-Clark with *Xanthomonas campestris* pv. *glycines*. Mixed aggregate culture control, A; virulent *X. c.* pv. *glycines* strain XP175 in small (B) or large (B′) aggregate cultures; avirulent *X. c.* pv. *glycines* strain S-9-8 in small (C) or large (C′) aggregate cultures; fungal elicitor in small (D) or large (D′) aggregate cultures.

Both size cell aggregates of Sb-Clark responded to each of the stressing agents producing similar levels of glyceollin with a marked but unequal decrease in the constitutive level of daidzein and genistein. Avirulent strain S-9-8 effected a virtual disappearance of these two isoflavonoids from the cell culture (Table 1).

Both large and small aggregates of Sb-4A when exposed to avirulent strain S-9-8 produced similar levels of glyceollin, whereas the virulent strain XP175 failed to do so with either size cell aggregates. Although the isoflavonoids declined in both size aggregates exposed to strain XP175, S-9-8 caused an almost total loss of daidzein and genistein (Table 1) as in the case of Sb-Clark.

Interaction of the large cell aggregates of either soybean cultivar with the fungal elicitor produced marked darkening, glyceollin accumulation, and a concurrent large decrease in daidzein and genistein. While the latter two phenomena occurred during the interaction with the small aggregates of both cultivars, neither small aggregate cell culture darkened (Fig. 1).

Both aggregate sizes of Sb-4A and the large aggregate culture of Sb-Clark became very viscous and gel-like when exposed to strain S-9-8. This was presumably due to copious production of bacterial exopolysaccharides accompanied by a concurrent increase in CFU per milliliter to 10^{10}. This was not observed in the other interactions. Where the challenged cultures became gel-like, a sharp decline in soybean cell viability occurred by 68 hr.

Table 1 Changes in the Isoflavonoids of Large and Small Cell Aggregates of Sb-Clark and Sb-4A (cv. Mandarin) upon Biotic Stressing (μg/g dry wt)

	Daidzein		Genistein		Glyceollin	
Treatment	Sb-Clark	Sb-4A	Sb-Clark	Sb-4A	Sb-Clark	Sb-4A
Small aggregates						
Control	2560	1573	5080	7343	0	0
Strain XP175[a]	576	354	940	1565	790	0
Strain S-9-8[b]	9	50	73	38	908	1395
P. infestans elicitor	1201	236	1180	870	880	975
Large aggregates						
Control	3240	2569	5866	8239	0	0
Strain XP175[a]	742	1837	515	3585	812	0
Strain S-9-8[b]	16	40	77	107	867	1140
P. infestans elicitor	1441	385	1109	980	912	1100

[a]Strain XP175 = *X. campestris* pv. *glycines* XP175.
[b]Strain S-9-8 = *X. campestris* pv. *glycines* S-9-8.

There was little change in the soybean cell viability in the other interactions (Table 2).

Following exposure to the stressing agent strain XP175, the larger cell aggregates of both soybean cultivars were found to have higher peroxidase levels than the small aggregates. The changes from the prestressed levels were of similar magnitude for both cell cultivars (Table 3).

Attachment of the bacterial strains to soybean cells of either aggregate size of Sb-Clark or Sb-4A cultures was not observed using phase contrast light microscopy and bacterial-induced clumping of cells of any of the cultures did not occur.

IV. DISCUSSION

In our earlier report (Zacharius and Kalan, 1990), data were presented which appeared to relate soybean suspension culture darkening with glyceollin induction during bacterial or fungal elicitor stress. A high level of cellular daidzein also seemed prerequisite to both phenomena. Concurrent with glyceollin induction, a large decline of the cellular daidzein and genistein usually occurred. In the present study, the small aggregates were able

Table 2 Cell Viability of Sb-Clark and Sb-4A (cv. Mandarin) Cultures and Bacterial Growth after 68 hr Interaction

Cell suspension	Aggregate size	Treatment	CFU/ml	% Viable soybean cells
Clark	Small	Control	0	89
	Large	Control	0	92
	Small	XP175[a]	2.20×10^8	89
	Large	XP175	1.12×10^9	85
	Small	S-9-8[a]	2.23×10^8	83
	Large	S-9-8	1.13×10^{10}	0
	Small	Fungal elicitor	0	88
	Large	Fungal elicitor	0	84
Sb-4A	Small	Control	0	85
	Large	Control	0	88
	Small	XP175	8.60×10^7	85
	Large	XP175	6.47×10^8	87
	Small	S-9-8	1.17×10^{10}	45
	Large	S-9-8	1.40×10^{10}	15
	Small	Fungal elicitor	0	85
	Large	Fungal elicitor	0	83

[a] *Xanthomonas campestris* pv. *glycines*.

Table 3 Changes in Peroxidase Levels in Cell Aggregates of Sb-Clark and Sb-4A (cv. Mandarin) on Challenge with *Xanthomonas campestris* pv. *glycines* strain XP175

		Absorbance change at 460 nm/mg protein	
Soybean cultivar	Aggregate size (μm)	Suspension	Suspension after XP175
Sb-Clark	<150	0.60	0.85
	>150	0.70	2.70
	Whole unsieved culture	0.68	2.50
Sb-4A	<150	0.65	0.85
	>150	0.75	2.70
	Whole unsieved culture	0.70	2.50

to accumulate glyceollin without observable darkening. When glyceollin accumulated in either size aggregates, the levels were virtually the same irrespective of whether the aggregates were darkened or not. In addition, in the case of Sb-4A exposed to virulent strain XP175, neither the small undarkened cell aggregates nor the large darkened cell aggregates accumulated glyceollin although the constituent isoflavonoids of the culture underwent a decline. This would suggest that metabolic consumption of these constitutive isoflavonoids need not be linked to glyceollin induction.

The propensity of the soybean culture to darken with biotic stressing was reflected in the size of the cell aggregates of the culture. Large aggregates darkened on exposure to any of the three stressing agents while the small ones did not. Earlier Zacharius and Kalan (1990) had reported that an unsieved cv. Mandarin culture of mixed size aggregates did not show observable darkening on stressing with fungal elicitor but the cultures had a low level of constitutive daidzein and genistein.

Cell darkening of suspension cultures with or without biotic stressing would appear to reflect the level of phenolic compounds present in the cells and medium. Interestingly, small cell aggregates used in this study had lower constitutive isoflavonoid levels than the larger cell aggregates. Siegel and Enns (1979) were able to prevent the discoloration and aggregation of soybean suspension cultures grown in B5 medium with 2,4-D by adsorbing the excess cellular polyphenols with either polyvinvlpyrrolidine or bovine serum albumin in the medium. Additionally, Singh et al. (1982) also associated higher levels of polyphenols in cowpea callus (*Vigna unguiculata* L.

Walp. subsp. *unguiculata*) with higher activities of peroxidase and polyphenol oxidase. Moreover, polyphenol content increased with 2,4-D concentration above 1 μg/ml while supplements of casein hydrolysate and coconut water produced the lowest polyphenol accumulation. In our study, soybean cultures were grown with 1 μg/ml of 2,4-D without casein hydrolysate and coconut water. Although we did not compare polyphenol levels, increased levels of peroxidase were associated with those cell aggregates which darkened markedly with addition of strain XP175.

Peroxidase levels were found by Verma and Van Huystee (1970) to be 2.5-fold greater in large peanut cell aggregates (2–4 mm) than in small ones (150 μm). It was also found that a cell mass less than 0.5 mm in diameter consists of undifferentiated uniform cells while cellular differentiation appears in large cell aggregates. Working with tobacco suspension cultures, Kuboi and Yamada (1976, 1978) found stimulation of the lignin biosynthetic pathway, during tracheid differentiation, occurs in large aggregates but low activities of shikimate dehydrogenase, cinnamic acid-4-hydroxylase, 5-hydroxyferulic acid-*O*-methyltransferase, and caffeic acid-*O*-methyltransferase inhibit differentiation in small aggregates. Both phenylalanine ammonia-lyase (PAL) and peroxidase exhibited activities in the small aggregates. Hahlbrock et al. (1974) found that the highest specific activity of PAL is associated with single cells and small aggregates, while the specific activities in large aggregates were considerably lower. This would explain finding greater secondary compound production in cultures of small aggregates and viable single cells. Kinnersley and Dougall (1980) succeeded in increasing the anthocyanin yield in *Daucus carota* L. cell cultures by screening for small cell aggregates and subculturing; a consequence of this selection was small aggregates with a lower level of endogeneous cytokinin (which reduce anthocyanin yield). Watts et al. (1984) found evidence to suggest that the presence of green, aggregated cells or low-temperature stress contributes to the ability of celery cell suspensions to synthesize secondary compounds. Nevertheless, the level of glyceollin accumulated in the large and small aggregates of soybean described here were quite similar.

Perhaps Apostol et al. (1989) provide in part the most rational explanation of our observations regarding the relationship of aggregate size, darkening, and glyceollin production. They found that elicitor-stimulated plant cell cultures responded with the rapid production of H_2O_2 which was subsequently used by extracellular peroxidases. Exogenous H_2O_2 alone stimulated phytoalexin production in the soybean cell suspension culture, and inhibition of elicitor stimulated glyceollin production was observed upon addition of catalase or other inhibitors of the oxidative burst. For inhibition to occur, the presence of catalase was necessary during elicitor addition.

Montillet and Degousée (1991) have reported much greater glyceollin-eliciting activity in soybean seedlings by two organic hydroperoxides than H_2O_2. H_2O_2 eliciting efficiency was comparable to the two organic hydroperoxides when tissue catalase activity was suppressed. Of further interest, Graham and Graham (1991) found that deposition of phenolic polymers in soybean cotyledon cell walls is an early and major response to treatment with fungal cell wall glucan. This is accompanied by a rapid and massive increase in activity of a specific group of anionic wall-bound peroxidases.

Analysis of our observations in light of the above suggests that the large aggregates of Sb-Clark and Sb-4A responded to biotic stress by severely darkening and at the same time a burst of H_2O_2 induced (as indicated by fourfold increase in peroxidase level) the glyceollin production. The small aggregates of both cultivars did not discolor because of a limited H_2O_2 burst indicated by only a slight increase in peroxidase level, but these aggregates presumably contained high activities of the early phytoalexin pathway enzymes, allowing for glyceollin production. The sharp decline in isoflavonoids in soybean suspension cultures treated with biotic elicitors (Zacharius and Kalan, 1990) almost certainly can be explained by the oxidative burst occurring during elicitation. Culture darkening and the level of destruction of the measured isoflavonoids following elicitation probably reflects firstly on the intensity of the oxidative burst and secondly on the induced level of the peroxidases present in the culture.

Strain S-9-8 is avirulent and strain XP175 is virulent on both cv. Clark and cv. Mandarin leaves. Fett (1984), however, found moderate glyceollin accumulation in leaves of cv. Clark inoculated with virulent strain XP175 and a weak hypersensitive response with no glyceollin accumulation following inoculation with avirulent strain S-9-8. Growth of strain S-9-8 was restricted in leaves cv. Clark 24 hr after inoculation when compared to growth of strain XP175. The leaf observations with cv. Clark were inconsistent with those described here for the cell cultures. Leaves of Sb-4A (cv. Mandarin) were not treated with either strain.

The level of glyceollin which accumulated in the inoculated large and small aggregate cultures did not appear to determine the bacterial cell populations attained (Table 2). Strain S-9-8 grew to much higher levels than strain XP175 in both the large and small aggregate Sb-4A cultures, while strain S-9-8 but not strain XP175 induced accumulation of appreciable levels of glyceollin. Both aggregate size cultures of Sb-Clark stressed with either strain of *X. campestris* yielded similar amounts of glyceollin (Table 1), yet strain S-9-8 grew 200 times greater in the large cell aggregations than in the small aggregates. Strain S-9-8 attained higher populations than did strain XP175 under all cell culture conditions except for the small aggregate sized Sb-Clark. Evidence offered here indicates little or no control by glyceollin on the growth of *X. campestris* in the cell suspension cultures. The percentage of viable soybean cells remaining after 68 hr exposure to the *X. campestris* agrees well with the bacterial populations attained.

Observations with phase microscopy did not indicate any attachment of strains S-9-8 or XP175 to Sb-Clark suspension cells of either aggregate type. This is of interest in view of the findings of Jones and Fett (1984) that strain S-9-8 is immobilized by electron-dense material in leaf intercellular spaces of cv. Clark while strain XP175 is not.

V. EPILOG

The use of plant cell suspension cultures to study the interaction of plants with biotic elicitors such as bacteria may lead to results that do not accurately reflect responses of the intact plant. This is evidenced by the result for glyceollin induction, bacterial growth, and bacterial immobilization presented here and in our earlier studies of the interaction of soybean cells with plant pathogenic bacteria (Fett, 1984; Fett and Zacharius, 1982, 1983). Careful considerations need to be given to the age of the cell suspension culture, the culture medium, and cell aggregate size distribution. The fact that plant cells in suspension cultures are continually bathed in large volumes of liquid whereas leaf intercellular spaces (where leaf-spotting bacteria reside and grow) are initially deficient in free water may preclude certain important interactions such as prolonged bacterial cell–plant cell contact from taking place. However, response of cell suspension culture to bacterial inoculation can accurately reflect certain in planta phenomena during plant–bacterial interactions as demonstrated by recent studies of Orlandi et al. (1992).

Future research should be directed at determining optimal cell suspension culture conditions and cell aggregate sizes to be used in order to more closely mimic in planta phenomena.

REFERENCES

Alves, L. M., Heisler, E. G., Kissinger, J. C., Patterson, J. M., and Kalan, E. B. (1979). Effects of controlled atmosphere on production of sesquiterpenoid stress metabolites by white potato tuber. Possible involvement of cyanide-resistant respiration. *Plant Physiol. 63*:359–362.

Apostol, I., Heinstein, P. F., and Low, P. S. (1989). Rapid stimulation of an oxidative burst during elicitation of cultured plant cells. *Plant Physiol. 90*:109–116.

Fett, W. F. (1984). Accumulation of isoflavonoids and isoflavone glucosides after inoculation of soybean leaves with *Xanthomonas campestris* pv. *glycines* and pv. *campestris* and a study of their role in resistance. *Physiol. Plant Pathol. 24*: 303–320.

Fett, W. F., and Zacharius, R. M. (1982). Bacterially induced glyceollin production in soybean cell suspension cultures. *Plant Sci. Lett. 24*:303–309.

Fett, W. F., and Zacharius, R. M. (1983). Bacterial growth and phytoalexin elicitation in soybean cell suspension cultures inoculated with *Pseudomonas syringae* pathovars. *Physiol. Plant Pathol. 22*:151–172.

Gamborg, O. L. (1975). Callus and cell culture. *Plant Tissue Culture Methods* (O. L. Gamborg and L. R. Wetter, eds.), National Research Council of Canada, Saskatoon.

Graham, M. Y., and Graham, T. L. (1991). Rapid accumulation of anionic peroxidases and phenolic polymers in soybean cotyledon tissues following treatment with *Phytophthora megasperma* f. sp. *glycinea* wall glucan. *Plant Physiol. 97*: 1445–1455.

Hahlbrock, K., Ebel, J., and Oaks, A. (1974). Determination of specific growth stages of plant cell suspension cultures by monitoring conductivity changes in the medium. *Planta 118*:75–84.

Jones, S. B., and Fett, W. F. (1985). Fate of *Xanthomonas campestris* infiltrated into soybean leaves: an ultrastructural study. *Phytopathology 45*:733–741.

Kinnersley, A. M., and Dougall, D. K. (1980). Increase in anthocyanin yield from wild carrot cell cultures by a selection system based on cell aggregate size. *Planta 149*:200–204.

Kuboi, T., and Yamada, Y. (1976). Caffeic acid-O-methyl transferase in a suspension of cell aggregates of tobacco. *Phytochemistry 15*:397–400.

Kuboi, T., and Yamada, Y. (1978). Regulation of the enzyme activities related to lignin synthesis in cell aggregates of tobacco cell culture. *Biochim. Biophy. Acta 542*:181–190.

Montillet, J. L., and Degousée, N. (1991). Hydroperoxydes induce glyceollin accumulation in soybean. *Plant Physiol. Biochem. 29*:689–694.

Orlandi, E. W., Hutcheson, S. W., and Baker, C. J. (1992). Early physiological responses associated with race-specific recognition in soybean leaf tissue and cell suspension treated with *Pseudomonas syringae* pv. *glycinea*. *Physiol. Mol. Plant Pathol. 40*:173–180.

Phillips, J. H. (1973). Dye exclusion tests for cell viability. *Tissue Culture Methods and Applications* (F. F. Kruse, Jr., and M. K. Patterson, Jr., eds.), Academic Press, New York, pp. 406–408.

Siegel, N. R., and Enns, R. K. (1979). Soluble polyvinvlpyrrolidine and bovine serum albumin adsorb polyphenols from soybean suspension cultures. *Plant Physiol. 63*:206–208.

Singh, B. D., Rao, G. S. R. L., and Singh, R. P. (1982). Polyphenol accumulation in callus cultures of cowpea (*Vigna sinensis*). *Indian J. Exp. Biol. 20*:387–389.

Verma, D. P. S., and van Huystee, R. B. (1970). Cellular differentiation and peroxidase isozymes in cell culture of peanut cotyledons. *Can. J. Bot. 48*:429–431.

Watts, M. J., Galpin, I. J., and Collin, H. A. (1984). The effect of growth regulators, light and temperature on flavour production in celery tissue cultures. *N. Phytol. 98*:583–591.

Worthington Biochemical Corporation (1972). *Enzymes and Enzyme Reagents*, Freehold, NJ, pp. 43–44.

Zacharius, R. M., and Kalan, E. B. (1984). Biotransformation of the potato stress metabolite, solavetivone, by cell suspension cultures of two solanaceous and three non-solanaceous species. *Plant Cell Rep. 3*:189–192.

Zacharius, R. M., and Kalan, E. B. (1990). Isoflavonoid changes in soybean cell suspensions when challenged with intact bacteria or fungal elicitors. *J. Plant Physiol. 135*:732–736.

7

Phytoalexins as a Factor in the Wilt Resistance of Cotton

M. H. Avazkhodjaev, S. S. Zeltser, and H. V. Nuritdinova
Institute of Experimental Biology of Plants, Academy of Sciences of Uzbekistan, Tashkent, Uzbekistan

Raviprakash G. Dani
Central Institute for Cotton Research, Nagpur, Maharashtra, India

I. INTRODUCTION

The *Verticillium* wilt of cotton (*Gossypium spp.*) always remained a serious problem in the cultivation of this crop. According to contemporary scientific thinking, the plant cell, like the animal cell, possesses an immunological control system, the function of which involves not only to defend against pathogenic microorganisms, but also to support the structural and functional integrity of the body. Disease resistance is significant in this context, for it is the result of interaction between the host plant genotype, the pathogen, and the surrounding environmental factors.

Among the principal causes of increase in susceptibility to wilt disease of cotton, the foremost are crop monoculture and planting of identical cultivars. Second comes the public's compulsive use of pesticides, different chemicals, and fertilizers. Third, there is the annual carryover of infected plant parts in the soil.

It is well known that in wild pathosystems a more or less rigid selection operates, thus upholding the balance of nature, whereas under cultural conditions two genotypes operate, i.e., that of the host plant and that of the parasite, which interact through metabolic changes within these two bodies. In contrast to animals, the plant body system is not endowed with highly specific antigen–antibody reactions. Immunity in plants is brought about by a whole series of less specific defense reactions. In the past 10–15 years, thanks to developments in molecular biology, particularly molecular genetics, it has been demonstrated that immunity involves not only a bodily reaction but a complex of several reactions, directed in support of its functional integrity, i.e., homeostasis. Considerable biochemical and molecular studies are being directed to disease resistance in cotton because of the unique tools of biotechnology that are now available for plant genetic improvement, involving the aspects of biochemical regulation of phytoalexin expression (Stewart, 1991) and selection at the cellular level for resistance to physiological stress (Dani, 1992), and so forth.

In the study of the physiological and biochemical processes of the host–pathogen interaction in the *Verticillium* wilt, two distinct phases are involved. The first phase consists of determination in which some very exclusive reactions occur, with transmission of signals, directly to the genetic apparatus. The second phase is the expression, characterized by the entire range of defense reactions.

It has been established through our long-term research that one of the fundamental wilt defense reactions in cotton consists of the so-called hypersensitivity reaction, which in itself embodies an entire array of parameters, starting with some nonspecific reactions (e.g., strengthening of oxidation recovery enzymes, formation of their isozymes, oxidation of polyphenols, formation of quinones) and culminating in the formation of phytoalexins.

Müller and Borger (1940) postulated a theory on the existence of phytoalexins, which has met with experimental confirmation over the past two to three decades. A wide range of phytoalexins have now been identified in beans, potatoes, orchids, beetroot, and other crops including cotton (Tomiyama, 1970; Cruickshank and Perrin, 1971; Keen et al., 1971; Metlitski et al., 1976; Fraile et al., 1982; Kuć and Rush, 1985; Bailey and Mansfield, 1985; Sun et al., 1989; Essenberg et al., 1990).

II. RESISTANCE REACTIONS OF COTTON

A. Phytoalexins

Studies such as those mentioned above provided impetus to further the genetic and biochemical investigations of phytoalexins. The series of investigations by Bell (1967, 1969, 1981) made a significant contribution to the understanding of phytoalexins in cotton. Many of his studies were concerned with flowering and boll initiation phases. During infection, in certain plants fungitoxicity of xylem vessels and the leaves supporting bolls often registered such an increase at flowering stage as to completely suppress the germination of fungal spores. It was shown that the fungitoxicity of the resulting products was due to certain ether-soluble phenolic compounds. Gossypol was primarily recognized. Besides gossypol, at least four other gossypol-related compounds were also isolated from xylem vessels of infected cottons. The complex of gossypol-like compounds akin to phytoalexins was termed "gossypol equivalents."

In the course of our own investigations on phytoalexins, attention was focused primarily on the method involving "drop diffusates," the fungitoxicity of which was used as an index for the activity of phytoalexins in cotton tissues. This method permits the recovery of phytoalexins without destruction of tissues (Metlitski et al., 1971; Muxamedova and Turakulov, 1974). Results of our research confirm that wilt resistance in cotton is closely related to the ability of the plant to produce phytoalexins in response to infection. As can be seen from the data presented in Table 1, high phytoalexin activity is characteristically found in the wilt-resistant wild relative of cotton, namely, subsp. *mexicanum*, while the lowest activity is noted in the wilt-susceptible *G. hirsutum* cvs. S 4727 and 1306 DB.

Thus a major reason for the high wilt resistance of the *G. hirsutum* cvs. Express and the chemo-mutants L3 and L4 would be their comparatively higher ability to produce phytoalexins in response to infection by the fungus *V. dahliae* Kleb., the causal organism of wilt. It was noted that the fungitoxicity of the diffusates increased with increasing density of suspended fungal conidia, attaining a maximum value in the concentration of 10^4–10^6 conidia/ml. However, with higher concentrations of the inoculum of the pathogen, the phytoalexin activity in the diffusates tended to register a sharp decline.

As indicated above, phytoalexin activity of the leaf-derived diffusates forms a clear criterion of wilt resistance of different cultivars of cotton. To understand the role of phytoalexins in the incompatibility reactions of cot-

Table 1 Fungitoxicity of Leaf Diffusates from Various Cotton Cultivars

Cultivar/form of cotton	% Suppression	
	Conidial germination	Hyphal growth
Tashkent-1	34.5	83.1
Tashkent-2	29.0	66.4
Tashkent-3	24.2	77.5
Chemomutant L3	29.7	75.9
L4	23.0	50.0
L-4727	24.0	60.9
1306 DV	4.0	7.4
Express 2	28.5	71.8

ton and the fungus *V. dahliae*, the structure of the substances that determine the fungitoxicity of the diffusates was resolved by means of IR, UV, nuclear magnetic resonance, and mass spectrometry (Sadykov et al., 1974; Karimdjanov et al., 1976). In the light of these investigations, and those carried out at the National Cotton Pathology Research Laboratory in the United States, the chemical structure of the cotton phytoalexins has been clearly established. These fall into the category of sesquiterpenoid and triterpenoid aldehydes, akin to gossypol (Fig. 1).

B. Location and Role of Phytoalexins

Our subsequent efforts were directed to investigations on the function and role of phytoalexins in disease-resistant cotton, and the difference in infection response reactions of the resistant and susceptible varieties, considering the localization of the processes in the infected tissues and the dynamics of their development. It was noted that a more complete picture of the defense reaction mechanism reflects at the initial determination phase of the wilt disease when the pathogen comes in contact with the host plant during the latent, symptomless period of the disease (from the moment of plant infection until the localization and cessation of artificially induced infection in the event of incompatibility or until the actual appearance of symptoms in the event of compatibility).

To determine the quantity of isohemigossypol in the xylem vessels of cotton varieties differing in their degree of wilt resistance, the method of Avazkhodjaev and Zeltser (1980) was adopted. The results revealed that production of phytoalexins was characteristic of both susceptible and resistant varieties. The difference lay in the speed of formation of isohemigossypol in response to infection. In wilt-resistant varieties, formation/accumulation of phytoalexins progresses at a faster rate than in susceptible ones. Corresponding with this, quantitative differences also manifest in the content of isohemigossypol during the early days of incubation period. In resistant plants, during the initial 5 days following infection, a more or less faster accumulation of phytoalexins was observed in the xylem vessels. In the resistant subsp. *mexicanum* and in varieties belonging to the class of Tashkent, as also in the case of the chemomutant L3, traces of isohemigossypol appeared on the chromatogram within 10–12 hr of infection. During this particular phase, phytoalexins were not detected in the tissues of the susceptible variety S-4727, while traces of isohemigossypol appeared only after 24–30 hr of incubation. Through such analyses of dynamics of phytoalexin formation, it is possible to judge the level of susceptibility or resistance of cotton to the *Verticillium* wilt. This property is also displayed by an array of new, promising, wilt-resistant *G. hirsutum* varieties recently

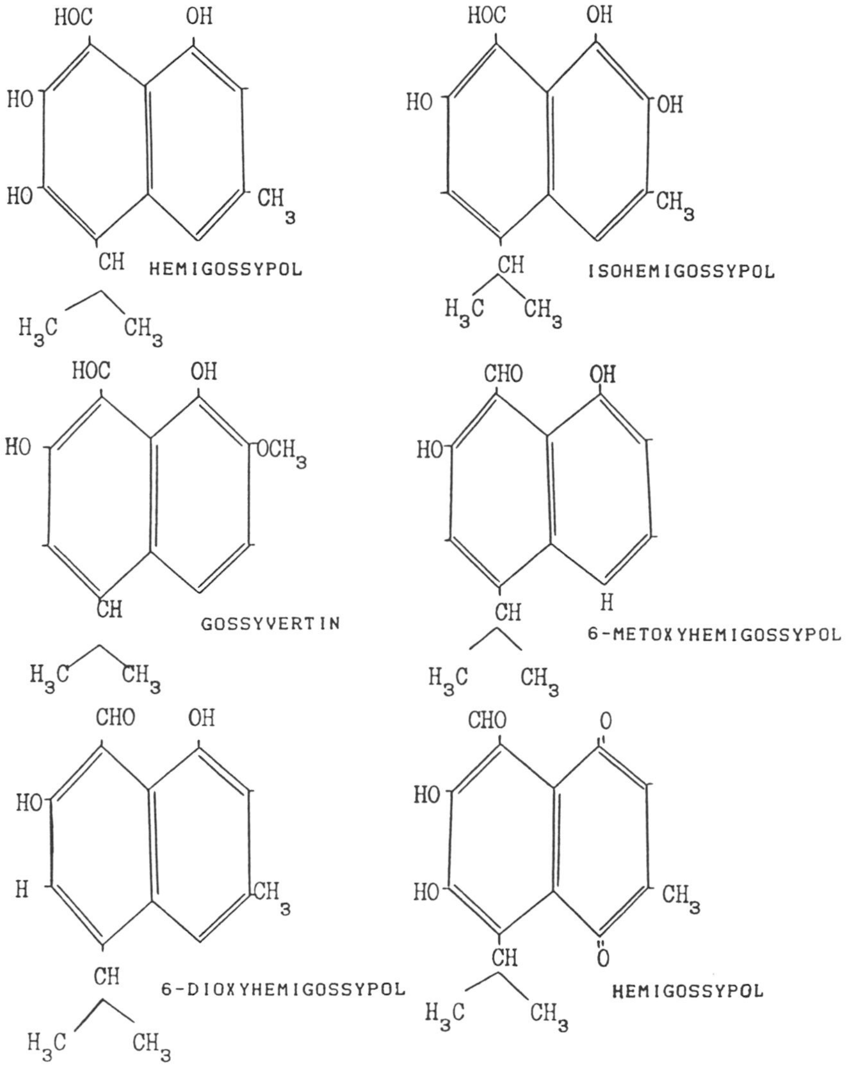

Figure 1 Chemical structure of phytoalexins of cotton.

released by the IEBP, including Uzbekistan 1 and 2, Express-2, Tashkent-6, AN-402, and several new lines in selection. All the experiments involving the tissues of resistant plants were marked by a much faster rate of formation and accumulation of isohemigossypol. It may be noted from the data represented in Fig. 2 that in the tissues of the susceptible cv. S-4727, in the

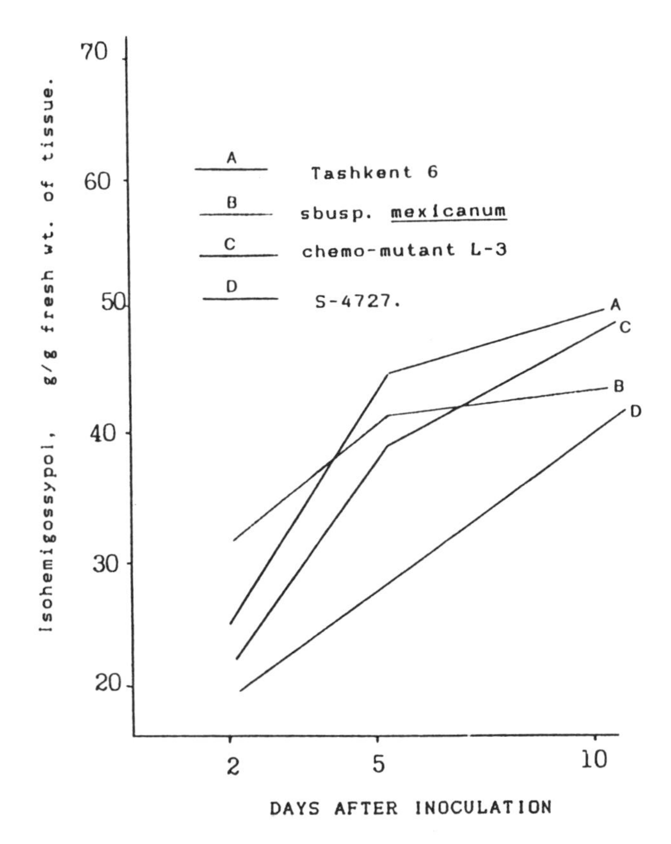

Figure 2 Dynamics of phytoalexin formation in xylem tissues of different cotton cultivars.

dynamics of the incubation period, a similar trend of quantitative increase in hemigossypol is registered, although the quantity of phytoalexins per se was less than that of the resistant ones, until the tenth day of the postinfection period.

At different stages of wilt, depending on environment and features of pathogenesis, the content of phytoalexins in the tissues of susceptible cottons may even be higher. However, this is not connected wit the incompatible reactions between the plant and the fungus *V. dahliae* but on the contrary characterises a progressive intensification of the wilt disease. At the beginning of the incubation period (2–5 days), wilt-resistant plants characteristically show fewer necrotized cells and consequently relatively higher doses of phytoalexins get accumulated in the infection spots, which may prove lethal to the pathogen. In tissues of susceptible plants, progression of the disease is faster since during these very days of latent infection a lesser quantity of phytoalexins is formed, although the number of necrotized cells is significantly higher. In other words, the rate of growth of the pathogen is faster than that of production of isohemigossypol and hence the fungitoxic compounds cannot completely block the development of infection. This is one of the characteristic properties of phytoalexins, determining their role in the wilt resistance of the cotton plant. A decisive moment in the event of infection is the speed of production of postinfectional inhibitors in tissues of host plant in response to infection and their accumulation in lethal doses in the sites of infection.

As was revealed from our subsequent studies, phytoalexins are formed in the cells surrounding the xylem vessels, and thus in comparison to other plant species the functioning of this defense reaction in cotton is unique. In those spots where infective structures of pathogen are present, primary cell walls of visceral parenchyma are destroyed inside the vessels, with the formation of tangentially isolated sectors (Fig. 3). A kind of "isolation" of the diseased sectors from the healthy ones occurs, which can be clearly seen under methyl blue stain. At times, these are in such large numbers that they completely cover parts of vessel segments, as in the natural formation of tyloses in several wood species (Yatsenko-Xmelevski, 1954). Results of our experiments also confirm that phytoalexins accumulate in the spherical tyloses of xylem vessels of infested plants and are clearly visible when their preparations are stained with antimony trichloride. Thus, in the tissues of cotton plant during infection with *V. dahliae*, certain mechanical barriers are formed, by way of blocking of vessel segments or their portions, with simultaneous localization of fungitoxic phytoalexins in them, which may completely suspend the development of the disease. In the susceptible cv. S-4727, within 48 hr of infection a larger number of necrotized cells were obtained than those in the cv. Tashkent-6. However, the concentration of phytoalexins was obviously not adequate for a complete suppression of infection. Consequently, despite the fact that phytoalexins are produced in response to infection in the susceptible plant, the pathogen appeared to continue to settle in xylem vessels, leaving behind a certain number of dead parenchyma cells. On the other hand, in the resistant cultivar, during incubation period, cells are necrotized in lesser numbers in response to infection, and higher doses of phytoalexins are accumulated in the zone of infection, inhibiting the subsequent spread of the pathogen in the plant body. Thus, the necrotized cells of the cotton plant may serve as barriers in the process of ramification of the pathogenic fungus, only if they contain lethal doses of phytoalexins.

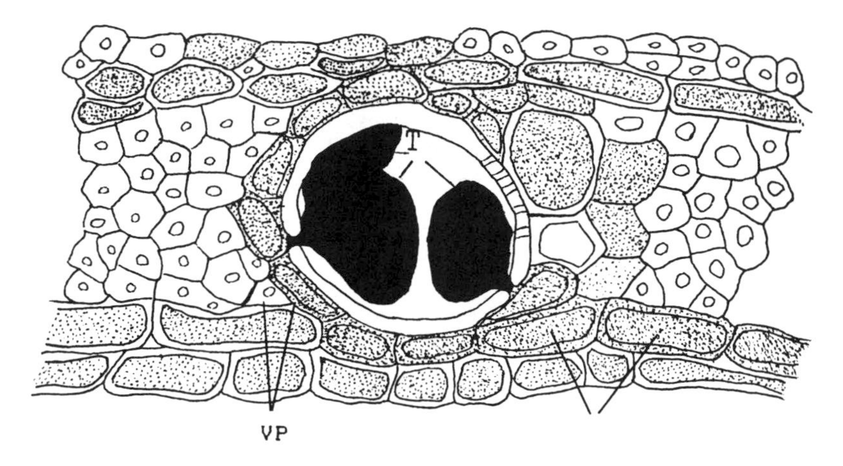

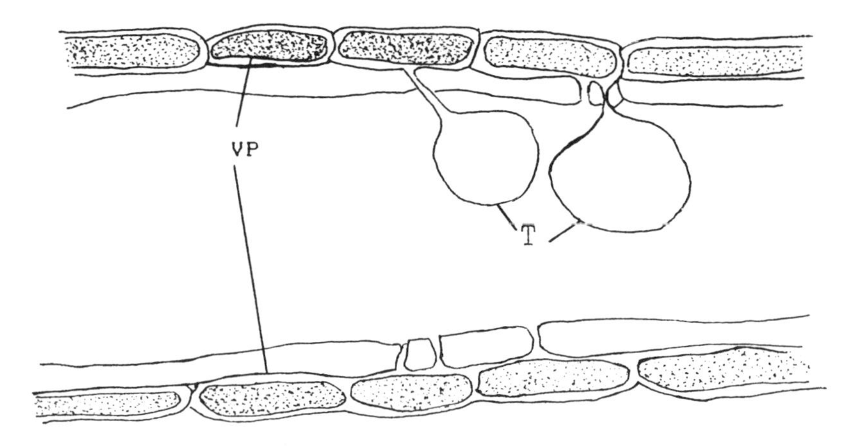

Figure 3 Pathological changes in xylem of cotton infected with *Verticillium* wilt.

C. Search for Inducers from Fungal Cell Wall

Albersheim and Anderson-Prouty (1975) stipulated a theory, according to which the mechanism of phytoalexin formation relates to the interaction of surface molecules of the interacting partners: the receptors, products of genes of the resistant host plant, and the pathogen molecules, products of the avirulence genes.

It has been established that not only the pathogen but certain cultural fluids containing inducers, extracts of fungal mycelia, and several high molecular weight metabolites of the pathogen, after their contact with the plant can selectively act as inducers, resulting in the formation of phytoalexins in the plant. Analogous behavior is displayed by an array of chemical substances, such as salts of heavy metals, antibiotics, inhibitors of ferments, fungicides, etc.

On the basis of reports in the literature on the nature of plant defense reactions (Metlitski et al., 1976; Dyakov, 1983; Albersheim and Valent, 1978; Zaki et al., 1972), we initiated a search for the inducer metabolites among various fractions isolated from mycelia and cultural fluids of the pathogen. Initially, a certain protein–lipopolysaccharide complex (PLPC) and its components were studied (Malysheva and Zeltser, 1968). Our experiments established the phytoalexin activity of phytotoxic metabolites of *V. dahliae*, isolated by the scientists of the Institute of Microbiology of the Uzbek Academy of Sciences (Borodin, 1978). The metabolites included di-2-ethylhexylphthalate, transaminic acid, an oligosaccharide, a polypeptide, a pigment, a common protein, and lipids from fungal mycelia and other sources.

Our subsequent efforts were directed to the study of pathogen cell wall derivatives. As per contemporary thinking, first in order of importance, considering the interaction of higher plants and their specific pathogens in the initial stage of infection, come host membrane and membranolytic agents of the inducer pathogen.

Subsequent progress of the disease would be mainly decided by the status of the plant membrane sensitivity toward the specific pathogen metabolites that interact with them. This interaction presupposes the presence of receptor areas on the host membranes as also the presence of the specific factors complementary to them on the pathogen envelopes. This theory is reflected in the series of reports concerning certain fungal, bacterial, and viral diseases (Hadwiger and Schwochan, 1969; Dyakov, 1976, 1983; Metlitski, 1976; Metlitski and Ozeretskovskaya, 1985). In this context, investigations by Albersheim and his group (Anderson-Prouty and Albersheim, 1975; Albersheim and Valent, 1978) are of considerable interest. Utilizing the methods of isolation and fractionation of cell walls with some modifications, we were able to isolate and identify the nature of the most active phytoalexin-inducing components of cell walls, differing in the degree of virulence of the races of *V. dahliae* (race 2 and the avirulent radiomutant R-177). Results of our experiments showed that comparatively high phytoalexin–inducing activity is exhibited by alkaline and lipid fractions of the cell walls of the fungus *V. dahliae*.

Spectral data characterized the active phytoalexin-inducing components of the cell walls as substances with typical polysaccharide structure, with characteristic functional group and bonds (see Avazkhodjaev et al., 1984a,b; Avazkhodjaev, 1985).

Analyses of the elicitor isolated from cell walls have indicated that the elicitor, complex in structure, is principally composed of carbohydrates (over 50%) and proteins (up to 7%), with traces of K and Ca cations and about 1% phosphorus. In the hydrolysates of the elicitors, the carbohydrate components are predominantly galactose, glucose, and mannose, while the amino acids include aspartic and glutamic acids and alanine. IR spectroscopic data confirmed the complexity of the elicitors and their glycoprotein nature (Figure 4).

D. Induction of Phytoalexins by Elicitors

The induction of phytoalexins by elicitors was subsequently studied, using the relatively resistant cv. Tashkent-6 and the susceptible cv. S-4727 (Table 2). Dynamics of quantitative contents of phytoalexins in xylem tissues was monitored. The results of thin-layer chromatography in plates of silufol-254 (Czechoslovakia) (Avazkhodjaev et al., 1984a,b) revealed that, with the introduction of elicitor, a small quantity of phytoalexin appeared as early as on the second day in stems of both cultivars. Within 2–5 days, larger quantities of phytoalexins were found accumulated in the relatively resistant cv. Taskhent-6, especially against the elicitor from the avirulent race. However, at a much later stage, the reverse was found to be the case: the content of phytoalexins was higher in the xylem of the susceptible variety, especially as against the inducer from the virulent race.

In some of our recent studies, further observations were made on the nature of changes in xylems in different cotton varieties varying in the degree of their wilt resistance, under the influence of the elicitor from *V.*

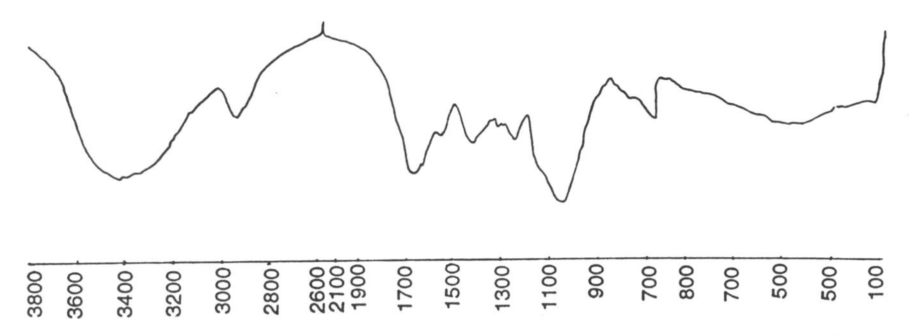

Figure 4 IR spectrum of the elicitor of the fungus *Verticillium dahliae*.

Table 2 Induction of Phytoalexins of Cotton, Isohemigossypol (IHG), and Gossypol Equivalents (GE) with the Help of Elicitor Isolated from Mycelia of the Fungus *Verticillium dahliae* (μg/g fresh wt of tissue)

		Days after infection														
		2			3			5			10			15		
Cotton cv.	Fungal race	IHG	GE	SUM	IHG	GE	SUM	IHG	GE	SUM	IHG	GE	SUM	IHG	GE	SUM
Tashkent-	R-177	8.7	10.2	18.9	31.3	68.7	100.0	36.7	76.3	113.0	37.8	83.2	121.0	45.3	36.7	82.6
6	2	4.3	7.9	12.2	29.8	57.2	88.0	31.7	70.7	102.4	40.4	100.0	140.0	100.2	105.3	205.5
S-4727	R-177	7.4	5.8	13.2	25.4	62.0	87.4	30.0	64.9	94.9	78.7	108.6	187.3	60.4	46.7	106.8
	2	3.6	t	3.6	22.3	44.0	66.3	23.3	63.7	87.0	146.7	285.4	138.7	160.0	116.7	276.7

dahliae (see Avazkhodjaev et al., 1990a). The pathological changes that occurred were the formation of tyloses and the choking of vessels, with subsequent formation of phytoalexins (Fig. 5). During the first 10 days after the introduction of the elicitor in the stems, there was a gradual development of these processes, from the lower portion of the stem to the upper regions. In the susceptible cv. S-4727, under the influence of the elicitor from the virulent races, such reactions proceeded more slowly than those in the relatively resistant cv. Tashkent-6. The type of response of the

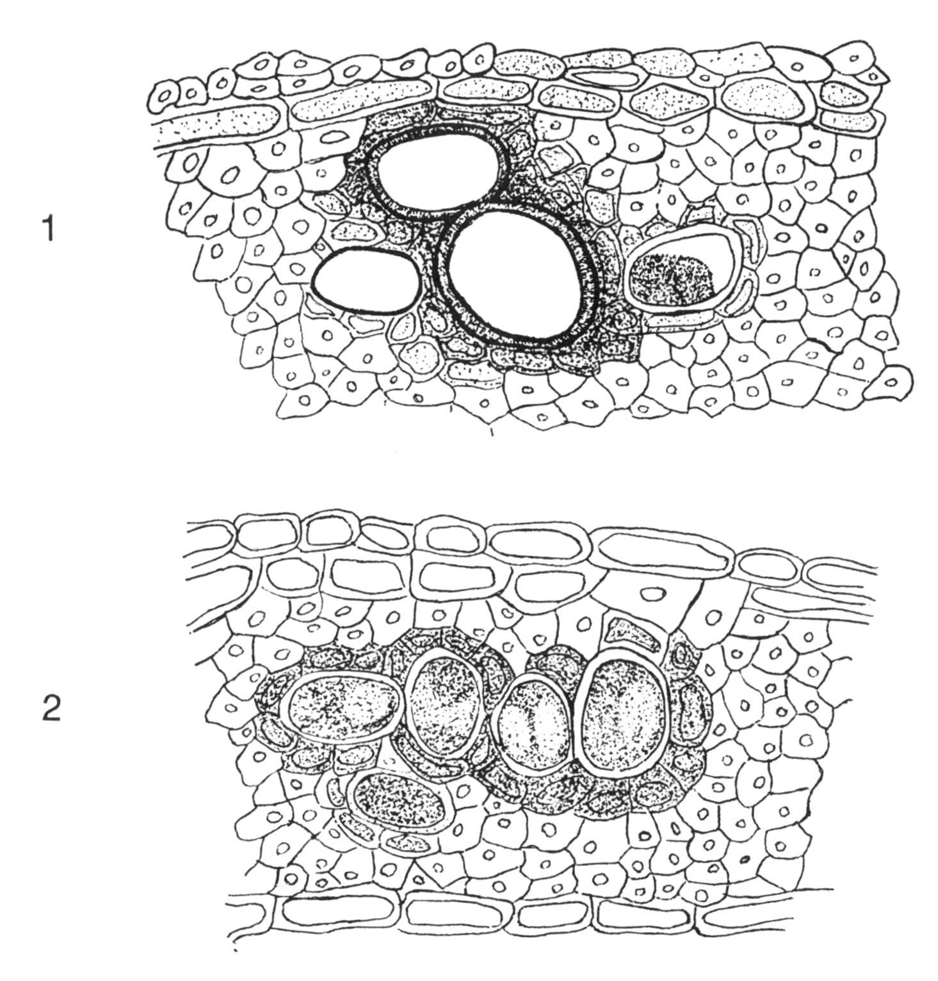

Figure 5 Pathological changes in xylem of cotton cv. S-4727 (1) and Tashkent-6 (2) in the spot of isolation of the elicitor of the fungus (after 48 hr).

given plant variety was determined by the speed of induction of defense reaction, which is also influenced by its metabolite, the inducer of phytoalexin formation in the plant, in doses lethal to the pathogen.

It is evident from the results that in the very initial phase following introduction of elicitor in xylem of resistant plant, higher amounts of phytoalexins are accumulated, as compared to those in the susceptible varieties. This leads to the confinement of infection to the lower portions of the stem, i.e., in the primary infection spot and obstructs the further growth of the fungus in the plant.

Observations on the inducible character of the defense reaction of cotton have been used as the basis for a laboratory-devised method of defense against wilt, on the lines of artificial intensification of phytoalexin formation in plant tissues by various means, including the use of chemical preparations that influence the natural mechanisms which limit or block infection. The function of such preparations relates to the induction of phytoalexin production in tissues. High ability to activate the defense reaction was shown by ionophore cyclopolyether derivatives, i.e., Dibenzo-18 and Crown-6, effecting two- to threefold lowering of mortality in field experiments (Avazkhodjaev et al., 1984a,b).

A more effective antiwilt immunizer was the preparation Biosol-2, which was synthesized in collaboration with chemists at the Bashkir branch of the Academy of Sciences of the erstwhile USSR. Preparations of the type Bisol have been introduced in cotton cultivation, since incidence of wilt under their influence is lowered by 50–70%, while the fiber yield increases by up to three to four quintals per hectare. Purpose-oriented screening and subsequent testing of the preparations is in progress.

During the recent years, researchers of phytoimmunity, having resolved the nonspecific character of indiction, have turned to compositional analysis of the metabolites from pathogens, which condition the specificity response of the host plant. Metlitski et al. (1986), considering phytoimmunity from a biological viewpoint wherein incompatibility is a natural rule and compatibility is almost exception, felt that specificity of the response reaction of the host plant has to be associated with various products that condition compatibility of the participants. Such a compatibility is the consequence of the lengthy process of natural evolution of the host and the pathogen, in which new genes for virulence appear and then overcome the genes that condition resistance in the host plant. Specificity thus may be controlled by products of the susceptibility gene of the plant and that of its complimentary parasite gene.

Several hypothetical models propose that in addition to nonspecific inducers (elicitors) conditioning the induction of phytoalexins in association with the receptor and the resistant plant, the parasites produce species-specific and race- and cultivar-specific suppressors, which obstruct the action of the elicitors (Terekhova and Dyakov, 1980; Heath, 1981; Bushnell and Rowell, 1981). Specificity of suppression is conditioned through a series of corresponding areas on suppressor and receptor, which must be complementary to each other, and thus effectively ensure the suppression of the effects of the elicitor by rendering the cell incapable of switching on the defense reaction.

Our experience points to the fact that the high varietal breakdown in cotton is largely due to the strength of the virulent race 2 of *V. dahliae* to inhibit the synthesis and accumulation of lethal doses of phytoalexins in the infected tissues.

Subsequent studies have underlined the significant role of suppressor metabolites in races of *V. dahliae*, blocking the defense reaction of cotton in the initial stage of *Verticillium* wilt. Suppression of phytoalexin formation is race-specific, since it is not displayed under influence of the avirulent mutant R-177, which apparently is one of the means to overcome the cultivar resistance of cotton (Table 3; Fig. 6; see also Avazkhodjaev et al., 1984a,b). As can be seen from the Fig. 6, in the experiment involving the introduction of metabolite (analogous to the suppressor of the virulent race 2) from the avirulent mutant R-177, no suppression of induction of phytoalexin formation was noticeable: the xylem vessels of the host plant showed the presence of lethal doses of the phytoalexin isohemigossypol. Suppression of the defense reaction—the formation of phytoalexin—by the suppressor of virulent race 2 was noticed in the host plant only in the initial determinative stages of the disease, since the process of phytoalexin formation was restored in the infected tissues on the fifth day of the experiment.

Table 3 Induction and Blocking of Phytoalexin Formation in Cotton Tissues

Experimental variant	Isohemigossypol content after 48 hr (μg/g dry wt of tissue)
1. Control (intact tissues)	Absent
2. Injection of avirulent mutant P-177 race	38.4 ± 1.3
3. Injection of virulent race 2 (2.5 ml spores/ml)	21.8 ± 0.9
4. Influence of metabolite (avirulent suppressor of race 2)	49.7 ± 1.2
5. Influence of suppressor of race 2 + after a day, inductor of phytoalexins	Absent
6. Introduction of phytoalexin inductor	46.2 ± 2.1

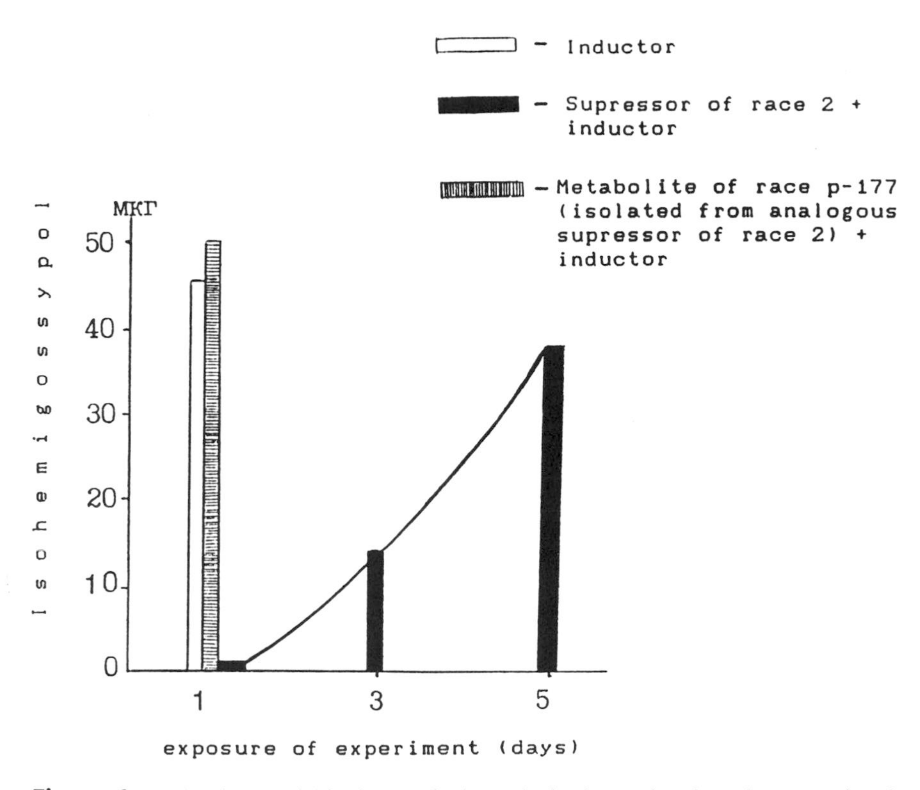

Figure 6 Induction and blockage of phytoalexin formation in xylem vessels of the cotton cv. Tashkent-6.

The results confirm that the high incidence of breakdown in wilt resistance of cotton is related to the ability of the race to suppress the formation and accumulation of lethal doses of phytoalexins. Since the suppression is race-specific, it is presumably one of the ways to surmount the cultivar resistance. On the basis of the collected information, our laboratory has proposed a method for selection of the wilt-resistant forms of cotton (Anonymous, 1986), which is used by plant breeders for obtaining an inducer of the *Verticillium* wilt of the selection material, which is resistant to the race prevalent at the given time.

Results of chemical analysis of the suppressor metabolite isolated from virulent race 2 of *V. dahliae*, as well as the phytoalexin-inducing preparations from the cell walls of the pathogen, indicate that the suppressor metabolite from virulent race 2 of *Verticillium dahliae* is glycoproteinaceous in nature and contains 34.4% carbohydrates and 56.3% common proteins. It was established with the help of IR spectroscopy, that the suppressor metabolite relates to lipoglycoproteins, with their characteristic complex bonds of functional groups.

In the hydrolysate of the suppressor, the following amino acids were identified: aspartic acid, glutamic acid, asparagine, aminobutyric acid, alanine, serine, lysine, threonine, histidine, leucine, isoleucine, arginine, tyrosine, cysteine and oystine, proline, tryptophan, phenylalanine, valine, and methionine. The sugars detected were raffinose, lactose, glucosamine, galactose, glucose, mannose, and rhamnose. It was also shown that the suppressor molecule had a lower molecular weight (of the order 8000–10,000) than the phytoalexin inducers.

III. RECEPTOR SUBSTANCES

In the light of the fact that the process of recognition of pathogen inducers is accompanied by the participation of the cell wall components of the host plant receptors, we were able to isolate certain lectin-like substances from the cell walls of the host plants, possessing the specificities of receptors. According to earlier reports (Sequeira, 1978; Lyutsik et al., 1981), lectins are proteins, possessing the ability to associate in a complementary manner with pathogen cell wall inducers of polysaccharide character.

It has been known that lectins participate in the formation of glycoprotein and polysaccharide complexes and in the process of intercellular recognition, which take place during host plant and pathogen interaction (Lyutsik et al., 1981; Markov and Xavkin, 1983; Faye and Crispeels, 1987). Results of our research on their physicochemical constitution indicate that lectins are mainly proteins forming a complex with trace amounts of carbohydrates (Nuritdinova, 1988).

In the course of our investigations, two fractions of lectins were traced in the helium columns, marked with quantitative differences of carbohydrate content. The amount of carbohydrate in the first fraction (11.7 mg/g dry wt) was about half the quantity of the second (22.8 mg/g dry wt). In the hydrolysates of the two fractions, the sugars galactose, glucose, mannose, and ramnose were present in a ratio of 4:3:1:2. Lectins of the first fraction contained about 31.4% proteins whereas those of the second contained 78.8%. Some quantitative variation in the intensity of absorption bands in the IR absorption spectra of the two fractions of lectins was observed, although the spectral data characterized the lectins as substances that are mainly proteinaceous in character, with functional groups and bonds inherent to proteins. Also, the IR spectrum displayed insignificant absorption bands characteristic of carbohydrates, which permits the lectins to be seen as a complex union of proteins with polysaccharides with a preponderance of protein structures, and thus confirms the data obtained

Table 4 Analysis of Lectin Fractions for Content (%) of Cations of Various Elements

Lectin	C	H	S	No	Cu	Ca	Al	Fe	Si	Mg	Mn
Fr. I	21.97	4.33	1.66	3.27	0.009	0.0031	0.054	0.16	0.26	t	t
Fr. II	27.22	5.11	1.17	4.42	0.003	0.0261	0.018	0.11	0.11	t	t

in earlier analysis (Avazkhodjaev et al., 1987). The lectin fractions contained small amounts of cations such as Cu, Ca, Al, Fe, Si, Mg, and Mn. (Table 4 shows the percentage content of the various elements.) The two fractions of lectins differed not only in terms of their proteins, carbohydrates, amino acids, and sugars but also in their individual physiological actions.

A. Lectins, Biological Properties, Function, and Immobilization

Data on the influence of the two fractions of lectins in the formation of phytoalexins are presented in Table 5 and Fig. 7. It is seen that the two fractions are different in their induction of phytoalexins (Fig. 7). The first fraction suppressed or inhibited the induction by about 53.1%, whereas the second strengthened it by about 131.8%.

Detailed studies were conducted on the biological properties of the lectins, i.e., their ability to react to the infectious structures and metabolites of the pathogen, forming determinant complexes through specially designed in vitro experiments. The lectin and its two fractions can react with *V. dahliae*, influencing germination of conidia and hyphal growth (Table 6). Inhibition of growth of pathogen is most evident during the reaction of lectins with an avirulent race of the fungus. By producing incubation mixtures of the complete lectin with the inductor isolated from *V. dahliae*, it has been established that lectins are concerned with the fungal metabolite—the elicitor of phytoalexins in the plant also.

Table 5 Influence of Lectin on the Formation of Phytoalexins in Hypocotyls of Cotton Plants (μg/g wt)

Lectin fraction	Isohemigossypol	Gossypol equivalent	Sum of phytoalexins	% of control
Control	36.0 ± 1.2	35.6 ± 0.9	71.6 ± 4.2	100.0
Fr. I	17.0 ± 0.9	21.0 ± 0.5	38.0 ± 2.1	53.1
Fr. II	50.4 ± 3.2	44.0 ± 2.3	94.4 ± 5.4	131.8

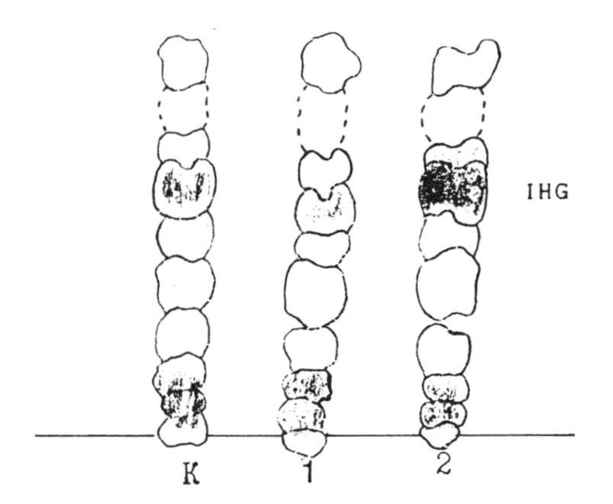

Figure 7 Chromatogram of chloroform extracts from hypocotyls of cotton plant treated with lectine fractions.

During subsequent research on the sites of contact of lectin with the components of the elicitor of pathogen (which immobilize the inducement of the disease) it was found that there was a discretionary contact of lectin with the sugars glucose, galactose, and mannose—basic components of the hydrolysate of the carbohydrate portion of the elicitor of *V. dahliae*. Such components were termed as heptanes (Avazkhodjaev et al., 1989a,b).

The ability of the lectin fractions isolated from the hypocotyls of cotton plant to agglutinate the conidia of the two pathogenic races of *Verticillium* differing in virulence (race 2 and the avirulent mutant R-177) as well as the influence of the heptane sugars on this process were investigated. The results showed that addition of heptanes to the incubation mixture inhibited

Table 6 Phytopathological Characteristics of the Experiment on the Influence of Lectins of Cotton Hypocotyls and Its Fractions on the Pathogen of *Verticillium* Wilt

	Control		Lectin I		Lectin II		Sum of Lectins	
	Hyphal growth	Conidial germination	%		Suppression			
Fungal race	(a)	(b)	(a)	(b)	(a)	(b)	(a)	(b)
R-177	100	100	85.0	52.0	71.2	33.6	81.1	41.0
Race 2	100	100	80.0	38.3	28.1	11.7	58.8	29.4

the agglutinization of conidia of both races by lectins and their fractions (Table 7). Thus, the introduction of heptane sugars individually to the incubation mixture lowers the intensity of agglutination of conidia of virulent race 2 by almost four times and that of avirulent race by five times as compared to control (without the addition of sugars), and three and four times, respectively, as compared to the introduction of xyloses and arabinoses in the mixture, taken as control. The aggregate of heptanes in the same concentration as that of the individual heptanes inhibited the agglutination of conidia of both the races caused by total lectin, even more significantly (8–10 times) (Avazkhodjaev et al., 1989a,b).

The difference in the agglutination responses shown by the races 2 and R-177 was noticed also when the first and second fractions of lectins were introduced into the incubation mixture. The aggregate of lectins, like its individual fractions, agglutinized the conidia of the avirulent race R-177 more intensively than the conidia of virulent race 2. Also, the first fraction of lectins displayed a higher ability to agglutinize the conidia of the avirulent race. It was concluded that the presence of the sugar heptanes in the incubation mixture inhibited the agglutination in all variants of the experiment significantly while the addition of xylose and arabinose in the form of control was obviously ineffective. The aggregate of sugar heptanes more strongly inhibited agglutination than each sugar taken individually. This

Table 7 Agglutinization of Conidia of *V. dahliae* with Lectin of Cotton and Its Fractions (in Relative Units, Average Aggregates of Agglutinized Conidia in 0.01 ml of Incubation Mixture)

Fungal race	2	R-177	2		R-177	
	Aggregate of lectins		Lectin fraction			
Variant			I	II	I	II
Sugar heptanes						
Glucose	9.1	8.7	14.9	11.0	7.7	11.1
Galactose	11.1	10.7	14.2	17.0	11.7	10.8
Mannose	9.8	9.9	17.4	15.3	9.7	7.3
Aggregate	6.0	6.3	6.1	7.4	0.8	5.0
Control						
Xylose	33.0	41.6	32.9	36.8	74.3	57.3
Arabinose	31.2	42.6	32.3	36.4	74.2	57.8
Buffer	45.0	51.1	40.9	46.6	77.1	58.8

was most evident in the variant with the first fraction of lectin and the conidia of R-177.

The experimental results thus pointed to the fact that agglutination is a specific indicator of the interaction of lectin with components of sugars from the surfaces of conidial cell walls of the fungus *V. dahliae*. Specific inhibition of agglutination of the conidia by lectins and by its fractions—sugar heptanes (glucose, galactose, mannose)—indicates that lectins form glycoconjugates with these sugars, i.e., they are components of the fungal cell walls, reacting with the lectins of the cotton plant.

Thus the property of the lectins to participate in the twin processes of recognition and the interaction of cells in the cotton plant fungus *V. dahliae*, which emphasizes their defensive role and ability to immobilize the pathogen, has been ably established from the results of our investigations. The degree of agglutination of the fungal conidia may serve as a test for the resistance of the plants and the virulence of the initiator of wilt.

B. Receptor Functions of Plasmalemmal and Microsomal Fractions

Since surface metabolites localized in the membranes of both the host plant and the pathogen are known to interact, it was expected that the lectins participating in the defense mechanisms would be concentrated in the plasmalemma and in membranes of the microsomal fraction of plants. We investigated the receptor function of the plasmalemma and microsomal fractions of the cotton plant, with the intention to gain an insight into the relation of its components with the membrane-active fungal metabolites—the elicitor of phytoalexin formation. The chemical composition of the hydrolysates of plasmalemma of the unaffected cells is known to be akin to that of lectin. Model studies on the composition of incubation mixtures of plasmalemma and microsomal fractions with elicitor showed that the incubating components possessed the ability to form conjugates. In the presence of microsomal fractions and plasmalemma, we observed a strong agglutination of conidia, especially in the avirulent race of *V. dahliae* (see Avazkhodjaev et al., 1990b). We have thus established that a property of metabolites of plant and pathogen (elicitors, immunodepressors, lectins) is that when present on the cell membrane surface they participate in the host–pathogen interaction, resulting in realization of either the markers of resistance of the host plant or the markers of virulence of the pathogen. With the help of the elicitor of phytoalexin formation, isolated from the fungal cell wall, specific sites of reaction have been identified in the receptor sections of microsomal fractions of cotton cells.

IV. EFFECT OF INOCULUM LOAD ON PHYTOALEXIN FORMATION

It has been noted that the ability of microsomal fractions to form glycoconjugates with elicitor (and, consequently, phytoalexins in plant tissues in response to induction) may be affected by certain unfavorable factors of the surrounding environment. One such important factor is the load of pathogen, i.e., quantity of the inoculum per unit surface of the plant. Data obtained on this factor (Table 8) reveal that the infection load significantly influences the rate of accumulation of lethal doses of phytoalexins. Against the background of a relatively small load of infection (infection with a weaker intensity), the plants displayed high concentrations of isohemigossypol in the lower portions of the stem, ranging between 42.7 and 45.2 $\mu g/g$ of the tissue, as early as on the fifth day following inoculation. An increase in the dose of artificially induced infection by a factor of 5 yielded a lesser quantity of phytoalexin during the same initial 5-day period (between 29.2 and 31.4 g), wherein isohemigossypol was also detected in the central portion of the infected stem. Thus, in this particular variant of the experiment, the infection zone was larger, and the content of the phytoalexins per unit infected tissue would be less than lethal to pathogen. This equally applies to phytoalexin accumulation patterns in the infected stems of cotton during the 10-day period of incubation: the high infection load caused a faster growth of the disease and the zone of infection was larger, i.e., the rate of proliferation of infection is higher than the rate of accumulation of lethal doses of phytoalexins.

Besides known external factors that influence the phytoalexin formation (temperature, humidity, pH of the medium, etc.), the race composition of the fungus *V. dahliae* present in the soil is also a significant influential factor. Results of experiments on the dynamics of phytoalexin formation in cotton, under the influence of races differing in virulence, are shown in Fig. 8. These indicate that the virulent race 2 of *V. dahliae* has the ability to delay the accumulation of high doses of phytoalexins in the tissues of resistant plants, which get strongly affected by the disease caused by this pathogen. Introduction of the avirulent mutant R-177 into the xylem vessels of cotton plant, on the contrary, induces a much faster formation and accumulation of phytoalexins in the sites of infection.

Virulent forms of pathogens are endowed with the ability to circumvent the defense mechanism of the host plant, rendering it inactive (Hadwinger and Schwochan, 1969; Metlitski and Ozeretksovskaya, 1973, 1985). This is one of the means of adaptation of the pathogen to the disease-resistant host plant, in which the defense mechanism is not destroyed but remains unstimulated. In this context, we conducted some experiments concerning the so-called cross-protection phenomenon.

Table 8 Isohemigossypol Content in the Different Parts of Stem of Cotton Plant at the Incubation Period of Wilt Disease (μg/g fresh wt)

Cotton cv.	Conc. of inoculum (million conidia/ml)	Days of incubation period/stem sections								
		5			10			15		
		Lower	Middle	Upper	Lower	Middle	Upper	Lower	Middle	Upper
Uzbekistan-1	2.5	29.2	7.4	–	89.9	8.75	t	61.2	13.4	7.1
	0.5	45.2	–	–	66.3	4.5	–	69.5	8.0	–
Tashkent-1	2.5	31.4	6.3	–	77.8	9.4	–	66.2	14.5	8.1
	0.5	42.7	–	–	59.6	5.3	–	73.3	10.4	–

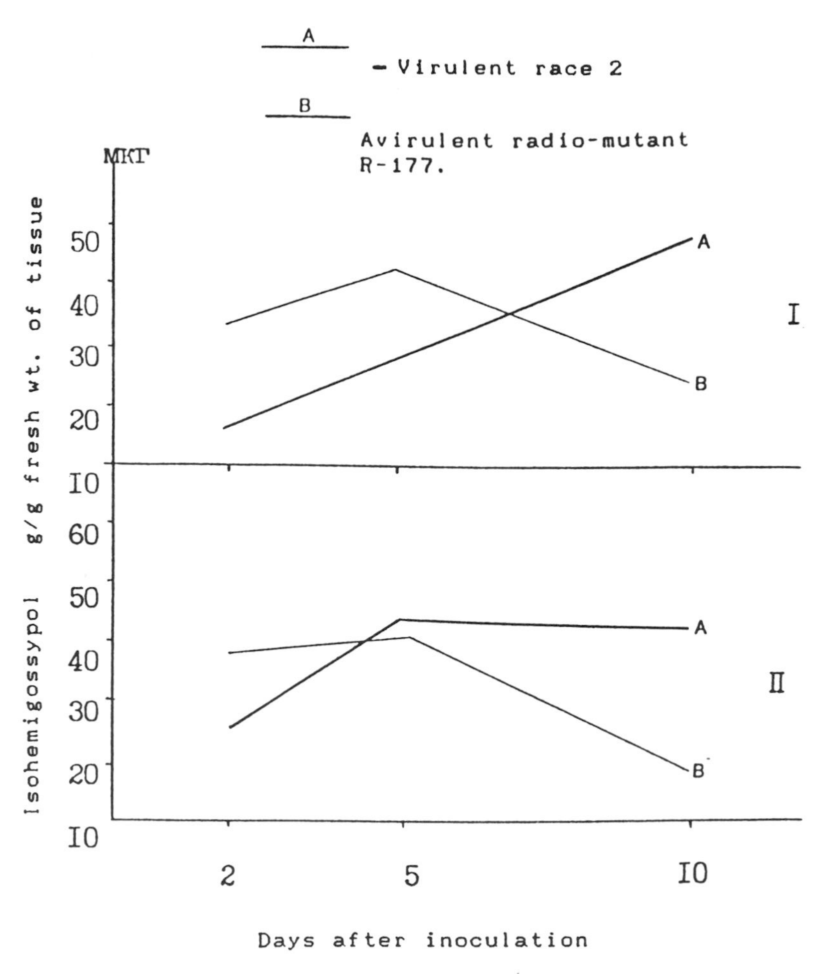

Figure 8 Dynamics of phytoalexin formation under the influence of races of *Verticillium dahliae* differing in virulence in the cotton cultivars S-4727 (I) and Uzbekistan (II).

We made some interesting observations on cotton plants in the tissues of which spore suspensions of avirulent mutant R-177 were first introduced and after 1–3 days the plants were reinoculated with a virulent race of the wilt fungus. Tissues of xylem vessels of the plant of the resistant cvs. Uzbekistan and S-4727 showed a remarkable production of isohemigossypol within 48 hr (35–40 μg/g) in response to infection by the avirulent mutant R-177. It can be assumed that such fast formation of high doses of phytoalexins delayed the development of artificial infection by the virulent race 2. This was confirmed from the growth patterns of wilt in xylem vessels as well as from observations made during 60 days of the postinfection period (Table 9). In the control variant involving infection with the avirulent mutant R-177, the plant was apparently healthy for about 50 days, without any signs of wilt disease, while in the control variant involving infection with race 2 all the inoculated plants displayed strong symptoms of the disease and perished within 18–20 days.

Stimulation of phytoalexin formation through injection of avirulent race increased the resistance to the strongly virulent race 2 of *V. dahliae*. Plants of the cv. S-4727 suffered higher wilt infection than those of the cv. Uzbekistan. The differences were evident in the speed of growth of the wilt as well as in the degree of infection of the experimental plants. However, a complete death of any plant, as was the case with the control variants, was not noticed. On the contrary, when reinoculated after 2–3 days, the degree of wilt infection was found reduced in both kinds of plants. In the cv. Uzbekistan, the plants remained practically healthy, especially 3 days after reinoculation with race 2.

Table 9 Infection of Cotton with Wilt in the Experiment with "Cross-Protection" (Infection Load 2.5 million Conidia/ml)

	% of infected plants		
Experimental variant	With high intensity	With low intensity	% Healthy plants
cv. S-4727			
Control (injection of race 2)	[a]	[a]	—
Avirulent mutant + reinoculation with race 2 after 1 day	50	50	—
Avirulent mutant + reinoculation with race 2 after 2 days	45	55	—
Avirulent mutant + reinoculation with race 2 after 3 days	30	70	—
cv. Uzbekistan-1			
Control (injection of race 2)	[a]	[a]	
Avirulent mutant + reinoculation with race 2 after 1 day	10	80	10
Avirulent mutant + reinoculation with race 2 after 2 days	—	40	60
Avirulent mutant + reinoculation with race 2 after 3 days	—	10	90

[a]All plants died, i.e., 100% mortality.

It was further observed that if the period between inoculation and reinoculation is increased up to 10 days, mass infection results, since the initial induction of phytoalexin by the avirulent race lasts only for a short duration and fades in time. Thus the fast rate of formation and accumulation of lethal doses of phytoalexins in response to avirulent mutant R-177 checks further development of the pathogen and subsequently the postinfectional inhibitors were not produced. Virulent races of *V. dahliae* cotton plant tissues; hence they settle rapidly in the xylem vessels, striking new sectors. Molecular mechanisms behind such responses have not been clearly understood, but their foreplay, with the resultant synthesis of inhibitors of the phytoalexin, is fully known.

V. EXPERIMENTS INVOLVING INTERACTION OF DISEASES

Occurrence of cross-protection was noticed when various other pathogenic microorganisms, such as those causing gummosis and root rot, were allowed to infect a cotton plant prior to its infection by *Verticillium*. The effect was either congenial or adverse depending on the reaction of variety and species to the particular pathogen and its level of virulence. Thus the cross-contact of the plant with the inducers of root rot, gummosis, and *Verticillium* wilt may lead either to the stimulation of its defense reactions or to the decline of its immunosystems and death. In the first case, if the variety turns out to be resistant to the inducer of black rot and gummosis, there may be nonspecific synthesis of phytoalexins in the plant tissues, which, with subsequent infection with *Verticillium*, would suppress the development of the fungus. In the susceptible cultivar, with escalation of disease the plant either perishes or is unable to withstand the subsequent wilt infection consequent to increased sensitivity to the pathogen.

It has been confirmed in our studies that the inducers of root rot (*Rhizoctonia solani*) and gummosis (*Xanthomonas malvacearum*) bring about the formation of phytoalexins in the tissues of seedlings of cotton, depending on the level of varietal resistance against these microorganisms (Table 10). Plants in the early phase of development were artificially infected by the inducers of gummosis and root rot. Those plants which survived were infected with *Verticillium* during the boll initiation stage. In all variants of the experiment with the preliminary infection of gummosis and root rot, a similar tendency for increase in the induction of phytoalexin formation was observed. In cv. Uzbekistan, following infection with *Verticillium*, the content of isohemigossypol was 56.8 μg/g fresh wt, while in those showing primary infection with the bacteria and root rot, the corresponding values were found to be 68.0 and 104.6 μg/g, respectively. In cv.

Table 10 Quantitative Analysis of Phytoalexins of Cotton in Hypocotyl Tissues Infected with the Bacteria *Xanthomonas malvacearum* and the Fungus *Rhizoctonia solani* (μg/g)

Cultivar											
Uzbekistan				White-gold				108 F			
Fungus		Bacteria		Fungus		Bacteria		Fungus		Bacteria	
IHG	GE	IHG	GE	IHG	GE	IHG	GE	IHG	GE	IHG	GE
10.0	39.8	7.3	19.2	15.4	32.4	13.0	21.0	20.2	80.5	7.4	54.0

IHG, isohemigossypol; GE, gossypol equivalent.

108 F, the corresponding values were 36.0, 50.0, and 72.0 μg/g fresh wt, respectively.

VI. XENOBIOTICS IN THE IMMUNOSYSTEM OF COTTON

Considerable influence on phytoalexin formation is exerted by various chemical preparations, such as fungicides, pesticides, and defoliants. Our data indicate that certain preparations under test (e.g., the defoliants Butifos and Butylcaptaks) substantially increased the disease tolerance of the cotton plant. When applied in low concentrations, the intensity of wilt disease was also low in a majority of experimental plants. Inoculation of the plant with *V. dahliae* following the introduction of the test preparations in xylem strengthened the formation of phytoalexins in the unaffected tissues of the plant, depending on the concentration of the defoliant. It was further noted that it is possible to increase the resistance of the plant to wilt by using the preparations for enhancing the overall immunity of the plant, but only with a precise control of timing, with a short duration of application and proper concentration for their action in the plant, and also taking care to minimize their harmful influence, if any, on the surrounding environment.

Pesticides at high concentrations may cause elevated phytotoxicity relative to the *Vericillium* wilt inducer and the plant, suppressing the hypersensitive reaction and phytoalexin formation in the tissues of the cotton plant, due to the consequent short period of latent infection and the high wilt incidence.

Certain unfavorable environmental factors may reduce the disease resistance of the plant. Presently, thanks to developments in molecular biology and molecular genetics, it has been established that immunity is not only the resistance of the body but also a complex set of reactions directed in support of its functional integrity, i.e., homeostasis.

One of the modes of control of the immunosystem in cotton, subject to the compounding influence of various anthropogenic factors (xenobiotics such as pesticides, herbicides, fertilizers, as well as ontogenic xenobiotics, such as pathogen toxins, phytoalexins, quinones, etc., that are induced in the plant tissues during disease incidence), is the elucidation of activity of the enzyme system, metabolizing or expelling the foreign elements from the plant body. Endogenic xenobiotics, such as phytoalexins, accumulating in the tissues in high concentrations, display toxic effects not only toward the pathogen, but toward the plant itself. It has been noted by us that one such system, very effective in detoxification and utilization of both exogenic and endogenic xenobiotics, is the system of glutathione–glutathione-*S*-transferase (Avazkhodjaev et al., 1991). It has been established that under the influence of this enzyme the phytoalexin isohemigossypol is metabolized to 60–70%. Varietal differences have been detected in the activity of this system in various cotton cultivars and species, which leads to the recommendation for its use as a physiological test for the desired selection of varieties with overall resistance to diseases and pesticides.

VII. EPILOG

Results of our research have elucidated the phenomenon of phytoalexins in considerable details. One of the characteristics of phytoalexin formation in the cotton plant differing from other host–pathogen systems is its production in the cells surrounding infected tissues into which the fungus does not penetrate. Defense reactions, relating to the incompatibility of the plant and the wilt fungus, are developed in the parenchyma, cell walls of which subsequently collapse from within, and the gaps are sealed due to infected segments of the xylem vessels. Mechanical and chemical barriers are simultaneously created in the path of infection by the cells since the latter are saturated with fungitoxic phytoalexins. It is demonstrated that the process of phytoalexin formation, in response to races of *V. dahliae* differing in virulence, is a feature of all varieties of cotton regardless of the status of their wilt resistance. A major factor limiting the disease resistance is the rate of formation and accumulation of lethal (to pathogen) doses of phytoalexins. Induction of phytoalexin formation in the cotton plant, similar to that in many other crops, is polygenic in character. Various metabolites of *V. dahliae* which induce phytoalexin formation (toxins, elicitors, proteins, their different fractions, etc.), as well as metabolites that block the induction of phytoalexin formation in cotton and suppressors that display high race specificity, have been identified. Suppression of phytoalexin formation and partial metabolization are two of the means employed by the virulent races of the wilt fungus to overcome cotton plant resistance (Avazkhodjaev et al., 1991).

The characteristics of receptor substances of the surface structures of the cells of the cotton plant, which display the ability to bind complementarily with the pathogen inductors such as lectins, have been enlisted. We have identified two fractions of the cotton plant lectins, differing with respect to chemical composition as well as in their physiological actions: the first fraction blocks phytoalexin formation, whereas the second induces it. Also elucidated is the ability of lectins to immobilize infectious structures of the fungus *V. dahliae*, hyphal growth, and the association of inductor with phytoalexin formation.

It may be noted in conclusion that for resolving important problems related with wilt disease of cotton it is imperative that systems of an all-inclusive and objective assessment of realization of immunological control of the plant be exploited. Ascertaining the realistic criteria of immunological control in plant permits not only the stimulation of activity of the protector metabolites but also the realization of simultaneous selection for resistance to diseases, vermins, pesticides, etc. In confirmation of the same, the data obtained by us on the activity of glutathione–glutathione-*S*-transferase system in the cotton plant and its role in the biotransformation of xenobiotics should prove useful. In the course of research on wild and cultivated new forms of cotton, some differences were detected, both in the phytoalexin-inducing ability and in the glutathione-*S*-transferase activity. Several of these would be useful in selection as initial materials for constitution of complex-resistant varieties of cotton.

ACKNOWLEDGMENTS

The authors acknowledge several friends and colleagues for tireless help and active interest. Thanks are especially due to our numerous colleagues in the Institute of Microbiology and the Bashkir branch of the erstwhile USSR, and to Dr. R. W. Jayashankar of IEBP, for valuable assistance. We are also grateful to Dr. Ram Das Akella, Senior Lecturer in Russian, Nagpur University, for a review of the English version of the text and valuable advice on translation. The fourth author (R. G. D.) thanks the Ministry of Human Resource Development, Government of India, and the Indian Council of Agricultural Research, for award of a postdoctoral fellowship in the erstwhile USSR. He also acknowledges the help by way of useful personal discussions with Dr. Margaret Essenberg and her colleagues at the Oklahoma State University, and with Dr. Alois Bell of the National

Cotton Pathology Research Laboratory, Texas A&M University, College Station, Texas.

REFERENCES

Albersheim, P., and Anderson-Prouty, A. J. (1975). Carbohydrates, proteins, cell surfaces and the biochemistry of pathogenesis. *Ann. Rev. Plant Physiol. 26*: 31–52.

Albersheim, P., and Valent, B. (1978). Host-pathogen interactions in plants. *J. Cell Biol. 6*(3):627–643.

Anderson-Prouty, A. J., and Albershiem, P. (1975). Isolation of a pathogen-synthesized fraction rich in glycon, that elicits a defence response in the pathogen host. *Plant Physiol. 56*(2):286–291.

Anonymous (1986). Patent No. 1507264, 19-12-1986. Institute of Experimental Biology of Plants, Tashkent (in Russian).

Avazkhodjaev, M. X. (1985). Physiological basis of wilt-resistance of cotton plant and the paths of its induction. D. Sc. dissertation, Academy of Sciences, USSR, Baku. (in Russian).

Avazkhodjaev, M. X., and Zeltser, S. S. (1980). *Physiological Factors of Wilt-Resistant Cotton Plant*, FAN, Academy of Sciences, Uzbekistan (in Russian).

Avazkhodjaev, M. X., Jumaniyazov, I., Tashmuxamedov, B. A., and Gagelgans, A. (1984). Phytoalexin inducing properties of cyclopolyethers. *Proc. of Second All-Union Conf. Macrocycles*, 1984 (in Russian).

Avazkhodjaev, M. X., Zeltser, S. S., and Adilova, A. N. (1984a). Characteristics of a metabolite of the fungus *Verticillium dahliae*, inducer of phytoalexins in the cotton plant. *Uzbekskii Biologickeskii Z. 1*:3–5 (in Russian).

Avazkhodjaev, M. X., Zeltser, S. S., and Adilova, A. N. (1984b). Induction and blockage of phytoalexin formation in the cotton plant. *Doklady Academii Nauk. Uzbek. SSR 8*:260–266 (in Russian).

Avazkhodjaev, M. X., Nuritdinova, X. V., Zeltser, S. S., and Adilova, A. N. (1987). Some properties of lectines of cotton plant in relation to the *Verticillium* wilt. *Doklady Academii Nauk. Uzbek. SSR 1*:56–58 (in Russian).

Avazkhodjaev, M. X., Nuritdinova, X. V., Zeltser, S. S., and Madaminova, L. M. (1989a). Formation of determinate complexes of lectine of the cotton plant, with components of the elicitor of the *Verticillium* wilt pathogen. *Uzbekskii Biologickeskii Z. 3*:6–8 (in Russian).

Avazkhodjaev, M. X., Nuritdinova, X. V., Zeltser, S. S., and Madaminova, L. M. (1989b). Receptor function of plasmalemma in the wilt disease of cotton. *Doklady Academii Nauk. Uzbek. SSR 11*:54–56 (in Russian).

Avazkhodjaev, M. X., Nuritdinova, X. V., and Zeltser, S. S. (1990a). Dynamics of pathological changes in xylem of the cotton plant under the influence of the fungus *Verticillium dahliae. Micologia i Fitopatologia 24*(3):240–244 (in Russian).

Avazkhodjaev, M. X., Zeltser, S. S., Nuritdinova, X. V., and Musaev, X. A. (1990b). Receptor function of plasmalemma in the wilt disease of the cotton plant. *Abstr.* All-Union Convention "*Physiological and biochemical aspects of plant immunity to diseases and injury.*" Ufa, Bashkiria 1990, pp. 5 (in Russian).

Avazkhodjaev, M. X., Musaev, X. A., Zeltser, S. S., and Nuritdinova, X. V. (1991). Role of glutathione glutathione-S-transferases in the detoxication of xenobiotics. *Uzbekskii Biologickeskii Zurnal 4*:59–60 (in Russian).

Bailey, J. A., and Mansfield, J. B. (1985). *Phytoalexins.* Naukova Dumka Publ. Kiev. Ukr. SSR. 320 p. (in Russian).

Bell, A. A. (1967). Formation of gossypol in infected or chemically irritated tissues of *Gossypium* Species. *Phytopathology 57*(7):759–764.

Bell, A. A. (1969). Phytoalexin production and *Verticillium* wilt resistance in cotton. *Phytopathology 59*(8):1119–1127.

Bell, A. A. (1981). Biochemical mechanisms of disease resistance. *Annual Review of Plant Physiology 32*:1221–1281.

Borodin, G. I. (1978). *Physiological-biochemical basis of pathogenesity of the fungus* Verticillium dahliae, *the inducer of* Verticillium *wilt of cotton.* D. Sc. Dissertation, Academy of Sciences. Uzbek SSR, Tashkent 1978. (in Russian).

Bushnell, W. R., and Rowell, J. B. (1981). Suppressors of defence reactions: a model for roles in specificity. *Phytopathology, 71*:1012–1014.

Cruickshank, J. A., and Perrin, D. R. (1971). Studies of phytoalexins. XI. The induction antimicrobial spectrum and chemical assay of phaseollin. *Phytopathology, 70*(3):209–229.

Dani, R. G. (1992). Biotechnological research of cotton- two decades in Soviet retrospection. *Advances in Plant Sciences* (*India*), *5*(2): (In press).

Dyakov, Y. T. (1976). Membrane aspects of phytopathology. *Obzor. MSX, VNIISE ISX, M.*2. (in Russian).

Dyakov, Y. T. (1983). Physiological-biochemical mechanisms of resistance of plants to fungal diseases. *Itogi Nauka i Texniki VNIITI, Zashita Rasyenyii,* 1983, pp. 5–90. (in Russian).

Essenberg, M., Grover, P. M., and Cover, E. C. (1990). Accumulation of antibacterial sesquiterpenoids in bacterially inoculated *Gossypium* leaves and cotyledons. *Phytochemistry 29*:(10)3107–3113.

Faye, L., and Crispeels, M. (1987). Transport and processing of the glycosylated precursor of concanavalin. A. in Jack-bean. *Planta 170*(2):217–224.

Fraile, A., Garsia-Arenal., J., Garsia-Serrano, J. J., Sagastra, L. (1982). Toxicity of phaseollin, phaseolliniso-flavan and kievitone to *Botrytis cinerra. Phytologia Journal 102*(2):161–169.

Hadwiger, L., and Schwochan, M. (1969). Host-resistance responses—an induction hypothesis. *Phytopathology 59*(2):223–227.

Heath, M. C. (1981). A generalized concept of host-parasite specificity. *Phytopathology, 71*:1121–1123.

Karimdjanov, A. K., Ismailov, A. I., Abdullaev, Z. S., Islambekov, S. Y., Kamaev, F. T., and Sadikov, A. S. (1976). Structure of gossyvertin—a new phytoalexin of cotton. *Ximiya Prip. Socd. 2*:238–242.

Keen, N. T., Sims, J. J., Erwin, D. C., Rice, E., and Patridge, J. E. (1971). 6-α-Hydorxyphaseollin: an antifungal chemical induced in soyabean hypocotyls by *Phytophtora megasperma* var. *soyae. Phytopathology, 61*(9):1084–1089.

Kuć, J., and Rush, Y. S. (1985). Phytoalexins. *Archieves Biochemistry et Biophysics* *236*(2):455–472.

Lyutsik, M. H., Panasyuk, E. N., and Lyutsik, A. D. (1981). *Lectines*. Lyvov Pub., Moscow, p. 153.

Malysheva, K. M., and Zeltser, S. S. (1968). Protein-lipid-polysaccharide complex cultural fluid and the mycelii of the fungus *Verticillium dahliae*, inducer of *Verticillium* wilt of cotton. *Doklady Akademi Nauk, SSSR, 179*(1):231-234 (in Russian).

Markov, E. Y., and Xavkin, E. V. (1983). Plant lectines: proposed functions. *Phisiologia Rastyenii, 30*(5):852–865. (in Russian).

Metlitski, L. V. (1976). *Phytoimmunity. Molecular mechanisms*. Nauka Publishers, Moscow. (in Russian). 50 pp.

Metlitski, L. V., and Ozeretskovskaya, O. L. (1973). *Phytoalexins*. Nauka Publishers, Moscow. (in Russian). 176 pp.

Metlitski, L. V., and Ozeretskovskaya, O. L. (1985). *How plants defend themselves against diseases*. Nauka Publishers, Moscow. P 114. (in Russian).

Metlitski, L. V., Lyubimova, N. V., and Muxamedova, R. A. (1971). Phytoalexin activity and wilt tolerance of cotton. *Izvestii Akademy Nauk, SSSR, Seria Biologia 2*:72–75.

Metlitski, L. V., Ozeretskovskaya, O. L., Yurganova, A. L., Savelyeva, O. H., Chalova, L. I., and Dyakov, Y. T. (1976). Phytoalexin induction in potato tubers with a metabolite of the fungus *Phytopthra infestans* (Mont) de Bary. *Doklady Akadmii Nauk, SSSR, 226*(5):1217-1220. (in Russian).

Müller, K. O., and Borger, H. (1940). Experimentalle untersuchungen uber die phytophora-Resistenz der kartoffel. *Arb. Biol. Reichsanst. Land Forstwirt.*, Berlin Dehlem *23*(2):189–231. (in German).

Muxamedova, R. A., and Turakulov, Y. X. (1974). Phytoalexin activity of cotton and its wilt tolerance. *Uzbeksi Bilogicheski Zhurnal 6*:5–7 (in Russian).

Nuritdinova, X. V. (1988). *Functional properties of exoosomose of amino acids and lectines during the expression of wilt disease of cotton*. Ph.D. Dissertation, Academy of Sciences. Uzbek SSR, Tashkent. (in Russian).

Sadykov, A. S., Metlitsky, L. V., Karimdjanov, A. K., Ismailov, A. I., Muxamedova, P. A., Avazkhodjaev, M. X., and Kamaev, F. G. (1974). Isohemigossypol: phytoalexin of cotton. *Doklady Akademy Nauk, SSSR 281*(6):1274-1276 (in Russian).

Sequeira, L. (1978). Lectins and their role in host-pathogen specificity. *Ann. Rev. Phytopathol. 16*:453–481.

Stewart, J. McD. (1991). *Biotechnology of Cotton* ICAC Review articles on cotton production, *No. 3*. CAB International, Wallingford.

Sun, J. T., Essenberg, M., and Melcher, U. (1989). Phytoactivated DNA nicking, enzyme inactivation, and bacterial inhibition by sesquiterpenoid phytoalexins from cotton. *Mol. Plant Microbe Interact. 2*(3):139–147.

Terekhova, V. A., and Dyakov, Y. T. (1980). Isolation of metabolites of *Phytopthora infestans* participating in the reaction of the parasite with potato under the influence of osmotic shock. *Bull. Moskovskova Obshyestva isp. Prirody, Otdel Biologii 85*(5):68–77 (in Russian).

Tomiyama, K. (1970). Phytoalexins. *Chem. Regul. Plants* *5*(2):105–115.
Yatsenko-Xmelevski, A. A. (1954). *Aspects and Methods of Anatomical Research of Wood*. Academy of Sciences, USSR (in Russian).
Zaki, O. J., Keen, N. T., and Erwin, D. C. (1972). Implication of vergosin and hemigossypol in the resistance of cotton to *Verticillium albo-atrum*. *Phytopathology* *62*(12):1406–1407.

8

Cotton (*Gossypium hirsutum*) Strategies of Defense Expression

Accumulation of Unique *Aspergillus flavus* Control Substances

H. J. Zeringue, Jr.
Southern Regional Research Center, Agricultural Research Service, U.S. Department of Agriculture, New Orleans, Louisiana

I. INTRODUCTION

Toxigenic strains of the fungi *Aspergillus flavus* Link ex Fries and *Aspergillus parasiticus* Speare produce secondary toxic metabolites called aflatoxins (Detroy et al., 1971). These toxins, particularly aflatoxin B_1, have the distinction of being the most carcinogenic of all natural products (Groopman et al., 1981). Much research effort has been focused on aflatoxins since elevated levels of the toxins are found in pre- and postharvest crops. In the United States a guideline of 20 ppb is the minimum aflatoxin allowed by the Food and Drug Administration (FDA) for interstate shipment of foods and feeds.

Currently, good farming practices have produced some positive results in control of aflatoxin contamination of cottonseed in the field. These include 1.) irrigation during periods of drought (the fungus seems to thrive during drought conditions), 2.) use of insecticides at marked stages of cotton boll development to prevent fungal entry through insect wounds, and 3.) the use of crop varieties that exhibit some resistance to *A. flavus* infection or aflatoxin accumulation.

The basic plan for elimination or control of aflatoxin in cottonseed has involved research on natural products in the cotton plant that inhibit *A. flavus* or toxin production. Identification of natural inhibitors and their mechanisms of production can be used to enhance aflatoxin inhibitory traits in cotton plants through either classical plant breeding or contemporary molecular engineering techniques.

II. METHODOLOGY

A. Fungi and Elicitor Preparation

In most cases, *A. flavus* SRRC 1000 was maintained on potato dextrose agar plates at 29°C. Fungal spores were inoculated on a defined medium of Adye and Mateles (1984) contained in 2.8 liter Fernback culture flasks. Inculated media were kept at 29°C without shaking for 10–14 days. Mycelia were washed with distilled H_2O, homogenized with 0.1 M phosphate buffer, filtered, defatted with $CHCL_3$-MeOH (1:1 v/v), washed with Me_2CO, and air-dried (Zeringue et al., 1982). Mycelia were resuspended in H_2O, autoclaved for 3 hr at 121°C, filtered on filter paper, and the filtrate was concentrated by vacuum distillation, dialyzed against H_2O at 2°C, and lyophilized. The hot-water–soluble mycelial extract elicitor was prepared from a 10-day-old culture and contained 44.6% protein and 37.5% carbohydrate.

B. Plants and Treatments of Leaf Disks and Cotton Boll Disks and TLC Quantitation

Acala SJ-2 cotton plants were grown under greenhouse conditions and were at least 2 months old at the time of leaf treatment. Both leaves and cotton boll surfaces were slightly abraded with 6- to 8-mm diameter wounds produced by carborundum disks or by a scalpel blade. Two areas of fifth or sixth true leaves per plant were scratched in areas midway between the outer margin of the leaf to the medium. Similar wounds were produced on developing cotton boll carpel surfaces in areas between the suture lines of the boll. Usually 10 μg of the mycelial extract in 10 μl sterile, distilled H_2O was applied to the wounded areas, and 2 days after treatment wounded areas were excised with 14- to 18-mm-diameter cork borers. Leaf disks (20 disks at a time) were placed in 40 ml 50% aqueous EtOH and were vacuum-infiltrated; the flasks containing the disks were subsequently placed on a reciprocating shaker for 12 hr at room temperature. Cotton boll disks (24 disks) were vacuum-infiltrated with 40 ml methyl alcohol–acetone (4:1 v/v). The disks were removed from the alcohol extract and the extract was concentrated to dryness by rotary evaporation under vacuum at 50°C. The residue was dissolved in MeOH to produce a 5% (w/v) solution which was spotted on silica gel thin-layer chromatography (TLC) plates and developed in CH_3CN-$CHCl_3$ (1:4 v/v). Plates were spotted with treated leaf or boll extracts and standards and scanned densitometrically at 365 nm with a Schoeffel spectrodensitometer (Model SD 3000) to obtain a semiquantitative analysis of the fluorescent spots. The R_f' values of the compounds in plant extracts were determined in three different solvent systems (Table 1).

Table 1 TLC of Plant Extracts and Standards in Three Solvent Systems

Band color (365 nm)	R_f values		
	System I[a]	System II[b]	System III[c]
Y	0.26 (0.26)[d]	0.22 (0.22)[d]	0.28 (0.28)[d]
B	0.38 (0.39)[e]	0.35 (0.34)[e]	0.10 (0.11)[e]
Y	0.48	0.46	0.42
Y	0.58 (0.58)[f]	0.55 (0.56)[f]	0.45 (0.46)[f]
Y-G	0.64	0.65	0.55

[a]$CHCl_3$-Me_2CO-HCOOH (80:19:1).
[b]MeCN-$CHCl_3$ (1:4).
[c]Hexane-EtOAc-MeOH (60:40:1).
[d]Lacinilene C standard.
[e]Scopoletin standard.
[f]Lacinilene C 7-methyl ether standard.
Y, yellow; B, blue; Y–G, yellow-green.
Source: Data from Zeringue (1984).

C. Collection and Identification of Major Volatiles Emitted from Cotton Leaves

Volatiles were trapped on glass Tenax columns (0.1-g Tenax GC 60–80 mesh) packed between glass wool in tubes (3/8 × 3 in.). Air streams were passed over the plant material and volatiles were collected for 30 min. The Tenax tube was loaded into an external inlet apparatus (Scientific Instrument Service, River Ridge, LA) interfaced with a Finnegan MAT GC/MS 400 Series instrument (Legendre et al. 1979). The volatiles were heat-desorbed onto a 50-m SE-54 column. The temperature program used was 3 min from −30°C to 30°C (to trap the volatiles at the head of the column) followed by 30–150°C at 2.5°/min and then 150–250°C at 10°/min. The peaks on the reconstructed ion chromatogram were tentatively identified by comparison with the Finnegan NBS library and quantitated using the ARICQ program.

III. PHYTOALEXINS

A. Induction of Phytoalexins

Early research at the Southern Regional Research Center (SRRC) showed that five cotton autofluorescent (365 nm illumination) phytoalexins could be induced in the cotton leaf or the developing cotton boll by a cell-free, hot-water-soluble, mycelial extract of *A. flavus* (Zeringue, 1984), including

1.) the sesquiterpenoid phytoalexins (Fig. 1), 2,7-dihydroxycadalene (DHC) and 2-hydroxy-7-methoxycadalene (DHMC); 2.) their oxidation products lacinilene C (LAC) and lacinilene C 7-methyl ether (LACME); and 3.) the coumarin phytoalexin scopoletin (SCOP). The denoted substances were induced when cell-free mycelial extracts were applied to artificially wounded areas on the cotton leaf or on the surfaces of the developing cotton boll (Zeringue, 1984, 1987, 1988). Fungal extracts appeared to elicit the production of the phytoalexins by the redirection of the terpenoid biosynthetic pathways in the cotton plant. A similar modification of terpenoid metabolism was demonstrated in potato tubers (Shih et al., 1973; Shih and Kuć, 1973).

A 10-day, time-course study was conducted on the induction of DHC, DHMC, LAC, and LACME in Acala SJ-2 cotton leaves that were slightly abraded by a 6-mm-diameter fine, carborundum sandpaper disk and treated with 10 μg of a cell-free hot-water–soluble extract of a toxigenic strain of *A. flavus* (Table 2) (Zeringue, 1987a). Initial accumulation of the compounds occurred 2 days after treatment with elevated levels on day 6 posttreatment. A gradual decline of the compounds occurred after the peak periods. Control wounded leaves also showed increases of the compounds on day 6 with a gradual decline toward the end of the test period. To characterize the distribution of induced phytoalexins in reference to the 6-mm-diameter wounded, treated areas, cork borers of various diameters were used to excise rings encircling the wounded areas (Table 3). After the second day of treatment, the greatest concentration of induced compounds

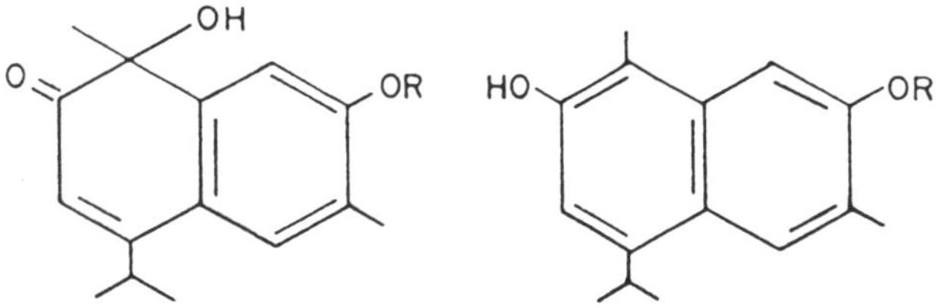

(1) R = H (2) R = H
(3) R = Me (4) R = Me

Lacinilene C (1), 2,7-Dihydroxycadalene (2),
Lacinilene C 7-methyl ether (3),
and 2-Hydroxy-7-methoxycadalene (4)

Figure 1 Lacinilene C (1), 2,7-dihydroxycadalene (2), lacinilene C 7-methyl ether (3), and 2-hydroxy-7-methoxycadalene (4).

Table 2 Quantitation by TLC Fluorodensitometry of the Induced Components in Extracts from Leaf Disks Taken from Cotton Leaves Treated with a Cell-Free Mycelial Extract of *A. flavus*

	Amount in (μg/g fresh wt)									
Component	1d	2d	3d	4d	5d	6d	7d	8d	9d	10d
2,7-Dihydroxycadalene	tr	13.4	3.1	1.0	tr	28.1	15.0	12.4	12.2	6.3
	(0)[a]	(0)	(tr)[b]	(tr)	(6.1)	(3.2)	(9.0)	(5.1)	(5.0)	(3.5)
Lacinilene C	tr	38.4	32.0	12.1	8.6	51.3	30.5	17.3	11.1	8.2
	(0)	(0)	(tr)	(tr)	(8.2)	(18.1)	(20.1)	(10.3)	(10.2)	(6.6)
2-Hydroxy-7-methoxycadalene	tr	25.5	12.1	8.3	3.3	36.4	28.3	20.2	20.0	18.1
	(0)	(0)	(tr)	(tr)	(tr)	(11.3)	(9.3)	(11.1)	(9.2)	(3.2)
Lacinilene C 7-methyl ether	0.5	23.1	3.2	2.6	5.2	38.1	26.2	23.8	21.5	17.3
	(0)	(tr)	(tr)	(tr)	(2.2)	(9.6)	(5.4)	(3.4)	(2.9)	(1.6)

[a]Wounded leaves treated with distilled water only.
[b]tr = trace < 1 μg.
Source: Data from Zeringue (1987a).

Table 3 Percentage Distribution After 2 and 5 Days of Induced Components in 6-mm-diameter Wounded/Fungal-Treated Areas and in Areas Encircling the 6-mm Areas in the Cotton Leaf

	Outer diameter (mm)					
Components	6	9	11	14	16	21
2,7-Dihydroxycadalene	29.1[a]	58.3	12.6	0	0	0
	86.2[b]	13.8	0	0	0	0
Lacinilene C	18.9[a]	77.0	4.1	0	0	0
	92.6[b]	7.4	0	0	0	0
2-Hydroxy-7-methoxycadalene	34.5[a]	58.0	4.6	2.1	0.9	0
	81.5[b]	18.5	0	0	0	0
Lacinilene C 7-methyl ether	42.7[a]	53.1	4.2	0	0	0
	94.1[b]	5.9	0	0	0	0

[a]Two days after treatment.
[b]Five days after treatment.
Source: Data from Zeringue (1987a).

was localized in a 3-mm area around the 6-mm wounded/fungal-treated area (Table 3). The 6-mm-diameter area was brownish in color and the adjacent 3-mm area had a light yellow–green appearance. Five days after treatment the four induced compounds were found predominately in the 6-mm-diameter wounded/treated area; this area plus the adjacent encircling 3-mm area were both brownish in color. The results are in agreement with Hargreaves and Baily (1978) who demonstrated that the phytoalexin phaseollin from the french bean (*Phaseolus vulgaris*) and other isoflavonoid phytoalexins are probably synthesized in tissues around necrotic cells and are absorbed and accumulate in the dead tissue.

B. Phytotoxic Effects

To understand the cyclic production of the induced phytoalexins in cotton leaves, a sensitive bioassay was developed that utilized the aquatic macrophyte *Lemna minor* L. (Einhelling et al., 1985; Zeringue, 1987a). LAC and LACME from excised leaf disks exhibited phytotoxic properties in the *L. minor* assay (Table 3). Both lacinilenes inhibited the growth of *L. minor*, but LAC produced the greatest inhibition. Both compounds produced root discoloration and some frond bleaching occurred with LAC. The phytotoxic properties of the two lacinilenes may explain some of the results described in Table 4. It is possible that cell death in the cotton leaf releases constitutive material (an endogenous elicitor) which initiates phytoalexin biosynthesis and the accumulation of the compounds shown in Table 1

Table 4 Separated Components[a] of Cotton Leaf Disks and Controls Assayed After 1 Week of Incubation with *L. minor*

Component	No. of leaf disks extracts[b]	No. of fronds after 1 week of incubation		Dry wt of *L. minor* as percentage of control
		Treated	Control	
Lacinilene C	1	24[c]	32	77
	2	11[c]	28	41
	3	8[d]	27	27
	4	11[d]	32	28
	5	9[d]	30	27
Lacinilene C 7-methyl ether	1	28	30	84
	2	22[c]	28	81
	3	20[c]	32	63
	4	20[c]	30	62
	5	18[c]	27	64

[a]Separated by preparative TLC.
[b]Aliquots prepared from 50-leaf disks to represent the amount of component in 1–5 leaf disks.
[c]Roots have a brownish discoloration.
[d]Some bleaching of fronds resulted.
Source: Data from Zeringue (1987a).

(Hargraves and Bailey, 1978). This reasoning could also explain the induction of the compounds in wounded control leaves on days 5–6 after wounding. The synergistic effect produced by a fungal extract and an endogenous elicitor may explain the peak accumulations of the compounds in treated leaf extracts on days 2 and 6 (Table 1). Synergistic elicitor effects have been observed between the β-glucan elicitor of *Phytophora megasperma* and the endogenous elicitor of soybean tissue (Darvill and Albersheim, 1984). Similar synergism has been observed between β-glucan and fatty acid elicitors in potato (Block and Kuć, 1983; Kuranty and Osman, 1983; Preisig and Kuć, 1983).

The lacinilene and cadalene precursors are induced in the *green* tissues of the cotton plant in contrast to the terpenoid phytoalexins desoxyhemigossypol and hemigossypol, together with their methyl esters, which are formed in germinating seed, young roots, hypocotyls, xylem vessels, cambial tissues, and cotton boll endocarp (Halloin and Bell, 1979; Veech, 1978; Hunter et al., 1978; Bell and Stipanovic, 1977). Varied levels of *A. flavus* inhibition were observed in direct bioautography on TLC plates that contained the five separated induced compounds (unpublished). Essenberg et al. (1982) demonstrated that DHC and LAC are bacteriostatic and accumulate in leaves and cotyledons of blight-resistant lines of cotton when the tissues were treated with a bacterial suspension of *Xanthomonas campestris* pv. *malvacearum*. Halloin and Greenblatt (1982) found accumulation of lacinilenes and cadalenes in cotton boll tissues after puncture inoculation of bolls with *Diplodia gossypina*; the bolls exhibited a bright, yellowish fluorescence under long-wave UV light and also resistance to infection by *Diplodia gossypina*.

C. Wound Response and Fungal Extract Treatment

Initial true leaves of Acala SJ-2 cotton plants wounded by gentle abrasion resulted in accumulations of ferulic acid (4-hydroxy-3-methoxycinnamic acid) detected on the third and fourth days after treatment. When similar wounds were treated with cell-free extracts of *A. flavus*, elevated levels of scopoletin (6-methoxy-7-hydroxycoumarin) were produced after 1 day of treatment and continuing for 7 days (Zeringue et al., 1985). Elevated levels of scopoletin were not detected in leaves that were only wounded, and enhanced quantities of ferulic acid were not observed in leaf tissue after wounding and exposure to the fungal extract (Table 5).

Increased ferulic acid in wounded cotton leaves and scopoletin in the

Table 5 Ferulic Acid and Scopoletin Concentrations[a] in Leaf Disks from the First True Leaf of Acala SJ-2 Cotton Plants That Were Wounded and Wounded-Plus-Treated with Cell-Free *A. flavus* Extract

Day after wounding	Wounded			Wounded plus treated with fungal extract	
	Ferulic acid				
	cis	*trans*	Scopoletin	Ferulic acid	Scopoletin
1	t[b]	t	2.4 ± 0.12	t	179.6 ± 2.64
2	t	t	2.1 ± 0.20	t	61.8 ± 3.28
3	262 ± 5.4[c]	1268 ± 6.2	1.3 ± 0.30	t	53.5 ± 2.73
4	47 ± 2.3	472 ± 6.7	1.0 ± 0.03	t	36.2 ± 1.66
5	t	t	t	t	27.4 ± 4.84
6	1	1	t	t	18.1 ± 2.13
7	1	1	t	t	14.3 ± 2.55

[a]Data reported, in μg/g fresh wt, are from analyses of extracts of 80 leaf disks for each day (20 wounded leaves and 20 wounded-plus-treated leaves). Each leaf disk was 15 mm in diameter, including a wounded area 6 mm in diameter. These data represent the mean of three test runs.
[b]Trace amounts, less than 1 μg/g fresh wt.
[c]Mean ± SE.
Source: Data from Zeringue et al. (1985).

fungal extract–treated, wounded leaves indicates either an increased synthesis of the compounds or an active transport of the substances from cells surrounding the wounded area. Apparently ferulic acid in the wounded leaves is derived from cinnamic acid and other intermediates of the phenylpropanoid pathway leading to the formation of lignin. It is possible that the rapid accumulation of scopoletin after a day of wounding and fungal extract treatment results from a redirection of cinnamic acid in the phenylpropanoid metabolic pathway via *ortho*-hydroxyl derivatives and lactonization to the production of the coumarin scopoletin (Rhodes and Wooltorton, 1978). Accumulation of scopoletin and its glycone scopolin have been shown in potatoes and tobacco infected with fungi, bacteria, and viruses (Mayr et al., 1963; Tanguy and Martin, 1972; Sequeria, 1969; Huges and Swain, 1960). Scopoletin acts as a stimulator of IAA oxidase (Imbert and Wilson, 1970), and peroxidase activities (Schafer et al., 1971). Accumulated scopoletin also acts as an inhibitor of catalase activity in tobacco infected by *Peronospora tabacina* (Hochberg and Cohen, 1977). The fact that scopoletin can alter the host enzyme systems may be of significance in the plant's defense system.

D. Carpel Wall Phytoalexins in the Developing Cotton Boll

Developing cotton bolls of Deltapine 61 that had been grown under controlled conditions were artificially wounded on the carpel surfaces and subsequently treated with the hot water–soluble mycelial extract of *A. flavus* at weekly intervals for 8 weeks postanthesis. Two days after treatment, the bolls were harvested and disks containing the treated surfaces were excised and extracted to determine induction of DHC, DHMC, LAC, LACME, and SCOP (Zeringue, 1988). Concentrations of the lacinilenes and cadalenes increased over the 8-week testing period and peaked either at the seventh or eighth week postanthesis boll age (Table 5). However, scopoletin concentrations were just the opposite with higher concentrations in extracts of younger bolls that peaked at 3 weeks postanthesis.

The highest concentrations of lacinilene and cadalene phytoalexins occurred during the period of boll opening (41–49 days postanthesis; Table 6). Lee (1988) inoculated cotton boll sutures with *A. flavus* spores at the initiation of boll opening, harvested the bolls after 2 or 4 weeks, and found no aflatoxin contamination in the seeds of the treated bolls. It is possible that induced phytoalexins could be translocated to deeper tissues within the boll either passively or actively. The lacinilene and cadalene phytoalexins are slightly water-soluble (Stipanovic and Wakelyn, 1974), and a penetrating carpel wall wound could allow dew or rain to transfer the phytoalexins into deeper tissues of the boll to effect a more widespread defense.

Table 6 Quantitation by TLC Fluorodensitometry of the Induced Components in Extracts from Cotton Boll Disks Treated with Cell-Free Mycelia Extracts of *A. flavus* at Weekly Intervals for an 8-Week Postanthesis Period

		Amount in (μg/g fresh wt of excised disks)							
	Days	1–7	8–14	15–21	22–28	29–35	36–42	43–49	50–56
Component	Weeks	1	2	3	4	5	6	7	8
2,7-Dihydroxycalalene		4.4	18.6	18.2	26.3	34.7	38.6	53.2	3.4
		(2.6)[a]	(1.5)	(2.6)	(1.5)	(tr)[b]	(tr)	(–)[d]	(–)
Lacinilene C		8.3	12.5	10.1	26.3	22.3	25.4	42.8	38.4
		(–)	(–)	(–)	(tr)	(3.7)	(3.7)	(5.6)	(4.3)
2-Hydroxy-7-methoxycadalene		–	tr	2.6	7.4	7.7	11.8	24.3	21.5
		(–)	(–)	(–)	(2.5)	(2.0)	(2.8)	(7.2)	(7.6)
Lacinilene C 7-methyl ether		7.2	10.1	12.5	14.8	14.3	15.2	34.3	26.4
		(–)	(–)	(–)	(–)	(tr)	(2.5)	(3.1)	(3.8)
Scopoletin		41.4	44.6	53.2	26.3	17.2	15.3	8.3	3.4
		(2.6)	(1.5)	(2.6)	(1.5)	(tr)	(tr)	(–)	(–)
Water content (%) of excised carpel disks		87.0[c]	87.1	87.8	87.5	83.9	74.8	58.9	49.2

[a]μg/g fresh wt in wounded bolls treated with sterile distilled H_2O only.
[b]Trace <1 μg.
[c]Determined by heating at 110°C until constant wt.
[d]None detected.
Note: (1) Nonwounded controls contained <1 μg components/g fresh wt. (2) Boll opening occurred 41–49 days.
Source: Zeringue (1988).

E. Environmental Stress Effects on Phytoalexin Production

The concentrations of DHC, DHMC, LAC, LACME, and SCOP were determined in cotton leaves treated with the hot water-soluble mycelial extract of *A. flavus* under a light regime of 14 hr light/10 hr dark and temperature cycles of 25/20°, 30/25°, 35/30°, and 40/35°C (Zeringue, 1990). The highest concentrations of the sesquiterpenoid and coumarin phytoalexins were produced at temperature cycles of 30/25° and 2 days after treatment with the fungal elicitor (Table 7). Temperature regimes lower than 30/25°C resulted in a 35% reduction in induced phytoalexins while the highest temperature cycle tested (40/35°C) yielded a 43% reduction.

Concentrations of the individual phytoalexins were measured in cotton leaves treated with the fungal elicitor and maintained at different levels of plant moisture stress (PMS). PMS values less than −11.7 bars resulted in decreased amounts of the five induced compounds (Table 8). Even slight leaf wilt (−15.5 bars) at the time of fungal elicitor treatment after 2 days (−24.3 bars) resulted in 45–58% reduction of the five phytoalexins.

From these findings it is apparent that an "ideal" set of environmental conditions of temperature and moisture is necessary for the maximum induction of the cotton leaf phytoalexins. Cotton plants grown in hot, dry, desert regions may not favor maximum production of the phytoalexins. Inability of host plants to produce defensive chemicals quickly when attacked by *A. flavus* may be related to the elevated incidence of aflatoxin contamination of cottonseed grown in desert regions.

IV. VOLATILE ELICITORS AND GASEOUS PHYTOALEXINS

Baldwin and Schultz (1983) damaged leaves of potted poplar (*Populous* X *euroamericana*) ramet and sugar maple (*Acer saccharum*) seedlings and demonstrated increased concentrations of phenolic compounds in the damaged leaves as well as in the leaves of undamaged plants sharing the same enclosure. Rhodes (1982) found that Stika willow (*Salix*) trees attacked by tent caterpillars (*Malacosoma california pluviale*) and nearby unattacked control trees exhibited altered leaf quality. Fall webworms (*Hyphantria cunea*) fed on the attacked leaves grew more slowly than those fed on the leaves from unattacked willows. These studies suggested that an airborne signal from damaged tree tissues stimulated biochemical changes in neighbouring undamaged trees and influenced the feeding and growth of phytophagous insects. Initial SRRC tests with volatiles involved interactions between *A. flavus* and host cotton plants and an assessment of a cotton plant's defense system relative to airborne signals. Zeringue (1987b) exposed cotton leaves for 7 days to volatile chemicals originating from 1.) *A. flavus*-infected cotton leaves, 2.) *A. flavus* cultures, or 3.) mechanically

Table 7 Temperature Effect[a] on the Elicitation[b] of Cotton Leaf Phytoalexins[c]

Diurnal temp. regime, °C, light/dark[d]	Phytoalexins[e] induced (μg/g fresh wt leaf disks)				
	DHMC	LACME	DHC	SCOP	LAC
25/20	21.3 ± 1.5	14.0 ± 1.3	8.1 ± 0.3	5.6 ± 0.8	23.1 ± 2.5
	(0)[f] ± (0)	(0.4) ± (0)	(0) ± (0)	(0.8) ± (0.1)	(tr)
30/25	27.6 ± 2.6	17.8 ± 2	13.2 ± 2.1	12.5 ± 1.3	38.5 ± 1.7
	(tr)[g]	(0.6) ± (0.3)	(tr)	(1.8) ± (0.4)	(1.6) ± (0.4)
35/30	22.7 ± 2.9	17.6 ± 1.6	8.6 ± 0.6	9.5 ± 0.8	21.6 ± 1.2
	(tr)	(0.8) ± (0.2)	(tr)	(1.2) ± (0.3)	(1.4) ± (0.8)
40/35	11.3 ± 1.8	8.1 ± 1.3	8.6 ± 1.2	3.7 ± 0.1	13.0 ± 1.0
	(tr)	(tr)	(tr)	(0.8) ± (0.2)	(tr)

[a]Means ± SD of two experimental runs with each run containing at least 15 leaf disks from different leaves of simular maturity and position harvested from different plants.
[b]Wounded leaves elicited by a cell-free mycelia extract of *A. flavus*.
[c]Determined 48 hr after elicitation.
[d]14-hr light/10-hr dark cycles.
[e]DHMC, 2-hydroxy-7-methoxycadalene; LACME, lacinilene C 7-methyl ether; DHC, 2,7-dihydroxycalalene; SCOP, scopoletin; LAC, lacinilene C.
[f]Wounded leaves treated with sterile, deionized water only.
[g]Trace <0.4.
Source: Data from Zeringue (1990).

Table 8 Plant Moisture Stress (PMS) Effect[a] on the Elicitation[b] of Cotton Leaf Phytoalexins[c]

Two-day range of PMS (bars)	Leaf appearance and degree of wilt	Phytoalexins induced (μg/g fresh wt leaf disks)				
		DHMC	LACME	DHC	SCOP	LAC
−8.1	Normal					
−11.7	Normal	26.2 ± 2.7	18.3 ± 2.9	11.3 ± 0.5	14.6 ± 1.7	40.6 ± 1.3
−15.5	Wilt+					
−24.3	Wilt++	15.2 ± 1.2	8.6 ± 0.4	5.3 ± 1.0	8.2 ± 0.9	18.1 ± 2.2
−22.5	Wilt+++					
−31.5	Wilt++++	3.1 ± 0.7	2.2 ± 0.1	4.0 ± 0.2	3.8 ± 0.5	6.1 ± 0.3

[a]Means ± SD of two experimental runs at 30/25°C with each run containing at least 15 leaf disks from 15 different leaves of simular maturity and position harvested from different plants.
[b]Wounded leaves were induced by a cell-free mycelia extract of *A. flavus*.
[c]Determined 48 hr after elicitation.
Source: Data from Zeringue (1990).

damaged cotton leaves. It was observed that volatiles from *A. flavus*-infected leaves caused significant increases (52% and 34%) in phloroglucinol-reactive compounds, expressed as "gossypol equivalents," in wounded and undamaged leaves, respectively (Table 9). Bell (1967) induced accumulation of phloroglucinol-reactive compounds in tissues of cotton by inoculating with conidia of *Verticillium alboatrum* or with sporangiospores of *Rhizopus nigricans*. He described the accumulation of the phloroglucinol-reactive phenolic compounds in terms of gossypol equivalents and discussed their accumulation relative to that of phytoalexin production. Subsequent studies at SRRC demonstrated that heliocide H_2 (C_{25} terpenoid aldehyde, a natural cotton insecticide) was one of the predominant products formed in the volatile recipient (wounded) and nonwounded cotton leaves. Heliocide H_2 is formed naturally by a Diels–Alder addition of hemigossypolone and myrcene in the pigment glands of the cotton plant, Stipanovic and Bell (1977). Zeringue and McCormick (1989) found that myrcene represents 17.7% of the total volatiles in nonwounded Acala SJ-2 leaves and 24.8% of the total volatiles in wounded Acala SJ-2 leaves. Leaf damage caused by penetrating hyphae of *A. flavus* might have released myrcene from the pigment glands, and its subsequent reaction with hemigossypolone could have increased the heliocide H_2 level in the volatile recipient leaves.

To elucidate the effects of cotton leaf volatiles on *A. flavus* cultures, microbial-free compressed air was passed continuously (2 and 7 days)

Table 9 Effects of Volatile Chemicals Emitted by Infected or Wounded Cotton Leaves or by *A. flavus* Cultures on the Gossypol Equivalent Content of Receptor Cotton Leaves after 7 Days Exposure

		Treatment condition means[a] in nm/mol gossypol equivalents/g dry leaf tissue		
Volatile source	Volatile receptor	Source leaves	Receptor leaves	Control leaves[b]
Cut leaves	Normal leaves	337 a[c]	225 b	197 b
A. flavus inoculated leaves	Wounded leaves	1003 c	702 c	337 d
A. flavus inoculated leaves	Normal leaves	481 e	339 f	225 g
A. flavus cultures	Normal leaves		295 h	229 h
A. flavus cultures	Wounded leaves		401 i	317 i

[a]There were at least three replicates of each treatment condition, and the data represent the means of the analyses of three subsamples in each treatment.
[b]Control leaves treated the same as receptor leaves, except they received only filtered compressed air.
[c]Means in separate columns in each treatment followed by the same letter are not significantly different at P = 0.05 according to Duncan's Multiple Range Test.
Source: Data from Zeringue (1987b).

Table 10 Effects of 2- and 7-Day Incubation of *A. flavus* in Contact with Cotton Leaf Volatiles

Cotton cultivar	Time	Dry wt as a percent of control
	2 days	
8160 NW[a]		90.7 ± 9.1[c]
8160 W[b]		32.4 ± 8.6
SJ-2 NW		115.3 ± 6.7
SJ-2 W		32.4 ± 3.2
	7 days	
8160 NW		100.6 ± 3.3
8160 W		96.6 ± 4.0
SJ-2 NW		179.3 ± 5.6
SJ-2 W		178.3 ± 1.9

[a]Nonwounded.
[b]Wounded.
[c]Standard error of mean of three separate experiments.
Source: Data from Zeringue and McCormick (1989).

through enclosed systems containing wounded or nonwounded leaves of glandular or glandless cotton; the emitted volatiles were bubbled through liquid cultures of *A. flavus* (Zeringue and McCormick, 1989). After 2 days of incubation (Table 10), volatiles from wounded, glandular, and glandless cotton leaves retarded the growth of *A. flavus*. After 7-day incubations (Table 9) fungal growth was stimulated by volatiles from wounded or nonwounded glanded cotton leaves, but not from either type of glandless cotton leaves. The results showed that wounded cotton leaves release antimicrobial volatiles from both wounded SJ-2 and 8160 after 2 days. In addition, SJ-2 wounded and SJ-2 nonwounded volatiles stimulated fungal growth after 7 days; the substances apparently were produced from lysigenous pigment glands in the leaves since the stimulatory effect was not observed in cultures receiving volatiles derived from glandless 8160 leaves.

To characterize the active components from wounded leaves, volatiles were trapped on Tenax collection tubes and volatile profiles were determined by direct injection/gas chromatography/mass spectral analysis at 2- and 7-day periods (Zeringue and McCormick, 1989). Over 90 compounds were identified by this method. Purified compounds of selected identified cotton leaf–derived volatiles were assayed in culture to determine their bioactivity. Of the individual volatile components tested, C_6–C_9 alkenals, especially *trans*-2-hexenal, exhibited the maximum inhibitory effect on the growth of fungus (Table 11). Stimulatory effects of the individual volatile

Table 11 Radial Growth of *A. flavus* as a Percentage of Control After 2 Days in Contact with Some Selected Volatiles.

Volatile component	Level of tested component (μl)				Concentration (μmol/μl)
	1	3	5	10	
Alcohols					
3-Methyl-1-butanol	91 ± 4[a]	88 ± 3	80 ± 6	55 ± 10	9.1
3-Methyl-2-butanol	95 ± 6	95 ± 4	78 ± 3	80 ± 6	9.1
2-Buten-1-ol	100 ± 3	100 ± 2	100 ± 3	100 ± 3	1.5
2-Butoxy alcohol	96 ± 1	90 ± 1	86 ± 2	86 ± 3	7.6
1-Pentanol	109 ± 2	100 ± 3	91 ± 2	90 ± 1	9.2
4-Penten-1-ol	98 ± 2	98 ± 6	89 ± 5	89 ± 3	9.6
cis-2-Hexene-1-ol	98 ± 2	93 ± 4	98 ± 6	98 ± 3	8.4
cis-3-Hexene-1-ol	85 ± 2	80 ± 8	74 ± 5	69 ± 6	8.4
1-Heptanol	136 ± 3	114 ± 5	73 ± 3	73 ± 4	7.0
3-Hepten-1-ol	100 ± 3	89 ± 2	80 ± 4	73 ± 3	7.0
1-Nonanol	136 ± 3	122 ± 3	128 ± 4	132 ± 3	5.7
1-Decanol	96 ± 2	92 ± 2	91 ± 2	91 ± 5	5.2
Aldehydes					
Hexanal	84 ± 5	76 ± 3	76 ± 2	0 ± 0	8.3
trans-2-Hexenal	0 ± 0	0 ± 0	0 ± 0	0 ± 0	8.6
2,4-Hexadienal	53 ± 3	0 ± 0	0 ± 0	0 ± 0	9.0
2-Hexenal, diethylacetal	98 ± 2	0 ± 0	0 ± 0	0 ± 0	4.9
Heptanal	67 ± 5	58 ± 3	49 ± 6	0 ± 0	7.4
trans-2-Hepenal	82 ± 3	0 ± 0	0 ± 0	0 ± 0	7.6
Octanal	114 ± 7	88 ± 5	50 ± 3	46 ± 3	6.5
trans-2-Octenal	77 ± 3	0 ± 0	0 ± 0	0 ± 0	6.7
Nonyl aldehyde	75 ± 4	60 ± 3	46 ± 3	0 ± 0	5.8
trans-2-Nonenal	82 ± 2	0 ± 0	0 ± 0	0 ± 0	6.0
N-Decyl aldehyde	96 ± 2	92 ± 8	91 ± 3	91 ± 3	5.3
Dodecyl aldehyde	136 ± 5	112 ± 4	104 ± 3	104 ± 2	4.5

Ketones					
2-Pentanone	133 ± 8	116 ± 3	116 ± 3	116 ± 5	9.4
3-Pentanone	111 ± 3	121 ± 3	118 ± 2	116 ± 2	9.4
Cyclohexanone	109 ± 6	107 ± 5	109 ± 3	91 ± 4	10.0
2-Heptanone	80 ± 2	84 ± 2	76 ± 3	71 ± 4	7.1
3-Heptanone	100 ± 1	96 ± 5	91 ± 4	91 ± 3	9.6
3-Octanone	82 ± 4	82 ± 3	82 ± 1	82 ± 3	6.3
2-Nonanone	94 ± 1	77 ± 2	77 ± 2	61 ± 3	5.9
Others					
Myrcene	85 ± 5	90 ± 5	88 ± 3	86 ± 3	5.9
Ocimene	95 ± 9	95 ± 3	88 ± 1	85 ± 2	5.8
Limonene	90 ± 6	85 ± 3	83 ± 3	90 ± 3	6.1
Camphene	131 ± 6	111 ± 5	111 ± 3	111 ± 2	6.2
α-Pinene	114 ± 5	114 ± 4	114 ± 4	114 ± 1	5.9
β-Pinene	114 ± 2	114 ± 3	114 ± 1	114 ± 2	6.3
Caryophyllene	93 ± 2	93 ± 2	93 ± 2	93 ± 2	4.4
4-Pentenoic acid	114 ± 3	108 ± 2	96 ± 3	86 ± 1	9.8
Ethyl acetate	100 ± 4	95 ± 1	91 ± 4	86 ± 3	10.2

[a]Mean ± SD for 3 replicates/tested level.

Source: Data from Zeringue and McCormick (1989).

components tested were not as pronounced as the inhibitory effects. Unbranched C_8–C_{12} alkanals, 2- and 3-C_5 alkanones, and α- and β-pinene stimulated the growth of *A. flavus*.

The bioactive alkenals are breakdown products of the unsaturated fatty acids linoleic and linolenic acids with their concentrations being greatly increased by mechanical damage of plant tissue (Lyr and Banasiak, 1983). The key enzyme in alkenal biosynthesis is a membrane-bound lipoxygenase which is believed to be located both in chloroplasts and mitochondria (Sekiya et al., 1979; MacLeod and Pikk, 1979; Hatanaka et al., 1978). Because the alkenals are highly inhibitory to *A. flavus* growth, they may be considered as gaseous phytoalexins because of their mode of production and antifungal effects.

Since C_6–C_{10} alkenals are released from damaged cotton leaves, it seems logical to ask if newly released volatiles could effect a defense expression in the same plant or in neighboring plants. In order to address the question pertinent experiments were conducted. Microbial-free, compressed air was designed to carry individual C_6, C_7, C_8, C_9, and C_{10} alkenals and alkanals into enclosed systems containing artificially wounded and nonwounded Acala SJ-2 developing cotton bolls (Zeringue, 1991). Two days after treatment, disks were excised from the treated cotton boll surfaces and extracted to determine the induction of DHC, DHMC, LAC, LACME, and SCOP. Results showed that all the alkenals produced elevated levels of the five induced phytoalexins in artificially wounded, developing cotton bolls. All tested volatile alkanals produced slightly elevated levels of SCOP in the artificially wounded cotton boll when compared to wounded controls. Wounding or tissue damage appears necessary for C_6–C_{10} alkenals to function as active "volatile elicitors" for phytoalexin production.

Results of SRRC studies demonstrate that C_6–C_{10} alkenals function both as volatile elicitors and as gaseous phytoalexins in the cotton plant. Airborne signals consisting of C_6, C_7, C_8, C_9, and C_{10} alkenals emitted from wounded cotton bolls or wounded cotton leaves can induce the production of the lacinilene, cadalene, and scopoletin phytoalexins on the carpel surfaces of the cotton bolls on the same plant or in bolls of nearby plants (Zeringue, 1991). It also has been demonstrated that damage of cotton leaves or cotton bolls can initiate production of the C_6–C_{10} alkenals, the fungitoxic gaseous phytoalexins. Volatile compounds appear to represent a specialized niche in basic cotton plant defense.

V. EPILOG

Research at SRRC will continue to focus on plant-derived metabolites that inhibit the growth of toxigenic strains of *Aspergillus* sp. or inhibit aflatoxin biosynthesis. This brief chapter has described several strategies that the cotton plant may utilize in defense against *Aspergillus* sp. It is apparent that induction of a combination of defense mechanisms in cotton plants will protect cottonseed from aflatoxin contamination. Elucidation of the defense systems will provide a basis for development of commercial cotton cultivars with enhanced resistance to *A. flavus* and aflatoxin.

REFERENCES

Adye, J., and Mateles, R. I. (1964). Incorporation of labelled compounds into aflatoxins. *Biochimica and Biophysica Acta 86*:418–420.

Baldwin, I. T., and Schultz, J. C. (1983). Rapid changes in tree leaf chemistry induced by damage: Evidence for communication between plants. *Science 221*: 277–279.

Bell, A. A. (1967). Formation of gossypol in infected or chemically irritated tissues of Gossypium Species. *Phytopathology 57*:756–764.

Bell, A. A., and Stipanovic, R. D. (1977). Biochemistry of disease and pest resistance in cotton. *Mycopathologia 65*:91–106.

Block, G. B., and Kuć, J. (1983). Elicitation of phytoalexins by arachidonic and eicosapentaenoic acids: a host survey. 75th Annual Meeting American Phytopathological Society, Ames, IA, Paper No. A486.

Darvill, A. G., and Albersheim, P. (1984). Phytoalexins and their elicitors—a defense against microbial infection in plants. *Ann. Rev. Plant Physiol. 35*:243–275.

Detroy, R. W., Lillehoj, E. B., and Ciegler, A. (1971). Aflatoxin and related compounds. *Microbial Toxins*, Vol. 6. (A. Ciegler, S. Kadis, S. J. Ajl, eds.), Academic Press, New York, pp. 3–178.

Einhelling, F. A., Leather, G. R., and Hobbs, L. L. (1985). Use of *Lemma minor* L. as a bioassay in allelopathy. *J. Chem. Ecol. 11*:65–72.

Essenberg, M., Doherty, M. d'A., Hamilton, B. K., Henning, V. T., Cover, E. C., McFaul, S. J., and Johnson, W. M. (1982). Identification and effects on *Xanthomonas campestris* pv. *malvacearum*, of two phytoalexins from leaves and cotyledons of resistant cotton. *Phytopathology 72*:1349–1356.

Groopman, J. D., Groy, R. G., and Wogan, G. N. (1981). In vitro reactions of aflatoxin B_1-aducted DNA. *Proc. Natl. Acad. Sci. USA 78*:5445.

Halloin, J. M., and Bell, A. A. (1979). Production of nonglandular terpenoid aldehydes within diseased seeds and cotyledons of *Gossypium hirsutum*. *J. Agric. Food Chem. 27*:1407–1409.

Halloin, J. M., and Greenblatt, G. A. (1982). *Abstract, Beltwide Cotton Production Research Conference*, January 3–7, 1982, Las Vegas, p. 46.

Hargreaves, J. A., and Bailey, J. A. (1978). Phytoalexin production by hypocotyls of *Phaseolus vulgaris* in response to constitutive metabolites release by damaged bean cells. *Physiol. Plant Pathol. 13*:89–100.

Hatanaka, A., Sebiya, J., and Kayiwara, T. (1978). Distribution of an enzyme system producing *cis*-3-hexenal and n-hexanal from linolenic and linoleic acid in some plants. *Phytochemistry 17*:869–872.

Hochberg, M., and Cohen, Y. (1977). Scopoletin-induced catalase inhibition in tobacco leaves infected by *Peronospora tabacina* Adam. *Israel Journal of Botany 26*:48–52.

Hughes, J. C., and Swain, T. (1960). Scopolin production in potato tubers infected with *Phytophthora infestans*. *Phytopathology 50*:398–402.

Hunter, R. E., Halloin, J. M., Veech, J. A., and Carter, W. W. (1978). Terpenoid accumulation in hypocotyls of cotton seedlings during aging and after infection by *Rhizoctonia solani*. *Phytopathology 68*:347–350.

Imbert, M. P., and Wilson, L. A. (1970). Stimulatory and inhibitory effects of scopoletin in IAA oxidase preparations from sweet potato. *Phytopathology 9*: 1787–1794.

Kuranty, M. J., and Osman, S. F. (1983). Class distribution, fatty acid composition and elicitor activity of *Phytophthora infestans* mycelial lipids. *Physiological Plant Pathology 22*:363–370.

Lee, L. S. (1988). Aflatoxin in Arizona cottonseed: lack of toxin formation following *Aspergillus flavus* inoculation at sutures. *Journal of the American Oil Chemists' Society 65*:127–128.

Legendre, M. G., Fisher, G. S., Schuller, W. H., Dupuy, H. P., and Rayner, E. T. (1979). Novel technique for the analysis of volatiles in aqueous and nonaqueous systems. *Journal of the American Oil Chemists Society 56*:552–555.

Lyr, H., and Banasiak, L. (1983). Alkenals, volatile defense substances in plants, their properties and activities. *Acta Phytopathology Academy Science Hungary*. *18*:3–12.

MacLeod, A. I., and Pikk, H. E. (1979). Formation of (E)-hex-2-enal and (Z)-hex-3-en-1-ol by fresh leaves of *Brassica oleracea*. *Journal of Agriculture and Food Chemistry 27*:469–475.

Mayr, H. H., Diskus, A., and Beck, W. (1963). Accumulation of scopoletin in tissue of *Nicotiana tabacum* L. infected by *Peronospora tabacina* Adams. *Phytopathologie Zeitschuft 47*:95–97.

Preisig, C., and Kuć, J. (1983). Regulations of the sesquiterpenoid eliciting activity of C-20 fatty acids in potato by carbohydrate isolated from *Phytophthora infestans*. *75th. Annual Meeting American Phytopathological Society, Ames, IA., USA. Paper No. A511.*

Rhodes, J. M., and Wooltorton, L. S. C. (1978). The biosynthesis of phenolic compounds in wounded plant storage tissues. In *Biochemistry of wounded plant tissues* (Gunter Kohl ed.) Walter de Gruyter, Berlin and New York.

Schafer, P., Wender, S. H., and Smith, E. C. (1971). Effect of scopoletin on two anodic isoperoxidases isolated from tobacco tissue culture W 38. *Plant physiology 48*:232–233.

Sekiya, J., Kajiwara, J. T., and Hatanaka, A. (1979). Volatile C_6-aldehyde formation via hydroperoxide from C_{18}-unsaturated fatty acids in etiolated alfalfa and cucumber seedings. *Agriculture Biological Chemistry 43*:969–980.

Sequeira, L. (1969). Synthesis of scopolin and scopoletin in tobacco plants infected by *Pseudomonas solanacearum*. *Phytopathology 63*:826–829.

Shih, M., Kric, J., and Williams, E. B. (1973). Suppression of steroid glycoalkaloid accumulation as related to rishitin accumulation in potato tubers. *Phytopathology 63*:821–826.

Shih, M., and Kuć, J. (1973). Incorporation of ^{14}C from acetate and mevalonate into rishitin and steroid glycoalkaloids by potato tuber slices inoculated with *Phytophtora infestans. Phytopathology 63*:826–829.
Stipanovic, R. D., Bell, A. A., O'Brien, D. H., and Lukefar, M. J. (1977). Heliocide H_2: an insecticidal sesquiterpenoid from cotton (*Gossypium*). *Tetrahedron Letters 6*:567–570.
Stipanovic, R. D., and Wakelyn, P. J. (1974). *Proceedings ACGIH Cotton Dust Symposium*. Atlanta, GA, November 1974, 225.
Tanguy, J., and Martin, C. (1972). Phenolic compounds and the hypersensitivity reaction in *nicotiana tabacum* L. infected with tobacco mosaic virus. *Phytochemistry 11*:19–28.
Veech, J. A. (1978). An apparent relationship between methoxy-substituted terpenoid aldehydes and the resistance of cotton to *Meloidogyne incognita. Nematologica 24*:81–87.
Zeringue, H. J., Jr., Neucere, J. N., and Parrish, F. W. (1982). Hemagglutinins of some *Aspergilli. Biochemical Systematics and Ecology 10*:217–220.
Zeringue, H. J., Jr. (1984). The accumulation of five fluorescent compounds in the cotton leaf induced by cell-free extracts of *Aspergillus flavus. Phytochemistry 23*(11):2501–2503.
Zeringue, H. J., Jr., Conkerton, E. J., and Chapital, D. C. (1985). *Canadian Journal of Botany 63*:2470–2472.
Zeringue, H. J., Jr. (1987a). A possible relationship between phytoalexin production in the cotton leaf and a phytoxic response. *Phytochemistry 26*:975–978.
Zeringue, H. J., Jr. (1987b). Changes in cotton leaf chemistry induced by volatile elicitors. *Phytochemistry 26*:1357–1360.
Zeringue, H. J., Jr. (1988). Production of carpel wall phytoalexins in the developing cotton boll. *Phytochemistry 27*:3429–3431.
Zeringue, H. J., Jr., and McCormick, S. P. (1989). Relationships between cotton leaf-derived volatiles and growth of *A. flavus. Journal of the American Oil Chemists Society 66*:581–585.
Zeringue, H. J., Jr. (1990). Stress effects on cotton leaf phytoalexins elicited by cell-free-mycelia extracts of *Aspergillus flavus. Phytochemistry 29*:1789–1791.
Zeringue, H. J., Jr. (1992). Effects of C_6–C_{10} alkenals and alkanals on eliciting a defense response in the developing cotton boll. *Phytochemistry 31*:2305–2308.

9

Sesquiterpenoid Phytoalexins Synthesized in Cotton Leaves and Cotyledons During the Hypersensitive Response to *Xanthomonas campestris* pv. *malvacearum*

Margaret Essenberg and Margaret L. Pierce
Oklahoma State University, Stillwater, Oklahoma

I. INTRODUCTION

Our work has focused on evaluation of the role of phytoalexins in heritable resistance of cotton to bacterial blight. The most thorough studies of this type by other workers have been with fungal diseases, and mostly with isoflavonoid phytoalexins (Mansfield, 1982; Hahn et al., 1985). This study evaluates the role of sesquiterpenoid phytoalexins in a bacterial leafspot disease. Cotton (*Gossypium* spp.) and *Xanthomonas campestris* pv. *malvacearum* are a genetically interesting system because of the race/cultivar specificity due to a number of major host genes for resistance, each of which confers resistance to a different set of races (Brinkerhoff, 1970). The pathogen's race phenotype is determined by avirulence genes (DeFeyter and Gabriel, 1991), each of which contains several 102-nucleotide-base-pair repeats of unknown function (D. W. Gabriel, personal communication). Upland cotton (*Gossypium hirsutum* L.) lines possessing several resistance genes have been developed whose resistance to bacterial blight is so high that no macroscopic symptoms or only pinpoint lesions are produced after inoculation with most races of the pathogen (Brinkerhoff et al., 1984).

Objectives of our work have been to determine whether phytoalexins contribute significantly to the high resistance of those cotton lines and to learn how the phytoalexins are biosynthesized. In evaluating the role of phytoalexins, we have taken an analytical approach by asking whether the pathogen is exposed to inhibitory doses of the phytoalexins at the time it becomes inhibited during incompatible interactions. Results of the study yielded information on whether phytoalexin accumulation is sufficient to account for resistance. The difficult task in such a study is determining the local phytoalexin concentration in contact with the pathogen. Our discovery that the cellular localization of sesquiterpenoid phytoalexins in cotton leaves and cotyledons could be directly observed by fluorescence microscopy of fresh, unstained tissue offered the opportunity to determine such local concentrations.

II. MATERIALS AND METHODS

The various plant lines and bacterial strains are described in the articles cited in the text. Cotton plants (*Gossypium hirsutum* L.) were grown in a growth chamber with a 14-hr light/10-hr dark cycle with 30°C maximum/19°C minimum temperatures (Pierce and Essenberg, 1987). *Xanthomonas campestris* pv. *malvacearum* Smith (Dye) was cultured, inocula were prepared as bacterial suspensions in sterile saturated $CaCO_3$ solution, and cotyledons were inoculated by infiltration from a syringe (no needle) as described by Pierce and Essenberg (1987). Entire leaves were infiltrated with a pressure-drive spray of inoculum through stomata of their abaxial surfaces. Bacterial growth trends in planta were determined by diluting and plating tissue homogenates (Essenberg et al., 1979a).

Sesquiterpenoids were extracted from frozen foliar tissues, worked up, resolved by reverse-phase high-performance liquid chromatography (HPLC), and quantitated by several methods (Essenberg et al., 1982, 1990; Pierce and Essenberg, 1987). Bioassays of their antibacterial activities were preformed on logarithmically growing cultures of *X. campestris* pv. *malvacearum* in defined medium in the dark (Essenberg et al., 1982) or with 300–700 nm radiation at 70 μEinsteins/m^2/sec (Steidl, 1988). Photoactivated nicking of DNA plasmids and inactivation of DNAse I and malate dehydrogenase by 2,7-dihydroxycadalene (DHC) or lacinilene C (LC) were performed with 300–700 nm radiation at 630 μEinsteins/m^2/sec and detected as described by Sun et al. (1989). Inactivation of cauliflower mosaic virus was performed similarly, was detected by inoculation of turnip leaves, and was characterized by polyacrylamide gel electrophoresis (PAGE) of viral polypeptides after denaturation with sodium dodecylsulfate and 2-mercaptoethanol and by alkaline agarose gel electrophoresis of viral DNA (Sun et al., 1988).

The yellow-green fluorescent cells of inoculated foliar cotton tissues were observed by fluorescence microscopy with filter sets designed for detection of fluorescein (450–490 nm excitation, 520–560 nm emission) (Essenberg et al., 1992b). Brightly fluorescent cells were separated from weakly fluorescent cells with a fluorescence-activated cell sorter, using filters with the same excitation and emission ranges as employed for fluorescence microscopy (Pierce and Essenberg, 1987). Necrotic cells were stained in 0.25-cm^2 leaf discs with an aqueous solution of Evan's blue. Exhaustive extraction of phytoalexins was performed by vacuum infiltration with ethanol/water (70:30) and sonication for 30–45 min at 20–30°C. The relative fluorescence intensities of cells before and after tissue extraction were viewed with a video camera fitted to the fluorescence microscope, amplified, digitized, and analyzed by computer (Essenberg et al., 1992b). Alkaline hydrolysis of materials remaining after extraction was performed with 1 N NaOH at room temperature for 24 hr. Details of the method for determining average phytoalexin concentration in the yellow-green fluorescent cells of foliar tissue were described by Essenberg et al. (1992a). Important in that method was determination of the fraction of the leaf tissue water that was in the hypersensitively responding (i.e., yellow-green fluorescent and/or brown) cells. The fraction of leaf tissue water that was in all mesophyll cells was determined by stereological analysis of cross-sections. The fraction of mesophyll cells that were hypersensitively responding was determined in fresh, unsectioned leaf samples by counting cells that were fluorescent and/or Evan's blue-stainable, as well as total mesophyll cells.

Incorporation of [1,2-^{13}C]acetate into DHC was by infiltration of inoculated OK1.2 cotyledons with a 20 mM solution at 52 hr postinoculation, followed by harvest at 70 hr (Essenberg et al., 1985). Incorporation of [5-^{3}H]mevalonolactone into DHC and 2-hydroxy-7-methoxycadalene (HMC) was by infiltration of inoculated cotyledons with a 30 μCi/ml solution at 26 hr postinoculation, followed by harvest at 43 hr. Degradation was by oxidation with a ruthenium-containing catalyst, followed by esterification of the resulting labeled isobutyric acid and HPLC of the ester to constant specific radioactivity (Davis et al., 1991). δ-Cadinene was identified by comparison of its proton nuclear magnetic resonance and mass spectra with spectra of authentic δ-cadinene isolated from cade oil and by cochromatography with authentic δ-cadinene (Davis, 1993). Cell-free preparations of the sesquiterpene cyclase were obtained by homogenizing inoculated, glandless cotyledons in cold buffer containing enzyme-stabilizing additives and then preparing from the homogenate a 27,000*g* supernatant (Munck and Croteau, 1990). Cyclase activity was assayed by incubating an aliquot of the supernatant (10 μg protein per assay) with 0.3 μCi [1-^{3}H]FPP for 30 min at 30°C, followed by hexane extraction of the reaction mixture, passage of the hexane through silica to remove polar compounds, and liquid scintillation counting of the eluate. Cochromatography of the eluate with authentic δ-cadinene by normal phase HPLC verified the identity of the predominant product as δ-cadinene (Davis, 1993).

III. RESULTS AND DISCUSSION

A. Association of Phytoalexins with Resistance to Bacterial Blight of Cotton

A search for organic solvent-extractable substances produced during the hypersensitive response of leaves and cotyledons led to isolation and identification of the sesquiterpene phenols DHC and LC (Essenberg et al., 1982). Methyl ethers of both compounds were also found: HMC and lacinilene C 7-methyl ether (LCME) (Essenberg et al., 1990).

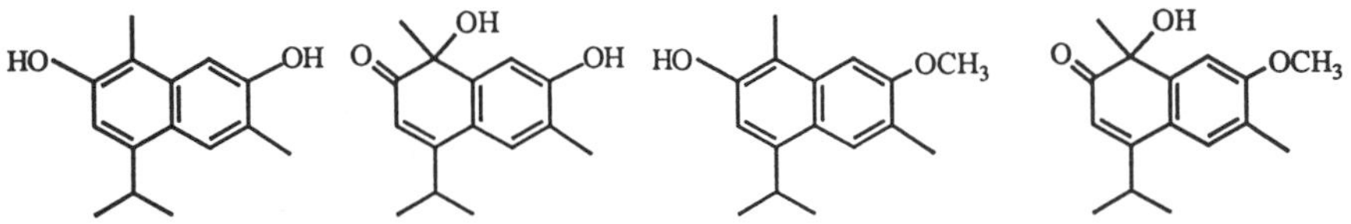

HO OH O OH OH HO OCH_3 O OH OCH_3

DHC LC HMC LCME

In three highly resistant cotton lines, phytoalexin accumulation occurred during the hypersensitive, resistant response. In inoculated leaves or cotyledons, the four sesquiterpenoids accumulated to levels of 100–500 nmol/g dry wt by the time bacterial multiplication was inhibited (3–4 days after inoculation) (Essenberg et al., 1982, 1990). Only very small amounts (1–30 nmol/g dry wt) were found in mock-inoculated resistant plants or in inoculated susceptible plants at this time. In the highly resistant lines, the four compounds continued to accumulate for several days after bacterial inhibition, each reaching levels of 1000–5000 nmol/g dry wt before declining.

Each highly resistant cotton line possesses two or three major genes for bacterial blight resistance plus a complex of "polygenes" (Essenberg et al., 1982; Pierce and Essenberg, 1987). Also of interest were a collection of lines of related genetic background, each of which possesses a different, race-specific resistance gene (Hunter and Brinkerhoff, 1961). Three of these lines, as well as the related, fully susceptible line Ac44, were tested for phytoalexin production after inoculation with each of six strains of *X. campestris* pv. *malvacearum* (of three different race phenotypes). In all 14 incompatible (resistant) interactions, higher amounts of all four of the sesquiterpenoids shown above were found at the time of bacterial inhibition than in all 10 of the compatible (susceptible) interactions (Shevell, 1985). Different resistance genes evidently control accumulation of the same phytoalexins. Furthermore, Ac44, which possesses no known genes for resistance to bacterial blight, accumulated high amounts of the four compounds during its nonhost response to *X. campestris* pv. *campestris*, a pathogen of crucifers (Essenberg et al., 1990). Thus in the many compatible and incompatible race/cultivar interactions and in the one heterologous pathovar/cotton interaction that we have analyzed, sesquiterpenoid phytoalexin accumulation correlated invariably with effective resistance to bacterial disease. Correlation of phytoalexin production with incompatibility has also been observed within differential race/cultivar sets of *Pseudomonas syringae* pv. *pisi* and pea (Hadwiger and Webster, 1984) and of *P. syringae* pv. *glycinea* and soybean (Long et al., 1985).

The abnormal behavior of inoculated cotton plants held in continuous darkness after inoculation gave us another opportunity to observe phytoalexin levels in relation to resistance (Morgham et al., 1988). In continuous darkness, inoculated leaves of both susceptible and highly resistant lines underwent confluent necrosis. However, the sesquiterpenoid phytoalexins did not accumulate in either line. In both lines, the bacteria multiplied to high population densities typically found in susceptible plants maintained in a normal light/dark cycle. Thus light was required for phytoalexin production in cotton leaves, and in its absence rapid tissue necrosis was insufficient to inhibit bacterial multiplication. These observations are consistent with the hypothesis that phytoalexin accumulation is responsible for bacterial inhibition under normal light/dark conditions, although it is possible that other light-dependent functions were instead responsible.

B. Antibacterial Activities of the Sesquiterpenoids

Bioassays of the sesquiterpenoids' effects on growth of *X. campestris* pv. *malvacearum* in vitro were at first conducted in darkness to avoid photooxidation, to which all four compounds are sensitive. DHC was the most potently antibacterial (Table 1) (Essenberg et al., 1982, 1990). HMC, its methyl ether, was not detectably active, perhaps because it is almost insoluble in water. The *S* enantiomer of LC was more active than the *R* enantiomer; however, the two enantiomers of LCME had comparable activities. Although the water solubilities of the phytoalexins in planta may differ somewhat from those reported in Table 1, it appears likely that the phytoalexins are soluble enough to diffuse through the apoplast in concentrations that in combination are bacteriostatic.

It was reported in 1983 that phaseollin and several other isoflavonoid phytoalexins were activated by irradiation with ultraviolet light to produce free radicals and to cause inactivation of the enzyme glucose-6-phosphate dehydrogenase in an in vitro assay system (Bakker et al., 1983). This report stimulated us to test the effects of irradiation on the biological activities of the cotton phytoalexins. In vitro studies showed that when irradiated with

Table 1 Antibacterial Concentrations, Water Solubilities, and Average Cellular Concentrations In Planta[a] of Cotton Leaf Phytoalexins

	Antibacterial concentrations (mM)[b]			Local concentration (mM)[d]	
Phytoalexin	ED_{50}	ED_{90}	Solubility (mM)[c]	Day 3	Day 4
DHC	0.35	0.5	0.6	6.2	6.3
LC	~1.5	~1.8	1.8	4.2	4.9
LCME	~0.6	–	1.2	1.1	1.7
No. HR necrotic cells per bacterial colony[e]:			day 3: 0.61	day 4: 4.5	
Avg. no. HR cells per cluster:			day 3: 2.4	day 4: 4.8	
Ratio of HR cell clusters to bacterial colonies:			day 3: 0.25	day 4: 0.93	

[a]Leaves of the highly resistant line OK1.2 were spray-infiltrated with 3 × 10^6 cells/ml of *X. campestris* pv. *malvacearum*.
[b]ED_{50} and ED_{90} were the phytoalexin concentrations which permitted 50% and 10% as many bacterial generations, respectively, as in an uninhibited control culture growing logarithmically in defined liquid medium at 30°C in the dark (Essenberg et al., 1982, 1990). The LC and LCME used for these bioassays were isolated from inoculated Im 216 cotyledons and were predominantly the *R* enantiomers. LCME did not exhibit an ED_{90} within its solubility range.
[c]Solubility in water at 30°C (mM) (Essenberg et al., 1990).
[d]Concentrations in hypersensitively responding (HR) cells, determined as described in the text with the minor modification that Evan's blue stain was used to facilitate the detection of responding cells that fluoresced little. Phytoalexin concentrations are means of determinations from two plants (Essenberg, Pierce, Cover, Richardson and Scholes, unpublished) and corrected for recovery rates of phytoalexin standards in the extraction and separation procedures: 33%, 24%, and 36% for DHC, LC, and LCME, respectively.
[e]Fluorescent and/or Evan's blue-stainable cell numbers per cm^2 of leaf corrected for cells wounded by the inoculation procedure, determined from mock-inoculated leaves. The number of bacterial colonies per cm^2 in inoculated leaves was assumed to be equal to the initial number of bacteria per cm^2 (Essenberg et al., 1979a,b).

white light (300–700 nm), DHC and LC can induce single-strand breaks in DNA, inactivate enzymes, and destroy the infectivity of cauliflower mosaic virus, apparently by crosslinking the viral DNA to coat protein (Sun et al., 1988, 1989). During exposure of DHC to light and oxygen, free radicals were detected by spin trapping with phenyl-*tert*-butylnitrone (Steidl, 1988). DHC, LC, and LCME were more inhibitory to liquid cultures of *X. campestris* pv. *malvacearum* when they were exposed to white light (300–700 nm) than in the dark (Steidl, 1988; Sun et al., 1989; Samad and Essenberg, unpublished). In the light, 0.10 mM DHC inhibited bacterial growth completely. HMC remained inactive and therefore does not qualify as a phytoalexin. The broadly destructive effects observed from photoactivated DHC suggest that during the hypersensitive response in sunlit cotton plants, these phytoalexins probably damage many components of the infecting bacterial cells. The only other report of phototoxic phytoalexins of which we are aware is on thiophenes elicited in *Tagetes* by *Fusarium oxysporum* (Kourany et al., 1988). However, the growing numbers of plant species discovered to contain phototoxins (Downum et al., 1991) suggest that other phytoalexins may have unrecognized phototoxicity.

C. Localization of the Phytoalexins in and Around All Hypersensitively Necrotic Cells

When foliar tissues of resistant cotton lines are inoculated with *X. campestris* pv. *malvacearum* by infiltration with a suspension of fewer than 10^7 bacteria/ml, the tissue does not undergo confluent collapse and necrosis, but instead develops clusters of dark brown, necrotic cells. Each cluster is the site of a bacterial colony which has grown from a single bacterial cell deposited during inoculation in the adjacent intercellular space. Our early work demonstrated that in resistant plants, growth of each colony is inhibited by a local resistant response (Essenberg et al., 1979a) manifested as a small bacteriostatic zone around the colony (Essenberg et al., 1979b). Fluorescence microscopy revealed that all of the brown, necrotic cells exhibit yellow-green fluorescence that is spectrally similar to that of the phytoalexins LC and LCME (Pierce and Essenberg, 1987; Essenberg et al., 1992b). Fluorescent materials may also be present in the apoplast, as suggested by apparent fluorescence of the cell walls of the necrotic cells and often also of neighboring cells.

To determine whether the most potent phytoalexin, DHC, is also localized in the fluorescent cells, we undertook the physical separation of yellow-green fluorescent cells from symptomless cells so that they could be analyzed directly for their phytoalexin contents. Macerating enzymes were used to obtain suspensions of palisade and spongy mesophyll cells from inoculated resistant cotyledons, and the cells were quickly subjected to fluorescence-activated cell sorting. From the phytoalexin contents determined in the brightly fluorescent and less fluorescent fractions and from the numbers of brightly fluorescent and symptomless, less fluorescent cells in each fraction, we calculated the average phytoalexin content of each of these two cell types (Pierce and Essenberg, 1987). A brightly fluorescent cell contained 10–25 times as much LC and 40 times as much DHC as a less fluorescent cell. When the relative numbers of brightly fluorescent and symptomless, less fluorescent cells in the cotyledonary tissue were taken into account, it was calculated that more than 90% of the DHC and more than 75% of the LC and LCME were associated with the fluorescent cells at infection centers (Table 2).

Since the bacterial colonies are in the intercellular spaces rather than inside the necrotic cells, it was important to know whether the phytoalexins can penetrate the host plasma membrane to contact the bacteria. Evan's blue, a dye that stains only cells with damaged plasma membranes (Taylor and West, 1980), was used to assess membrane integrity. Of the fluorescent cells observed in leaf samples 2, 3, and 4 days after inoculation, 98 ± 2% were stained by Evan's blue and/or were dark brown (Pierce and Essenberg, unpublished). We know that the dark brown cells have dysfunctional membranes because they are collapsed (Essenberg et al., 1979a, 1992b). We conclude that all of the fluorescent cells have damaged membranes permeable to small molecules.

An efflux experiment demonstrated that the phytoalexins were able to diffuse from the fluorescent cells. The abaxial epidermis was peeled from strips of inoculated resistant cotyledons, and the strips were gently stirred in a buffered 0.4 M mannitol solution (Pierce et al., unpublished). The phytoalexins LC, LCME, and DHC diffused into the medium with first-order kinetics and half-times of 3 hr, 6 hr, and 16 hr, respectively.

However, not all of the yellow-green fluorescent material in hypersen-

Table 2 Distribution of Phytoalexins Between Brightly Fluorescent Cells and Symptomless, Less Fluorescent Cells per Unit Area of Cotyledon[a]

	Fraction of mesophyll tissue (%)	Phytoalexins (pmol/mm^2)			Contribution to total (%)[b]		
		DHC	LC	LCME	DHC	LC	LCME
Experiment 1							
Brightly fluorescent	23	68.0	86	62	93	88	100
Less fluorescent	77	5.3	11	0	7	12	0
Experiment 2							
Brightly fluorescent	23	110.0	140	71	92	75	79
Less fluorescent	77	9.6	47	18	8	25	21

[a]Cells were isolated from resistant OK1.2 cotyledons three days after infiltration with 5 × 10^6 cells/ml of *X. campestris* pv. *malvacearum* and subjected to fluorescence-activated cell sorting.

[b]It was assumed that the phytoalexin amounts lost during isolation and sorting had the same distribution between brightly fluorescent and less fluorescent cells in the intact tissue as the amounts recovered in the sorted cells.

Source: Reproduced with permission from Pierce and Essenberg, 1987. Copyright 1987, Academic Press, Ltd.

sitively responding cotton foliar tissue was extractable. After exhaustive extraction with ethanol/water (70:30), cotyledons still contained fluorescent cells. They were the same cells that fluoresced in fresh tissue: of 507 fluorescent cells in resistant cotyledons harvested 3–4 days after inoculation and examined before and after extraction, 503 (99%) had visually detectable residual fluorescence (Essenberg et al., 1992a). Alkaline hydrolysis partially removed this residual fluorescence, indicating its probable origin in both wall-bound phenolic monomers and phenolic polymers.

This finding that essentially all of the fluorescent cells possessed nonextractable fluorescence raised the question of whether all or only some of them also contained the fluorescent phytoalexins. To answer this question, fluorescence intensity (520–560 nm) of individual yellow-green fluorescent palisade cells was measured with a videodensitometer before and after tissue extraction. Extraction removed at least 35% of the fluorescence from all cells, and the most frequent loss was 85–90%. Thus all fluorescent cells of hypersensitively responding tissue possessed extractable as well as nonextractable fluorescence. When an extract of hypersensitively responding tissue was fractionated, LC and LCME accounted for 75% of the yellow-green fluorescence in the set of fractions (Essenberg et al., 1992b). These results strongly suggest that in inoculated resistant cotton foliar tissue, all hypersensitively responding cells accumulate both free phytoalexins and covalently bound fluorescent phenolics.

Autofluorescence or absorbance that is spectrally similar to that of phytoalexins has also been used to localize phytoalexins in leaves of broadbean (Mansfield et al., 1974) and grapevine (Langcake and Pryce, 1976) in response to *Botrytis* infection, in resistant oat infected with *Pyricularia* or *Puccinia* (Mayama and Tani, 1982), and in juvenile sorghum leaves infected with *Colletotrichum* (Snyder et al., 1992). In broadbean and grapevine, healthy cells adjacent to necrotic cells fluoresced most intensely, whereas in oat and sorghum, as in cotton, the collapsed cells were the sites of fluorescence or pigmentation. Histochemistry has also been used to locate phytoalexins at or near the invading pathogen during expression of resistance in cotton stems, bean leaves and hypocotyls, flax leaves, and tomato roots and stems (reviewed in Pierce and Essenberg, 1987).

D. Phytoalexin Concentrations in the Hypersensitively Necrotic Cells

The findings described above, that the phytoalexins are predominantly localized in cells which can be identified by their autofluorescence, enabled us to determine the average phytoalexin concentrations in those cells. Tissue water content was determined as the difference between fresh and dry weights. Other samples of the same leaves were examined by fluorescence microscopy so that the fraction of mesophyll cells exhibiting yellow-green fluorescence and/or brown pigmentation could be determined by counting. The fraction of the leaf tissue water that was in those hypersensitively responding cells was computed. The phytoalexin content of a given amount of leaf tissue was then divided by the computed volume of its hypersensitively necrotic cells to give average millimolar concentration in those cells (Essenberg et al., 1992a).

Results of applying this method to highly resistant line OK1.2 are shown in Table 1. Each of the phytoalexins accumulated in the hypersensitively necrotic cells to a concentration higher than its water solubility (Essenberg et al., 1990). At both 3 and 4 days after inoculation, DHC and LC concentrations were greater than their ED_{90} levels in the dark. However, at 3 days there was only about one cluster of detectably fluorescent and/or Evan's blue-stainable (necrotic) cells for every four bacterial colonies. By day 4, numbers of necrotic cells had increased so that there were 4.5 necrotic cells per bacterial colony. Since the average cluster consisted of 4.8 cells, there was nearly one (0.93) cluster per bacterial colony. This correspondence agrees well with the observation that at 4 days bacterial multiplication in these leaves was inhibited.

Cellular phytoalexin concentrations high enough to account for bacteriostasis on the day it was observed have also been determined for the highly resistant cotton lines Im 216 and WbM(0.0) (Pierce et al., unpublished results). On the same day, in the genetically related susceptible line WbM(4.0), phytoalexin levels were too low to be inhibitory, which is consistent with the fact that bacteria were still multiplying.

Cellular phytoalexin concentrations were also determined in several of the Ac44 lines possessing single resistance genes (Shevell et al., unpublished results). At the time of bacterial inhibition, inhibitory phytoalexin concentrations were found in leaves infected with incompatible bacterial strains, but in leaves infected with compatible strains, cellular phytoalexin concentrations were too low to be inhibitory.

In summary, the phytoalexins are localized in the right place to be effective—in the leaky, dead, or dying mesophyll cells closest to the intercellular bacterial colonies. In the various cotton lines studied, which have different bacterial blight resistance genes and different levels of resistance, the phytoalexins have been found at concentrations higher than in vitro bioassays indicate that they are needed to fully inhibit multiplication of the bacterial pathogen. These phytoalexin-loaded cells become numerous enough to inhibit essentially all bacterial colonies on the day when bacteriostasis was observed. We conclude that our determinations of cellular phytoalexin concentrations provide strong evidence that phytoalexin accumulation is sufficient to account for resistance of cotton foliar tissue to bacterial blight.

Studies by other investigators have yielded similar evidence for a few plant/fungal pathogen interactions. One approach has been to cut thin tissue slices and to subject alternate or replicate slices either to staining for presence of the pathogen or to quantitative analysis for phytoalexins. Phytoalexin concentration is computed by dividing the slice's phytoalexin content by its water content. This approach was applied to the soybean hypocotyl/*Phytophthora megasperma* var. *sojae* system by Yoshikawa et al. (1978), using 0.25-mm-thick slices, and at high resolution by Hahn et al. (1985), using 15-μm-thick slices. Results support the hypothesis that the phytoalexin glyceollin I has an important early role, but may not be solely responsible for resistance. Metlitskii et al. (1985) employed the tissue slice approach in studying potato tuber inoculated on a cut surface with compatible and incompatible races of *Phytophthora infestans* and found antifungal concentrations of rishitin at the arrested front of incompatible fungal growth. During the resistant response of juvenile sorghum to *Colletotrichum graminicola*, epidermal cells respond to fungal appressoria by accumulating orange–red deoxyanthocyanidin phytoalexins. Snyder et al. (1991) obtained an estimate of 150 mM phytoalexin in subcellular vesicle-like inclusions by microspectrophotometry. Although dilution occurs when these vesicles release the phytoalexin within the cell, it seems likely that the concentration is still higher than the 9 μM that inhibits fungal growth in vitro. Thus in the plant/fungal pathogen systems in which local phytoalexin concentrations at sites of pathogen growth have been experimentally estimated, as well as in the plant/bacterial pathogen system that we study, results indicate that phytoalexins can contribute significantly to observed resistance.

E. Biosynthetic Pathways

The structures of phytoalexins produced in cotton leaves and cotyledons suggest that they are terpenoid in origin, and isotope incorporation into DHC and HMC from labeled acetate and mevalonolactone has confirmed this (Essenberg et al., 1985; Davis et al., 1991). Several things have been learned about how *trans, trans*-farnesylpyrophosphate (FPP), a common intermediate of terpene biosynthesis, enters the specialized pathway to the cotton leaf phytoalexins. When acetate containing ^{13}C at 90% enrichment in both positions was fed to cotyledons during phytoalexin biosynthesis, the positions of adjacent carbon atoms labeled from the same acetate molecule were deduced from the ^{13}C nuclear magnetic resonance spectrum of DHC (Essenberg et al., 1985). This information showed which of the three possible folding patterns of the farnesyl precursor occurs during formation of the bicyclic ring structure of DHC. The same cyclization pattern has subsequently been demonstrated in gossypol, a bisesquiterpene from cotton that functions in resistance to herbivores (Masciadri et al., 1985).

During the biosynthesis of DHC in planta, a tritium atom from [5-^{3}H]mevalonolactone was transferred to the methine carbon of the isopropyl side chain (Davis et al., 1991). It seems likely that this transfer occurs as a 1,3-hydride shift from C-1 of the farnesyl precursor following its cyclization to a 10-membered ring cation. The retention of tritium on the isopropyl group throughout the pathway to DHC suggested further that if [1-^{3}H]FPP were used as the substrate with cell-free preparations from phytoalexin-synthesizing tissues, intermediates of the pathway would be tritium-labeled. To simplify the mixture of labeled compounds produced, a cotton line which lacks pigment glands was used. Its leaves and cotyledons have only very low amounts of gossypol and the other terpenoids stored in pigment glands of other cotton lines (Elzen et al., 1985), but it produces high levels of DHC, HMC, LC, and LCME after bacterial infection. Cell-free preparations from its inoculated cotyledons cyclized [1-^{3}H]FPP or unlabeled FPP to δ-cadinene, an unsaturated, bicyclic hydrocarbon with carbon skeleton of DHC (Davis, 1993).

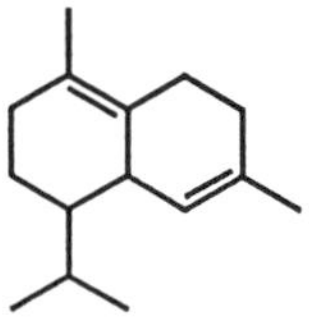

δ-Cadinene

This enzymic activity is strongly infection-induced. Thus it appears that biosynthesis of the phytoalexins from FPP begins with cyclization and is followed by hydroxylation and aromatization. In this, it follows the pattern that has been found for biosynthesis of other cyclic sesquiterpenes (Cane, 1990).

IV. EPILOG

We will continue our exploration of the foliar tissue's response to individual bacterial colonies. The observation that cells become Evan's blue–stainable (i.e., necrotic) as they become visibly fluorescent raises the question of where biosynthesis of these fluorescent phytoalexins occurs. Is it in a short burst just before death? Or in living, neighboring cells followed by transport to the dead or dying cells? Or in both places? Our biosynthetic studies are providing tools for answering these questions. We are currently performing pulse label experiments to gain information about time and place of biosynthesis relative to host cell death. We are also purifying the sesquiterpene cyclase described above. If it is proven to catalyze a reaction on the biosynthetic pathway to the phytoalexins, then an immunocytochemical stain for it or an in situ hybridization probe for its messenger RNA can reveal the cells in which phytoalexin biosynthesis is induced.

We are also working to elucidate the rest of the biosynthetic pathway. Several putative intermediates have been identified. Isotopic incorporation and dilution experiments will be conducted to show whether these compounds are biosynthetic intermediates between FPP and the phytoalexins. Assays for later enzymes of the pathway may then be developed, so that these enzymes may be isolated.

Preparation and identification of cDNA clones for biosynthetic enzymes will give us the opportunity to block these steps in plants by transforming them with antisense constructs of the genes. This strategy is now being applied by other investigators to legumes to test the role of isoflavonoid phytoalexins in disease resistance. Such manipulation of genes affecting phytoalexin concentration or activity offers promise of powerful tests of whether phytoalexins are *necessary* for resistance. The analyses of phytoalexin concentrations in planta that we have described in this chapter have, by comparison, provided evidence concerning whether phytoalexin accumulation is *sufficient* to account for observed disease resistance.

The cDNA clones for biosynthetic enzymes may be used as probes for identification of genomic clones. The cadalene and lacinilene phytoalexins that we study are only four of a large group of terpenoids in cotton that have the same carbon skeleton and function variously as constitutive defenses against herbivory, as attractants to pollinating insects, or as infection-induced defenses against fungal and bacterial pathogens (Bell, 1986). Isolation of genes for the biosynthetic enzymes, along with their variously regulated promoters, will open up possibilities of crop improvement by enhancing or repressing synthesis of these compounds in particular organs of the plant or by introducing genes for foreign defense proteins under the regulation of cotton's own stress-responsive promoters.

ACKNOWLEDGMENTS

We thank our colleagues Ellen C. Cover, Paul E. Richardson, Vernon E. Scholes, Roushan A. Samad, and Judith L. Shevell for permission to describe their unpublished work, and Oral Roberts University for use of the microspectrophotometer and the cell sorter. The work described here was supported by a grant from the Herman Frasch Foundation, by the Cooperative State Research Service, U.S. Department of Agriculture, under Agreement No. 5901-0410-9-0236-0, by the National Science Foundation under Grants Nos. PCM-8117015, PCM-8316759, and DMB-8616650, and by the Oklahoma Agricultural Experiment Station, of which this is journal article J-6312.

REFERENCES

Bakker, J., Gommers, F. J., Smits, L., Fuchs, A., and De Vries, F. W. (1983). Photoactivation of isoflavonoid phytoalexins: involvement of free radicals. *Photochem. Photobiol. 38*(3): 323–329.

Bell, A. A. (1986). Physiology of secondary products. *Cotton Physiology* (J. R. Mauney and J. McD. Stewart, eds.), Cotton Foundation, Memphis, pp. 597–621.

Brinkerhoff, L. A. (1970). Variation in *Xanthomonas malvacearum* and its relation to control. *Annu. Rev. Phytopathol. 8*: 85–110.

Brinkerhoff, L. A., Verhalen, L. M., Johnson, W. M., Essenberg, M., and Richardson, P. E. (1984). Development of immunity to bacterial blight of cotton and its implications for other diseases. *Plant Dis. 68*(2): 168–173.

Cane, D. E. (1990). Enzymatic formation of sesquiterpenes. *Chem. Rev. 90*(7): 1089–1103.

Davis, G. D. (1993). δ-Cadinene synthase from hypersensitively responding cotton cotyledons: identification of substrate and product and partial purification of the enzyme. PhD thesis, Oklahoma State University, Stillwater.

Davis, G. D., Eisenbraun, E. J., and Essenberg, M. (1991). Tritium transfer during biosynthesis of cadalene stress compounds in cotton. *Phytochemistry, 30*(1): 197–199.

De Feyter, R., and Gabriel, D. W. (1991). At least six avirulence genes are clustered on a 90-kilobase plasmid in *Xanthomonas campestris* pv. *malvacearum*. *Mol. Plant–Microbe Interact. 4*(5): 423–432.

Downum, K. R., Swain, L. A., and Faleiro, L. J. (1991). Influence of light on plant allelochemicals: a synergistic defense in higher plants. *Arch. Insect Biochem. Physiol. 17*(4): 201–212.

Elzen, G. W., Williams, H. J., Bell, A. A., Stipanovic, R. D., and Vinson, S. B. (1985). Quantification of volatile terpenes of glanded and glandless *Gossypium hirsutum* L. cultivars and lines by gas chromatography. *J. Agric. Food Chem. 33*(6): 1079–1082.

Essenberg, M., and Pierce, M. L. (1994). Role of phytoalexins in resistance of cotton to bacterial blight. *Bacterial Pathogenesis and Disease Resistance* (S.-D. Kung and D. D. Bills, eds.), World Scientific, Singapore, pp. 303–304.

Essenberg, M., Cason, E. T. Jr., Hamilton, B., Brinkerhoff, L. A., Gholson, R. K., and Richardson, P. E. (1979a). Single cell colonies of *Xanthomonas malvacearum* in susceptible and immune cotton leaves and the local resistant response to colonies in immune leaves. *Physiol. Plant Pathol. 15*(1): 53–68.

Essenberg, M., Hamilton, B., Cason, E. T. Jr., Brinkerhoff, L. A., Gholson, R. K., and Richardson, P. E. (1979b). Localized bacteriostasis indicated by water dispersal of colonies of *Xanthomonas malvacearum* within immune cotton leaves. *Physiol. Plant Pathol. 15*(1): 69–78.

Essenberg, M., Doherty, M. d'A., Hamilton, B. K., Henning, V. T., Cover, E. C., McFaul, S. J., and Johnson, W. M. (1982). Identification and effects on *Xanthomonas campestris* pv. *malvacearum* of two phytoalexins from leaves and cotyledons of resistant cotton. *Phytopathology 72*(10): 1349–1356.

Essenberg, M., Stoessl, A., and Stothers, J. B. (1985). The biosynthesis of 2,7-dihydroxycadalene in infected cotton cotyledons: the folding pattern of the farnesol precursor and possible implications for gossypol biosynthesis. *J. Chem. Soc. Chem. Commun. 9*: 556–557.

Essenberg, M., Grover, P. B. Jr., and Cover, E. C. (1990). Accumulation of antibacterial sesquiterpenoids in bacterially inoculated *Gossypium* leaves and cotyledons. *Phytochemistry 29*(10): 3107–3113.

Essenberg, M., Pierce, M. L., Cover, E. C., Hamilton, B., Richardson, P. E., and Scholes, V. E. (1992a). A method for determining phytoalexin concentrations in fluorescent, hypersensitively necrotic cells in cotton leaves. *Physiol. Mol. Plant Pathol. 41*(2): 101–109.

Essenberg, M., Pierce, M. L., Hamilton, B., Cover, E. C., Scholes, V. E., and Richardson, P. E. (1992b). Development of fluorescent, hypersensitively necrotic cells containing phytoalexins adjacent to colonies of *Xanthomonas campestris* pv. *malvacearum* in cotton leaves. *Physiol. Mol. Plant Pathol. 41*(2): 85–99.

Hadwiger, L. A., and Webster, D. M. (1984). Phytoalexin production in five cultivars of peas differentially resistant to three races of *Pseudomonas syringae* pv. *pisi*. *Phytopathology 74*(11): 1312–1314.

Hahn, M. G., Bonhoff, A., and Grisebach, H. (1985). Quantitative localization of the phytoalexin glyceollin I in relation to fungal hyphae in soybean roots infected with *Phytophthora megasperma* f. sp. *glycinea*. *Plant Physiol. 77*(3): 591–601.

Hunter, R. E., and Brinkerhoff, L. A. (1961). Report of the Bacterial Blight Committee, Oklahoma, *Proc. Cotton Dis. Council 21*: 7–8.

Kourany, E., Arnason, J. T., and Schneider, E. (1988). Accumulation of phototoxic thiophenes in *Tagetes erecta* (Asteraceae) elicited by *Fusarium oxysporum*. *Physiol. Mol. Plant Pathol. 33*(2): 287–297.

Langcake, P., and Pryce, R. J. (1976). The production of resveratrol by *Vitis vinifera* and other members of the Vitaceae as a response to infection or injury. *Physiol. Plant Pathol. 9*(1): 77–86.

Long, M., Barton-Willis, P., Staskawicz, B. J., Dahlbeck, D., and Keen, N. T. (1985). Further studies on the relationship between glyceollin accumulation and the resistance of soybean leaves to *Pseudomonas syringae* pv. *glycinea*. *Phytopathology, 75*(2): 235–239.

Mansfield, J. W. (1982). The role of phytoalexins in disease resistance. *Phytoalexins* (J. A. Bailey and J. W. Mansfield, eds.), Blackie, London, pp. 253–288.

Mansfield, J. W., Hargreaves, J. A., and Boyle, F. C. (1974). Phytoalexin production by live cells in broad bean leaves infected with *Botrytis cinerea*. *Nature* *252*(5481): 316–317.

Masciadri, R., Angst, W., and Arigoni, D. (1985). A revised scheme for the biosynthesis of gossypol. *J. Chem. Soc. Chem. Commun.* *22*: 1573–1574.

Mayama, S., and Tani, T. (1982). Microspectrophotometric analysis of the location of avenalumin accumulation in oat leaves in response to fungal infection. *Physiol. Plant Pathol.* *21*(2): 141–149.

Metlitskii, L. V., Ozeretskovskaya, O. L., Chalenko, G. I., Karavayeve, K. A., and Vasiukova, N. I. (1985). Methodological approaches to studying potato phytoalexins. *Mikologia i Fitopathologia* *19*(5): 429–435.

Morgham, A. T., Richardson, P. E., Essenberg, M., and Cover, E. C. (1988). Effects of continuous dark upon ultrastructure, bacterial populations and accumulation of phytoalexins during interactions between *Xanthomonas campestris* pv. *malvacearum* and bacterial blight susceptible and resistant cotton. *Physiol. Mol. Plant Pathol.* *32*(1): 141–162.

Munck, S. L., and Croteau, R. (1990). Purification and characterization of the sesquiterpene cyclase patchoulol synthase from *Pogostemon* cablin. *Arch. Biochem. Biophys.* *282*(1): 58–64.

Pierce, M., and Essenberg, M. (1987). Localization of phytoalexins in fluorescent mesophyll cells isolated from bacterial blight-infected cotton cotyledons and separated from other cells by fluorescence-activated cell sorting. *Physiol. Mol. Plant Pathol.* *31*(2): 273–290.

Shevell, J. L. (1985). Phytoalexin production in congenic cotton lines challenged with races of *Xanthomonas campestris* pv. *malvacearum*, M.Sc. thesis, Oklahoma State University, Stillwater.

Snyder, B. A., Leite, B., Hipskind, J., Butler, L. G., and Nicholson, R. L. (1991). Accumulation of sorghum phytoalexins induced by *Colletotrichum graminicola* at the infection site. *Physiol. Mol. Plant Pathol.* *39*(6): 463–470.

Steidl, J. R. (1988). Synthesis of 2,7-dihydroxycadalene, a cotton phytoalexin; photo-activated antibacterial activity of phytoalexins from cotton; effect of reactive oxygen scavengers and quenchers on biological activity and on two distinct degradation reactions of DHC. PhD thesis, Oklahoma State University, Stillwater.

Sun, T. J., Essenberg, M., and Melcher, U. (1989). Photoactivated DNA nicking, enzyme inactivation, and bacterial inhibition by sesquiterpenoid phytoalexins from cotton. *Mol. Plant–Microbe Interact.* *2*(3): 139–147.

Sun, T. J., Melcher, U., and Essenberg, M. (1988). Inactivation of cauliflower mosaic virus by photoactivatable cotton phytoalexin. *Physiol. Mol. Plant Pathol.* *33*(1): 115–126.

Taylor, J. A., and West, D. W. (1980). The use of Evan's blue stain to test the survival of plant cells after exposure to high salt and high osmotic pressure. *J. Exp. Bot.* *31*(121): 571–576.

Yoshikawa, M., Yamauchi, K., and Masago, H. (1978). Glyceollin: its rôle in restricting fungal growth in resistant soybean hypocotyls infected with *Phytophthora megasperma* var. *sojae*. *Physiol. Plant Pathol.* *12*(1): 73–82.

10

Chemistry, Biology, and Role of Groundnut Phytoalexins in Resistance to Fungal Attack

P. V. Subba Rao*
Centre de Coopération Internationale en Recherche Agronomique pour le Développement (CIRAD), Montpellier, France

Richard N. Strange
University College London, London, England

I. INTRODUCTION

A. The Groundnut Crop

Groundnuts (*Arachis hypogaea* L.), or "peanuts" as they are known in the United States, are an important oilseed crop belonging to the Fabaceae (Leguminosae). They are cultivated in most of the tropical regions of the world, where rainfall is adequate, lying between the latitudes 40°N and 40°S (McDonald, 1984). The total area devoted to groundnut cultivation is approximately 20 million ha (FAO, 1990; Wynne et al., 1991) and the average annual world production is estimated at 23.1 million t of pods (FAO, 1990).

The growth habit of the plant is unusual in that the pods are borne underground. After fertilization, the gynophore (or peg) elongates toward the ground and, on penetrating the soil, begins to form a pod. Once the kernels have differentiated the pod tissue becomes woody and dies, giving rise to the familiar pitted structure of the mature groundnut.

B. Production Constraints

Only a third of the world production of groundnuts comes from large commercial plantations, the remainder being produced by small holdings in developing countries where yields are extremely low, averaging 780 kg/ha.

**Current affiliation*: University College London, London, England

Low yields are caused by variety of constraints, among which pests and diseases are the most significant (Gibbons, 1980). In contrast, yields are far higher in developed countries, averaging 3000 kg/ha, and some cultivars are capable of yields as high as 5000 kg/ha under ideal growing conditions (McDonald, 1984).

Groundnuts are host to approximately 50 genera of fungi, 1 bacterium, 15 viruses, 16 nematodes, and 2 phanerogamic parasites, as well as a number of insect pests; they are also susceptible to physiological disorders such as deficiency diseases and injuries caused by climatic factors (Jackson and Bell, 1969; Feakin, 1973; COPR, 1981; Porter et al., 1984; Subrahmanyam and Ravindranath, 1988). Many of these agents are major constraints to groundnut production worldwide and reduce yields as well as quality of the crop substantially.

Two of the more serious disorders of groundnuts are leaf spot diseases and infection of kernels by fungi of the *Aspergillus flavus* group. The main causal agents of leaf spot diseases are *Cercospora arachidicola* (early leaf spot), *Phaeoisariopsis personata* (late leaf spot), and rust, *Puccinia arachidis*, which can cause losses of up to 50% (Subrahmanyam and McDonald, 1983, 1987). In addition, web blotch, caused by *Phoma arachidicola*, is important in some African countries (Cole, 1982). The *A. flavus* group of fungi consists of *A. flavus* itself and the closely related *A. parasiticus*. They produce powerful mycotoxins which can have serious consequences if consumed (see below).

C. Disease Control Strategies

Effective chemical control measures are available for leaf spot diseases but not for infection of kernels by the *A. flavus* group of fungi, which often occurs in the soil. Use of chemical methods to control leaf spot diseases is limited in developing countries for several reasons including lack of money to pay the high costs of the chemicals, lack of local availability of the chemicals themselves and the means of applying them, and lack of knowledge about their use. Hence, control measures have been directed to selecting and breeding disease-resistant cultivars.

Although disease control by breeding resistant cultivars can be very effective and is particularly appropriate for developing countries, the process is painfully long. Moreover, resistance genes for each disease and pest have to be introduced separately into agronomically acceptable cultivars. One way in which this process might be accelerated is to enhance the plant's defense mechanisms.

In this chapter we review one defense mechanism, the phytoalexin response, and suggest ways in which it might be exploited to increase the resistance of groundnuts to its more important parasites.

II. PHYTOALEXINS

A. Definition and Role in Defense

Phytoalexins are low molecular weight antimicrobial compounds that are synthesized by and accumulated in plants in response to microbial challenge (Paxton, 1981). Although the role of these compounds in resistance is not included in this definition, there is now ample evidence to suggest that rapid accumulation of sufficient concentration of phytoalexin in the vicinity of the challenging parasite is inhibitory for its growth and is crucial for defense (reviewed by Mansfield, 1982; Strange, 1992). Such evidence comes from two types of experiments: biochemical and genetic.

Biochemical evidence demonstrates that phytoalexins accumulate to inhibitory concentrations at the time that the parasite ceases to grow. For example, Hahn et al. (1985) showed, using a radioimmune assay, that within 8 hr of inoculation the concentration of the soybean phytoalexin, glyceollin I, exceeded the amount required to inhibit the in vitro growth of *Phytophthora megasperma* f. sp. *glycinea* in roots of a resistant cultivar but not in those of a susceptible cultivar. Similar results were also obtained for interactions of soybean roots and the soybean cyst nematode, *Heterodera glycines*; again using a radioimmune assay, Huang and Barker (1991) found that concentrations of glyceollin I in the head region of the nematode reached 0.3 μmol/ml by 24 hr after inoculation in a resistant cultivar whereas none was found in a susceptible cultivar.

Genetic evidence for the role of phytoalexins in resistance has been reviewed by VanEtten et al. (1989) with particular reference to the pea parasite, *Nectria haematococca*. In brief, only strains of the fungus that were tolerant of the pea phytoalexin pisatin and were able to degrade it by demethylation were virulent. When these were crossed with strains that were avirulent and sensitive to pisatin, the progeny generally fell into the two parental phenotypes, i.e. 1.) those that were tolerant of pisatin, were able to degrade it, and were virulent (Pda^+), and 2.) those that were sensitive to pisatin, could not degrade it, and were avirulent (Pda^-).

The importance of pisatin demethylase, the enzyme responsible for the degradation of pisatin, has been corroborated by experiments in which a Pda^- isolate was transformed with a fragment of DNA containing a gene which conferred the ability to demethylate pisatin (Ciufetti et al., 1988). Three of the transformants were more tolerant of pisatin and two were significantly more pathogenic to pea.

Recently, VanEtten's team (Miao et al., 1991) mapped one *Pda* gene (*Pda6*) to a small, meiotically unstable chromosome which was dispensable for normal growth. They point out that these properties are characteristic of B chromosomes and suggest that such genetic elements might be an important means by which variation could be generated in pathogenic fungi. This would be of particular significance if such genetic elements were laterally transmissible and could confer phytoalexin tolerance to other fungi in a manner analogous to R plasmids in bacteria (Miao et al., 1991).

B. Chemistry of Groundnut Phytoalexins

When a phytoalexin is first suspected to be involved in the defense of a plant against parasitic attack it is normally necessary to be able to extract it and to detect it in the extract. Unless the compound is already known, detection inevitably involves bioassay with a suitable test microorganism. The next step is to isolate the active compound and determine its chemical characteristics. Once this has been achieved, physicochemical techniques or radioimmune assays may be used for quantifying it in plant extracts.

Extraction and Partial Purification

Both groundnut kernels (cotyledons) and groundnut leaves present problems for the extraction of phytoalexins. Cotyledons are rich in oil and this can interfere with the subsequent chromatography if it is not removed. Chlorophyll can also interfere with chromatographic procedures if too much is extracted when making preparations from leaves.

Arora and Strange (1991) compared three methods for extracting cotyledons. The cotyledons were either 1.) vacuum-infiltrated with acetonitrile and incubated for 48 hr at 25°C, 2.) homogenized in 80% acetone, or 3.) homogenized in 95% ethanol.

Leaves were extracted by vacuum infiltration of either 50% methanol (Subba Rao et al., 1988a, 1990, 1991) or 60% ethanol (Edwards and Strange, 1991). Higher concentrations of the alcohols resulted in the extraction of too much chlorophyll.

The phytoalexins were partially purified by either solid phase extraction or solvent partitioning. In solid phase extraction, the acetonitrile extracts of cotyledons were diluted to 25% acetonitrile with water and applied to a cartridge containing ODS silica. After washing with 25% acetonitrile, phytoalexins were eluted from the cartridge in 100% acetonitrile. Ethanol extracts of leaves were treated similarly but with this material the original 60% ethanol concentration of the extractant was reduced to 25% by film evaporation before application to the cartridge.

When phytoalexins were partially purified by solvent partitioning, the organic solvent was first removed from extracts by film evaporation. The resulting aqueous preparations were partitioned three times against hexane to remove oils and then three times against ethyl acetate (Subba Rao et al.,

1988a, 1990, 1991; Edwards and Strange, 1991; Arora and Strange, 1991). Ethyl acetate extracts were usually dried over anhydrous sodium sulfate before reduction of the volume by film evaporation. (See Fig. 1 for a summary of these techniques.)

Bioassay

Phytoalexins are often detected initially by their antifungal activity. The most widely used technique is the thin layer chromatography (TLC) bioassay (Homan and Fuchs, 1970). Solutions of phytoalexins extracted from the plant are spotted on a TLC plate and chromatographed. After removal of the solvent, the plate is sprayed with a suspension of spores of a darkly pigmented fungus in a nutrient solution and the plate incubated at high humidity and suitable temperatures for growth of the fungus for 48–72 hr. The phytoalexins appear as white spots devoid of fungal growth whereas the remainder of the plate is covered by a dark mat of fungal mycelium (Fig. 2). Suitable test fungi are *Cladosporium cucumerinum* (Aguamah et al., 1981; Edwards and Strange, 1991), *C. herbarum* (Ingham, 1976), and *C. cladosporioides* (Subba Rao et al., 1988a, 1990, 1991).

Chemical and Physical Detection

Chromogenic Reagents. Many phytoalexins are phenolic and therefore form colored reaction products with reagents such as diazotized *p*-nitroaniline (DPN), diazotized sulfanilic acid (DSA), and Gibbs reagent. These color reactions can be an aid to identification, particularly when combined with Rf data (Ingham, 1982). Other reagents such as iodine vapor and antimony chloride ($SbCl_3$), which detect lipoidal phytoalexins, have also been used (Subba Rao et al., 1988a).

UV Properties. Some phytoalexins appear as green or light or dark blue spots on chromatograms when illuminated with long-wavelength UV light but many do not. The UV absorption properties of solutions of pure compounds can provide useful information for identification (Ingham, 1982).

Identification

Once pure preparations of phytoalexins have been obtained, mass and nuclear magnetic resonance spectrometry are powerful means for their identification and generally provide enough data to propose a structure if the compound is new. For example, these techniques were used to deduce the structure of the arachidins from groundnut cotyledons (Keen and Ingham, 1976; Aguamah et al., 1981; Cooksey et al., 1988).

AGUAMAH ET AL., 1981

Groundnut kernels soaked in H_2O

↓

Cut into 1-2 mm thick slices

Incubated 48 h in dark at 25°C

↓

Homogenized in 95% EtOH

↓

Filtered and evaporated at 40°C

↓

Partitioned three times with petroleum ether b.p. 40-60°C

↓

Aqueous fraction | Petroleum fraction (discarded)

↓

Partitioned three times with EtOAc

↓

EtOAc combined fractions | Aqueous fraction (discarded)

↓

Preparative HPLC on Si column (mobile phase, Petroleum ether b.p. 40-60°C:EtOAc - 7:3 v/v)

↓

Active fractions dissolved in acetonitrile

↓

Analytical HPLC ODS column (mobile phase MeCN-H_2O 1:1 V/V)

Figure 1 Methods for the extraction, separation, and quantitative analysis of groundnut phytoalexins.

SUBBA RAO ET AL. 1988a, 1990, 1991

Infected leaves ground in 50% MeOH

↓

Incubated at room temperature in dark for 48 h

↓

Filtered and reduced volume to ¼ *in vacuo* at 40°C

↓

Centrifuged 10 min (15 000g)

↓ Pellet (discarded)

↓ Aqueous fraction

↓

Partitioned three times with hexane

↓ Hexane fraction

↓ Aqueous fraction

↓

Partition three times with EtOAc

↓ Aqueous phase (discarded)

↓ EtOAc fractions combined

↓

Open column chromatography on Si gel (mobile phase $CHCl_3$-MeOH 50:1 and 50:10 or Hexane:EtOAc:MeOH - 50:50:5 (V/V))

↓

Analytical HPLC on a Si column (mobile phase Hx-EtOAc 65:35 V/V or a methanol gradient in Hexane)

↓

Analytical HPLC on a C_{18} column (mobile phase a gradient of methanol in 1% acetic acid from 50-99% in 20 min).

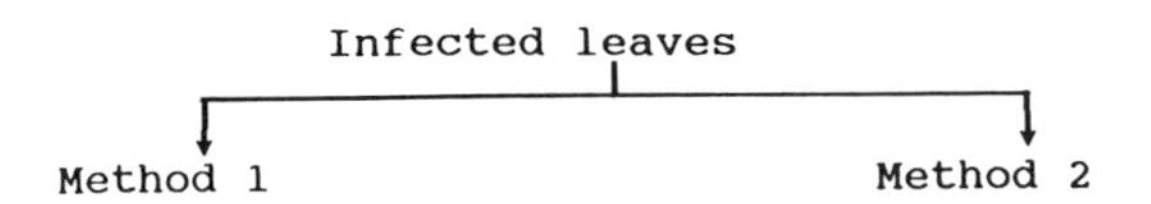

Method 1

Air-dried samples vacuum infiltrated with 60% EtOH

↓

EtOH removed *in vacuo* at 40°C

↓

Aqueous solution partitioned three times with EtOAc. Aqueous phase discarded

↓

Combined EtOAc fractions evaporated to dryness and dissolved in $CHCl_3$

↓

Flash chromatography on a silica gel column: phytoalexins eluted in a stepwise gradient of EtOAc in hexane

↓

Semi-preparative HPLC on a C_{18} column: phytoalexins eluted in acetonitrile-H_2O 1:1 (V/V)

Method 2

Samples vacuum infiltrated with 60% EtOH

↓

EtOH diluted to 25%

↓

Solid phase cartridge (500 mg) Techoprep C_{18}: phytoalexins eluted in acetonitrile

Methods 1 and 2 combined

↓

Analytical HPLC on a C_{18} column: phytoalexins eluted in a gradient of acetonitrile in 1% acetic acid

Figure 1 Continued.

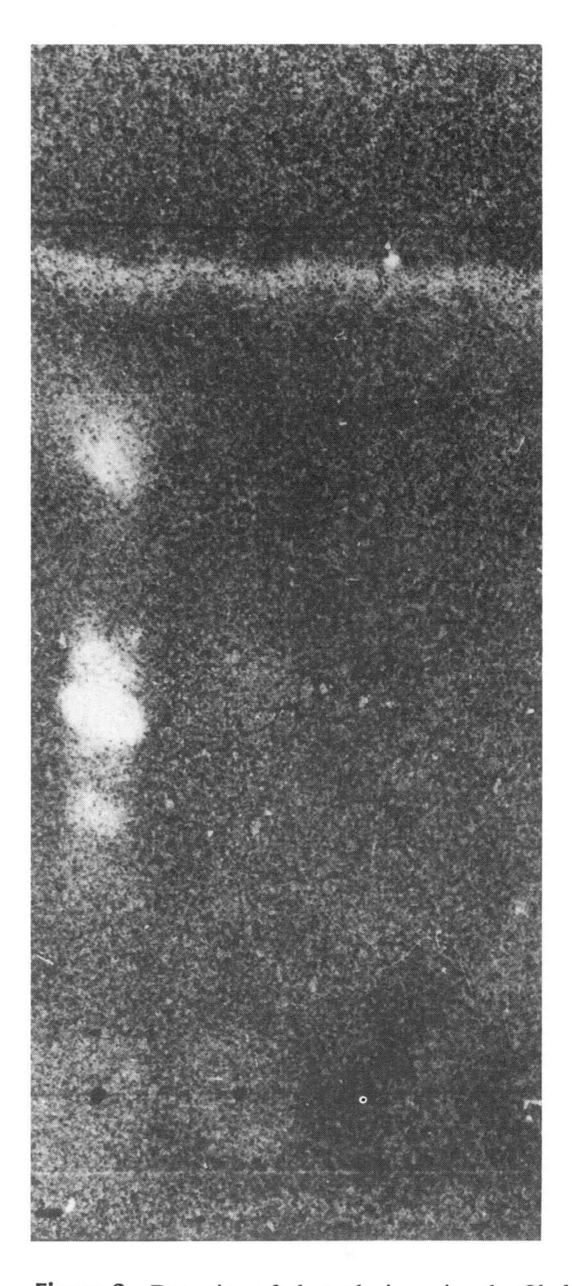

Figure 2 Detection of phytoalexins using the *Cladosporium* TLC test. White areas indicate the location of phytoalexins where fungal growth was inhibited.

Table 1 Phytoalexins Isolated from Groundnut

Compound	Tissue	Method of elicitation	Ref.
Stilbenes			
1. 3,5,4′-Trihydroxystilbene (resveratrol)	Cotyledons	(i) *Helminthosporium carbonum*	Ingham, 1976; Cooksey et al., 1988
		(ii) Wounding	
2. Arachidin I (4-(3-methylbut-1-enyl)-3,5,4′-tetrahydroxystilbene)	″	Wounding and natural microflora	Aguamah et al., 1981
3. Arachidin II (4-(3-methylbut-2-enyl)-3,5,4′-trihydroxystilbene)	″	Wounding and natural microflora	Keen and Ingham, 1976; Aguamah et al., 1981
4. Arachidin III (4-(3-methylbut-1-enyl)-3,5,4′-trihydroxystilbene)	″	Wounding and natural microflora	Aguamah et al., 1981
5. Arachidin IV (3-isopentadienyl-4,3′,5′-trihydroxystilbene)	″	(i) Wounding under sterile conditions	Cooksey et al., 1988
	Leaves	(ii) *Puccinia arachidis*	Subba Rao, 1987
Flavonoids			
6. Medicarpin	″	(i) *Cercospora arachidicola*	Strange et al., 1985
	″	(ii) *Phoma arachidicola*	″
	″	(iii) *Puccinia arachidis*	Strange, 1987b
			Subba Rao, 1987

7. Demethylmedicarpin	″	*Cercospora arachidicola*	Edwards and Strange, 1991
8. Daidzein	″	″	Edwards and Strange, 1991
9. Formononetin	″	″	Edwards and Strange, 1991
10. 7,4′-Dimethoxy-2′-hydroxy-isoflavanone	″	″	Edwards and Strange, 1991
11. 7,2′-Dihydroxy-4′-methoxy-isoflavanone	″	″	Edwards and Strange, 1991
Fatty acid–related compounds			
12. Nonyl phenol	″	*Puccinia arachidis*	Subba Rao et al., 1988b
13–15. Alkyl *bis* phenyls	″	″	″
16. Methyl linolenate	″	″	″
17. 13-Hydroxyoctadecadiene-9,11-methyloate	″	″	″
18. 9-Hydroxyoctadecadiene-10,12 methyloate	″	″	″
19. 1,2,3(2-Acetyloxy)tricarboxylic propanic acid	″	″	″

Quantitative Determination

For many phytoalexins, high-performance liquid chromatography (HPLC) is the method of choice (Strange, 1987; Subba Rao et al., 1988a, 1990, 1991; Edwards and Strange, 1991; Arora and Strange, 1991). The compounds are usually separated by reverse phase chromatography using an ODS silica column and quantitated by reference to internal standard compounds and/or external standards of the authentic phytoalexins.

Although no work has yet been achieved with a radioimmunoassay of groundnut cotyledons, this sensitive technique will undoubtedly be of importance for establishing whether phytoalexins accumulate at the right time and in the right place to explain the cessation of growth of invading microorganisms (Hahn et al., 1985; Huang and Barker, 1991).

Structures and Chemical Characteristics

The compounds so far identified as phytoalexins from groundnut fall into three main classes: stilbenes, flavonoids, and compounds related to long-chain fatty acids (Table 1 and Fig. 3). Chromatographic data and their reactions to chromogenic reagents are given in Table 2. Spectral characteristics are found in Table 3.

C. Biology of Groundnut Phytoalexins

Elicitation

Wounding of sterile tissue caused elicitation of stilbene phytoalexins in groundnut cotyledons (Aguamah et al., 1981; Narayanaswamy and Mahadevan, 1983; Arora and Strange, 1991). Subba Rao (1987) also found that the wounding caused by detaching leaflets from plants elicited phytoalexins.

Although salts of heavy metals and UV light have commonly been used to elicit phytoalexins in other plants, Edwards and Strange (1991) found that these techniques only resulted in low levels of accumulation in groundnut leaflets. Accordingly, these and other workers have used fungal challenge as a means of elicitation. For example, Ingham (1976) elicited phytoalexins in groundnut leaflets by inoculation with the nonpathogen *Helminthosporium carbonum*, and similarly, Narayanaswamy and Mahadevan (1983) used the nonpathogens *Curvularia spicata* and *Helminthosporium oryzae*. Variation in phytoalexin yield occurred according to the challenging organism. For example, *C. spicata* induced 33 μg of *trans*-resveratrol/ml of diffusate, whereas *H. oryzae* induced 667 μg/ml (Narayanaswamy and Mahadevan, 1983).

Substantial yields of phytoalexins have been obtained from plants that were naturally infected with major fungal pathogens such as *Cercospora arachidicola* (Edwards and Strange, 1991), *Phoma arachidicola* (Strange et

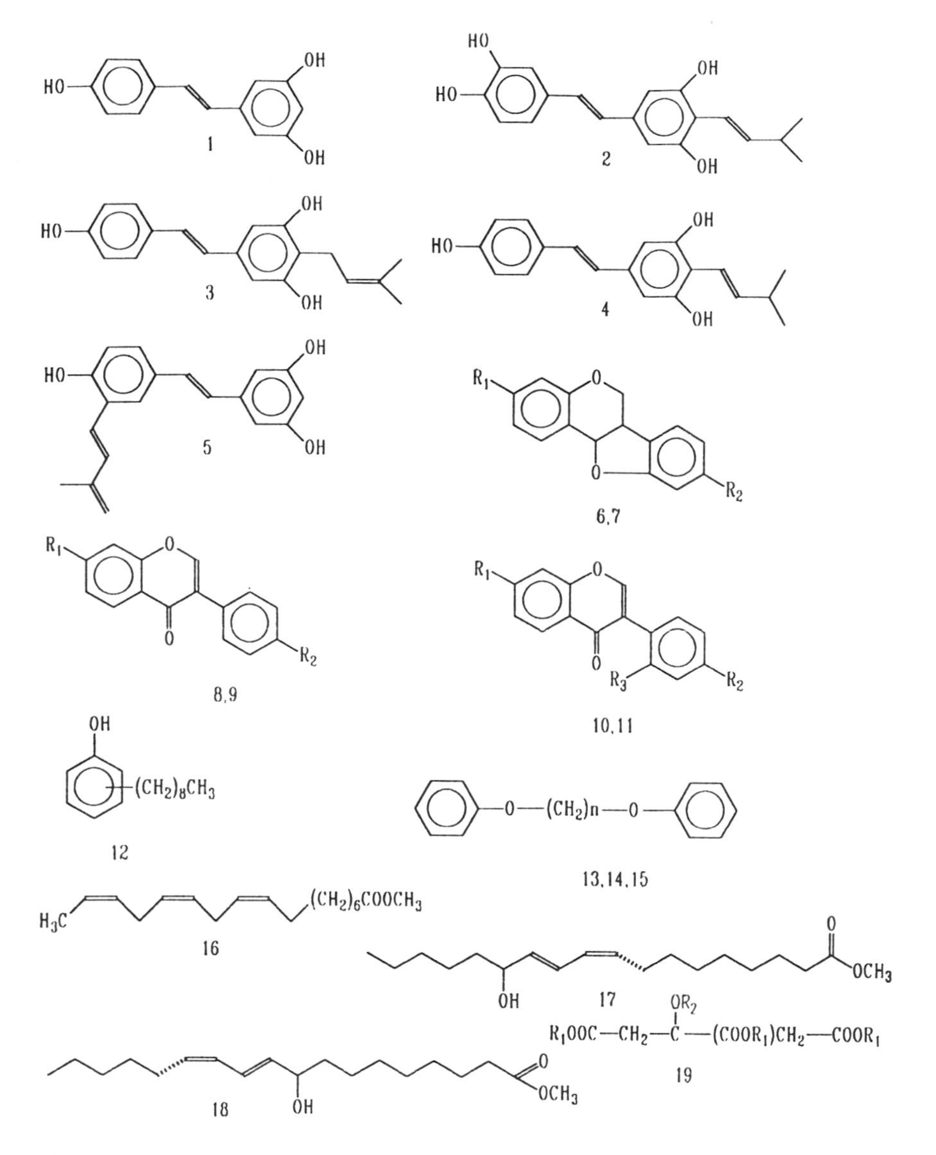

Figure 3 Structures of groundnut phytoalexins. 1, Resveratrol. 2, Arachidin I (4-(3-methyl-but-1-enyl)-3,5,3′4′-tetrahydroxystilbene). 3, Arachidin II (4-(3-methyl-but-2-enyl)-3,5,4′-trihydroxystilbene). 4, Arachidin III (4-(3-methyl-but-1-enyl)-3,5,4′-trihydroxystilbene. 5, Arachidin IV (3-isopentadienyl-4,3′,5′-trihydroxystilbene). 6, R_1 = OH, R_2 = OCH_3 medicarpin. 7, R_1 = R_2 = OH demethylmedicarpin. 8, R_1 = R_2 = OH daidzein. 9, R_1 = OH, R_2 = OCH_3 formononetin. 10, R_1 = R_2 = OCH_3, R_3 = OH 7,4′-dimethoxy-2′-hydroxyisoflavanone. 11, R_1 = R_3 = OH, R_2 = OCH_3 7,2′-dihydroxy-4′-methoxyisoflavanone. 12, Nonyl phenol. 13–15, Alkyl *bis* phenyls (n = 11, 12, or 13). 16, Methyl linolenate. 17, 13-Hydroxyoctadecadiene-9,11-methyloate. 18, 9-Hydroxyoctadecadiene-10,12-methyloate. 19, R_1 = C_4H_9, R_2 = $COCH_3$ 1,2,3(2-acetyloxy)tricarboxylic propanic acid.

Table 2 Chromatographic Characteristics of Groundnut Phytoalexins and Their Reactions with Chromogenic Reagents

Compound	R_f	Solvents for Si-gel TLC	Retention time on HPLC and chromatographic conditions	Reaction to:		
				$SbCl_3$	DPN	DSA
Stilbenes						
1. *cis*- and *trans*-3,5,4′-Trihydroxy-stilbene (resveratrol) (Ingham, 1976; Cooksey et al., 1988)	0.11	$CHCl_3$:MeOH (100:04 v/v)	NR: Spherisorb ODS (250 × 4.6 mm i.d.), gradient 40–65% acetonitrile in acetic acid	NT	Orange	NT
2. Arachidin I (Aguamah et al., 1981)			8.0 min: Hypersil ODS (250 × 4.6 mm i.d.), acetonitrile–water 1:1 (v/v), FR: 1 ml/min	NT	NT	NT
3. Arachidin II (Aguamah et al., 1981)			9.3 min: Conditions as for Arachidin I	NT	NT	NT
cis- and *trans*-Arachidin II	0.14	$CHCl_3$:MeOH (100:03 v/v)		NT	NT	Bright yellow
(Narayanaswamy and Mahadevan, 1983)	0.28	″		NT	NT	″
4. Arachidin III (Aguamah et al., 1981)			10.5 min: Conditions as for Arachidin I, FR: 4 ml/min	NT	NT	NT
5. Arachidin IV			NR: Conditions as for resveratrol			

Flavonoids						
6. Medicarpin	0.55	$CHCl_3$:MeOH (50:01 v/v)	7.2 min: Spherisorb ODS (250 × 4.6 mm i.d.)	NT	Yellow	NT
	0.62	Benzene:MeOH (9:1 v/v)	CH_3CN–H_2O (1:1 v/v)			
	0.71	EtOAc:Cyclo Hx (1:1 v/v)	21.90 min: Spherisorb ODS (250 × 4.6 mm i.d.), acetonitrile from 35% to 40% in 12 min and to 75% in 23 min in 1% acetic acid, FR: 1.5 ml/min	NT	Pale orange	NT
7. Demethylmedicarpin	0.54	EtOAc:Cyclo Hx (1:1 v/v)	11.20 min: Conditions as for medicarpin	NT	Bright orange	NT
8. Daidzein	–	–	9.10 min: Conditions as for medicarpin	NT	NT	NT
9. Formononetin	0.42	As above	19.02 min: Conditions as for medicarpin	NT	Pale orange	NT
10. Isoflavanone 1 (7,4′-dimethoxy-2′-hydroxyisoflavanone)	0.46	EtOAc:Cyclo Hx (1:1 v/v)	21.00 min: Conditions as for medicarpin	NT	Yellow–orange	NT
11. Isoflavanone 2 (7,2′-dihydroxy-4′-methoxyisoflavanone)	–	–	13.38 min: Conditions as for medicarpin	NT	NT	NT

Table 2 Continued

Compound	R_f	Solvents for Si-gel TLC	Retention time on HPLC and chromatographic conditions	Reaction to:		
				$SbCl_3$	DPN	DSA
Fatty acid related compounds						
12. Nonyl phenol (Subba Rao, 1987)	0.85	Hx:EtOAc:MeOH (60:40:01 v/v)	18 min: Silica (300 × 9.5 mm i.d.), (Hx-EtOAC 95:05 v/v), FR:1 ml/min	Brown	NT	NT
	0.63	Hx:EtOAc (3:1 v/v)				
13–15. Alkyl *bis* phenyls Subba Rao et al., 1991)	0.22	$CHCl_3$:MeOH (50:02 v/v)	55 min: Silica (300 × 9.5 mm i.d.), Hx followed by EtOAc, gradient 0–30% in Hx: Fr: 1 ml/min	Yellow	NT	NT
16. Methyl linolenate (Subba Rao et al., 1988 a)	0.90	Hx:EtOAc:MeOH (50:50:05 v/v)	15.2 min: Silica C_{18} (300 × 9.5 mm i.d.), MeOH–H_2O–CH_3COOH (50:50:01 to 99:0:01 v/v) in 20 min, FR: 1 ml/min	Brown–yellow		NT
	0.44	$CHCl_3$:MeOH (50:02 v/v)				
17–18. Dienols (Subba Rao et al., 1990)	0.87–0.90	Hx:EtOAc:MeOH (50:50:05 v/v)		Dark brown	NT	NT
	0.48–0.52	Hx:EtOAc (5:1 v/v)				
	0.65–0.67	″				
19. 1,2,3-(2- acetyloxy) tricarboxylic propanic acid (Subba Rao, 1987)	0.45	Hx:EtOAc (3:1 v/v)	16 min: silica (300 × 9.5 mm i.d.), Hx-EtOAc (100:20 v/v)	Dark brown	NT	NT

DPN, diazotized *p*-nitroaniline; DSA, diazotized sulfanilic acid; FR, flow rate; Hx, hexane; NR, not recorded; NT, not tested.

Table 3 Absorption Spectra of Groundnut Phytoalexins

Compound	Absorption (γ_{max}, nm)
Stilbenes	
1. 3,5,4′-Trihydroxystilbene (resveratrol)	281, 297 (sh), 306, 320, 336 (sh)
2. Arachidin I (4-(3-methylbut-1-enyl)-3,5,3′,4′-tetrahydroxystilbene)	220, 245 (sh), 310 (sh), 346 (sh)
3. Arachidin II (4-(3-methylbut-2-enyl)-3,5,4′-trihydroxystilbene)	220, 295 (sh), 307, 324, 340 (sh)
4. Arachidin III (4-(3-methylbut-1-enyl)-3,5,4′-trihydroxystilbene)	219, 241 (sh), 327 (sh), 331, 346 (sh), 364 (sh)
5. Arachidin IV (3-isopentadienyl-4,3′,5′-trihydroxystilbene)	296
Flavonoids	
6. Medicarpin	282 (sh), 287
7. Demethylmedicarpin	282 (sh), 287
8. Daidzein	212, 238 (sh), 249, 262 (sh), 305
9. Formononetin	210, 238 (sh), 249, 262 (sh), 305
10. 7,4′-Dimethoxy-2′-hydroxyisoflavanone	277, 311
11. 7,2′-Dihydroxy-4′-methoxyisoflavanone	275, 311
Fatty acid–related compounds	
12. Nonyl phenol	–
13–15. Alkyl *bis* phenyls	–
16. Methyl linolenate	230, 286
17. 13-Hydroxyoctadecadiene-9, 11-methyloate	–
18. 9-Hydroxyoctadecadiene-10,12-methyloate	–
19. 1,2,3(2-Acetyloxy)tricarboxylic propanic acid	–

al., 1985), *Puccinia arachidis* (Subba Rao et al., 1988a, 1990, 1991), and *Rhizoctonia bataticola* (Narayanaswamy and Mahadevan, 1983).

Two other factors that profoundly affect phytoalexin elicitation and accumulation are the genotype of the plant and environmental conditions. These topics will be discussed later along with the role of phytoalexins in defense.

Biosynthesis and Degradation

Elicitor treatment causes the derepression of the biosynthetic pathway leading to phytoalexin synthesis. Once synthesized the phytoalexin may be subject to degradation by either host or parasite (Yoshikawa, 1983; VanEtten et al., 1989).

The phytoalexins of groundnuts so far described are stilbenes, flavonoids, and compounds related to long-chain fatty acids (Table 1 and Fig. 3). A considerable amount of information about the biosynthetic pathways for flavonoid and stilbene phytoalexins is available and the reader is referred to the reviews of Hahlbrock and Griesbach (1975, 1979), Stoessl (1982), Dewick (1982), Dixon et al. (1983), Ebel and Hahlbrock (1983), Kindl (1985), Ebel (1986), Smith and Banks (1986), Keen (1986), and VanEtten et al. (1989).

In brief, isoflavonoids and stilbenes are formed by the shikimic-polymalonate acid route. The first step is the synthesis of cinnamic acid from phenylalanine catalyzed by phenyl alanine ammonia-lyase. This is then converted to 4-coumaroyl-CoA by the action of cinnamate hydroxylase and 4-coumarate:CoA ligase. Reaction of 4-coumaroyl-CoA with three units of malonyl-CoA gives an intermediate which can cyclize in two different ways leading to two series of compounds. Isoflavonoids are found in one of these series and stilbenes in the other. Medicarpin, an important isoflavonoid phytoalexin of groundnut leaves, is probably synthesized via daidzein and formononetin, compounds that have also been found in leaves infected by *Cercospora arachidicola* (Edwards and Strange, 1991). The alternative cyclization leads to the biosynthesis of stilbene phytoalexins of groundnut via decarboxylation and the appropriate hydroxylation and prenylation of the 4-coumaroyl-CoA–malonyl-CoA intermediate.

At present there is little information concerning the biosynthesis of the other groundnut phytoalexins such as methyl linolenate, the dienols, and alkyl *bis* phenyl ethers. Presumably, methyl linoleate is formed according to the normal route of fatty acid synthesis, i.e., by the progressive condensation of two-carbon units derived from malonyl-CoA. Ravisé (personal communication) suggested that the dienols are also formed via this route and that the alkyl *bis* phenyl ethers are derived from acetyl-CoA.

Mechanisms for the degradation of groundnut phytoalexins are virtually unknown, although Edwards and Strange (1991) suggested that the occurrence of demethylmedicarpin in leaves infected with *Cercospora arachidicola*, by analogy with the demethylation of pisatin by *Nectria hamatococca* (VanEtten et al., 1989), was a degradation product. Demethylmedicarpin was not detected in groundnut leaves infected by rust.

Biological Activity

The spectrum of biological activity of groundnut phytoalexins has hardly begun to be explored and so far has been confined to antifungal activity, especially against parasites of groundnut (Table 4). The lowest ED_{50} value

Table 4 Antifungal Activity of Some Groundnut Phytoalexins

Phytoalexin	ED_{50} values ($\mu g/ml$) for the test fungi									
	H. carbonum		*Cladosporium* spp.		*C. arachidicola*		*A. flavus*		*P. arachidis*	
	SG	HE	SG	HE	SG	HE	SG	HE	SG	HE
1. Resveratrol	NT	50.0	NT	NT	NT	NT	NT	NT	NT	NT
2. Arachidin I	NT	NT	3.6	4.3	11.5	21.0	12.8	4.9	NT	NT
3. Arachidin II	NT	NT	7.6	22.1	25.1	63.0	12.7	6.8	NT	NT
4. Arachidin III	NT	NT	4.9	13.0	17.0	36.3	8.9	9.7	NT	NT
5. Arachidin IV	NT	NT	NT	NT	NT	NT	14.0	11.3	NT	NT
6. Medicarpin	NT	25.0	17.0	NT	5.0	V	NT	NT	NT	NT
7. Demethylmedicarpin	NT	NT	140.0	NT	92.0	NT	NT	NT	NT	NT
10. 7,4′-Dimethoxy-2′-hydroxyisoflavanone	NT	NT	NT	NT	16.0	NT	NT	NT	NT	NT
13–15. Alkyl *bis* phenyl ethers	NT	NT	NT	NT	NT	NT	NT	NT	40.0	35.0[a]
16. Methyl linolenate	NT	NT	NT	NT	NT	NT	NT	NT	37.5	NT
18, 19. Dienols	NT	NT	NT	NT	NT	NT	NT	NT	3.8	2.2[a]

[a]Tested for inhibition of germ tube growth only as *P. arachidis* is an obligate parasite and does not grow beyond appressorium formation outside the host.

H. carbonum, *Helminthosporium carbonum*; *C. arachidicola*, *Cercospora arachidicola*; *A. flavus*, *Aspergillus flavus*; *P. arachidis*, *Puccinia arachidis*; SG, spore germination; HE, hyphal extension; NT, not tested; V, variable. Compounds are numbered as in Table 1 and Fig. 3.

Source: Data taken from Cooksey et al. (1988), Edwards and Strange (unpublished), Ingham (1976), Subba Rao (1988a, 1990, 1991), and Wotton and Strange (1985, 1987).

(2.2 μg/ml) was that of the two isomeric dienols for the inhibition of germ tube growth of uredospores of *Puccinia arachidis*, and the highest (140 μg/ml) was that of demethylmedicarpin for inhibition of spore germination of *Cladosporium* spp. (Table 4).

Role in Defense

Attempts to assess the role of phytoalexins in resistance have been made mainly with leaf spot fungi and fungi of the *A. flavus* group. In order to provide a context for evaluating these data, it is worth reviewing the criteria that have been proposed to establish whether a phytoalexin can be considered to play a role in defense. These are as follows:

1. The compound must accumulate in response to infection.
2. The compound must be inhibitory to the invading organism.
3. The compound must accumulate to inhibitory concentrations in the vicinity of the parasite at the time the parasite ceases growing.
4. Varying the rate of accumulation of the phytoalexin causes a corresponding variation in the resistance of the plant.
5. Varying the sensitivity of the invading organism causes a corresponding variation in its virulence (Strange, 1992).

Of these criteria, the first three are mandatory and the remaining two provide corroborative evidence.

Resistance to Toxigenic *Aspergillus* Species. Groundnuts are frequently infected by the mycotoxigenic fungi *A. flavus* and *A. parasiticus*. Strains of *A. flavus* normally synthesize aflatoxin B_1 and G_1 whereas strains of *A. parasiticus* usually produce aflatoxin B_1, B_2, G_1, and G_2. In addition to their toxigenicity, these compounds are reported to be carcinogenic, mutagenic, and teratogenic (Hayes, 1980). Aflatoxin B_1 is the most toxic of the four compounds and is also a powerful liver carcinogen (Heathcote and Hibbert, 1978). Infection of groundnuts usually takes place before harvest from soil infested by the fungi but may also occur during harvest and storage (Williams and McDonald, 1983; Strange, 1991).

In attempts to find a simple method for screening genotypes for resistance to mycotoxigenic *Aspergillus* spp. an in vitro technique has been established. Groundnut cotyledons are hydrated to 20% moisture and rolled in a spore suspension of the fungi before incubation. Results of such tests have shown that cultivars which are resistant in the in vitro screen are also resistant in the field (Mehan et al., 1981; Zambettakis, 1983).

Susceptibility to *Aspergillus* spp. is influenced by the environment (Strange, 1991). In particular, drought stress and high temperatures increase susceptibility (Hill et al., 1983; Cole et al., 1985; Wotton and Strange, 1985, 1987).

The evidence that phytoalexins play a role in resistance to *Aspergillus* spp. will be reviewed according to the five criteria set out in the preceding section.

1. Stilbene phytoalexins accumulated in groundnut cotyledons in response to challenge by *A. flavus* using the in vitro technique described above (Wotton and Strange, 1987; Arora and Strange, unpublished). After a lag of 2 days phytoalexins accumulated rapidly, reaching >140 μg/g fresh wt of kernels (Wotton and Strange, 1987).

2. Wotton and Strange (1985, 1987) found that arachidins I, II, and III (Fig. 3) inhibited both spore germination and hyphal extension of *A. flavus* in vitro with ED_{50} values ranging from 8.9 to 12.8 μg/ml and from 4.9 to 9.7 μg/ml, respectively (Table 4).

3. Wotton and Strange (1987) found that the fungus essentially ceased to grow after 2 days in the in vitro seed testing technique. This coincided with the time that phytoalexins began to accumulate rapidly, going from ~12 μg/g fresh wt to >50 μg/g fresh wt. This latter figure represents a concentration that is about four times the ED_{50} value.

4. Wotton and Strange (1987) and Arora and Strange (1991, and unpublished results) found both genotypic and environmentally induced variation in phytoalexin accumulation. When inoculated with *A. flavus*, the more susceptible cultivar, TMV2, accumulated resveratrol and arachidin IV, whereas a more resistant cultivar, J11, accumulated both these compounds and additionally arachidin III. Moreover, total phytoalexin in the more resistant cultivar was >300 μg/g fresh wt 120 hr after inoculation by the in vitro technique compared with <90 μg/g fresh wt for the susceptible cultivar. These differences were reflected in the accumulation of aflatoxins which reached ~150 ppb in the more resistant cultivar but >12,000 ppb in the susceptible cultivar.

Wotton and Strange (1987) confirmed that drought stress enhanced the susceptibility of groundnut kernels to *A. flavus* when tested by the in vitro technique. Reduction in water supply also inhibited the phytoalexin response of such kernels.

Since phytoalexins are elicited in groundnut kernels by wounding in the absence of pathogenic challenge it was possible to evaluate the phytoalexin response without the complications consequent on the presence of the pathogen in infection experiments. In time course studies of three cultivars, phytoalexins accumulated within 24 hr of wounding kernels and reached maxima in 96–120 hr, after which they began to decline (Wotton and Strange, 1985). Cultivars differed in their speed of response, the susceptible cultivar producing the least amount of phytoalexin at 24 hr but outstripping the other cultivars at 96 hr. In other experiments, phytoalexin accumulation, measured in a set of 10 groundnut cultivars at 24 hr after wounding, ranged from 28 to 935 μg/g fresh wt and was negatively correlated with susceptibility in the in vitro screen (see above). No such correlation was obtained with phytoalexin accumulation at 96 hr after wounding (Wotton and Strange, 1985). It seems, therefore, that resistance to *A. flavus* may be related to early accumulation of phytoalexin rather than the final amounts found after prolonged incubation. Similar conclusions have been reached by workers investigating other systems (Mansfield, 1982).

5. At present there are no data reporting variation in the sensitivity of *A. flavus* to groundnut phytoalexins.

Resistance to Leaf Spot Fungi. The possibility that phytoalexins play a role in limiting infection by the leaf spot fungi, *Cercospora arachidicola* (early leaf spot), *Phaeoisariopsis personata* (late leaf spot), *Phoma arachidicola* (web blotch), and *Puccinia arachidis* (rust) will be discussed in the context of the five criteria previously stated.

1. Cole (1982) reported that early colonization of groundnut leaflets by *C. arachidicola* prevented their subsequent parasitism by *P. arachidicola*. Juice from the *C. arachidicola*-infected leaflets, but not healthy leaflets, contained a compound that was inhibitory to *P. arachidicola*. The compound was isolated by Strange et al. (1985) and identified as medicarpin. These results suggested that the accumulation of medicarpin prevented the colonization of early leaf spot–infected leaves with web blotch.

In similar experiments Subba Rao (1987) found that the in vitro toxicity of crude extracts from resistant genotypes infected with rust were always more toxic toward uredospore germination of the fungus than those of susceptible ones. Moreover, the higher toxicity levels were maintained throughout the latent period (Subba Rao, 1987).

In further experiments Subba Rao et al. (1988b) isolated 12 antifungal compounds from rust-infected leaves and determined the structures of eight of them. Methyl linolenate, two isomeric dienols, namely 13-hydroxyoctadecadien-9,11-methyloate and 9-hydroxyoctadecadien-10,12-methyloate, a nonyl phenol, 1,2,3(2-acetyloxy)tricarboxylic propanic acid, and three alkyl *bis* phenyl ethers with aliphatic chains corresponding to 11, 12, and 13 carbon atoms were reported as phytoalexins for the first time. Medicarpin and arachidin II were also isolated but in very low concentrations (Subba Rao, 1987).

In contrast, Edwards and Strange (1991) found that concentrations of both formononetin and medicarpin exceeded 120 μg/g fresh wt in leaves infected by *Cercospora arachidicola*. These workers, in addition, found demethylmedicarpin (>50 μg/g fresh wt) a possible degradation product of medicarpin. Cole et al. (unpublished) also found medicarpin and low concentrations of demethylmedicarpin in the healthy upper leaves of plants infected naturally in the field with *C. arachidicola* or *P. arachidicola*. These leaves did not develop symptoms of disease when detached and incubated under conditions that were conducive to fungal growth. It is not known whether the phytoalexins were translocated from the lower infected leaves or whether they were synthesized in situ in response to a signal from the infected part of the plant.

2. An important component of the phytoalexin response to leaf spot diseases was generally medicarpin although this was not true for all cultivars. The ED_{50} for medicarpin when tested for inhibition of spore germination of *C. arachidicola* was 5 μg ml and for demethylmedicarpin, a probable degradation product, 92 μg ml (Table 4; Edwards, 1992). It was difficult to obtain an ED_{50} value for medicarpin against hyphal extension since the fungus degraded it to demethylmedicarpin (Edwards, 1992). Spore germination of *C. arachidicola* was also inhibited by 7,4′-dimethoxy-2′-hydroxyisoflavanone, a prominent component of the phytoalexin response of some cultivars to infection by this fungus (Table 4; Edwards, 1992).

The inhibitory activity of methyl linolenate, a mixture of the dienol isomers and a mixture of the alkyl *bis* phenyl ethers, was estimated against uredospores of *P. arachidis* in vitro. The ED_{50} values for uredospore germination were 37.5, 3.8, and 40 μg/ml for methyl linolenate, the dienols, and alkyl *bis* phenyl ethers, respectively. Similarly, the ED_{50} values for germ tube growth were 2.2 and 35 μg/ml for dienols and alkyl *bis* phenyl ether, respectively (Table 4).

3, 4, and 5. No adequate evaluation of the role of phytoalexins in resistance to leaf spot diseases has been carried out according to these criteria.

III. STRATEGIES FOR INCREASING PLANT RESISTANCE BY EXPLOITING THE PHYTOALEXIN RESPONSE

The evidence reviewed above for phytoalexin involvement in the resistance of groundnut to fungal diseases is circumstantial. A major requirement is to show that phytoalexin accumulation slows the progress of infection. As previously discussed, such evidence may be biochemical or genetic. In the biochemical approach, the concentrations of the different compounds should be assessed at infection sites in order to ascertain the most important. These could then be assayed routinely by HPLC. Alternatively, if HPLC is not sufficiently sensitive to detect inhibitory concentrations of the phytoalexins in the vicinity of the parasite, radioimmune assays could be developed.

One genetic approach to establishing phytoalexin involvement in resistance hinges on the discovery, or the development by mutation, of strains of pathogens that differ markedly in their sensitivity. A correlation of virulence with insensitivity to phytoalexin is evidence for a role of the phytoalexin in resistance (VanEtten et al., 1989). If the mechanism of tolerance is phytoalexin degradation, as seems likely in the case of *Cercospora arachidicola*, which appears to demethylate medicarpin, then transformation of nondegrading isolates with the gene for medicarpin demethylase should enhance the virulence of the recipients (cf. VanEtten et al., 1989). Another approach is to prevent the production of phytoalexins by transforming plants with an antisense gene to one of the key biosynthetic enzymes. If the phytoalexin is important to disease resistance, such transgenic plants will be more susceptible than control plants.

Once phytoalexins have been demonstrated unequivocally to be crucial in defense against infection, then it will be important to exploit the response in order to combat disease. This may be done in three ways:

1. Plants may be selected for a rapid phytoalexin response to challenge by parasites resulting in the accumulation of inhibitory concentrations of the compounds before the parasite is able to grow away from the infected area. It will be important that this response occurs under conditions that are likely to obtain in the field. Critically, with regard to infection of groundnut cotyledons by *Aspergillus* spp., the response should occur even if the plant is drought-stressed (see above; Wotton and Strange, 1987).

2. Plants may develop systemic acquired resistance to infection. For example, as reported above, leaves from the tops of groundnut plants whose lower leaves were infected by leaf spot fungi remained healthy and contained medicarpin. One possibility is that the compound was synthesized in situ in response to a signal. Recently, salicylic acid has been shown to function as such a signal for a number of defense reactions in tobacco and cucumbers (Malamy et al., 1990; Métraux et al., 1990). It would be interesting to know if salicylic acid can act in this way in groundnuts and if the response is the production of medicarpin. Such a finding would prompt experiments in determining whether the application of salicylic acid might promote resistance without seriously compromising the yield of the crop. If yields were not significantly depressed, the application of salicylic acid through a sprinkler system might be worth investigating as a possible control measure.

3. Isoflavone reductase is a key enzyme in the biosynthesis of medicarpin. In groundnuts it leads to the formation of (+)-medicarpin (Strange et al., 1985) whereas in alfalfa the enantiomer (−)-medicarpin is formed. It is unlikely that parasites that rely on the degradation of one enantiomer of medicarpin for virulence would be able to degrade the other. This could be tested with groundnut pathogens. If (−)-medicarpin is not degraded by, for example, *C. arachidicola* or *P. arachidicola* then it would be important to genetically engineer plants to produce this form of the phytoalexin by

transforming them with the isoflavone reductase from alfalfa. Such plants could then be tested for their resistance to the leaf spot fungi.

IV. EPILOG

We now have some knowledge of the chemistry of the phytoalexins produced by groundnuts. However, it is likely that this is still incomplete. Much work remains to be done in establishing the variation in the phytoalexin response among *Arachis* spp. as well as the mechanisms by which they are elicited. A further area of research is the effect of environmental conditions on the phytoalexin response, particularly drought stress.

Research aimed at assessing the role of phytoalexins in resistance is urgently needed. The evidence so far is circumstantial. Critical biochemical and genetic experiments along the lines outlined in this chapter now need to be done.

If, as seems likely, the phytoalexin response is found to play a crucial role in the resistance of groundnuts to infectious disease, then this response should be exploited by conventional plant breeding and, perhaps, by genetic engineering.

ACKNOWLEDGMENTS

The authors thank Dr. A. Ravisé (Director of Research Retd, ORSTOM, Paris) for his comments.

REFERENCES

Aguamah, G. E., Langcake, P., Leworthy, D. P., Page, J. A., Pryce, R. J., and Strange, R. N. (1981). Two novel stilbene phytoalexins from *Arachis hypogaea*. *Phytochemistry 20*: 1381–1383.

Arora, M. K., and Strange, R. N. (1991). Phytoalexin accumulation in groundnuts in response to wounding. *Plant Sci. 78*: 157–163.

Ciufetti, L. M., Welltring, K.-M., Turgeon, B. G., Yoder, O. C., and VanEtten, H. D. (1988). Transformation of *Nectria haematococca* with a gene for pisatin demethylating activity, and the role of pisatin detoxification in virulence. *J. Cell Biochem. 12C*: 278 (Abstr).

Cole, D. L. (1982). Interactions between *Cercospora arachidicola* and *Phoma arachidicola*, and their effects on defoliation and kernel yield of groundnut. *Plant Pathol. 31*: 355–362.

Cole, R. J., Sanders, T. H., Hill, R. A., and Blankenship, P. D. (1985). Geocarposphere temperatures that induce preharvest aflatoxin contamination of peanuts under drought stress. *Mycopathologia 91*: 41–46.

Cooksey, C. J., Garratt, P. J., Richards, S. E., and Strange, R. N. (1988). A dienyl stilbene phytoalexin from *Arachis hypogaea*. *Phytochemistry 27*: 1015–1016.

COPR (Centre for Overseas Pest Research) (1981). *Pest Control in Tropical Grain Legumes*, Overseas Development Administration, London.

Dewick, P. M. (1982). Isoflavonoids, *The Flavonoids: Advances in Research* (J. B. Harborne and T. J. Mabry, eds.), Chapman and Hall, London, pp. 535–640.

Dixon, R. A., Dey, P. M., and Lamb, C. J. (1983). Phytoalexins: enzymology and molecular biology. *Adv. Enzymol. Rel. Areas Mol. Biol. 55*: 1–136.

Ebel, J. (1986). Phytoalexin synthesis: the biochemical analysis of the induction process. *Annu. Rev. Phytopathol. 24*: 235–264.

Ebel, J., and Hahlbrock, K. (1982). Biosynthesis. *The Flavonoids: Advances in Research* (J. B. Harborne and T. J. Mabry, eds.), Chapman and Hall, London, pp. 641–679.

Edwards, C. (1992). Investigation of the Phytoalexin Response of *Arachis hypogaea* L. and its possible involvement in resistance to pathogens, PhD thesis, University of London.

Edwards, C., and Strange, R. N. (1991). Separation and identification of phytoalexins from leaves of groundnut (*Arachis hypogaea*) and development of a method for their determination by reversed-phase high-performance liquid chromatography. *J. Chromatogr. 547*: 185–193.

FAO. (1990). *Production Year Book, 44*: 109.

Feakin, S. D. (ed.) (1973). Pest control in groundnuts, PANS Manual No. 2, Overseas Development Natural Resources Research Institute, ODA, London.

Gibbons, R.W. (1980). Groundnut improvement research technology for the semiarid tropics, *Proceedings of the International Symposium on Development and Transfer of Technology for Rainfed Agriculture and the SAT Farmer*, August 28–September 1, 1979, International Crops Research Institute for the Semi-Arid Tropics, Patancheru, India, pp. 27–37.

Hahlbrock, K., and Griesbach, H. (1975). Biosynthesis of flavonoids. *The Flavonoids: Advances in Research* (J. B. Harborne and T. J. Mabry, eds.), Chapman and Hall, London, pp. 866–915.

Hahlbrock, K., and Griesbach, H. (1979). Enzymic controls in biosynthesis of lignin and flavonoids. *Annu. Rev. Plant Physiol. 30*: 105–130.

Hahn, M. G., Bonhoff, A., and Griesbach, H. (1985). Quantitative localization of the phytoalexin glyceollin I in relation to fungal hyphae in soybean roots infected with *Phytophthora megasperma* f. sp. *glycinea*. *Plant Physiol. 77*: 591–601.

Hayes, A. W. (1980). Mycotoxins: a review of biological effects and their role in human diseases. *Clin. Toxicol. 17*: 45–83.

Heathcote and Hibbert (1978). Production of aflatoxins. *Developments in Food Science. 1. Aflatoxins: Chemical and Biological Aspects*, Elsevier, Amsterdam, pp. 26–29.

Hill, R. A., Blankenship, P. D., Cole, R. J., and Sanders, T. H. (1983). Effects of soil moisture and temperature on preharvest invasion of peanuts by the *Aspergillus flavus* group and subsequent aflatoxin development. *Appl. Environ. Microbiol. 45*: 628–633.

Homans, A. L., and Fuchs, A. (1970). Direct bioautography on thin layer chromatograms as a method for detecting fungitoxic substances. *J. Chromatogr. 51*: 327–329.

Huang, J. S., and Barker, K. R. (1991). Glyceollin-I in soybean-cyst nematode interactions: spatial and temporal distribution in roots of resistant and susceptible soybeans. *Plant Physiol. 96*: 1302–1307.

Ingham, J. L. (1976). 3,5,4′-Trihydroxystilbene as a phytoalexin from groundnut (*Arachis hypogaea*). *Phytochemistry 15*: 1791–1793.

Ingham, J. L. (1982). Phytoalexins from the Leguminosae. *Phytoalexins* (J. A. Bailey and J. W. Mansfield, eds.), Blackie, Glasgow, pp. 21–80.

Jackson, C. R., and Bell, D. K. (1969). Diseases of peanut (groundnut) caused by fungi. *Research Bulletin, 56*, Univ. of Georgia, College of Agriculture, Experiment Station.

Keen, N. T. (1986). Phytoalexins and their involvement in plant disease resistance. *Iowa State J. Res. 60*: 477–499.

Keen, N. T., and Ingham, J. L. (1976). New stilbene phytoalexins from American cultivars of *Arachis hypogaea*. *Phytochemistry 15*: 1794–1795.

Kindl, H. (1985). Biosynthesis of stilbenes. *Biosynthesis and Biodegradation of Wood Components* (T. Highchi, ed.), Academic Press, New York, pp. 345–377.

Malamy, J., Carr, J. P., Klessig, D. F., and Raskin, I. (1990). Salicylic acid: a likely endogenous signal in the resistance response of tobacco to viral infection. *Science 250*: 1002–1004.

McDonald, D. (1984). The ICRISAT Groundnut Program, *Proceedings of the Regional Groundnut Workshop for Southern Africa*, March 26–29, 1984, Lilongwe, Malawi, International Crops Research Institute for the Semi-Arid Tropics, Patancheru, India.

Mansfield, J. W. (1982). The role of phytoalexins in disease resistance. *Phytoalexins* (J. A. Bailey and J. W. Mansfield, eds.), Blackie, Glasgow, pp. 133–180.

Mehan, V. K., McDonald, D., Nigam, S. N., and Lalitha, B. (1981). Groundnut cultivars with seed resistance to invasion by *Aspergillus flavus*. *Oléagineux 36*: 501–505.

Métraux, J. P., Signer, H., Ryals, J., Ward, E., Wyssbenz, M., Gaudin, J., Raschdorf, K., Schmid, E., Blum, W., and Inverardi, B. (1990). Increase in salicylic acid at the onset of systemic acquired resistance in cucumber. *Science 250*: 1004–1006.

Miao, V. P., Covert, S. F., and VanEtten, H. D. (1991). A fungal gene for antibiotic resistance on a dispensable (B) chromosome. *Science 254*: 1773–1776.

Narayanaswamy, P., and Mahadevan, A. (1983). Isolation of phytoalexins from germinating seeds of groundnut (*A. hypogaea*). *Acta Phytopathologica Academiae Scientiarum Hungaricae 18*: 33–36.

Paxton, J. (1981). Phytoalexins: a working redefinition. *Phytopathol. Z. 101*: 106–109.

Porter, D. M., Smith, D. H., and Rodriquez-Kabana, R. (eds.) (1984). Compendium of peanut diseases. *Am. Phytopathol. Soc.*

Smith, D. A., and Banks, S. W. (1986). Biosynthesis, elicitation and biological activity of isoflavonoid phytoalexins. *Phytochemistry 25*: 979–995.

Stoessl, P. (1982). Biosynthesis of phytoalexins. *Phytoalexins* (J. A. Bailey and J. W. Mansfield, eds.), Blackie, Glasgow, pp. 133–180.

Strange, R. N. (1987). HPLC of phytoalexins. *Modern Methods of Plant Analysis, New Series 5* (H. F. Linskens and J. F. Jackson, eds.), Springer-Verlag, Berlin, pp. 121–148.

Strange, R. N. (1991). Natural occurrence of mycotoxins in groundnuts, cottonseed, soya and cassava. *Mycotoxins and Animal Feed Stuffs; Natural Occurrence, Toxicity and Control* (J. E. Smith and R. S. Henderson, eds.), CRC Press, Boca Raton, pp. 341–362.

Strange, R. N. (1992). Resistance: the role of the hypersensitive response and phytoalexins. *Plant Response to Foliar Pathogens* (P. G. Ayres, ed.), Biosis Scientific, Oxford.

Strange, R. N., Ingham, J. L., Cole, D. L., Cavill, M. E., Edwards, C., Cooksey, C. J., and Garatt, P. J. (1985). Isolation of the phytoalexin, medicarpin from leaflets of *Arachis hypogaea* and related species of the tribe Aeschynomeneae. *Z. Naturforschung. 40c*: 313–316.

Subba Rao, P. V. (1987). La rouille de l'arachide: étude de quelques mécanismes de défense de l'hôte, Thése de Doctorat, Université de Paris XI, Orsay, France.

Subba Rao, P. V., Geiger, J. P., Einhorn, J., Malosse, C., Rio, B., Nicole, M., Savary, S., and Ravisé, A. (1988a). Isolation of methyl linolenate, a new antifungal compound from *Arachis hypogaea* L. leaves infected with *Puccinia arachidis* Speg. *Oléagineux 43*: 173–177.

Subba Rao, P. V., Geiger, J. P., Einhorn, J., Rio, B., Malosse, C., Nicole, M., Savary, S., and Ravisé, A. (1988b). Host defence mechanisms against groundnut rust. *Int. Arachis Newslett. (IAN) 4*: 16–18.

Subba Rao, P. V., Einhorn, J., Geiger, J. P., Malosse, C., Rio, B., and Ravisé, A. (1990). New dienol phytoalexins from *Arachis hypogaea* L. infected with *Puccinia arachidis* Speg. *Oléagineux 45*: 225–227.

Subba Rao, P. V., Einhorn, J., Geiger, J. P., Malosse, C., Rio, B., and Ravisé, A. (1991). Alkyl bis-phenyl ethers, new phytoalexins produced by *Arachis hypogaea* L. infected with *Puccinia arachidis* Speg. *Oléagineux 46*: 501–507.

Subrahmanyam, P., and McDonald, D. (1983). Rust disease of groundnut. ICRISAT Information Bulletin No. 13, International Crops Research Institute for the Semi-Arid Tropics, Patancheru, India.

Subrahmanyam, P., and McDonald, D. (1987). Groundnut rust disease: epidemiology and control, *Groundnut Rust Disease*. Proceedings of a Discussion Group Meeting Held at ICRISAT Center, Patancheru, India. Sept. 24–28, 1984, pp. 23–39.

Subrahmanyam, P., and Ravindranadh, V. (1988). Diseases of groundnut caused by fungi and nematodes. *Groundnut* (P. S. Reddy, ed.), ICAR, New Delhi, pp. 453–507.

VanEtten, H. D., Matthews, D. E., and Matthews, P. S. (1989). Phytoalexin detoxification: importance for pathogenicity and practical implications. *Annu. Rev. Phytopathol. 27*: 143–164.

Williams, R. J., and McDonald, D. (1983). Grain molds in the tropics; problems and importance. *Annu. Rev. Phytopathol. 21*: 153–178.

Wotton, H. R., and Strange, R. N. (1985). Circumstantial evidence for phytoalexin involvement in the resistance of peanuts to *Aspergillus flavus*. *J. Gen. Microbiol. 131*: 487–494.

Wotton, H. R., and Strange, R. N. (1987). Increased susceptibility and reduced phytoalexin accumulation in drought stressed peanut kernels challenged with *Aspergillus flavus*. *Appl. Environ. Microbiol. 53*: 270–273.

Wynne, J. C., Beute, M. K., and Nigam, S. N. (1991). Breeding for disease resistance in peanut (*Arachis hypogaea* L.). *Annu. Rev. Phytopathol. 29*: 279–303.

Yoshikawa (1983). Macromolecules, recognition and the triggering of resistance. *Biochemical Plant Pathology* (J. A. Callow, ed.), Wiley, New York, pp. 267–298.

Zambettakis, C. (1983). Results of research on selected groundnut hybrids to limit *Aspergillus flavus infection. Comptes Rendu de Séances de l'Academie d'Agriculture de France 69*: 44–50.

11
Phytoalexins from the Crucifers

Thierry Rouxel, Albert Kollman, and Marie-Hélène Balesdent
Institut National de la Recherche Agronomique, Versailles, France

I. INTRODUCTION

A. General Information

Crucifers are distributed all around the world and are present in almost every available kind of environment from the Arctic Circle to the tropics (Hedge, 1976). However, the greatest number of genera are found in temperate regions of the Northern Hemisphere.

Cruciferous crops are grown as 1.) oilseed crop for seed-based condiments and oils, including oils of industrial use, 2.) forage and fodder crops for animal feeds, and 3.) vegetables for human consumption. Many of the cruciferous crops, especially within the genus *Brassica*, have been cultivated in the neolithic age and have been described by the ancient Greeks, Romans, Indians, and Chinese (Prakash and Hinata, 1980). Age-old breeding has selected a wide range of crop types within each of the species, and the extensive variation within species has caused considerable taxonomic confusion (Crisp, 1976; Prakash and Hinata, 1980; Williams and Hill, 1986). Each of the main cultivated *Brassica* species thus contains numerous varieties which are grown for oil, vegetables, or animal feeds. However, varieties of different species, i.e., *B. napus*, *B. juncea*, and *B. rapa*, grown for oil are often referred to by the same common name, oilseed rape. Rapeseed oil is the fourth commonly traded oil in the world. Most northern European countries produce rapeseed, i.e., *B. napus* var. *oleifera*, as their main edible oil crop (Williams and Hill, 1986). Canada, India, and the European Economic Community (EEC) are currently the main oilseed rape producers.

The six major cultivated *Brassica* species are genetically closely related. Three diploid species, *B. nigra* (bb), *B. rapa* syn. *campestris* (aa), and *B. oleracea* (cc), are the natural genitors of the three allotetraploid species *B. napus* (aacc), *B. juncea* (aabb), and *B. carinata* (bbcc) (Fig. 1) (U, 1935).

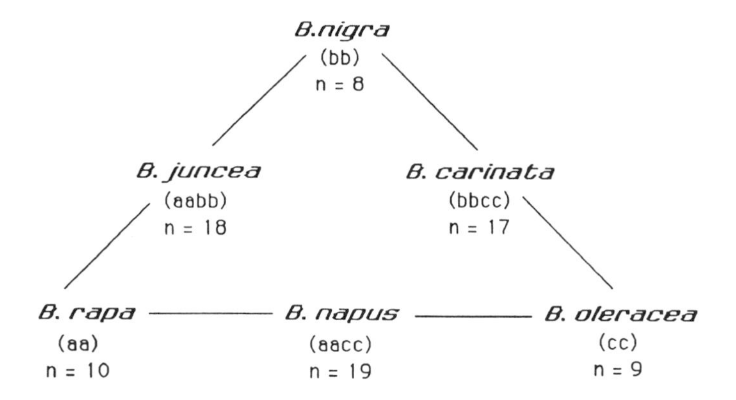

Figure 1 Genetic relationships of the cultivated *Brassica* species. (From U, 1935.)

B. *Brassica napus* and Blackleg Disease of Crucifers

Diseases of *B. napus* are mainly due to fungal pathogens, most of them having only a low incidence on yield. One of the major diseases of oilseed rape is blackleg of crucifers caused by *Leptosphaeria maculans* (Desm.) Ces. et de Not. [imperfect stage *Phoma lingam* (Tode ex Fr.) Desm.] (for a review, see Gabrielson, 1983). The disease was the limiting factor for the growth of most *Brassica* crops worldwide in the 1970s and the early 1980s. The breeding of *B. napus* cultivars displaying adult stage resistance, i.e., field resistance, has provided fairly good control of the disease since then.

The life cycle of the pathogen (Gabrielson, 1983; Regnault et al., 1987) is characterized by a very long symptomless development inside the host plant. The ascospores, discharged from the previous year's stubbles, are the main source of primary inoculum (Brunin and Lacoste, 1970). The pathogen penetrates the leaves or cotyledons of the plantlets through stomata or wounds. These primary infections of leaves have been shown to be a prerequisite for stem infection (Hammond et al., 1985). After entering the plant, the pathogen grows profusely between the epidermis and palisade layer, forming a sheet of hyphae. Branches behind the hyphal front grow downward into the spongy mesophyll. The pathogen then causes the typical primary symptom of the disease on leaves, stems, or cotyledons, a greyish green tissue collapse on which pycnidiae differentiate. Pycniospores can cause secondary infection of the diseased or surrounding plants. During the appearance of the necrotic symptom, hyphae ahead of the necrotic zone eventually reach minor veins and, using the intercellular spaces of the cortex beneath the veins, initiate a systemic growth through the petiole into the main stem (Hammond et al., 1985). During this systemic growth, the fungus is biotrophic, and no macroscopic symptom can be observed. Eventually, upon reaching the basis of the stem, the pathogen turns necrotrophic and causes the stem canker responsible for the lodging of the plants. The sexual stage of the fungus grows as a saprophyte on *Brassica* stubbles, and it can thus survive in the soil for 4–5 years, still remaining infectious (Alabouvette and Brunin, 1970).

Brassica with the B genome, i.e., *B. juncea*, *B. nigra*, and *B. carinata* (Fig. 1), display a complete resistance to *L. maculans*, expressed as a hypersensitive response, at all growth stages of the plants. In contrast, *B. napus* displays a field resistance only, which is incomplete and only expressed at the adult stage. This latter resistance could be due to the development of morphological barriers restricting the systemic infection of the plants (Hammond and Lewis, 1986).

According to Roy (1978), genetic determinants of adult stage resistance could be localized in the A genome of *B. napus* and be polygenic in nature. In contrast, the hypersensitive resistance of the mustards could be located in the B genome and be oligo- or monogenic (Roy, 1978). Recent findings by Keri et al. (1990) confirmed that the hypersensitive response of *B. juncea* to two *L. maculans* isolates was controlled by two nuclear genes with dominant recessive epistatic action.

Adult stage resistance allows a reduction in yield losses. The plants remain infected, but a small percentage of them develop a stem canker causing the lodging at the moment of the harvest. In France, the development of a "new" disease, the premature ripening of oilseed rape, in which *L. maculans* may be involved (Brun and Jacques, 1990), along with an increased aggressiveness of *L. maculans* to the double-low *B. napus* cultivars in Europe, Canada, and Australia, questions the adequacy of the adult stage resistance to control the disease.

As a consequence, one of the main challenges of the breeders is now to introduce the complete resistance of *Brassica* with the B genome into *B. napus*. Until now, diverse approaches have been tested, using interspecific hybrid progeny *B. napus* × *B. juncea* (Roy, 1978, 1984; Sacristan and Gerdemann, 1986; Rouxel et al., 1990a), *B. napus* × *B. nigra* interspecific hybrids (Sjodin and Glimelius, 1989), *B. napus*–*B. nigra* addition lines (Jahier et al., 1987; Rouxel et al., 1990c), or *B. napus* × *B. insularis* interspecific hybrids (Mithen and Lewis, 1988). Most of this work is still in progress today.

C. Objectives

The aim of our study of *Brassica* defense responses was, first, to have a better knowledge of the biochemical events associated with resistance to *L.*

maculans and, second, to know whether it was possible to screen for disease resistance genes at an early stage using a qualitative and quantitative analysis of plant phytoalexin response. Breeding programs to transfer resistance genes from the B genome in a *B. napus* background were mainly interested in this work. Third, we wished to test whether this host–pathogen system could be a suitable model for further research of the molecular and genetic determinants of plant–pathogen interaction, and for a study of the earliest events leading to resistance or susceptibility.

II. MATERIALS AND METHODS

A. Microorganism Cultures

The virulent isolates (1-144, 2-3, and single-ascospore isolate IIa1) and the nonvirulent isolate SV1 (Badawy and Hoppe, 1989; Koch et al., 1989; Rouxel et al., 1989, 1990b; Hassan et al., 1991) of *L. maculans* were used. They were maintained on malt-agar or V8-agar medium in the conditions previously described (Rouxel et al., 1989).

Erwinia chrysanthemi pv. *dahliae*, *E. carotovora* spp. *atroseptica*, and *Xanthomonas campestris* pv. *campestris* were grown on YDA agar medium, in the dark, at 30°C.

B. Plant Material and Inoculations

The interspecific hybrids progeny were issued from a cross between *B. napus* and *B. juncea*. These plants were self-pollinated twice and screened for resistance to *L. maculans* at S0 and S1 generations, according to a cotyledon inoculation test. Cultivars or breeding lines of *B. carinata*, *B. nigra*, *B. juncea*, *B. rapa* syn. *campestris*, *B. napus*, *B. oleracea*, and progeny issuing from an interspecific cross *B. napus* × *B. juncea* were grown in the greenhouse conditions previously described (Rouxel et al., 1989).

The results of the interactions between plants and *L. maculans* were evaluated according to a cotyledon inoculation test (Williams and Delwiche, 1979). Cotyledons of plantlets at growth stage 1 (Harper and Berkenkamp, 1975) were wounded with a needle and a 10-μl droplet of *L. maculans* pycniospore suspension (10^6 spores/ml) was placed over the wound. Incubation took place in a growth chamber at 18°C. Ten to 14 days after the inoculation, symptoms were assessed using the Williams and Delwiche rating system (Williams and Delwiche, 1979). This scale is divided in 10 classes, with 1 representative of the hypersensitive resistance and 9 representative of a highly susceptible interaction.

C. Abiotic Elicitation of Indole Phytoalexins

Except for time-course studies whereby whole plants were elicited and incubated under greenhouse conditions, abiotic elicitations were performed on detached leaves or leaf discs floating on water. Plant samples ranged from 200–400 mg fresh wt (one leaf) to 2 g. Leaves were elicited by spraying 10 mM $CuCl_2$ or $AgNO_3$ in 5 mg/liter Tween 80. Incubation took place in the growth chamber, at 18°C, under continuous white fluorescent light, for 48 hr.

UV elicitation was obtained by placing the detached leaves 25 cm under a 25-W germicidal lamp, and irradiating them for 2 hr or more. Incubation lasted 48 hr as described previously.

D. Effect of Synthesis Inhibitors and Ethylene on Phytoalexin Accumulation

B. juncea leaf discs were floated for 3 hr on solutions of actinomycin D, cycloheximide, or sirodesmin PL. Leaf discs were either maintained on the inhibitor solution or rinsed and floated on water. Elicitation with copper chloride was performed as described above and incubation took place in the growth chamber, at 18°C, under continuous illumination, for 12 hr.

Low concentrations of ethylene were applied for 12 hr on *B. napus* or *B. juncea* leaf discs maintained on water, in closed transparent plastic boxes. $CuCl_2$ elicitation of ethylene-treated leaves and incubation took place as described above.

E. Biotic Elicitation of Indole Phytoalexins

L. maculans was used as a 10^6 per ml pycniospore suspension in water.

Twenty-four-hour-old bacteria were suspended in water and the concentration adjusted to 5×10^7 cells per ml, by measuring the absorbance of the suspension at 600 nm.

Mycelial wall extracts or culture filtrates of *L. maculans* were obtained from 14-day- and 20-day-old cultures in modified Fries medium (Férézou et al., 1977).

Mycelial walls were washed, homogenized, and further purified according to Ayers et al. (1976). Acid or base hydrolysis of the purified cell walls was performed according to Ricci (1986). Purified mycelial walls and extracts were freeze-dried and maintained at −20°C. For the elicitation they were diluted in sterile twice-distilled water (500 mg dry wt per ml).

In all cases of biotic elicitation, leaves were surface-sterilized, rinsed with sterile water, and wounded using a multineedle device. Elicitors were applied by spraying the solutions or suspensions on leaves. Incubation took place in the growth chamber conditions described above, under a 16-hr photoperiod.

F. HPLC Apparatus

A Waters multisolvent delivery system equipped with a Waters U6K injector was used for high-performance liquid chromatography (HPLC). Chromatographic and spectral data from the eluate were acquired with a Waters 990 photodiode array detector. Data were computed by the Waters 990+ software loaded on NEC APC III.

G. Extraction and Purification of Brassilexin and Cyclobrassinin Sulfoxide

Detached *B. juncea* leaves (10–12 g fresh weight) were elicited using copper chloride as described above.

Phytoalexins were extracted either by macerating the elicited tissues in ethanol (95% v/v) with a Waring blender for 1–2 min, at room temperature, or by immersing the samples in 80°C ethanol (95% v/v) for 15 min.

The ethanolic extract was filtrated and evaporated to dryness at 40°C under reduced pressure. The dried extract was partitioned between twice-distilled water and diethyl ether (100 ml/50 ml, three times) and the combined diethyl ether phases were evaporated to dryness. The residue was taken up in 1 ml ethanol and the sample was applied to a column of silica gel 40 (35–70 mesh) and eluted with ethyl acetate–hexane (1:1). The thin layer chromatography (TLC) *Cladosporium* bioassay (Rouxel et al., 1989) permitted the detection of fungitoxic fractions which were collected, combined, dried (40°C, reduced pressure), and dissolved in a small volume of ethanol. The phytoalexins were further purified by HPLC. Fifty-milliliter samples were injected on a Brownlee C18 column (4.6 × 220 mm, 5-mm particle size) eluted with methanol–water (1:1 v/v, 1.5 ml/min). The fungitoxic fractions were combined, dried, recovered in a small volume of dichloromethane, and purified on a Brownlee Spheri-5 silica column (4.6 × 220 mm, 5-mm particle size) eluted with dichloromethane (1.5 ml/min). Further purification and characterization of the phytoalexins are described in Devys et al. (1988, 1990).

H. Determination of Indole Phytoalexins by HPLC

Elicited leaves or leaf discs were extracted and the hexane samples were cleaned up using Spe-ed octadecyl cartridges (Applied Separations, Inc.) as described in Section III (Results).

The colorless eluates obtained after clean-up were dried under reduced pressure (40°C) and recovered in dichloromethane. The solvent was evaporated at 50°C under a stream of nitrogen. The residue was finally taken up in 100 μl 95% (v/v) ethanol and 5- to 20-μl samples were chromatographed through a Brownlee 5-mm particle size C18 column (4.6 × 220 mm). The solvent delivery consisted of a linear gradient from 50% methanol v/v in water to 100% methanol in 5 min which was then maintained constant for 10 min. The flow rate was constant at 1.5 ml/min. Pure samples of the phytoalexins were used as standards.

III. RESULTS

A. First Reports

The first evidence of the accumulation of antifungal compounds, synthesized de novo by *Brassica* plants, was obtained in parallel by our team and that of Dr. Takasugi, in Japan (Takasugi et al., 1986; Sarniguet, 1986; Rouxel et al., 1987). Using the standard *Cladosporium* TLC bioassay, we showed that elicited *B. juncea* extracts displayed a large fungitoxic spot, which was not detected in control plants, when plants were elicited using the abiotic elicitor silver nitrate (Rouxel et al., 1989). The fungitoxic spot was localized at R_f 0.55 when silica gel TLC plates were developed using ethyl acetate. *B. napus* plants challenged by the abiotic elicitor displayed a much weaker fungitoxic spot at the same R_f. Such a weak fungitoxic spot was also observed when *B. juncea* plants were inoculated with *L. maculans*.

B. Chemical Structures

The phytoalexin was extracted from elicited *B. juncea* leaves by either macerating the tissues in ethanol with a Waring blender or immersion of the samples in hot ethanol. It was then purified using column chromatography and HPLC.

The chemical structure of the compound was established on the basis of physicochemical data, i.e., UV, IR, high-resolution mass spectroscopy (HR MS), ^{13}C and ^{1}H NMR, and nuclear Overhauser enhancement (nOe) difference experiments (Devys et al., 1988). These data confirmed that the new compound, brassilexin, was a sulfur-containing indole sharing structural similarities with brassinin, cyclobrassinin, or methoxybrassinin, previously characterized from other *Brassica* species (Takasugi et al., 1988) (Fig. 2).

Using similar elicitation, purification, and analysis procedures, an additional phytoalexin, cyclobrassinin sulfoxide, was obtained from *B. juncea* (Fig. 2) (Devys et al., 1990).

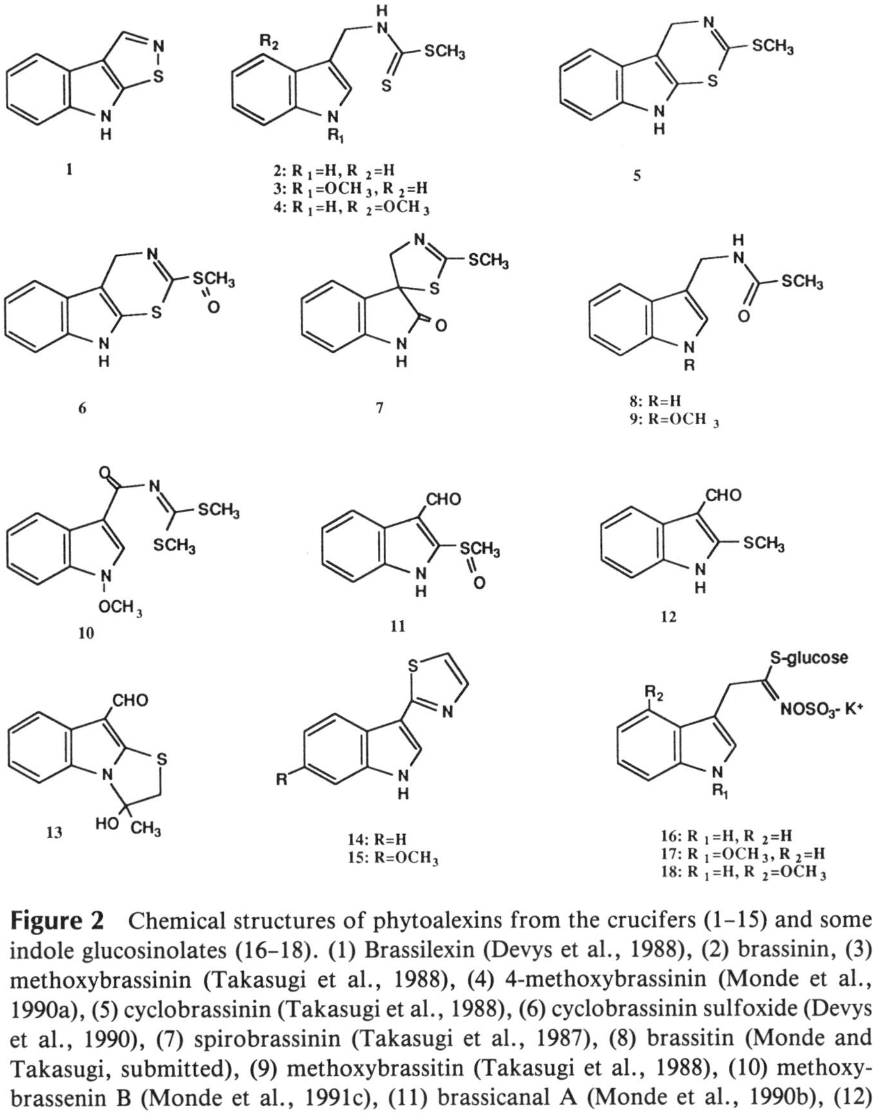

1

2: R_1=H, R_2=H
3: R_1=OCH_3, R_2=H
4: R_1=H, R_2=OCH_3

5

6

7

8: R=H
9: R=OCH_3

10

11

12

13

14: R=H
15: R=OCH_3

16: R_1=H, R_2=H
17: R_1=OCH_3, R_2=H
18: R_1=H, R_2=OCH_3

Figure 2 Chemical structures of phytoalexins from the crucifers (1–15) and some indole glucosinolates (16–18). (1) Brassilexin (Devys et al., 1988), (2) brassinin, (3) methoxybrassinin (Takasugi et al., 1988), (4) 4-methoxybrassinin (Monde et al., 1990a), (5) cyclobrassinin (Takasugi et al., 1988), (6) cyclobrassinin sulfoxide (Devys et al., 1990), (7) spirobrassinin (Takasugi et al., 1987), (8) brassitin (Monde and Takasugi, submitted), (9) methoxybrassitin (Takasugi et al., 1988), (10) methoxybrassenin B (Monde et al., 1991c), (11) brassicanal A (Monde et al., 1990b), (12) brassicanal B (Monde et al., 1990b), (13) brassicanal C (Monde et al., 1991b), (14) camalexin (Browne et al., 1991; Tsuji et al., 1992), and (15) methoxycamalexin (Browne et al., 1991); (16) glucobrassicin, (17) neoglucobrassicin, (18) 4-methoxyglucobrassicin.

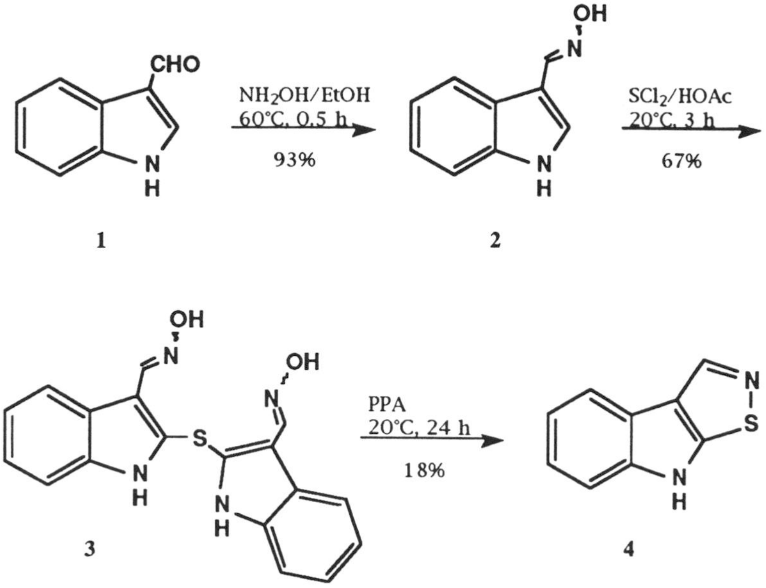

Figure 3 Synthesis of brassilexin starting from 3-indolecarbaldehyde. (From Devys and Barbier, 1990a.)

Apart from these two compounds, 15 other phytoalexins have been characterized from *Brassica* species, i.e., *B. rapa* and *B. oleracea* (Takasugi et al., 1988; Monde et al., 1990a, 1990b, 1991b), or other cruciferous plants, i.e., *Raphanus sativus* (Takasugi et al., 1987), *Camelina sativa* (Browne et al., 1991) and *Arabidopsis thaliana* (Tsuji et al., 1992) (Fig. 2). Up to now, the phytoalexins from cruciferous plants have been reported to possess an indole or oxindole nucleus with the appendage containing one or two sulfur atoms. No departure from this scheme has yet been reported.

C. Synthesis

Brassilexin has been synthesized with an overall 11% yield, starting from 3-indolecarbaldehyde (Fig. 3) (Devys and Barbier, 1990a). In addition, brassilexin was obtained in 30% yield by periodate-induced degradation of cyclobrassinin (Devys and Barbier, 1990b).

Three other phytoalexins have been successfully synthesized:

1. Brassinin was obtained from 3-(aminomethyl)indole treated with carbon disulfide in the presence of pyridine and triethylamine. The dithiocarbamate salt thus obtained was methylated with methyl iodide to give brassinin in 66% yield (Takasugi et al., 1988).

2. Cyclobrassinin was obtained, in 35% yield, by bromination of brassinin with pyridinium bromide perbromide, followed by dehydrobromination (Takasugi et al., 1988).
3. A simple synthetic method was developed for methoxybrassinin starting from indole-3-carboxaldehyde (Somei et al., 1992). The compound was obtained in seven steps, in 12% overall yield.

D. Toxicology Studies

Brassilexin inhibited both the germination of *L. maculans* pycniospores in water and its hyphal growth. The germination was completely suppressed at 12.5 mg/liter (75 μM), and a significant reduction in hyphal growth was obtained for concentrations as low as 1.5–3.12 mg/liter (9–18 μM) (Rouxel et al., 1989). A similar effect on *Alternaria brassicae* conidia or mycelium was observed (Rouxel, 1988).

These data can be compared with those reported for other phytoalexins from various plant families, i.e., toxicity values ranging from 10 to 100 μM (Smith, 1982).

Since it was observed that the compound may have a fungistatic effect rather than a lethal one (Rouxel et al., 1989), the possible metabolization of brassilexin by *L. maculans* has been assessed in liquid growth medium (Fig. 4). Pycniospores of the virulent isolate IIa1 and the nonvirulent one SV1 were grown in liquid Fries medium, in the presence of increasing concentrations of the phytoalexin. After 21 days of growth, the mycelium was freeze-dried and weighed, and the phytoalexin was extracted from the medium and determined. Under these conditions, the growth of both isolates was totally suppressed for concentrations above 12 mg/liter (70 μM). However, the virulent isolate IIa1 displayed a highly heterogeneous sensitivity to the phytoalexin. In a few cases it was even able to grow in the presence of 16 mg/liter brassilexin, which was not the case for the nonvirulent isolate SV1. Except for lethal concentrations, only a low rate of the initial concentration of brassilexin could be extracted from the medium thus suggesting, in both the virulent and nonvirulent isolate, an ability to metabolize the compound (Fig. 4).

Similar experiments, performed with brassinin, showed that this phytoalexin was metabolized into methyl(3-indolylmethyl)dithiocarbamate *S*-oxide, which is metabolized two to three times faster than brassinin (Soledade et al., 1991). The carboxylic acid that eventually resulted from this degradation was at least 10 times less toxic to the fungus than brassinin. Moreover, recent findings by Taylor et al. (1991b) showed that virulent isolates were able to specifically and rapidly metabolize brassinin, as compared to nonvirulent isolates.

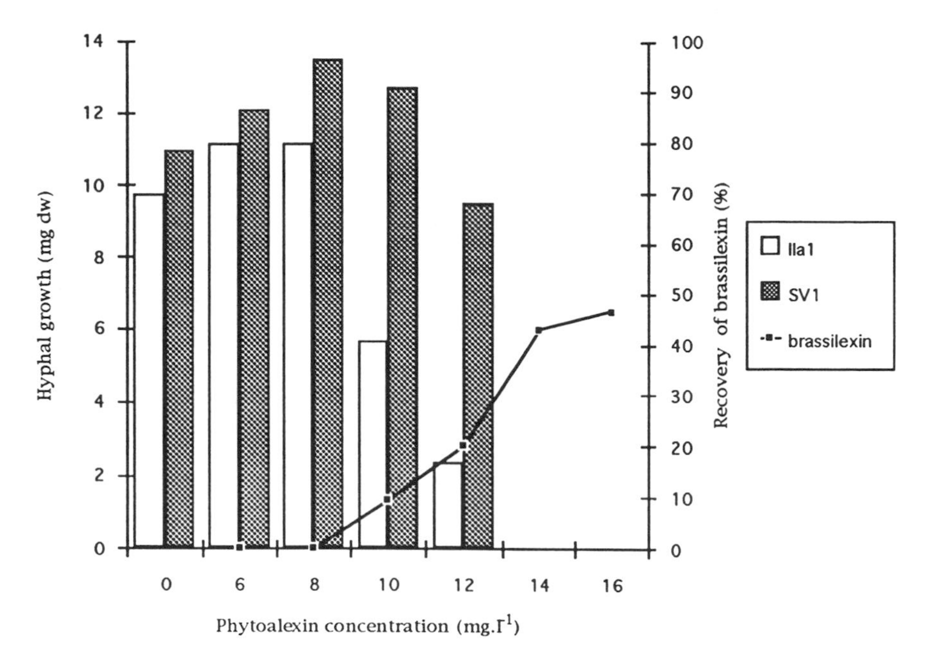

Figure 4 Toxicity of brassilexin to *L. maculans* and degradation of the compound by the fungus. The toxicity was assessed on the virulent isolate, IIa1, and the nonvirulent one, SV1, by weighing their freeze-dried mycelium after 21 days of growth in the presence of the phytoalexin. Brassilexin degradation by isolate SV1 is expressed as a percentage of the concentration of phytoalexin initially added to the medium. Each value represents the mean of five replications. The experiment was repeated twice.

Up to now, only a few other data are available on the toxicology of phytoalexins from *Brassica*. As it is the case for most of the phytoalexins isolated today, they would seem to act as general toxicants (Smith, 1982). Most of these compounds have been reported to be toxic to numerous fungal species including *L. maculans* (Takasugi et al., 1988; Dahiya and Rimmer, 1988a; Monde et al., 1990a, 1991b; Browne et al., 1991).

The toxicity of brassilexin to human KB cancer cells and human normal cells has been compared to that of cyclobrassinin and two other synthetic sulfur-containing indoles. Under these conditions, the LD_{50} was found to be 47 μM (8 mg/liter) for brassilexin and 94 μM (22 mg/liter) for cyclobrassinin. The toxic concentrations were similar for KB culture or normal cells. In addition, the toxicity of brassilexin to plants has also been investigated. It suppressed the germination of watercress seeds at 300 μM (50 mg/liter) (Tempête et al., 1991).

E. Analytical Conditions

Most of the quantitative analyses of phytoalexin production by cruciferous plants were performed using HPLC procedures (Dahiya and Rimmer, 1989; Rouxel et al., 1989; Monde and Takasugi, 1992). However, using chlorophyllous parts of plants, these analyses were usually complicated by the presence of pigments, waxes, and other interfering compounds. Moreover, using aqueous gradients, e.g., water–methanol, these compounds were also highly damaging to HPLC equipment and columns. The authors usually bypassed these problems using nonchlorophyllous parts of the plants, or callus tissues (Dahiya and Rimmer, 1988b; Monde et al., 1991a). The extracts could also be passed through silica and C18 cartridges prior to analysis, to remove highly polar or nonpolar compounds (Monde et al., 1991a). As an alternative, diffusates from elicited plant tissues and etiolated stems or leaves were analyzed (Dahiya and Rimmer, 1988a, 1989).

Since our aim was an investigation of the early interaction between the plants and a leaf pathogen, we chose to study the response of nonetiolated elicited leaves. The use of an unusual clean-up procedure, prior to analysis, allowed interfering compounds to be removed with retention of the phytoalexins (Kollmann et al., 1989; Rouxel et al., 1991) (Fig. 5). Elicited tissues were homogenized in 95% (v/v) ethanol, evaporated to dryness under reduced pressure, and the residues taken up in hexane. Hexane samples were passed through dried, unsolvated C18 cartridges and the sorbent washed with hexane until elution of a colorless solution occurred. The cartridges were then eluted with methanol-water (50:50) and the eluate, colorless or slightly colored, contained the phytoalexins. The yield of the method was very high for brassilexin, brassinin, and methoxybrassinin, since more than 80% of a known amount of phytoalexin was recovered after extraction, clean-up, and HPLC analysis. It was lower for cyclobrassinin (61–73%), but elution of the cartridges with methanol–water (65:35) increased the yield of this compound.

In addition, due to the characteristic UV spectra of these phytoalexins (Rouxel, 1988), the use of a photodiode detector (PAD Waters 990) permitted quantitative analyses. The three-dimensional display of the UV spectra of the compounds during the HPLC, along with a possible further computation of the spectra and chromatograms, allowed an unequivocal recognition of each of the phytoalexins, even when weak responses to the elicitation were obtained or when small plant samples (<1 g fresh wt) were analyzed.

A major improvement of reverse phase HPLC analysis of indole phytoalexins has recently been proposed by Monde and Takasugi (1992; see also Monde et al., 1991a). The authors used a complex acetonitrile–methanol–

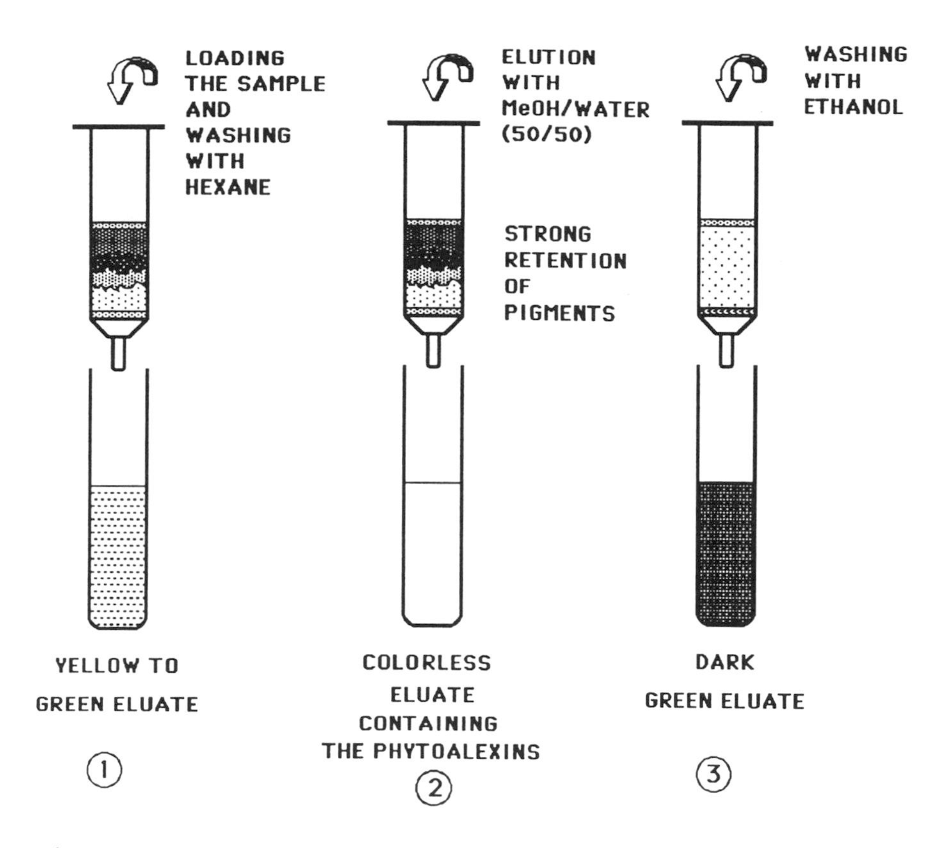

Figure 5 Sample clean-up protocol using reverse phase cartridges. (From Kollmann et al., 1989.)

water gradient system (Fig. 6) which allowed them to obtain a well-resolved chromatogram of 13 phytoalexins and three related indole metabolites. They finally suggested that a combination of Kollmann's protocol to clean up colored plant extracts with their HPLC analysis method could be the "standard analytical procedure of cruciferous phytoalexins."

F. Induction of Phytoalexin Accumulation

The efficiency of various elicitors was tested on *B. juncea*. Abiotic elicitors like silver nitrate, copper chloride, or UV light allowed the accumulation of higher amounts of phytoalexins than *L. maculans* (Table 1). The fungus, extracts of its cell walls, or most of the chromatographic fractions of its culture filtrate were also slow and low inducers of phytoalexin accumulation. The only fraction that allowed both the development of visible reac-

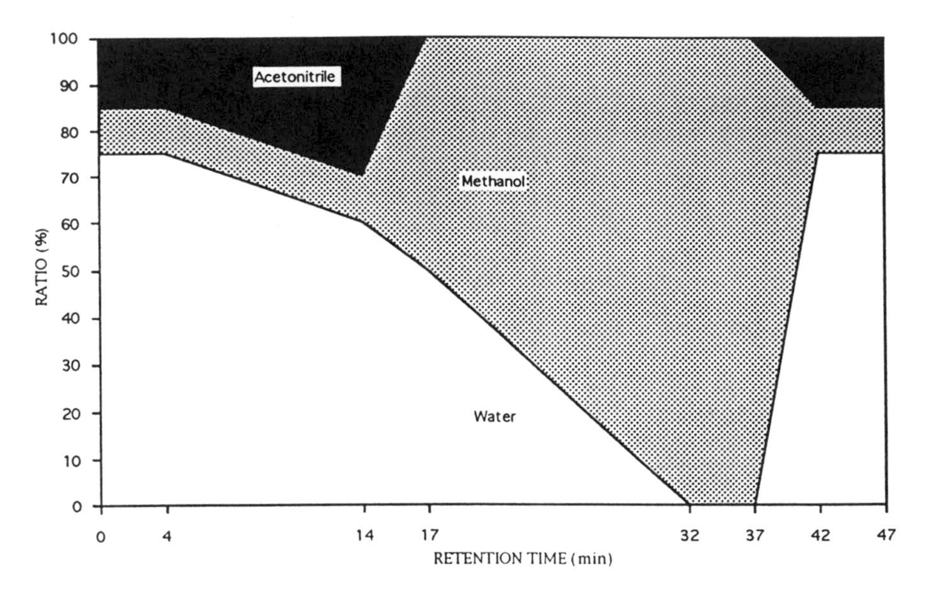

Figure 6 Complex gradient system used for an optimal resolution of cruciferous phytoalexins during reverse phase HPLC analysis. (From Monde and Takasugi, 1992.)

tions and phytoalexin accumulation, fraction 4, was also the one that contained the main toxic activity. Even though sirodesmin PL, the main toxin produced in vitro, did not induce a phytoalexin response in *Brassica* plants (see below), we did not succeed in separating the toxic from the eliciting activity using liquid chromatography or HPLC procedures. It is thus unclear whether uncharacterized elicitors or toxic compounds originated the response. So far, the best biotic inducers we tested were various nonhost bacteria which consistently induced a visible hypersensitive response and a rapid and seemingly specific accumulation of phytoalexins (Table 1).

Similar results were obtained by Dahiya and Rimmer (1989) when comparing the efficiency of chemicals to that of *L. maculans* to induce cyclobrassinin and methoxybrassinin accumulation in *Brassica* plants. $AgNO_3$ and, to a lesser extent, $CuCl_2$ and $HgCl_2$ were found to induce the highest accumulation of both methoxybrassinin and cyclobrassinin. The authors also observed that "the effectiveness of a chemical agent as a phytoalexin elicitor appears to be specific for a plant species as well as for plant organs." Dahiya and Rimmer (1989) pointed out the importance of the age of the plants or temperature conditions for the accumulation of phytoalexins. They showed that plant tissues from 35-day-old plants accumulated more

Table 1 Effect of Various Elicitor Treatments to Induce Phytoalexin Accumulation in *B. juncea* cv. Aurea[a]

		Brassilexin	Cyclobrassinin sulfoxide	Methoxybrassinin	Cyclobrassinin
Abiotic elicitors	$CuCl_2$ 10 mM (48 hr)	7.77 ± 3.01	16.06 ± 5.22	0	9.57 ± 3.63
	$AgNO_3$ 10 mM (48 hr)	5.60 ± 3.07	ND	ND	ND
	UV light (48 hr)	2.90 ± 0.76	ND	ND	ND
L. maculans	*L. maculans* IIa1 (4 d)	0.03 ± 0.01	0.07 ± 0.04	0	0.17 ± 0.03
	L. maculans SV1 (4 d)	0.30 ± 0.11	0.26 ± 0.07	0	0.67 ± 0.15
L. maculans cell wall extracts	SV1 total cell wall (3 d)	0	0	0	0.07 ± 0.04
	SV1 acid hydrolysis (3 d)	0.16 ± 0.07	0	0	0.54 ± 0.35
L. maculans culture filtrate fractions	Culture medium, fr. 4 (3 d)	0.07 ± 0.02	0	0	0.19 ± 0.10
	IIa1 20 days, fr. 4 (3 d)	0.48 ± 0.03	0.54 ± 0.20	0	0.53 ± 0.46
Nonhost bacteria	*X. campestris* (48 hr)	0.04 ± 0.04	0	0	0.29 ± 0.17
	E. carotovora (48 hr)	0.18 ± 0.16	0	0	2.12 ± 1.66
	E. chrysanthemi (48 hr)	0.71 ± 0.25	0	0	0.84 ± 0.45

[a]Copper chloride and silver nitrate aqueous solutions were sprayed on detached leaves. For UV elicitation, detached leaves were placed under a 25-W germicidal lamp and irradiated for 2 hr. In all cases of abiotic elicitation, incubation took place at 18°C under continuous illumination for 48 hr. In all other cases, detached leaves were wounded and sprayed with suspensions of the microorganism, extracts of the cell wall, or fractions of the culture filtrate. Incubation took place at 18°C under 16 hr photoperiod for 3–4 days. Values are mean ± SD of three to five replicates. They are expressed as μg/g fresh wt. Experiments were repeated two or three times. ND, not done.

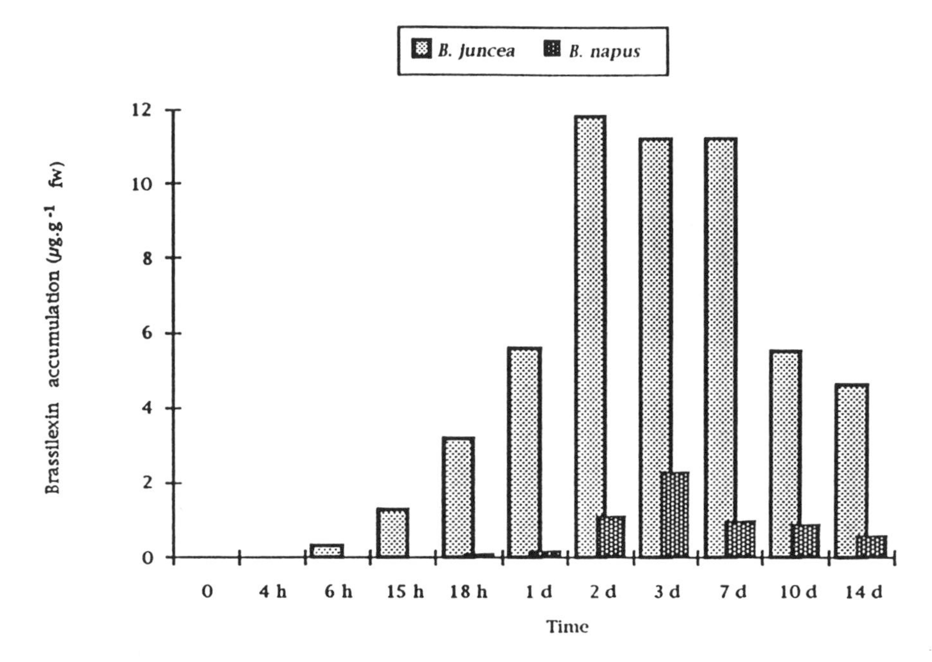

Figure 7 Time-course study of brassilexin accumulation in *B. juncea* cv. Aurea and *B. napus* cv. Brutor. Plants in the greenhouse were sprayed with a 10 mM aqueous copper chloride solution. Incubation took place under greenhouse conditions. Each value is the mean of three experiments with five replications per experiment. (From Rouxel et al., 1989.)

phytoalexin than tissues from 20-, 25-, and 40-day-old plants. Plants inoculated by *L. maculans* accumulated an optimal amount of phytoalexins at 20 and 25°C. Higher incubation temperatures (i.e., 30, 35, or 40°C) resulted in a drastic decrease of phytoalexin synthesis.

G. Time Course Studies on Phytoalexin Accumulation

Due to their greater efficiency, we used abiotic elicitors to compare the kinetics of brassilexin accumulation in a plant susceptible to *L. maculans*, *B. napus* cv. Brutor, and a hypersensitive one, *B. juncea* cv. Aurea (Fig. 7). Under these conditions, the first traces of brassilexin were detected 6 hr after the elicitation in *B. juncea*, and 12 hr later in *B. napus*. Moreover, *B. juncea* always accumulated more brassilexin than *B. napus*.

A more elaborate study was recently performed by Monde et al. (1991a). Using turnip roots, the authors studied the time-course accumulation of four indole phytoalexins (brassinin, methoxybrassinin, cyclobrassinin, and spirobrassinin) and that of nine glucosinolates following UV irradiation (Fig. 8). Under these conditions, the first traces of brassinin, methoxybrassinin, and cyclobrassinin were observed 8 hr after the elicitation. In contrast, the first traces of spirobrassinin were only observed 2 days after elicitation, and it then became the main part of the phytoalexin response. The increase in spirobrassinin accumulation was paralleled by a decrease in cyclobrassinin.

H. Effect of Synthesis Inhibitors on the Accumulation of Brassilexin

To assess whether brassilexin was indeed synthesized de novo, the effects of cycloheximide as a protein synthesis inhibitor and that of actinomycin D as an inhibitor of transcription were evaluated by floating *B. juncea* leaf discs for 3 hr on a solution of the inhibitor, prior to abiotic elicitation. In addition, the effect of sirodesmin PL, a toxin produced by the fungus in high amounts in culture medium (Férézou et al., 1977; Rouxel et al., 1988), was assessed under the same conditions. Both actinomycin D and cycloheximide were potent inhibitors of brassilexin accumulation, since concentrations of 16 μM for actinomycin D, and 35 μM for cycloheximide nearly totally suppressed phytoalexin accumulation (Table 2). These data thus suggested de novo synthesis. In contrast, sirodesmin PL only moderately inhibited

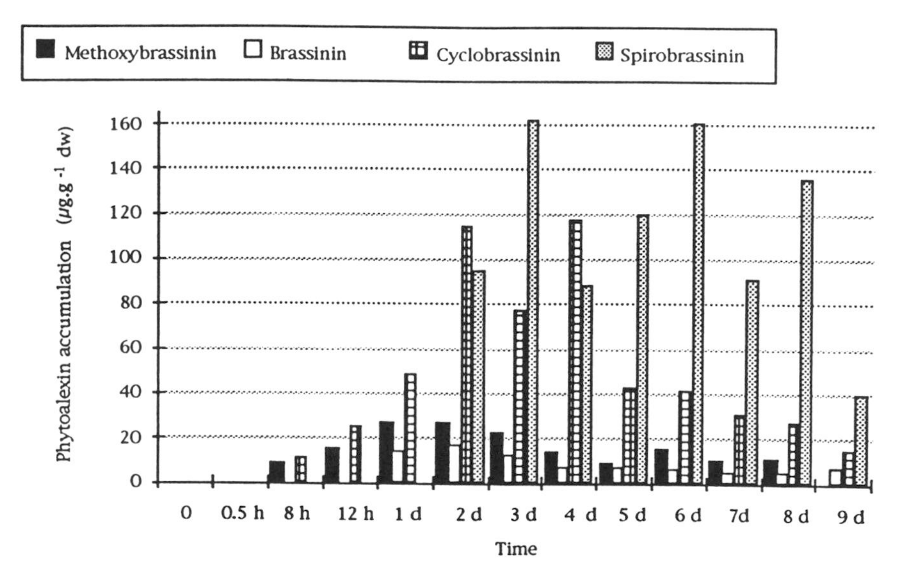

Figure 8 Time-course study of indole phytoalexin accumulation in *B. rapa* root tissue, following UV irradiation. (From Monde et al., 1991a.)

Table 2 Effect of Synthesis Inhibitors and Sirodesmin PL on Brassilexin Accumulation by *B. juncea*[a]

		Brassilexin accumulation (μg/g fresh wt)	
		Unelicited	Elicited
Actinomycin D	Control	0	4.97 ± 1.85
	2 μM	0	2.56 ± 0.62
	8 μM	0	1.56 ± 0.18
	16 μM	0	0.33 ± 0.12
Cycloheximide	Control	0	4.08 ± 1.66
	18 μM	0	0.66 ± 0.03
	35 μM	0	0.22 ± 0.13
	35 μM/rinsed	0	0
Sirodesmin PL	Control	0	5.65 ± 1.05
	10 μM	0	2.46 ± 1.46
	20 μM	0	3.36 ± 0.32
	40 μM	0	3.28 ± 0.34
	20 μM/rinsed	0	2.29 ± 1.46

[a]Leaf discs were floated for 3 hr on a solution of the inhibitor prior to the abiotic elicitation. They were either rinsed and floated on water, or maintained on the inhibitor solution during the elicitation. Each data is the mean ± SD of three experiments with five replications per experiment. 0, not detected.

brassilexin accumulation (40–59% of the control plant), whatever the tested concentration (Table 2). In addition, under our experimental conditions, sirodesmin PL did not by itself elicit the plant defense response. However, the weak efficiency of sirodesmin PL on brassilexin accumulation brought into question its possible involvement as a suppressor of plant defense responses (Rouxel et al., 1988).

I. Effect of Ethylene on the Accumulation of Phytoalexins

Ethylene was produced in high amounts in response to an abiotic elicitation, either by *B. juncea* or *B. napus* (Bouchenak et al., 1990). Exogenously applied ethylene never induced phytoalexin accumulation. However, low concentrations of exogenously applied ethylene (0.1–1 μl/liter) highly enhanced brassilexin, cyclobrassinin, or methoxybrassinin accumulation, in *B. napus* or *B. juncea*, when leaf discs were treated for 12 hr with the growth regulator prior to abiotic elicitation (Fig. 9). From these preliminary results, it was suggested that ethylene may act as a secondary messenger to enhance defense responses.

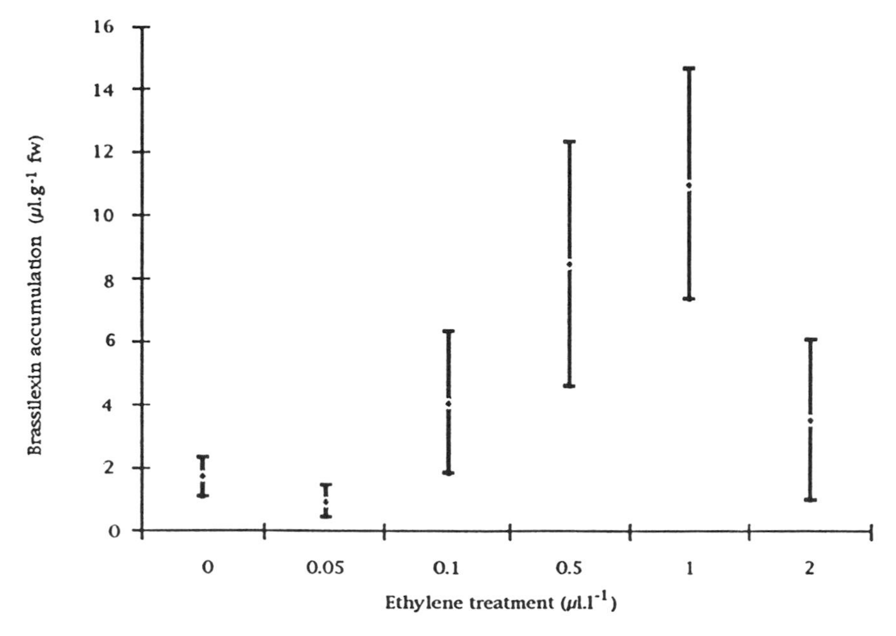

Figure 9 Effect of exogenously applied ethylene on brassilexin accumulation by *B. juncea*. Each datum represents the mean ± SD of three replications. The experiment was repeated three times.

J. Species-Specific Accumulation of Phytoalexins

To compare phytoalexin accumulation in *Brassica* species, we chose to use a nonspecific abiotic elicitation and a short time of incubation, i.e., 48 hr, which has been shown to allow an optimal accumulation of brassilexin in *B. juncea* (Fig. 7). Using copper chloride, we elicited 42 cultivars or breeding lines of *Brassica* belonging to the U triangle plus populations or cultivars of related species, *B. adpressa*, *Raphanus sativus*, and *Sinapis alba* (Rouxel et al., 1991). Five phytoalexins, brassilexin, brassinin, methoxybrassinin, cyclobrassinin, and cyclobrassinin sulfoxide were analyzed. Under these conditions, three of the phytoalexins, i.e., brassilexin, cyclobrassinin, and cyclobrassinin sulfoxide, were found within at least some lines of all species, whereas brassinin was only detected in *B. oleracea* and *B. napus* and methoxybrassinin within these two species and *B. rapa* and *B. carinata*. Brassinin, however, was always a minor part of the response. None of the five indole phytoalexins could be found in *Raphanus sativus* or *Sinapis alba*, although compounds displaying related UV spectra at different retention times were observed during the HPLC analysis of the extracts.

The profiles obtained for representative lines within a species (Fig. 10) showed 1.) an overall low accumulation of these phytoalexins within *B. oleracea*; 2.) the important part of brassilexin plus cyclobrassinin sulfoxide accumulation in the total response of plant with the B genome (see Fig. 1), as compared to species lacking it; 3.) the importance of methoxybrassinin in *B. oleracea* and *B. napus*.

B. oleracea lines usually accumulated the five phytoalexins in low amounts, even though a few lines displayed an important response, due to their high ability to rapidly accumulate methoxybrassinin. Only traces of brassilexin, brassinin, and cyclobrassinin sulfoxide were detected.

Low amounts of brassilexin and cyclobrassinin sulfoxide were also produced by *B. rapa* and *B. napus*. In the former case, however, one line displayed a quite unusual pattern of phytoalexin accumulation (see below). The production of cyclobrassinin was highly variable from one line to the other. However, both *B. napus* and *B. rapa* produced high amounts of cyclobrassinin. *B. napus* also consistently accumulated high levels of methoxybrassinin.

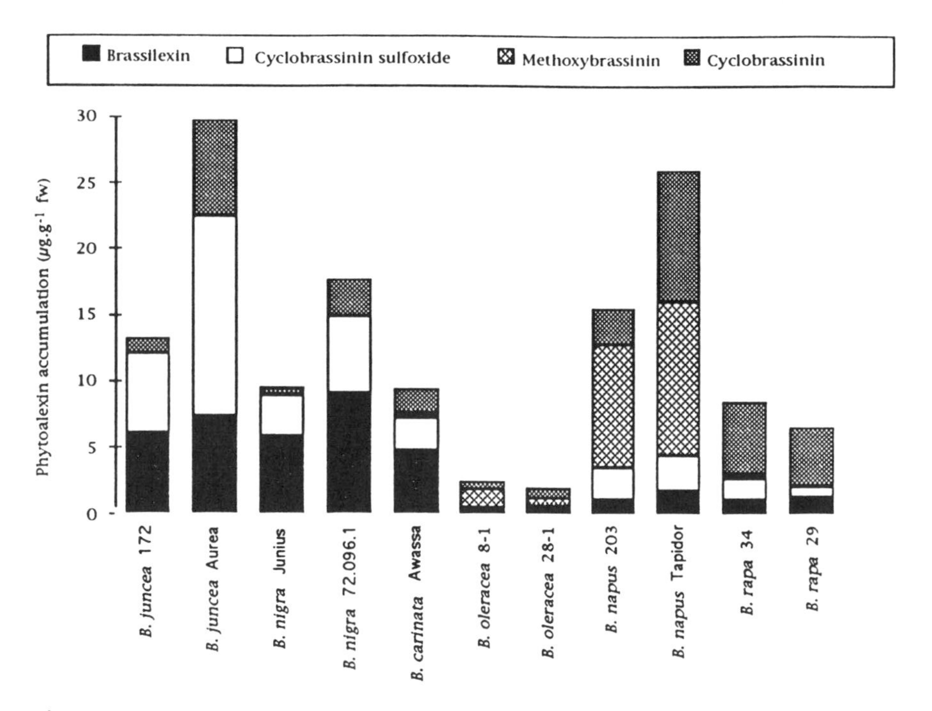

Figure 10 Phytoalexin profiles of representative *Brassica* lines following a $CuCl_2$ elicitation. (From Rouxel et al., 1991.)

B. juncea and *B. nigra* accumulated no brassinin or methoxybrassinin and high levels of brassilexin plus cyclobrassinin sulfoxide. One *B. nigra* line, however, or individuals within other lines only produced low levels of brassilexin. *B. nigra* usually accumulated low amounts of cyclobrassinin, whereas the production of this compound by *B. juncea* was either very low or as high as that of *B. napus*, depending on the line.

Finally, the one cultivar of *B. carinata* that we tested mainly accumulated brassilexin and, to a lesser extent, cyclobrassinin and cyclobrassinin sulfoxide. Only traces of methoxybrassinin were found within that species.

When considering a pool of all of the cultivars within a species, and calculating the accumulation of a given phytoalexin as a percentage of the total accumulation of each species (Table 3) it can be noticed that each of the diploid species produced different major phytoalexins and the profiles of the allotetraploid species were intermediate between their two diploid genitors (Table 3). Hence methoxybrassinin and, to a lesser extent, cyclobrassinin are mainly accumulated by plants with the C genome, cyclobrassinin by plants with the A genome, and brassilexin and cyclobrassinin sulfoxide by plants with the B genome. The A and C genomes are supposed to be phylogenetically more closely related to each other than either are to the B genome (Prakash and Hinata, 1980), and the phytoalexin accumulations of these two genomes are more similar to each other than to the B genome.

Most of the cultivars within a species could not be clearly discriminated according to their phytoalexin profiles. Some of them differed from the average pattern in the production of one given phytoalexin. A few cultivars, or lines, nonetheless displayed unusual profiles departing from the profile of the species.

K. Phytoalexin Accumulation and Hypersensitive Resistance to *L. maculans*

There was a strong correlation between resistance to *L. maculans*, according to a cotyledon inoculation test, and the accumulation of brassilexin (Fig. 11). No susceptible plant accumulated amounts of brassilexin comparable to that of the majority of lines possessing the B genome when challenged by the nonspecific elicitor, and no line with a high ability to accumulate brassilexin was susceptible to the disease (Fig. 11). A lesser correlation was observed to occur for cyclobrassinin sulfoxide, and low or no correlation for the three remaining phytoalexins (Rouxel et al., 1991). Finally, the total accumulation of the five indole phytoalexins could not be correlated with hypersensitive resistance to the pathogen either.

While there is a high correlation coefficient between the ability to accumulate brassilexin following an abiotic elicitation and disease resistance,

Table 3 Accumulation of Phytoalexins by the Different *Brassica* Species[a]

Species	Genome	Brassilexin	Cyclobrassinin sulfoxide	Brassinin	Methoxybrassinin	Cyclobrassinin
B. oleracea	cc	9.0 ± 13.4	5.8 ± 9.5	3.8 ± 5.5	56.0 ± 27.8	25.4 ± 2.62
B. napus	aacc	9.0 ± 7.6	10.5 ± 2.8	1.3 ± 1.8	47.3 ± 11.8	31.9 ± 12.9
B. rapa	aa	15.4 ± 6.9	13.9 ± 6.9	0	9.7 ± 9.3	61.0 ± 29.1
B. juncea	aabb	32.8 ± 9.9	44.5 ± 16.0	0	0	22.7 ± 21.2
B. nigra	bb	44.7 ± 14.4	43.4 ± 9.9	0	0	11.9 ± 9.3
B. carinata	bbcc	51.5	25.5	0	3.8	19.2

[a]Except for *B. carinata* (only one cultivar), 4–16 cultivars or breeding lines were analyzed for their phytoalexin profile following an abiotic elicitation. Each value is expressed as a percentage of the total production of indole phytoalexins within the considered species. Data are mean ± SD.
Source: From Rouxel et al. (1991).

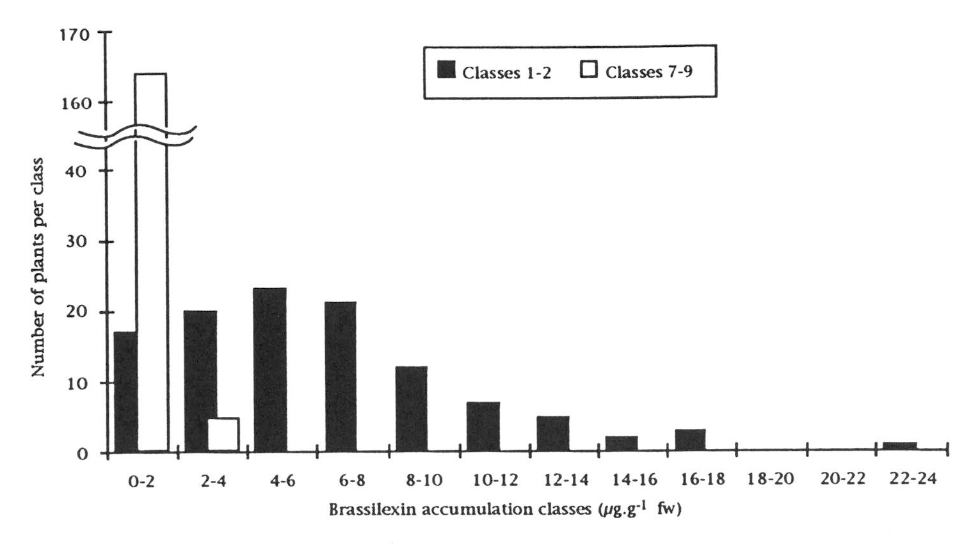

Figure 11 Distribution of *Brassica* plants, for a given set of lesion class when inoculated by *L. maculans*, as a function of their belonging to a class of brassilexin accumulation. Lesion classes 1–2 are representative of high-level resistance. Classes 7–9 are representative of susceptibility. Intermediate disease ratings were not considered for the analysis. Forty-two accessions of *Brassica* species and three lines of interspecific hybrid progeny *B. napus* × *B. juncea* are depicted on the graph. (Modified from Rouxel et al., 1991.)

and between the presence of these two variables and the B genome, there are some exceptions to this relationship, which means that the data should be interpreted with care. 1.) One *B. nigra* line, obtained from wild plants collected in Turkey, and individuals within other lines with the B genome, which consistently gave a hypersensitive reaction to the pathogen, accumulated only low amounts of brassilexin. 2.) Two *B. rapa* lines, 76-1 and 75-1, which were obtained from wild plants collected in Sicily and Algeria and lack the B genome, gave a hypersensitive reaction to inoculation. However, only 75-1 accumulated amounts of brassilexin similar to that of plants possessing the B genome. Recent findings that 76-1 segregates for resistance suggest that the genetic determinants of hypersensitive resistance could be different from that of species with the B genome (Mithen, personal communication). 3.) The accessions of *S. alba* and *R. sativus*, which are very closely related to *Brassica*, developed a superficially similar hypersensitive reaction to the pathogen but did not accumulate any of these five indole phytoalexins. Closely related compounds have nonetheless been observed to be elicited in *R. sativus* (Takasugi et al., 1987), *S. alba*, *Camelina sativa*, and *Capsella bursa-pastoris* (Conn et al., 1988; Browne et al., 1991; Tewari, personal communication) (see Fig. 2) and could be involved in defense responses.

L. Use of Phytoalexin Quantitative Analysis in Screening for Disease Resistance Genes

Using copper chloride as a nonspecific elicitor, we studied the brassilexin production of interspecific hybrid progeny between *B. napus* and *B. juncea* (Roy, 1978, 1984) and compared it to the response of both parent plants (Table 4). In parallel, the plants were assessed for resistance to *L. maculans* according to a cotyledon inoculation test. Three different interspecific progeny, termed 85-2, 85-3, and 85-4, were analyzed. 85-4 displayed a high level of resistance. It also accumulated a high level of brassilexin similar to that of species with the B genome. 85-3 was very susceptible and its response to the inoculation did not differ significantly from that of *B. napus*. Finally, 85-2 showed very heterogeneous responses to the pathogen. Both 85-2 and 85-3 accumulated an amount of brassilexin which was comparable with the one of *B. napus* (Table 4). The most heterogeneous interspecific

Table 4 Disease Severity and Brassilexin Accumulation in Various *Brassica* Species and Interspecific Hybrid Progeny[a]

			Mean disease severity class		Brassilexin accumulation (μg/g)	
Brassica	*B. juncea*	cv. Aurea	1.3 ± 0.5	a	5.9 ± 1.9	b
	B. nigra	cv. Junius	1.1 ± 0.4	a	4.3 ± 0.5	b
	B. carinata	cv. Awassa 67	1.0 ± 0.4	a	4.6 ± 2.0	b
	B. napus	cv. Brutor	7.8 ± 0.5	c	1.7 ± 0.3	a
		cv. Primor	7.9 ± 0.6	c	1.8 ± 0.7	a
		cv. Jet Neuf	7.2 ± 0.7	c	2.0 ± 0.8	a
		cv. Bienvenu	7.5 ± 0.7	c	2.5 ± 0.7	a
Interspecific hybrid progeny		85-2	4.9 ± 2.7	bc	2.4 ± 0.6	a
		85-3	7.7 ± 0.8	c	1.9 ± 0.7	a
		85-4	2.5 ± 0.7	b	5.3 ± 1.7	b

[a]Disease severity was assessed according to a cotyledon inoculation test (Williams and Delwiche, 1979). The scale is divided into 10 classes, with the higher ratings being representative of susceptibility. Brassilexin accumulation was quantitated in individual plants following an abiotic elicitation. Values are mean ± SD. In each column, values followed by different letters are significantly different (P = 0.01).

Source: From Rouxel et al (1990a).

hybrid progeny, 85-2, was further separated into five classes according to the result of the cotyledon inoculation test and its ability to accumulate brassilexin. Each of the plants were self-pollinated twice again and individuals belonging to each of the classes were assessed for resistance to *L. maculans*. All of the families displayed moderate to high susceptibility to the pathogen, and none accumulated more brassilexin than the parent plants. The lowest susceptibility of line 85-2-4 was not associated with a higher accumulation of brassilexin. However, plants of line 85-2-4 accumulated higher amounts of cyclobrassinin than the other lines. This difference was of low significance due to the highly variable accumulation of this compound from one plant to the other (data not shown).

To assess the usefulness of this methodology, we are currently undertaking a similar work with *B. napus–B. nigra* addition lines (Jahier et al., 1987; Rouxel et al., 1990c).

IV. DISCUSSION

The data presented here demonstrate that crucifers, like most plant families, accumulate fungitoxic compounds when challenged by microorganisms or abiotic stress. According to Paxton's definition (1981), these compounds are phytoalexins: 1.) they are toxic to *Brassica* pathogens as well as to other microorganisms; 2.) they are low molecular weight compounds; 3.) they are absent, or nondetected, in healthy plants and are synthesized de novo; 4.) they accumulate during a plant–microorganism interaction.

As compared to other plant families, e.g., the Fabaceae (Leguminoseae) (Ingham, 1982), in which more than 100 phytoalexins have been characterized, only limited numbers of phytoalexins from cruciferous plants have been described. However, the first characterizations of these compounds are quite recent and up to now only a few cruciferous plants have been assessed for phytoalexin production.

As has been described for other plant families, all the phytoalexins from the crucifers display structural similarities. As compared to the usual phytoalexins from other plant species, these compounds are characterized by their structural originality, i.e., the presence of an indole or oxindole system linked with one or more sulfur atoms. Cruciferous plants all contain a large group of sulfur-containing glycosides, the glucosinolates. These compounds may be hydrolyzed, e.g., following tissue damage, by the endogenous enzyme myrosinase. Among the glucosinolates some, like glucobrassicin (Fig. 2), contain an indole moiety, thus suggesting close biochemical relationships between these compounds and indole phytoalexins. Indole phytoalexins have been shown not to be degradation products of indole glucosinolates (Takasugi et al., 1988), but the hydrolysis products of indole glucosinolates are nonetheless toxic to *L. maculans* (Mithen et al., 1986). Time-course studies of phytoalexins and indole glucosinolate accumulation, following UV irradiation of plant tissues, showed that a decrease in the phytoalexin cyclobrassinin paralleled an increase in indole glucosinolate accumulation. Finally, the close biochemical relationship of both classes of products is also supported by deuterium-labeled tryptophan incorporation in both brassinin and cyclobrassinin (Monde and Takasugi, 1991). However, the control of the synthesis of both classes of products is different since indole glucosinolate accumulation is strongly induced on wounding of plant tissues, which is not the case for phytoalexins. Further studies are thus still needed to "elucidate the exact nature of the biosynthetic relationship, if any, between these two classes of biologically active compounds" (Monde et al., 1991a).

Using an abiotic elicitor, a species-specific accumulation of phytoalexin was observed. Devys and Barbier (1992) recently demonstrated the biosynthetic sequence cyclobrassinin to brassilexin through cyclobrassinin sulfoxide. In this respect, the high correlation coefficient observed between the ability to accumulate brassilexin and, to a lesser extent cyclobrassinin sulfoxide, and blackleg resistance, and between the presence of these two variables and the B genome suggests that the presence, or efficiency, of a few enzymes could be responsible for the observed differences. However, these data only represent part of the plant response, since the complete phytoalexin response of cruciferous plants, or most of the biochemical relationships that may exist between the phytoalexins, is still unknown. Moreover, even the high-resolution HPLC analyses do not permit the quantitative analysis of all the phytoalexins in an extract (Monde and Takasugi, 1992). It also remains difficult to compare our results with that obtained by the Takasugi team or Dahiya and Rimmer. Actually, each of the teams used different *Brassica* species or cultivars, analyzed different organs of the plants (and even different phytoalexins), and used different inducers of phytoalexin accumulation and quite different incubation conditions. As an illustration, the overall amount of phytoalexin quantified by Dahiya and Rimmer (1989) was much higher than obtained by us or the Takasugi group, whatever the tested elicitor. This result is not surprising, however, since these authors only analyzed the reacting parts of the plant and usually incubated the tissues for longer times than we did. To suppress such discrepancies, a critical assessment of the differential responsiveness of plant tissues as a function of the inducer, and an assessment of the effect of physical factors, e.g., light conditions, on the response of the plants remain to be done. The mastering of these variables and a better knowledge of indole phytoalexin biosynthetic pathways will then allow, using sophisticated HPLC analysis procedures, minimization of plant-to-plant variations and better differentiation of species-specific or cultivar-specific accumulation.

The quantification of one phytoalexin, brassilexin, permitted us to differentiate interspecific hybrid progeny resistant to *L. maculans* from susceptible ones. The technique was assessed using small-plant samples (200–400 mg fresh wt) so that a single leaf from a single plant could be examined and the plant still be used for further breeding. Its objective was to screen for an intrinsic ability which may be a property of the part of the B genome responsible for complete resistance to the pathogen, without using the parasite. Along with molecular biology characterization, phytoalexin "fingerprints" of single plants, using sophisticated HPLC methodologies, will allow a better differentiation between individuals and may be incorporated into breeding programs.

A strong correlation was observed between resistance to *L. maculans* and the ability to accumulate high levels of brassilexin, together with the presence of the B genome. However, recent findings cast doubts on the significance of the macroscopic reactions usually associated with resistance (Gugel et al., 1990). Moreover, correlations do not imply that the compounds really play a role in resistance. A comprehensive study of the variability of the pathogen, along with the development of methods allowing an unequivocal quantification of the result of the interaction between the plant and the pathogen, are now needed to confirm these correlations. Genetic studies of phytoalexin detoxification, or specificity of interaction isolate/cultivar, are now at hand and will hopefully unequivocally determine the involvement of phytoalexins in the resistance responses of cruciferous plants.

V. EPILOG

According to Hill and Williams (1988) and to the data presented here, *L. maculans* in combination with *Brassica* has considerable potential as a model host–pathogen system for the study of the genetic and molecular bases of plant–microorganism interactions and plant disease resistance.

The main interests of this system are 1.) the ease of culture of both host and pathogen; 2.) the development of a methodology allowing a reproductive induction of the sexual state of the pathogen in culture along with the ability to develop tetrad analyses (Mengistu et al., 1990); 3.) the high pathogenic variability of the pathogen in natural populations along with the finding that cultivar-isolate type–specific interactions occur in this system; 4.) the broad genetic basis of *Brassica* crops, like *B. napus*, and the ease with which the interaction can be studied due to 5.) the rapid and intense production of pycniospores which can easily be quantified, 6.) the rapid evaluation of the result of the interaction using a seedling test which allows a differentiation of specific resistance vs. susceptibility (Williams and Delwiche, 1979; Rouxel et al., 1990a), 7.) the possible use of rapid-cycling *Brassica* to undertake genetic analyses. Moreover, 8.) the recent development of powerful biochemical and molecular tools for the study of microorganisms now allows an unequivocal characterization of *L. maculans* isolates (Koch et al., 1991; Taylor et al., 1991a; Balesdent et al., 1992). 9.) Both the host and the pathogen are amenable to cellular and molecular biology techniques. In this respect, it is important to keep in mind that *Arabidopsis thaliana*, one of the most studied model plants, is a crucifer. Finally, 10.) the data presented here, along with similar studies performed by other teams, give a new insight into *Brassica* defense reactions, thus providing new tools to develop a quantification of the interaction and favoring a quest for defense genes or disease resistance genes.

ACKNOWLEDGMENTS

This work was supported by a grant from the Institut National de la Recherche Agronomique (AIP, 1986). The authors wish to thank H. H. Hoppe (Gesamthochsule Kassel, Fachbereich Landwirtschaft, Witzenhausen, F.R.G.) for single-ascospore lines of *L. maculans*; M. Takasugi (Department of Chemistry, Faculty of Science, Hokkaido University, Sapporo, Japan), M. Devys and M. Barbier (Chimie des Substances Naturelles, CNRS, Gif-sur-Yvette) for authentic samples of indole phytoalexins; N. N. Roy (Department of Agriculture, Perth, Australia) for interspecific hybrid progeny; A.-M. Chèvre, M. Renard (Amélioration des Plantes, INRA, Le Rheu), L. Boulidard (Amélioration des Plantes, INRA, Versailles), and R. Mithen (John Innes Centre for Plant Science Research, Colney Lane, Norwich NR4 7UH, UK) for crucifer seeds and breeding lines; J. F. Bousquet (Pathologie Végétale, INRA, Versailles) who permitted this research to take place in his lab and then allowed the development of a fruitful collaboration; N. T. Keen and D. Cooksey (University of California, Plant Pathology, Riverside, CA 92521, USA) for review of the manuscript. Special thanks are due to A. Sarniguet (Pathologie Végétale, INRA, Le Rheu) who initiated this work. We are very grateful to C. Sutre, A. Delaunay, and F. Bouchenak, whose help to perform part of this work was greatly appreciated.

REFERENCES

Alabouvette, C., and Brunin, B. (1970). Recherches sur la maladie du colza due à *Leptosphaeria maculans* (Desm.) Ces. et De Not. I. Rôle des restes de culture dans la conservation et la dissémination du parasite. *Ann. Phytophathol. 2*(3): 463–475.

Ayers, A. R., Ebel, J., Valent, B., and Albersheim, P. (1976). Host-pathogen interactions. X. Fractionation and biological activity of an elicitor isolated from the mycelial walls of *Phytophthora megasperma* var. *sojae*. *Plant Physiol. 57*: 760–765.

Badawy, H. M. A., and Hoppe, H. H. (1989). Production of phytotoxic sirodesmins by aggressive strains of *Leptosphaeria maculans* differing in interactions with oilseed rape genotypes. *J. Phytopathol. 127*: 146–157.

Balesdent, M. H., Gall, C., Robin, P., and Rouxel, T. (1992). Intraspecific variation in soluble mycelial protein and esterase patterns of *Leptosphaeria maculans* French isolates. *Mycol. Res. 96*: 677–686.

Bouchenak, F., Rouxel, T., Kollmann, A., Touraud, G., and Bousquet J.-F. (1990). Ethylene: phytopathological significance in the hypersensitive response of *Brassica juncea* leaves to *Leptosphaeria maculans*. *Plant Physiol. 93*(1): 53 (abst).

Browne, L. M., Conn, K. L., Ayer, W. A., and Tewari, J. P. (1991). The camalexins: new phytoalexins produced in the leaves of *Camelina sativa* (Cruciferae). *Tetrahedron 47*(24): 3909–3914.

Brun, H., and Jacques, M. A. (1990). Le dessèchement prématuré des pieds de colza. Quelques symptômes et agents pathogènes associés. *La Déf. Vég. 262*: 7–12.

Brunin, B., and Lacoste, L. (1970). Recherches sur la maladie du colza due à *Leptosphaeria maculans* (Desm.) Ces. et De Not. II. Pouvoir pathogène des ascospores. *Ann. Phytopathol. 2*(3): 477–488.

Conn, K. L., Tewari, J. P., and Dahiya, J. S. (1988). Resistance to *Alternaria brassicae* and phytoalexin elicitation in rapeseed and other crucifers. *Plant Sci. 56*: 21–25.

Crisp, P. (1976). Trends in the breeding and cultivation of cruciferous crops, *The Biology and Chemistry of the Cruciferae* (J. G. Vaughan, A. J. Macleod, and B. M. G. Jones, eds.), Academic Press, London, pp. 69–118.

Dahiya, J. S., and Rimmer, S. R. (1988a). Phytoalexin accumulation in tissues of *Brassica napus* inoculated with *Leptosphaeria maculans*. *Phytochemistry 27*: 3105–3107.

Dahiya, J. S., and Rimmer, S. R. (1988b). High-performance liquid chromatography of phytoalexins in stem callus tissues of rapeseed. *J. Chromatogr. 448*: 448–453.

Dahiya, J. S., and Rimmer, S. R. (1989). Phytoalexin accumulation in plant tissues of *Brassica* spp. in response to abiotic elicitors and infection with *Leptosphaeria maculans*. *Bot. Bull. Academia Sinica 30*: 107–115.

Devys, M., Barbier, M., Loiselet, I., Rouxel, T., Sarniguet, A., Kollmann, A., and Bousquet, J. F. (1988). Brassilexin, a novel sulphur-containing phytoalexin from *Brassica juncea* L., (Cruciferae). *Tetrahedron Lett. 29*: 6447–6448.

Devys, M., Barbier, M., Kollmann, A., Rouxel, T., and Bousquet, J. F. (1990). Cyclobrassinin sulphoxide, a sulphur-containing phytoalexin from *Brassica juncea*. *Phytochemistry 29*: 1087–1088.

Devys, M., and Barbier, M. (1990a). A convenient synthesis of isothiazolo[5,4*b*]indole (brassilexin) via a polyphosphoric acid initiated ring closure. *Synthesis 3*: 214–215.

Devys, M., and Barbier, M. (1990b). Oxidative ring contraction of the phytoalexin cyclobrassinin: a way to brassilexin. *J. Chem. Soc. Perkin Trans.* 2856–2857.

Devys, M., and Barbier, M. (1992). In vitro demonstration of a biosynthetic sequence for the Cruciferae phytoalexins. *Z. Naturforschung. C 47c*: 318–319.

Férézou, J. P., Riche, C., Quesneau-Thierry, A., Pascard-Billy, C., Barbier, M., Bousquet, J. F., and Boudart, G. (1977). Structures de deux toxines isolées des cultures du champignon *Phoma lingam* Tode: la sirodesmine PL et la desacetylsirodesmine PL. *Nouv. J. Chim. 1*: 327–334.

Gabrielson, R. L. (1983). Blackleg disease of crucifers caused by *Leptosphaeria maculans* (*Phoma lingam*) and its control. *Seed Sci. Technol. 11*: 749–780.

Gugel, R. K., Seguin-Swartz, G., and Petrie, G. A. (1990). Pathogenicity of three isolates of *Leptosphaeria maculans* on *Brassica* species and other crucifers. *Can. J. Plant Pathol. 12*: 75–82.

Hammond, K. E., Lewis, B. G., and Musa, T. M. (1985). A systemic pathway in the infection of oilseed rape plants by *Leptosphaeria maculans*. *Plant Pathol. 34*: 557–565.

Hammond, K. E., and Lewis, B. G. (1986). Ultrastructural studies of the limitation of lesions caused by *Leptosphaeria maculans* in stems of *Brassica napus* var. *oleifera*. *Physiol. Mol. Plant Pathol. 28*: 251–265.

Harper, F. R., and Berkenkamp, B. (1975). Revised growth-stage key for *Brassica campestris* and *B. napus*. *Can. J. Plant Sci. 55*: 657–658.

Hassan, A. K., Schulz, C., Sacristan, M. D., and Wostemeyer, J. (1991). Biochemical and molecular tools for the differenciation of aggressive and non-aggressive isolates of the oilseed rape pathogen, *Phoma lingam*. *J. Phytopathol. 131*: 120–136.

Hedge, I. C. (1976). A systematic and geographic survey of the old world cruciferae. *The Biology and Chemistry of the Cruciferae* (J. G. Vaughan, A. J. Macleod, and B. M. G. Jones, eds.), Academic Press, London, pp. 1–45.

Hill, C. B., and Williams, P. H. (1988). *Leptosphaeria maculans*, cause of blackleg of crucifers. *Advances in Plant Pathology, Vol. 6, Genetics of Plant Pathogenic Fungi* (D. S. Ingram and P. H. Williams, eds.), Academic Press, London, pp. 169–174.

Ingham, J. L. (1982). Phytoalexins from the Leguminosae. *Phytoalexins* (J. A. Bailey and J. W. Mansfield, eds.), Blackie, Glasgow, pp. 21–80.

Jahier, J., Tanguy, A.-M., Chevre, A.-M, Tanguy, X., and Renard, M. (1987). Extraction of disomic addition lines *B. napus–B. nigra* and introduction of *B. nigra* type *Phoma lingam* resistance to rapeseed. *Proceedings of the 7th International Rapeseed Congress*, Poznan, Poland, pp. 445–450.

Keri, M., Rimmer, S. R., and van den Berg, C. G. J. (1990). The inheritance of resistance of *Brassica juncea* to *Leptosphaeria maculans*. *Can. J. Plant Pathol. 12*(3): 335 (abst).

Koch, E., Badawy, H. M. A., and Hoppe, H. H. (1989). Differences between aggressive and non-aggressive single spore lines of *Leptosphaeria maculans* in cultural characteristics and phytotoxin production. *J. Phytopathol. 124*: 52–62.

Koch, E., Song, K., Osborn, T. C., and Williams, P. H. (1991). Relationship between pathogenicity and phylogeny based on restriction fragment length polymorphism in *Leptosphaeria maculans*. *Mol. Plant–Microbe Interac.* *4*: 341–349.

Kollmann, A., Rouxel, T., and Bousquet, J. F. (1989). Efficient clean up of non-aqueous plant extracts using reversed-phase cartridges. Applications to the determination of phytoalexins from *Brassica* spp. by high-performance liquid chromatography. *J. Chromatogr.* *473*: 293–300.

Mengistu, A., Rimmer, S. R., and Williams, P. H. (1990). In vitro induction of pseudothecia, ascospore release and variation in fertility among geographic isolates of *Leptosphaeria maculans*. *Phytopathology* *80*: 1007 (abst).

Mithen, R. F., Lewis, B. G., and Fenwick, G. R. (1986). In vitro activity of glucosinolates and their products against *Leptosphaeria maculans*. *Trans. Br. Mycol. Soc.* *87*(3): 433–440.

Mithen, R. F., and Lewis, B. G. (1988). Resistance to *Leptosphaeria maculans* in hybrids of *Brassica oleracea* and *Brassica insularis*. *J. Phytopathol.* *123*: 253–258.

Monde, K., Sasaki, K., Shirata, A., and Takasugi, M. (1990a). 4-Methoxybrassinin, a sulphur-containing phytoalexin from the chinese cabbage *Brassica campestris* L. ssp. *pekinensis*. *Chem. Lett.* *2*: 209–210.

Monde, K., Katsui, N., Shirata, A., and Takasugi, M. (1990b). Brassicanal-A and brassicanal-B, novel sulphur-containing phytoalexins from *Brassica oleracea*. *Phytochemistry* *29*(5): 1499–1500.

Monde, K., Takasugi, M., Lewis, J. A., and Fenwick, G. R. (1991a). Time-course studies of phytoalexins and glucosinolates in UV-irradiated turnip tissue. *Z. Naturforschung C* *46c*: 189–193.

Monde, K., Sasaki, K., Shirata, A., and Takasugi, M. (1991b). Brassicanal C and two dioxindoles from cabbage. *Phytochemistry* *30*(9): 2915–2917.

Monde, K., Sasaki, K., Shirata, A., and Takasugi, M. (1991c). Methoxybrassenins A and B, sulphur-containing stress metabolites from *Brassica oleracea* var. *capitata*. *Phytochemistry* *30*(12): 3921–3922.

Monde, K., and Takasugi, M. (1991). Biosynthesis of cruciferous phytoalexins: the involvement of a molecular rearrangement in the biosynthesis of brassinin. *J. Chem. Soc. Chem. Commun.* 1582–1583.

Monde, K., and Takasugi, M. (1992). High-performance liquid chromatographic analysis of cruciferous phytoalexins using a ternary mobile phase gradient. *J. Chromatogr.* *598*: 147–152.

Paxton, J. D. (1981). Phytoalexins-a working redefinition. *Phytopathol. Z.* *101*: 106–109.

Prakash, S., and Hinata, K. (1980). Taxonomy, cytogenetics and origin of crop brassicas, a review. *Opera Botanica* *55*: 1–57.

Regnault, Y., Laville, J., and Penaud, A. (1987). *Cahiers techniques: Les maladies du colza d'hiver*, CETIOM, Paris, France.

Ricci, P. (1986). Etude des relations hôte-parasite dans l'interaction compatible entre *Dianthus caryophyllus* L. et *Phytophthora parasitca* Dastur. Nature et intervention d'un mécanisme de défense élicitable. Thèse de l'Université de Paris XI, Orsay, France.

Rouxel, T. (1988). Pouvoir pathogène de *Leptosphaeria maculans* et réaction hypersensible de *Brassica* spp. Intervention d'une pathotoxine, la sirodesmine PL, et d'une phytoalexine, la brassilexine. Thèse de l'Université de Paris XI, Orsay, France.

Rouxel, T., Sarniguet, A., Kollmann, A., and Bousquet, J. F. (1987). Identification d'une phytoalexine dans *Brassica juncea* (moutarde brune) et *Brassica napus* (colza) en relation avec la résistance à *Phoma lingam*. In *1er Congrès de la Société Française de Phytopathologie*, Nov. 19-20, 1987, SFP, Rennes, France (abst).

Rouxel, T., Chupeau, Y., Fritz, R., Kollmann, A., and Bousquet, J. F. (1988). Biological effects of sirodesmin PL, a phytotoxin produced by *Leptosphaeria maculans*. *Plant Sci. 68*: 77-86.

Rouxel, T., Sarniguet, A., Kollmann, A., and Bousquet, J. F. (1989). Accumulation of a phytoalexin in *Brassica* spp. in relation to hypersensitive reaction to *Leptosphaeria maculans*. *Physiol. Mol. Plant Pathol. 34*: 507-517.

Rouxel, T., Renard, M., Kollmann, A., and Bousquet, J. F. (1990a). Brassilexin accumulation and resistance to *Leptosphaeria maculans* in *Brassica* spp. and progeny of an interspecific cross *B. juncea* × *B. napus*. *Euphytica 46*: 175-181.

Rouxel, T., Kollmann, A., and Bousquet, J. F. (1990b). Zinc suppresses sirodesmin PL toxicity and protects *Brassica napus* plants against the blackleg disease caused by *Leptosphaeria maculans*. *Plant Sci. 68*: 77-86.

Rouxel, T., Kollmann, A., Sutre, C., Chèvre, A.-M., Renard, M., and Boulidard, L. (1990c). Induction abiotique de l'accumulation de phytoalexines soufrées chez les *Brassica* et résistance hypersensible à *Leptosphaeria maculans*. *2éme Congrès de la Société Française de Phytopathologie*, Nov. 28-30, 1990, SFP. Montpellier, France (abst.)

Rouxel, T., Kollmann, A., Boulidard, L., and Mithen, R. (1991). Abiotic elicitation of indole phytoalexins and resistance to *Leptosphaeria maculans* within Brassiceae. *Planta 184*: 271-278.

Roy, N. N. (1978). A study on disease variation in the populations of an interspecific cross of *Brassica juncea* L. × *Brassica napus* L. *Euphytica 27*: 145-149.

Roy, N. N. (1984). Interspecific transfer of *Brassica juncea*-type high blackleg resistance to *Brassica napus*. *Euphytica 33*: 295-303.

Sacristan, M. D., and Gerdemann, M. (1986). Different behavior of *Brassica juncea* and *B. carinata* as sources of *Phoma lingam* resistance in experiments of interspecific transfer to *B. napus*. *Plant Breed. 97*: 304-314.

Sarniguet, A. (1986). Accumulation d'une phytoalexine chez *Brassica juncea* en relation avec la résistance hypersensible à *Phoma lingam*. Diplôme d'Etude Approfondie en Phytopathologie, Université de Paris-Sud-Orsay, Paris, France.

Sjodin, C., and Glimelius, K. (1989). "*Brassica naponigra*" a somatic hybrid resistant to *Phoma lingam*. *Theor. Appl. Genet. 77*: 651-656.

Smith, D.A. (1982). Toxicity of phytoalexins. *Phytoalexins* (J. A. Bailey and J. W. Mansfield, eds.), Blackie, Glasgow, pp. 218-252.

Soledade, M., Pedras, C., and Taylor, J. L. (1991). Metabolic transformation of the phytoalexin brassinin by the "blackleg" fungus. *J. Org. Chem. 56*: 2619-2621.

Somei, M., Kobayashi, K., Shimizu, K., and Kawasaki, T. (1992). A simple synthesis of a phytoalexin, methoxybrassinin. *Heterocycles 33*: 77–80.

Takasugi, M., Katsui, N., and Shirata, A. (1986). Isolation of three novel sulphur-containing phytoalexins from the chinese cabbage *Brassica campestris* L. ssp. *pekinensis* (Cruciferae). *J. Chem. Soc. Chem. Commun.* 1631-1632.

Takasugi, M., Monde, K., Katsui, N., and Shirata, A. (1987). Spirobrassinin, a novel sulphur-containing phytoalexin from the daikon *Raphanus sativus* L. ssp. *hortensis* (Cruciferae). *Chem. Lett.* 1631-1632.

Takasugi, M., Monde, K., Katsui, N., and Shirata, A. (1988). Novel sulphur-containing phytoalexins from the chinese cabbage *Brassica campestris* L. ssp. *pekinensis* (Cruciferae). *Bull. Chem. Soc. Jpn. 61*: 285–289.

Taylor, J. L., Borgmann, I., and Séguin-Swartz, G. (1991a). Electrophoretic karyotyping of *Leptosphaeria maculans* differenciates highly virulent from weakly virulent isolates. *Curr. Genet. 19*: 273–277.

Taylor, J. L., Soledade, M., Pedras, C., and Morales, V. M. (1991b). Molecular genetic and chemical characterization of the rapeseed "blackleg" causal organism. *3rd Congress of the International Society for Plant Molecular Biology: Molecular Biology of Plant Growth and Development* (R. B. Hallick, ed.), Tucson, Arizona, October, 6–11, 1991 (abst).

Tempête, C., Devys, M., and Barbier, M. (1991). Growth inhibition of human cancer cell cultures with the indole sulphur-containing phytoalexins and their analogues. *Z. Naturforschung. C 46c*: 706–707.

Tsuji, J., Jackson, E. P., Gage, D. A., Hammerschmidt, R., and Somerville, S. C. (1992). Phytoalexin accumulation in *Arabidopsis thaliana* during the hypersensitive reaction to *Pseudomonas syringae* pv. *syringae. Plant Physiol. 98*: 1304-1309.

U.N. (1935). Genome-analysis in *Brassica* with special reference to the experimental formation of *B. napus* and peculiar mode of fertilization. *Jap. J. Bot. 7*: 389-452.

Williams, P. H., and Delwiche, P. A. (1979). Screening for resistance to blackleg of crucifers in the seedling stage, *Proceedings of a Eucarpia Conference on the Breeding of Cruciferous Crops*, Wageningen, Netherlands, pp. 164–170.

Williams, P. H., and Hill, C. B. (1986). Rapid-cycling populations of *Brassica. Science 232*: 1385–1389.

12

Scoparone (6,7-Dimethoxycoumarin), a *Citrus* Phytoalexin Involved in Resistance to Pathogens

Uzi Afek
The Volcani Center, Bet Dagan, Israel

Abraham Sztejnberg
The Hebrew University of Jerusalem, Rehovot, Israel

I. INTRODUCTION

Not much work has been done on induced resistance of *Citrus* to pathogens. Studies with *Phytophthora citrophthora* (Smith and Smith) Leonian indicated that morphological exclusion of the fungus in mature cells cannot be responsible for *Citrus* resistance to this pathogen, and biochemical or physiological factors may also be involved in resistant and susceptible reactions. Mature cells of resistant *Citrus* species may possess inhibitors, or hypersensitivity reaction may be induced by substances produced by the plant cells after infection (Broadbent, 1969).

Two fungitoxic compounds were found by Hartmann and Nienhaus (1974a, b) in bark of *Citrus limon* (L.) Burm. infected by *P. citrophthora* and *Hendersonula toruloidea* Natrass. One of these compounds was identified as xanthoxylin (2-hydroxy-4,6-dimethoxyacetophenone). No xanthoxylin could be demonstrated in healthy bark or following mechanical wounding or chemical treatments. Musumeci and Olivera (1975, 1976) reported about total phenol increase in sweet orange [*Citrus sinensis* (L.) Osbeck] and sour orange (*C. aurantium* L.) following inoculation with *P. citrophthora*. However, the increase in the resistant species (sour orange) was higher than in the susceptible species (sweet orange). These investigators pointed to compound 1 as a phytoalexin accumulated in *Citrus* tissue after inoculation with *P. citrophthora*.

Xanthyletin (6,7-dimethylpyranocoumarin) (Kahn et al., 1985) and seselin (7,8-dimethylpyranocoumarin) (Vernenghi et al., 1987) were isolated from *Citrus* tissue infected with *P. citrophthora* and *P. parasitica* Dastur. Both xanthyletin and seselin showed an inhibitory activity against these pathogens in vitro. Ismail et al. (1978) reported that the synthesis of umbelliferone (7-hydroxycoumarin) was greatly enhanced during healing of injured grapefruit. Ben-Yehoshua et al. (1987, 1988) isolated several antifungal substances from pomelo fruit, some of which are coumarin derivatives.

Several studies reported that scoparone is involved in defense mechanisms of *Citrus* against pathogens such as *P. citrophthora* (Afek and Sztejnberg, 1986, 1988a), *Guignardia citricarpa* Kiely (De Lange et al., 1976), *Penicillum digitatum* Sacc. (Kim et al., 1991; Rodov et al., 1992), and *Diaporthe citri* (Faw.) (Aritmo et al., 1986). Treatments such as γ irradiation (Riov, 1971; Dubery and Schabort, 1987; Afek and Sztejnberg, 1993), high temperature (Afek and Sztejnberg, 1988b, 1993; Aritmo and Homma, 1988; Kim et al., 1991), UV illumination (Rodov et al., 1992), and fosetyl-Al and phosphorous acid (Afek and Sztejnberg, 1989) were found to stimulate scoparone production in *Citrus*.

This chapter examines the role of scoparone as a phytoalexin involved in *Citrus* resistance to pathogens. Different chemical and physical treatments that may increase scoparone concentration in *Citrus* tissue and an attempt to understand the biosynthesis of scoparone from phenylalanine are discussed.

II. MATERIALS AND METHODS

Fungal and Plant Material

Experiments were done with the fungus *Phytophthora citrophthora* (Smith and Smith) Leonian (isolate C-16) and with 3-month-old *Citrus* branches and mature fruits. Twenty-five to 30-cm-long and 7- to 10-mm-thick branches were pruned from 3-year-old *Citrus* seedlings; *Citrus sinensis* (L.) Osbeck (Shamouti); *C. aurantium* L. (sour orange); *C. jambhiri* Lush. (rough lemon); *Poncirus trifoliata* Raf. (trifoliate orange); *C. reticulata* Blanco × *C. sinensis* (Niva); and *C. macrophylla* Webster (Macrophylla). These species were chosen because one group (Macrophylla, trifoliate orange, and sour orange) is resistant to *P. citrophthora* and the other group (rough lemon, Shamouti, and Niva) is susceptible to *P. citrophthora*.

Mature *Citrus* fruits of *C. paradisi* Macfadden (grapefruit), *C. sinensis* (Valencia), Shamouti, and sour orange were picked from a 12-year-old orchard.

A. Inoculation

Incisions were made in the bark of branches and agar discs were cut from an actively growing zone of mycelia of *P. citrophthora* on PDA medium and placed over the incisions, fungal side downward. The inoculated branch sections were incubated in a humid chamber at 20°C and 28°C in darkness (Afek and Sztejnberg, 1988a). Fruits were inoculated by removing pieces of flavedo with a 3-mm-diameter cork borer to a depth of 0.2–0.5 mm at four to five sites around the equatorial plane of the fruit. A 3-mm-diameter disc, cut from an actively growing PDA culture of *P. citrophthora*, was placed on the fruit wounds and the inoculated fruits were incubated in a humid chamber at 24°C in darkness.

B. Extraction, Purification, Identification, and Quantification of Scoparone

Slices of inoculated, necrotic bark, cut from the outer edge of the wounds, were extracted with distilled water. The antifungal component extracted from the inoculated bark was concentrated and was eluted with EtOAc/ petroleum ether (1 : 1, v/v) and partitioned with $CHCl_3$ from which it crystallized upon evaporation, as colorless needles with mp 146–147°C.

Ultraviolet (UV) spectrophotometry, infrared (IR) analysis, ^{1}H NMR (nuclear magnetic resonance) and ^{13}C NMR spectra indicated that the active antifungal agent that was extracted and purified from *Citrus* bark and fruit inoculated with *P. citrophthora* was scoparone (Fig. 1).

C. Scoparone Quantification

Solutions of *Citrus* samples were analyzed spectrofluorometrically with excitation at 340 nm and emission reading at 430 nm. Concentration of scoparone in the *Citrus* tissue was calculated by a comparison with the standard curve (Afek and Sztejnberg, 1988a; Afek et al., 1986).

D. γ Irradiation

γ Ionizing irradiation was applied with cobalt radiant (^{60}Co). Branches and fruits were irradiated with 0-, 100-, 200-, 300-, and 400-krad dose. Inoculation of the irradiated branch and fruit with *P. citrophthora* was done 24 hr later.

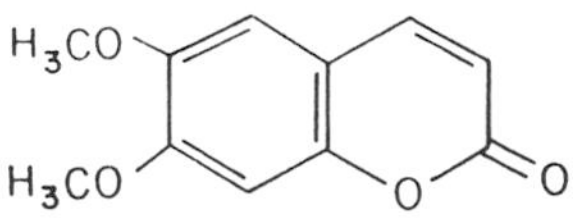

H_3CO H_3CO O O

Figure 1 Scoparone.

E. Fosetyl-Al and Phosphorus Application and Bioassay

Citrus branches were immersed in aqueous solutions containing fosetyl-Al (0–800 μm/ml) or H_3PO_3 (0–400 μm/ml) for 3 hr. The branches were then washed with tap water and inoculated with *P. citrophthora*. Advance of the pathogen (lesion length) and scoparone concentration were measured 4 days later.

F. Bioassay

ED_{50} values were determined by adding increasing concentrations of scoparone or fosetyl-Al or H_3PO_3 to cooled molten PDA immediately before it was poured into Petri plates. Disc of *P. citrophthora*, taken from an actively growing colony on PDA, was placed fungal side downward in the center of each plate. Plates were incubated in darkness at 25°C for 8 days (Afek and Sztejnberg, 1988a, 1989).

G. Labeling and Aminooxyacetic Acid Application

Excised branches (1 cm length and 0.5 cm diameter) were immersed in a solution containing [^{14}C]phenylalanine (3.14×10^5 dpm) or in 10 mM aminooxyacetic acid (AOA) (in sterile deionized water) for 3 hr, inoculated with *P. citrophthora*, and incubated at 24°C. The labeled scoparone, concentration of scoparone, and lesion length were measured 4 days later (Afek and Sztejnberg, 1988a).

III. SCOPARONE PRODUCTION IN *CITRUS*

A. Accumulation of Scoparone in Various Plant Parts and in Different Varieties of Citrus

Small concentrations of scoparone naturally exist in healthy *Citrus* barks and fruit peel (Afek and Sztejnberg, 1988a; Tatum and Berry, 1977). However, its concentration in *Citrus* tissue increases following inoculation with *P. citrophthora*. The production of scoparone and the lesion length in the bark of 3-month-old *Citrus* branches, resistant (macrophylla, trifoliate, and sour orange) and susceptible (rough lemon, Shamouti, and Niva) to *P. citrophthora*, were measured daily during incubation at 20°C and 28°C for 8 days after inoculation with the pathogen. Scoparone was induced in both groups of *Citrus*, but the concentration was higher and increased more rapidly in the resistant species. Twenty-four hours after the inoculation, concentration of scoparone in bark of the resistant species was about three times more than in the susceptible species. High concentration of the phytoalexin within the first 24–48 hr is critical in stopping the advance of the pathogen in vivo. However, scoparone concentration continued to increase and reached the maximum at 20°C, 4 days after the inoculation. The concentrations were 440, 415, and 250 μg/g fresh wt in macrophylla, trifoliate, and sour orange, and 42, 31, and 28 μg/g fresh wt in rough lemon, Shamouti, and Niva, respectively (Fig. 2).

The length of lesions caused by *P. citrophthora* in the bark 4 days after inoculation was 2.5 mm in macrophylla, 3.2 mm in trifoliate, 5.0 in sour orange, 11.0 mm in rough lemon, 15.5 mm in Shamouti, and 17.0 mm in Niva (Fig. 2).

Scoparone showed inhibitory activity against various phytopathogenic fungi in vitro (Table 1) and ED_{50} value for inhibition growth of *P. citrophthora* was 97 μg/ml (Fig. 3).

These results indicate that scoparone is involved in resistance and not a result of resistance (necrosis). Lesion length in the branches caused by *P. citrophthora* infection is inversely proportional to the increase in the phytoalexin concentration. As the lesion increases the concentration of scoparone decreases. Lesion length, 4 days after inoculation with *P. citroph-*

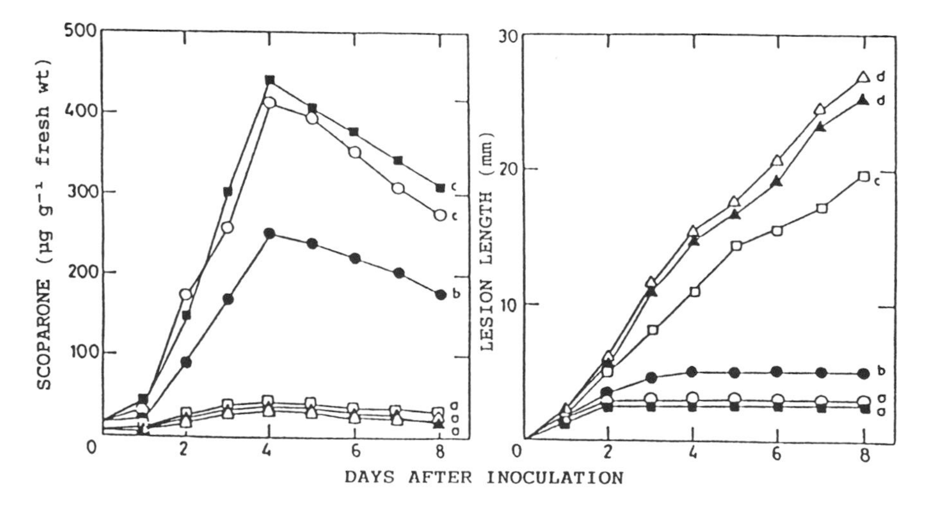

Figure 2 Accumulation of scoparone and lesion length in *Citrus* bark of the resistant species macrophylla (■), trifoliate orange (○), sour orange (●), and the susceptible species rough lemon (□), Shamouti (△), Niva (▲), after inoculation with *Phytophthora citrophthora* at an incubation temperature of 20°C. Significant differences were indicated by different letters within each time period according to Duncan's multiple range test ($p = 0.05$). Statistical analysis of scoparone concentration was done starting on the second day and lesion length on the third day (Afek and Sztejnberg, 1988a).

Table 1 Effective Dose of Scoparone for 50% Inhibition (ED_{50}) of Mycelial Growth of *Phytophthora citrophthora* Compared with Conidial Germination Inhibition of Six Other Phytopathogenic Fungi In Vitro

Fungal species	ED_{50} of scoparone (μg/ml)
Phytophthora citrophthora	97
Verticillium dahliae	61
Penicillium digitatum	64
Penicillium italicum	60
Colletotrichum gloeosporioides	54
Hendersonula toruloidea	90
Botryiodiplodia (Diplodia) natalensis	85

Source: Data from Afek et al. (1986).

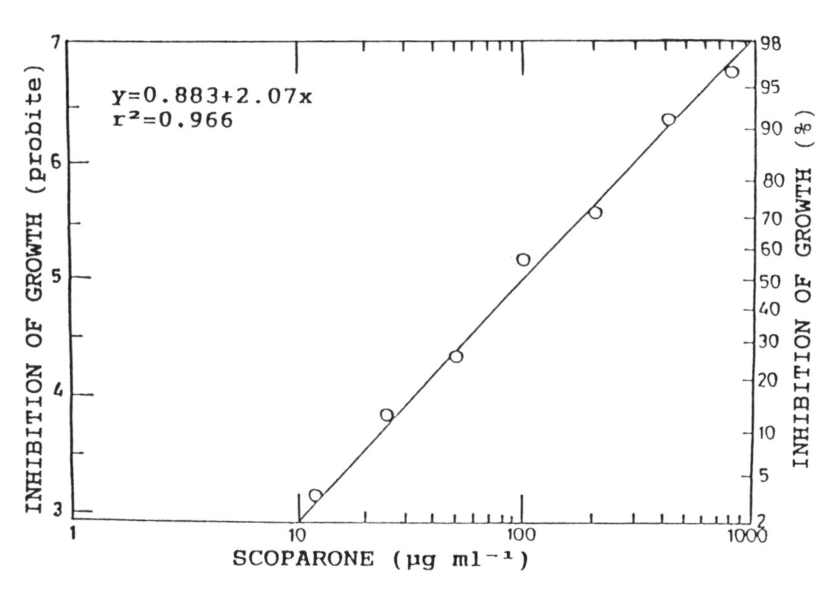

Figure 3 Dosage-response of *Phytophthora citrophthora* mycelial growth to log concentration of scoparone, expressed as a linear regression (Afek and Sztejnberg, 1988a).

thora at incubation temperatures of 20°C and 28°C, was negatively correlated with increased accumulation of scoparone in vivo (Fig. 4). Unlike Hartmann and Nienhaus (1974b) the authors did not find any correlation between the degree of *Citrus* resistance and the accumulation of xanthoxylin in the bark after infection with *P. citrophthora*. Accumulation of xanthoxylin appeared to be a result of the necrotic reaction and seemed to play a part in eliminating the pathogen after it had been inhibited by other defense mechanisms of the plant.

Arimoto et al. (1986) reported similar results with the fungus *Diaporthe citri* and suggested that scoparone was produced by the *Citrus* in response to infection by this pathogen.

Musumeci and Olivera (1975, 1976) described compound 1 as a phytoalexin accumulated in *Citrus* after inoculation with *P. citrophthora*. Because the final structural formula of compound 1 is still unknown, this compound cannot be related to one of the known *Citrus* phytoalexins. However, according to the description, it seems that compound 1 belongs to the coumarin group. Other coumarins such as xanthyletin (Khan et al., 1985) and seselin (Vernenghi et al., 1987) may also play a part as phytoalexins in *Citrus* resistance to pathogens.

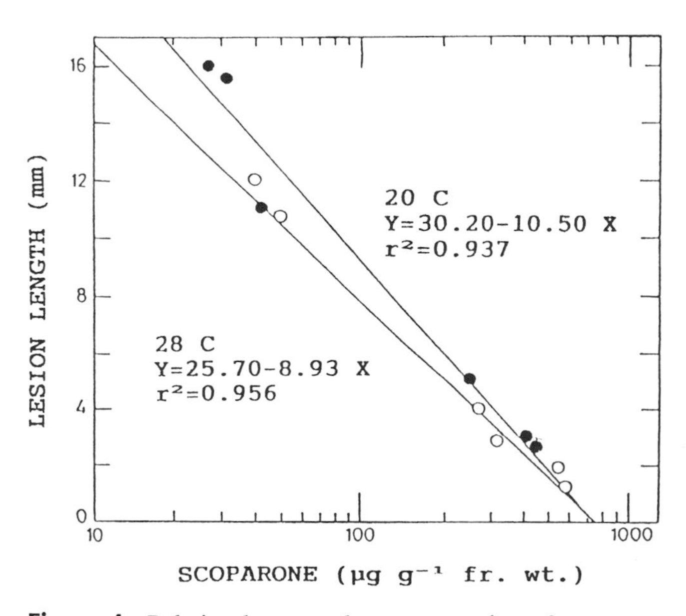

Figure 4 Relation between the concentration of scoparone and lesion length in *Citrus* bark 4 days after inoculation with *Phytophthora citrophthora* at an incubation temperature of 20°C (●) and 28°C (○) (Afek and Sztejnberg, 1988b).

B. Temperature Effect

In all the tested species, scoparone concentration, after inoculation with *P. citrophthora*, was higher while the lesion length was shorter at 28°C (Figs. 2 and 5). It was especially expressed in rough lemon, which showed resistant reaction. The concentration of scoparone increased up to 290 μg/g fresh wt (as compared to 42 μg/g fresh wt at 20°C after 4 days of incubation). The lesion length in the bark of rough lemon, at 28°C 4 days after the inoculation, was 5.2 mm. However, in the other species scoparone concentrations in the same period were 587, 515, and 326 μg/g fresh wt in macrophylla, trifoliate, and sour orange, and 53 and 39 μg/g fresh wt in Shamouti and Niva, respectively. Lesion length in the bark in these conditions was 1.3 mm in macrophylla, 1.9 mm in trifoliate, 3.5 mm in sour orange, 10.8 mm in Shamouti, and 11.7 mm in Niva (Fig. 5).

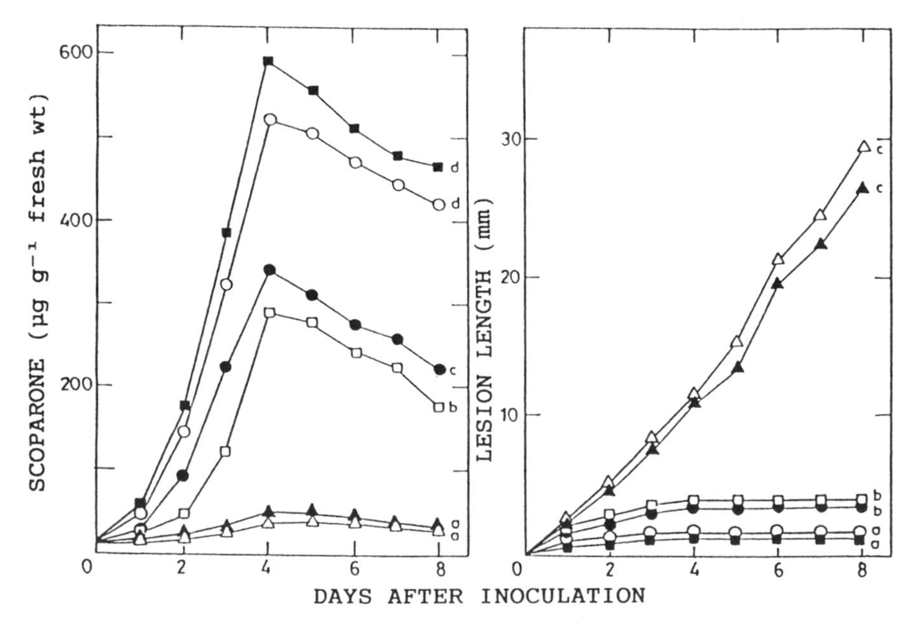

Figure 5 Accumulation of scoparone and lesion length in *Citrus* bark of the resistant species macrophylla (■), trifoliate orange (○), sour orange (●), and the susceptible species rough lemon (□), Shamouti (△), Niva (▲), after inoculation with *Phytophthora citrophthora* at an incubation temperature of 28°C. Significant differences were indicated by different letters within each time period according to Duncan's multiple range test ($p = 0.05$). Statistical analysis of scoparone concentration was done starting on the second day and lesion length on the third day (Afek and Sztejnberg, 1988b).

In the noninoculated control the concentration of scoparone was 12–18 μg/g fresh wt (at both temperatures) and wounding had no effect on scoparone production.

Temperature effect on *Citrus* resistance was discussed by Hartmann and Nienhaus (1974a). Lemon trees, which were found to be susceptible to *P. citrophthora* at 20°C, were resistant at 28°C. The optimal temperature for mycelial growth of this pathogen in vitro is 28°C. Disease inhibition in *Citrus* bark at 26–28°C seems to be caused by defense mechanisms of the host tissue. Results of this study confirm that the elevated temperature (28°C) increases resistance of *Citrus* to *P. citrophthora* by stimulating scoparone production (Fig. 5). This may explain why several *Citrus* rootsocks were found to be more resistant to *P. citrophthora* in the summer than in the winter (Hartmann and Nienhaus, 1974a; Broadbent, 1969).

In both temperatures scoparone concentration reached the peak on the fourth day after inoculation and then started to decline. Such a pattern of accumulation and degradation is typical for phytoalexins in plants (Bailey and Mansfield, 1982).

Kim et al. (1991) tested the temperature effect on lemon fruit resistant to *Penicillium digitatum*. It was found that greater amounts of scoparone accumulated in lemon peel when inoculated with *P. digitatum* and incubated at 36°C as compared to 17°C. Heat treatment at 36°C prevented decay development while treatment at 17°C had no effect. However, since the optimum temperature is 24°C for *P. digitatum* in vitro it is possible that both heat treatment at 36°C and scoparone accumulation have an inhibitory effect on *P. digitatum* in vivo (Fig. 6). The effect of temperature on scoparone production was studied by Arimoto and Homma (1988), who found that the largest quantity of this compound was produced in *Citrus* melanose spots and scars at 25°C, with progressively lower levels at lower temperatures until no scoparone was detected at 10°C or 5°C. Similarly, results of Afek and Sztejnberg (1988b) indicated that the high-temperature effect was strongly expressed in rough lemon. This species is considered to be susceptible to *P. citrophthora*. Rough lemon branch that gave a susceptible reaction at 20°C reacted as a resistant species at 28°C. In this case scoparone concentration in the bark was seven times higher at 28°C than at 20°C, in parallel with the increase in the resistance of *Citrus* against *P. citrophthora*.

C. Effect of γ Irradiation and UV Illumination

Citrus branch and fruit were inoculated with *P. citrophthora* 24 hr after treatment with different γ-irradiation doses. Scoparone concentration and lesion length in the branches were measured 4 days after the inoculation

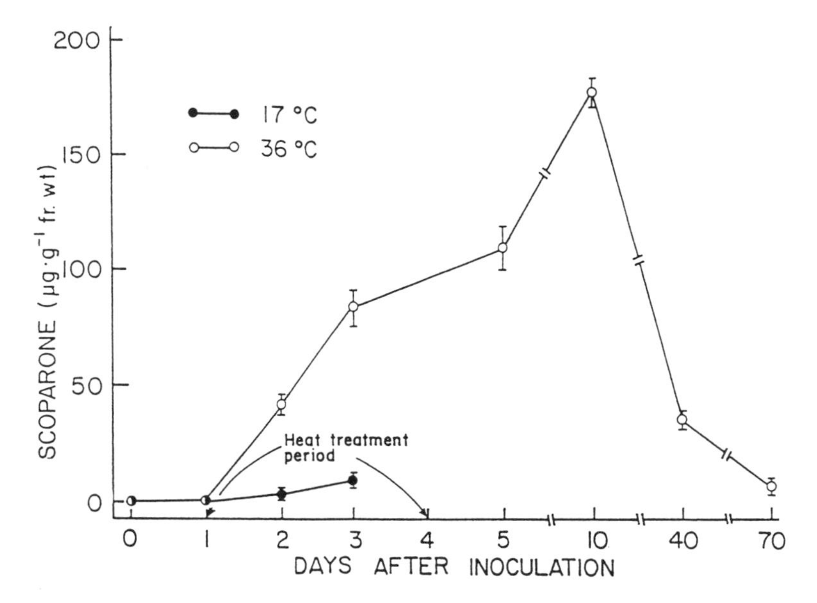

Figure 6 Effect of heat treatment on accumulation of scoparone in lemon flavedo inoculated with *Penicillium digitatum* (Kim et al., 1991).

whereas scoparone concentration and infected area of the fruits were measured after 7 days. Maximum scoparone concentration in the inoculated branch was achieved after treatment with 400 krad γ irradiation and reached up to 970 and 530 μg/g fresh wt in macrophylla and sour orange, and 100 and 82 μg/g fresh wt in rough lemon and Shamouti, respectively (Table 2).

In the noninoculated, irradiated (400-krad) branch, scoparone concentration was much higher (about seven times) in the resistant species and moderately (two to three times) in susceptible species as compared to the nonirradiated control. However, γ irradiation significantly increased scoparone concentration in both inoculated and noninoculated branches only when treated with a 300- or 400-krad dose. This treatment also resulted in significant decrease of lesion length in all the tested species (Table 2).

In fruits (sour orange, Valencia, Shamouti, and grapefruit), scoparone concentration after treatment with 400 krad and subsequent inoculation with *P. citrophthora* was about two to three times greater than in the nonirradiated fruit. Maximum concentration of 94.7 μg/g fresh wt was achieved in infected grapefruit (Table 3). However, as in branches, a significant increase in scoparone concentration occurred only after treatment with 300 and 400 krad γ irradiation.

Table 2 Scoparone (scop) Concentration (μg/g fresh wt) and Lesion Length (les leng) (mm) in Bark of the Resistant *Citrus* Species Macrophylla and Sour Orange, and the Susceptible Species Rough Lemon and Shamouti Noninoculated (non) and 96 hr After Inoculation (ino) with *Phytophthora citrophthora* at Incubation Temperatures of 24°C Treated with 0, 100, 200, 300, and 400 krad γ-Irradiation Dose ($p = 0.05$)

Dose (krad)	Macrophylla			Sour orange			Rough lemon			Shamouti		
	scop		les leng	scop		les leng	scop		les leng	scop		les leng
	non	ino	ino	non	ino	ino	non	ino	ino	non	ino	ino
0	18	451	2.6[a]	17	242	5.3	17	38	10.0	14	32	13.0
100	17	443	2.7	16	257	4.5	16	42	11.5	15	28	12.1
200	20	467	2.4	18	253	5.0	17	45	11.0	15	35	14.2
300	80	630	1.0	83	380	2.5	35	73	6.0	24	50	9.0
400	127	970	0.8	115	530	2.0	57	100	4.0	34	82	6.0
LSD	22	132	0.6	21	73	1.2	11	19	1.6	8	14	2.2

[a]Each number in the table is an average of five replicates.

Table 3 Scoparone (scop) Concentration ($\mu g/g$ fresh wt) and Lesion Area (les area) (mm^2) in Fruit Peel of Valencia, Sour Orange, Grapefruit, and Shamouti Noninoculated (non) and 7 days after Inoculation (ino) with *Phytophthora citrophthora* at Incubation Temperatures of 24°C Treated with 0, 100, 200, 300, 400 krad γ-Irradiation Dose ($p = 0.05$)

Dose (krad)	Valencia			Sour orange			Grapefruit			Shamouti		
	scop		les area	scop		les area	scop		les area	scop		les area
	non	ino	ino	non	ino	ino	non	ino	ino	non	ino	ino
0	8.6	15.0	196[a]	8.6	13.1	171	9.5	25.9	175	9.6	16.4	183
100	10.3	16.2	179	8.4	16.6	170	11.2	28.1	171	9.0	18.6	186
200	10.5	17.5	162	9.0	14.0	191	9.8	27.4	163	10.7	18.1	160
300	19.2	31.3	167	17.3	33.3	167	26.7	77.2	83	17.8	25.3	154
400	18.7	30.4	162	18.1	35.8	163	31.0	94.7	50	19.0	28.7	150
LSD	4.3	7.2	37	4.7	6.6	41	7.0	15.0	31	3.9	6.1	43

[a]Each number in the table is an average of five replicates.

γ Irradiation affected the size of lesions only in grapefruit (Table 3). Infected areas of grapefruit significantly decreased from 175 mm^2 to 83 and 50 mm^2 after treatment with 300 and 400 krad, respectively.

Riov (1971) and Dubery and Schabort (1987) had reported that scoparone accumulated in *Citrus* fruit peel after treatment with γ irradiation. These investigators did not detect scoparone in nonirradiated tissue. Results of this research show that a small amount of scoparone naturally exists in branches and fruits of various species of *Citrus* (Tables 2 and 3).

Furthermore, this study suggests that a correlation exists among scoparone concentration, lesion length, and resistance of *Citrus* tissue. γ Irradiation (300 and 400 krad) increased scoparone concentration in both inoculated (*P. citrophthora*) and noninoculated branches and fruits (Tables 2 and 3). The level of scoparone in the inoculated, irradiated bark, both resistant and susceptible, and grapefruit fruit peel was sufficient to inhibit the growth of the fungus in vivo while in sour orange, Shamouti, and Valencia fruits scoparone concentration was low and hence the fungal growth was not inhibited.

Riov et al. (1968) reported that γ irradiation increased the activity of phenylalanine ammonia-lyase (PAL), a key enzyme in the biosynthesis of coumarins and phenols in plants (Hanson and Havir, 1981; Legrand, 1983; Jones, 1984). Afek and Sztejnberg (1988a) also supported this finding and related it to scoparone accumulation. They showed that AOA, a competitive inhibitor of PAL, suppressed scoparone production in *Citrus* and this was accompanied with decreased resistance.

Kim et al. (1991) measured the concentrations of scoparone in lemon fruit illuminated with UV light. The concentrations were 27, 77, and 120 μg/g fresh wt when treated with 1.5, 3.0, and 4.5 $\times$ 10^4 erg/mm, respectively, after 7 days of incubation. Later, scoparone concentration declined (Fig. 7). Rodov et al. (1992) tested the effect of UV light in kumquat. The quantity of the phytoalexin reached a peak of 530 μg/g fresh wt 11 days after treatment with UV dose of 1.5 $\times$ 10^3 J/m^2 and then declined rapidly to trace levels after a month (Fig. 8).

Chalutz et al. (1992) reported that grapefruit exposed to 30–60 sec of UV illumination and subsequently inoculated with *P. digitatum* showed reduction in incidence of green mold. A process of induced resistance in grapefruit seemed to develop gradually with time after treatment. Maximum response was observed between 24 and 48 hr after the treatment. The activity of PAL in the peel of UV-treated grapefruit increased within 24 hr following treatment and remained high at 48 hr, compared with the low and unchanged level of PAL activity in the peel of the nontreated control. These results suggest a possible involvement of an induced mechanism of resistance. This may be due to an increase of antifungal compounds in vivo.

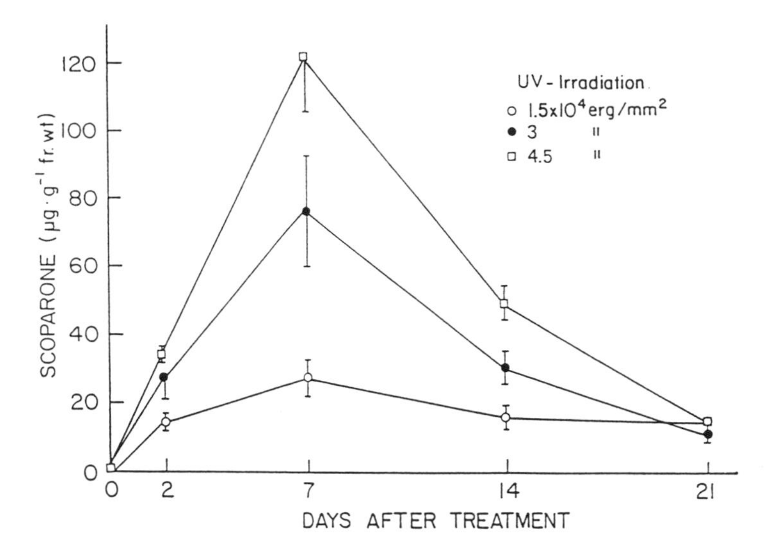

Figure 7 Effect of UV dose on accumulation of scoparone in lemon fruits (Kim et al., 1991).

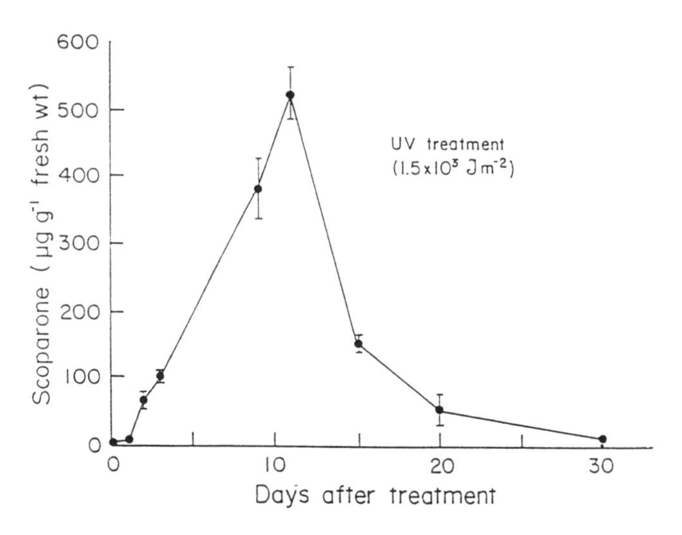

Figure 8 Time course of scoparone accumulation and depletion in UV-treated kumquat fruit (Rodov et al., 1992).

However, the mechanism of resistance suggested by Chalutz et al. (1992) has to be stimulated within the first 24–48 hr after treatment and infection with the pathogen.

We now can assume that the concentrations of scoparone required for inhibition of growth of *P. citrophthora* in vitro and in vivo are similar. A comparison between scoparone concentrations that inhibit the advance of *P. citrophthora* in vivo and its growth in vitro showed that the advance of the pathogen in marcophylla and trifoliate terminates 2 days after the inoculation, when the concentration of the phytoalexin is about 150 $\mu g/g$ fresh wt (Fig. 2). This concentration is equivalent to ED_{65} for *P. citrophthora* in vitro (Fig. 3). In the susceptible *Citrus* species when scoparone concentrations remain low the advance of the pathogen is not inhibited.

D. Effect of Fosetyl-Al and Phosphorous Acid

Fosetyl-Al and phosphorous acid were found to induce resistance in *Citrus* against *P. citrophthora* by increasing scoparone accumulation in the tissue (Afek and Sztejnberg, 1989). Studies have shown that fosetyl-Al (trade name Aliette, Rhone-Poulenc Sanitaire, Lyon, France) has little effect on mycelial growth of oomycetes in vitro (Sanders et al., 1983; Bompeix and Saindrenan, 1984), but has the ability to control diseases of plants caused by these pathogens (Farih et al., 1981a, b; Sanders et al., 1983). Coffey and Bower (1984), Fenn and Coffey (1984, 1985), and Ouimette and Coffey (1989) suggested that fosetyl-Al is degraded into phosphorous acid (H_3PO_3) in vivo and this compound acts directly against the pathogen as a fungicide. Alternatively, fosetyl-Al may act indirectly against pathogens by activating host defense mechanisms (Bompeix et al., 1980; Guest, 1984a, b; Vernenghi and Ravise, 1985; Khan et al., 1986).

The accumulation of scoparone and the advance of *P. citrophthora* (lesion length) in the bark of 3-month-old *Citrus* branches were measured after treatment with fosetyl-Al or H_3PO_3 and inoculation with *P. citrophthora*. In all the *Citrus* species tested except for Niva (very susceptible to the pathogen), concentration of scoparone in inoculated bark treated with 300 $\mu g/ml$ fosetyl-Al or 125 $\mu g/ml$ H_3PO_3 was two to four times greater than in the controls (inoculated and nontreated branches). Treatment with higher concentration of these compounds decreased scoparone concentration (Figs. 9 and 10).

Lesion length in macrophylla, sour orange (resistant to *P. citrophthora*), and rough lemon (susceptible to *P. citrophthora*) decreased more rapidly than in Niva after treatments with 0–300 $\mu g/ml$ of fosetyl-Al or 0–125 $\mu g/ml$ of H_3PO_3. In Niva, treatments with fosetyl-Al and H_3PO_3 had no effect on scoparone concentrations, and lesion length sharply decreased

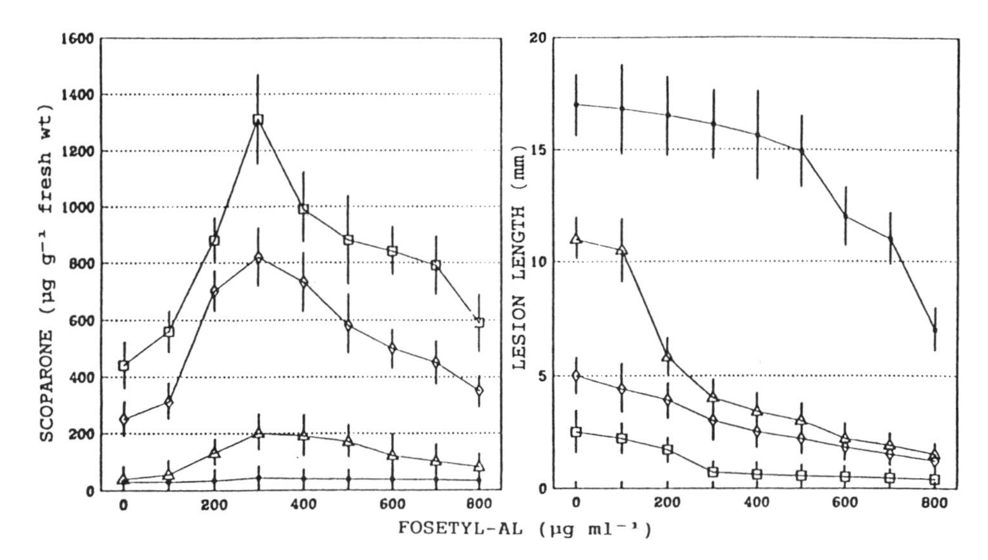

Figure 9 Accumulation of scoparone (*left*) and lesion length (*right*) in *Citrus* bark of the resistant species macrophylla (□) and sour orange (◇), and the susceptible species rough lemon (△), and Niva (■), 96 hr after inoculation with *Phytophthora citrophthora* at an incubation temperature of 20°C. Branches were treated with fosetyl-Al 3 hr before inoculation. Vertical bars are standard errors. (Afek and Sztejnberg, 1989).

only when treated with more than 500 µg/ml fosetyl-Al or 200 µg/ml H_3PO_3 (Figs. 9 and 10). Fosetyl-Al and H_3PO_3 did not induce scoparone production in healthy tissue in any of the species tested. The regression analysis showed that inhibition of mycelial growth area of *P. citrophthora* was significantly correlated with increasing concentrations of fosetyl-Al (r^2 = 0.960, $P < 0.01$) and H_3PO_3 (r^2 = 0.949, $P < 0.01$). The ED_{50} values of fosetyl-Al and H_3PO_3 for mycelial growth were 55 and 7 µg/ml, respectively.

Probably the argument as to whether fosetyl-Al and H_3PO_3 act directly against pathogens in vivo (Coffey and Bower, 1984; Fenn and Coffey, 1984, 1985; Ouimette and Coffey, 1989) or act indirectly against pathogens in vivo by increasing resistance (Bompeix et al., 1980; Guest, 1984a, b; Khan et al., 1985, 1986; Vernenghi and Ravise, 1985) will remain open. Results of the present study support both opinions for direct and indirect effects of fosetyl-Al and H_3PO_3 on *P. citrophthora* in vivo. Similarly, Smillie et al. (1989) presented evidence for both direct and indirect modes of action of phosphite on *Phytophthora* spp. causing disease in plants.

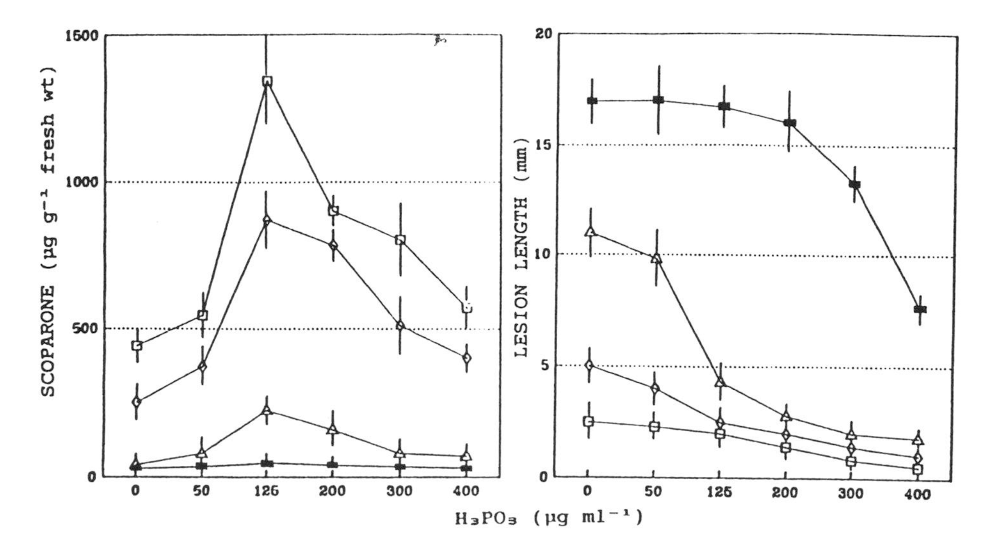

Figure 10 Accumulation of scoparone (*left*) and lesion length (*right*) in *Citrus* bark of the resistant species macrophylla (□) and sour orange (◇), and the susceptible species rough lemon (△), and Niva (■), 96 hr after inoculation with *Phytophthora citrophthora* at an incubation temperature of 20°C. Branches were treated with H_3PO_3 3 hr before inoculation. Vertical bars are standard errors (Afek and Sztejnberg, 1989).

Fosetyl-Al and H_3PO_3 induce resistance in *Citrus* by increasing concentrations of scoparone in macrophylla, sour orange, and rough lemon. However, when fosetyl-Al penetrates the plant it is degraded to H_3PO_3, which is much more toxic to *Phytophthora* spp. than fosetyl-Al and inhibits the pathogen as a fungicide (Coffey and Bower, 1984; Fenn and Coffey, 1984, 1985). Results of this study show that the concentration of H_3PO_3 required for the same effect as that of fosetyl-Al is 40%. This is a good indication that fosetyl-Al is degraded into about 40% H_3PO_3 in vivo.

This research has led to the suggestion that fosetyl-Al and H_3PO_3 affect *P. citrophthora* in vivo in two ways. Apparently, at low-level treatments it stimulates the host defense mechanisms, while at higher levels it acts directly as a fungicide. In the *Citrus* species macrophylla, sour orange, and rough lemon, maximum production of scoparone is stimulated by 300 μg/ml fosetyl-Al or 125 μg/ml H_3PO_3, but the concentration of scoparone in Niva remains low (Figs. 9 and 10). This is explained by the "potential of resistance" (the particular selection of *Citrus* to produce scoparone) that these three species have and Niva has not.

Scoparone, which is involved in *Citrus* resistance, and the fungicide both have an effect on *P. citrophthora* in *Citrus* plants with the potential of resistance, whereas in Niva only the fungicide has an effect on the fungus. Thus, the concentrations of fosetyl-Al and H_3PO_3 that are required to stop the advance of the pathogen in macrophylla, sour orange, and rough lemon are lower than those in Niva.

To support this hypothesis, we assume that in vitro and in vivo ED_{50} values are similar and that lesion length in vivo reasonably reflects fungal growth in vitro. From Figs. 9 and 10 it appears that 50% lesion length in Niva (on which fosetyl-Al and H_3PO_3 directly affected the pathogen without the involvement of scoparone) occurred after application of fosetyl-Al (750 μg/ml) or H_3PO_3 (350 μg/ml). In vitro, the ED_{50} values for fosetyl-Al or H_3PO_3 were 55 and 7 μg/ml, respectively. It suggests that only 55 μg/ml fosetyl-Al and 7 μg/ml H_3PO_3 are present when ED_{50} value is reached in vivo. If one assumes from this that only 2% of the applied H_3PO_3 or 7.3% of fosetyl-Al enter the tissue, it means 22 μg/ml fosetyl-Al and 2.5 μg/ml H_3PO_3, which are equivalent to only ED_{20} and ED_{25} values, respectively, in vitro. Probably, these concentrations in macrophylla, sour orange, and rough lemon cannot stop the advance of *P. citrophthora* without the involvement of scoparone.

Furthermore, treatments of inoculated macrophylla, sour orange, and rough lemon with 800 μg/ml fosetyl-Al or 400 μg/ml H_3PO_3 did not increase scoparone concentration in vivo. The explanation for this is that treatments with high concentrations of these compounds inhibit the physiological activities of *P. citrophthora*. When the physiological activities of the pathogen are completely inactivated, the effect of fosetyl-Al and H_3PO_3 on scoparone production is similar to their effect on noninfected tissue. Fosetyl-Al or H_3PO_3 alone cannot stimulate scoparone production in healthy tissue.

E. Probable Hypothesis for the Biosynthetic Pathway of Scoparone in *Citrus*

Cinnamic acid was reported to be a precursor of several coumarins in plants (Sequeira, 1969; Fritig, 1972; Clarke and Baines, 1976). Other studies show that the amino acid phenylalanine is the precursor of cinnamic acid in plants (Amerhein and Zenk, 1977; Grisebach, 1977; Gross, 1978). The results of Afek and Sztejnberg (1988a) indicate that the precursor of scoparone is apparently phenylalanine. The authors tried to prove this theory in two different ways: measurement of the distribution of radioactive scoparone in the bark 4 days after treatment with the isotope [^{14}C]phenylalanine and inoculation with *P. citrophthora*. The results showed that scoparone

```
                PAL
        AOA---->¦
                ¦
                V
PHENYLALANINE---->CINNAMIC ACID---->P-COUMARIC ACID---->

---->4-METHOXY O-COUMARIC ACID---->7-METHOXYCOUMARIN
```

Figure 11 Biosynthesic pathway of 7-methoxycoumarin from the amino acid phenylalanine (Brown, 1978; Legrand, 1983).

was labeled with ^{14}C. Total incorporation was 11% and 15% in the resistant species, sour orange, and macrophylla, respectively, as compared to 1.5% in the susceptible species, Shamouti and Niva.

Phenylalanine is deaminated to *trans*-cinnamic acid by the enzyme PAL (Hanson and Havir, 1981; Legrand, 1983; Jones, 1984).

It is known that AOA is a competitive inhibitor of PAL (Amerhein, 1978; Hoagland and Duke, 1982). The precursor of simple coumarin and 7-methoxycoumarin, whose chemical structure is similar to scoparone (6,7-dimethoxycoumarin), is cinnamic acid (Brown, 1978). We can assume that the biosynthetic pathways of scoparone and 7-methoxycoumarin are similar. Inhibition of PAL activity by AOA may inhibit the production of scoparone in the biosynthesis pathway (Fig. 11). Results of the present research show that resistant species responded as susceptible species following treatment with 10 mM AOA. Scoparone concentrations in the bark of the resistant species treated with 10 mM AOA 4 days after inoculation with *P. citrophthora* was 32.5–43.4 μg/g fresh wt as compared to 295–472 μg/g fresh wt in the control (inoculated and nontreated branches).

Lesion length in this group, 4 days after the inoculation varied from 9.2 to 12.0 mm as compared to 2.8 to 5.1 in the nontreated control (Fig. 12). The effect of AOA on scoparone concentration and on lesion length in the susceptible species was insignificant.

IV. EPILOG

We believe that the present research establishes the role of scoparone as a phytoalexin conferring resistance of *Citrus* to *P. citrophthora*. This chapter confirms the hypothesis of Broadbent (1969) that morphological exclusion of *P. citrophthora* in *Citrus* tissue cannot be responsible for the resistance to this pathogen, and that biochemical or physiological differences must exist between resistant and susceptible species. Our study shows differences between resistant and susceptible *Citrus* species following inoculation with *P. citrophthora*. Scoparone accumulates in both groups but the concentration in the resistant species rapidly increases and reaches 10–15 times higher

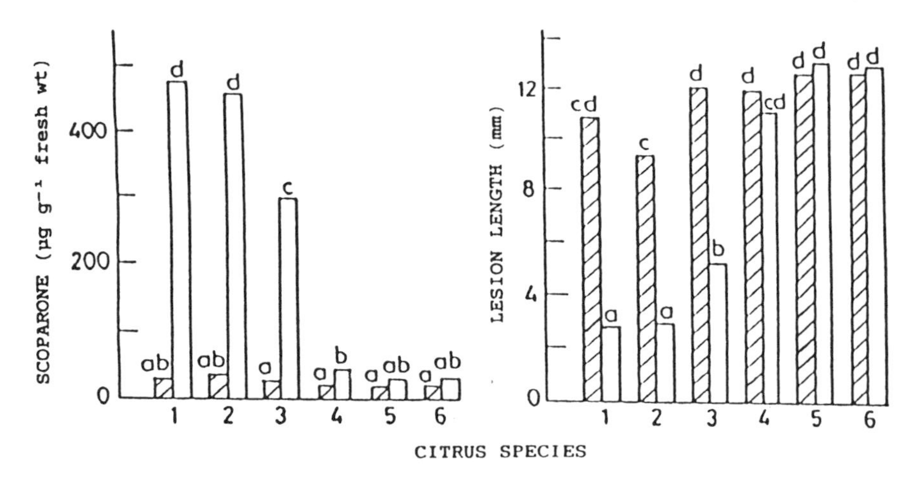

Figure 12 The concentration of scoparone (*left*) and lesion length (*right*) in the bark of the following *Citrus* species 96 hr after inoculation with *Phytophthora citrophthora*: (1) macrophylla (resistant); (2) trifoliate orange (resistant); (3) sour orange (resistant); (4) rough lemon (susceptible); (5) Shamouti (susceptible); (6) Niva (susceptible), treated (▨) and nontreated (□) with aminooxyacetic acid (AOA). The incubation temperature after the inoculation was 20°C. Different letters indicate significant differences according to Duncan's multiple range test (p = 0.05) (Afek and Sztejnberg, 1988a).

than that in the susceptible species. The advance of the pathogen (lesion length) at that time is 2–6 times greater in the susceptible species as compared to the resistant species.

The effect of physical treatments, such as temperature, γ irradiation, and UV illumination, and of chemical treatments, such as fosetyl-Al and phosphorous acid, on *Citrus* resistance is expansively discussed in this chapter. Difference between the high- and low-temperature effect on *Citrus* can give a good explanation for seasonal influences on *Citrus* susceptibility. Furthermore, nonchemical approaches, such as high temperature and γ-irradiation treatments, applied to increase resistance against diseases may replace, or at least reduce, the use of chemicals against pathogens.

Results of fosetyl-Al and phosphorous acid experiments give another point of view for the argument whether these compounds act directly on pathogens in vivo, similar to the effect of fungicides, or indirectly by increasing resistance. Probably, in *Citrus*, these compounds act against *P. citrophthora* in both ways.

An understanding of the biosynthetic pathway of scoparone in *Citrus* can give a good opportunity to understand the mechanism of resistance.

Future studies should focus on the involvement of new approaches, such as UV treatments as well as controlled and modified atmosphere treatments, on the increased *Citrus* resistance to pathogens.

REFERENCES

Afek, U., and Sztejnberg, A. (1988a). Accumulation of scoparone, a phytoalexin associated with resistance of *Citrus* to *Phytophthora citrophthora*. *Phytopathology 78*: 1678–1682.

Afek, U., and Sztejnberg, A. (1988b). The involvement of scoparone (6,7-dimethoxycoumarin) in resistance of *Citrus* rootstocks against *Phytophthora citrophthora*. *Proceeding of the Sixth International Citrus Congress, 2*. Margraf Publishers, Weikersheim, Germany. pp. 779–785.

Afek, U., and Sztejnberg, A. (1989). Effect of fosetyl-Al and phosphorous acid on scoparone, a phytoalexin associated with resistance of *Citrus* to *Phytophthora citrophthora*. *Phytopathology 79*:736–739.

Afex, U., and Sztejnberg, A. (1993). Temperature and gamma irradiation effects on scoparone, a citrus phytoalexin conferring resistance to *Phytophthora citrophthora*. *Phytopathology 83*: 753–758.

Afek, U., Sztejnberg, A., and Carmely, S. (1986). 6,7-Dimethoxycoumarin, a *Citrus* phytoalexin conferring resistance against *Phytophthora* gummosis. *Phytochemistry 25*: 1855–1856.

Amerhein, N. (1978). Novel inhibitors of phenylpropanoid metabolism in higher plants. *Regulation of Secondary Product and Plant Hormone Metabolism* (M. Luckner and K. Schreiber, eds.), Pergamon Press, Oxford. pp. 173–182.

Amerhein, N., and Zenk, M. H. (1977). Metabolism of phenylpropanoid compounds. *Physiologie Vegetale 15*: 251–260.

Aritmo, Y., and Homma, Y. (1988). Studies on *Citrus* melanose and *Citrus* stem-end rot by *Diaporthe citri* (Faw.) Wolf. 9. Effect of light and temperature on the self-defense reaction of *Citrus* plants. *Ann. Phytophatol. Soc. Jpn. 54*: 282–289.

Aritmo, Y., Homma, Y., and Ohsawa, T. (1986). Studies on *Citrus* melanose and *Citrus* stem-end rot by *Diaporthe citri* (Faw.) Wolf. 5. Identification of phytoalexin in melanose spot. *Ann. Phytopathol. Soc. Jpn. 52*: 620–625.

Bailey, J. A., and Mansfield, J. W. (1982). *Phythoalexins*. Wiley. New York.

Ben-Yehoshua, S., Shapiro, B., and Moran, R. (1987). Individual seal-packing enables the use of curing at high temperatures to reduce decay and heal injury of *Citrus* fruits. *HortScience 22*: 777–783.

Ben-Yehoshua, S., Shapiro, B., Kim, J. J., Sharoni, J., Carmeli, S., and Kashman, Y. (1988). Resistance of *Citrus* fruit to pathogens and its enhancement by curing. *Proceedings of the Sixth International Citrus Congress, 3*. Margraf Publishers, Weikersheim, Germany. pp. 1371–1379.

Bompeix, G., and Saindrenan, P. (1984). In vitro antifungal activity of fosetyl-Al and phosphorous acid on *Phytophthora* species. *Fruits 39*: 777–785.

Bompeix, G., Ravise, A., Raynal, G., Fettouche, F., and Durand, M. C. (1980).

Method of obtaining necrotic blocking zones on detached tomato leaves using aluminum tris. Hypotheses as to its mode of action in vivo. *Annales de Phytopathologie. 12*: 337–351.

Broadbent, P. (1969). Observation on mode of infection of *Phytophthora citrophthora* in resistant and susceptible *Citrus* roots. *Proceedings of the First International Citrus Symposium*. International Society of Citriculture. *3*: 1207–1210.

Brown, S. A. (1978). Biochemistry of the coumarins. *Rec. Adv. Phytochem. 12*: 249–285.

Chalutz, E., Droby, S., Wilson, C. L., and Wisniewski, M. E. (1992). UV-induced resistance to postharvest diseases of *Citrus* fruit. *J. Photochem. Photobiol. 15*: 367–374.

Clarke, D. D., and Baines, P. S. (1976). Host control of scopolin accumulation in infected potato tissue. *Physiol. Plant Pathol. 9*: 199–203.

Coffey, M. D., and Bower, L. A. (1984). In vitro variability among isolates of eight *Phytophthora* species in response to phosphorous acid. *Phytopathology 74*: 738–742.

DeLange, J. H., Vincent, A. P., Du Plessis, L. M., Van Wyk, P. J., and Ackerman, L. G. J. (1976). Scoparone (6,7-dimethoxycoumarin) induced in *Citrus* peel by black spot, *Guignardia citricarpa* Kiely. *Phytophylactica 8*: 83–84.

Dubery, I. A., and Schabort, J. C. (1987). 6,7-Dimethoxycoumarin: a stress metabolite with antifungal activity in gamma-irradiated *Citrus* peel. *Suid-Afrikaanse Tydskrif vir Wetenskap 83*: 440–441.

Farih, A., Menge, J. A., Tsao, P. H., and Ohr, H. D. (1981a). Metalaxyl and efosite aluminum for control of *Phytophthora* gummosis and root rot on *Citrus*. *Plant Dis. 65*: 654–657.

Farih, A., Tsao, P. H., and Menge, J. A. (1981b). Fungitoxic activity of efosite aluminum on growth, sporulation, and germination of *Phytophthora citrophthora* and *P. parasitica*. *Phytopathology 71*: 934–936.

Fenn, M. E, and Coffey, M. D. (1984). Studies on the in vitro and in vivo of fosetyl-Al and phosphorous acid. *Phytopathology 74*: 606–611.

Fenn, M. E., and Coffey, M. D. (1985). Further evidence for the direct mode of action of fosetyl-Al and phosphorous acid. *Phytopathology 75*: 1064–1068.

Fritig, B., Hirth, L., and Ourisson, G. (1972). Biosynthesis of phenolic compounds in healthy and diseased tobacco plants and tissue cultures. *Hoppe-Seyler's Z. Physiol. Chem. 353*: 134–135.

Grisebach, H. (1977). Biochemistry of lignification. *Naturwissenschaften 65*: 619–625.

Gross, G. G. (1978). Recent advances in the chemistry and biochemistry of lignin. *Rec. Adv. Phytochem. 12*: 177–220.

Guest, D. I. (1984a). The influence of cultural factors on the direct anti-fungal activities of fosetyl-Al, propamocarb, metalaxyl, SN 75196 and Dowco 444. *Phytopathol. Z. 111*: 155–164.

Guest, D. I. (1984b). Modification of the defense responses in tobacco and capsicum following treatment with fosetyl-Al [Aluminum tris (O-ethyl phosphonate)]. *Physiol. Plant Pathol. 25*: 123–124.

Hanson, K. R., and Havir, E. A. (1981). Phenylalanine ammonia-lyase. *The Biochemistry of Plants*, Vol. 7 (P. K. Stumpf and E. E. Conn, eds.), Academic Press, New York, pp. 577–622.

Hartmann, G., and Nienhaus, F. (1974a). Gummosis pathogen *Phytophthora citrophthora* (Smith and Smith) Leonian and *Hendersonula toruloidea* Nattrass on *Citrus limon* in Lebanon. II. Seasonal influence. *Z. Pflanzenkrankheit. Pflanzenschutz 81*: 433–457.

Hartmann, G., and Nienhaus, F. (1974b). The isolation of xanthoxylin from bark of *Phytophthora* and *Hendersonula*-infected *Citrus limon* and its fungitoxic effect. *Phytopathol. Z. 81*: 97–113.

Hoagland, R. E., and Duke, S. O. (1982). Effect of glyphosate on metabolism of phenolic compounds. VIII. Comparison of the effects of aminooxyacetate and glyphosate. *Plant Cell Physiol. 23*: 1081–1088.

Ismail, M. A., Rouself, R. L., and Brown, G. A. (1978). Wound healing in *Citrus*. Isolation and identification of 7-hydroxycoumarin (umbelliferone) from grapefruit flavedo and its effect on *Penicillium digitatum* Sacc. *HortScience 13*: 358.

Jones, D. H. (1984). Phenylalanine ammonia-lyase: regulation of its induction and its role in plant development. *Phytochemistry 23*: 1349–1350.

Khan, A. J., Vernenghi, A., and Ravise, A. (1986). Incidence of fosetyl-Al and elicitors on the defense reactions of *Citrus* attacked by *Phytophthora* spp. *Fruits 41*: 587–595.

Khan, A. J., Kunesch, G., Chuilon, S., and Ravise, A. (1985). Biological activity of xanthyletin, a new phytoalexin of *Citrus*. *Fruits 40*: 807–811.

Kim, J. J., Ben-Yehoshua, S., Shapiro, B., Henis, Y., and Camely, S. (1991). Accumulation of scoparone in heat-treated lemon fruit inoculated with *Penicillium digitatum* Sacc. *Plant Physiol. 97*: 880–885.

Legrand, M. (1983). Phenylpropanoid metabolism and its regulation in disease. *Biochemical Plant Pathology* (J. A. Callow, ed.), Wiley, New York, pp. 367–384.

Musumeci, M. R., and Olivera, A. R. (1975). Accumulation of phenols and phytoalexins in citrus tissue inoculated with *Phytophthora citrophthora* (Smith and Smith) Leonian. *Summa Phytopathologica 1*: 275–282.

Musumeci, M. R., and Olivera, A. R. (1976). Accumulation of phenols and phytoalexins in *Citrus* tissues inoculated with *Phytophthora citrophthora. Summa Phytopathologica 2*: 27–31.

Ouimette, D.G., and Coffey, M. D. (1989). Phosphonate levels in avocado (*Persea americana*) seedlings and soil following treatment with fosetyl-Al or potassium phosphonate. *Plant Dis. 73*: 212–215.

Riov, J. (1971). 6,7-Dimethoxycoumarin in the peel of gamma irradiated grapefruit. *Phytochemistry 10*: 1923.

Riov, J., Monselise, P. S., and Kahan, S. R. (1968). Effect of gamma radiation on phenylalanine ammonia-lyase activity and accumulation of phenolic compound in *Citrus* fruit peel. *Environ. Exp. Bot. 8*: 463–466.

Rodov, V., Ben-Yehoshua, S., Kim, J. J., Shapiro, B., and Ittach, Y. (1992). Ultraviolet illumination induces scoparone production in kumquat and orange fruit and improves decay resistance. *J. Am. Soc. Hort. Sci. 117*: 788–792.

Sanders, P. L., Houser, W. I., and Cole, H. (1983). Control of *Pythium* spp. and *Pythium* blight of turfgrass with fosetyl aluminum. *Plant Dis. 67*: 1382–1383.
Sequeira, L. (1969). Synthesis of scopolin and scopoletin in tobacco plants infected with *Pseudomonas salanacearum*. *Phytopathology 59*: 473–478.
Smillie, R., Grant, B. R., and Guest, D. (1989). The mode of action of phosphite: evidence for both direct and indirect modes of action on three *Phytophthora* spp. in plants. *Phytopathology 79*: 921–926.
Tatum, J. H., and Berry, R. E. (1977). 6,7-Dimethoxycoumarin in the peels of *Citrus*. *Phytochemistry 16*: 1091–1092.
Vernenghi, A., and Ravise, A. (1985). Stimulation de la reaction de la tomate contre le *Phytophthora parasitica* par le phosetyl-Al par des eliciteurs fongiques. *Fruits 40*: 495–502.
Vernenghi, A., Ramiandrasoa, F., Chuilon, S., and Ravise, A. (1987). Phytoalexines des *Citrus*: seselin proprietes inhibitrices et modulation de synthese. *Fruit 42*: 103–111.

13
Stilbene Phytoalexins and Disease Resistance in *Vitis*

Wilhelm Dercks
Fachhochschule Erfurt, Erfurt, Germany

L. L. Creasy and C. J. Luczka-Bayles
Cornell University, Ithaca, New York

I. INTRODUCTION

A. General Information on Stilbene Compounds

Stilbenes have been found in a number of plant families (Ingham, 1976, 1982; Ingham and Harborne, 1976). Their antifungal nature has been implicated in preventing wood decay (Hart and Shrimpton, 1979; Hart, 1981) and in disease resistance of plants against different pathogens (Ward et al., 1975; Ingham, 1976; Hart, 1981; Aguamah et al., 1981). Biosynthetically, stilbenes are derived from the shikimic-polymalonic acid pathway. Stilbene synthase, which is the key enzyme in the biosynthesis of stilbenes in groundnuts (*Arachis hypogaea*) and also in *Vitis* spp. (Fritzemeier and Kindl, 1981; Melchior and Kindl, 1991), converts one molecule of *p*-coumaroyl-CoA and three molecules of malonyl-CoA into 4,3′,5′-trihydroxystilbene, commonly known as *trans*-resveratrol (Ingham, 1976; Rupprich and Kindl, 1978). In UV-irradiated grapevine leaves, phenylalanine was a good but tyrosine a poor precursor (Langcake and Pryce, 1977b), indicating that the 4′-hydroxyl is probably introduced at the cinnamic acid stage. *trans*-Resveratrol (hereafter resveratrol) appears to be the most widely distributed stilbene in several plant families.

B. Phytopathological Relevance of Stilbenes in *Vitis* Species: An Overview on Initial Research Work

The production of stilbenes by *Vitis* spp. as a response to fungal infection was first reported by Langcake and Pryce (1976). They observed a blue fluorescence in the zone of apparently healthy cells adjacent to the advancing margin of lesions in leaves of *Vitis vinifera* infected with *Botrytis cin-*

erea when the leaves were examined under long-wavelength ultraviolet radiation (365 nm). The compound responsible was identified as resveratrol. It was not detectable in healthy leaves but accumulated locally to high levels at the sites of infection in diseased leaves. Resveratrol was also found to be a major constituent of lignified stem tissue (Langcake and Pryce, 1976; Pool et al., 1981). In addition to the predominant resveratrol, other compounds considered oligomers of resveratrol and termed viniferins were also synthesized. The most important ones were α- and ϵ-viniferin (Langcake and Pryce, 1977a,c; Pryce and Langcake, 1977). These three compounds were produced sequentially in the leaves following infection suggesting the occurrence of a polymerization process. A similar compound, pterostilbene, was also identified, but was only found in very small amounts (Langcake et al., 1979). The structural formulas and chemical names of these four stilbenes are given in Fig. 1.

These initial findings stimulated research projects in several different groups around the world. Whereas each group focused on specific aspects, the overall goal, for a long time, has been the same: the elucidation of the involvement of stilbenes in disease resistance of *Vitis* spp. Only recently has attention been directed to other aspects, e.g., the significance for human health of the contents of resveratrol in grapes. The purpose of this chapter is to highlight the major achievements made with emphasis on the results generated in our laboratory and to indicate some potential future prospects of stilbene research.

C. Antifungal Properties of Stilbenes

The EC_{50} values of various stilbenes derived from orchinol ranged from 10 to 50 μg/ml in a number of fungi including *Phytophthora infestans*, *Monilinia fructicola*, and *Venturia inaequalis* (Ward et al., 1975). The compounds were more active against spore germination than mycelial growth. With *Helminthosporium carbonum*, an EC_{50} value of 50 μg/ml for resveratrol against mycelial growth was obtained (Ingham, 1976). EC_{50} values of resveratrol for germination of the grapevine pathogens *Plasmopara viticola* and *B. cinerea* ranged from 71 to over 200 μg/ml. This was also the case with mycelial growth of the latter fungus. The EC_{50} values for ϵ-viniferin ranged from 19 to over 200 μg/ml (Langcake and Pryce, 1976, 1977a; Langcake, 1981; Luczka, 1982; Stein, 1984; Stein and Blaich, 1985).

Compared with antibiotic agents or fungicides this activity does not appear very potent, but there is little doubt that the quantities found in diseased leaf tissue contribute effectively to cessation of the parasite's growth in the host tissue. Concentrations of 400–600 μg/ml fresh wt, dependent on *Vitis* spp. and stilbene tested, have been measured in grapevine

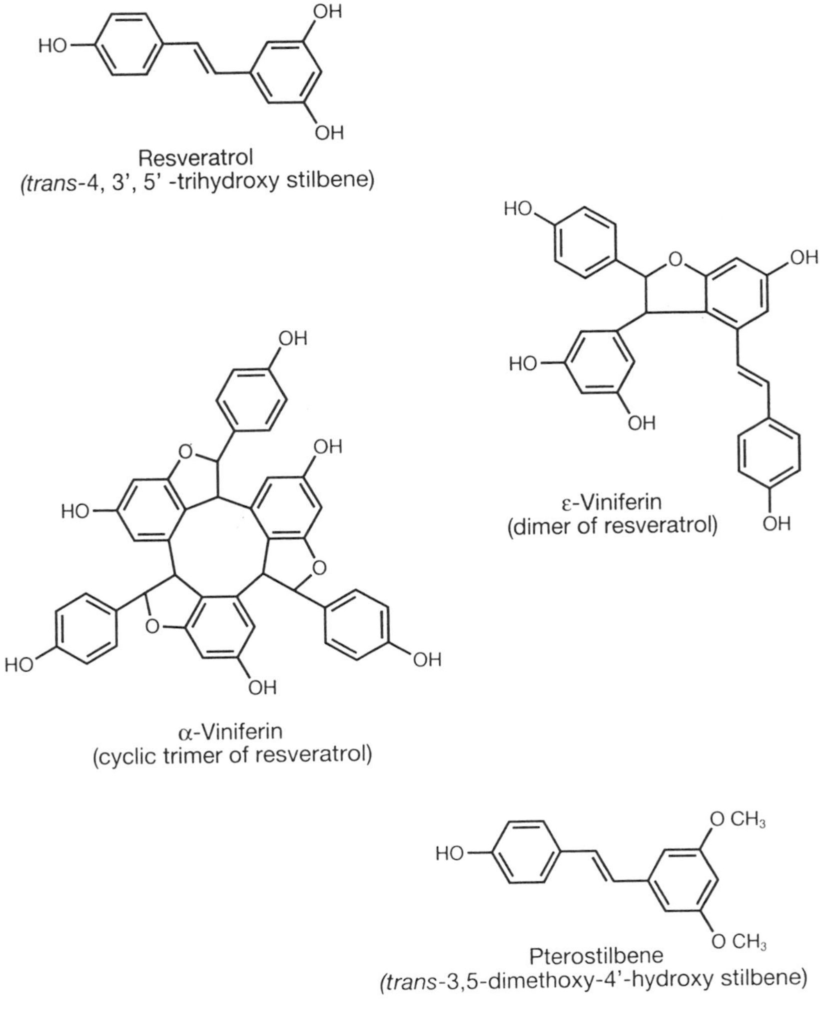

Resveratrol
(*trans*-4, 3', 5' -trihydroxy stilbene)

ε-Viniferin
(dimer of resveratrol)

α-Viniferin
(cyclic trimer of resveratrol)

Pterostilbene
(*trans*-3,5-dimethoxy-4'-hydroxy stilbene)

Figure 1 Stilbene compounds identified in *Vitis* species.

leaves (Langcake and Pryce, 1976; Langcake, 1981). This suggests that amounts at the actual infection sites may be even higher.

Little is known about the mode of action of stilbenes in fungi, but recent work suggests an interference of pterostilbene with the functionality of membrane proteins (Pezet and Pont, 1988b, 1990; Pont and Pezet, 1991). Structure–activity studies with substituted *trans*-hydroxystilbenes derived from resveratrol and pterostilbene showed that the effects on membrane proteins were related to the electronic character, lipophilicity, and molecular weight of the substances (Pont and Pezet, 1990). An involvement of benzylic or allylic radicals with fungicidal mode of action was implicated by experiments with structural analogs of stilbenes (Arnoldi et al., 1989).

D. Objectives of the Investigations

The studies summarized herein have served to find answers to different questions. In the following, the objectives of all investigations described in the following sections are explicitly listed to clarify the flow of logic in this chapter.

1. *Determination of basic features of stilbene synthesis in leaves and berries of* Vitis *spp.* These studies were conducted to evaluate the methods of stilbene induction to study patterns of synthesis, the capacity of various *Vitis* spp. to synthesize stilbenes (stilbene production potential), and the role of endogenous and exogenous plant conditions for stilbene synthesis.
2. *Aspects of host–parasite interactions between* Vitis *spp. and selected pathogens in leaves. B. cinerea*: These experiments were carried out to study the rate and level of accumulation following inoculation and the relation between stilbene production potential and disease resistance. *P. viticola*: The purpose of these investigations was to determine the stilbene production potential with respect to resistance, the role of inoculum density in elicitation of stilbenes, the plant's response during the early stages of infection, and the comparative relevance of resveratrol and ϵ-viniferin.
3. *Occurrence of stilbenes in foods derived from grapes.* The adoption of new cultivars selected for disease resistance raises questions on whether new compounds with toxicological properties will be introduced in the food supply. In one well-documented case, a disease-resistant cultivar of potato (Akeley et al., 1968) was introduced only to be later withdrawn from production when it was found that the increased disease resistance was due to exceptionally high concentrations of mutagenic glycoalkaloids (Zitnak and Johnston, 1970). Studies were conducted on both grape berries and wines to quantify the stilbenes present in them.
4. *Involvement of stilbene phytoalexins in the mode of action of the fungicide fosetyl-Al against* P. viticola *in* Vitis *spp.* These studies were conducted to determine whether or not fosetyl-Al can induce stilbene synthesis in leaves of *Vitis* spp. and a potential fosetyl-Al induced accumulation of stilbenes is significant enough to be considered an important aspect of its mode of action.

II. MATERIALS AND METHODS

A. Plants of *Vitis* Species Used and Selection of Leaf and Berry Material for Experiments

The plants used were rooted cuttings taken from the *Vitis* species collection at the New York State Agricultural Experiment Station, Geneva, New York 14456, USA. The vines were maintained in the greenhouse except for yearly 8-week rest periods in a cold room. The second to sixth fully expanded leaf was selected for experiments because very young and very old leaves do not synthesize high concentrations of stilbenes (Pool et al., 1981; Luczka, 1982; Stein, 1984).

To study the phytoalexin production potential of grape berries over the course of the season, berries were harvested from field- or greenhouse-grown plants at sampling times considered appropriate (Creasy and Coffee, 1988).

B. Culture of and Experiments with Pathogens

Plasmopara viticola was maintained on detached grapevine leaves incubated on moist filter paper in Petri dishes. Subsequent inoculation of leaves or leaf discs which had been cut with a cork borer for specific experiments were carried out with sporangia suspensions. Stilbene detoxification tests were conducted with sporangia in aqueous suspension.

Botrytis cinerea was maintained on V-8 juice agar. Spore germination tests for assessment of antifungal activity of stilbenes were carried out in malt extract broth incubated in multiwell tissue culture plates. Mycelium growth inhibition tests were conducted on solid Czapek Dox agar. Inoculation experiments were carried out with drops of spore suspensions placed on the lower surface of leaves or leaf discs.

C. Induction of Stilbenes

To assess the general capacity for phytoalexin synthesis (stilbene production potential) in *Vitis* spp., the lower surfaces of leaves were exposed to 8 min UV irradiation (254 nm) from a spectroline UV lamp (0.6 mW/cm^2). UV irradiation is known to cause rapid transcription of defense genes in plants (Chappel and Hahlbrock, 1984). After 48 hr in darkness (to avoid photochemical alternations of stilbenes: Battersby and Greenock, 1961; Blackburn and Timmons, 1969; Blaich and Bachmann, 1980) leaf discs were cut, blotted, weighed, and stored at $-20°C$ until extracted.

To assess fungal elicitation of stilbenes, leaves were inoculated as described below and samples collected at appropriate times following inoculation (see below).

D. Stilbene Purification and Quantitation

Stilbenes are present in lignified stem tissue of most *Vitis* spp. (Langcake and Pryce, 1977a; Pool et al., 1981). External standards for analysis were extracted from canes in 70% methanol. Upon removal of the methanol, the aqueous phase was extracted with ethyl acetate. Purified compounds were obtained by high-performance liquid chromatography (HPLC). Stilbenes were identified by their UV spectra in ethanol (Langcake and Pryce, 1976; Ingham, 1982) as well as by mass spectrometry. Concentrations were determined spectrophotometrically.

E. Stilbene Extraction and Measurement

Stilbenes were extracted from tissue in methanol and partitioned in ethyl acetate (Langcake and Pryce, 1976; Pool et al., 1981). Ethyl acetate was distilled off in vacuo. All these procedures were carried out in dim light to avoid photochemical alterations of stilbenes (see above). Initial measurements (Pool et al., 1981) were carried out with thin layer chromatography (TLC) and gas chromatography (GC), for which the trimethylsilyl derivatives were made. Later (Luczka, 1982; Creasy and Coffee, 1988; Dercks and Creasy, 1989a, b), concentrations of stilbenes were measured by HPLC on a silicic acid column with 4% methanol/96% methylene chloride as eluent. Wine analyses required two HPLC separations, the first on silicic acid with 3% methanol in methylene chloride and the second after complete isomerization to *cis*-resveratrol on a reverse phase column (Siemann and Creasy, 1992).

F. Application of the Fungicide Fosetyl-Al

Leaf discs were cut and placed upper surface down on fungicide-soaked filter paper in Petri dishes. In time–course studies on stilbene accumulation discs were sampled as found appropriate (Dercks and Creasy, 1989b; see also above).

III. BASIC FEATURES OF STILBENE SYNTHESIS IN LEAVES AND BERRIES OF *VITIS* SPECIES

A. Induction and Patterns of Stilbene Production

Stilbenes, especially resveratrol, are present in lignified stem tissue of most *Vitis* spp. (Langcake and Pryce, 1977a; Pool et al., 1981), but either are not detectable in healthy leaves and berries or are present at only very low levels. There are, however, several ways of inducing their synthesis (Table 1).

Table 1 Overview of Inducing Principles of Stilbene Formation in *Vitis* spp.

Inducing principles	Examples	Refs.
Chemicals	Phenyloxyalkyl-carbonic acids; 2-desoxy-D-glucose	Blaich and Bachmann (1980)
	Sugar solutions; mucic acic	Stein and Hoos (1984)
	Fosetyl-Al	Raynal et al. (1980) Stein and Hoos (1984) Dercks and Creasy (1989b)
UV irradiation	–	Langcake and Pryce (1976, 1977b) Pool et al., (1981) Luczka (1982) Barlass et al. (1987) Creasy and Coffee (1988) Dercks and Creasy (1989a) Jeandet et al. (1991)
Mechanical injury	–	Langcake and Pryce (1976)
Inoculation with fungal pathogens	*B. cinerea*	Langcake and Pryce (1976) Langcake and McCarthy (1979) Pool et al. (1981) Luczka (1982) Stein and Blaich (1985)
	P. viticola	Langcake (1978, 1981) Langcake and Lovell (1980) Raynal et al. (1980) Barlass et al. (1987) Dercks and Creasy (1989a,b)

The unspecific nature of the inducing principles led Langcake (1981) to term the viniferins stress metabolites rather than phytoalexins. Whereas inoculation studies are indispensable to determine the involvements of stilbene, in disease resistance (see below), UV irradiation has been instrumental in the study of production patterns in *Vitis* spp. In all our studies we have never detected either α-viniferin or pterostilbene. Therefore, the focus has been on resveratrol and ϵ-viniferin. The synthesis of these compounds as induced by irradiation with short-wavelength (254 nm) UV radiation is species specific. These results of Luczka (1982) will be discussed in more detail because they exemplify basic general patterns. *Vitis* spp. differed in the quantity and quality of stilbenes synthesized. UV irradiation–induced resveratrol and ε-viniferin concentrations over time were compared for four species (Fig. 2). *V. rupestris* B-38 produced the highest concentrations of stilbene compounds (400 nmol/g fresh wt) while *V. vinifera* cv. "Chardonnay" produced the lowest concentrations (90 nmol/g fresh wt). *V. doaniana* and *V. riparia* B-50 were two of the species between these two extremes.

The time course of production of resveratrol and ε-viniferin found by Langcake and Pryce (1977b) was very similar to the one shown by *V. riparia* B-50. In this species, resveratrol accumulated more quickly in the tissue than ε-viniferin. As polymerization proceeded, resveratrol concentration declined while ε-viniferin continued to build. After 3 days, levels of both compounds began to decline. *V. rupestris* B-38 and *V. doaniana* showed similar curves. In *V. vinifera* cv. "Chardonnay," however, ε-viniferin accumulated more quickly than resveratrol and was the predominant compound. The relative proportions of the two compounds in each species were similar. Species which produced high amounts of resveratrol also produced high amounts of ε-viniferin and vice versa. The actual ratios between the two compounds depended on when they were measured. In addition, the time required to reach maximum total stilbene content depended somewhat on the amount produced. It took longer for *V. rupestris* B-38 to reach its peak (3.5 days) than *V. vinifera* cv. "Chardonnay" (1.5 days).

B. Stilbene Production Potential of *Vitis* Species

From the above results it is obvious that each *Vitis* spp. may have its own characteristic response to induction. Some species are able to synthesize high concentrations of stilbenes, whereas others attain medium or low levels. With regard to the correlation between accumulated stilbene levels and resistance to pathogens, attempts were made to rank *Vitis* spp. according to their capacity to synthesize stilbenes (hereafter referred to as stilbene production potential). To assess this general potential, plants of different *Vitis* spp. were induced by UV irradiation and *B. cinerea* in 1980–81 (Table 2) and again in 1985–86 by UV light (Table 3). Resveratrol production potential of grape berries was measured in six cultivars by UV induction in 1988 (Table 4).

A direct comparison between the data of Tables 2 and 3 is not possible (different plants, different conditions, different calculations: amounts of stilbenes per cm^2 leaf area or mg fresh wt, respectively). The independent sets of data, however, had crucial aspects in common. Whereas *Vitis* spp. with "intermediate" stilbene production potential showed variable results and were not easy to define, the order of species with high or low potential, respectively, was never reversed. For instance, "Castor" or *V. rupestris* B-38 always exhibited a high potential whereas *V. vinifera* spp. always showed

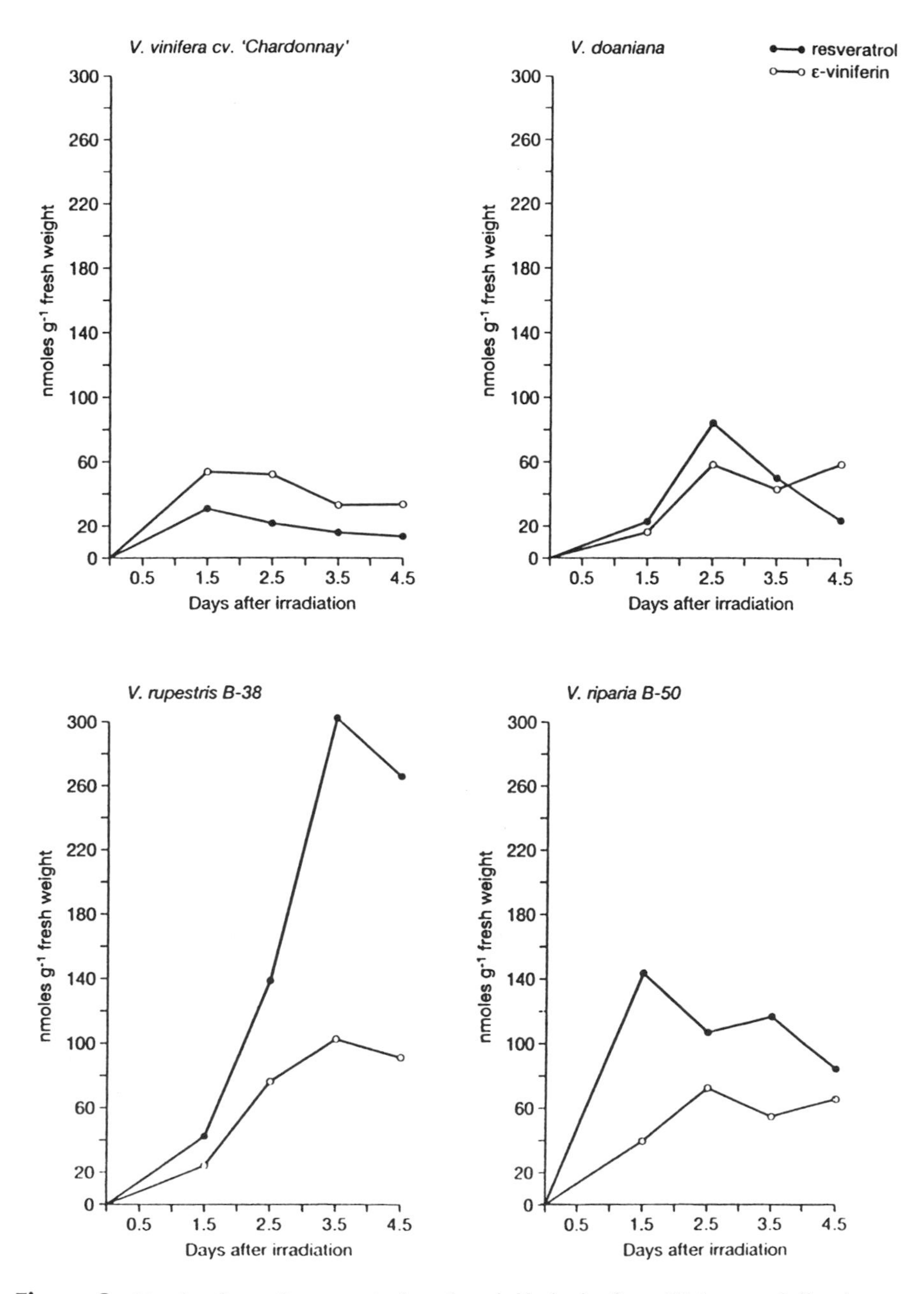

Figure 2 Production of resveratrol and ϵ-viniferin in four *Vitis* spp. following irradiation with short-wavelength (254 nm) UV radiation.

Table 2 Total Stilbene Production (Resveratrol plus ϵ-Viniferin)[a] in Leaves of 10 *Vitis* spp. in 1980–81[b]

Vitis spp.	nmol stilbenes/ cm^2 leaf area	Stilbene production potential
V. rupestris B-38	44	High
V. longii	34	↑
V. rupestris 6544	29	
V. treleasei	28	
V. champini CH-3-48	27	
V. riparia B-50	25	
V. andersonii	20	
V. cinerea I-66	18	
V. vinifera cv. "Chardonnay"	16	
V. argentifolia GBC-17	15	Low

[a]Mean of three tests (one test induced by UV irradiation, two tests induced by inoculations with *B. cinerea*).
[b]Greenhouse plants.
Source: Data from Luczka (1982).

a low capacity for stilbene synthesis. Another common feature was that resveratrol was the predominant stilbene in native American *Vitis* spp. whereas ϵ-viniferin was more common in European *V. vinifera* spp. In summary, UV irradiation has made it possible to select *Vitis* spp. with high and low phytoalexin potential to study specific questions like, for instance, the relation between stilbene contents and disease resistance (see below).

In berries, no stilbenes other than resveratrol were found in significant amounts. The resveratrol production potential changed during the season. Greenhouse-grown berries had highest potential after set but before maturation. In field samples the highest potential was reached near véraison but it decreased after late August, independent of maturity (Creasy and Coffee, 1988). Similar results were obtained by Jeandet et al. (1991). The resveratrol production potential was also found to be species-specific (Table 4).

C. Role of Endogenous and Exogenous Plant Conditions for Stilbene Synthesis

Many pitfalls in the initial work on grapevine stilbenes were caused by the enormous variability of stilbene production in grapevines. This pertains to variability not only between different species but also between:

Table 3 Total Stilbene Production (Resveratrol plus ε-Viniferin)[a] in Leaves of 17 *Vitis* spp. in 1985–86[b]

Vitis spp.	nmol stilbenes/ g fresh wt	Stilbene production potential
Interspecific cv. "Castor" (B-7-2)	2551	High
V. riparia B-50	200	↑
V. rupestris B-38	176	
V. doaniana	173	
V. andersonii	160	
Interspecific cv. "Pollux" (B-6-18)	139	
V. acerifolia	132	
V. argentifolia GBC-17	123	
Interspecific cv. "Vignoles"	118	
V. rupestris 6544	105	
V. vinifera cv. "Riesling"	90	
V. vinifera cv. "Chardonnay"	83	
V. champini CH-3-48	65	
V. cinerea I-66	63	
V. vinifera cv. "Bacchus"	51	
V. vinifera cv. "Müller-Thurgau"	45	
V. treleasei	25	Low

[a]Mean of 2 tests (induced by UV irradiation).
[b]Greenhouse plants.
Source: Data from Dercks and Creasy (1989a).

Table 4 Resveratrol Production Potential[a] in Berry Skin of Six *Vitis* spp. Near Véraison in 1988[b]

Vitis spp.	nmol resveratrol/ cm^2 berry skin
V. labrusca cv. "Concord"	49
V. vinifera cv. "Cabernet Sauvignon"	49
V. vinifera cv. "Riesling"	22
Interspecific cv. "Chancellor"	22
V. labrusca cv. "Catawba"	18
Interspecific cv. "Cayuga White"	18

[a]Mean of two tests (induced by UV irridation).
[b]Field-grown plants.
Source: Data from Creasy and Coffee (1988).

Table 5 Relationship Between Origin of Leaf Material and Variability of Stilbene Production in *Vitis* spp.

Origin	Variability of stilbene production
Greenhouse	Intermediate to high
Field	Extreme

Source: Data from Luczka (1982); Stern (1984); Dercks and Creasy (unpublished).

Different clones of the same species
Different plants of the same clone
Different leaves of the same plant
Differences between leaf halves of the same leaf

This set of complications is further magnified by plant vigor and leaf age as well as by origin (greenhouse or field leaves). Tables 5 and 6 summarize the experience of independent researchers (Langcake and McCarthy, 1979; Pool et al., 1981; Luczka, 1982; Stein and Blaich, 1985; Barlass et al., 1987; Bavaresco and Eibach, 1987; Dercks and Creasy, 1989a, b).

To obtain reproducible and reliable quantitative results it has become customary to exclude field-grown leaves from specifically designed induction experiments (for instance, in Section VI) which, although high in stilbene production potential, exhibit a degree of variability detrimental to any quantitative analysis. Picking midshoot leaves from greenhouse plants grown under the same conditions provides the best chance to obtain homogenous leaf material. This, of course, does not pertain to studies where specific aspects of stilbene production in field leaves or berries are ad-

Table 6 Relationship Between Leaf Position, Development Stage, and Stilbene Production Potential in Leaves of *Vitis* spp.

Leaf position	Development stage	Stilbene production potential
Shoot base	Old	Low
Shoot center	Mature	High
Shoot tip	Young	Intermediate

Source: Data from Luczka (1982); Stern (1984); Dercks and Creasy (unpublished).

dressed, as in the study of production patterns during the course of the season (Creasy and Coffee, 1988; Jeandet et al., 1991).

IV. ASPECTS OF HOST–PARASITE INTERACTIONS BETWEEN *VITIS* SPECIES AND SELECTED PATHOGENS IN LEAVES

A. *Botrytis cinerea*

Rate and Level of Stilbene Accumulation Following Inoculation

The concentration of both resveratrol and ϵ-viniferin increased 1 day after inoculation in *V. riparia* B-50. This augmentation continued for another 2–3 days and was very similar to the one described for *V. riparia* B-50 following induction by UV as shown in Fig. 2. These results have been described in detail elsewhere (Pool et al., 1981). Contrary to the response induced by UV irradiation, the concentrations did not decline until the last testing date (6 days after inoculation). This may be attributed to the fact that in contrast to the brief UV irradiation period, the inducing principle remained present in the leaves. Very similar studies conducted by Blaich et al. (1982) and Stein and Hoos (1984) showed exactly the same pattern for nine *Vitis* spp. of different stilbene production potential and corresponding level of resistance to *B. cinerea*. The relative ranking of the species (eight of which were the same as in our studies) according to their stilbene production potential was very similar to ours (Tables 2 and 3). These studies clearly demonstrate that stilbenes accumulate at a faster rate and to a higher level in *Vitis* spp. with high phytoalexin potential than in species with low phytoalexin potential.

Relationship Between Stilbene Production Potential and Disease Resistance

In the *Vitis* spp. used by Stein and Hoos (1984) in the study discussed above, there was a close positive correlation between stilbene production potential and resistance to *B. cinerea*. When the authors analyzed this correlation in 95 *Vitis* spp., they found three different groups (Table 7).

This is the same pattern as found in the work of our laboratory. The results are described in detail elsewhere (Luczka, 1982). Of 10 *Vitis* spp. tested, the following ones fell into the respective groups:

Group 1: *V. cinerea* I-66, *V. champini* CH-3-48, *V. argentifolia* GBC-17
Group 2: *V. rupestris* B-38, *V. longii*
Group 3: *V. vinifera* cv. "Chardonnay," *V. andersonii*, *V. riparia* B-50, *V. rupestris* 6544, *V. treleasei*. (The occurrence of *V. vinifera* cv. "Chardonnay" in this group cannot be easily explained. Normally it is very susceptible to *B. cinerea*.)

Table 7 Groups of Relationships Between Stilbene Production Potential[a] and Level of Resistance to *B. cinerea* in 95 *Vitis* spp. (according to Stein, 1984[b] and Stein and Hoos[c])

Group	Stilbene production potential	Level of resistance to *B. cinerea*	Relative frequency of *Vitis* spp.
1	Low	Low	High
2	Intermediate to high	High	High
3	Low	High	Low

[a]Induced by inoculations with *B. cinerea*.
[b]Description of results including identification of *Vitis* spp.
[c]Description of results only.

The existence of genotypes in the third group suggests that resistance may, in several cases, be associated with factors other than stilbene phytoalexins, which cannot be surprising since many of them are already known (Fregoni, 1983; Stein, 1984; Jeandet and Bessis, 1989). There is a clear message from the above results, i.e., high phytoalexin potential is never associated with susceptibility. These findings heavily support the important role of stilbenes in resistance of *Vitis* spp. to *B. cinerea*. Further substantiation stems from the results of Langcake and McCarthy (1979) and Langcake (1981). They determined the amounts of resveratrol around the actual lesions and found an inverse relationship between stilbene concentration and susceptibility to *B. cinerea*. *B. cinerea* is a more important parasite on grape berries than on leaves (unfortunately, berries are much more difficult to work with) and it is interesting to note that there is a good correlation between the resistance of leaves and maturing berries (Stein and Hoos, 1984; Stein and Blaich, 1985). Midage berries are usually very resistant to *B. cinerea* at the time they have the highest potential for synthesizing resveratrol although the resveratrol potential decreases after véraison (Creasy and Coffee, 1988; Jeandet et al., 1991).

B. *Plasmopara viticola*

Relationship Between Stilbene Production Potential and Disease Resistance

When 17 *Vitis* spp. were analyzed (Dercks and Creasy, 1989a), four different groups were found (Table 8). Although the number of species tested was smaller than for *B. cinerea*, and it cannot be said that all species will follow the same pattern, the overall picture appears to be the same. Resistance can be mediated by factors other than stilbenes, but high concentrations of stilbenes are never linked with susceptibility. Thus, for downy

Table 8 Groups of Relationships Between Stilbene Production Potential[a] and Level of Resistance to *P. viticola* in 17 *Vitis* spp.

Group	Stilbene production potential	Level of resistance to *P. viticola*	*Vitis* spp.
1	low to intermediate	low	*V. acerifolia, V. argentifolia* GBC-17
			V. treleasei
			V. vinifera cv. 'Müller-Thurgau'
			V. vinifera cv. 'Chardonnay'
			V. vinifera cv. 'Riesling'
2	intermediate	intermediate	*V. andersonii*
			V. rupestris B-38, *V. rupestris* 6544
			Interspecific cultivar 'Vignoles'
3	high	high	*V. doaniana, V. riparia* B-50
			Interspecific cultivar 'Castor'
			Interspecfiic cultivar 'Pollux'
4	low	high	*V. cinerea* I-66, *V. champini* CH-3-48
			V. vinifera cv.'Bacchus'

[a]Induced by UV irradiation (see Table 3).
Source: Data from Dercks and Creasy (1989a).

mildew, the conclusions of stilbenes being important factors in the resistance of *Vitis* spp. (similar results were obtained by Bavaresco and Eibach, 1987) are in line with findings from *B. cinerea*.

Role of Inoculum Density in the Elicitation of Stilbene Synthesis and Plant Response During Early Stages of Infection

Since phytoalexins are only locally synthesized at the attempted site of infection and the number of infection sites may be low if a leaf is challenged by only a few fungal spores, the overall response of the leaf may be difficult to measure or even detect on a per-leaf base. To test whether or not the stilbene response per sample unit can be optimized for quantitative research by modifying the inoculum density, comparative studies were carried out with *Vitis* spp. of different stilbene production potential and corresponding levels of resistance to *P. viticola* (Dercks and Creasy, 1989a). It was shown that by increasing the number of infection sites it is possible to enhance stilbene accumulation (Table 9). Furthermore, it became obvious that more sporangia are produced (i.e., the attack is higher) when the inoculum den-

Table 9 Effect of Increasing the Inoculum Density[a] of *P. viticola* on Accumulation of Stilbenes[b] and Sporangia Production in Leaves of Three *Vitis* spp.

Vitis spp.	Stilbene production and level of resistance to *P. viticola*	Level of stilbenes actually reached in experiment	Effect[c] of increased inoculum density on:	
			Stilbene accumulation	Sporangia production
V. vinifera cv. "Riesling"	Low	Low	+	+[d]
V. rupestris 6544	Intermediate	Intermediate	+[d]	+
Interspecific cv. "Castor"	High	High	+	+

[a]Inoculum densities tested (sporangia/ml): 0; 30,000; 90,000; 150,000.
[b]Tested before inoculation; 2, 4, 6, 8, and 10 hr after inoculation; 1, 2, 3, 4, 5 days after inoculation.
[c]+, positive; −, negative.
[d]Reduced effect again with the highest inoculum density tested.
Source: Data from Dercks and Creasy (1989a).

sity is increased. Thus, it was even possible to break resistance of the interspecific cultivar "Castor." With 150,000 sporangia/ml, reproduction was obtained whereas this was not possible with lower inoculum densities.

These time course studies also revealed that the most pronounced stilbene synthesis occurred during the first 2–10 hr following inoculation and again later around the time of sporulation (Fig. 3). The absolute levels reached depended on the *Vitis* spp. tested. They were highest in "Castor," intermediate in *V. rupestris* 6544, and lowest in "Riesling," reflecting the stilbene production potential of the respective *Vitis* spp. (Dercks and Creasy, 1989a).

The implications of these findings are twofold. First, they demonstrate how important it is to take into account the severity of fungal challenge when assessing the comparative degree of phytoalexin response and corresponding level of disease resistance in *Vitis* spp. This factor has frequently been overlooked in studies with leaves from vines in the field where disease pressure is seldom homogenous. Second, the results show that phytoalexin synthesis can occur very rapidly in plants. In most pathosystems this was observed after days; here after hours. When drawing conclusions one should not only look at the plant's internal "rhythm" of synthesis, but also at the challenging pathogen's development characteristics. The infection process and colonization of host tissue by *P. viticola* are extremely rapid.

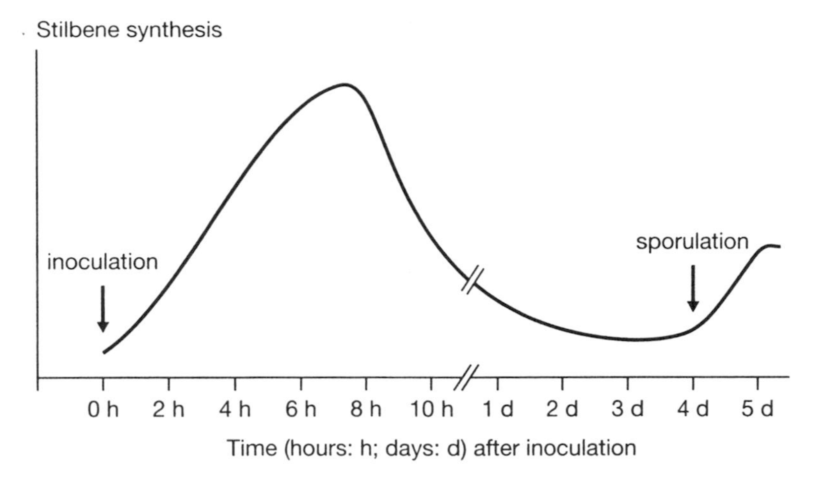

Figure 3 Schematic representation of stilbene synthesis in three *Vitis* spp. *V. vinifera* cv. "Riesling"; *V. rupestris* 6544; Interspecific cv. "Castor" (B-7-2) following inoculation with *P. viticola*.

Hypersensitive reactions were seen after only a few hours by light and electron microscopic studies, and there is strong evidence that stilbenes are involved in this reaction (Langcake and Lovell, 1980; Langcake, 1981). Our findings may have implications for studies with other pathogens which develop as rapid as *P. viticola*.

Comparative Relevance of Resveratrol and ϵ-Viniferin

The antifungal properties of stilbenes in general have already been reviewed above. There is a lot of heterogeneity in these data concerning pathogens of *Vitis* spp. from several research groups. One concept has been that resveratrol should not be considered a phytoalexin because of its low fungitoxicity compared with that of ϵ-viniferin (Langcake, 1981). In retrospect, it appears possible that the variability of the data can be attributed to the fact that most studies were carried out with concentrations beyond the limits of solubility of the compounds. When soluble concentrations of the two substances were used (60–70 μg/ml) resveratrol was twice as inhibitory as ϵ-viniferin toward germination of sporangia of *P. viticola* (Dercks and Creasy, 1989a). These would, however, be equimolar concentrations. Former comparisons had been made on a weight basis. In addition, ϵ-viniferin was found to be less stable than resveratrol in water. In the presence of *P. viticola* sporangia, a rapid degradation of ϵ-viniferin, but not of resveratrol, was observed (Table 10). These results, together with the fact that ϵ-viniferin has never been detected by us in leaf tissue colonized by the

Table 10 Characteristic Features of Resveratrol and ϵ-Viniferin with Regard to Phytopathological Significance

Feature	Resveratrol	ϵ-Viniferin
Short-term degradation in water	–[a]	+[b]
Degradation by sporangia of *P. viticola*	–	+
Presence in diseased leaves of *Vitis* spp.		
B. cinerea	+	+
P. viticola	+	–
Antifungal activity[c]		
B. cinerea	=ϵ-Viniferin	=Resveratrol
P. viticola	>ϵ-Viniferin	<Resveratrol

[a] –, not observed.
[b] +, observed.
[c] Assessed on a molar basis.
Source: Data from Luczka (1982) and Dercks and Creasy (1989a,b).

P. viticola isolate used throughout our studies, point to an underestimation of resveratrol's regulatory role in the interaction between *P. viticola* and *Vitis* spp. by several researchers. The significance of ϵ-viniferin appears to be more related to endo- and exogenous plant conditions than to active resistant responses of the host. In *B. cinerea*, resveratrol was equal to ϵ-viniferin in terms of antifungal activity (Luczka, 1982).

V. OCCURRENCE OF STILBENES IN GRAPE PRODUCTS

A. Grapes and Juice

Resveratrol is found in all parts of the grape cluster (Creasy and Coffee, 1988). The component parts of "Riesling" berries, analyzed at harvest for their residual resveratrol content, showed great differences. The highest concentration was found in the seeds which, although contributing only 7% of berry weight, had 73% of the total berry resveratrol. The skin and pulp, composing half of berry weight, contained 18% of the resveratrol. The small percentage remaining was found in the juice (Siemann and Creasy, unpublished data).

B. Wines

Wines should reflect the grapes from which they were made. Wine analyses revealed unexpected large differences in resveratrol concentration and we found that wine-making techniques greatly influenced the transfer of resveratrol into wine (e.g., fermentation on or off skins) or removal from wine (i.e., use of resins for fining; Siemann and Creasy, 1992). Research wines made from a disease-resistant variety had a higher concentration of resveratrol than one made from a more susceptible variety. In the same experiment, wines from both varieties produced from vineyard plots not treated with fungicides had higher concentrations than those from the fungicide-sprayed plots (Siemann and Creasy, unpublished data).

VI. INVOLVEMENT OF STILBENE PHYTOALEXINS IN THE MODE OF ACTION OF THE FUNGICIDE FOSETYL-AL AGAINST *P. VITICOLA* IN *VITIS* SPECIES

Fosetyl-Al (aluminum ethyl phosphite) is a systemic fungicide active against fungi belonging to the Oomycetes (Cohen and Coffey, 1986). Its active metabolite in plant tissue is phosphite (H_3PO_3), also referred to as phosphonate, phosphonic, or phosphorous acid (Fenn and Coffey, 1984, 1985; Luttringer and de Cormis, 1985; Saindrenan et al., 1985). Both fosetyl-Al and phosphite have a direct effect on fungal metabolism and cell wall synthesis (Dercks and Buchenauer, 1987; Dunstan et al., 1990; Smillie et al., 1990) but the primary mode of action in fungi is still unknown. Recent investigations suggest an interference with metabolism of phosphates (Barchietto et al., 1990) and specifically with synthesis of the nucleotide adenylate (Griffith et al., 1990). There are also a number of reports which show that accumulation of phenolic compounds and phytoalexins in plant tissue is associated with the action of phosphite in plants (e.g., Bompeix et al., 1980; Durand and Sallé, 1980; Raynal et al., 1980; Dercks and Buchenauer, 1986; Guest, 1984, 1986; Saindrenan et al., 1988; Afek and Sztejnberg, 1989; Nemestothy and Guest, 1990). These findings have led to the assumption that there may be a combination of direct and indirect modes of action of phosphite (Dercks and Creasy, 1989b; Smillie et al., 1989, 1990). The following is a brief description of the principles of fosetyl-Al–induced accumulation of stilbenes in *Vitis* spp. It is based on the work of Dercks and Creasy (1989b), where the results are described in detail.

Two basic ideas were pursued throughout the studies. One was to measure stilbenes at critical times during the infection process to see whether stilbenes are specifically synthesized against *P. viticola.* Sooner or later phytoalexins will always be synthesized unspecifically because of massive cell death in any kind of reaction, but this need not necessarily be related to induction by fosetyl-Al. The other idea was to make a distinction of the fosetyl-Al–induced effect in *Vitis* leaf tissue between species of low, intermediate, and high phytoalexin potential and corresponding levels of resistance to *P. viticola* to see if a stimulation of phytoalexin synthesis in susceptible species might lead to a response equivalent to that in resistant species. The results are schematically summarized in Table 11.

In noninoculated leaf tissue, fosetyl-Al caused only a moderate induction of stilbene synthesis irrespective of the stilbene production potential of the *Vitis* spp. tested. In inoculated tissue, the induced response depended strongly on the inherent capacity for phytoalexin synthesis, i.e., the response of *Vitis* spp. with low stilbene production potential was not as high as that in species with higher potential. Furthermore, the phytoalexin response was greater with postinfectional than with preinfectional treatments, probably because of the many more cells invaded by the pathogen and, hence, involved in the synthesis of stilbenes which occurs following cell damage or death.

Despite the weak induction of stilbenes in *Vitis* spp. with low phytoalexin production potential, fosetyl-Al had a pronounced effect on the reduction of sporangia formation following preinfectional treatments. This was even stronger in species with higher potential. The effect of postinfectional fungicide treatments at the same concentrations was weaker, but depended heavily on the amounts of resveratrol synthesized. These were higher in *Vitis* spp. with high than in species with low phytoalexin production potential.

The results show that fosetyl-Al must have a direct influence on the pathogen because reproduction of *P. viticola* was reduced despite the lack of significant concentrations of stilbenes being produced. This was obvious with preinfectional treatments. On the other hand, it cannot be denied that stilbenes are involved in the action of the fungicide. This became obvious with postinfectional treatments where the efficacy was correlated with accumulated levels of stilbenes which, in turn, were closely linked with phytoalexin production potential and level of inherent resistance to *P. viticola* of the *Vitis* spp. tested.

These findings are evidence for a combination of direct and indirect modes of action of fosetyl-Al against *P. viticola* in *Vitis* spp. They are consistent with the hypothesis that the fungicide first interferes with metabolism of the fungus. This interference could then result in alterations of the physiology of the pathogen–host interaction and thus lead to a response of resistance in the host in which phytoalexins are involved. This view has distinctly evolved in the more recent literature (e.g., Dunstan et al., 1990; Smillie et al., 1989, 1990). On the other hand, claims of a primary indirect mode of action of fosetyl-Al via stimulation of natural defense mechanisms in plants which were made initially when the potent direct antifungal activity of phosphite had not yet been realized (e.g., Bompeix et al., 1981) are no longer sustained.

Table 11 Schematic Overview on the Effect[a] of Fosetyl-Al[b] (Pre- and Postinfectional Treatments) on the Accumulation of Stilbenes in Noninoculated and Inoculated Leaf Tissue[c] as Well as on the Reduction of Sporangia Formation in Inoculated Leaf Tissue of Three *Vitis* spp.

Vitis spp.	Stilbene production potential and level of resistance to *P. viticola*	Accumulation of stilbenes in:			Reduction of sporangia formation in inoculated tissue:	
			Inoculated tissue			
		Noninoculated tissue	Fosetyl-Al preinfectional	Fosetyl-Al postinfectional	Fosetyl-Al preinfectional	Fosetyl-Al postinfectional
V. vinifera cv. "Riesling"	Low	+	+ +	+ + +	+ + +	+
V. rupestris 6544	Intermediate	+	+ + +	+ + + +	+ + + + +	+ + +
Interspecific cv. "Castor"	High	+	+ + +	+ + + + +	+ + + + +	+ + + + +

[a] +, moderate; + +, noticeable; + + +, pronounced; + + + +, strong; + + + + +, very strong.
[b] Fungicide concentrations tested: 200 and 400 μg/ml. The higher concentration was always more effective with regard to every parameter.
[c] Inoculated with *P. viticola*.
Source: Data from Dercks and Creasy (1989b).

VII. EPILOG

A. Stilbenes and Disease Resistance in *Vitis* Species

There is no doubt today that stilbene phytoalexins are very important factors in the resistance of *Vitis* spp. to *B. cinerea* and *P. viticola* (Fregoni, 1983; Stein, 1984; Jeandet and Bessis, 1989). A leaf high in concentrations of these compounds will successfully ward off a challenge by these pathogens. There is evidence that this might also be the case with berries (Pezet and Pont, 1988a; Jeandet et al., 1991), but less information is available for these organs, largely because berries are experimentally more difficult to work with than leaves. With regard to *B. cinerea*, however, more data on berries are urgently needed because this pathogen is eminently more important on fruit than on leaves. The realization that at some developmental stages there is a good correlation between the resistance of berries and leaves, and that the conclusions from studies with leaves may be extended to berries, certainly needs more substantiation (Stein and Hoos, 1984; Stein and Blaich, 1985).

With regard to *Uncinula necator* there is virtually no information in the literature on the relationship between stilbene contents and resistance of leaves or berries. The reason might be that researchers may not have found a good correlation between resveratrol concentrations and disease resistance but never reported on their negative findings. It is interesting to note that stilbenes can only be induced by UV irradiation on the abaxial leaf surfaces (Langcake and Pryce, 1977b; Pool et al., 1981) and that powdery mildew is found more frequently on the adaxial than on the abaxial leaf surfaces, although it is principally able to attack all green parts of the plant (Bulit and Lafon, 1978). Furthermore, *U. necator* only penetrates the epidermal cells and not the deeper cell layers (Heintz and Blaich, 1990). If it were generally true that a large part of the resveratrol synthesis is located not just in but below the epidermal cells, which is suggested by the findings of Blaich and Bachmann (1980), then it would also be conceivable that only insufficient amounts of the antifungal compound might come into physical contact with the pathogen to account for inhibition of fungal growth. This is certainly different from *B. cinerea* and *P. viticola*, which both penetrate the outer cell layers and continue to grow inside the tissue of the inner cell layers.

The antifungal properties of stilbenes initially stirred hopes that these compounds might be used as biochemical markers of resistance in breeding programs. Before this question can be finally resolved it is necessary to overcome the difficulties discussed in Section III.C. A phenotype high in stilbene production potential at any given point in time may not remain so throughout its lifetime. Not all leaves on a plant are the same in potential;

neither are different plants within a clone, nor different clones within a species. Because of this, stilbenes have failed to serve as qualitative markers of resistance in breeding and screening efforts (Luczka, 1982; Barlass et al., 1987). Luczka's work (1982) has also shown the difficulty of determining traits of inheritance of high stilbene accumulation through crosses between parents of known stilbene production potential. None of the progeny followed an obvious pattern and phenotypes belonging to all groups of stilbene production potential were found irrespective of the parents' potential. It is of the utmost importance to characterize the genetics of stilbene synthase expression in *Vitis* spp. to finally elucidate the involvement of stilbenes in disease resistance throughout the host's life cycle.

In view of this, hopes for future disease resistance is some plant species by introducing a gene from groundnut (*Arachis hypogaea*) which codes for stilbene synthase should meet with cautious enthusiasm (Hain et al., 1990). Even though the gene was expressed in transgenic tobacco plants and resveratrol was identified, it is difficult to see why the basic features of stilbene production (see above) would be any more predictable or usable in transgenic plants than in its native location.

B. Detoxification of Stilbenes by Fungal Parasites of *Vitis* Species

Accumulation of stilbenes is the result of synthesis, turnover of these toxic metabolites by plant tissue (Hoos and Blaich, 1988), and active degradation by the invading pathogen. Our work has shown that *P. viticola* may detoxify ε-viniferin. The question whether this feature may be important for pathogenicity, as found in some other parasite–host interactions (Van Etten et al., 1989), deserves further study. At the moment, it is not known as to whether detoxification of ε-viniferin is peculiar to some or common to all isolates of *P. viticola*. However, Langcake's observations (1978) contrasted with ours, make it likely that the former may be the case. It would also be interesting to see whether or not some isolates are able to degrade resveratrol. It has been reported that resveratrol and pterostilbene are subject to detoxification by *B. cinerea* (Hoos and Blaich, 1990; Pezet et al., 1991). Further research in this area is needed.

C. Comparative Relevance and Distribution of Stilbene Compounds in Different Plant Organs

Our investigations have produced evidence that resveratrol is more important for the regulation of the interaction between *Vitis* spp. and *P. viticola* than ε-viniferin. We do not know much about α-viniferin and pterostilbene. In which organs (leaves, berries, shoots, canes) are they heavily concentrated? Are they all phytoalexins or are some just products of woody plant parts (which seems to be the case for ϵ-viniferin) with little relevance for disease resistance to pathogens that attack green organs? Also, there is little information on the location of stilbenes within the cell and cell layers apart from Blaich and Bachmann's cytological studies (1980) which showed that resveratrol appears to be deposited within small areas of the cytoplasm or in the periplasm, probably near plasmodesmata, mostly in and below the epidermal cells. Is this commonly the case or were the deeper cell layers just not sufficiently challenged in this particular experiment?

D. Mode of Action of Stilbenes

Despite the studies of Pezet and Pont (1988b, 1990) and Pont and Pezet (1991) (see also chapter 14 by Pezet and Pont), this field is virtually a blank. If we are to understand more about the overall significance of stilbenes, more information on their mode of action at the biochemical and molecular, in addition to the histological, level is needed.

E. Can Synthesis of Stilbenes Be Induced Systemically?

Almost all research to date has been devoted to the elucidation of mechanisms of local stilbene accumulation. To the best of our knowledge, nobody has looked deeply into the question of whether synthesis of stilbenes can also be part of a process of systemic acquired resistance, the signal for which is translocatible in the plant. If so, the upper leaves of the plant could be "immunized" by challenging the lower leaves with an appropriate agent or compound.

ACKNOWLEDGMENTS

This work was supported by the New York State College of Agriculture and Life Sciences and the Deutsche Forschungsgemeinschaft.

The authors thank J. M. Babcock, J. A. Becker, and J. O. Becker for typing and editorial assistance with the manuscript. We are also grateful to R. Pezet for providing a paper in press and R. Blaich for valuable discussions.

REFERENCES

Afek, U., and Sztejnberg, A. (1989). Effects of fosetyl-Al and phosphorous acid on scoparone, a phytoalexin associated with resistance of *Citrus* to *Phytophthora citrophthora*. *Phytopathology 79*: 736–739.

Aguamah, G. E., Langcake, P., Leworthy, D. P., Page, J. A., Pryce, R. J., and

Strange, R. N. (1981). Isolation and characterization of novel stilbene phytoalexins from *Arachis hypogaea*. *Phytochemistry 20*: 1381–1383.
Akeley, R. V., Mills, W. R., Cunningham, C. E., and Watt, J. (1968). Lenape: a new potato variety high in solids and chipping quality. *Am. Potato J. 45*: 142–145.
Arnoldi, A., Farina, G., and Merlini, L. (1989). Relazione struttura: attività antifungina di fitoalessine isoflavonoidi. *Notiziario sulle Malattie delle Piante 110*: 171–187.
Barchietto, T., Saindrenan, P., and Bompeix, G. (1990). Effects of phosphonate on phosphate metabolism in *Phytophthora citrophthora*. *Pest. Sci. 30*: 365–366.
Barlass, M., Miller, R. M., and Douglas, T. J. (1987). Development of methods for screening grapevines for resistance to infection by downy mildew. II. Resveratrol production. *Am. J. Enol. Viti. 38*: 65–68.
Battersby, A. R., and Greenock, A. I. (1961). *cis*- and *trans*-3,3′, 4,4′-Tetramethoxystilbenes. *J. Chem. Soc.* (part 2): 2592–2593.
Bavaresco, L., and Eibach, R. (1987). Investigations of the influence of N-fertilizer on resistance to powdery mildew (*Oidium tuckeri*), downy mildew (*Plasmopara viticola*), and on phytoalexin synthesis in different grapevine varieties. *Vitis 26*: 192–200.
Blackburn, E. V., and Timmons, C. J. (1969). The photocylisation of stilbene analogues. *Q. Rev. Chem. Soc. London 23*: 482–503.
Blaich, R., and Bachmann, O. (1980). Die Resveratrolsynthese bei Vitaceen; Induktion und zytologische Beobachtungen. *Vitis 19*: 230–240.
Blaich, R., Bachmann, O., and Stein, U. (1982). Causes biochemiques de la résistance de la vigne à *Botrytis cinerea*. *Eur. Medit. Plant Prot. Org.* (EPPO) *Bull. 12*: 167–170.
Bompeix, G., Fettouche, F., and Saindrenan, P. (1981). Mode d'action du phoséthyl-al. *Phytiatrie-Phytopharmacie 30*: 257–272.
Bompeix, G., Ravisé, A., Raynal, G., Fettouche, F., and Durand, M. C. (1980). Modalités de l'obtention des nécroses bloquantes sur feuilles detachées de tomate par l'action du tris-o-éthyl phosphonate d'aluminium (phoséthyl d'aluminium), hypothèses sur son mode d'action in vivo. *Annales de Phytopathologie 12*: 337–351.
Bulit, J., and Lafon, R. (1978). Powdery mildew of the vine. *The Powdery Mildews* (D. M. Spencer, ed.), Academic Press, London, pp. 525–548.
Chappel, J., and Hahlbrock, K. (1984). Transcription of plant defense genes in response to UV light or fungal elicitor. *Nature 311*: 16–18.
Cohen, Y., and Coffey, M. D. (1986). Systemic fungicides and the control of oomycetes. *Ann. Rev. Phytopathol. 24*: 311–338.
Creasy, L. L., and Coffee, M. (1988). Phytoalexin production potential of grape berries. *J. Am. Soc. Hort. Sci. 113*: 230–234.
Dercks, W., and Buchenauer, H. (1986). Untersuchungen zum Einfluss von Aluminiumfosetyl auf den pflanzlichen Phenolstoffwechsel in den Pathogen-Wirt-Beziehungen *Phytophthora fragariae*-Erdbeere und *Bremia lactucae*-Salat. *J. Phytopathol. 115*: 37–55.
Dercks, W., and Buchenauer, H. (1987). Comparative studies on the mode of action

of aluminium ethyl phosphite in four *Phytophthora* species. *Crop. Prot. 6*: 82–89.

Dercks, W., and Creasy, L. L. (1989a). The significance of stilbene phytoalexins in the *Plasmopara viticola*-grapevine interaction. *Physiol. Mol. Plant Pathol. 34*: 189–202.

Dercks, W., and Creasy, L. L. (1989b). Influence of fosetyl-Al on phytoalexin accumulation in the *Plasmopara viticola*-grapevine interaction. *Physiol. Mol. Plant Pathol. 34*: 203–213.

Dunstan, R. H., Smillie, R. H., and Grant, B. R. (1990). The effects of subtoxic levels of phosphonate on the metabolism and potential virulence factors of *Phytophthora palmivora*. *Physiol. Mol. Plant Pathol. 36*: 205–220.

Durand, M. C., and Sallé, G. (1981). Éffet du tris-o-éthyl phosphonate d'aluminium sur le couple *Lycopersicum esculentum* Mill.-*Phytophthora capsici* Léon. Étude cytologique et cytochimique. *Agronomie 1*: 723–732.

Fenn, M. E., and Coffey, M. D. (1984). Studies on the in vitro and in vivo antifungal activity of fosetyl-Al and phosphorous acid. *Phytopathology 74*: 606–611.

Fenn, M. E., and Coffey, M. D. (1985). Further evidence for the direct mode of action of fosetyl-Al and phosphorous acid. *Phytopathology 75*: 1064–1068.

Fregoni, M. (1983). Fattori genetici ed agronomici predisponenti alla *Botrytis*. *Vignevini 10*: 35–42.

Fritzemeier, K. H., and Kindl, H. (1981). Coordinate induction by ultraviolet light of stilbene synthase, phenylalanine ammonia-lyase and cinnamate-4-hydroxylase in leaves of Vitaceae. *Planta 151*: 48–52.

Griffith, J. M., Smillie, R. H., and Grant, B. R. (1990). Alterations in nucleotide and pyrophosphate levels in *Phytophthora palmivora* following exposure to the antifungal agent potassium phosphonate (phosphite). *J. Gen. Microbiol. 136*: 1285–1291.

Guest, D. I. (1984). Modification of defense response in tobacco and capsicum following treatment with fosetyl-Al [aluminium tris (o-ethyl phosphonate)]. *Physiol. Plant Pathol. 25*: 125–134.

Guest, D. I. (1986). Evidence from light microscopy of living tissues that fosetyl-Al modifies the defense response in tobacco seedlings following inoculations by *Phytophthora nicotianae* var. *nicotianae*. *Physiol. Plant Pathol. 29*: 251–261.

Hain, R., Bieseler, B., Kindl, H., Schröder, G., and Stöcker, R. (1990). Expression of a stilbene synthase gene in *Nicotiana tabacum* results in synthesis of the phytoalexin resveratrol. *Plant Mol. Biol. 15*: 325–335.

Hart, J. H. (1981). Role of phytostilbenes in decay and disease resistance. *Ann. Rev. Phytopathol. 19*: 437–458.

Hart, J. H., and Shrimpton, D. M. (1979). Role of stilbenes in resistance of wood to decay. *Phytopathology 69*: 1138–1143.

Heintz, C., and Blaich, R. (1990). Ultrastructural and histochemical studies on interactions between *Vitis vinifera* L. and *Uncinula necator* (Schw.) Burr. *New Phytologist 115*: 107–117.

Hoos, G., and Blaich, R. (1988). Metabolism of stilbene phytoalexins in grapevines: oxidation of resveratrol in single-cell cultures. *Vitis 27*: 1–12.

Hoos, G., and Blaich, R. (1990). Influence of resveratrol on germination of conidia and mycelial growth of *Botrytis cinerea* and *Phomopsis viticola*. *J. Phytopathol. 129*: 102–110.

Ingham, J. L. (1976). 3,5,4-Trihydroxystilbene as a phytoalexin from groundnuts (*Arachis hypogaea*). *Phytochemistry 15*: 1791–1793.

Ingham, J. L. (1982). Phytoalexins from the leguminosae. *Phytoalexins* (J. A. Bailey and J. W. Mansfield, eds.), Wiley, New York, pp. 21–80.

Ingham, J. L., and Harborne, J. B. (1976). Phytoalexin induction as a new dynamic approach to the study of systematic relationships among higher plants. *Nature 260*: 241–243.

Jeandet, P., and Bessis, R. (1989). Une réflexion sur les mécanismes morphologiques et biochimiques de l'interaction vigné-*Botrytis*. *Bull. l'Office International de la Vigne et du Vin 703–704*: 637–657.

Jeandet, P., Bessis, R., and Gauthéron, B. (1991). The production of resveratrol (3,5,4′-trihydroxystilbene) by grape berries in different developmental stages. *Am. J. Enol. Vitic. 42*: 41–46.

Langcake, P. (1978). Phytoalexin production by grapevines in relation to infection by *Plasmopara viticola* (Abstr). Abstract of papers, 3rd International Congress of Plant Pathology, Munich, p. 248.

Langcake, P. (1981). Disease resistance of *Vitis* spp. and the production of the stress metabolites resveratrol, ε-viniferin, α-viniferin and pterostilbene. *Physiol. Plant Pathol. 18*: 213–226.

Langcake, P., Cornford, C. A., and Pryce, R. J. (1979). Identification of pterostilbene as a phytoalexin from *Vitis vinifera* leaves. *Phytochemistry 18*: 1025–1027.

Langcake, P., and Lovell, P. (1980). Light and electron microscopical studies of the infection of *Vitis* spp. by *Plasmopara viticola*, the downy mildew pathogen. *Vitis 19*: 321–337.

Langcake, P., and McCarthy, W. V. (1979). The relationship of resveratrol production to infection of grapevine by *Botrytis cinerea*. *Vitis 18*: 244–253.

Langcake, P., and Pryce, R. J. (1976). The production of resveratrol by *Vitis vinifera* and other members of the vitaceae as a response to infection or injury. *Physiol. Plant Pathol. 9*: 77–86.

Langcake, P., and Pryce, R. J. (1977a). A new class of phytoalexins from grapevines. *Experientia 33*: 151–152.

Langcake, P., and Pryce, R. J. (1977b). The production of resveratrol and the viniferins by grapevines in response to ultraviolet irradiation. *Phytochemistry 16*: 1193–1196.

Langcake, P., and Pryce, R. J. (1977c). Oxidative dimerisation of 4-hydroxystilbenes in vitro: production of a grapevine phytoalexin mimic. *J. Chem. Soc. Chem. Commun. 7*: 208–210.

Luczka, C. J. (1982). Stilbene phytoalexins and susceptibility to *Botrytis cinerea* in *Vitis*. M.S. Thesis, Cornell University, Ithaca, NY.

Luttringer, M., and De Cormis, L. (1985). Absorption, dégradation et transport du phoséthyl-Al et de son métabolite chez la tomate (*Lycopersicon esculentum* Mill.). *Agronomie 5*: 423–430.

Melchior, F., and Kindl, H. (1991). Coordinate and elicitor-dependent expression of stilbene synthase and phenylalanine ammonia-lyase genes in *Vitis* cv. Optima. *Arch. Biochem. Biophys. 288*: 552–557.

Nemestothy, G. S., and Guest, D. I. (1990). Phytoalexin accumulation, phenylalanine ammonia lyase activity and ethylene biosynthesis in fosetyl-Al treated resistant and susceptible tobacco cultivars infected with *Phytophthora nicotianae* var. *nicotianae. Physiol. Mol. Plant Pathol. 37*: 207–219.

Pezet, R., and Pont, V. (1988a). Mise en évidence de ptérostilbène dans les grappes de *Vitis vinifera. Plant Physiol. Biochem. 26*: 603–607.

Pezet, R., and Pont, V. (1988b). Activité antifongique dans les baies de *Vitis vinifera*: effets d'acides organiques et du ptérostilbène. *Revue Suisse de Viticulture, d'Arboriculture, et d'Horticulture 20*: 303–309.

Pezet, R., and Pont, V. (1990). Ultrastructural observations of pterostilbene fungitoxicity in dormant conidia of *Botrytis cinerea* Pers. *J. Phytopathol. 129*: 19–30.

Pezet, R., Pont, V. and Hoang-Van, K. (1991). Evidence for oxidative detoxification of pterostilbene and resveratrol by a laccase-like stilbene oxidase produced by *Botrytis cinerea. Physiol. Mol. Plant Pathol. 39*:441–450.

Pont, V., and Pezet, R. (1990). Relation between the chemical structure and the biological activity of hydroxystilbenes against *Botrytis cinerea. J. Phytopathol. 130*: 1–8.

Pont, V., and Pezet, R. (1991). Un moyen de défense naturel des végétaux: les phénols. *Revue Suisse d'Agriculture 23*: 237–241.

Pool, R. M., Creasy, L. L., and Frackelton, A. S. (1981). Resveratrol and the viniferins, their application to screening for disease resistance in grape breeding programs. *Vitis 20*: 136–145.

Pryce, R. J., and Langcake, P. (1977). α-Viniferin: an antifungal resveratrol trimer from grapevines. *Phytochemistry 16*: 1452–1454.

Raynal, G., Ravisé, A., and Bompeix, G. (1980). Action du tris-o-éthylphosphonate d'aluminium (phoséthyl d'aluminium) sur la pathogénie de *Plasmopara viticola* et sur stimulation des réactions de défense de al vigne. *Annales de Phytopathologie 12*: 163–175.

Rupprich, N., and Kindl, H. (1978). Stilbene synthases and stilbenecarboxylate synthases, I. Enzymatic synthesis of 3,5,4′-trihydroxystilbene from ρ-coumaroyl coenzyme A and malonyl coenzyme A. *Hoppe-Seyler's Z. Physiol. Chem. 359*: 165–172.

Saindrenan, P., Barchietto, T., Avelino, J., and Bompeix, G. (1988). Effects of phosphite on phytoalexin accumulation in leaves of cowpea infected with *Phytophthora cryptogea. Physiol. Mol. Plant Pathol. 32*: 425–435.

Saindrenan, P., Darakis, G., and Bompeix, G. (1985). Determination of ethyl phosphite, phosphite, and phosphate in plant tissues by anion-exchange high performance liquid chromatography and gas chromatography. *J. Chromatogr. 347*: 267–273.

Siemann, E. H., and Creasy, L. L. (1992). Concentration of the phytoalexin resveratrol in wine. *Am. J. Enol. Vitic. 43*:49–52.

Smillie, R. H., Dunstan, R. H., Grant, B. R., Griffith, J. M., Iser, J., and Niere, J. O. (1990). The mode of action of the antifungal agent phosphite. *Eur. Med. Plant Prot. Org.* (EPPO) *Bull. 20*: 185–192.

Smillie, R.H., Grant, B. R., and Guest, D. I. (1989). The mode of action of phosphite: evidence for both direct and indirect modes of action on three *Phytophthora* spp. in plants. *Phytopathology 79*: 921–926.

Stein, U. (1984). Untersuchungen über biochemische und morphologische Merkmale der Botrytisresistenz bei Vitaceen. PhD Thesis, University of Karlsruhe, FRG.

Stein, U., and Blaich, R. (1985). Untersuchungen über die Stilbenproduktion und Botrytisanfälligkeit bei *Vitis*-Arten. *Vitis 24*: 75–87.

Stein, U., and Hoos, G. (1984). Induktions- und Nachweismethoden für Stilbene bei Vitaceen. *Vitis 23*: 179–194.

Van Etten, H. D., Mathews, D. E., and Mathews, P. S. (1989). Phytoalexin detoxification: importance for pathogenicity and practical implications. *Ann. Rev. Phytopathol. 27*: 143–164.

Ward, E. W. B., Unwin, C. H., and Stoessl, A. (1975). Postinfectional inhibitors from plants. XV. Antifungal activity of the phytoalexin orchinol and related phenanthrenes and stilbenes. *Can. J. Bot. 53*: 964–971.

Zitnak, A., and Johnston, G. R. (1970). Glycoalkaloid content of B5141-6 potatoes. *Am. Potato J. 47*: 256–260.

14
Mode of Toxic Action of Vitaceae Stilbenes on Fungal Cells

Roger Pezet and Vincent Pont
Swiss Federal Agricultural Research Station of Changins, Nyon, Switzerland

I. INTRODUCTION

Though the toxicity of phytoalexins is well documented, their mechanism of action on fungal cells is poorly understood. However, many fragmentary reports are available and it is possible to propose a mechanism of fungitoxic mode of action of the hydroxystilbenes. These compounds belong to the large family of plant phenolics from which numerous other phytoalexins are recognized. According to their basic common chemical structure, it is expected that all phenolics act on similar biochemical pathways and that their fungicidal mode of action is identical.

In a well-documented review, Smith (1982) discussed the toxicity of phytoalexins. We can assume that extensive membrane damages occur soon after fungi are exposed to phytoalexins and suppression of exogenous respiration may reflect insufficient uptake of substrate due to membrane damages. More precise is the toxic effect of rishitin to membranes; it acts by increasing their permeability to small molecular weight compounds (Lyon, 1980).

Both phaseollin and pisatin have been shown to be toxic to plant and fungal cells in a similar manner by disrupting either the structure or the functioning of the plasma membrane (Hargreaves, 1980). Laks and Prunner (1989) suggest that pisatin and maackiain, as well as other flavonoid phytoalexin analogues, function primarily against fungi as uncouplers of oxidative phosphorylation. According to O'Neil and Mansfield (1982), the antifungal activity of flavonoids and isoflavonoids depends on some common physiochemical attributes, perhaps lipophilicity, and on the ability to penetrate fungal cell wall and membranes rather than on a common structure.

One may be surprised that so many biocide molecules are phenols possessing important conjugated systems. In such molecules, electrons of chemical bonds are not immobile. They move inside the molecules, and this movement of electrons is due to alternate single and double bonds (Fig. 1).

Chemical structure of stilbenes constitutes a conjugated system with aromatic character as described above. Hart (1981) reports the effects of stilbenes on respiration of fungal cells. He suggests that they act as uncoupling agents and they could form protein–phenol complexes. Uncoupling of electron transport and photophosphorylation were also observed on isolated chloroplasts of *Spinacia oleracea* in the presence of different stilbenes (Gorham and Coughlan, 1980). The conjugated system of hydroxystilbenes plays an important role in the formation of charge transfer complexes (CTCs), as described by Slifkin (1980), favoring contact and affinity with proteins. Like hydroxystilbenes, the aromatic hydrocarbons, a group of synthetic molecules, are able to constitute CTC with proteins too. Most of these molecules are synthetic fungicides and for this reason their mechanism of action was better known than that of phenolic phytoalexins. All these informations are useful to explain how stilbenes kill fungal cells.

II. EFFECT OF HYDROXYSTILBENES ON FUNGAL CELLS

Natural hydroxystilbenes that we have studied are phytoalexins produced by Vitaceae, resveratrol (3,5,4′-trihydroxystilbene) and pterostilbene (3,5-dimethoxy-4′-hydroxystilbene). Other hydroxystilbenes with different physicochemical properties were synthesized in our laboratory in order to explain relations between the chemical structure and the biological activity of these compounds (Table 1) (Pont and Pezet, 1990).

A. Effect on Respiration

The first observed effect of hydroxystilbenes on conidia of *Botrytis cinerea* is a decrease of oxygen uptake. This effect may be slight or important, depending on the substituents of the hydroxystilbenic basic structure (Table 2). Pterostilbene is the best inhibitor of conidial respiration. We observed that oxygen uptake was interrupted some minutes after the addition of pterostilbene (Fig. 2).

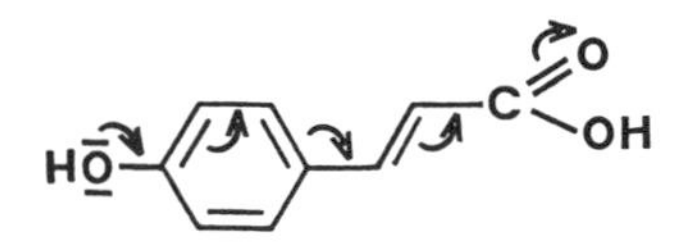

Figure 1 Conjugated system of *p*-coumaric acid. Arrows represent electron displacement within the molecule (electronic delocalization).

Table 1 Structure of *trans*-Hydroxystilbenes Tested on the Conidia of *Botrytis cinerea*

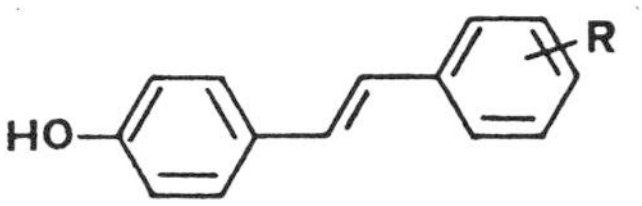

R		mp observed (°C)	Ref.
H	(1)	–	
4-CH_3	(2)	211–212	Veschambre, et al. (1967)
3,4-$(OCH_3)_2$	(3)	184–186	–
4-Cl	(4)	184–185	Massarani (1957)
			Veschambre et al. (1967)
3,5-$(OCH_3)_2$	(5)	86–86.5	King et al. (1953)
			Späth and Schläger (1940)
3,5-$(OH)_2$	(6)	254–254.5	Nonomura et al. (1963)
			Rupprich et al. (1980)
3-Cl	(7)	130–131	Veschambre et al. (1967)
3,4-Cl_2	(8)	142–143	–
3,5-Cl_2	(9)	148–150	–

Table 2 Oxygen Uptake After 10 min (nmol O_2 min) by 10^7 6-day-old Conidia of *Botrytis cinerea* Treated by the Compounds of Table 1 at 5×10^{-5} M[a]

R		nmol O_2
H	(1)	8.33
4-CH_3	(2)	7.53
3,4-$(OCH_3)_2$	(3)	7.71
4-Cl	(4)	5.40
3,5-$(OCH_3)_2$	(5)	2.79
3,5-$(OH)_2$	(6)	6.74
3-Cl	(7)	5.51
3,4-Cl_2	(8)	2.92
3,5-Cl_2	(9)	4.85
Conidia only: 10.4 nmol O_2/min		

[a]Cell respiration was measured polarographically with a Clark electrode (PO_2 analyzer, Bachofer) at 25 ± 0.01°C.

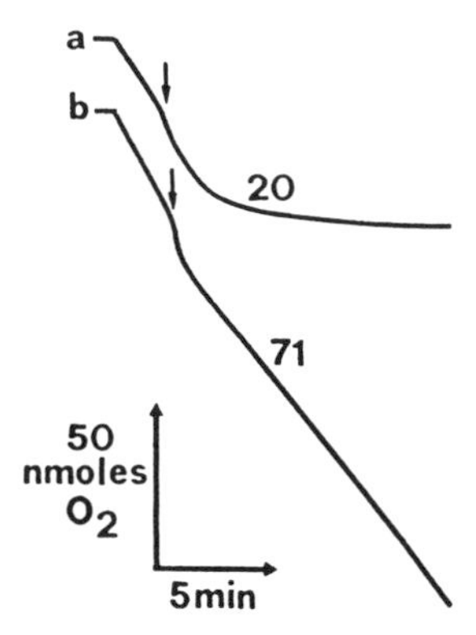

Figure 2 Effect of pterostilbene on respiration of conidia of *Botrytis cinerea*. a, Pterostilbene treated conidia. b, Control. Pterostilbene (50 μl/ml of an ethanolic solution at 2.56 mg/ml) or ethanol for the control (50 μl/ml) are added some minutes (*arrows*) after the suspension of conidia (2.5×10^7/ml) was placed into a vessel of a Clark-type oxygen electrode (PO_2 Analyzer, Bachofer) maintained at 25 $\pm$ 0.01°C.

B. Visible Effects on Conidia

When conidia of *B. cinerea* are placed on media containing different concentrations of hydroxystilbenes, germination is more or less affected depending on the chemical structure of the hydroxystilbenes. The more dramatic effect is a total inhibition of germination and the destruction of all organelles and cell membranes (Pezet and Pont, 1990). Conidia appear under a light microscope to be structurally transformed (Fig. 3). We call this effect the transformation of conidia. Their size decreased from a mean of 10×8.5 μm to 8×6 μm and we observed a withdrawal of the cytoplasm from the wall. An outflow of cytoplasmic matter (*arrow*, Fig. 3) was sometimes observed.

Figures 4 and 5 show the effects of concentration of different hydroxystilbenes on the inhibition of germination and on the transformation of *B. cinerea* conidia. The most toxic compounds are pterostilbene and three chlorinated hydroxystilbenes. Resveratrol remains totally inactive, even at a concentration of 10^{-3} M. Both inhibition of germination and transformation of conidia and the effect on cellular respiration design the pterostilbene as the known most toxic natural hydroxystilbene produced by the Vitaceae.

C. Effect on Cellular Structures

Fungitoxic action of pterostilbene was studied through ultrastructural observations. Dormant conidia of *B. cinerea* are round to elliptical and relatively smooth. The ultrastructure of healthy dormant conidia was similar to

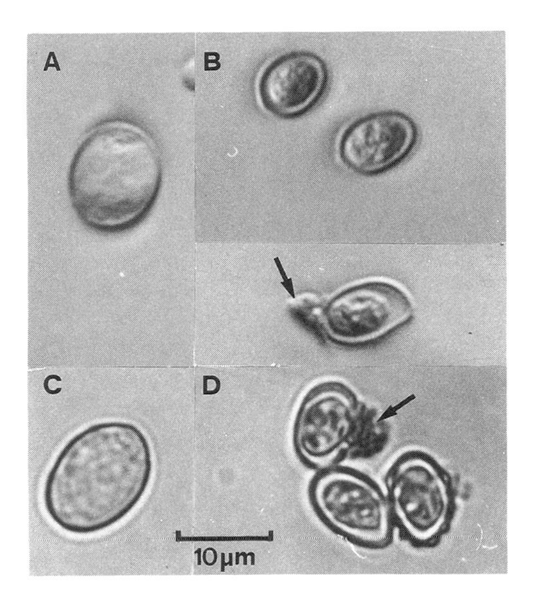

Figure 3 Effect of pterostilbene on dormant conidia of *Botrytis cinerea* visible under light microscopy. A and C, Untreated conidium. B and D, Pterostilbene treated conidia (5×10^{-4} M). Observed under differencial interference contrast (DIC) (A and B). Observed under light microscope (C and D). Note the cytoplasmic withdrawal from the spore wall. Some conidia present outflow of cytoplasm (*arrows*).

that previously described (Buckley et al., 1966; Gull and Trinci, 1971). Figure 6 shows that healthy conidia possess several nuclei; mitochondria are usually round to ovoid with many cristae; numerous vacuoles, probably containing glycogen, are located in the cytoplasm. Few round lipid bodies are visible and the plasma membrane appears well defined with its three layers—the external, dense to the electron, and the two internal layers, more clear.

Application of pterostilbene at 5×10^{-4} M on dormant conidia induces very rapid modifications of their ultrastructure. Chronological observations, after 1–10 min following the addition of pterostilbene, show that mitochondria and nuclear membrane are very rapidly (5 min) affected by the phytoalexin. These membranes become thicker and many electron-dense lipid bodies are visible in the cytoplasm. Soon after endoplasmic reticula are disorganized and ribosomes tend to disappear. Before a complete disorganization of cytoplasmic organelles and a disruption of the cell membranes, large electron-dense lipid bodies are formed close to the mitochondrial and nuclear membranes. The ultimate stage, where cyto-

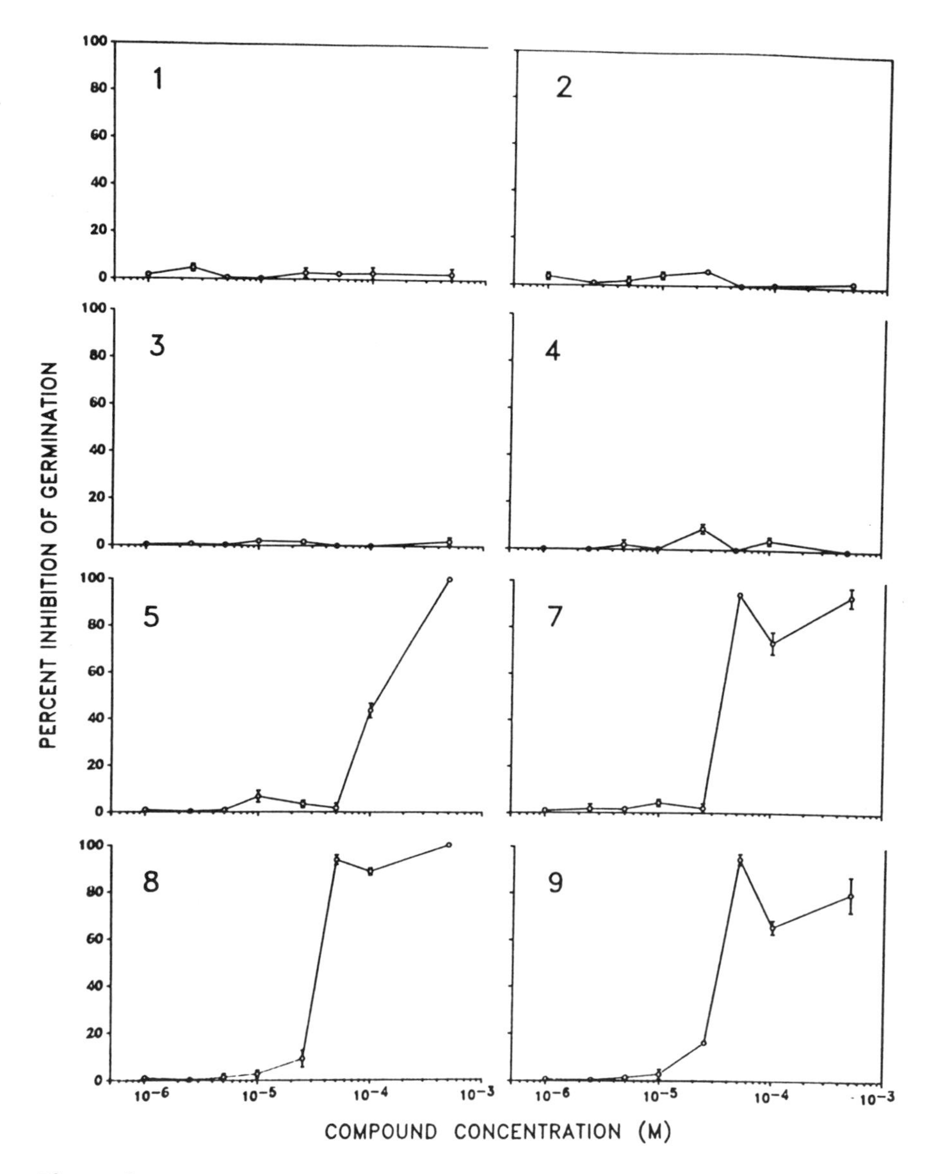

Figure 4 Effect of concentration of hydroxystilbenes of Table 1 on the inhibition of germination of 6-day-old conidia of *Botrytis cinerea* (Pont and Pezet, 1990). Inactive compound 6 (resveratrol) is not represented.

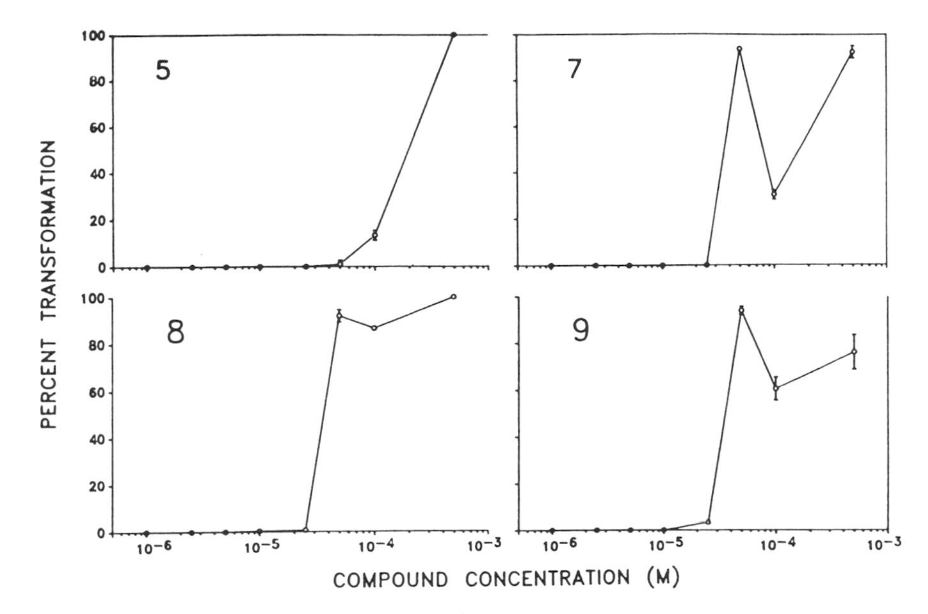

Figure 5 Effect of concentration of hydroxystilbenes of Table 1 on the transformation of the aspect of 6-day-old conidia of *Botrytis cinerea* (Pont and Pezet, 1990). Compounds 1, 2, 3, 4, and 6 were inactive and are not represented.

plasmic material seems to be dissolved and coagulated, corresponds to the lethal transformation stage of conidia (Fig. 7). The degree of effect on membranes appears to occur in the following order: mitochondrial membrane ≥ nuclear membrane > endoplasmic reticulum > cytoplasmic membrane.

The rapid modification of mitochondrial membranes corresponds to the dramatic decrease of O_2 uptake of conidia in the presence of pterostilbene. Large electron-dense lipid bodies visible in mitochondrial and nuclear membranes are probably phospholipids released from disorganized membranes. The presence of such lipid bodies in nuclear and mitochondrial membranes of *Sclerotinia fructigena* has been described by Najim and Turian (1979) and associated to phospholipids because of the high affinity of osmic acid to this class of lipids.

III. PROPOSED MECHANISM OF ACTION

We have seen that hydroxystilbenes with different substituents have different fungitoxicity levels. The benzenic substituents of these compounds can be described by the σ values characterizing the electron-attracting or donat-

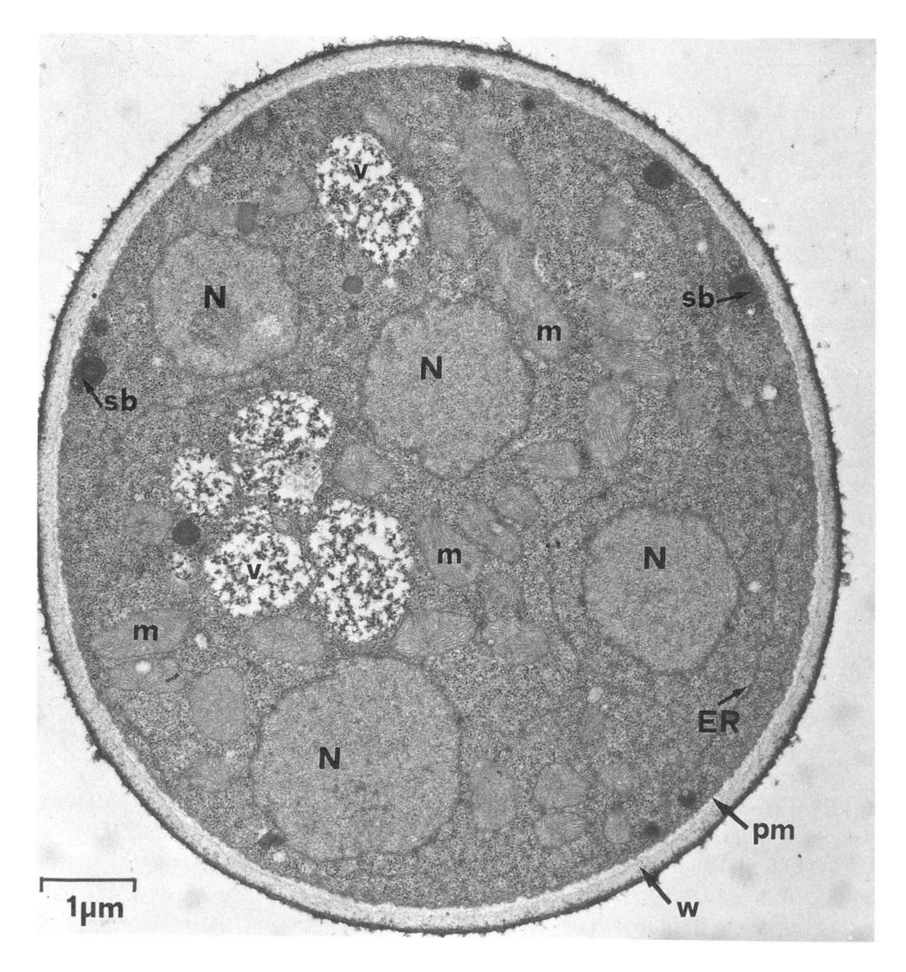

Figure 6 Healthy dormant conidia of *Botrytis cinerea*. N, nucleus; m, mitochondria; v, vacuoles; ER, endoplasmic reticulum; w, wall; pm, plasma membrane; sb, storage bodies; lb, lipid bodies.

ing power. A good linear relationship (r = 0.93) can be calculated between the σ value of 10 monosubstituted styryl groups of hydroxystilbenes (σ_{st}) given by Veschambre et al. (1967) and the σ value of benzenic substituents. In consequence σ_{st} values of disubstituted styryl groups of hydroxystilbenes are deduced from these values.

The antifungal efficiency of a compound may be determined also by its capacity to invade lipophilic membrane sites. In this area, the relation between the biocidal character of flavonoids, phytoalexins, and their lipophilicity were related by Laks and Pruner (1989). The parameter of hydrophobicity, R_m, of hydroxystilbenes was calculated from R_f obtained by thin layer chromatography (TLC) on reverse phase C_{18} utilizing the solvent system MeOH–H_2O (8 : 2) as eluent (Rittich et al., 1980) according to the formula $R_m = \log(1/R_f - 1)$. Finally, the method described by Bondi (1964) permitted a calculation of the van der Waals volumes V_m. These chemical parameters of the tested hydroxystilbenes are given in Table 3.

In view of the inhibition of germination and respiration, and transformation of conidia of *B. cinerea*, it can be concluded that the biological activity of the tested hydroxystilbenes depended largely on the σ values. It tended to increase with the electron-attracting power of the substituents. In this view, the σ_{st} values were in better agreements with the apparent small differences in the biological effects, for example, among the chloro compounds 3-Cl[7], 3,4,-Cl_2[8], and 3,5-Cl_2[9]. The inhibition of germination and the transformation of conidia for these compounds did not always reach 100% beyond the "critical concentration" of 2.5×10^{-5} M. At high concentrations, these lipophilic compounds crystallized partly in the agar during the experiments. This fact resulted in a loss of activity at the appropriate concentration of 10^{-4} M. However, for the respiration, the difference of activity between compounds with substituents 3,4-Cl_2[8] and 3,5-Cl_2[9] remained unexplained.

The weak activity of resveratrol[6], in spite of a σ of $+0.16$, may be the consequence of its hydrophilic character ($R_m = 0.60$), which renders it difficult to dissolve in the membranes.

The real toxicity of pterostilbene is always higher than its theoretical toxicity calculated from its electronic characteristics. On the other hand, the important volume of the methoxy groups of this stilbene can perturb the intermolecular bonds more than the other substituents. Methoxy groups give the possibility to form hydrogen bonding with their oxygen atoms. This factor, including the possible formation of CTCs (Slifkin, 1980) and hydrogen bonding with the phenolic-OH (Looms and Battaile, 1966), favors the contact of pterostilbene with membranous proteins. It is clearly accepted that amino acids, particularly aromatic amino acids, form CTCs with electron acceptors (Slifkin, 1980). The delocalization of electrons in the conjugated system represented by hydroxystilbenes enhances the polarization of the molecules when the substituents are electron-attracting groups. Generally, the more electron-attracting are the stilbenes, the more toxic are their effects. The affinity of hydroxystilbenes for proteins could be the consequence of formation of CTCs between these compounds and amino acids of the proteins.

Inhibition of respiration by pterostilbene is probably indirect. We have described the effects of this stilbene on mitochondrial and nuclear mem-

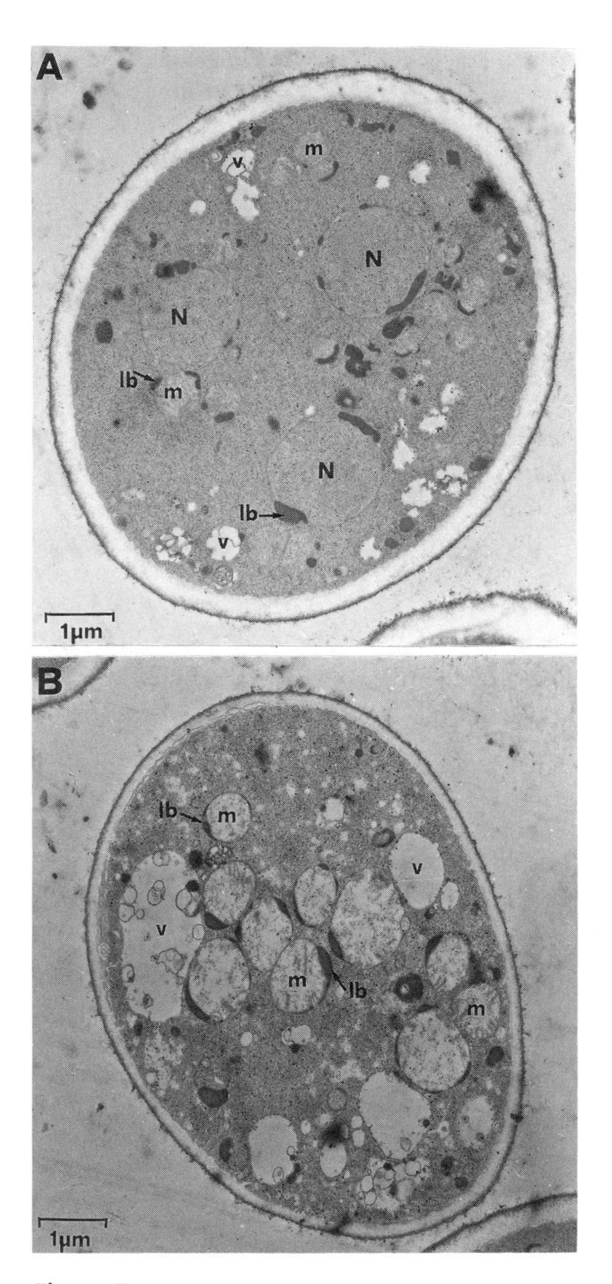

Figure 7 Aspect of dormant conidia of *Botrytis cinerea* induced by pterostilbene (5×10^{-4} M) after different times of incubation. A, 5 min; B, 15 min; C, 30 min; D, 3 hr. (See legends in Figure 6.)

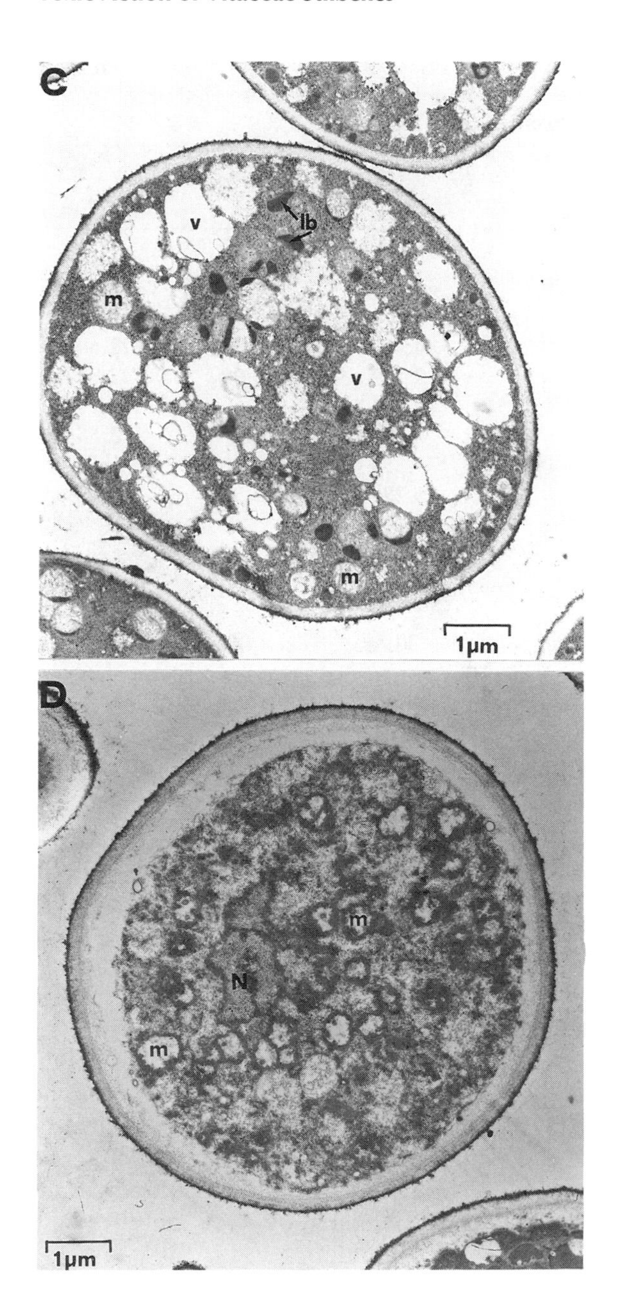
C
v
lb
m
v
m
1µm
D
m
N
m
1µm

Table 3 Aromatic Substituent Constants (σ, σ_{St}) of Substituted Styryl Groups, Parameter of Hydrophobicity (R_m), and Van der Waals Volumes (V_w) of Substituted Hydroxystilbenes Presented in Table 1

R		σ	σ_{St}	R_m	V_w (cm^3/mol)
H	(1)	0	0.172	+0.07	114.16
4-CH_3	(2)	−0.17	0.153	+0.23	125.31
3,4-$(OCH_3)_2$	(3)	−0.12	0.162	−0.09	142.85
4-Cl	(4)	+0.23	0.216	+0.23	123.64
3,5-$(OCH_3)_2$	(5)	+0.05	0.189	+0.09	142.85
3,5-$(OH)_2$	(6)	+0.16	0.205	−0.60	125.20
3-Cl	(7)	+0.37	0.236	+0.27	123.64
3,4-Cl_2	(8)	+0.52	0.255	+0.43	133.12
3,5-Cl_2	(9)	+0.75	0.286	+0.55	133.12

branes and on endoplasmic reticulum. These effects on membranes are identical to those reported by Lyr (1987) in connection with the mechanism of action of aromatic hydrocarbon fungicides (AHFs). These compounds possess electron-conjugated systems as stilbenes do. The σ value was estimated for one of them, dichloromethoxyphenol [DCMP], as +0.33. This indicates that DCMP possesses an electron-attracting power as important as 3-chloro-4′-hydroxystilbene[7].

Lyr (1987) proved that the primary toxic effect of AHFs is an induced lipid peroxidation, especially in the mitochondrial and nuclear membranes and in the endoplasmic reticulum. The reason is the interaction of AHFs with flavin enzymes, such as cytochrome C reductase, or other monooxygenases, located within the membrane system. AHF blocks the electron transport from flavin to the substrate and induces, by generation of free radicals, a pathological peroxidation of membranous phospholipids. Comparing AHF-resistant and sensitive strains of *Mucor*, Werner (1980) found that the mitochondrial proteins of resistant strains contained a low concentration of tyrosine 0.2% of the total amino acids, compared to 2.9% in sensitive strains. Addition of antioxidant and anti–free radicals as α-tocopherol acetate to a culture of *Mucor*, inhibited by AHF, nullifies lipid peroxidation and counteracts the growth inhibition.

Amino acid tyrosine is an important site where electrons can enter cytochrome to reach the iron-active center (Salemme et al., 1973). Aromatic structure of tyrosine could favor CTCs (Slifkin, 1980) with all molecules possessing conjugated system and electron-attracting capacities. Pterostilbene and other hydroxystilbenes having positive σ belong, as do AHFs, to this category of compounds.

IV. EPILOG

Many natural phenolics like hydroxystilbenes possess an important conjugated system. This electronic character, associated with the nature of the substituents, determine the σ value relating to the biological effects of hydroxystilbenes. We found among the high conjugated phenols not only many phytoalexins but other active compounds, such as fungitoxins and bactericidal compounds as well (Pont and Pezet, 1991).

Pterostilbene is produced in a very low concentration by leaves and immature berries of grapevines comparing to resveratrol. The low fungitoxicity of this last compound, even at high concentration, is explained by its hydrophilic character incompatible with the lipophilicity of biological membranes. Its direct role in the defense mechanism of Vitaceae against fungal attacks is probably very poor. Pterostilbene, even at low concentrations, displays, when associated to glycolic acid, a high toxicity toward *B. cinerea* (Pezet and Pont, 1988). This organic acid, at a relatively high concentration in immature berries, is toxic to *B. cinerea* conidia where it provokes important damages to the organelle membranes. Microorganisms can oxidize glycolic acid to glyoxylic acid (Corpe and Stone, 1960), probably as it has been described for plants, by glycolate oxidase with a concomitant production of hydrogen peroxide (Metzler, 1977).

The synergistic effects of pterostilbene and glycolic acid may increase peroxide production and consequent damages to the biological membranes.

All these informations suggest that the mechanism of action of hydroxystilbenes involves important lipid peroxidation by blocking flavin enzymes such as cytochrome c reductase and similar monooxygenases.

ACKNOWLEDGMENTS

We are grateful to Paul Parey, Editor for authorization to use previously published figures and tables from the *Journal of Phytopathology*, Vols. 129 and 130.

REFERENCES

Bondi, A. (1964). Van der Waals volumes and radii. *J. Phys. Chem. 68*:441–451.

Buckley, P. M., Sjaholm, V. E., and Sommer, N. F. (1966). Electron microscopy of *Botrytis cinerea* conidia. *J. Bacteriol. 91*(5):2037–2044.

Corpe, N. A., and Stone, R. W. (1960). Oxidation of glycolic acid by *Penicillium crysogenum*. *J. Bacteriol. 80*:452–456.

Gorham, J., and Coughlan, S. J. (1980). Inhibition of photosynthesis by stilbenoids. *Phytochemistry 19*:2059–2064.

Gull, K., and Trinci, P. J. (1971). Fine structure of spore germination of *Botrytis cinerea*. *J. Gen. Microbiol. 68*:207–220.

Hargreaves, J. A. (1980). A possible mechanism for the phytotoxicity of the phytoalexin phaesollin. *Physiol. Plant Pathol. 16*:351–357.

Hart, J. H. (1981). Role of phytostilbenes in decay and disease resistance. *Ann. Rev. Phytopathol. 19*:437–458.

King, F. E., Cotterill, C. B., Godson, D. H., Jurd, L., and King, T. J. (1953). The chemistry of extractive hardwood. XIII. Colourless constituents of *Pterocarpus* species. *J. Chem. Soc.* 3693–3697.

Laks, P. E., and Pruner, M. S. (1989). Flavonoid biocides: structure/activity relations of flavonoid phytoalexin analogues. *Phytochemistry 28*(1):87–91.

Loomis, W. D., and Battaile, J. (1966). Plant phenolic compounds and the isolation of plant enzymes. *Phytochemistry 5*:423–438.

Lyon, G. D. (1980). Evidence that the toxic effect of rishitin may be due to membrane damage. *J. Exp. Bot. 31*(123):957–966.

Lyr, H. (1987). Mechanism of action of aromatic hydrocarbon fungicides. *Modern Selective Fungicides* (H. Lyr, ed.), Longman, pp. 75–89.

Massarani, E. (1957). Ulteriori ricerche sopra derivati stilbenici e difeniletanici. *Il Farmaco-Ed. Sc 12*(5):380–386.

Metzler, D. E. (1977). Light in biology. *Biochemistry: The Chemical Reactions of Living Cells*. Academic Press, New York, p. 787.

Najim, L., and Turian, G. (1979). Ultrastructure de l'hyphe végétatif de *Sclerotinia fructigena*. *Can. J. Bot. 57*:1299–1313.

Nonomura, S., Kanagawa, H., and Makimoto, A. (1963). *Chem. Abst. 60*:4240c (Abstr)

O'Neil, T. M., and Mansfield, J. W. (1982). Antifungal activity of hydroxyflavans and other flavonoids. *Trans. Br. Mycol. Soc. 79*(2):229–237.

Pezet, R., and Pont, V. (1988). Activité antifongique dans les baies de *Vitis vinifera*: effets d'acides organiques et du ptérostilbène. *Rev. Suisse Viti. Arbor. Hort. 20*:303–309.

Pezet, R., and Pont, V. (1990). Ultrastructural observations of pterostilbene fungitoxicity in dormant conidia of *Botrytis cinerea* Pers. *J. Phytopathol. 129*:19–30.

Pont, V., and Pezet, R. (1990). Relation between the chemical structure and the biological activity of hydroxystilbenes against *Botrytis cinerea*. *J. Phytopathol. 130*:1–8.

Pont, V., and Pezet, R. (1991). Un moyen de défense naturel des végétaux: les phénols. *Revue Suisse Agric 23*(4):237–241.

Rittich, B., Polster, M., and Kralik, O. (1980). Reversed-phase high-performance liquid chromatography. I. Relationship between R_M values and Hansch's pi parameters for a series of phenols. *J. Chromatogr. 197*:43–50.

Rupprich, N., Hildebrand, H., and Kindl, K. (1980). Substrate specificity in vivo and in vitro in the formation of stilbenes. Biosynthesis of rhaponticin. *Arch. Biochem. Biophys. 200*:72–78.

Salemme, F. R., Krant, J., and Kaman, M. D. (1973). Structural bases for function in cytochrome c. *J. Biol. Chem. 248*(22):7701–7716.

Slifkin, M. A. (1980). The significance of charge transfer interactions in biology. *Molecular Interactions*, Vol. 2. H. Ratajezak and W. J. Orville-Thomas, eds.), Wiley, London, pp. 271–304.

Smith, D. A. (1982). Toxicity of phytoalexins. *Phytoalexins* (J. A. Bailey and J. W. Mansfield, eds.), Blackie, Glasgow, pp. 218–252.

Spath, E., and Schlager, J. (1940). Über die Inhaltstoffe des roten Sandelholzes. II. Mitteil.: *Die Konstitution des Pterostilbens. Ber. 8*:881–884.

Veschambre, H., Dauphin, G., and Kerkomard, A. (1967). Transmission des effets électroniques dans les molécules de trans-hydroxy-4 stilbènes et d'hydroxy-4 tolanes. *Bull. Soc. Chim. 8*:2846–2854.

Werner, P. (1980). Zum Wirkungsmechanismus des systemischen Fungicides Chloroneb und zu den möglischen Ursachen erzeugter Resistenz gegenüber *Mucor mucedo* L. *Fres. Dissertation Martin-Lüther-University*, Halle-Wittemberg, 1–142.

15

Inducible Compounds in *Phaseolus*, *Vigna*, and *Dioscorea* Species

S. A. Adesanya
Obafemi Awolowo University, Ile-ife, Nigeria

M. F. Roberts
London University, London, England

I. INTRODUCTION

Phytoalexins are inducible compounds first observed by Müller and Borger in 1940 and defined by Müller (1958) as antibiotics produced as a result of biochemical interactions between a host plant and a parasite. This definition has been modified by several workers (Harborne, 1977; Deverall, 1982). Essentially, they are chemical compounds that accumulate in the living hypersensitive tissues around the infection sites (Bailey, 1973; Rahe, 1973; Mansfield et al., 1974). The cause of hypersensitive reaction may be infection due to microorganisms or to the toxic effects of their breakdown products and metabolites (biotic inducers), or chemical and physical factors (abiotic inducers) (Cruickshank and Perrin, 1968; Hadwiger and Schwochau, 1971; Bailey et al., 1980). There are indications of different modes of activity for these two groups of inducers. Biotic inducers appear to activate defense genes in the host, which leads to enzyme synthesis for the production of the compounds (Darvill and Albersheim, 1984; Dhawale et al., 1989; Ellis et al., 1989; Preisig et al., 1991); the mode of action of abiotic inducers is not clear since phytoalexins have been known to accumulate in tissues treated with agents blocking genetic transcription and translation processes (Yoshikawa, 1978). There are indications to suggest that on infection constitutive compounds in the host cause the parasite to release elicitors which activate the phytoalexin production genes (Kiraly et al., 1972; Keen et al., 1983; Ersek and Kiraly, 1986). The induced phytoalexins demonstrate toxicity against nonpathogens as well as pathogens; pathogens, however, usually have a way of avoiding their toxic effects by catabolism or the formation of conjugates with glycosides (Tani and Mayama, 1982; Weltring et al., 1982; Willeke and Barz, 1982; Denny and Van Etten, 1983; Tahara et al., 1987).

Plants which rapidly accumulate phytoalexins are known to be more resistant to infection. There are, however, doubts about their primary role in disease resistance (Ersek and Kiraly, 1986).

The exploitation of phytoalexins relies on their inherent properties. The genetic factor leads to the possibility of evolution of peculiar biosynthetic pathways and compounds with restricted distribution that can be used in chemotaxonomy. Interfamilial differences such as the production of isoflavonoids in the Fabaceae (Leguminosae) and terpenoids in Solanaceae have been observed together with intrageneric variations such as the production of medicarpin (**47**), and maackiain (**58**) by some *Trigonella* species, and medicarpin and vestitol (**38**) by others (Ingham and Harborne, 1976; Ingham, 1981a; Ingham, 1990a; Kuć, 1982). This type of study has been used in clarifying the taxonomy of Trifolieae and recently Phaseolinae species (Ingham and Harborne, 1976; Ingham, 1981b, 1990a,b) and intrageneric variations in the genus *Glycine* (Keen, 1986).

The major biological property attributed to phytoalexins is antifungal and antimicrobial against a range of organisms. It is noteworthy that such chemotoxicities principally involve basic interference with physiological processes and therefore possible activity against other life forms; for example, in a test of several isoflavonoids against zoopathogens it was found that phaseollinisoflavan (**43**) a phytoalexin in several *Phaseolus* species, was very active (Gordon et al., 1980; Smith, 1982). Other biological activities of isoflavonoid phytoalexins have been well documented in a review by Smith and Banks (1986).

It has been established that previous inoculation of plants with a nonpathogen protects the plant against subsequent infection by pathogens (Müller and Börger, 1940; Bell and Pressley, 1969; Mansfield, 1982). This cross-protection property relies on the ability of the plant to accumulate high levels of its phytoalexins more rapidly on infection and is the key property for disease resistance (Mansfield, 1982). Thus, monitoring the levels of phytoalexins in induced plants can assist in the selection and breeding of resistant strains for breeding.

In the present work, some aspects of elicited isoflavonoids in *Phaseolus* and *Vigna* species and dihydrostilbenes in *Dioscorea* species will be described. These two groups of compounds have common biosynthetic origins in that both groups are formed from either cinnamoyl-CoA or *p*-coumaroyl-CoA and malonyl-CoA.

A. *Phaseolus–Vigna* Species

These species belong to the family Fabaceae (Leguminoseae, subfamily Papilionoideae), tribe Phaseolae, and subtribe Phaseolinae. They constitute the largest group and are the most important source of food in the subtribe

(Kay, 1979; Duke, 1981). The two dominant genera, *Vigna* (80 species) and *Phaseolus* (28 species), are difficult to separate taxonomically. Many species in one genus often have synonyms in the other, such as *P. aureus* (Roxb.) syn. *V. radiata* (L.) R. Wilczek. This is complicated by their widespread distribution in temperate and tropical zones which might have led to the evolution of regional characteristics and subsequent classification into Asiatic and tropical groups (Maekawa, 1955; Verdecourt, 1970; Marechal et al., 1978). This confusion provided a reason for the use of chemical characteristics for a chemotaxonomic solution. Earlier work on the protein and hybridization properties of *Phaseolus* species indicated intergeneric differences in the *Phaseolus–Vigna* complex that was largely in conformity with the classification by Marechal et al. (1978) and also shows that *P. lunatus* is possibly less evolved than other members of the genus (Kloz, 1962; Kloz and Klozova, 1974).

Production of phytoalexins involves the synthesis of individual enzymes for the several steps in their biosynthesis. Various intermediates as well as final products will accumulate and differences in these factors may permit the separation of close species chemotaxonomically. The inducible constituents of only a few species in the *Phaseolus–Vigna* complex have been investigated. These are *P. vulgaris* (20 compounds) (Woodward, 1979a, 1979b, 1980; Ingham, 1982; Biggs et al., 1983), *P. lunatus* (2), *P. aureus* (2), *V. unguiculata* (7) (Ingham, 1982), *P. mungo* (none) (Smith, 1971), and *Voandozeia subterranea* (syn. *Vigna subterranea*) (1) (Ingham, 1982). Our work has extended the investigation and data in four species, *P. lunatus*, *P. aureus*, *P. mungo*, and *P. coccineus* (Adesanya et al., 1984a, 1985; O'Neill et al., 1983, 1986), while recent work by Ingham (1990b) has provided additional data for *V. subterranea*, *V. angularis*, and *V. umbellata* (Table 1). The large number of known and novel isoflavonoids isolated (Fig. 1) provided an opportunity for a detailed analysis of their biosynthetic pathways and their comparative toxicities toward the fungi *Aspergilus niger* and *Cladosporium cucumerinum*.

B. *Dioscorea* Species

The major genus of the Dioscoreaceae is *Dioscorea*, which is divided arbitrarily into several sections with several species spread over all the sections. Other minor genera include *Tamus* and *Rajania* (Coursey, 1967). Most of the species are typified by twining stems and large storage tubers which in some are reduced to rhizomes and in others grow on the aerial part as bulbils. The tuberous species have a high carbohydrate content and are cultivated in the tropics as food crops (Coursey, 1967). The nonpoisonous species such as *D. rotundata* Poir., *D. alata* L., and *D. cayanensis* L. are widely consumed; the poisonous ones such as *D. dumentorum* Kunth. Pax.

and *D. hispida* L. are only safe after the removal of the poisonous alkaloid dioscorine.

These tubers deteriorate rapidly in storage due to microbial infection, with an estimated loss of over half of the yearly harvests in Nigeria (Coursey, 1967). Thus, concerted efforts are being made to find ways of increasing their shelf life (dormancy period) and also of preventing microbial attack.

The peel of *D. rotundata* has been shown to contain the constitutive antifungal phenanthrenes batatasin I (**77**), and hircinol (**81**) (Coxon et al., 1982), and that of *D. decipiens*, 2,7-dihydroxy-1,3,5-trimethoxy-9,10-dihyrophenanthrene (**80**) (Sunder et al., 1978). Constitutive compounds responsible for dormancy in the bulbils of *D. opposita (D. batatas*) have been isolated and characterized as the phenanthrene batatasin I and the dihydrostilbenes (bibenzyls) batatasin II, III (**72**), IV (**71**), and V (**73**) (Fig. 1) (Hashimoto et al., 1972; Hashimoto and Tajima, 1978). The distribution of these substances in *Dioscorea* species varies, with some species having none (Ireland et al., 1981). Chemical investigations in breeding experiments have linked resistance to anthracnose disease in *D. alata* to the level of phenolics (Alozie et al., 1987) indicating a possible role for inducible compounds in the resistance to infection. A subsequent study of tissues of *D. batatas* inoculated with bacteria led to the isolation of dihydropinosylvin (**68**) as the major compound, together with batatasin I, batatasin IV, 3-hydroxy-5-methoxybibenzyl (**69**), 6,7-dihydroxy-2,4-dimethoxy-9,10-dihydrophenanthrene (**78**), and 2,7-dihydroxy-4,6-dimethoxy-9,10-dihydrophenanthrene (**79**) (Fig. 1) (Takasugi et al., 1987). This study has been extended to the inducible compounds in *D. rotundata*, *D. alata*, *D. dumentorum*, *D. bulbifera*, and *D. mangenotiana*, with the aim of evaluating their potential in prolonging dormancy, disease resistance, crop protection, and chemotaxonomy in the various sections.

II. METHODOLOGY

Plant Materials and Microorganisms

Seeds of *Phaseolus vulgaris* L. var. Prince, *P. aureus* Roxb. (synonym, *Vigna radiata* (L.) R. Wilczek), *P. mungo* L. (synonym *V. mungo* (L.) Hepper.), and *P. lunatus* L. were purchased from Thompson and Morgan Ltd., London and those of *P. coccineus* L., var. Scarlet Emperor, were obtained from Northrop King Seeds (Minneapolis, MN). Seeds of *Sorghum bicolor* (L.) Mench. and tubers of *Dioscorea rotundata* Poir., *D. alata* L., *D. dumentorum* (Kunth.) Pax. were brought at the market in Ile-ife, Nigeria. Wild bulbils of *D. bulbifera* L. and tubers of *D. mangenotiana* Meige. were collected at various sites around Ile-ife.

ISOFLAVONES

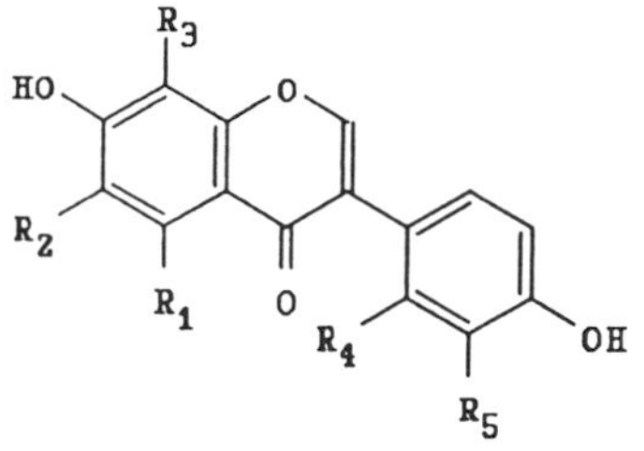

1. $R_1 = R_2 = R_3 = R_4 = R_5 = H$ — Diadzein
2. $R_2 = R_3 = R_4 = R_5 = H$, $R_1 = OH$ — Genistein
3. $R_2 = R_3 = R_4 = R_5 = H$, $R_1 = OCH_3$ — Isoprunetin
4. $R_1 = R_2 = R_3 = R_5 = H$, $R_4 = OH$ — 2'-hydroxydiadzein
5. $R_2 = R_3 = R_5 = H$, $R_1 = R_4 = OH$ — 2'-hydroxygenistein
6. $R_2 = R_3 = R_5 = H$, $R_4 = OCH_3$, $R_1 = OH$ — 2'-methoxygenistein
7. $R_2 = R_3 = R_5 = H$, $R_1 = OCH_3$, $R_4 = OH$ — 2'-hydroxyisoprunetin
8. $R_2 = R_5 = H$, $R_1 = R_3 = R_4 = OH$ — 8,2'-dihydroxygenistein
9. $R_1 = R_2 = R_5 = H$, $R_3 = R_4 = OH$ — 8,2'-dihydroxydiadzein
10. $R_2 = R_5 = H$, $R_1 = R_4 = OH$, $R_3 = CH_2CH{=}C(CH_3)_2$ — 2,3-dehydrokievitone
11. $R_2 = R_5 = H$, $R_1 = R_4 = OH$, $R_3 = CH_2CH{=}C(CH_3)CH_2OH$ — 2,3-dehydrokievitol
12. $R_2 = R_3 = H$, $R_1 = R_4 = OH$, $R_5 = CH_2CH{=}C(CH_3)_2$ — Phaseoluteone
13. $R_3 = R_5 = H$, $R_1 = R_4 = OH$, $R_2 = CH_2CH{=}C(CH_3)_2$ — Luteone

ISOFLAVONONES

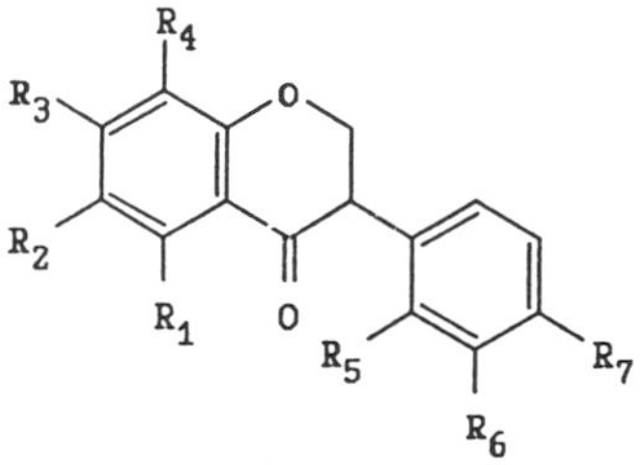

14. $R_1 = R_2 = R_4 = R_6 = H$, $R_3 = R_5 = R_7 = OH$ — 2'-hydroxydihydrodiadzein
15. $R_1 = R_2 = R_6 = H$, $R_3 = R_4 = R_5 = R_7 = OH$ — 8,2'-dihydroxy dihydrodiadzein
16. $R_2 = R_4 = R_6 = H$, $R_1 = R_3 = R_5 = R_7 = OH$ — Dalbergioidin
17. $R_2 = R_4 = R_6 = H$, $R_3 = R_7 = OH$, $R_1 = R_5 = OCH_3$ — 5,2'-dimethoxydalbergioidin
18. $R_2 = R_4 = R_6 = H$, $R_1 = R_3 = R_7 = OH$, $R_5 = OCH_3$ — Isoferreirin
19. $R_1 = R_2 = R_6 = H$, $R_3 = R_5 = R_7 = OH$, $R_4 = CH_2\text{-}CH{=}C(CH_3)_2$ — 5-deoxykievitone
20. $R_1 = R_2 = R_6 = H$, $R_3 = R_5 = R_7 = OH$, — 5-deoxykievitone hydrate

Figure 1 Structure of known and novel isoflavonoids, including isoflavones, isoflavonones, cyclo-derivatives of isoflavones and isoflavonones, isoflavans, pterocarpans, furano-pterocarpans, coumestans, stilbenes, dihydrostilbenes, and dihydrophenanthrenes.

$R_4 = CH_2\text{-}CH_2\text{-}C(CH_3)_2OH$

21. $R_1 = R_2 = R_6 = H, R_3 = R_5 = R_7 = OH,$ $R_4 = CH_2\text{-}CH{=}C(CH_3)CH_2OH$ — 5-deoxykievitol
22. $R_2 = R_6 = H, R_1 = R_3 = R_5 = R_7 = OH,$ $R_4 = CH_2\text{-}CH{=}C(CH_3)_2$ — Kievitone
23. $R_2 = R_6 = H, R_1 = R_3 = R_5 = R_7 = OH,$ $R_4 = CH_2\text{-}CH_2\text{-}C(CH_3)_2OH$ — Kievitone hydrate
24. $R_2 = R_6 = H, R_1 = R_3 = R_5 = R_7 = OH,$ $R_4 = CH_2\text{-}CH{=}C(CH_3)CH_2OH$ — Kievitol
25. $R_2 = R_6 = H, R_1 = R_3 = R_5 = OH, R_7 = OCH_3,$ $R_4 = CH_2\text{-}CH{=}C(CH_3)_2$ — 4'-methoxykievitone
26. $R_2 = R_6 = H, R_1 = OH, R_3 = R_5 = R_7 = OCH_3$ $R_4 = CH_2\text{-}CH{=}C(CH_3)_2$ — Trimethoxykievitone
27. $R_2 = R_6 = H, R_1 = R_3 = R_5 = R_7 = OCH_3,$ $R_4 = CH_2\text{-}CH{=}C(CH_3)_2$ — Tetramethoxykievitone
28. $R_2 = H, R_1 = R_3 = R_5 = R_7 = OH,$ $R_4 = R_6 = CH_2\text{-}CH{=}C(CH_3)_2$ — 3'-(γ,γ-dimethylallyl) kievitone
29. $R_4 = H, R_1 = R_3 = R_7 = OH, R_5 = OCH_3$ $R_2 = R_6 = CH_2\text{-}CH{=}C(CH_3)_2$ — Isosophoranone
30. $R_2 = R_4 = R_6 = H, R_1 = R_7 = OH, R_3 = R_5 = OCH_3$ — Cajanol
31. $R_2 = R_4 = H, R_1 = R_3 = R_7 = OH, R_5 = OCH_3,$ $R_6 = CH_2\text{-}CH{=}C(CH_3)_2$ — Sophoraisoflavanone A

CYCLO-DERIVATIVES OF ISOFLAVONES AND ISOFLAVONONES

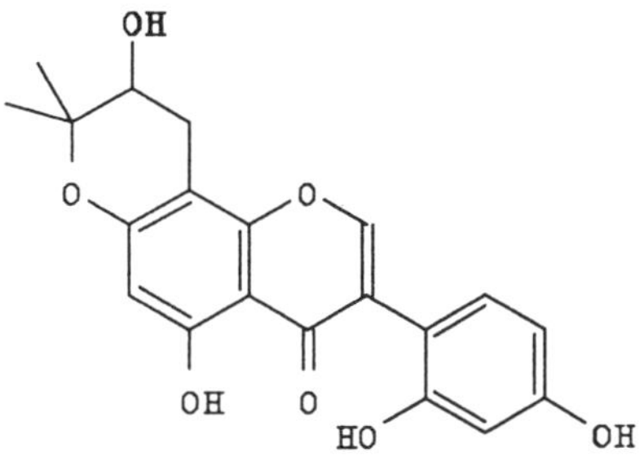

32. Cyclo-2,3-dehydrokeivitone hydrate

Figure 1 Continued.

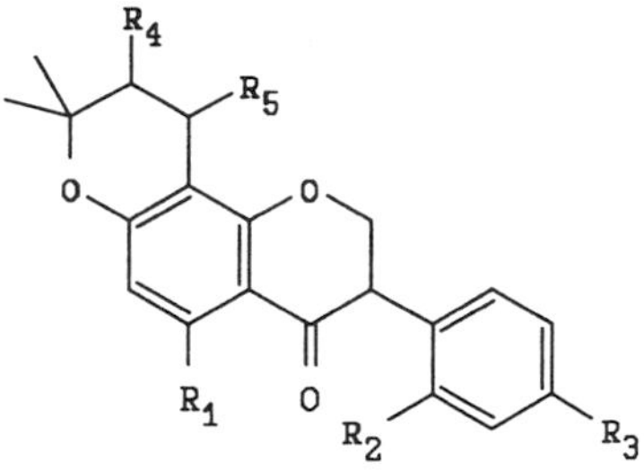

33. $R_4 = R_5 = H_2$, $R_1 = R_2 = R_3 = OH$	Cyclokievitone
34. $R_4 = R_5 = H$, $R_1 = R_2 = R_3 = OCH_3$	Trimethoxykievitone
35. $R_4 = R_5 = H$, $R_1 = R_2 = R_3 = OH$	1",2"-dehydrocyclokievitone
36. $R_1 = R_2 = R_3 = H$, $R_5 = H_2$, $R_4 = OH$	Cyclokievitone hydrate

ISOFLAVANS

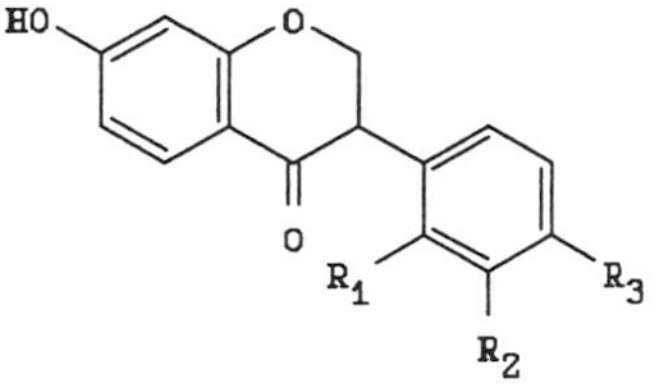

37. $R_2 = H$, $R_1 = R_3 = OH$	Demethylvestitol
38. $R_2 = H$, $R_1 = OH$, $R_3 = OCH_3$	Vestitol
39. $R_2 = H$, $R_3 = OH$, $R_1 = OCH_3$	Isovestitol
40. $R_2 = H$, $R_1 = R_3 = OCH_3$	Sativan
41. $R_3 = OH$, $R_1 = R_2 = OCH_3$	Laxifloran
42. $R_3 = OH$, $R_1 = OCH_3$, $R_2 = CH_2CH{=}C(CH_3)_2$	2'-methoxyphaseollidin isoflavan

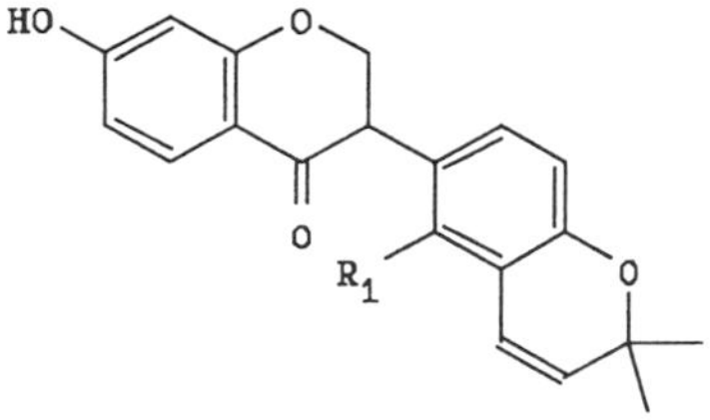

43. $R_1 = OH$	Phaseollin isoflavan
44. $R_1 = OCH_3$	2'-methoxyphaseollin isoflavan

PTEROCARPANS

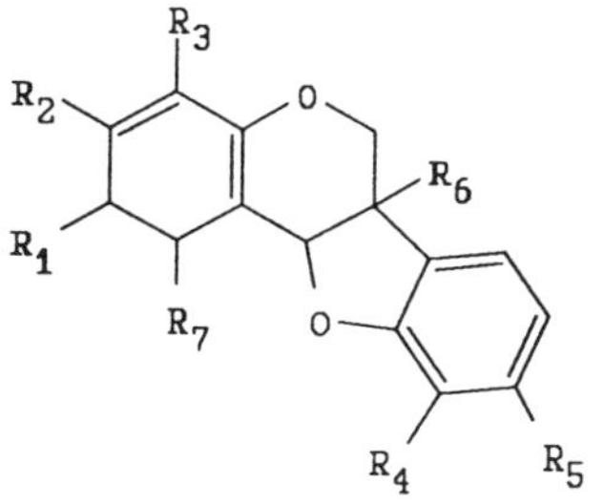

45. $R_1 = R_3 = R_4 = R_6 = R_7 = H$, $R_2 = R_5 = OH$ — Demethylmedicarpin
46. $R_1 = R_3 = R_4 = R_6 = R_7 = H$, $R_5 = OH$, $R_2 = OCH_3$ — Isomedicarpin
47. $R_1 = R_3 = R_4 = R_6 = R_7 = H$, $R_2 = OH$, $R_5 = OCH_3$ — Medicarpin
48. $R_1 = R_3 = R_4 = R_7 = H$, $R_2 = R_5 = R_6 = OH$ — Glycinol
49. $R_1 = R_3 = R_6 = R_7 = H$, $R_2 = R_5 = OH$, $R_4 = CH_2CH{=}C(CH_3)_2$ — Phaseollidin
50. $R_1 = R_3 = R_6 = H$, $R_2 = R_5 = OH$, $R_7 = OCH_3$, $R_4 = CH_2CH{=}C(CH_3)_2$ — 1-methoxyphaseollidin
51. $R_3 = R_6 = R_7 = H$, $R_2 = R_5 = OH$, $R_1 = R_4 = CH_2CH{=}C(CH_3)_2$ — 2(γ,γ-dimethylallyl) phaseollidin
52. $R_1 = R_6 = R_7 = H$, $R_2 = R_5 = OH$, $R_3 = R_4 = CH_2CH{=}C(CH_3)_2$ — 4(γ,γ-dimethylallyl) phaseollidin
53. $R_3 = R_7 = H$, $R_2 = R_5 = R_6 = OH$, $R_1 = R_4 = CH_2CH{=}C(CH_3)_2$ — 2,10-(dimethylallyl) glycinol
54. $R_1 = R_3 = R_7 = H$, $R_2 = R_6 = OH$, $R_5 = OCH_3$, $R_4 = CH_2CH{=}C(CH_3)_2$ — Cristacarpin
55. $R_1 = R_3 = R_6 = R_7 = H$, $R_2 = R_5 = OH$, — Dolichin A

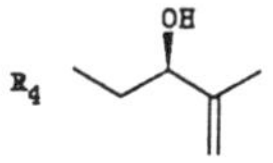

56. $R_1 = R_3 = R_6 = R_7 = H$, $R_2 = R_5 = OH$, — Dolichin B

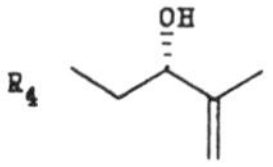

Figure 1 Continued.

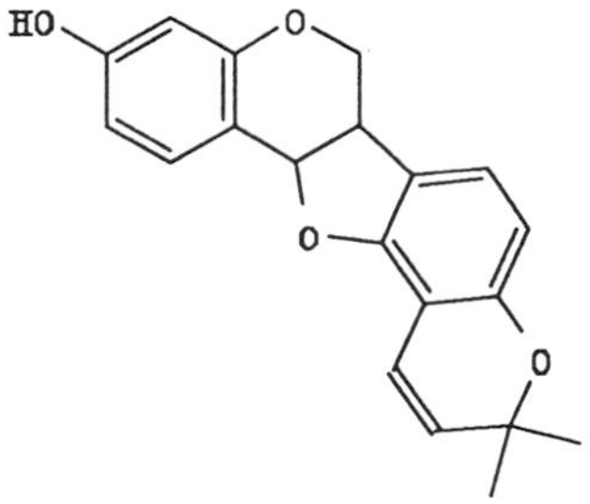

HO O O O

57. Phaseollin

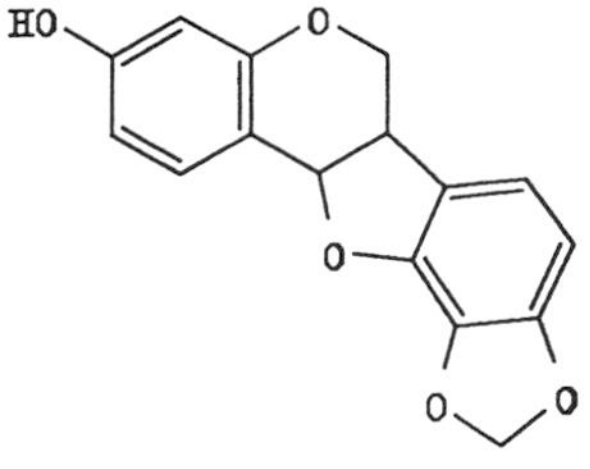

HO O O O O

58. Maackian

FURANO-PTEROCARPANS

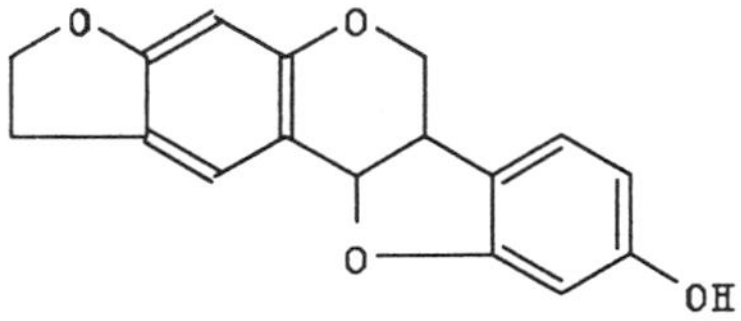

O O O OH

59. Neodunol

COUMESTANS

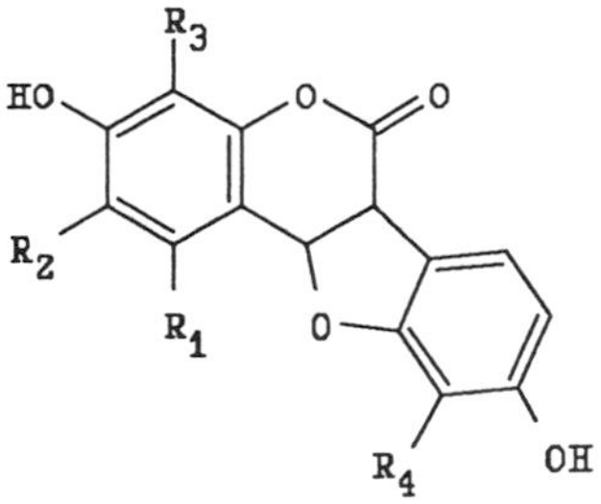

R_3 HO O O R_2 R_1 O R_4 OH

60. $R_1 = R_2 = R_3 = R_4 = H$ Coumesterol
61. $R_2 = R_3 = R_4 = H$, $R_1 = OH$ Aureol
62. $R_1 = R_3 = R_4 = H$, $R_2 = CH_2CH{=}C(CH_3)_2$ Psoralidin
63. $R_1 = R_2 = R_4 = H$, $R_3 = CH_2CH{=}C(CH_3)_2$ Phaseol
64. $R_1 = R_2 = R_3 = H$, $R_4 = CH_2CH{=}C(CH_3)_2$ Isosojagol

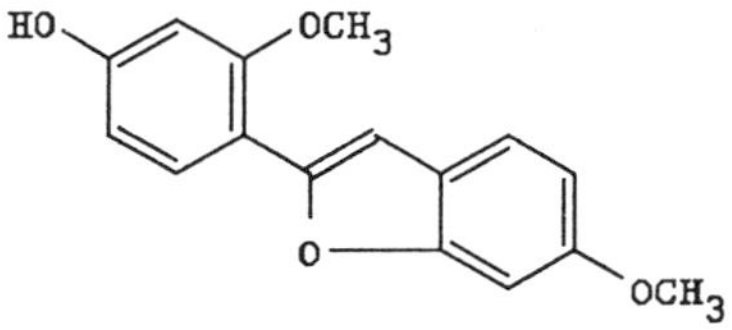

65. Vignafuran

STILBENES

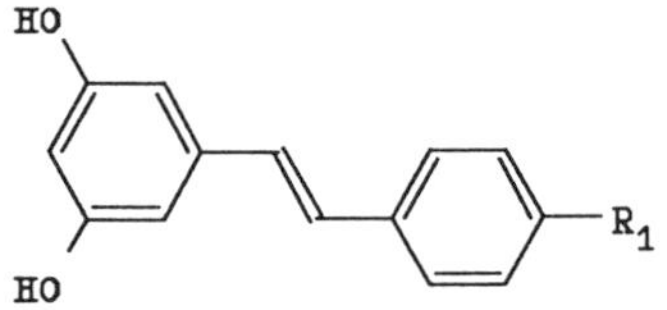

66. $R_1 = H$ Pinosylvin
67. $R_1 = OH$ Trans-resveratrol

DIHYDROSTILBENES

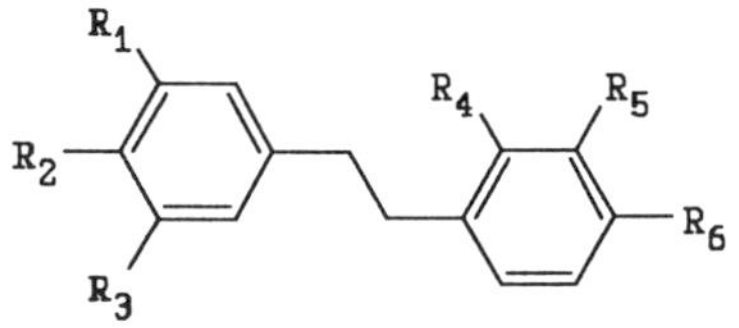

68. $R_2 = R_4 = R_5 = R_6 = H, R_1 = R_3 = OH$ Dihydropinosylvin
69. $R_2 = R_4 = R_5 = R_6 = H, R_1 = OH, R_3 = OCH_3$ 3-hydroxy-5-methoxy dihydrostilbene
70. $R_2 = R_5 = R_6 = H, R_1 = R_3 = R_4 = OH$ Demethylbatatasin IV
71. $R_2 = R_5 = R_6 = H, R_1 = R_4 = OH, R_3 = OCH_3$ Batatasin IV
72. $R_2 = R_4 = R_6 = H, R_1 = R_5 = OH, R_3 = OCH_3$ Batatasin III
73. $R_5 = R_6 = H, R_4 = OH, R_1 = R_2 = R_3 = OCH_3$ Batatasin V
74. $R_2 = R_5 = R_6 = H, R_4 = OH, R_1 = R_3 = OCH_3$ 4-hydroxy-3,5-dimethoxy-dihydrostilbene
75. $R_5 = R_6 = H, R_2 = R_4 = OH, R_1 = R_3 = OCH_3$ 2,4-dihydroxy-3,5-dimethoxy bibenzyl
76. $R_2 = R_4 = R_5 = H, R_1 = R_3 = R_6 = OH$ Dihydroresveratrol

Figure 1 Continued.

DIHYDROPHENANTHRENES

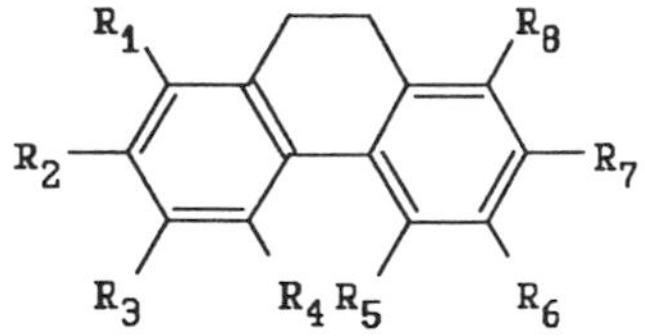

77. $R_1 = R_3 = R_5 = R_8 = H$, $R_6 = OH$, $R_2 = R_4 = R_7 = OCH_3$ Batatasin I
78. $R_1 = R_3 = R_5 = R_8 = H$, $R_6 = R_7 = OH$, $R_2 = R_4 = OCH_3$
79. $R_1 = R_3 = R_5 = R_8 = H$, $R_2 = R_7 = OH$, $R_4 = R_6 = OCH_3$
80. $R_4 = R_6 = R_8 = H$, $R_2 = R_7 = OH$, $R_1 = R_3 = R_5 = OCH_3$
81. $R_1 = R_3 = R_6 = R_7 = R_8 = H$, $R_2 = R_5 = OH$, $R_4 = OCH_3$

Cultures of bacteria, *Bacillus cereus, Staphylococcus aureus, Pseudomonas aeruginosa,* and *Escherichia coli* were obtained from the Department of Microbiology, Obafemi Awolowo University, Ile-ife. The fungi *Cladosporium cucumerinum* Ell. and Arth., and *Aspergillus niger* Van Teigh. were obtained from the Commonwealth Mycological Institute, Kew, London, while *C. cladosporoides, A. niger, A. flavus* (clinical isolate), *Tricophyton mentagyrophytes* (clinical isolate), and *Botryodiplodia theobromae* Pat. were obtained from the Department of Microbiology, Obafemi Awolowo University, Ile-ife.

B. Induction, Extraction, Detection, and Isolation of Phytoalexins

Sterilized seeds of the *Phaseolus* species were germinated in running water and grown for about 11 days in sterilized vermiculite before induction by immersing their roots in aqueous 3 mM solution of $CuCl_2$ for 19 hr (control seedlings in water alone). They were further grown for 4 days before analysis (Adesanya, 1984).

Peeled tubers or bulbils were cut into small discs and washed with sterile water before dipping in mycelia suspensions of *B. theobromae* or 3 mM solution of $HgCl_2$ for 5 min (control specimens in water) before incubation for 4 days (Fagboun et al., 1987).

Induced and control specimen were extracted into EtOAc, and the extracts subjected to thin layer chromotography (TLC) bioassay using *C. cucumerinum* (*Phaseolus* species) or *C. cladosporoides* (*Dioscorea* species). The anitfungal zones appearing as white on a green background was correlated with duplicate TLC plates sprayed with vanillin/H_2SO_4 or Fast Blue B salt solution (Adesanya, 1984).

For isolation of compounds in induced tissues, large-scale EtOAc extracts of induced tissues were fractionated by column chromatography and purified by preparative TLC using various solvents. The pure compounds, were characterized using their ultraviolet (UV), infrared (IR), mass spectrometry (MS), and nuclear magnetic resonance (NMR) spectroscopic data (Adesanya, 1984; Fagboun et al., 1987; Takasugi et al., 1987).

C. Assay of Induced Compounds and Enzymes

Phaseollin (**57**) and kievitone (**22**) in *P. vulgaris*, phaseollidin (**49**) and kievitone in *P. aureus*, and kievitone in *P. mungo* were isolated by preparative TLC and quantified by high-performance liquid chromatography (HPLC), while dihydropinosylvin and demethylbatatasin IV (**70**) and batatasin IV were similarly isolated and quantified by UV spectroscopy. Sampling was done over 120 hr in induced and control specimens, respectively, and amounts produced extrapolated from standard curves obtained with authentic compounds (Adesanya, 1984; Cline et al., 1989). The enzymes phenylalanine ammonia-lyase (PAL), tyrosine ammonia-lyase (TAL), and polyphenol oxidase (PPO) in *D. alata* were assayed simultaneously in induced and control tubers as were the phytoalexins. The assays were done using standard procedures (Cline et al., 1989).

D. Antimicrobial Assay

Paper discs were loaded with various graded concentrations of the dihydrostilbenes and placed on bacteria-seeded agar plates. Zones of inhibition were measured for the active compounds after a 24-hr incubation period (Takasugi et al., 1989).

E. Antifungal Assay

Graded concentrations of the isoflavonoids were dispensed into multiwell assay trays. The volume in each well was made up to 50 μl by the addition of ethanol and then 950 μl of a spore suspension in Czapek Dox liquid nutrient added. Controls without isoflavonoids and test preparations were incubated in the dark for 3 days. MIC was taken as a range between the last well with mycelia development and the next concentration without growth (Adesanya et al., 1986).

The dihydrostilbenes were assayed using agar plates with holes in a circle containing various graded concentrations of the compounds and water only in the control. A mycelial plug was placed in the middle and zones of limitation of mycelial growth measured from the edge of each well. The MIC was taken as a range between the well showing the least zone of inhibition and the next concentration without a zone of inhibition (Adesanya et al., 1989).

F. Seed Germination Tests

Twenty surface-sterilized seeds of *S. bicolor* were put into sterile petri dishes. Fifty-milliliter aqueous solutions of dihydrostilbenes containing a drop of Tween 20 were sterilized by membrane filtration and 5-ml portions of each concentration dispensed into separate Petri dishes. Controls contained water and Tween 20 only. Samples and controls were incubated in the dark and the number of germinated seeds counted at 24-hr intervals. Experiments were performed in quadruplicate and the average percent germination calculated (Cline et al., 1989).

G. Seedling Root Elongation Tests

Twenty seedlings of *S. bicolor* grown in sterile distilled water and having about 2 cm root length were distributed into sterile Petri dishes and 5-ml test solutions prepared as in seed germination tests added to the Petri dishes. The increase in root length of all seedlings in each dish was measured and the average readings from four dishes per concentration was calculated (Cline et al., 1989).

III. RESULTS AND DISCUSSION

A. Induction of Phytoalexins

Fabaceae: Phaseolae: Subtribe Phaseolinae

Earlier reports on *P. vulgaris* had shown that the major phytoalexins, kievitone and phaseollin, accumulate after induction of the enzyme PAL in tissues elicited with either biotic or abiotic elicitors, with kievitone reaching higher levels than phaseollin (Bailey, 1982; Adesanya, 1984). These compounds similarly accumulated in *P. aureus,* while only kievitone was elicited in *P. mungo* seedlings treated with $CuCl_2$ solution, contrary to the report of Smith (1971) that *P. mungo* seed pods contain no phytoalexin diffusate (Adesanya, 1984). These two compounds have been demonstrated to accumulate after enzyme induction by an abiotic elicitor. Only the constituents of infected tissues of other species of the Phaseolinae subfamily have been analyzed (Ingham, 1982); these analyses have led to the isolation of biosynthetically related compounds in different plants.

The compounds isolated from large-scale $CuCl_2$ treated seedlings of *P. vulgaris*, *P. coccineus*, *P. aureus*, *P. lunatus*, and *P. mungo* (Adesanya et al., 1984a, 1985; O'Neill et al., 1983, 1986) together with those reported in *P. vulgaris* and *V. unguiculata* (Ingham 1982) and *V. subterranea*, *V. angularis*, and *V. umbellata* (Ingham, 1990b) are presented in Table 1. Other species of the Phaseolinae investigated so far are *Macroptilium atropurpureum* and *M. marti* whose leaf diffusates contain genistein (**2**), 2′-

Table 1 Inducible Constituents of *Phaseolus* and *Vigna* Species

Compounds	Yield [a,b,c]								
	Vigna spp.				*Phaseolus* spp.				
	angularis[1]	*subterranea*[1]	*umbellata*[1]	*unguiculata*[2]	*vulgaris*[3]	*coccineus*	*aureus*[1]	*lunatus*	*mungo*
5-Deoxy Compounds									
Diadzein 1	–	–	–	–	+	14	–	48	–
2′-Hydroxydiadzein 4	–	–	–	–	+	–	–	25	70
8,2′-Dihydroxydiadzein 9	–	–	–	–	–	–	–	23	–
2′-Hydroxydihydro diadzein 14	–	–	–	–	+	22	–	–	10
8,2′-Dihydroxydihydro diadzein 15	–	–	–	–	–	–	–	6	–
5-Deoxykievitone 19	–	–	–	–	+	–	1300	13	90
5-Deoxykievitone hydrate 20	–	–	–	–	–	–	–	–	90
5-Deoxykievitol 21	–	–	–	–	–	–	–	19	–
Demethylvestitol 37	–	–	–	+	+	61	–	–	40
2-Methoxyphaseollidin isoflavan 42	–	–	–	+	–	–	–	–	–
Phaseollinisoflavan 43	–	–	–	–	+ +	30	–	–	–
2-Methoxyphaseollin isoflavan 44	–	–	–	–	+	–	–	–	–
Glycinol 48	–	–	–	–	–	5	–	–	10
Demethylmedicarpin 45	–	+	+	–	+	–	+	–	–
Medicarpin 47				+	–	–	–	–	–
Phaseollidin 49	–	+	+	+ +	+ +	24	2000	4	–
2-(γ,γ-Dimethylallyl)phaseollidin 51	–	–	–	–	–	–	–	4	–
4-(γ,γ-Dimethylallyl)phaseollidin 52	–	–	–	–	–	–	–	6	–
2,10-Di(γ,γ-dimethylallyl)glycinol 53	–	–	–	–	–	–	–	11	–
Phaseollin 57	–	–	–	+	+ + +	172	–	–	–
Coumestrol 60	–	–	–	+	–	26	–	6	–
Psoralidin 62	–	–	–	–	–	–	–	9	–
Phaseol 63	–	–	–	–	–	–	900	–	–
Isosojagol 64	–	–	–	–	–	18	–	–	–
Vignafuran 65	–	–	–	+ + +	–	–	–	–	–

5-Hydroxy Compounds									
Genistein 2	+	+	–	–	++	99	5700	56	3600
Isoprunetin 3	–	–	–	–	–	7	–	–	–
2′-Hydroxygenistein 5	+	+	+	–	++	61	1600	51	340
2′-Methoxygenistein 6	–	–	–	–	–	–	–	9	–
2′-Hydroxyisoprunetin 7	–	–	–	–	–	11	–	–	–
8,2′-Dihydroxygenistein 8	–	–	–	–	–	–	–	12	–
2,3-Dehydrokievitone 10	–	–	–	–	+	–	1600	8	–
2,3-Dehydrokievitol 11	–	–	–	–	–	–	–	2	–
Phaseoluteone 12	–	–	–	–	+	32	–	–	–
Luteone 13	–	–	–	–	–	–	–	7	–
Dalbergioidin 16	+	+	+	–	++	–	150	–	560
5,2′-Dimethoxy dalbergioidin 17	–	–	–	–	–	13	–	–	–
Isoferrein 18	–	–	–	–	–	14	–	23	200
Kievitone 22	+	+	+	+++	+++	110	3600	832	1980
Kievitone hydrate 23	–	–	–	–	–	–	97	–	–
Kievitol 24	–	–	–	–	–	–	–	213	–
4′-Methoxykievitone 25	–	–	–	–	–	–	–	–	100
3′(γ,γ-Dimethylallyl)kievitone 28	–	–	–	–	–	–	–	47	–
Cyclo-2,3-dehydro kievitone 32	–	–	–	–	–	–	–	10	–
1″,2″-Dehydrocyclo kievitone 35	–	–	–	–	+	12	1200	5	120
Cyclokievitone hydrate 36	–	–	–	–	–	–	–	47	100
Aureol 61	–	–	–	–	–	31	1500	–	210

[a]Numerical yield values $10^2 \times \mu g/g$ fresh weight as reported by Adesanya (1980).

[b]+ Relative presence of compound as reported in literature.

[c]– Compound absent.

[1]Ingham, 1990.

[2]Ingham, 1982; Bailey, 1973; Preston et al., 1975; Lampard, 1974.

[3]Woodward, 1979a, 1979b, 1980; Adesanya, 1980.

hydroxygenistein (**5**), dalbergioidin (**16**), kievitone, demethylmedicarpin (**45**), phaseollin, and phaseollidin. *M. bracteatum* and *M. lathyroides* contain **2**, **5**, **16**, and **22** and *Macrotyloma axillare* leaves accumulate **5**, **16**, **57**, and **22** (Ingham, 1982, 1990b). *Dolichos biflorus* produces genistein, 2′-hydroxygenistein, dalbergioidin, isoferrierin (**18**), dolichin A (**55**) and B (**56**), *Lablab niger* produces 2′-hydroxygenistein, dalbergioidin, kievitone, phaseollidin (**49**), demethylvestitol (**37**), isovestitol (**39**), laxifloran (**41**), and vignafuran (**65**), while *Psophocarpus tetragonolobus* gives phaseollidin, isomedicarpin (**47**), 1-methoxylphaseollidin (**50**), demethylmedicarpin, and cristacarpin (**54**) (Ingham, 1982). *Neorautanea mitis* produces **45**, neodunol **59**, **49**, and **37**, while *Oxyrhynchus volubilis* has only glycinol (**48**). *Stenophylis angustifolia* and *S. stenocarpa* have the same compounds as *M. atropurpureum* except for **2** and **57**. *Stophostyles helvola* and *Dipogon lignosus* produce the same phytoalexins as *M. atropurpureum* with the addition of demethylvestitol (**37**) (Ingham, 1982, 1990b). Kievitone appears to be the most common inducible compound in this subtribe, produced in the greatest quantity, followed by phaseollidin, although their concentrations differ in different species. Exceptions have been found in *P. coccineus*, which produces larger amounts of phaseollin than kievitone (Adesanya, 1984), *D. lignosus* and *S. angustifolia* with higher amounts of phaseollidin, and *V. subterranea*, which is devoid of kievitone (Ingham, 1990b).

Dioscoreaceae

D. opposita (*D. batatas*), the first plant investigated in this family, produces the phenanthrenes batatasin I (**77**), 6,7-dihydroxy-2,4-dimethoxy-9,10-dihydrophenanthrene (**78**), 2,7-dihydroxy-4,6-dimethoxy-9,10-dihydrophenanthrene (**79**), and dihydrostilbenes, dihydropinosylvin (**68**), 3-hydroxy-5-methoxybibenzyl (**69**), and batatasin IV (**71**) in response to the bacterium *Pseudomonas cichorri* (Takasugi et al., 1987). Other compounds isolated in our investigation of other *Dioscorea* species infected with the fungi *Botryodiplodia theobromae* are presented in Table 2 (Fagboun et al., 1987; Adesanya et al., 1989; 1989; Cline et al., 1989; Kaganda and Adesanya, 1990). Dihydropinosylvin (**68**) appears to be the principal induced compound in the section Enantiophylum. On induction of one of the species, *D. alata*, by the fungi, dihyropinosylvin, demethylbatatasin IV, and batatasin IV accumulate in the infected yams (Fig. 1). By contrast, in tubers treated with $HgCl_2$, only dihydropinosylvin shows appreciable increase (Fig. 2). Although the lower amounts of **68** accumulate in those induced with $HgCl_2$, in both cases increase in PAL activity precedes the onset of dihydrostilbene formation with less increase with the $HgCl_2$ treatment. TAL activity was somewhat lower in both treatments than in the controls. These results sug-

Table 2 Induced Constituents of *Dioscorea* Species

Compounds	Presence in *Dioscorea* species[a,b]					
	rotundata[1] (tuber)	*alata*[2] (tuber)	*magentiana*[3] (tuber)	*opposita*[4] (bulbils)	*bulbifera*[5] (bulbils)	*dumentorum*[5] (tuber)
Dihydrostilbenes						
Dihydropinosylvin 68	+++	+++	+	+++	–	–
3-Hydroxy-5-methoxybibenzyl 69	–	–	–	+	–	–
Demethylbatatasin IV 70	+	+	–	–	++	–
Batatasin IV 71	+	+	–	+	–	–
Batatasin III 72	–	+	–	–	–	–
2′-Hydroxy-3,5-dimethoxy bibenzyl 74	–	–	–	+	–	–
2′,4′-Dihydroxy-3,5-dimethoxy bibenzyl 75	–	–	++	–	–	–
Dihydroresveratrol	–	–	–	–	–	++
Dihydrophenanthrenes						
Batatasin I 77	–	–	–	+	–	–
6,7-Dihydroxy-2,4-dimethoxy phenanthrene 78	–	–	–	+	–	–
2,7-Dihydroxy-4,6-dimethoxy phenanthrene 79	–	–	–	+	–	–

D. rotundata, *D. alata*, *D. mangentiana*, and *D. opposita* (*D. batatas*) in section Enantiophylum; *D. bulbifera* in section Opsophyton; *D. dumentorum* in section Lasiophyton.

[a] + Relative quantity in a specie, as reported in literature.

[b] – Compound not present in induced specie.

[1] Fagboun et al., 1987. [2] Cline et al., 1989. [3] Kaganda and Adesanya, 1990. [4] Takasugi et al., 1987. [5] Adesanya et al., 1989.

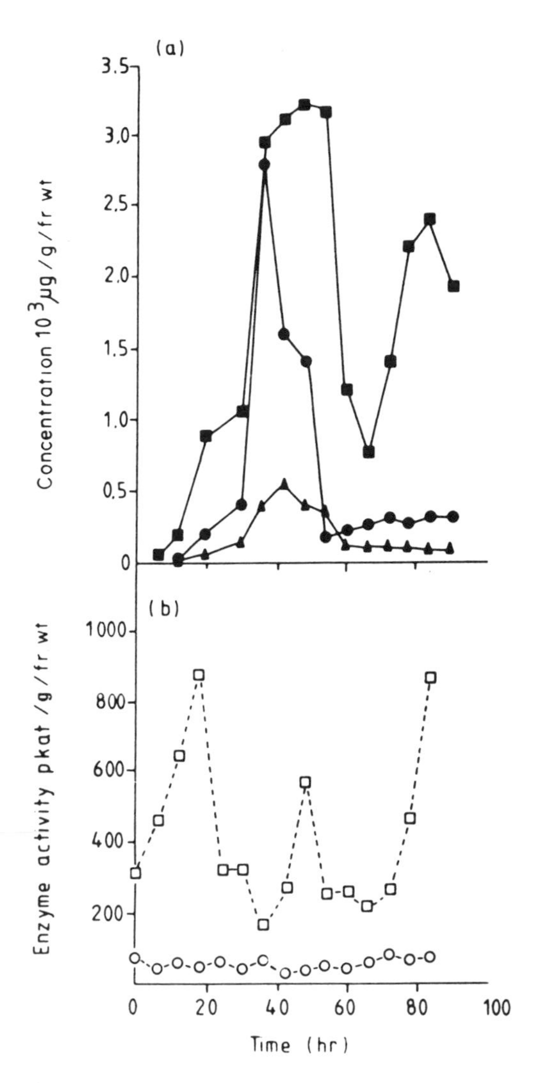

Figure 2 Enzyme activities in relation to the accumulation of (a) dihydrostilbenes **1–3** in *B. theobromae*-induced tubers. ■-■, dihydropinosylvin (**1**); ●-●, demethylbatatasin IV (**2**); ▲-▲, batatasin IV (**3**); (b) □-□, phenylalanine ammonia-lyase (PAL); ○-○, tyrosine ammonia-lyase (TAL).

gest the synthesis of the dihydrostilbenes is via a route involving phenylalanine as demonstrated for the stilbenes (Rudloff and Jorgensen, 1963; Schöppner and Kindl, 1979; Fritzemeier and Kindl, 1981; Stoessl, 1982). Since these compounds are antimicrobial, they can be regarded as phytoalexins except batatasin IV which has been found, without deliberate induction, in *D. batatas* (Hashimoto et al., 1972; Hashimoto and Tajima, 1978).

The lower levels of dihydropinosylvin resulting from $HgCl_2$ elicitation may also relate directly to enzyme inhibition and rapid onset of cell death (Hargreaves, 1979). However, the results obtained did not appear to support the suggestion that abiotic inducers do not affect phytoalexin synthesis but only inhibit their degradation. PPO activity (Fig. 2) did not show any correlation with the progressive browning which was observed in experiments with controls or in tubers treated with $HgCl_2$. Its transient increase in $HgCl_2$ induced tubers supports the proposal for an altered role for PPO in stressed systems where its availability may assist in phenolic deposition in cell walls as a mechanism to limit infections (Fearman and Diamond, 1967; Yoshikawa, 1978; Mayer and Hard, 1979; Cline et al., 1987).

B. Characterization of Isolated Compounds

The *Phaseolus* species investigated produced 42 isoflavonoids of the following types: isoflavones, isoflavanones, isoflavans, pterocarpans, and coumestans (Table 1). In the *Dioscorea* species, six dihydrostilbenes were isolated (Table 2). The identity of these compounds was established from their UV, IR, MS, and ^{1}H and ^{13}C NMR spectral data, together with their color reactions on TLC plates and by comparison with literature values (Adesanya, 1984; Adesanya et al., 1984a,b, 1985, 1986; O'Neill, 1983; O'Neill et al., 1983, 1984, 1986). These data conformed with the established analytical profile for flavonoids (Mabry et al., 1970) except in a few cases. The collection of a large number of isoflavonoids provided an opportunity for a comparative study of their spectral characteristics in line with an earlier study of spectral properties of flavonoids (Mabry et al., 1970). Diagnostic differences were observed in the UV spectra of the isoflavones and isoflavanones, examples of which are given in Table 3.

C. UV Spectroscopy of Isoflavonoids

The UV spectrum of most isoflavones in either EtOH or MeOH have principal absorption maxima between 245 and 270 nm as established in earlier reports (Mabry et al., 1970; Ingham, 1982). However, isoflavones with three or more oxygen atoms on the A ring absorb at 260–270 nm, while those with fewer oxygen atoms have their maxima at 245–255 nm. Methylation of the 5-OH group or its absence causes a hypochromic shift of 5–10

Table 3 Some Examples of the Effect of Shift Reagents on UV Spectrum of Isoflavones and Isoflavanones

	Oxygenation pattern										
	A Ring				B Ring				Shift (nm)		
Compound	5	6	7	8	2′	3′	4′	λ_{max}	NaOMe	NaOAc	$AlCl_3$
Isoflavone											
Daidzein **1**			OH				OH	249	10	7	–
Genistein **2**	OH		OH				OH	261	14	13	12
Isoprunetin **3**	OCH_3		OH				OH	255	10	10	–
2′-Hydroxydaidzein **4**			OH		OH		OH	248	11	11	7
2′-Hydroxygenistein **5**	OH		OH		OH		OH	262	12	10	7
2′-Methoxygenistein **6**	OH		OH		OCH_3		OH	261	10	9	9
2′-Hydroxyisoprunetin **7**	OCH_3		OH		OH		OH	255	17	13	12
8,2′-Dihydroxygenistein **8**	OH		OH	OH	OH		OH	264	17	13	10
8,2′-Dihydroxydaidzein **9**			OH	OH	OH		OH	251	6	3	4
2,3-Dehydrokeivitone **10**	OH		OH	X	OH		OH	265	10	10	10
Phaseoluteone **12**	OH		OH		OH	X	OH	254	9	8	3
Luteone **13**	OH	X	OH		OH		OH	265	15	10	11
Isoflavanone											
2′-Hydroxydihydrodaidzein **14**			OH		OH		OH	277	60	60	–
8,2′-Dihydroxydihydrodaidzein **15**			OH	OH	OH		OH	285	55	55	–
Dalbergioidin **16**	OH		OH		OH		OH	289	35	35	31
5,2′-Dimethoxydalbergioidin **17**	OCH_3		OH		OCH_3		OH	286	36	36	–
Isoferreirin **18**	OH		OH		OCH_3		OH	289	36	35	26
5-Deoxykievitone **19**			OH	X	OH		OH	287	55	53	–
Kievitone **22**	OH		OH	X	OH		OH	293	42	41	22
4′-Methoxykievitone **25**	OH		OH	X	OH		OCH_3	293	42	42	24
3′-(γ,γ-Dimethylallyl)kievitone **28**	OH		OH	X	OH	X	OH	295	42	43	22
Isosophoranone[1] **29**	OH	X	OH		OCH_3	X	OH	293	38	36	–
Cajanol[2] **30**	OH		OCH_3		OCH_3		OH	287	71*	ND	23
Sophoraisoflavanone[3] **31**	OH		OH		OCH_3	X	OH	291	ND	40	21

ND, not determined; * = (NaOH); X = $CH_2CH{=}C(CH_3)_2$.
[1]Delle Monache et al. (1977); [2]Ingham (1976). [3]Komatsu et al. (1978).

nm due to the absence of the normal hydrogen bonding between the 5-OH and carbonyl groups (Mabry et al., 1970). In the isoflavones the principal maxima occurs at 285–300 nm with the 5-hydroxy types and 277–287 nm with the 5-deoxy types. The formation of the benzofuran ring in pterocarpans produces an extended molecular conjugation resulting in characteristic twin peaks or a peak with a shoulder at 280–290 nm.

Mabry et al. (1970) and Voirin (1983) reported subtle differences in the effect of shift reagents on the UV characteristics of flavones such that it was possible to use these characteristics to differentiate the different classes of flavonoids. A later study shows that isoflavones may be classified in a similar manner by some shift reagents. NaOMe ionizes all phenolic groups on the flavonoid and isoflavonoid nucleus leading to a bathochromic shift of 10–20 nm from the principal absorption maximum for isoflavones and 35–60 nm for isoflavanones and pterocarpans. $AlCl_3$ reagent forms a complex between the 5-OH and the carbonyl groups that results in a bathochromic shift for 5-hydroxyflavonoids and expectedly none for the 5-deoxy types. Similarly, shifts of 6–15 nm and 20–30 nm were observed for 5-hydroxyisoflavones and 5-hydroxyisoflavanones, respectively (Table 3) (Mabry et al., 1970; Adesanya, 1984). Although no shift was observed for 5-deoxy-2′-deoxyisoflavonoid derivatives, differences were observed when free 2′-hydroxy substituents are present. Table 3 shows that in isoflavones, but not in isoflavanones, when the 5-OH is absent or derivatized, a free 2′-OH does form a weak complex with the carbonyl causing a shift of 4–7 nm. Thus an observed shift with $AlCl_3$ for isoflavones could indicate either free 5-OH or 2′-OH in compounds where the 5-OH is absent in contrast to the report for flavones (Mabry et al., 1970; Voirin, 1983).

Although Mabry et al. (1970) did not report any data for flavonoids with isopentenyl side chains, the presence of such side chains at various positions on the isoflavone molecule did not affect the $AlCl_3$-induced shifts. In the isoflavanones, no shifts were observed when the side chain is at C-6 (Table 3). Delle Monache et al. (1977) reported no $AlCl_3$ shift for isosophoranone (**29**), a 5-hydroxy-6,3′-diprenylated isoflavanone. Similar observations were reported for acetophenones prenylated at various positions on the benzene ring (de Lima et al., 1975). Thus it appears that the presence of a bulky group at C-6 of isoflavanones ortho to the chelating hydroxyl group on a benzene ring prevents $AlCl_3$ complexation and consequently results in no shift of the absorption maxima.

D. Biosynthetic Considerations

Isoflavonoids

Enzymic studies and radiolabeling experimentation have helped to elucidate biosynthetic pathways and to establish a relationship between protein synthesis and phytoalexin formation. Important in this sequence of events is the overall increase in the rate of formation of RNA in plants following infection. Results suggest de novo mRNA synthesis is required for an effective defense response which enables the plant to resist infection (Bell et al., 1984; Ryder et al., 1984; Schmelzer et al., 1985).

The biosynthetic pathways for flavonoids and isoflavonoids are well established (Dewick, 1992 and references therein) and many of the required enzymes have been isolated and studied in relation to phytoalexin elicitation. The modification of phenylalanine to cinnamoyl-CoA requires two important enzymes, (PAL) and cinnamoyl-CoA ligase, and increases in both these enzymes are observed in response to elicitation (Hahlbrock et al., 1980; Whitehead et al., 1982; Dixon et al., 1983; Robbins et al., 1985).

The formation of isoflavonoids has been comprehensively studied by Grisebach and collaborators, who demonstrated the biosynthesis of the isoflavone skeleton from precursors of the general flavonoid pathway. The key steps in isoflavonoid biosynthesis involve a 1,2-aryl shift of the B ring which takes place after the formation of the C-15 chalcone intermediate, by chalcone synthase and an NADPH reductase, and its subsequent conversions to the isomeric flavanone by chalcone isomerase (Hagmann and Grisebach, 1984; Hakim and Dewick, 1984; Koch and Grisebach, 1986; Bless and Barz, 1988; Dewick, 1988; Hakamatsuka et al., 1991). All nine enzymes required for the biosynthesis of prenylated pterocarpans, i.e., glyceollins, have now been identified in elicited soybean cells and, as a result of this and other studies, the pathways to the various groups of isoflavonoids are well established (Grisebach et al., 1989; Welle and Grisebach, 1991).

The differentiation of the 5-hydroxy- and 5-deoxyisoflavonoids has also been suggested to occur at the chalcone stage (Woodward, 1980). This led to the proposal of different pathways for the 5-hydroxy- and 5-deoxyisoflavonoids isolated from *P. vulgaris* (Woodward, 1979b, 1980). These pathways can be extended for the compounds isolated from various *Phaseolus* species.

The apparent absence of some of the proposed intermediates in *P. aureus*, *V. unguiculata* (Table 1), and other Phaseolae species is due to the analysis of either small amounts of infected tissues, compounds at zones of antifungal activities in TLC bioassays, or diffusates from induced tissues (Smith and Ingham, 1980; Ingham, 1982), which invariably leads to small number of compounds. However the absence of phaseollin, a major phytoalexin in the proposed 5-deoxy pathway in *P. aureus*, is most likely due to the plant's inability to synthesize this compound. Such deficiency was also observed for *P. lunatus* and *P. mungo* (Table 1). On the other hand, the absence of some key intermediates like 2′-hydroxydiadzein (**4**) and 2′-

hydroxydihydrodiadzein (**14**) of the 5-deoxy pathway in *P. coccineus* and *P. lunatus*, respectively, and dalbergioidin (**16**) of the 5-hydroxy pathway in the two plants may be due to rapid turnover of these compounds during biosynthesis of the final products.

In the 5-deoxy pathway, the primary products from the chalcone appear to be diadzein (**1**), 2′-hydroxydiadzein (**4**), and 2′-hydroxydihydrodiadzein (**14**) progressively to give all the related isoflavonoids except vignafuran (**65**), which may arise from the chalcone through a different route (Martin and Dewick, 1979; Stoessl, 1982). The 8-hydroxylated products, 8,2′-dihydroxydiadzein (**9**) and 8,2′-dihydroxydihydrodiadzein (**15**), found only in *P. lunatus*, can be derived directly from the 2′-hydroxy precursors. The major pathways to the isoflavones, isoflavanones, isoflavans, pterocarpans, and coumestans may be found. Some of these constituents are prenylated, followed by the cyclization of the prenyl group. *P. mungo* appears to prenylate only isoflavanones, while *P. lunatus* may not possess the ability to cyclize its prenylated products (Table 1).

Following similar analysis, the primary products in the 5-hydroxy pathway are genistein (**2**), 2′-hydroxygenistein (**5**), and dalbergioidin (**16**). These metabolites can give the methylated compounds, isoprunetin (**3**), 2′-methoxygenistein (**6**), 2′-hydroxyisoprunetin (**7**), 5,2′-dimethoxydalbergioidin (**17**), and isoferreirin (**18**) found in *P. coccineus* and *P. lunatus* together with 8,2′-dihydroxygenistein (**8**) of *P. lunatus* (Table 1). This path seems to produce only isoflavones and isoflavanones (Table 1). The position of the 1-hydroxylated coumestan, aureol (**61**), is not clear, as it could be derived from this path following the corresponding steps for the synthesis of coumestans in the 5-deoxy pathway (Stoessl, 1982) or by hydroxylation of an appropriate coumestan in the 5-deoxy pathway. However, prenylation of isoflavanones, especially to kievitone, the major phytoalexin in most of the species, seems common while the prenylation of isoflavones appears to be restricted to *P. lunatus* (Adesanya et al., 1986).

The hydration of the isopentenyl side chain of kievitone or phaseollidin was first observed as a result of fungal detoxification in cultures fed with these compounds (Bailey et al., 1977; Kuhn et al., 1977; Smith et al., 1980). Similar hydrates, 5-deoxykievitol (**21**) and 5-deoxykievitone hydrate (**20**), derivable from 5-deoxykievitone (**19**); kievitol (**24**) and kievitone hydrate (**23**), derivable from kievitone, 2,3-dehydrokievitol (**11**), derivable from 2,3-dehydrokievitone (**10**), and cyclokievitone hydrate (**36**), derivable from cyclokievitone (**33**), were isolated from $CuCl_2$-induced tissues (Adesanya, 1984; O'Neill et al., 1986). Clearly, the host plant also has the capability to carry out these conversions, which in all instances reduces the fungitoxicity of the constituent and thus could be considered as a detoxifying process by the host cells. Toxicity of phytoalexins to host cells and ability of the cells to detoxify phytoalexins when applied at sublethal doses have been demonstrated (Smith, 1982; Van Etten et al., 1982). These results also support the probable existence of a phytoalexin-detoxifying capability in *P. mungo*, *P. lunatus*, and possibly other *Phaseolus* species. Enzymes involved in detoxification reactions have also been characterized in fungi and induced plants; for example, kievitone hydratase has been isolated from the cell-free culture filtrate of *Fusarium solani* f. sp. *phaseoli* and also from bean tissues inoculated with the fungi (Kuhn et al., 1977b; Smith et al., 1984).

Dihydrostilbenes

Stilbenes have also been shown to arise from phenylalanine and like flavonoids there is a requirement for the action of PAL to give cinnamic acid (Rudloff and Jorgensen, 1963; Langcake and Price, 1977; Schöppner and Kindl, 1979; Stoessl, 1982). This enzyme was also activated in *D. alata* tissues induced with *B. theobromae* and $HgCl_2$ indicating a similar role in the Dioscoreaceae for this enzyme. Cinnamic acid may be converted to *p*-coumaric acid and these two acids, as their CoA derivatives react with malonyl-CoA to give either pinosylvin or resveratrol. In recent years, the enzyme responsible for the formation of pinosylvin from malonyl-CoA and cinnamoyl-CoA has been isolated from *Pinus sylvestris* (Schöppner and Kindl, 1979). The partially purified enzyme from *Rheum rhaponticum* was found to catalyze the formation of 3,5,4′-trihydroxystilbene (resveratrol **67**) (Rupprich et al., 1980). Stilbene synthases, which occur in a small number of genera, have been grouped into two types: those which utilize *p*-coumaroyl-CoA and malonyl-CoA and form resveratrol (Schöppner and Kindl, 1984; Liswidowati et al., 1991), and the *Pinus* or Vitaceae (Fitzmeier and Kindl, 1981) enzyme, which utilizes cinnamoyl-CoA and malonyl-CoA and leads to pinosylvin accumulation (Gehlert et al., 1990). It is likely that the dihydrostilbenes dihydropinosylvin (**68**) and dihydroresveratrol (**76**) may arise from pinosylvin (**66**) and resveratrol (**67**), respectively, by the reduction of the bridge double bond and other dihydrostilbenes derived from these products or appropriate stilbene precursors (Stoessl, 1982). It has also been suggested that the biosynthesis of phenanthrenes and dihydrophenanthrenes proceeds via dihydro-*m*-coumaric acid in Combretaceae and Dioscoreaceae (Fritzmeier and Kindl, 1983); it is possible that the 2′-hydroxydihydrostilbenes arise from this route to give demethylbatatasin IV (**70**), which is subsequently metabolized to batatasin III, IV, and V in *Dioscorea* species (Fig. 1).

The induced formation of pinosylvin synthase is an early, selective, and sensitive process elicited by fungi or environmental stress. Young pine seedlings, especially, respond quickly to the attack of various fungi by synthesizing the enzymes required for pinosylvin accumulation (Gehlert et al., 1990). Pine stilbene synthase cDNAs have been isolated from *P. sylvestris* in young seedlings challenged with *B. cinerea* and the full-length cDNA encoding pinosylvin-forming stilbene synthase has been sequenced and a comparison made with resveratrol-forming stilbene synthases and chalcone-forming synthases; there were found to be significant differences in both instances (Schröder et al., 1988; Lanz et al., 1990; Melchoir and Kindl, 1991). Low concentrations of fungal spores during periods of high humidity activate the expression of substantial amounts of stilbene synthase. The quick response of young seedlings to fungal attack seems to be an important means of defense (Ebel, 1989; Hain et al., 1990). Enhanced concentrations of ozone cause a substantial increase in stilbene synthase activity (Roseman et al., 1991) and in this respect the level of pinosylvin synthase mRNA is an excellent indicator of environmental stress (Schwekendiek et al., 1992).

E. Chemotaxonomic Considerations

Phaseolus–Vigna *Complex*

The distinction between *Vigna* and *Phaseolus* species has long been a matter of taxonomic interest. Analysis of the morphological characteristics, protein content, and enzymic activities has indicated that the Asiatic beans which had previously been classed as *Phaseolus* species are better placed in the *Vigna* genus (Kloz, 1962; Casimir and LeMarchand, 1966; Kloz and Klozova, 1974; Chrispeels and Baumgartner, 1978). Marechal's (1978) classification of the *Phaseolus–Vigna* complex had also placed *P. aureus*, an Asiatic sample, in *Vigna* subsection *ceratotropis*. Comparison of the chemical constituents of the *Phaseolus* and *Vigna* species shown in Table 1 with those which have been reported for other species in the Leguminoseae indicates some differences (Ingham, 1982, 1990b). In general, most of the compounds, including those with isoflavan structures that occur in the genera *Phaseolus*, *Vigna*, *Lablab*, and *Erythrina*, are of little chemotaxonomic importance. Recent investigations clearly showed that the ability to produce methoxylated derivatives, once thought to be restricted to *Vigna*, is far more widespread (Table 1) (Lampard, 1974; Preston, 1975; Ingham, 1981b, 1990b). Phaseollinisoflavan (**43**), found only in the typical *Phaseolus* species *P. vulgaris* and *P. coccineus*, may be of chemotaxonomic importance.

It is interesting to note that *P. lunatus* produced no cyclized derivative of its many prenylated compounds and has the novel 8-hydroxylation pathway. These observations confirm earlier observations that *P. lunatus* is different from other *Phaseolus* species in its protein characteristics (Kloz, 1962; Kloz and Klozova, 1974).

Dioscorea *Species*

Ireland et al. (1981) attempted to separate different sections of the genus unsuccessfully by studying the distribution of naturally occurring dormancy inducing batatasin I-V in tubers and bulbils of 13 tropical and 2 temperate *Dioscorea* species. The major phytoalexins isolated from *Dioscorea* species are shown in Table 2. The major phytoalexin in *D. batatas* (Takasugi et al., 1987), *D. rotundata* (Fagboun et al., 1987), *D. alata* (Cline et al., 1989), and minor in *D. mangenotiana* (Kaganda and Adesanya, 1990) in the section Enantiophyllum is dihydropinosylvin. It was not found in *D. bulbifera* (section Opsophyton) and *D. dumentorum* (section Lasiophyton). Dihydroresveratrol has only been found in *D. dumentorum* (Adesanya et al., 1989).

F. Biological Properties

Antimicrobial Activities

Isoflavonoids. The isolation of significant amounts of isoflavones, isoflavanones, pterocarpans, and isoflavans allowed a further study of the structure–activity requirements of isoflavonoids. Perrin and Cruickshank (1969) had earlier suggested that a common three-dimensional shape was important, but further studies did not confirm this hypothesis (Van Etten, 1967). Harborne and Ingham (1978), as a result of work done with isoflavone luteone (**13**), suggested that a lipophilic side chain was important for fungitoxicity and a similar suggestion was proffered for kievitone with an additional requirement for phenolic hydroxyl groups by Smith (1978). Light microscopic examination revealed the gross effects of active isoflavonoids on the development of hyphae as disorganized growth patterns involving swollen hyphal tips, cellular granulation, and the production of septal plugs (Smith, 1978; Adesanya et al., 1986). Comparative antifungal studies using *Aspergillus niger* and *Cladosporium cucumerinum* were done on all compounds with good yield, their chemical derivatives, together with demethylmedicarpin (**45**), medicarpin (**47**), and isomedicarpin (**46**) isolated from *Trifolium repens* (Table 4).

Differential sensitivity of the two fungi used was observed and this supports Van Etten's (1967) observation in the test of pterocarpans and related isoflavonoids against *Aphanomyces euteiches* Drechs. and *Fusarium solani* Mart. Sacc. f.sp. *cucurbitae* (Burk, F. R. Jones) Synd. and Hans. The results in Table 4 show that pterocarpans (**45, 46, 47, 49**) and isoflavans (**37, 43**) are more fungitoxic than other classes of isoflavonoids and support the result obtained by Van Etten (1967). He suggested that the comparable fungitoxic activities of (−) vestitol (**38**), (−)sativan (**40**) and (+)3-hydroxy-9-methoxypterocarpan (**47**) were related to the fact that the isoflavans can

Table 4 Inhibition of Spore Germination in *Cladosporium cucumerinum* and *Aspergillus niger* by Isoflavonoids

	Minimum inhibitory concentration[a] (μg/ml)	
Compound	*C. cucumerinum*	*A. niger*
Isoflavones		
1 Daidzein	50–75	50–75
2 Genistein	50–75	50–75
3 Isoprunetin	50–75	25–50
4 2′-Hydroxydaidzein	25–50	50–75
5 2′-Hydroxygenistein	50–75	25–50
6 2′-Methoxygenistein	50–75	25–50
7 2′-Hydroxyisoprunetin	25–50	50–75
10 2,3-Dehydrokievitone	25–50	25–50
11 2,3-Dehydrokievitol	>100	>100
12 Phaseoluteone	10–25	10–25
Isoflavonones		
16 Dalbergioidin	75–100	50–75
18 Isoferreirin	10–25	10–25
19 5-Deoxykievitone	25–50	50–75
22 Kievitone	10–25	25–50
23 Kievitone hydrate	>100	>100
24 Kievitol	>100	>100
26 2′,4′,5-Trimethyl kievitone	25–50	10–25
27 2′,4′,5,7-Tetramethyl kievitone	>100	>100
28 3′(γ,γ-Dimethylallyl)kievitone	25–50	25–50
31 Cyclokievitone	50–75	50–75
32 3′,4′,5-Trimethylcyclokievitone	>100	>100
33 1″,2″-Dehydrocyclokievitone	25–50	50–75
Pterocarpans		
45 Demethylmedicarpin	25–50	50–75
46 Isomedicarpin	10–25	50–75
47 Medicarpin	10–25	25–50
49 Phaseollidin	10–25	40–50
57 Phaseollin	25–50	50–60
Isoflavans		
37 Demthylvestitol	25–50	50–75
43 Phaseollinisoflavan	10–25	50–75
Coumestans		
61 Aureol	25–50	50–75

[a]Minimum inhibitory concentration = the range of isoflavonoid concentrations between which fungal spore (approx 2×10^6 spores/ml) germination at 25°C in Czapeck Dox liquid nutrient ceases. All isoflavonoids tested had shown fungitoxic activity in the TLC bioassay.

assume a conformation similar to the three-dimensional shape of pterocarpans. Although the coumestans have been generally considered not to be fungitoxic (Smith, 1982), the planar coumestan, aureol **(61)**, isolated from *P. aureus* had significant fungitoxic activity.

The isoflavone phaseoluteone **(12)** is more active than its nonprenylated analog 2′-hydroxygenistein **(5)**; the isoflavanone kievitone is more active than dalbergioidin and the pterocarpan phaseollidin **(49)** is more active than demethylmedicarpin **(45)**, showing that the dimethylallyl side chain enhances fungitoxicity. This was noted earlier by Harborne and Ingham (1978) and Smith (1978) using a more limited range of compounds, and it was suggested that the side chain improves lipophilicity and hence aids penetration of the fungal cell wall. However, the presence of a second side chain as in 3′-(γ,γ-dimethyallyl)kievitone **(28)** did not further enhance fungitoxicity when compared to kievitone and cyclization of the side chain shows a reduction in activity when cyclokievitone **(33)** and phaseollin **(57)** are compared to kievitone and phaseollidin, respectively (Table 4).

The oxidative product of kievitone, kievitone hydrate **(23)**, isolated earlier as a product of fungal metabolism of kievtone is not fungitoxic (Kuhn et al., 1977). Similar reduction in activity was observed in the series even when the side chain was oxidized to a primary alcohol as in kievitol **(24)**, and 2,3-dehydrokievitol **(11)** (Adesanya et al., 1986). Fungal oxidative metabolites of pterocarpan phytoalexins such as phaseollidin and phaseollin also have similar reduced activity (Ingham, 1976; Bailey et al., 1977; Smith et al., 1980). However, phenolic hydroxyl groups appear important for activity, since the tetramethylated derivative of kievitone (2′,4′,5,7-tetramethylkievitone,**27**) was inactive, while 2′,4′,7-trimethylkievitone **(26)** was active. This indicates that at least one phenolic group is required for fungitoxicity. In the isoflavone series, compounds with C-7, C-4′ hydroxylation have fungitoxic activity and further hydroxylation at C-2′ causes only slight changes in toxicity (Table 4). This is in agreement with the results of Ingham et al. (1977) with *C. cucumerinum* and *Phytophthora megasperma*. He suggests that it is the presence of oxygen at these positions that is important, since methoxyl substitution of these hydroxyls does not radically alter fungitoxic activity. The lack of difference in activity between diadzein **(1)** and genistein **(2)** suggests that hydroxylation at C-5 is relatively unimportant in isoflavones. However, some differences in fungitoxic activity are observed for kievitone and 5-deoxykievitone **(19)** in the isoflavanone series. Perrin and Cruickshank (1969) suggested that hydroxyl groups at C-3 and C-9 are important for activity in pterocarpans, but this is not confirmed by this work or by other workers (Van Etten, 1967). It would be of interest to evaluate the contribution to activity of oxygenation at C-7 and C-4 (C-3 and C-9 for pterocarpans). Isoflavonoids lacking this oxygenation pattern are rare in nature.

The significant toxicity of isoferreirin (**18**) when compared to dalbergioidin (**16**) might seem to show that methylation increases both lipophilicity and activity, but this is not so in the isoflavone and pterocarpan series, since 2′-methoxygenistein (**6**) has an activity similar to that of 2′-hydroxygenistein (**5**), while medicarpin (**47**) and isomedicarpin (**46**) are only slightly more fungitoxic than demethylmedicarpin (**45**) (Table 4).

Dihydrostilbenes. In this class, too few compounds have been tested for antimicrobial activity to provide a detailed comparative study. However, Takasugi et al. (1987) have shown that induced phenanthrenes and dihydrostilbenes in *D. batatas* (Table 2) are generally more effective against fungi than bacteria, and the dihydrostilbenes more active than the phenanthrenes. Our results are in agreement with this observation and also show that dihydropinosylvin and batatasin IV are significantly more toxic (<25 μg/ml, respectively) to *Aspergillus fumigatus* than demethylbatatasin IV (**70**) (>100 μg/ml), and dihydroresveratrol (**76**) was also not active against *A. niger* and *A. flavus* (Adesanya, unpublished; Adesanya et al., 1989). This might suggest the need for at least two free hydroxyl groups for activity. The role of demethylbatatasin IV in disease resistance is not yet clear. It has only transient accumulation in fungal-induced tubers and no significant one in $HgCl_2$-treated ones (Figs. 2 and 3), suggesting that it could be a product of fungal/plant detoxification process, or a phytoalexin with yet-to-be-defined toxicity. It is also noteworthy that the stilbene analog of **68** and **76**, pinosylvin (**66**), and *trans*-resveratrol (**67**) with antimicrobial activities have been isolated from induced *Pinus* and *Arachis* species, respectively (Ingham, 1976; Schöppner and Kindl, 1979).

Growth Inhibitory Activity

The growth inhibitory activities of batatasin IV and batatasin III in lettuce seed germination, hypocotyl elongation, and wheat coleoptile section elongation tests have been established (Hasimoto and Tajima, 1978). Batatasin IV also showed strong persistent inhibition of the gemination of seeds of *Sorghum bicolor*, and by comparison dihydropinosylvin (**68**) showed a similar activity at a higher concentration, with the effect wearing off after 24 hr (Fig. 4) (Cline et al., 1989). Demethylbatatasin IV was not active (Fig. 4). Batatasin IV also showed a strong and persistent inhibition of the elongation of roots of *S. bicolor* seedlings, with dihydropinosylvin having a similar activity and demethylbatatasin IV showing a lesser action, but was still an effective inhibitor of root elongation (Fig. 5). In all experiments the roots did not recover with time and root tips, at high concentrations of

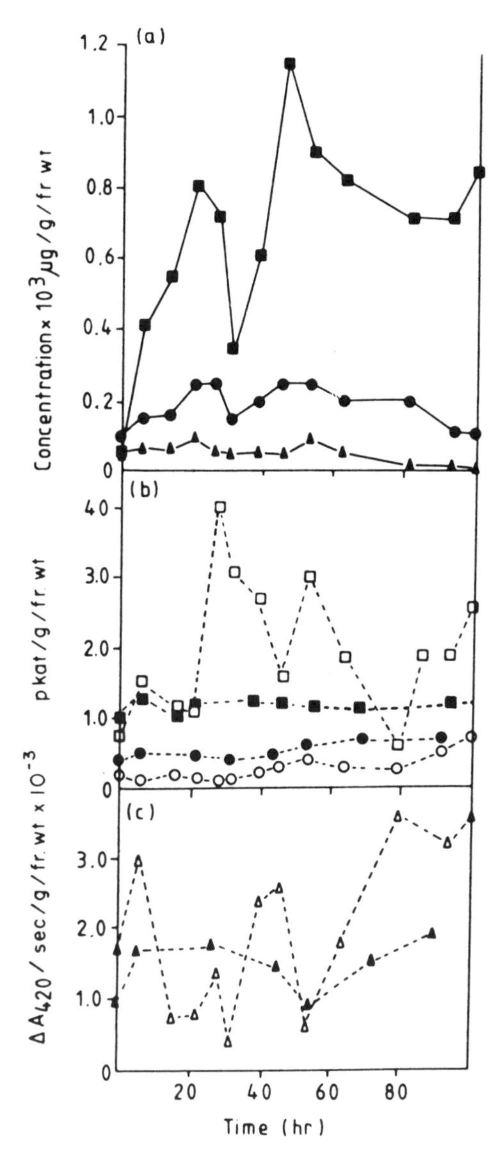

Figure 3 Enzyme activities in relation to the accumulation of dihydrostilbenes **1**–**3** in $HgCl_2$ induced tubers. (a) ■-■, dihydropinosylvin (**1**); ●-●, demethylbatatasin IV (**2**); ▲-▲, batatasin IV (**3**). (b) After elicitation: □-□, PAL; ○-○, TAL; in controls: ■-■, PAL; ●-●, TAL. (c) After elicitation: △-△, PPO; in controls: ▲-▲, PPO.

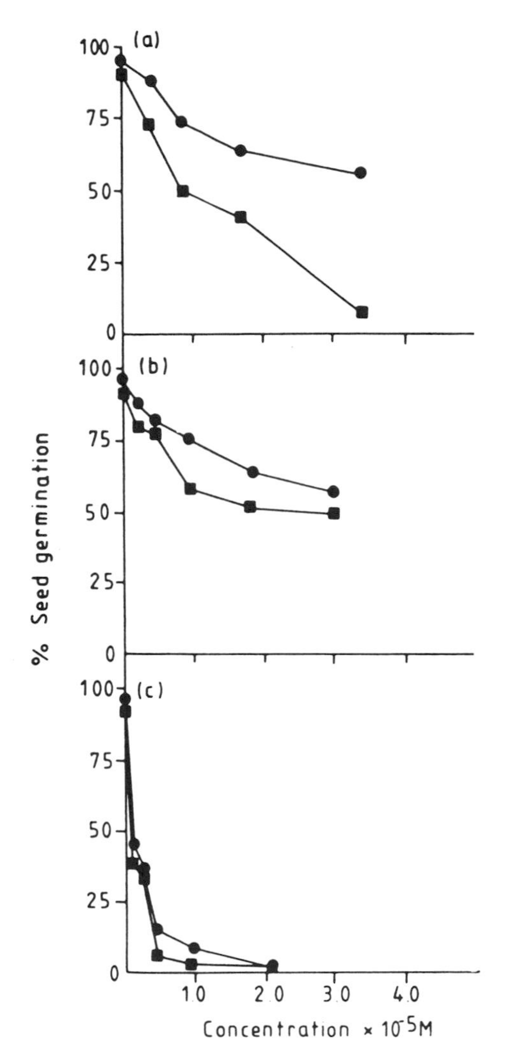

Figure 4 Inhibitory activities of dihydrostilbenes **1–3** on germination of *S. bicolor* seeds. ●-●, 48 hr incubation; ■-■, 24 hr incubation.

dihydropinosylvin and batatasin IV collapsed presumably due to an uncontrolled leakage of metabolites and electrolytes from the cell due to severe effects on the cell membrane (Tahara et al., 1972). The result suggests that dihydropinosylvin and demethylbatatasin IV do not possess dormancy-inducing characteristics already established for batatasin IV and other batatasins (Hashimoto et al., 1972), since they cannot prevent seed germination,

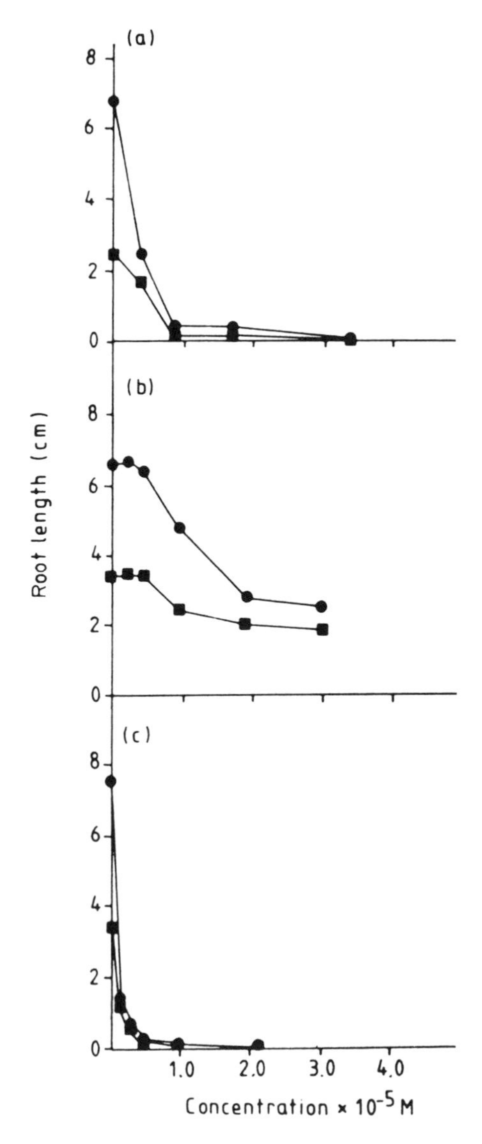

Figure 5 Inhibitory activities of dihydrostilbenes **1–3** on root elongation of *S. bicolor* seedlings. ●-●, 48 hr incubation; ■-■, 24 hr incubation.

but like other phytoalexins are toxic to the parasites as well as host cells at high concentrations. The role of batatasin IV in disease resistance is not clear, since it accumulates to a low level in *Dioscorea* species, including *D. batatas* (Takasugi et al., 1987; Cline et al., 1989), and also occurs naturally in *D. batatas* (Hashimoto et al., 1972).

IV. EPILOG

Selected species in the Phaseoleae and Dioscoreaceae have been investigated for their inducible compounds. Large-scale and detailed analysis of $CuCl_2$ induced tissues of *Phaseolus* species led to the isolation of many isoflavonoids that provided an insight into their possible biogenesis, structure–antifungal activity relationships, and chemotaxonomy. The analysis of fungitoxic spots in the TLC bioassay of fungus induced tissues of *Dioscorea* species gave only a few dihydrostilbenes by comparison. Thus a detailed large-scale analysis of induced tissues has more analytical potential than isolation of fungitoxic compounds alone.

Although it was possible to speculate on the biosynthetic pathway to most of the isoflavonoids and dihydrostilbenes considered, the correct pathway still needs to be established by isolating and characterizing the terminal enzymes. This is being done for some compounds in other species (Preisig et al., 1991). Phytoalexins have obvious chemotaxonomic potential in these two families and so detailed analysis of induced compounds in more species could assist in the taxonomy of these families. Such investigations could also yield more novel compounds and new biological activities. The bioassay of the dihydrostilbenes was extended to human fungal parasites. *A. fumigatus* and *Tricophyton mentagyrophytes*, in order to evaluate their usefulness in treating human infections. Similar investigations have shown that phaseollinisoflavan is very active against zoopathogens (Gordon et al., 1980; Smith, 1982). Such screenings against other test systems could expose other potential uses of these compounds.

The role of many of these inducible compounds in disease resistance needs to be established as they vary in composition, number, and amounts in various species. In *P. lunatus*, *P. mungo*, and *P. vulgaris*, the largest group of antifungal isoflavonoids was the isoflavanones, whereas in *P. aureus* and *P. coccineus* there is a more even distribution of isoflavones, isoflavanones, and pterocarpans. Only *P. coccineus* produces significant levels of isoflavans, and the only fungitoxic coumestan isolated, aureol, was found in both *P. aureus* and *P. mungo*.

Although the work on the *Dioscorea* species is as yet not comprehensive, variations are also noticeable in the type of major compounds produced, the seemingly inactive demethylbatatasin IV was the major inducible compound in *D. bulbifera*, while *D. dumentorum* contains dihydroresveratrol and a yet-to-be-characterized compound, the other species produce dihydropinosylvin as the major compound, except *D. mangenotiana*, which is more prone to rot than the other species considered. It is likely that the plant uses all the active compounds it produces in a concerted effort to combat fungal invasion, with each compound having different functions (Mansfield, 1982; Smith, 1982).

The work on the species in the Dioscoreaceae, particularly the economic yams, is aimed at breeding resistant species and finding ways of protecting yams under storage from microbial attack while increasing their dormancy period, thus increasing their shelf life. The results obtained so far indicate that batatasin IV and to a lesser extent dihydropinosylvin might be good subjects for further studies that would include the evaluation of their toxicity to humans and metabolism in both plants and humans.

ACKNOWLEDGMENTS

The authors with to thank the librarians at the School of Pharmacy for help in checking references and Mr. B. C. Homeyer for help with the compilation of the figures and tables.

REFERENCES

Adesanya, S. A. (1984). Inducible and constitutive isoflavonoids from *Phaseolus* and *Vigna* species. PhD Thesis, University of London.

Adesanya, S. A., Ogundana, S. K., and Roberts, M. F. (1989). Dihydrostilbene phytoalexins from *Dioscorea bulbifera* and *D. dumentorum*. *Phytochemistry 28*(3):773–774.

Adesanya, S. A., O'Neill, M. J., and Roberts, M. F. (1984a). Induced and constitutive isflavonoids in *Phaseolus mungo*. *Z. Naturforschung. C, Biosciences 39*: 881–893.

Adesanya, S. A., O'Neill, M. J., and Roberts, M. F. (1984b). Isoflavonoids from *Phaseolus coccineus*. *Phytochemistry 24*(11):2699–2702.

Adesanya, S. A., O'Neill, M. J., and Roberts, M. F. (1986). Structure related fungitoxicity of isoflavonoids. *Physiol. Mole. Plant Pathol. 29*:95–103.

Alozie, S. O., Nwankiti, A. O., and Oti, E. (1987). Source of resistance in anthracnose blotch disease of water yam (*Dioscorea alata*) caused by *Colletotrichum gloeosporioides* Penz. 2. Relationship between phenol content and resistance, *Beitraege zur Tropischen Landwirtschaft und Veterinaermedizin 25H.1*:55–59.

Arnoldi, A., and Merlini, L. (1990). Antifungal activity relationships for some isoflavonoid phytoalexins. *J. Agric. Food Chem. 38*:834–838.

Bailey, J. A. (1973). Production of antifungal compounds in cowpea (*Vigna sinen-*

sis) and pea (*Pisum sativum*) after virus infection. *J. Gen. Microbiol. 75*:119–123.

Bailey, J. A. (1982). Mechanism of phytoalexin accumulation. *Phytoalexins* (J. A. Bailey and J. W. Mansfield, eds.), Blackie, London, pp. 289–378.

Bailey, J. A. (1984). The relationship between symptom expression and phytoalexin concentration in hypocotyls of *Phaseolus vulgaris* infected with *Collectotrichum lindemuthianim. Physiol. Plant Pathol. 4*:477–488.

Bailey, J. A., Burden, R. S., Mynett, A., and Brown, C. (1977). Metabolism of phaseollin by *Septoria nodorum* and other nonpathogens of *Phaseolus vulgaris. Phytochemistry 16*:1541–1544.

Bailey, J. A., Rowell, P. M., and Arnold, G. M. (1980). The temporal relationship between host cell death, phytoalexin accumulation and fungal inhibition during hypersensitive reactions of *Phaseolus vulgaris* to *Colletotrichum lindemuthianum. Physiol. Plant Pathol. 17*:329–339.

Bell, A. A. (1981). Biochemical mechanisms of disease resistance. *Ann. Rev. Plant Physiol. 32*:21–81.

Bell, A. A., and Pressley, J. T. (1969). Heat inhibited or heat killed conidia of *Verticillium albo-atrum* induce disease resistance in cotton. *Phytopathology 59*: 1147–1151.

Bell, J. N., Dixon, R. A., Bailey, J. A., Rhowell, P. M., and Lamb, C. J. (1984). Differential induction of chalcone synthase mRNA activity at the onset of phytoalexin accumulation in compatible and incompatible plant pathogen interactions. *Proc. Nal. Acad. Sci. USA 91*:3384–3488.

Biggs, D. R., Shaw, J. G., and Yates, M. K. (1983). Identification of 41,7-dihydroxy-2′,3′-dimethoxyisoflavan in bean roots. *J. Nat. Prod. 46*(5):742–744.

Bless, W., and Barz, W. (1988). Isolation of pterocarpan synthase, the terminal enzyme of pterocarpan phytoalexin biosynthesis in cell suspension culture of *Cicer arietinum. Fed. Eur. Biol. Soc. Lett. 235*:47–50.

Casimir, J., and Le Machand, L. (1966). *Bulletin Jardin Botanique de L'Etat a Bruxelles. 36*:233.

Chrispeels, M. J., and Baugartner, B. (1978). Seriological evidence confirming the assignment of *Phaseolus aureus* and *P. mungo* to the genus *Vigna. Phytochemistry 17*:125–126.

Cline, E. I., Adesanya, S. A., Ogundana, S. K., and Roberts, M. F. (1989). Induction of PAL activity and dihydrostilbene phytoalexins in *Dioscorea alata* and their plant growth inhibitory properties. *Phytochemistry 28*(10):2621–2625.

Coursey, D. G. (1967). *Yams*. Longmans, London.

Coxon, D. T., Ogundana, S. K., and Dennis, C. (1982). Antifungal phenathrenes in yam tubers. *Phytochemistry 21*(6):1389–1392.

Cruickshank, I. A. M., and Perrin, D. R. (1968). The isolation and partial characterization of monilicolin A, a polypeptide with phaseollin inducing activity from *Monilinia fructicola. Life Sci. 7*:449–458.

Darvill, A. G., and Albersheim, P. (1984). Phytoalexins and their elicitors: a defense against microbial infections in plants. *Ann. Rev. Plant Physiol. 35*:243–275.

De Lima, O. G., De Mello, J. F., De Barros-Coeiho, J. S., De Andrade Lyra, F. G., De Albuquerque, M. M., Marini-Bettolo, G. B., Delle Monache, G., and Delle Monache, F. (1975). New prenylated chalcones from *Longicarpus neuroscapha* betth. *Il. Farmaco Ed. Sci. 30*:326–342.

Delle Monache, G., Delle Monache, F., Marini-Bettolo, G. G., De Albuquerque, M., De Mello, J., and De Lima, O. G. (1977). *Gazzette Chimica Italiana 107*: 189.

Denny, T., and Van Etten, H. D. (1983). Tolerance of *Nectria haematococca* MPVI to phytoalexin pisatin in the absence of detoxification. *J. Gen. Microbiol. 129*: 2893–2901.

Deverall, B. J. (1982). The concept of phytoalexins. Phytoalexins (J. A. Bailey and J. W. Mansfield, eds.), Blackie, London, pp. 1–20.

Dewick, P. M. (1992). Biosynthesis of shikimate metabolites. *Nat. Prod. Rep. 9*: 153–181.

Dewick, P. M. (1988). Isoflavonoids. *The Flavonoids: Advances in Research Since 1980* (J. B. Harborne, ed.). Chapman-Hall, New York, pp. 125–240.

Dhawale, S., Souciet, G., and Kuhn, D. N. (1989). Increase in chalcone synthetase, messenger RNA in pathogen inoculated soybeans with race specific resistance is different in leaves and roots. *Plant Physiol. 91*(3):911–916.

Dixon, R. A. (1986). The phytoalexin response: elicitation signaling and control of host gene expression. *Biol. Rev. 61*:239–291.

Duke, J. A. (1983). *Handbook of Legumes of World Economic Importance*. Plenum Press, London.

Ebel, J. (1986). Phytoalexin synthesis: the biochemical analysis of the induction process. *Ann. Rev. Phytopathol. 24*:235–264.

Ellis, J. S., Jennings, A. C., Edwards, L. A., Mavandad, M., Lamb, C. J., and Dixon, R. A. (1989). Defense gene expression in elicitor treated cell suspension cultures of french bean cultivar Imuna. *Plant Cell Rep. 8*(8): 504–507.

Ersek, T., and Kiraly, Z. (1986). Phytoalexins. Warding-off compounds in plants? *Physiol. Plantarum 68*:343–346.

Fearman, H., and Diamond, A. (1967). *Phytopathology 15*:279.

Fagboun, D. E., Ogundana, S. K., Adesanya, S. A., and Roberts, M. F. (1987). Dihydrostilbene phytoalexins from *Dioscorea rotundata*. *Phytochemistry 26*(12): 3187–3189.

Fritzemeier, K. H., and Kindl, H. (1981). Coordinate induction by UV light of stilbene synthetase, phenylalanine ammonia lyase and cinnamate-4-hydroxylase in leaves of Vitaceae. *Planta 151*:48–52.

Fritzemeier, K. H., and Kindl, H. (1983). Two different pathways leading to phenanthrene and 9,10-dihydrophenanthrenes of the genus *Dioscorea*. *Z. Naturforschung. C, Biosciences 39*:217–221.

Gehlert, R., Schöppner, A., and Kindl, H. (1990). *Mol. Plant-Microbe Interact. 3*: 444–449.

Gordon, M. A., Lappa, E. W., Fitter, M. S., and Lindsay, M. (1980). Susceptibility of zoopathogenic fungi to phytoalexins. *Antimicrob. Agents Chemother. 12*: 120–123.

Grisebach, H., Edelmann, L., Fischer, D., Kochs, G., and Welle, R. (1989). Biosynthesis of phytoalexins and nod-gene inducing isoflavones in soybean. *Signal Molecules in Plants and Plant-Microbe Interactions* (B. Lugtenberg, ed.), NATO ASI Series Vol H-36, Springer, Heidelberg, pp. 57–64.

Hadwiger, L. A., and Schwochau, M. E. (1971). Ultraviolet light induced formation of pisatin and phenylalanine ammonia lyase. *Plant Physiol. 47*:346–351.

Hagmann, M. L., Heller, W., and Grisebach, H. (1984). Induction of phytoalexin synthesis in soybean: Stereospecific 3,9-dihydroxypterocarpan 6a-hydroxalase from elicitor-induced soybean cell cultures. *Eur. J. Biochem. 142*:127–931.

Hahlbrock, K., and Scheel, D. (1989). Physiology and molecular biology of phenylprionoid metabolism. *Ann. Rev. Plant Physiol. Plant Mol. Biol. 40*:347–369.

Hain, R., Bieseler, B., Kindl, H., Schröder, G., and Stöcker, R. (1990). *Plant Mol. Biol. 15*:325–335.

Hakamatsuka, T., Noguchi, H., Ebizuka, Y., and Sankawa, U. (1988). Deoxychalcone synthase from cell suspension cultures of *Pueraria lobata. Chem. Pharm. Bull. 36*:4225–4228.

Hakim, A. M., Ani, A., and Dewick, P. M. (1984). Isoflavonoid biosynthesis: concerning the aryl migration. *J. Chem. Soc. Perkin Trans. I.* 2831–2838.

Harborne, J. B. (1977). *Introduction to Ecological Biochemistry.* Pergamon Press, London.

Harborne, J. B., and Ingham, J. L. (1978). Biochemical aspects of coevolution of higher plants with their fungal parasites. *Biochemical Aspects of Coevolution in Higher Plants* (J. B. Harborne, ed.), Academic Press, London, pp. 343–405.

Hargreaves, J. A. (1979). Investigation into the mercuric chloride stimulated phytoalexin accumulation in *Phaseolus vulgaris* and *Pisum sativum. Physiol. Plant Pathol. 15*:279–287.

Hashimoto, T., Hasegawa, K., and Kawarada, A. (1972). Batatasins: new dormancy inducing substances of yam bulbils. *Planta 108*:369–374.

Hashimoto, T., and Tajima, M. (1978). Structures and synthesis of growth inhibitors, batatasin IV and V and their physiological activities. *Phytochemistry 17*: 1179–1184.

Ingham, J. L. (1976). Fungal modification of pterocarpan phytoalexin from *Melilotus alba*, and *Trifolium pratense. Phytochemistry 15*:1489–1495.

Ingham, J. L. (1981a). Phytoalexin induction and its chemosystematic significance in the genus *Trigonella. Biochem. Syst. Ecol. 9*(4):275–281.

Ingham, J. L. (1981b). Phytoalexin induction and its taxonomic significance in the Leguminosae (subfamily Papillionoideae). *Advances in Legume Systematics.* Proceedings of the International Legume Conference, Kew, 1978. (R. M. Polhill, and P. G. Ravens, eds.), London, pp. 599–626.

Ingham, J. L. (1982). Phytoalexins in the leguminosae. *Phytoalexins* (J. A. Bailey and J. W. Mansfield, eds.), Blackie, London, pp. 21–80.

Ingham, J. L. (1990a). A further investigation of phytoalexin formation in genus *Trifolium. Z. Naturforschung C, Biosciences 45*(7–8):829–834.

Ingham, J. L. (1990b). Systematic aspects of the phytoalexin formation within the tribe Phaseolae of the Leguminosae, subfamily Papillionoideae. *Biochem. Syst. Ecol. 18*(5):329–344.

Ingham, J. L., and Harborne, J. B. (1976). Phytoalexin induction as a new dynamic

approach to the study of systematic relationships among higher plants. *Nature 260*:241–243.
Ingham, J. L., Keen, N. T., and Hymowitz, T. (1977). A new isoflavone phytoalexin from fungus inoculated stems of *Glycine wightii*. *Phytochemistry 16*: 1943–1946.
Ireland, C. R., Schwabe, W. W., and Coursey, D. G. (1981). The occurrence of batatasins in Dioscoraceae. *Phytochemistry 20*(7):1569–1571.
Kaganda, N. G., and Adesanya, S. A. (1990). A new dihydrostilbene phytoalexin from diseased *Dioscorea mangenotiana*. *J. Nat. Prod. 53*(5):1345–1346.
Kay, D. E. (1979). *Food, Legumes, Crop and Product Digest No. 3*. Tropical Product Institute, London.
Keen, N. T., Lyne, R. L., and Hymowitz, T. (1986). Phytoalexin production as a chemosystematic parameter within the Genus *Glycine*. *Biochem. Sys. Ecol. 14*: 418–486.
Keen, N. T., Yoshikawa, M., and Wang, M. C. (1983). Phytoalexin elicitor activity of carbohydrates from *Phytophthora megasperma* f. sp., *glycinea* and other sources. *Plant Physiol. 71*:466–471.
Kiraly, Z., Barma, B., and Ersek, T. (1972). Hypersensitivity as a consequence, not the cause of plant resistance to infection. *Nature 239*:456–472.
Kloz, J. (1962). An investigation of the protein character of four *Phaseolus* species with special reference to the question of their genus. *Biologia Plantarum 4*:85–90.
Kloz, J., and Lozova, E. (1974). Protein euphaseolin in Phaseolinae—a chemicotaxonomical study. *Biologia Plantarum 16*:290–300.
Koch, G., and Grisebach, H. (1986). Enzymic synthesis of isoflavones. *Eur. J. Biochem. 155*:311–318.
Koch, G., and Grisebach, H. (1989). Phytoalexin synthesis in soy bean: purification and reconstitution of cytochrome P-450: 3,9-dihydroxypterocarpan 6a-hydroxylase and separation from cytochrome P-240:cinnamaaate-4-hydroxylase. *Arch. Biochem. Biophys.*
Kuć, J. (1982). Phytoalexins in the Solanaceae. *Phytoalexins* (J. A. Bailey and J W. Mansfield, eds.), Blackie, London, pp. 21–80.
Kuhn, P. J., Smith, D. A., and Ewing, D. F. (1977). 5,7,2′,4′-Tetrahydroxy-8-(3”-hydroxy-3”-methyl-butyl)isoflavanone, a metabolite of kievitone produced by *Fusaium solani* f.sp. Phaseoli. *Phytochemistry 16*:296–297.
Kuhn, P. F., Smith, D. A., and Ewing, D. F. (1977b). *Physiol. Plant Pathol. 14*: 179.
Lampard, J. F. (1974). Demethylhomopterocarpin; an antifungal compound in *Canavalia ensiformis* and *Vigna unguiculata* following infection. *Phytochemistry 13*:291–292.
Langcake, P., and Pryce, R. J. (1977). The production of resveratrol and the viniferins by grapevines in response to ultraviolet radiation. *Phytochemistry 16*:1193–1196.
Lanz, T., Schröder, G., and Schröder, J. (1990). *Planta 181*:169–175.
Liswidowati, Melchior, F., Hohmann, F., Schwer, B., and Kindl, H. (1991). *Planta 183*:307–314.

Mabry, T. J., Markham, K. R., and Thomas, M. B. (1970). *The Systematic Identification of Flavonoids*. Springer, Berlin.

Maekawa, F. (1955). Topomorphological and taxonomical studies in Phaseoleae Legiminosae. *Jap. J. Bot. 15*:103–116.

Mansfield, J. W. (1982). The role of phytoalexins in disease resistance. *Phytoalexins* (J. A. Bailey and J. W. Mansfield, eds.), Blackie, London, pp. 253–288.

Mansfield, J. W., Hargreaves, J. A., and Boyle, F. C. (1974). Phytoalexin production by live cells in broad bean leaves infected with *Botrytis cinerea. Nature 252*:316.

Marechal, R., Mascharpa, J., and Stainer, F. (1978). Etude taxonomique d'un group complexe d'especes des agenies *Phaseolus* et *Vigna* (Papilionaceae) sur la base de donnees morphologiques et polliniques, traites par l'analyse informatique. *Boisseira 28*:1–273.

Martin, M., and Dewick, P. M. (1979). Biosynthesis of 2-arylbenzofuran phytoalexin, vignafuran in *Vigna unguiculata. Phytochemistry 18*:1309–1317.

Mayer, A. M., and Hard, E. (1979). Polyphenol oxidases in plants. *Phytochemistry 18*:193–215.

Melchior, F., and Kindl, H. (1991). *Arch. Biochem. 288*:552–557.

Müller, K. O. (1958). Studies on phytoalexins. 1. The formation and immunological significance of phytoalexins produced by *Phaseolus vulgaris* in response to infections with *Sclerotinia fructicola* and *Phytophthora infestans. Aust. J. Biol. Sci. 11*:275–300.

Müller, K. O., and Börger, H. (1940). Experimentelle utersuchungen uber die *Phytophthora* resistenz der kartoffel. *Arb. Biol. Anst. Reichsant 23*:189–231.

O'Neill, M. J. (1983). Aureol and Phaseol, two new coumestans from *Phaseolus aureus* Roxb., *Z. Naturforschung C, Biosciences 38*:698–700.

O'Neill, M. J., Adesanya, S. A., and Roberts, M. F. (1983). Antifungal phytoalexins from *Phaseolus aureus* Roxb., *Z. Natturforschung C, Biosciences 38*:693–697.

O'Neill, M. J., Adesanya, S. A., and Roberts, M. F. (1984). Isosojagol, a coumestan from *Phaseolus coccineus. Phytochemistry 23*(11):2704–2705.

O'Neill, M. J., Adesanya, S. A., Roberts, M. F., and Pantry, I. R. (1986). Novel inducible isoflavonoids from lima bean, *Phaseolus lunatus. Phytochemistry 25*: 1315–1322.

Perrin, D. R., and Cruickshank, I. A. M. (1969). The antifungal activity of pterocarpans towards *Monilinia fructicola. Phytochemistry 8*:971–978.

Preisig, C. C., Van Etten, H. D., and Mareau, R. A. (1991). Induction of 6a-hydroxymaackiain-3-O-methyl transferase activities during biosynthesis of pisatin. *Arch. Biochem. Biophys. 290*(2):468–473.

Preston, N. W. (1975). 2′-O-methylphaseollidinisoflavin from infected tissue of *Vigna unguiculata. Phytochemistry 14*:1131–1132.

Preston, N. W., Chamberlain, K., and Skipp, R. A. (1975). A 2-arylbenzofuran phytoalexin from cowpea (*Vigna unguiculata*). *Phytochemistry 14*:1843–1844.

Rahe, J. E. (1973). Occurrence and levels of the phytoalexin phaseollin in relation to delimitation at sites of infection of *Phaseolus vulgaris* by *Collectotrichum lindemuthinum. Can. J. Bot. 51*:2423–2430.

Robbins, M. P., Bkolwell, G. P., and Dixon, R. A. (1985). Metabolic changes in elicitor-treated bean cells. Selectivity of enzyme induction in relation to Phytoalexin accumulation. *Eur. J. Biochem. 148*:563–569.

Roseman, D., Heller, W., and Sandermann, H. (1991). *Plant Physiol. 97*:1280–1286.

Rudloff, E., and Jorgensen, E. (1963). The biosynthesis of pinosylvin in the sapwood of *Pinus resinosa* AIT. *Phytochemistry 2*:297–304.

Rupprich, N., Hildebrand, H., and Kindl, H. (1980). Substrate specificity in vivo and in vitro in the formation of stilbenes, biosynthesis of rhaponticin. *Arch. Biochem. Biophys. 200*:72–78.

Ryder, T. B., Cramer, C. C., Bell, J. N., Robbins, M. P., Dixon, R. A., and Lamb, C. J. (1984). Elicitor rapidly induces chalcone synthase mRNA in *Phaseolus vulgaris* cells at the onset of the phytoalexin defense response. *Proc. Nl. Acad. Sci. USA 81*:5724–5728.

Schmelzer, E., Som sich, I., and Hahlbarock, K. (1985). Coordinated changes in transcription and translation rates of phenylalanine ammonia-lyase and 4-coumarate:CoA ligase mRNAs in elicitor-treated *Petroselenium crispum* cells. *Plant Cell Rep. 4*:293–296.

Schöppner, A., and Kindl, H. (1979). Stilbene synthetase (pinosylvine synthase) and its induction by ultraviolet light. *Fed. Eur. Biochem Soc. Lett. 108*(2):349–352.

Schöppner, A., and Kindl, H. (1984). Purification and properties of a stilbene synthase from induced cell suspension cultures of peanut. *J. Biol. Chem. 259*: 6806–6811.

Schröder, G., Brown, J. W. S., and Schröder, J. (1988). *Eur. J. Biochem. 172*:161–169.

Schwekendiek, A., Pfeiffer, G., and Kindl, H. (1992). Pine stilbene synthase of cDNA, a tool for probing environmental stress. *Fed. Eur. Biochem. Soc. 301*: 41–44.

Smith, D. A. (1978). Observations on the fungitoxicity of kievitone. *Phytopathology 48*:81–87.

Smith, D. A. (1982). Toxicity of phytoalexins. *Phytoalexins* (J. A. Bailey and J. W. Mansfield, eds.), Blackie, London, pp. 218–252.

Smith, D. A., and Banks, S. W. (1986). Biosynthesis, elicitation and biological activity of isoflavonoid phytoalexins. *Phytochemistry 25*(5):979–995.

Smith, D. A., and Ingham, J. L. (1980). Legumes, Fungal Pathogens, and Phytoalexins. *Advances in Legume Sciences* (R. J. Summerfield, and A. H. Bonting, eds.), HMSO, London, pp. 207–222.

Smith, D. A., Kuhn, P. J., Bailey, J. A., and Burden, R. S. (1980). Detoxification of phaseollidin by *Fusarium solani* f.sp. Phaseoli. *Phytochemistry 19*:1673–1675.

Smith, D. A., Wheeler, H. E., Banks, S. W., and Cleveland, T. E. (1984). Association between lowered kievitone hydratase activity and reduced virulence to bean in variants of *Fusarium solani* f. sp. phaseoli. *Physiol. Plant Pathol. 25*:135–147.

Smith, J. M. (1971). *Physiological Plant Pathol. 1*:85.

Stoessl, A. (1982). Biosynthesis of Phytoalexins. *Phytoalexins* (J. A. Bailey and J. W. Mansfield, eds.), Blackie, London, pp. 218–252.

Sun, Y., Wu, Q., Van Etten, H. D., and Hrazdina, G. (1991). Stereoisomerism in plant disease resistance: induction and isolation of 7,2′-dihydroxy-4′5′-methylenedioxyisoflavone. *Arch. Biochem. Biophys. 284*:167–173.

Sunder, R., Rangaswami, S., and Reddy, G. C. (1978). A new dihydrophenanthrene from *Dioscorea decipiens. Phytochemistry 17*:1037.

Tahara, S., Ingham, J., Nakahara, S., Mizutami, J., and Harborne, J. B. (1984). Fungitoxic dihydrofuranisoflavones and related compounds in white lupin, *Lupinus albus. Phytochemistry 23*:1889–1990.

Takasugi, M., Kawashima, S., Monde, K., Katsui, N., Masamune, T., and Shirata, A. (1987). Antifungal compounds from *Dioscorea batatas* inoculated with *Pseudomonas cichorii. Phytochemistry 26*(2):371–375.

Tani, T., and Mayama, A. (1982). *Plant Infection: The Physiological and Biochemical Basis.* Japan Science Society Press, Tokyo.

Van Etten, H. D. (1967). Antifungal activity of pterocarpans and other selected isoflavonoids. *Phytochemistry 15*:655–659.

Van Etten, H. D., Matthews, D. E., and Smith, D. A. (1982). Metabolism of phytoalexins. *Phytoalexins* (J. A. Bailey and J. W. Mansfield, eds.), Blackie, London, pp. 181–217.

Van Etten, H. D., Matthews, D. E., Matthews, P., Miao, V., Maloney, A., and Straney, D. (1989). A family of genes for phytoalexin detoxification in the plant pathogen *Nectria haematococca. Signal Molecules in Plants and Plant Microbe Interactions* (B. J. J. Lugtenberg, ed.), NATO ASI Series H36, pp. 219–228.

Verdecourt, B. (1970). Studies in the Leguminosae Papilionoideae for the flora of tropical East Africa: IV. *Kew Bull. 24*:507–570.

Viorin, B. (1983). UV spectral differentiation of 5-hydroxy and 5-hydroxy-3-methoxy flavones with mono-(4′), di-(3′,4′) or tri(3′,4′,5′) substituted B-rings. *Phytochemistry 22*:2107–2145.

Welle, R., and Grisebach, H. (1991). Properties and solubilization of phenyltransferase of isoflavonoid phytoalexins biosynthesis in soybeans. *Phytochemistry 30*(2):479–484.

Weltring, K. M., Mackenbrock, K., and Barz, W. (1982). Demethylation, methylation and 3′-hydroxylation of isoflavones by *Fusarium* fungi. *Z. Naturforschung. 37c*:570–574.

Wengenmayer, H., Ebel, J., and Grisebach, H. (1974). *Eur. J. Biochem. 50*:135.

Whitehead, I. M., Dey, P. M., and Dixon, R. A. (1982). *Planta 154*:156.

Willeke, U., and Barz, W. (1982). Catabolism of 5-hydroxyisoflavones by fungi of the genus *Fusarium. Arch. Microbiol. 132*:266–269.

Woodward, M. D. (1979a). Phaseoluteone and other 5-hydroxyisoflavonoids from *Phaseolus vulgaris. Phytochemistry 18*:363–365.

Woodward, M. D. (1979b). New isoflavonoids related to kievitone from *Phaseolus vulgaris. Phytochemistry 18*:2007–2010.

Woodward, M. D. (1980). Phaseollin formation and metabolism in *Phaseolus vulgaris. Phytochemistry 19*:921–927.

Yoshikawa, M. (1978). Diverse modes of action of biotic and abiotic phytoalexin elicitors. *Nature 275*:546.

16

Involvement of Phytoalexins in the Response of Phosphonate-Treated Plants to Infection by *Phytophthora* Species

P. Saindrenan
Institut de Biologie Moléculaire des Plantes, C.N.R.S., Strasbourg, France

David I. Guest
University of Melbourne, Parkville, Victoria, Australia

I. INTRODUCTION

The rapid and localized pathogen-induced synthesis of phytoalexins is thought to play an important defensive rôle in the early response of many plants to infection by potential parasites (Dixon and Lamb, 1990). In susceptible hosts, the growth of biotrophic fungal pathogens is unrestricted, and a compatible host–parasite interaction may develop given favorable environmental conditions. Responses usually associated with resistance are not entirely absent, but rather are not expressed soon enough or with sufficient magnitude at the infection site to be effective (Kuć and Rush, 1985). Presumably the necessary elicitor signals are either not released in sufficient magnitude or are not recognized sufficiently rapidly by the plant.

In terms of strategies for plant disease protection, the ability to transform a compatible interaction into an incompatible interaction by modifying the pathogenic ability of the parasite (Kiraly et al., 1972) or the defense response of the host (Langcake, 1981) offers many exciting possibilities. A number of chemicals appear to have this ability. The antipenetrant tricyclazole acts by reducing the pathogenic competence of the parasite (Woloshuk et al., 1980). Probenazole (Sekizawa and Mase, 1981), 2,2-dichloro-3,3-dimethylcyclopropane carboxylic acid (Langcake et al., 1983), and methyl-2,6-dichloroisonicotinic acid (Métraux et al., 1991) alter the host physiology so that disease resistance is increased.

Phosphonates, including the phosphonate anion ($H_2PO_3^-$) and fosetyl-Al (aluminum tris-*O*-ethyl phosphonate, Aliette), are systemic compounds that control diseases caused by oomycetes, particularly *Phytophth-*

ora species (Bertrand et al., 1977; Pegg et al., 1985; Cohen and Coffey, 1986). Compared to other systemic fungicides, they exert little activity on mycelial growth in vitro at the submillimolar concentrations usually found in vivo in protected plants (Bompeix et al., 1981; Guest, 1984a). Conversely, their effects in vitro frequently do not match their performance in vivo. This paradox has been explained by a hypothesis proposing a complex mode of action, involving the active participation of the host defense reactions, mainly phytoalexin accumulation (Bompeix et al., 1981; Guest, 1984a). However, at higher application rates, and with certain host–parasite interactions, a direct mode of action is more plausible (Fenn and Coffey, 1984; Dercks and Buchenauer, 1986; Ye and Deverall, 1986).

Bompeix and coworkers (Bompeix et al., 1980; Vo-Thi et al., 1979) made the original observation that tomato leaves produced more phenolic compounds following inoculation with *P. capsici* when the leaves were floated on a solution of fosetyl-Al than they did when floated on distilled water. The response was associated with the formation of necrotic blocking zones and hypersensitive cell death (Durand and Sallé, 1981) similar to that observed in the incompatible interaction between *Cladosporium fulvum* and tomato (Lazarovits and Higgins, 1975a,b). Although the involvement of phenylpropanoid metabolism in the resistance response of tomato has not yet been clearly demonstrated (Kuć, 1982), it was suggested that enhanced production of phenolic compounds might be a component of the resistance mechanism of phosphonate-treated leaves. These observations were subsequently confirmed in other plant–pathogen interactions (Guest, 1984b; Khan et al., 1986; Saindrenan et al., 1988a,b; Afek and Sztejnberg, 1989; Smillie et al., 1989; Nemestothy and Guest, 1990).

In this chapter, we will discuss the effects of phosphonates (disodium phosphonate salt, $Na_2HPO_3.5H_2O$, and fosetyl-Al) on the stimulation of the plant's defense reactions via phytoalexin accumulation in two well-defined model systems, cowpea (*Vigna unguiculata* (L.) Walp. cv. Tvu 645)–*Phytophthora cryptogea* and tobacco (*Nicotiana tabacum* L.)–*Phytophthora nicotianae* var. *nicotianae*. Some other plant–pathogen interactions will be described briefly. Indeed, it has been shown that the protection given by submillimolar concentrations of phosphonates is higher in plants having highly active dynamic defense systems (Guest and Grant, 1991). Moreover, experimental treatments that suppress phytoalexin accumulation provide useful clues to assess the role of phytoalexins and other defense reactions in the mode of action of phosphonates. A range of biosynthetic inhibitors have been used to identify the metabolic pathways that are important to resistance and to the effect of phosphonate on these pathways.

II. PLANT–PATHOGEN INTERACTIONS

A. The Cowpea–Phytophthora Cryptogea Interaction

Phytoalexins are the best candidates as biochemical markers for the resistance of cowpea to fungal parasites (Lampard, 1974; Preston, 1975; Preston et al., 1975; Partridge and Keen, 1976). Cowpea phytoalexins are isoflavonoids, with the exception of the 2-arylbenzofuran phytoalexin, vignafuran (Preston et al., 1975). Bioassays on detached leaves of the susceptible cultivar Tvu 645 were used to determine whether local concentrations of phytoalexins at the infection site in compatible treated leaves reach inhibitory levels by the time the pathogen is inhibited. Inoculation of untreated leaves produced light brown expanding lesions and hyphae had ramified throughout the leaf tissues 48 hr after inoculation (Fig. 1). In leaves floated on buffered phosphonate solution (2.44 mM), dark brown necrotic lesions

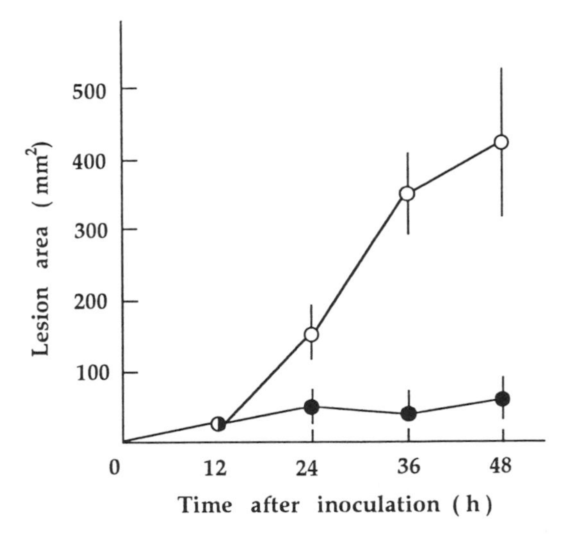

Figure 1 Lesion areas caused by *Phytophthora cryptogea* in cowpea leaves of the susceptible cultivar TVu645 untreated (○) or treated (●) with phosphonate (2.44 mM). Data are representative of three separate experiments; bars represent ±SE; 12 infected sites were observed in each experiment. The primary leaves were detached, inoculated on their adaxial side with a mycelial plug, and floated on a solution containing 40 mM MES buffer adjusted to pH 6.5 and 50 mg/liter benzimidazole with or without phosphonate. The assessment of the extent of infection was done after clearing leaf discs in chloroform–ethanol–acetic acid (30:60:10) and staining the tissues in a 0.002% Rose Bengal solution.

developed. Fungal growth ceased about 24 hr after inoculation. In leaves floating on buffer alone, small amounts of kievitone and phaseollidin were produced, while no vignafuran was detected (Fig. 2). As infection proceeded, the concentration of both phytoalexins decreased. In phosphonate-treated leaves, larger quantities of phytoalexins were produced in two different patterns of accumulation. Kievitone accumulation reached a maximum (134.5 μg/g fresh wt) 24 hr after inoculation, decreasing afterward. Phaseollidin and vignafuran accumulated gradually to levels of 160 μg/g fresh wt and 9 μg/g fresh wt, respectively. These different patterns of the appearance of the 5-hydroxylated isoflavone kievitone and the 5-deoxyisoflavonoid phaseollidin reflects a typical feature of induction of isoflavonoid phytoalexins in legumes (Whitehead et al., 1979; Robbins et al., 1985). Cessation of fungal growth coincides with levels of kievitone which exceed the ED_{90} value (100 μg/ml) for fungal growth (Saindrenan et al., 1988a), so that the involvement of kievitone in the restriction of *P. cryptogea* is likely. Phaseollidin reaches inhibitory concentrations (ED_{90} = 130 μg/ml) at a later stage of infection, while vignafuran, which has been reported to be induced specifically in cowpea leaves infected with *Colletotrichum lindemuthianum* (Preston et al., 1975), does not accumulate to a sufficient level in treated leaves and cannot on its own be involved in the restriction of fungal growth. Furthermore, cowpea leaves do not appear to produce the cyclo derivative of phaseollidin, phaseollin, in response to infection by *P. cryptogea*. This contrasts with cowpea hypocotyls infected by tobacco necrosis virus (Bailey, 1973).

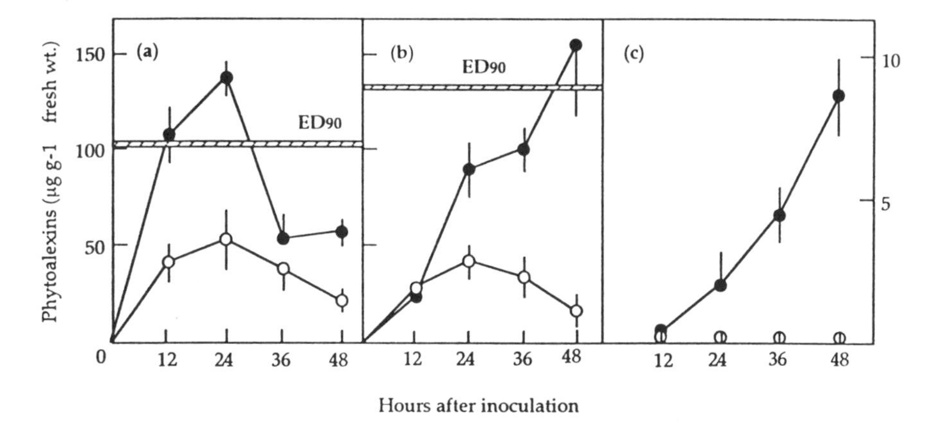

Figure 2 Time-course accumulation of kievitone (a), phaseollidin (b), and vignafuran (c) in untreated (○ and treated (●) cowpea leaves following inoculation with *P. cryptogea*. Hatched bars represent ED_{90} values of phytoalexins determined in vitro by radial growth bioassays.

From these results it was clear that phosphonate treatment enhanced the rate and magnitude of phytoalexin accumulation in the necrotic cells of the lesions. Whether a causal relationship existed between enhanced host defenses and disease protection still remained an open question. The first answers came from studies showing that the levels of phosphonate in lesions where the pathogen was completely restricted were only sufficient to inhibit mycelial growth by 25% in a medium containing a phosphate level similar to that found in infected tissues (Saindrenan et al., 1988a). Phosphate, by its structural analogy with phosphonate, has been shown to interfere with the activity of phosphonates in planta (Bompeix et al., 1980; Smillie et al., 1989). Moreover, the infection did not create a metabolic sink leading to the accumulation of the toxophore, which might then have caused a direct inhibition of the fungus. Indeed, although phosphonate is poorly inhibitory to *P. cryptogea* in vitro, especially in nonlimiting phosphate conditions (Bompeix and Saindrenan, 1984), as it is the case during the infection court in cowpea leaves where phosphate content is about 7 mM (Saindrenan et al., 1988a), accumulation of the oxyanion at the site of infection might be expected.

A second line of evidence was presented using inhibitors of biosynthetic pathways of phytoalexins. α-Aminooxyacetate (AOA) is a competitive inhibitor of phenylalanine ammonia-lyase (PAL; EC 4.3.1.5) (Amrhein et al., 1976), which is thought to play an important role in the regulation of isoflavonoid phytoalexin biosynthesis (Dixon et al., 1983). It was shown that 5 mM AOA annulled the effect of phosphonate in the detached leaf bioassay and that leaves exhibited symptoms of complete susceptibility (Saindrenan et al., 1988b). Using tritiated phosphonic acid, Fenn and Coffey (1985) argued that the aminooxy compound might interfere with the uptake of phosphonate by *P. capsici*, thus reducing the efficacy of the toxophore in tomato-treated plants. However, no such effects were found in either the mycelium of *P. cryptogea* or cowpea leaf (Saindrenan et al., 1988b). It was subsequently shown that under the experimental conditions used by Fenn and Coffey (1985) an exchange of the phosphorus-bound tritium of phosphonic acid with the hydrogen of water occurred, making it uncertain as to what was really measured (Bompeix et al., 1989). In a very simple way, Bompeix (1989) established unequivocally that AOA did not affect the activity of phosphonate against three different species of *Phytophthora* in axenic culture. Because the reversal effect of AOA was not as a result of the inhibition of phosphonate uptake, it might be expected that its effect was through the modification of host defense reactions. In vitro experiments showed that AOA inhibited PAL extracted from cowpea leaves and the inhibition closely paralleled the increase in susceptibility of leaves to *P. cryptogea* (Saindrenan et al., 1988b). Five millimolar AOA reduces

Table 1 Influence of α-Aminooxyacetate (AOA) on Phosphonate-Induced Accumulation of Phytoalexins and Levels of Extractable Phenylalanine Ammonia-lyase (PAL) in Infected Cowpea Leaves

Treatment[a]	Phytoalexins (μg/g fresh wt)[b]		PAL activity[c] (nmol cinnamate h^{-1} mg^{-1} protein)
	Kievitone	Phaseollidin	
Control	40.9 ± 13.9	75.1 ± 12.2	45.2 ± 15.5
Control + AOA (5 mM)	12.3 ± 8.4	7.3 ± 4.4	19.5 ± 2.4
Phosphonate (2.44 mM)	96.1 ± 24.5	90.1 ± 18.1	92.9 ± 20.9
Phosphonate (2.44 mM) + AOA (5 mM)	24.4 ± 6.2	10.4 ± 5.5	35.1 ± 16.4

[a]Leaves were floated for 4 hr in darkness on control or 5 mM AOA solutions before inoculation. The leaves were then inoculated and placed again on the same medium but with 2.44 mM phosphonate.
[b]Phytoalexin contents of infected area were determined 24 hr after inoculation. Data are the mean of two separate experiments ± SD.
[c]PAL activity was assayed radiometrically. Data are the mean of two separate experiments ± SD.
Source: From Saindrenan et al. (1988b).

kievitone and phaseollidin accumulation in phosphonate-treated leaves as well as in untreated leaves (Table 1). These results indicate that the decrease of phytoalexin accumulation may be caused by inhibition of PAL activity in vivo. Likewise, the reduction in phytoalexin content coincides with a decrease in PAL activity. Indeed, AOA does not cause an increase in extractable PAL activity as reported for another inhibitor L-α-aminooxy-β-phenylpropionic acid (AOPP) (Amrhein and Gerhardt, 1979), but rather a substantial decrease which may be the consequence of the inhibition of transaminase by AOA (John et al., 1978), resulting in lowered PAL synthesis (Havir, 1981). Alternatively, AOA may remain tightly bound to the enzyme during extraction (Hoagland and Duke, 1982).

The effect of AOA in reducing phytoalexin content is positively correlated with the increase in susceptibility of cowpea leaves. These results clearly demonstrate the active participation of the host defense reactions in the mode of action of phosphonate and, particularly in this model system, of the phenylpropanoids which play an important role in the resistance of some legumes to pathogens.

The primary target of phosphonate seems to be the pathogen. Indeed, it was shown in some cases that the in vivo performance of the compound paralleled its in vitro activity (Dolan and Coffey, 1988). Moreover, phosphonate was shown to enhance the capacity of *P. cryptogea* to overproduce molecules with elicitor activity in culture filtrates (Saindrenan et al., 1990) and therefore to alter indirectly the outcome of the plant–pathogen interaction through a modification of fungal pathogenicity. Barchietto et al. (1992) suggested that phosphonate may interact with phosphate for the catalytic site of phosphorylating enzymes, as well as for the regulation of several enzymes, inducing a physiological state similar to that induced by a limitation of phosphate, thus leading to the overproduction of elicitor compounds.

B. The Tobacco–*Phytophthora nicotianae* var. *nicotianae* Interaction

Resistant and susceptible cultivars of tobacco differ in the magnitude and timing of the phytoalexin response following infection by *P. nicotianae* var. *nicotianae* (Nemestothy and Guest, 1990). Sesquiterpenoid phytoalexins accumulate earlier and faster in the resistant cultivar (NC 2326) than they do in the susceptible cultivar (Hicks) (Fig. 3). Inhibition of sesquiterpenoid biosynthesis by the specific inhibitor of HMG CoA reductase, mevinolin

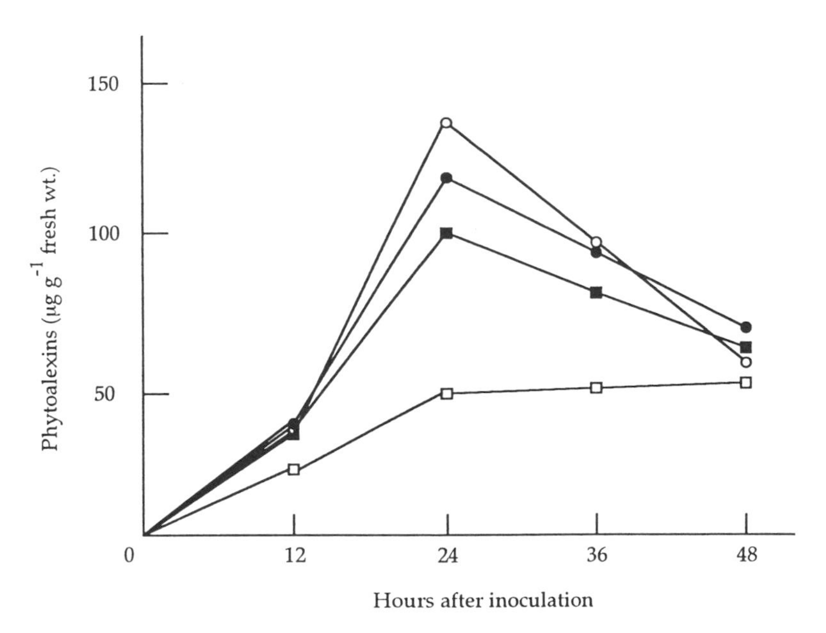

Figure 3 Dynamics of total sesquiterpenoid phytoalexin (capsidiol, phytuberin, rishitin, and phytuberol) accumulation in tobacco stems inoculated with zoospores of *Phytophthora nicotianae* var. *nicotianae*: cv. "Hicks" (□); cv. "Hicks" + 10 mg/plant fosetyl-Al (■); cv. "NC 2326" (○) and cv. "NC 2326" + 10 mg/plant fosetyl-Al (●).

Table 2 Effect of Fosetyl-Al (10 mg/plant) and Inhibitors on the Total Sesquiterpenoid Phytoalexin Accumulation in Tobacco Stems (mg/g fresh wt.) 24 hr After Inoculation with Zoospores of *Phytophthora nicotianae* var. *nicotianae*

Cultivar treatment	"Hicks"	"Hicks" + fosetyl	"NC 2326"	"NC 2326" + fosetyl
Control	50	107	134	114
Inhibitor:				
Mevinolin	9	8	6	8
AOA	46	48	52	60
AHPP	66	78	108	121
AVG	65	101	120	115
SED (6 replicates) = 6				

AOA, d-aminooxyacetate; AHPP, L-α-aminohydrazino-β-phenylpropionic acid; AVG, aminoethoxyvinylglycine.
Source: From Nemestothy and Guest (1990).

(Chappell and Nable, 1987), and the nonspecific inhibitor or aromatic biosynthesis, AOA (Amrhein et al., 1976; John et al., 1978) inhibits phytoalexin accumulation and increases susceptibility in a leaf disc bioassay of the resistant cultivar (Tables 2 and 3) (Nemestothy and Guest, 1990). These results provide strong circumstantial evidence that sesquiterpenoid phytoalexin biosynthesis and accumulation is involved in the expression of race-specific resistance of tobacco to this pathogen. Specific inhibition of PAL

Table 3 Lesion Diameters (mm) 60 hr After Inoculation with *Phytophthora nicotianae* var. *nicotianae* of Leaf Discs Floated on 40 mM MES Buffer Containing 1 mg/liter Benzimidazole, and, Where Indicated, Fosetyl-Al (282 μM), Mevinolin (10 μM), AOA (100 μM), AHPP (100 μM), or AVG (30 μM)[a]

Cultivar treatment	"Hicks"	"Hicks" + fosetyl	"NC 2326"	"NC 2326" + fosetyl
Control	27	12	16	4
Inhibitor:				
Mevinolin	28	25	35	12
AOA	29	24	29	14
AHPP	22	14	18	5
AVG	22	11	15	4
POx	25	25	32	16
SED (6 replicates) = 2				

[a]Propylene oxide (POx) treated discs were exposed to 1 ml/liter for 1 hr before inoculation.

by L-α-aminohydrazino-β-phenylpropionic acid (AHPP; Munier and Bompeix, 1985) or ethylene by aminoethoxyvinylglycine (AVG; Paradies et al., 1980) did not affect the susceptibility or resistance of leaf discs (Tables 2 and 3), suggesting at most a minor role for phenylpropanoids, such as lignin, or for ethylene, in the early stages of resistance of tobacco to this pathogen (Nemestothy and Guest, 1990).

We also demonstrated that, coinciding with its ability to protect a susceptible cultivar against the black shank pathogen, the phosphonate compound fosetyl-Al enhances sesquiterpenoid phytoalexin accumulation (Table 2) and other responses following inoculation (Guest, 1986; Guest et al., 1989; Nemestothy and Guest, 1990). Two pieces of evidence suggest a causal link between enhanced host defenses and disease protection. First, mevinolin and AOA inhibit sesquiterpenoid phytoalexin accumulation (Table 2) and reduce the efficacy of fosetyl-Al (Table 3) in the susceptible cultivar. AOA would also reduce lignin deposition (Hammerschmidt, 1984). Second, at the rates applied and at the phosphate concentrations present, the measured concentration of fosetyl-Al or its toxophore, phosphonate, at the infection site is only partially inhibitory to mycelial growth in vitro (Smillie et al., 1989).

Our results support the hypothesis that in this and some other host-parasite interactions, enhanced speed and magnitude of phytoalexin accumulation is involved, along with a direct obstruction of normal growth of the pathogen, in the complex mode of action of phosphonates (Saindrenan et al., 1988a; Smillie et al., 1989; Guest and Bompeix, 1990; Guest and Grant, 1991). The relative importance of the "direct" and "indirect" components of the mode of action varies with the characteristics of the particular host–parasite pair studied, the application rates, and possibly environmental factors (Guest and Grant, 1991). Our hypothesis is supported by the evidence that fosetyl-Al had no additive effect on the resistance, or on the phytoalexin, PAL, or ethylene responses of the resistant cultivar, "NC 2326." When phytoalexin biosynthesis was blocked in the resistant cultivar, lesion growth was still partially inhibited in fosetyl-Al–treated leaf discs, indicating that factors other than phytoalexins are involved in the total resistant response in this cultivar. There is little significant direct effect of phosphonate on fungal growth in planta, because there was no residual inhibition of lesion growth in leaf discs of cv. "Hicks" taken from fosetyl-Al–treated plants following treatment with mevinolin, or in leaf discs from treated plants subsequently killed by exposure to propylene oxide (Nemestothy and Guest, 1990).

A primary effect of phosphonates on the pathogen may be to interfere with its ability to elude phytoalexin elicitation (Guest and Bompeix, 1984), perhaps through alterations to water-soluble cell wall components of *Phy-*

tophthora (Dunstan et al., 1990). Many other elements of the resistant response may be triggered by a phosphonate-affected pathogen in a resistant host. By disturbing the pathogen's recognition–avoidance mechanisms, fosetyl-Al could enable the activation of an "idle capacity" in the susceptible cultivar, supporting the hypothesis that all plants have the potential to express resistance (Kuć, 1983).

Phosphonates enhance plant responses normally associated with stress and disease resistance. The suppression of sesquiterpenoid phytoalexin biosynthesis and, to a lesser degree, lignin deposition reduces the activity of phosphonates. We propose that rapid phytoalexin accumulation, and possibly lignin deposition, is involved in the mode of action of phosphonates against this pathogen in tobacco, in combination with, and following, the direct disturbance of the pathogen. Identification of the precise site(s) of metabolic disturbance caused by phosphonates may provide some insight into the regulation of the host–parasite interaction.

C. Other Interactions

Fosetyl-Al was shown to control *Phytophthora parasitica* infections of capsicum fruits (Guest, 1984b). The control was associated with enhanced hypersensitivity and increased capsidiol accumulation similar to that found in genetically incompatible interaction. The *Citrus–P. citrophthora* interaction is also well documented (Khan and Ravisé, 1985; Afek and Sztejnberg, 1986; Khan et al., 1986). It was found that low concentrations of fosetyl-Al and phosphonate act against *P. citrophthora* by increasing defense mechanisms against the pathogen and accumulation of the 6,7-dimethoxycoumarin phytoalexin, scoparone, while higher concentrations of phosphonates act directly as a fungistat (Afek and Sztejnberg, 1989). Dercks and Buchenauer's (1986) study of the strawberry–*P. fragariae* and lettuce–*Bremia lactucae* interactions led them to claim that increases in phenolic compounds in plants treated with fosetyl-Al were of secondary importance in disease control. However, the high application rates of fosetyl-Al used by these authors may explain the restriction of fungal growth via a direct action of the compound on the pathogen, possibly through the toxicity of the aluminum ion (Muchovej et al., 1980). Smillie et al. (1989) demonstrated that in lupin, which lacks a highly active dynamic defense system, phosphonate, even at high concentrations in the plant, is unable to stop the growth of *P. cinnamomi*.

III. EPILOG

The evidence that phytoalexins contribute to plant disease resistance comes from two types of investigations that provide either circumstantial and correlative evidence or experimental evidence.

There is a large body of circumstantial evidence that resistant plants accumulate more phytoalexins at the site of attempted infection, faster than susceptible plants (Dixon, 1986; Preisig and Kuć, 1987; Nemestothy and Guest, 1990). The most convincing evidence comes from studies of near isogenic cultivars of the same species such as soybeans, cowpea, or tobacco.

There are now several lines of experimental evidence that phytoalexins are central to disease resistance in some plants. Van Etten et al. (1989) demonstrated an absolute correlation between virulence and phytoalexin detoxification by the necrotrophic *Nectria haematococca*. They were able to transform avirulent fungi with the pisatin demethylase gene, which also conferred virulence.

Treatments that inhibit phytoalexin biosynthesis or accumulation, such as those described in this chapter, increase the susceptibility of the plant to disease (Moesta and Grisebach, 1982; Massala et al., 1987; Saindrenan et al., 1988b; Nemestothy and Guest, 1990). Conversely, treatments that enhance phytoalexin accumulation increase disease resistance (Ward, 1984; Saindrenan et al., 1988a; Nemestothy and Guest, 1990). A model for the mode of action of the phosphonates has been proposed (Guest and Grant, 1991; Barchietto et al., 1992). In this model, sublethal concentrations of phosphonate, such as those commonly encountered in treated plants, induce a stress physiology in oomycetes. One consequence of this is that factors that normally suppress the host recognition system are not produced, or fail to function, and that elicitors are either overproduced or become more active. The net result is that the plant recognizes the pathogen as an incompatible invader and invokes its normal array of defense responses as a consequence.

These experiments open several avenues of investigation. First, it will be instructive to learn the precise mechanism by which phosphonates disturb the virulence mechanisms of *Phytophthora* spp. The loose selectivity of phosphonates for the oomycetes suggest a distinctive metabolism, probably phosphate metabolism. The fact that interference with this metabolism affects the virulence of these pathogens suggests that further exploration may provide information about the virulence factors of these organisms, and thus the factors that control phytoalexin accumulation in the host. It may also open new avenues of disease control based on the exploitation of preexisting but dormant mechanisms.

Despite the accumulated evidence, phytoalexins are not a universal resistance mechanism, nor should they be viewed in isolation from other plant disease resistance mechanisms, with which they certainly act in synergy. Phytoalexins have not been found in some plants which have well-described disease resistance mechanisms, such as wheat or cucumber. Although there is often a strong link between gene-for-gene resistance and rapid phytoalexin accumulation, no mechanism for this linkage has been described. There has been no report of a gene encoding rapid phytoalexin accumulation, nor has there been a report of a "phytoalexin-minus" mutant plant that is consequently susceptible to disease.

REFERENCES

Afek, U., and Sztejnberg, A. (1986). A *Citrus* phytoalexin, 6,7-dimethoxycoumarin, as a defense mechanism against *Phytophthora citrophthora*, and the influence of Aliette and phosphorous acid on its production. *Phytoparasitica 14*:26.

Afek, U., and Sztejnberg, A. (1989). Effects of fosetyl-Al and phosphorous acid on scoparone, a phytoalexin associated with resistance of *Citrus* to *Phytophthora citrophthora*. *Phytopathology 79*:736–739.

Amrhein, N., and Gerhardt, J. (1979). Superinduction of phenylalanine ammonia lyase in gherkin hypocotyls caused by the inhibitor, L-α-aminooxy-β-phenylpropionic acid. *Biochimica and Biophysica Acta 583*:434–442.

Amrhein, N., Gödeke, K. H., and Kefeli, V. I. (1976). The estimation of relative intracellular phenylalanine ammonialyase (PAL) activities and the modulation in vivo and in vitro by competitive inhibitors. *Berichte des Deutschen Botanischen Gesellshaft 89*:247–259.

Bailey, J. A. (1973). Production of antifungal compounds in cowpea (*Vigna sinensis*) and pea (*Pisum sativum*) after virus infection. *J. Gen. Microbiol. 75*:119–125.

Barchietto, T., Saindrenan, P., and Bompeix, G. (1992). Physiological responses of *Phytophthora citrophthora* to a subinhibitory concentration of phosphonate. *Pest. Biochem. Physiol. 42*:151–166.

Bertrand, A., Ducret, J., Debourge, J. C., and Horrière, D. (1977). Etude des propriétés d'une nouvelle famille de fongicides: les monoéthyl phosphites métalliques. Caractéristiques physicochimiques et propriétés biologiques. *Phytiâtrie-Phytopharmacie 26*:3–18.

Bompeix, G., and Saindrenan, P. (1984). In vitro antifungal activity of fosetyl-Al and phosphorous acid on *Phytophthora* species. *Fruits 39*:777–786.

Bompeix, G., Ravisé, A., Raynal, G., Fettouche, F., and Durand, M-C. (1980). Modalités de l'obtention des nécroses bloquantes sur feuilles détachées de tomate par l'action du tris-O-ethylphosphonate d'aluminium (phoséthyl d'aluminium), hypothèses sur son mode d'action in vivo. *Annales de Phytopathologie 12*:337–351.

Bompeix, G., Fettouche, F., and Saindrenan, P. (1981). Mode d'action du phoséthyl-Al. *Phytiâtrie-Phytopharmacie 30*:257–272.

Bompeix, G. (1989). Fongicides et relations plantes-parasites: cas des phosphonates. *Comptes Rendus de l'Académie d'Agriculture 6*:183–189.

Bompeix, G., Burgada, R., d'Arcy-Lameta, A., and Soulié, M-C. (1989). Phosphonate hydrogen exchange in aqueous solutions of ^{2}H and ^{3}H-labelled phosphonic acid. *Pest. Sci. 25*:171–174.

Chappell, J., and Nable, R. (1987). Induction of sesquiterpenoid biosynthesis in tobacco cell suspension cultures by fungal elicitor. *Plant Physiol. 85*:469–473.

Cohen, Y., and Coffey, M. D. (1986). Systemic fungicides and the control of oomycetes. *Ann. Rev. Phytopathol. 24*:311–338.

Dercks, W., and Buchenauer, H. (1986). Investigations on the influence of aluminium ethyl phosphonate on plant phenolic metabolism in the pathogen–host interactions, *Phytophthora fragariae*-strawberry and *Bremia lactucae*-lettuce. *J. Phytopathol. 115*:37–55.

Dixon, R. A. (1986). The phytoalexin response: elicitation, signalling and control of host gene expression. *Biol. Rev. 61*:239–291.

Dixon, R. A., and Lamb, C. J. (1990). Molecular communication in interactions between plants and microbial pathogens. *Ann. Rev. Plant Physiol. Plant Mol. Biol. 41*:339–367.

Dixon, R. A., Dey, P. M., and Lamb, C. J. (1983). Phytoalexins: enzymology and molecular biology. *Adv. Enzymol. Rel. Areas Mol. Biol. 55D*:1–135.

Dolan, T. E., and Coffey, M. D. (1988). Correlative in vitro and in vivo behaviour of mutant strains of *Phytophthora palmivora* expressing different resistances to phosphorous acid and fosetyl-Na. *Phytopathology 78*:974–978.

Dunstan, R. H., Smillie, R. H., and Grant, B. R. (1990). The effects of sub-toxic levels of phosphonate on the metabolism and potential virulence factors of *Phytophthora palmivora. Physiol. Mol. Plant Pathol. 36*:205–220.

Durand, M.-C., and Sallé, G. (1981). Effet du tris-O-éthyl phosphonate d'aluminium sur le couple *Lycopersicum esculentum* Mill.-*Phytophthora capsici* Leon. Etude cytologique et cytochimique. *Agronomie 1*:723–731.

Fenn, M. E., and Coffey, M. D. (1984). Studies on the in vitro and in vivo antifungal activity of fosetyl-Al and phosphorous acid. *Phytopathology 74*:606–611.

Fenn, M. E., and Coffey, M. D. (1985). Further evidence for the direct mode of action of fosetyl-Al and phosphorous acid. *Phytopathology 75*:1064–1068.

Guest, D. I. (1984a). The influence of cultural factors on the direct antifungal activities of fosetyl-Al, Propamocarb, Metalaxyl, SN75196 and Dowco 444. *Phytopathol. Z. 111*:155–164.

Guest, D. I. (1984b). Modification of defense responses in tobacco and capsicum following treatment with fosetyl-Al. *Physiol. Plant Pathol. 25*:125–134.

Guest, D. I. (1986). Evidence from light microscopy of living tissues that fosetyl-Al modifies the defence response in tobacco seedlings following inoculation with *Phytophthora nicotianae* var. *nicotianae. Physiol. Mol. Plant Pathol. 29*:151–161.

Guest, D. I., and Bompeix, G. B. (1984). Fosetyl-Al as a tool in understanding the resistant response in plants. *Phytophthora Newslett. 12*:62–69.

Guest, D. I., and Bompeix, G. (1990). The complex mode of action of phosphonates. *Aust. Plant Pathol. 19*:113–115.

Guest, D. I., and Grant, B. R. (1990). The complex action of phosphonates in plants. *Biol. Rev. 66*:159–187.

Guest, D. I., Upton, J. C. R., and Rowan, K. S. (1988). Fosetyl-Al alters the respiratory response in *Phytophthora nicotianae* var. *nicotianae*-infected tobacco. *Physiol. Mol. Plant Pathol. 34*:257–265.

Hammerschmidt, R. (1984). Rapid deposition of lignin in potato tuber tissue as a response to fungi non-pathogenic on potato. *Physiol. Plant Pathol. 24*:33–42.

Havir, E. A. (1981). Modification of L-phenylalanine ammonia lyase in soybean cell suspension cultures by 2-aminooxyacetate and L-2-aminooxy-3-phenylpropionate. *Planta 152*:124–130.

Havir, E. A., and Hanson, K. R. (1968). Phenylalanine ammonia lyase. I. Purification and molecular size of the enzyme from potato tubers. *Biochemistry 7*: 1896–1899.

Hoagland, R. E., and Duke, S. E. (1982). Effects of glyphosate on metabolism of phenolic compounds. VIII. Comparison of the effects of aminooxyacetate and glyphosate. *Plant Cell Physiol. 23*:1081–1088.

John, R. A., Charteris, A., and Fowler, L. J. (1978). The reaction of aminooxyacetate with pyridoxal phosphate-dependent enzymes. *Biochem. J. 171*:771–779.

Khan, A. J., and Ravisé, A. (1985). Stimulation of defence reactions in Citrus by fosetyl-Al and fungal elicitors against *Phytophthora* spp. *Br. Crop Prot. Council Monogr. 31*:281–284.

Khan, A. J., Vernenghi, A., and Ravisé, A. (1986). Incidence of fosetyl-Al and elicitors on the defence reactions of Citrus attacked by *Phytophthora* spp. *Fruits 41*:587–595.

Kiraly, Z., Barna, B., and Ersek, T. (1972). Hypersensitivity as a consequence, not the cause, of plant resistance to infection. *Nature 239*:456–458.

Kuć, J. (1982). Phytoalexins from the Solanaceae. In *Phytoalexins* (J. A. Bailey and J. W. Mansfield, eds.), Blackie, London, pp. 81–105.

Kuć, J. (1983). Induced systemic resistance of plants to disease caused by fungi and bacteria. *The Dynamics of Host Defence*, (J. A. Bailey and B. J. Deverall, eds.), Academic Press, Sydney, pp. 191–221.

Kuć, J., and Rush, J. S. (1985). Phytoalexins. *Arch. Biochem. Biophys. 236*:455–472.

Lampard, J. F. (1974). Demethylhomopterocarpin: an antifungal compound in *Canavalia ensiformis* and *Vigna unguiculata* following infection. *Phytochemistry 13*:291–292.

Langcake, P. (1981). Alternative chemical agents for controlling plant disease. *Phil. Trans. R. Soc. London B 295*:83–101.

Langcake, P., Cartwright, D. W., and Ride, J. P. (1983). The dichlorocyclopropanes and other fungicides with an indirect mode of action. *Systemische Fungizide und Antifungale Verbindungen* (H. Lyr and C. Polter, eds.), Akademie-Verlag, Berlin, pp. 199–210.

Larazovits, G., and Higgins, V. J. (1975a). Histological comparison of *Cladosporium fulvum* race 1 on immune, resistant and susceptible tomato varieties. *Can. J. Bot. 54*:224–234.

Larazovits, G., and Higgins, V. J. (1975b). Ultrastructure of susceptible, resistant and immune reactions of tomato to races of *Cladosporium fulvum*. *Can. J. Bot. 54*:235–249.

Legrand, M., Fritig, B., and Hirth, L. (1976). Enzymes of the phenylpropanoid pathway and the necrotic reaction of hypersensitive tobacco to tobacco mosaic virus. *Phytochemistry 15*:1353–1359.

Massala, R., Legrand, M., and Fritig, B. (1987). Comparative effects of two competitive inhibitors of phenylalanine ammonia lyase on the hypersensitive resistance of tobacco to tobacco mosaic virus. *Plant Physiol. Biochem. 25*:217–225.

Métraux, J. P., Ahl-Goy, P., Staub, T., Speich, J., Steinemann, A., Ryals, J., and Ward, E. (1991). Induced systemic resistance in cucumber in response to 2,6-dichloro-isonicotinic acid and pathogens. *Advances in Molecular Genetics of Plant–Microbe Interactions*, Vol. 1 (H. Hennecke and D. P. S. Verma, eds.), Kluwer, Dordrecht, pp. 432–439.

Moesta, P., and Grisebach, H. (1982). L-2-Aminooxy-3-phenylpropionic acid inhibits phytoalexin accumulation in soybean with concomitant loss of resistance against *Phytophthora megasperma* f. sp. *glycinea*. *Physiol. Plant Pathol. 21*: 65–70.

Munier, R. L., and Bompeix, G. (1985). Inhibition de la L-phenylalanine ammonia lyase de *Rhodotorula glutinis* par des phenylalanines N-substituées. *Comptes Rendus de l'Académie des Sciences Paris 300*:203–206.

Muchovej, J. J., Maffia, L. A., and Muchovej, R. M. C. (1980). Effects of exchangeable soil aluminium and alkaline calcium salts on the pathogenicity and growth of *Phytophthora capsici* from green pepper. *Phytopathology 70*:1212–1214.

Nemestothy, G. N., and Guest, D. I. (1990). Phytoalexin accumulation, phenylalanine ammonia lyase activity and ethylene biosynthesis in fosetyl-Al treated resistant and susceptible tobacco cultivars infected with *Phytophthora nicotianae* var. *nicotianae*. *Physiol. Mol. Plant Pathol. 37*:207–219.

Paradies, I., Konze, J. R., Elstner, E. F., and Paxton, J. D. (1980). Ethylene: Indicator but not inducer of phytoalexin synthesis in soybean. *Plant Physiol. 66*:1106–1109.

Partridge, J. E., and Keen, N. T. (1976). Association of the phytoalexin kievitone with single-gene resistance of cowpeas to *Phytophthora vignae*. *Phytopathology 66*:426–429.

Pegg, K. G., Whiley, A. W., Saranah, J. B., and Glass, R. J. (1985). Control of Phytophthora root rot of avocado with phosphorous acid. *Aust. Plant Pathol. 14*:25–29.

Preisig, C. L., and Kuć, J. A. (1987). Phytoalexins, elicitors, enhancers, suppressors and other considerations in the regulation of R-gene resistance to *Phytophthora infestans* in potato. *Molecular Determinants of Plant Disease* (S. Nishimura et al., eds.), Japan Science Society Press/Springer-Verlag, pp. 203–221.

Preston, N. W. (1975). 2′-O-methoxyphaseollidinisoflavan from infected tissue of *Vigna unguiculata*. *Phytochemistry 14*:1131–1132.

Preston, N. W., Chamberlain, K., and Skipp, R. A. (1975). A 2-arylbenzofuran phytoalexin from cowpea (*Vigna unguiculata*). *Phytochemistry 14*:1843–1844.

Robbins, M. P., Bolwell, G. P., and Dixon, R. A. (1985). Metabolic changes in elicitor-treated bean cells. Selectivity of enzyme induction in relation to phytoalexin accumulation. *Eur. J. Biochem. 148*:563–569.

Saindrenan, P., Barchietto, T., Avelino, J., and Bompeix, G. (1988a). Effects of phosphite on phytoalexin accumulation in leaves of cowpea infected with *Phytophthora cryptogea*. *Physiol. Mol. Plant Pathol. 32*:425–435.

Saindrenan, P., Barchietto, T., and Bompeix, G. (1988b). Modification of the phosphite induced resistance response in leaves of cowpea infected with *Phytophthora cryptogea* by α-aminooxyacetate. *Plant Sci. 58*:245–252.

Saindrenan, P., Barchietto, T., and Bompeix, G. (1990). Effects of phosphonate on the elicitor activity of culture filtrates of *Phytophthora cryptogea* in *Vigna unguiculata*. *Plant Sci. 67*:245–251.

Sekizawa, Y., and Mase, S. (1981). Mode of controlling action of probenazole against rice blast disease with reference to the induced resistance mechanism in rice. *J. Pest. Sci. 6*:91–94.

Smillie, R. H., Grant, B. R., and Guest, D. I. (1989). The mode of action of the fungicide phosphite: evidence for both direct and indirect modes of action in plants. *Phytopathology 79*:921–926.

van Etten, H. D., Matthews, D. E., and Matthews, P. S. (1989). Phytoalexin detoxification: Importance for pathogenicity and practical implications. *Annu. Rev. Phytopathol. 27*:143–164.

Vo-Thi, H., Bompeix, G., and Ravisé, A. (1979). Rôle du tris-O-éthylphosphonate d'aluminium dans la stimulation des réactions de défense des tissus de tomate contre le *Phytophthora capsici*. *Comptes Rendus de l'Académie des Science Paris 288*:1171–1174.

Ward, E. W. B. (1984). Suppression of metalaxyl activity by glyphosate: evidence that host defence mechanisms contribute to metalaxyl inhibition of *Phytophthora megasperma* f. sp. *glycinea* in soybeans. *Physiol. Plant Pathol. 25*:381–386.

Whitehead, I. M., Dey, P. M., and Dixon, R. A. (1979). Differential patterns of phytoalexin accumulation and enzyme induction in wounded and elicitor-treated tissues of *Phaseolus vulgaris*. *Planta 154*:156–164.

Woloshuk, C. P., Sisler, H., Tokousbalides, M. C., and Dutky, S. R. (1980). Melanin biosynthesis in *Pyricularia oryzae*: site of tricyclazole inhibition and pathogenicity of melanin-deficient mutants. *Pest. Biochem. Physiol. 14*:256–264.

Ye, X. S., and Deverall, B. J. (1986). Effects of Aliette on *Brema lactucae* on lettuce. *Trans. Br. Mycol. Soc. 86*:597–602.

17

Phytoalexins in Forage Legumes

Studies on Detoxification by Pathogens and the Role of Glycosidic Precursors in Roots

V. J. Higgins and J. Hollands
University of Toronto, Toronto, Ontario, Canada

Dallas K. Bates
Michigan Technological University, Houghton, Michigan

I. INTRODUCTION

Until recently, our research on phytoalexins concentrated on various aspects of the role of the phytoalexins medicarpin (3-hydroxy-9-methoxypterocarpan)(**1**), or medicarpin and maackiain (3-hydroxy-8,9-methylenedioxypterocarpan(**2**), in foliage of alfalfa and in foliage and roots of red clover, respectively. With foliage, the initial objective was to determine whether phytoalexins might explain differences in host specificity and also whether they might be involved in limiting lesion size in leaf spot diseases. The work with roots arose initially out of our curiosity about the origins and relationships of phytoalexins in the roots as compared with the leaves. Unfortunately, because of our dissatisfaction with the genetics of these crops, our interest gradually turned to leaf mold of tomato leaving many interesting questions about these forage legume systems unanswered; however, in this chapter, we take the opportunity to briefly review some of that work and to present some as-yet unpublished results.

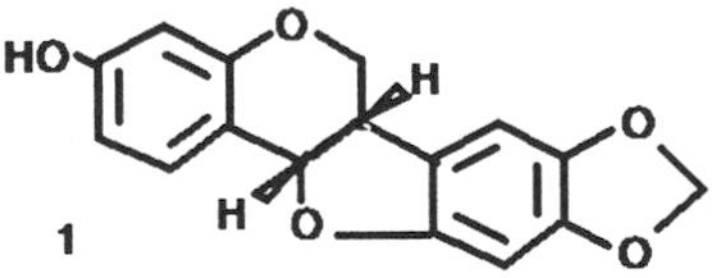

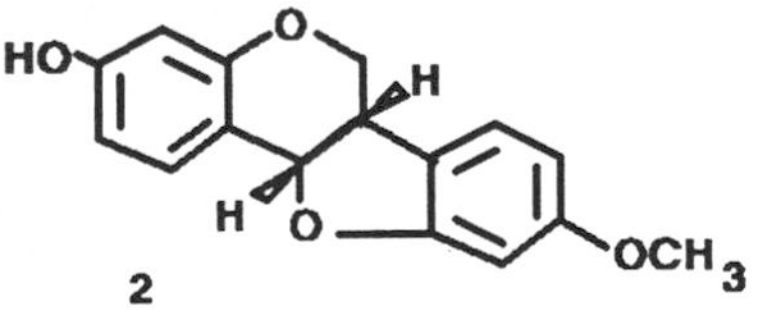

The general picture that emerged initially from work with alfalfa was that pathogens of alfalfa generally could rapidly convert medicarpin to nontoxic components (Higgins and Millar, 1969; Higgins, 1972) whereas nonpathogens lacked that ability or produced a derivative that was still toxic (Higgins and Millar, 1970). Further work with the alfalfa pathogen *Stemphylium botryosum* showed that it could also metabolize the phytoalexin maackiain from red clover, a host on which this fungus is only weakly virulent (Duczek and Higgins, 1976a), and pisatin and phaseollin from peas and beans, two nonhosts of *S. botryosum* (Heath and Higgins, 1973). Although on peas and beans the slower rate of phytoalexin metabolism appeared to allow enough of the compounds to accumulate to account for the inhibition of *S. botryosum*, this was not the case on red clover. Thus, medicarpin and maackiain did not appear from this work to be involved in the low virulence of this fungus on red clover. Interestingly, phaseollin, maackiain, and medicarpin were all reduced to their respective isoflavans (Higgins et al., 1974; Higgins, 1975) by a system, which at least for maackiain, could be induced by any of the three compounds.

As pointed out by Van Etten et al. (1982, 1989), any means of interfering with the pathogen's ability to detoxify phytoalexins could affect the pathogenicity of the pathogen. In this regard, we tested S-containing pterocarpans for their ability to inhibit the reduction of maackiain to dihydromaackian (DHMaa) (7,2′-dihydroxy-4′,5′-methylenedioxyisoflavan)(**3**).

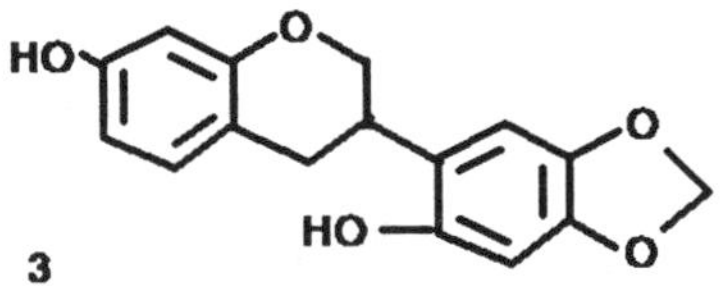

Our hypothesis was that reduction of the S-containing molecule by the pterocarpan oxidoreductase would create an -SH derivative which might bind irreversibly to the enzyme and thus act as a "suicide substrate" (Walsh, 1983). Alternatively, even if the S-containing pterocarpans simply proved to be good competitive inhibitors, they could be useful tools for studying the importance of phytoalexin detoxification in planta. As well, knowledge of such potential enzyme inhibitors might be of potential commercial value in disease control. The choice of this system was influenced, not only by the fact that we had a ready source of maackiain as the result of a gift of trifolirhizin from Albert Stoessl (formerly Agricultural Canada Research Institute, London, Ontario), but also because using the drop diffusate method with this fungus on clover leaflets would be a convenient means of eventually testing inhibition in planta. Some results from this project are discussed below.

An intriguing aspect of maackiain production in red clover is the fact that the glycoside of maackiain, trifolirhizin, occurs in roots in relatively high concentration yet no more than 1–2 $\mu g/g$ fresh wt of tissue was found in foliage (Duczek and Higgins, 1976a). Previously, we reported (McMurchy and Higgins, 1984) that trifolirhizin appeared to be a source of the phytoalexin maackiain in red clover roots infected with *Fusarium roseum*. The production of pterocarpanoid phytoalexins from preformed glycosides or other isoflavone conjugates has recently received renewed attention, particularly in soybean (Graham et al., 1990; Morris et al., 1991). Below are summarized results of a study designed not only to clarify the source of maackiain in red clover roots but to investigate the presence of a comparable glycoside of medicarpin.

II. METHODOLOGY

Fungal Metabolism of Phytoalexins by *Stemphylium botryosum*: Inhibition by S-Containing Pterocarpanoid

The chemicals tested as inhibitors of maackiain conversion to DHMaa were synthesized by D. K. Bates, D. A. Hay, and J. A. Fleming (Michigan Technological University, Houghton, Michigan) and were derivatives of 6a,11a-dihydrobenzo[3,2-c][1]benzopyrans (11a-thioptercarpans) including 11-thiapterocarpan (TP) (**4**), 3-methoxy-11-thiapterocarpan (3-OMe-TP) (**5**), 3-hydroxy-11-thiapterocarpan (3-OH-TP) (**6**), and dihydrobenzothiophene (DHBTP) (**7**).

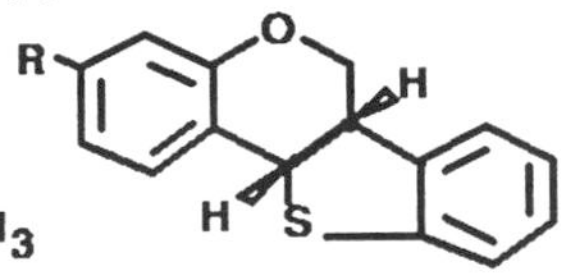

4) R = H
5) R = OCH_3
6) R = OH

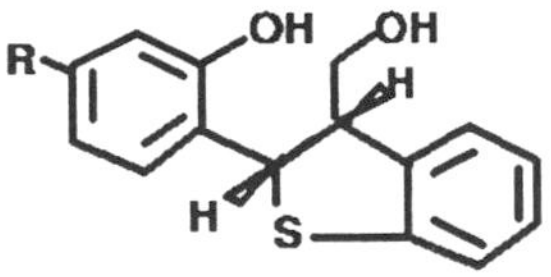

7

Maackiain conversion assays were basically carried out as in Higgins (1975). Maackiain, DHMaa, TP, and 3-OMe-TP were analyzed by high-performance liquid chromatopraphy (HPLC) using a C_{18} column and 70% acetonitrile in water. Bioassays of each compound were made on germ tube growth as described in Duczek and Higgins (1976b).

Phytoalexin Production from Preformed Glycosides

Maackiain and medicarpin were measured by HPLC as above, or by UV absorbance of compounds eluted from thin layer chromatography plates as described in McMurchy and Higgins (1984). Maackiain and medicarpin were removed from root extracts by partitioning with CCl_4, which does not remove the glycosides; the extracts were then treated with a β-glucosidase and the maackiain and medicarpin released by hydrolysis of the glycosides were extracted and measured.

III. METABOLISM OF PHYTOALEXINS BY *STEMPHYLIUM BOTRYOSUM*: INHIBITION BY S-CONTAINING PTEROCARPANOIDS

As all inhibitor tests were to be done in situ with germinated spores, this required that the S-pterocarpan derivatives tested have no or very little toxicity to fungal growth; this requirement eliminated 3-OH-TP and DHBTP which were as toxic as maackiain (Fig. 1). Fortunately, TP and 3-OMe-TP were less toxic (Fig. 1) and the toxic effect did not change as the concentration increased.

The effect of TP, the compound chosen for detailed study, on the conversion of maackiain (at 10 μg/ml) by germinated conidia of *S. botryosum* was monitored over an 8-hr period during which about 50% loss of maackiain normally occurred in untreated controls (Fig. 2). Treatment of conidia with TP generally reduced the maackiain loss to about 20–30% but again this effect was not significantly concentration-dependent (Fig. 2). Maackiain loss over the 8-hr period occurred at a relatively constant rate for untreated controls (Fig. 3) with DHMaa detectable by 2 hr. TP-treated conidia showed a less uniform loss of maackiain (Fig. 3) and

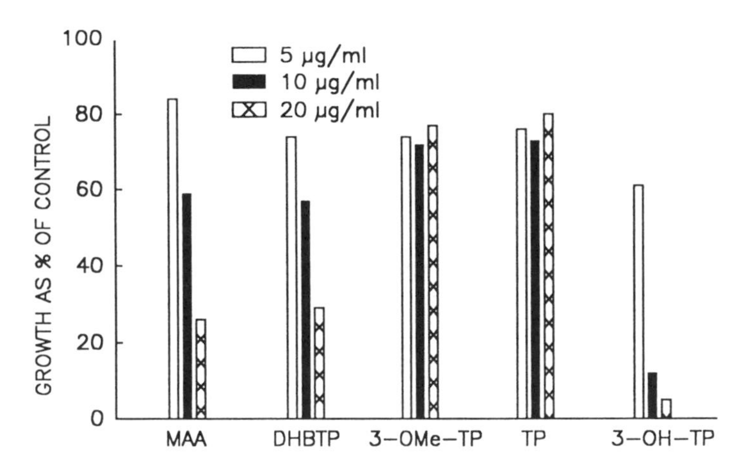

Figure 1 Effect of maackiain (Maa), dihydrobenzothiophene (DHBTP), 3-methoxy-11-thiapterocarpan (3-OMe-TP), 11-thiapterocarpan (TP), and 3-hydroxy-11-thiapterocarpan (3-OH-TP) on germ tube growth of *Stemphylium botryosum* conidia pregerminated before treatment with 5, 10, or 20 μg/ml of each compound. Germ tubes (n = 100) were measured after a 4-hr incubation and the average increase in growth is presented as a percentage of the untreated control.

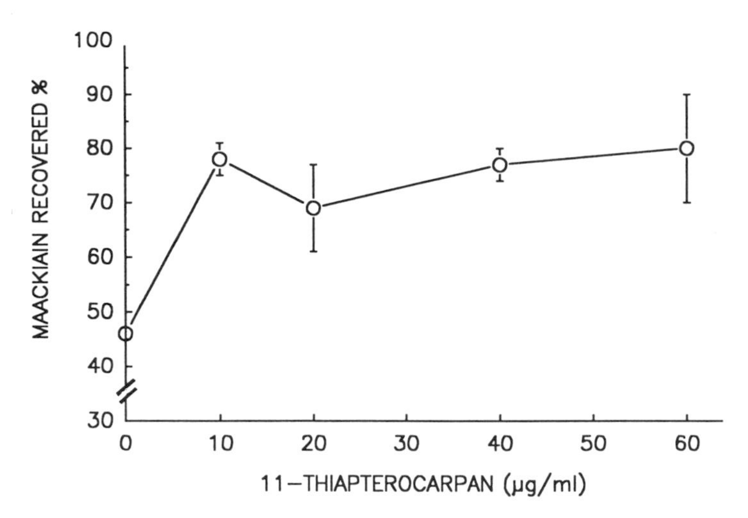

Figure 2 Effect of increasing concentrations of 11-thiapterocarpan (TP) on the conversion of maackiain to dihydromaackiain by *Stemphylium botryosum* after 8 hr incubation with 10 μg/ml of maackiain. Recovery of maackiain is shown as a percentage of that recovered from conidial suspensions immediately after maackiain addition (zero time).

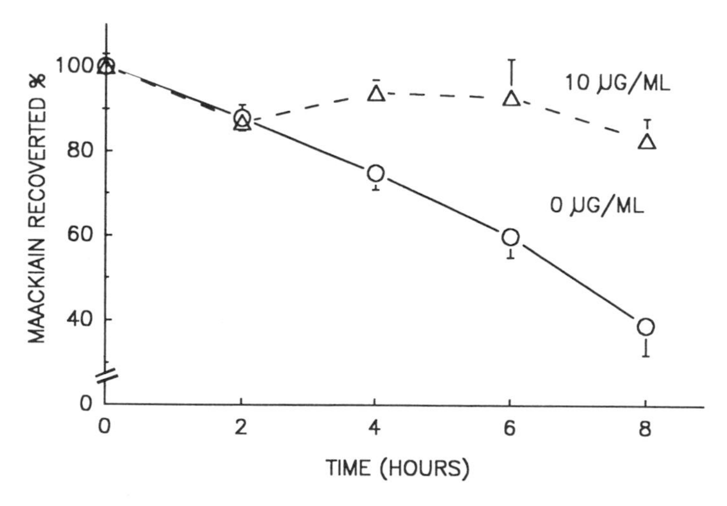

Figure 3 Kinetics of maackiain (added to give 10 μg/ml) conversion to dihydromaackiain by *Stemphylium botryosum* during an 8 hr period in the presence or absence of 11-thiapterocarpan (TP) at 10 μg/ml. Recovery of maackiain is shown as a percentage of the maackiain recovered immediately after maackiain addition (zero time).

DHMaa was first detected at 4 hr. (Other experiments indicated that DHMaa is further metabolized by this fungus; thus the kinetics of DHMaa concentration could not be used to monitor the rate of maackiain conversion.) About 90% of added TP was recovered from cultures throughout the incubation period indicating that TP was not metabolized by the fungus and that it was not, or at least not irreversibly, changed during the interaction with the enzyme. Thus, it does not appear to act as a typical suicide substrate.

In an in situ experiment, such as the above, it is impossible to differentiate between inhibition of enzyme synthesis vs. enzyme activity. One approach to answering this question took advantage of our knowledge about the inducibility of the *S. botryosum*-maackiain conversion system which is assumed to be due to induction of the synthesis of an oxidoreductase (Higgins, 1975). Maackiain (to give 10 μg/ml) or an equivalent volume of solvent was added to pregerminated spores which were incubated for a further 17 hr before the addition of more maackiain (to give 20 μg/ml) with or without TP. If TP inhibits enzyme synthesis rather than activity, the effect of TP on maackiain conversion should be decreased. Inhibition of maackiain by TP still occurred in induced conidia (Fig. 4) suggesting that enzyme

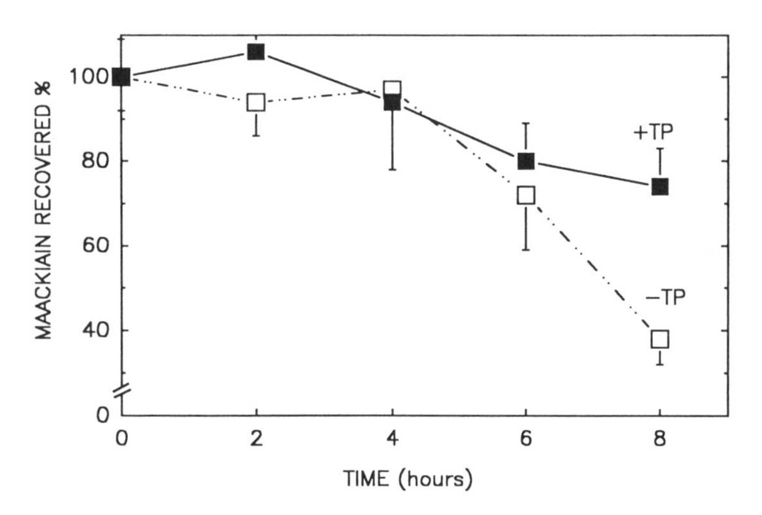

Figure 4 Effect of preinducing *Stemphylium botryosum* with maackiain on the ability of 11-thiapterocarpan (TP) to inhibit the conversion of maackiain to dihydromaackiain. Pregerminated conidia were incubated with maackiain (10 μg/ml) for 8 hr before the addition of more maackiain (20 μg/ml) and TP. Recovery of maackiain is shown as a percentage of the maackiain recovered immediately after the addition of the second maackiain treatment (zero time as graphed).

activity is affected; however, interpretation of these data are complicated by an array of factors including the inhibitory effects of both compounds on growth. Confirmation of a direct interaction of TP with the enzyme must await purification and characterization of the enzyme. A soluble constitutively produced pterocarpan:NADPH oxidoreductase isolated from *Ascochyta rabiei* has been shown to reduce both medicarpin and maackiain to their respective isoflavans (Hohl and Barz, 1987; Hohl et al., 1989). The enzyme from *S. botryosum*, although induced, is probably very similar in activity and merits further investigation.

Despite our lack of understanding of the mode of inhibition, some preliminary tests were made to determine if TP was effective in planta. Maackiain and medicarpin usually do not accumulate beyond a few μg/ml when red clover leaves are inoculated with *S. botryosum*, presumably because of the rapid metabolism of these compounds. Preliminary experiments to test the effect of TP on maackiain and medicarpin accumulation on *S. botryosum*-inoculated clover leaflets indicated that TP was not effective in planta, i.e., the levels of maackiain and medicarpin in drop diffusates containing TP and the fungus were not markedly different from the combined total for either alone. Such data must be treated with caution, however, because of the possibility that TP might also inhibit enzymes involved in the final steps of pterocarpanoid synthesis. One such candidate enzyme is NADPH:isoflavone oxidoreductase which has been isolated from several plants including chickpea (Tiemann et al., 1991). In situ inhibition of a plant enzyme could be explored using abiotic elicitors or fungi that do not metabolize maackiain.

IV. PHYTOALEXIN PRODUCTION FROM PREFORMED GLYCOSIDES

Challenge of roots of axenically grown red clover seedlings by the abiotic elicitor $CuCl_2$ was used to clarify the role of trifolirhizin, and a comparable glycoside of medicarpin, in the production of maackiain and medicarpin, respectively. The kinetics (Fig. 5A) of the loss of trifolirhizin and increase in maackiain in the 24-hr period following $CuCl_2$ treatment were comparable to those seen (McMurchy and Higgins, 1984) following inoculation with *F. roseum*. Although a glycoside of medicarpin was not isolated and characterized, techniques similar to those used to monitor trifolirhizin levels indicated the presence of such a glycoside. The kinetics (Fig. 5B) of glycoside loss and appearance of medicarpin following $CuCl_2$ treatment were very similar to those for trifolirhizin and maackiain.

To further test if maackiain and medicarpin were originating from their respective glycosides as a result of β-glucosidase activity in the injured tissue, as suggested by the kinetic data, glyphosate, an inhibitor of the

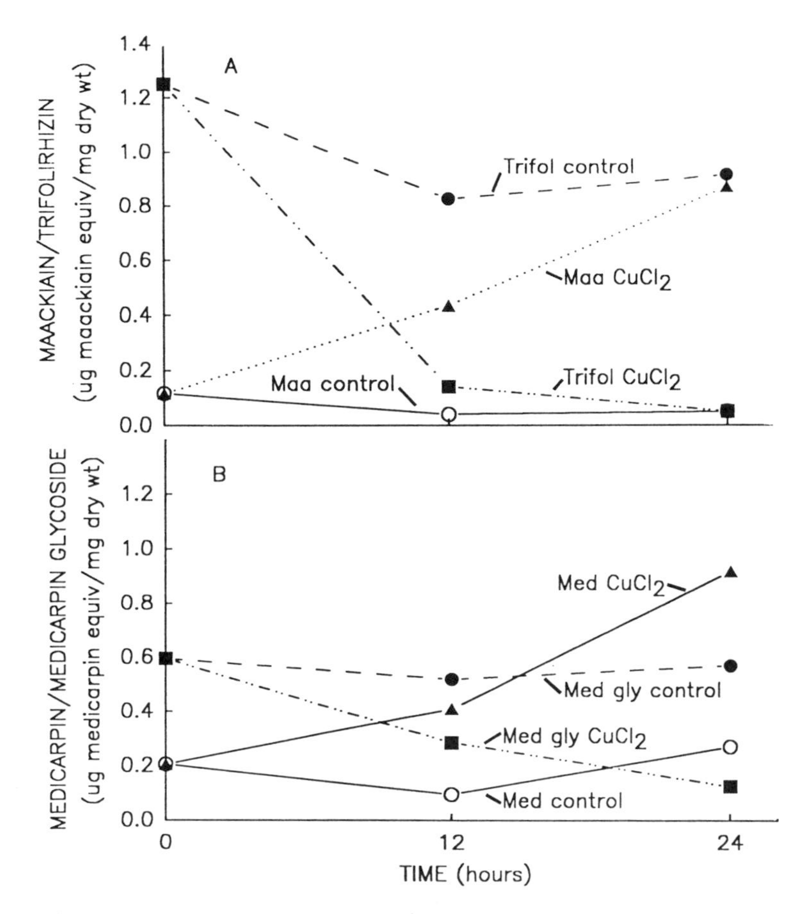

Figure 5 (A) Changes in trifolirhizin (measured as the aglycone maackiain) and maackiain concentration in red clover roots treated with $CuCl_2$ (10^{-3} M) and incubated 12 or 24 hr. (B) Comparable changes in a glycoside (Med gly) of medicarpin (measured as the aglycone) and medicarpin in the same roots.

shikimate pathway (Hollander and Amrhein, 1980), was used on $CuCl_2$-treated roots. If maackiain and medicarpin are formed by de novo synthesis, then treatments with glyphosate should show reduced amounts of these compounds by 24 hr. The data (Fig. 6A,B) for both maackiain and medicarpin indicate that the presence of glyphosate at 10 μg/ml did not affect the relative glycoside/phytoalexin ratio normally seen after $CuCl_2$ treatment. This suggests that the contribution of de novo synthesis to phytoalexin synthesis in the initial 24-hr period is at most very limited. Similar experiments with the phenylalanine ammonia lyase (PAL) inhibitor aminooxyace-

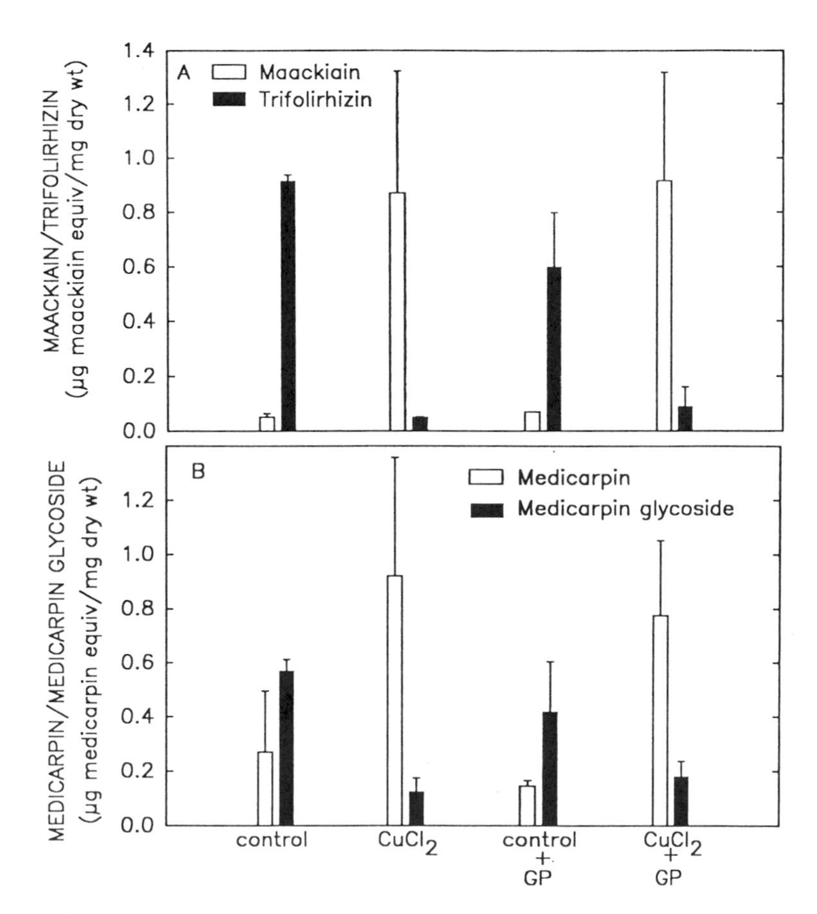

Figure 6 (A) Effect of the glyphosate (GP) (10 μg/ml) on $CuCl_2$-induced changes in the concentration of trifolirhizin (measured as the aglycone maackiain) and maackiain in red clover roots. (B) Comparable effects of glyphosate on concentrations of medicarpin glycoside and medicarpin in the same roots.

tic acid (AOA), in which only trifolirhizin/maackiain levels were monitored, gave similar results.

Interestingly, and although difficult to substantiate statistically because of the high variation in glycoside levels in unelicited plants, the level of glycoside in AOA- or glyphosate-treated controls appeared to decrease over the 24-hr incubation period. This suggests that there may be a constant turnover of glycoside and that in the presence of an inhibitor such as AOA or glyphosate this loss is not replaced and results in a diminishing pool of glycosides. Johal and Rahe (1990), after showing a strong positive correlation between the effects of glyphosate on loss of resistance of bean to *Colletotrichum lindemuthianum* and loss of phytoalexin production, expressed concerns about the possible effects on plant health of residues of preplant herbicides containing glyphosate. For plants such as red clover, in which glycosides appear to be an important initial source of phytoalexins following pathogen attack, such residues might also result in depletion of the glycoside pool and further reduce the resistance of the crop to root rot pathogens.

Increasing evidence shows that the role of phytoalexin glycosides in resistance is not unique to clover. The studies of Graham and coworkers (1990) suggest that in all organs of soybean seedlings glyceollin biosynthesis following infection by *Phytophthora* is not solely dependent on the induction of enzymes of the early phenylpropanoid pathway; instead glyceollin is partly produced from daidzein released by hydrolysis of daidzein conjugates. In contrast, in soybean leaves challenged with *Phytophthora*, such glycoside precursors did not appear to play a role in biosynthesis of glyceollin (Morris et al., 1991). More intriguing is the recent work of Mackenbrock and Barz (1991) who showed that maackiain and medicarpin are partially (30%) produced de novo by chickpea suspension cells despite high levels of the potential precursor formonetin 7-*O*-glycoside-6″-*O*-malonate (FGM); in contrast, cells treated with the PAL inhibitor L-α-aminooxy-β-phenylpropionic acid (AOPP) totally utilized FGM to produce these same phytoalexins. (Interestingly, the pool size of FGM, although not discussed by the authors, appeared to decline gradually in the presence of AOPP as would be expected if some turnover normally occurs as discussed above.) Obviously, the role of preformed precursors in the biosynthesis of phytoalexins deserves more study but the results can be expected to vary from plant to plant, organ to organ, and pathogen to pathogen.

V. EPILOG

Some of our past questions regarding the biosynthesis, role, and detoxification of phytoalexins will soon be answered by the use of molecular techniques. There is no doubt that antisense technology will be used to experimentally prevent phytoalexin detoxification by pathogens. This will be an important means of clarifying the importance of detoxification in those many host–pathogen systems which are not amenable to the types of genetic analysis used by Van Etten and coworkers for *Nectria* (see Van Etten et al., 1989). Nonetheless, commercial exploitation of antisense technology related to detoxification mechanisms in the fungus seems unlikely as does the use of chemical inhibitors of such enzymes. The latter are unlikely to be so specific that they will not also affect the biosynthetic enzymes used by the plant in the final stages of phytoalexin synthesis.

Also, our increasing knowledge of the enzymes involved in the final biosynthetic steps of phytoalexin biosynthesis will allow prevention of the final reactions via antisense techniques. Understanding the role of pools of glycosidic precursors in phytoalexin production will be very important in such experiments as the appropriate antisense gene may block the development of such pools. If these glycosidic compounds have a function other than defense, phenotypic changes may be detected in the transgenic plants. Most experiments which have relied on blocking PAL or other enzymes occurring early in the biosynthetic pathway of pterocarpanoid phytoalexins have generally recognized that many other plant functions rely on the same pathway, but have failed to adequately consider the effects of preformed precursors on the experiments.

Increasingly, as bioengineering of plants and fungi becomes more routine, we need to investigate the relative contribution of each of the plant's defense components to the final disease interaction and to understand the interactions between components. For example, it is possible that, in at least some plants, phytoalexins change the growth of the fungus just enough that the hyphal tips become amenable to total inhibition by the activity of chitinase and/or glucanases. Indeed, the effect of phytoalexins on fungal development is poorly researched. Duczek and Higgins (1976b) showed that maackiain and medicarpin inhibited formation of appressoria by *Cochliobolus* (*Helminthosporium*) *carbonum*; although appressorium formation is a developmental response that usually proceeds phytoalexin production, comparable developmental changes, e.g., haustorium formation in the rusts, that occur after penetration might be similarly affected.

Our reliance on in vitro bioassays of hyphal growth and our failure to reproduce the gradual buildup of phytoalexins that occurs in plants may prove to have mislead us about their relative importance in resistance. Similarly, our failure to treat plants with gradually increasing amounts of fungal elicitors have undoubtedly created misleading data on the role of many elicitors, e.g., in tomato, a cell wall-derived (nonspecific) elicitor from *Cladosporium fulvum* appears to be degraded by enzymes produced by infected or stressed plants (Peever and Higgins, 1989). As well, the elicitor activity is inhibited by other low molecular weight components of the apoplast (Lu and Higgins, unpublished). Under normal conditions of infection by *C. fulvum*, it is assumed that these "suppressors" build up at about the same rate as the nonspecific elicitor, so that nonspecific elicitation of the defense response never occurs. How many other elicitors under active study are likewise "artifacts" of our assay systems? Molecular dissections of defense responses will eventually reveal how much we have been led astray by our failure to properly mimic the pathogen and plant in our experimentation.

REFERENCES

Duczek, L. J., and Higgins, V. J. (1976a). The role of medicarpin and maackiain in the response of red clover leaves to *Helminthosporium carbonum*, *Stemphylium botryosum*, and *S. sarcinaeforme*. *Can. J. Bot.* *54*:2609–2619.

Duczek, L. J., and Higgins, V. J. (1976b). Effect of treatments with the phytoalexins medicarpin and maackiain on fungal growth in vitro and in vivo. *Can. J. Bot.* *54*:2610–2619.

Graham, T. L., Kim, J. E., and Graham, M. Y. (1990). Role of constitutive isoflavone conjugates in the accumulation of glyceollin in soybean infected with *Phytophthora megasperma*. *Mol. Plant–Microbe Interact.* *3*:157–166.

Heath, M. C., and Higgins, V. J. (1973). In vitro and in vivo conversion of phaseollin and pisatin by an alfalfa pathogen *Stemphylium botryosum*. *Physiol. Plant Pathol.* *3*:107–120.

Higgins, V. J. (1972). Role of the phytoalexin medicarpin in three leaf spot diseases of alfalfa. *Physiol. Plant Pathol.* *2*:289–300.

Higgins, V. J. (1975). Induced conversion of the phytoalexin maackiain by the alfalfa pathogen *Stemphylium botryosum*. *Physiol. Plant Pathol.* *6*:5–18.

Higgins, V. J., and Millar, R. L. (1969). Comparative ability of *Stemphylium botryosum* and *Helminthosporium turcicum* to induce and degrade a phytoalexin from alfalfa. *Phytopathology* *59*:1493–1499.

Higgins, V. J., and Millar. R. L. (1970). Degradation of alfalfa phytoalexin by *Stemphylium loti* and *Colletotrichum phomoides*. *Phytopathology* *60*:269–271.

Higgins, V. J., Stoessl, A., and Heath, M. C. (1974). Conversion of phaseollin to phaseollinisoflavan by *Stemphylium botryosum*. *Phytopathology* *64*:105–107.

Hohl, B., Arnemann, M., Schwenen, L., Stockl, D., Bringmann, G., Jansen, J., and Barz, W. (1989). Degradation of the pterocarpan phytoalexin (−)-maackiain by *Aschochyta rabei*. *Z. Naturforschung.* *44c*:771–776.

Hohl, B., and Barz, W. (1987). Partial characterization of an enzyme from the fungus *Ascochyta rabiei* for the reductive cleavage of pterocarpan phytoalexins to 2′-hydroxyisoflavans. *Z. Naturforschung.* *42c*:897–901.

Hollander, H., and Amrhein, N. (1980). The site of inhibition of the shikimate pathway by glyphosate I. Inhibition by glyphosate of phenylpropanoid synthesis in buckwheat (*Fagopyrum esculentum* Moench.). *Plant Physiol.* *66*:823–829.

Johal, G. S., and Rahe, J. E. (1990). Role of phytoalexins in the suppression of resistance of *Phaseolus vulgaris* to *Colletotrichum lindemuthianum* by glyophosate. *Can. J. Plant Pathol.* *12*:225–348.

Mackenbrock, U., and Barz, W. (1991). Elicitor-induced formation of pterocarpan phytoalexins in chickpea (*Cicer arietinum* L.) cell suspension cultures from constitutive isoflavone conjugates upon inhibition of phenylalanine ammonium lyase. *Z. Naturforschung.* *46c*:43–50.

McMurchy, R. A., and Higgins, V. J. (1984). Trifolirhizin and maackiain in red clover: changes in *Fusarium roseum* "Avenaceum"-infected roots and *in vitro* effects on the pathogen. *Physiol. Plant Pathol. 25*:229–238.

Morris, P. F., Savard, M. E., and Ward, E. W. B. (1991). Identification and accumulation of isoflavonoids and isoflavone glucosides in soybean leaves and hypocotyls in resistance responses to *Phytophthora megasperma* f. sp. *glycinea*. *Physiol. Mol. Plant Pathol. 39*:229–244.

Peever, T., and Higgins, V. J. (1989). Suppression of the activity of non-specific elicitor from *Cladosporium fulvum* by intercellular fluids from tomato leaves. *Physiol. Mol. Plant Pathol. 34*:471–482.

Tiemann, K., Inze, D., Van Montagu, M., and Barz, W. (1991). Pterocarpan phytoalexin biosynthesis in elicitor-challenged chickpea (*Cicer arietinum* L.) cell cultures: purification, characterization and cDNA cloning of NADPH:isoflavone oxidoreductase. *Eur. J. Biochem. 200*:751–757.

Van Etten, H. D., Matthews, D. E., and Smith, D. A. (1982). Metabolism of phytoalexins. *Phytoalexins* (J. A. Bailey and J. W. Mansfield, eds.), Wiley, New York, pp. 181–217.

Van Etten, H. D., Matthews, D. E., and Matthews, P. S. (1989). Phytoalexin detoxification: importance for pathogenicity and practical implications. *Annu. Rev. Phytopathol. 27*:143–164.

Walsh, C. T. (1983). Suicide substrates: mechanism based enzyme inactivators with therapeutic potential. *Trends Biochem. Sci. 8*(7):254–257.

18

Induction of Phytoalexin Synthesis in *Medicago sativa* (Lucerne)-*Verticillium albo-atrum* Interaction

Christopher J. Smith, J. Michael Milton, and J. Michael Williams
University of Wales, Swansea, Wales

I. INTRODUCTION

The first report of a wilt disease of *Medicago sativa* L. (lucerne, alfalfa) caused by *Verticillium albo-atrum* was from Sweden in 1918 (Hedlund, 1923), but during the period 1938–1950 the disease was reported in a number of European countries, including Germany (Richter and Klinkowski, 1938), Holland (Hansen and Weber, 1948), and France (Krietlow, 1962). In the United Kingdom, where there had been a large postwar increase in lucerne cultivation, 12,800 hectares in 1942 to 44,000 hectares in 1954, the first recorded outbreak was in 1952 (Noble et al., 1953). Following this first occurrence the disease spread rapidly, resulting in a decline in the area under cultivation to 15,000 hectares by 1970. Subsequently the disease was recorded in Canada in 1964 (Aube and Sackston, 1964) and in Washington State (Graham et al., 1977), and today the disease is of major significance in North America.

Isaac and coworkers carried out extensive studies on a number of aspects of the disease in the United Kingdom (Isaac, 1957a; Isaac and Lloyd, 1959; Isaac and Heale, 1961). From these studies it became apparent that a degree of specificity existed in the interaction of *V. albo-atrum* with the host plant. For example, from a large number of strains in the fungus isolated from a range of different host plants, only those strains isolated from lucerne were virulent when inoculated into this particular host. Such a reaction is unusual in *Verticillium* infections; the only other such cases that have been reported are from peppermint (Nelson, 1947) and Brussels sprout infection caused by *V. albo-atrum* (sic) (*V. dahliae*) (Isaac, 1957b).

In common with other plant–pathogen combinations, incompatibility between lucerne and *V. albo-atrum* is frequently associated with the hypersensitive response (HR). This expression of resistance, which is accompanied by induction of a number of active defense mechanisms, results in the formation of a barrier of defense-related molecules around the invading microorganism. In roots of lucerne, for example, synthesis of a lipid–suberin conjugate is induced (Newcomb and Robb, 1989), with a resulting limitation to the spread of the fungus, while synthesis of inhibitors of pathogen-secreted polygalacturonases has been observed in some lucerne tissues in response to *V. albo-atrum* (Degra et al., 1988).

No doubt such mechanisms have important influences on the development of infection under a variety of circumstances but there is now good evidence to indicate that in many plant–fungal interactions accumulation of phytoalexins is a significant determinant of resistance (Smith, 1994). For example, plants with an established resistance to a pathogen produce phytoalexins to higher concentrations than susceptible genotypes (Mayama et al., 1981), and phytoalexins accumulate to sufficiently high concentrations in the vicinity of the invading microorganism to inhibit its further growth (Pierce and Essenberg, 1987). Further evidence is provided by demonstrations that inhibition of phytoalexin synthesis can lead to loss of resistance (Waldmüller and Grisebach, 1987) as well as by related observations concerning the relationship between the ability of a microorganism to detoxify phytoalexins and its virulence (Van Etten et al., 1989). More comprehensive discussions of the evidence for the role of phytoalexins as determinants of disease resistance will be found elsewhere in this volume and in a number of reviews (see, for example, Dixon, 1986; Keen, 1990).

In lucerne, however, Higgins and Millar (1968) observed that when droplets of a suspension of spores from *Helminthosporium sativum*, a fungus nonpathogenic to lucerne (incompatible reaction), were placed on the upper surfaces of the leaf, the phytoalexins medicarpin, a pterocarpan, and sativan and vestitol, both of which are isoflavanols (Fig. 1), diffused into the droplets. These three compounds are the major phytoalexins that can be detected in tissues of lucerne in response to a variety of invading microorganisms. They arise from the reactions of the central phenylpropanoid pathway (Fig. 1) by which L-phenylalanine is converted to 4-coumaroyl-CoA, and which provides precursors for a number of different isoflavanoid phytoalexins in a variety of legumes (Hahlbrock and Scheel, 1989). 4-Coumaroyl-CoA is subsequently converted to medicarpin, vestitol, and sativan by a series of reactions that were elucidated in lucerne tissues by Dewick and Martin (Dewick, 1975; Dewick and Martin, 1976, 1979).

The interaction between *V. albo-atrum* and lucerne has been studied in our own laboratories for a number of years, and Khan and Milton (1975, 1978, 1979), using the drop diffusate technique, examined the response of

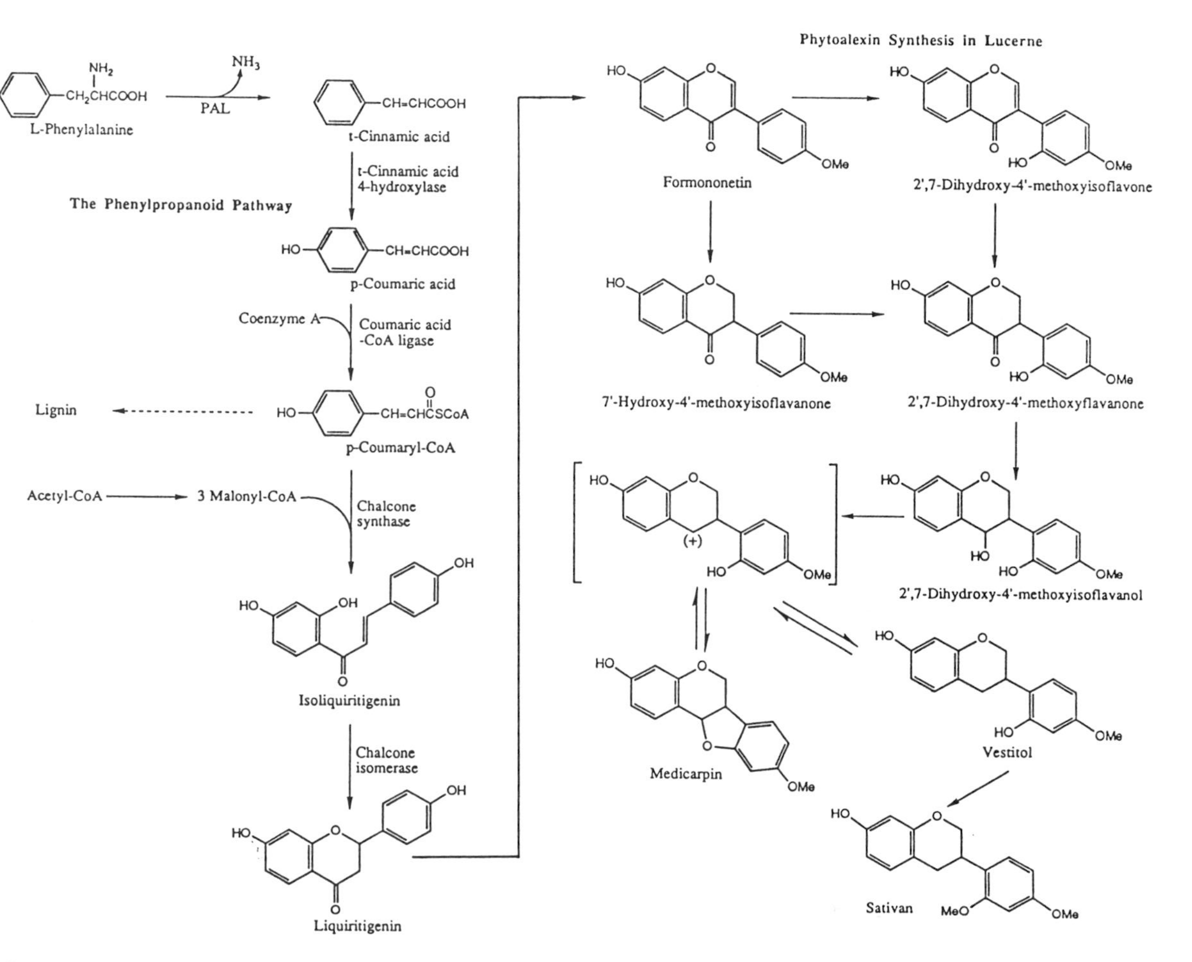

Figure 1 Biosynthetic route of phytoalexin formation in lucerne tissues.

lucerne leaf tissue to a range of different isolates of *V. albo-atrum* and *V. dahliae*. Their studies showed that the concentration of spores in the droplet, the length of exposure to spores, and the presence of light all contribute to the outcome of the plant–fungal interaction and hence influence the pathogenicity of the particular fungal isolate. Of particular significance was their conclusion that the differential pathogenicity of the isolates used in the study was closely correlated with the concentration of phytoalexin induced in and accumulated by the host tissues. *V. albo-atrum* isolated from tomato plants, for example, was not pathogenic to lucerne and was found to induce higher concentrations of phytoalexins in lucerne leaf tissue than did the fungus that had originally been isolated from lucerne. In addition, the lucerne isolate appeared to be relatively insensitive to the lower concentrations of phytoalexins it induced in lucerne leaf tissue.

At first glance it may appear strange that these studies were performed with leaf tissues and a fungus, the normal point of entry of which would be the root. Induction of a defense response in tissues that are remote from the point of infection is not an unusual occurrence, however, and phytoalexin accumulation in leaf tissues could play a significant role in preventing spread of the fungus throughout the rest of the plant. In any case the results of these studies clearly indicated the cellular interactions that occur between plant and fungus and have since been confirmed in the relatively undifferentiated tissues of suspension culture where morphological differentiation is not a problem.

This conclusion, that the differential pathogenicity observed was closely related to the concentration of phytoalexins accumulated in the host, is one that has been reached by other workers studying a variety of other host–pathogen interactions. It highlighted the potential of this particular defense response to be a determinant of resistance in the lucerne–*V. albo-atrum* interaction and gave rise to three related areas of investigation in our laboratories: 1.) determination of the kinetics of phytoalexin accumulation in the incompatible interaction; 2.) characterization of the elicitor component of the pathogen responsible for induction of phytoalexin accumulation in the host; and 3.) characterization of the intracellular mechanism regulating induction of the response in the host.

II. MATERIALS AND METHODS

A. Growth of Plants

Seeds of lucerne, cvs. Maris Kabul and Europe, were soaked overnight in running tap water prior to sowing in a peat-based compost. Once germinated, seedlings were transferred to a soil-based compost and maintained at

22 ± 2°C under a 16-hr photoperiod at an intensity of 12,000 lux, when plants were intended for use in bioassay, or transferred to a greenhouse when used in pathogenicity studies.

B. Development and Maintenance of Lucerne, Callus, Culture, and Cell Suspension Cultures

Leaves were removed from 6-week old plants of lucerne and surface-sterilized by washing in a 5% aqueous hypochlorite solution for 2 min. Squares, 1 × 1 cm, were cut aseptically and transferred to the surface of Murashige and Skoog (1962) agar medium. Growth of callus was allowed to develop for 6 weeks at 25 ± 1°C in the dark. Once established the undifferentiated callus tissue was maintained on the same medium by transference of a small (~0.5 cm^3) piece of callus to fresh medium at 6-week intervals.

Liquid cell suspension cultures of *M. sativa* were initiated from callus cultures maintained in this laboratory, by aseptic transference of 1 cm^3 solid callus to 100 ml Murashige and Skoog (1962) liquid medium at pH 6.7. Cultures were incubated in the dark at 25 ± 1°C on an orbital shaker (100 rpm) and subcultured every 2 weeks. For experiments, cultures were used 5 days after subculturing and additions were made aseptically by passing solutions through sterile membrane filters (Minisart, NML 0.2 μm, Sartorius, UK) from a sterile hypodermic syringe. At the end of the experimental incubation period cells were harvested by centrifugation for 2 min at 500*g* prior to further analysis.

C. Preparation of Elicitor

The isolate of *V. albo-atrum* pathogenic to *M. sativa* (designated V_1) was obtained from field-grown plants of *M. Sativa* and the nonpathogenic (designated V_2) from *Lycopersicon esculentum*. Elicitors from both were isolated from 6-week-old liquid cultures grown in Dox's medium. The medium was filtered to remove mycelia and spores, dialyzed exhaustively against distilled water, lyophilized, and the resulting powder dissolved in distilled water at a concentration of 4 mg/ml.

Elicitor activity was assessed by the ability to induce phytoalexin synthesis, or increases in phenylalanine ammonia-lyase (PAL) activity, in shoots of *M. sativa*, placed for 16 hr with their cut ends in 50 ml distilled water containing putative elicitor. Where cell suspension cultures were employed, addition of putative elicitor was carried out as described above, and phytoalexin content was determined at 16 hr and PAL activity after 4 hr.

D. Cut-Shoot Assay for Elicitation of Phytoalexin Accumulation and Phenylalanine Ammonia-lyase Activity

The ability of various components to elicit phytoalexin accumulation and increases in PAL activity in lucerne leaf tissue was determined using shoots of 6-week-old plants. Shoots were harvested and their cut ends placed in water. The lower 0.5 cm of each stem was removed while remaining under the surface of the water and four shoots were transferred to each conical flask containing 50 ml of a solution of the compound under test in water. A few drops of oil were added to the surface of the solution to reduce evaporation. Flasks were transferred to a controlled environmental chamber, 23 ± 2°C, 7000 lux, for 16 hr before leaves were removed for extraction of phytoalexins or for determination of PAL activity.

E. Extraction of Phytoalexins

Phytoalexins were extracted from leaf tissue using a modification of the method described by Keen (1978). Leaves were removed from the stem tissue of experimental shoots, weighed, and transferred to flasks containing 40% ethanol–water (v/v), (16 ml/g fresh wt tissue). Leaf tissue was subjected to vacuum infiltration for 5 min at the end of which time the vacuum was removed and the process repeated. Each flask was incubated in the dark at 25 ± 1°C, on an orbital shaker (100 rpm) for 2 hr. At the end of this time the leaf tissue was removed by filtration, washed (5 ml distilled water/g fresh wt), and the aqueous washings combined with the ethanolic filtrate. The resulting solution was partitioned three times against ethyl acetate or diethyl ether (0.5 ml solvent/ml extract). The solvent extracts were combined and the solvent was removed by rotary film evaporation at 30°C under reduced pressure until 1 ml remained. The residue was transferred to a 1-dram vial and the remaining solvent removed under a stream of filtered air at 40°C. Samples were stored at −20°C until required.

F. High-Performance Liquid Chromatography of Phytoalexins

Samples were dissolved in 100 μl of acetonitrile (AnalaR, BDH) containing 0.01% (w/v) dibutylphthalate (DBP), which was used as an internal standard. The samples were filtered (0.45 μm Acro LC3S, Gelman Sciences) to remove any particulate matter and the resultant filtrate was subjected to high-performance liquid chromatography (HPLC) using an instrument (Milton Roy, UK) fitted with an integrator (CI 10B, Milton Roy, UK). Typically, a 10-μl sample was injected, via a rheodyne valve, onto a reverse phase column (S5ODS, Spherisorb, 5 μm, 250 mm × 4.9 mm (i.d.), Hichrom, UK) and was eluted isocratically with acetonitrile–water (3 : 2 v/v) at a flow rate of 1.3 ml/min. Compounds eluting from the column were detected by their absorption at 240 nm. A standard ratio of retention times for each phytoalexin, compared to the retention time of DBP, was determined by the injection of authentic medicarpin, sativan, and vestitol together with 0.01% DBP in acetonitrile. Using this standard ratio, the retention time of each phytoalexin could be determined in unknown samples.

The quantity of medicarpin in a sample was determined using the ratio of the integrated areas of the phytoalexin and the DBP peaks, corrected for a response factor that adjusts for the difference in absorbances of the phytoalexins and DBP. Response factors were determined for each of the phytoalexins by injection of known quantities of each phytoalexin and the same quantity of DBP followed by comparison of the integrated areas.

G. Experimental Use of Cell Suspension Cultures and Extraction of Phytoalexins

Where suspension cultures were used to determine the effects of elicitor, or of various agonists and antagonists of the elicitation process, on the production of phytoalexins or PAL, the cultures were used 5 days after subculturing. Effectors in solution were added to the cultures aseptically by passage through a sterile membrane filter (Minisart, NML 0.2 μm, Sartorius, UK) from a syringe. The volume added never exceeded 2 ml and where it was necessary to use ethanol to solubilize the effector the final concentration in the culture was never greater than 0.18 M. At this concentration ethanol was found to have no effect on the phytoalexin content or the activity of PAL of cell cultures. Equivalent volumes of water or ethanol were used in control cultures. For extraction of phytoalexins at the end of the experimental treatment period, ethanol was added to the suspension culture to a final concentration of 40% (v/v). The culture was then subjected to vacuum infiltration and the extraction procedure described for leaf tissue.

H. Extraction and Assay of PAL

Cell-free homogenates were prepared from tissues of *M. sativa* by grinding in 50 mM Tris-HCl, pH 8.6, containing 10 mM ascorbic acid (3 ml buffer/g fresh wt), with a pestle and mortar and with the aid of a little acid-washed sand. The homogenate was filtered through two layers of Miracloth (Calbiochem, UK), centrifuged (20,000*g*, 20 min, 4°C), and the supernatant dialyzed against 50 mM Tris-HCl, pH 8.6 (20 ml buffer/ml sample) for 16 hr at 4°C.

The dialyzate was assayed spectrophotometrically for PAL activity (Bolwell et al., 1985) by the addition of 1 ml 30 mM L-phenylalanine in 50 mM Tris-HCl, pH 8.6 (final concentration in the assay 10 mM), to a 2-ml sample of dialyzate (protein content in the range 0.7–1.0 mg). Absorbance at 290 nm was measured after 2 hr incubation at 30°C, against a control incubation containing 10 mM D-phenylalanine in place of the L isomer.

III. PHYTOALEXIN ACCUMULATION IN THE LUCERNE–*V. ALBO-ATRUM* INTERACTION

A. Development of an HPLC Method for Quantitation of Lucerne Phytoalexins

In the earlier studies of Khan and Milton (1975), the drop diffusate technique of Cruickshank and Perrin (1960) was employed. In this method lucerne leaves were treated with a spore suspension prepared from *V. albo-atrum* and the phytoalexins that were synthesized by the leaf in response to the challenge accumulated in the droplets. Following collection of the droplets from leaf surfaces the phytoalexins were extracted into a suitable organic solvent, purified by thin layer chromatography (TLC) using silica gel plates developed with a chloroform–carbon tetrachloride solvent (3 : 1 v : v), and the areas corresponding to phytoalexins were scraped from the plate. The phytoalexins which were eluted from the adsorbent were quantitated from their absorbance at characteristic wavelengths. Using this approach, Flood et al. (1978) identified medicarpin, sativan, and vestitol as the major phytoalexins present and the presence of at least four other components with the properties of phytoalexins was indicated. A number of errors are inherent in these methods, however, and HPLC is now widely used in the analyses of isoflavanols and related components (Williams and Harborne, 1989). Until fairly recently, few HPLC methods had been designed specifically for phytoalexins, the majority of those in use being general methods for separation of isoflavanoid compounds with little adaptation for the specific phytoalexins and host tissues involved. As a prerequisite for our quantitative studies of phytoalexin accumulation, therefore, we undertook development of an efficient extraction and separation method, based on HPLC, for the phytoalexins of lucerne.

The current method in use in our laboratory is already described in, Section II, Materials and Methods, but some comments relating to general features and rationale of the approach will be made here. Except where studies are performed with the drop diffusate technique the extraction method that has been found to be the most satisfactory with tissues of lucerne, including cell suspension cultures, is a modification of the facilitated diffusion method (Keen, 1978). In terms of the efficiency of extraction of phytoalexins and the lower content of interfering compounds coextracted, this technique has advantages over homogenization of the tissue, and routinely as little as 1 gm fresh wt tissue can be handled, with the actual amount depending on such factors as source of the tissue, duration of exposure to the eliciting stimulus, sensitivity of the tissue to the stimulus, and lower limit of detection of the HPLC system. Where leaf or stem tissue is involved 40% ethanol is added directly to the tissue but in the case of cell suspension cultures absolute ethanol is added to the whole culture, including medium, until a concentration of 40% is reached. The growth medium is included in the extraction procedure because lucerne cultures secrete phytoalexins into the medium and these would be lost if the cells were removed from the medium before extraction takes place.

It is sometimes possible to quantitate the phytoalexin content of an ethanolic extract directly by HPLC, and this may be appropriate for cell suspension cultures with their lower content of phenolics and absence of chlorophyll. In our experience, however, a clearer separation of components is achieved and column life is extended if a partitioning step is included in the procedure. Routinely we have found ethyl acetate to be a more satisfactory solvent than the tetrachloromethane that was used originally with lucerne extracts (Higgins, 1972). However, more recently we have used diethyl ether because it is as efficient as ethyl acetate for extraction of medicarpin, but with its lower boiling point it can be removed with a stream of filtered air rather than by rotary film evaporation. This permits simultaneous handling of multiple samples.

Liquid chromatography on a column of reverse phase adsorbent offered the best possibility for resolution of the three phytoalexins vestitol, medicarpin and sativan, and trials were carried out using a 250 × 4.9 mm (i.d.) column of S5ODS Spherisorb silica (Hichrom, UK), 5-μm particle size and elution with combinations of acetonitrile and water. Gradient elution has been used to separate the phytoalexins extracted from a variety of sources by a number of workers, both before and since our initial experiments (see, for example, Edwards and Strange, 1991), and we found that a gradient from 47% to 85% aqueous acetonitrile over a 15-min period (1.3 ml/min flow rate) achieved a satisfactory separation of a mixture of the three authentic phytoalexins. The same conditions were used to achieve separation of an extract of lucerne leaf tissue that has been challenged for 16 hr with elicitor derived from *V. albo-atrum*. The three phytoalexins are separated from each other by this gradient; however, leaf extracts contain a number of components that are more polar than the phytoalexins and these are not fully resolved from vestitol, which is the first of the phytoalexins to be eluted.

Accurate quantitation of any phytoalexin requires a knowledge of the absorption coefficient at the monitoring wavelength, establishment of the linearity of the response of the detector, and inclusion in the extract of an internal standard so that peak areas may be expressed relative to the known amount of standard that has been added. In our system, absorption coefficients were determined by spectrophotometry of authentic samples of the phytoalexins and linearity of the response was established by injection of a series of standards ranging in quantity up to 10 μg.

Selection of an internal standard was initially more difficult and DBP was chosen because of its availability, its absorption spectrum in relation to that of the phytoalexins, and because out of the compounds tested it was the only one to be fully resolved from the phytoalexins and the other components present in extracts of lucerne. Because of the differences between the wavelength of maximum absorption of the three phytoalexins and of the internal standard, however, monitoring of the eluant was carried out at a "compromise" wavelength of 240 nm where absorption of the phytoalexins is reduced compared to absorption at their maxima, but absorption of the internal standard while not maximal is still significant.

Routinely, then, DBP has been included as an internal standard in the extracts, but, using gradient elution, analysis time was prolonged because of the time taken at the end of each separation to reequilibrate the column to the starting conditions. However, vestitol, the earliest eluting phytoalexin of those present in lucerne, has rarely been encountered in our studies and since application of a gradient is more critical to separation in this area of the chromatogram rather than for resolving medicarpin and sativan, isocratic elution was tested as a means of separating the phytoalexins of lucerne. Elution with 60% aqueous acetonitrile at a flow rate of 1.3 ml/min was found to give the optimum resolution with retention times, relative to DBP, of 0.25, 0.30, and 0.40, for vestitol, medicarpin, and sativan respectively. While separation of vestitol from other components is not entirely satisfactory with isocratic elution, it has rarely been detected in our samples and so we routinely use this method (Fig. 2a,b).

B. Chemical Synthesis of Medicarpin, Sativan, and Vestitol

Samples of the authentic phytoalexins were required for the development of the HPLC assay and racemic vestitol, sativan, and medicarpin were synthesized (Evans, 1986) as follows. The most efficient synthesis of these compounds involved, as the key step, the thallium(III)-induced oxidative rearrangement of the intermediate chalcones (**1**) to form the 1,2-diaryl-3,3-dimethoxypropan-1-ones (**2**). The latter were converted, without isolation, into isoflavones (**3**) (Farkas et al., 1974).

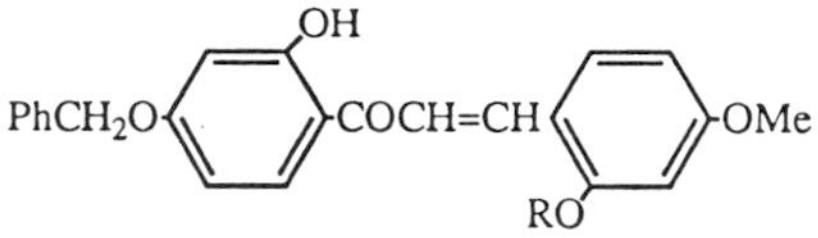

(1) R = Me or CH_2Ph

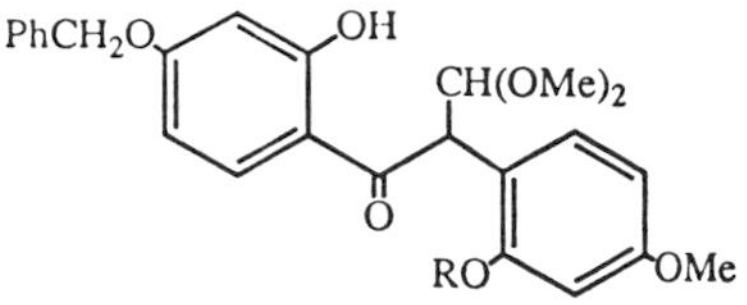

(2)

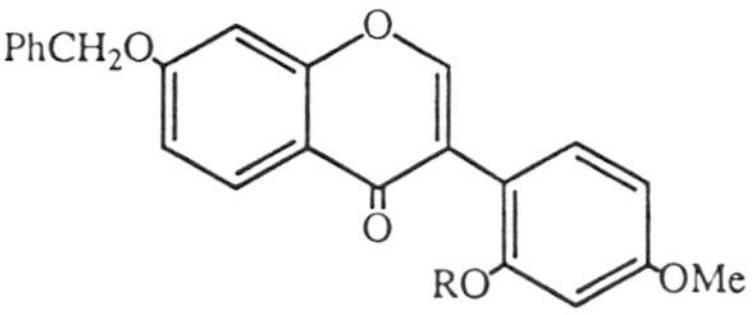

(3)

Sativan (Dewick and Martin, 1979) and vestitol (Farkas et al., 1974) have been synthesized by this route but full details were not given for sativan. The chalcone (**1**, R = Me), prepared by condensation of 4-benzyloxy-2-hydroxyacetophenone and 2,4-dimethoxybenzaldehyde, was converted to the isoflavone (**3**, R = Me), mp 137–139°C in 73% yield by reaction with thallium(III) nitrate followed by acid hydrolysis. Reaction of an acetone solution of the isoflavone with hydrogen at atmospheric pressure over 10% palladium on charcoal gave sativan (36% after recrystallization). The synthesis of medicarpin by the same route (Dewick, 1977) was complicated by the need to include a ^{14}C-labeled methyl group at C-4′, and the isoflavone (**3**, R = CH_2Ph) was not reported. This isoflavone was readily prepared in 70% yield (mp 134–135°C) from the known chalcone (**1**, R = CH_2Ph) and debenzylation by hydrogenolysis over 10% palladium on charcoal under controlled conditions (the reaction being terminated after 2 eq of hydrogen had been consumed) gave 7,2′-dihydroxy-4′-methoxyisoflavone, mp 218–220°C. Reduction of this isoflavone with sodium borohydride in ethanol followed by the action of 1 M hydrochloric acid gave medicarpin (mp 200–201°C), isolated by extraction into ethyl acetate.

An alternative route to medicarpin was also used. This had the advantage that the same isoflavone intermediate (**3**, R = Me) could be used to synthesize sativan and medicarpin. The route to medicarpin involved selective aluminum chloride–catalyzed demethylation of (**3**, R = Me) in refluxing acetonitrile, conditions which also removed the benzyl protecting group (Cocker et al., 1965). The resulting crude product containing 7,2′-dihydroxy-4′-methoxyisoflavone was converted as above to medicarpin,

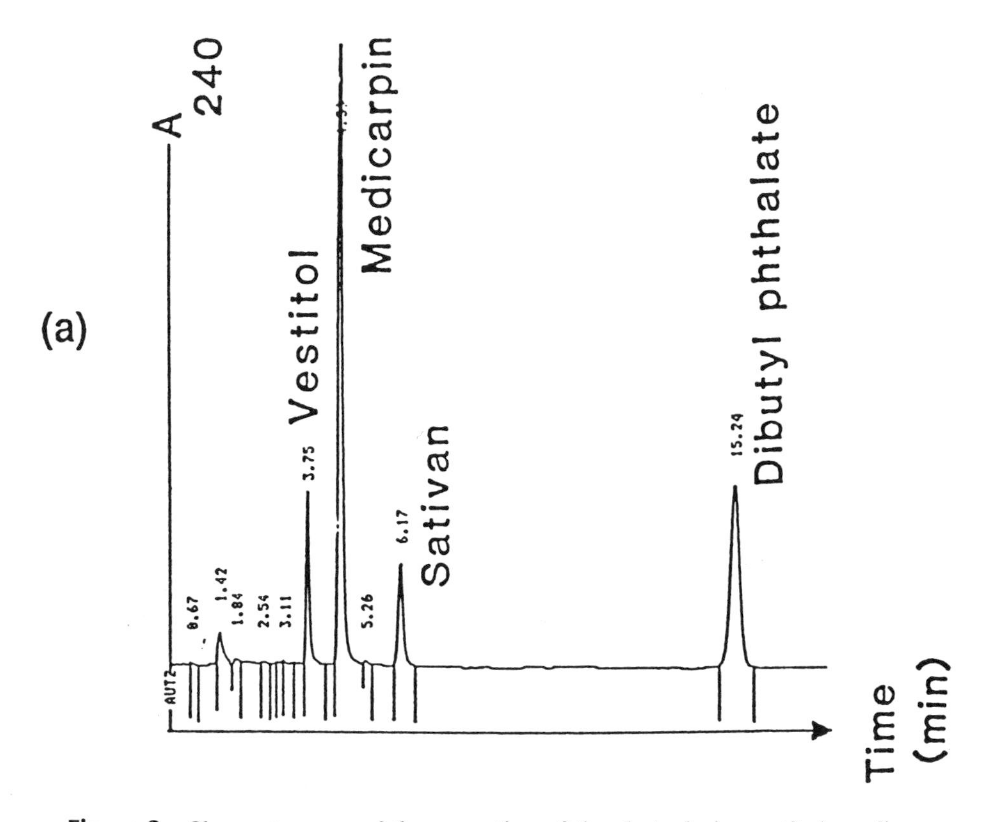

Figure 2 Chromatograms of the separation of the phytoalexins vestitol, medicarpin, and sativan by HPLC. (a) Authentic samples of the phytoalexins and the internal standard di-*n*-butylphthalate. (b) Separation of an extract of lucerne leaf tissue that has been challenged for 16 hr with the V_2 elicitor preparation. Details of conditions will be found in Section II, Materials and Methods.

which was isolated by chromatography on a chromatotron (silica gel, solvent: chloroform). Vestitol was prepared from isoflavone (**3**, R = CH_2Ph) by hydrogenation/hydrogenolysis over 10% palladium on charcoal in acetone (cf Farkas et al., 1974). All three phytoalexins were characterized by mass spectrometry.

C. Future Development on Methodology

Two improvements to the HPLC method are at present under consideration. The choice of DBP as an internal standard is not entirely satisfactory because of its relatively long retention time compared to sativan, which has

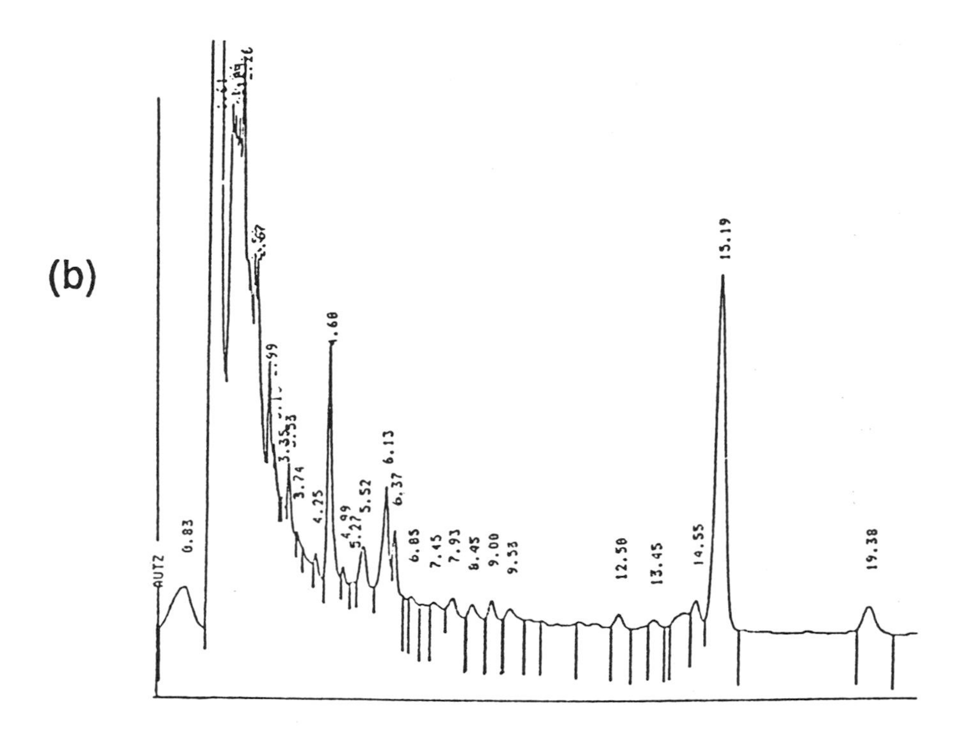

the longest retention time of the phytoalexins, and because it is necessary to monitor at 240 nm in order to detect DBP (λ_{max} 225 nm) while the λ_{max} of vestitol, medicarpin, and sativan are 282, 287, and 284 nm, respectively. The 4-alkoxyacetophenones are being examined as potential internal standards partially because their absorbance characteristics are more suited to this application, but also because the retention times of the different alkoxyacetophenones vary with the length of the alkyl chain. So, for instance, 4-butoxyacetophenone has a λ_{max} of 270 nm and a retention time of 1.32 times that of sativan on the ODS column (250 × 4.9 mm i.d.) when 80% aqueous methanol is used as eluant. Its performance as an internal standard in extracts of lucerne is currently being assessed.

The second area where an improvement to the current methodology may be achieved is the detection system. All three phytoalexins fluoresce in the region of 310 nm when excited at their absorption maxima, close to 280 nm (Williams, unpublished). Potentially this is a more selective method of monitoring because few of the other components present in lucerne extracts fluoresce at that wavelength in response to incident light of 280 nm. Depending on the circumstances and equipment available, a fluorescent method may have a far greater sensitivity than the absorbance method and the technique is in the process of evaluation.

D. Phytoalexin Accumulation in Response to Fungal Elicitors

Khan and Milton's observation that the differential pathogenicity of *V. albo-atrum* isolates was related to phytoalexin accumulation (1975) prompted us to examine the isolates to determine what elements of the fungi could affect the host in such a way that a difference in expression of the defense response resulted. The same host tissues, and thus the same metabolic potentials, were involved in each case, so that some aspect of the fungus must have been responsible for the different outcomes of the interactions. There are a number of possible explanations including differences in structure of the isolates such that the host is unable to detect the pathogen or does not detect it quickly enough to activate phytoalexin production, while the nonpathogen is detected at an early stage; secretion by the pathogen of suppressors of phytoalexin synthesis that prevent the host from producing phytoalexins; and a greater efficiency of some isolates to metabolize phytoalexins, so that in vivo their concentrations in the host do not reach antimicrobial levels.

Our approach has been to focus on the pathogen factor(s) of the fungus responsible for induction of phytoalexin synthesis in the host, and for this purpose two isolates were used, one recovered from lucerne (V_1) and the other from plants of tomato (*Lycopersicon esculentum*) (V_2). Pathogenicity studies had already established that the V_1 isolate is capable of successfully infecting lucerne plants cv. Europe (compatible interaction) while a high degree of resistance is shown to isolate V_2.

Phytoalexin accumulation can be induced not only by the living microorganism but also by components derived from it, termed *elicitors* (Keen 1990), molecules which by virtue of the structural information they contain or the functional activity they possess are characteristic of the organism. Mostly such elicitors are not race-specific, i.e., they are found to be produced by avirulent and virulent races of the pathogen alike and elicit phytoalexin synthesis in both resistant and susceptible cultivars of the host.

Nevertheless they provide a useful starting point to the investigation of the differential response of lucerne since at least one possibility, metabolism of phytoalexins by the pathogen, is eliminated and the complexity of the interaction is greatly reduced. Elicitors of phytoalexin synthesis have been previously isolated from *V. albo-atrum* by several workers. For example, Zaki et al. (1972) identified a protein–lipopolysaccharide component that induced formation of gossypol-related phytoalexins in stem tissue of cotton. A glycoprotein elicitor also is known to elicit synthesis of medicarpin in lucerne (Onuorah, 1987).

In our present studies we have used elicitors derived from the two isolates of *V. albo-atrum*, V_1 and V_2, prepared as described in Section II, Materials and Methods. In the initial experiments the elicitors were used in their crude state because there is some evidence to indicate that induction of synthesis of the phytoalexin phaseollin in *Phaseolus vulgaris*, by an elicitor from *Colletotrichum lindemuthianum*, relies on the presence of a number of related polysaccharides, none of which by themselves can induce synthesis of the phytoalexin (Hamdan and Dixon, 1989). Fractionation in the early stages may have led to the loss of a component without which elicitation of the phytoalexin response may not have taken place.

Leaf tissues respond to treatment with V_2 elicitor by production of phytoalexins (Fig. 3), accumulation being dependent on the time of treatment with the maximum concentration occurring 16 hr after the initial challenge with elicitor (data not shown). The concentration of phytoalexin accumulated also depends on the concentration of elicitor employed, though the concentration of the elicitor that achieves maximum synthesis of the phytoalexin is different for each of the phytoalexins. Only low concentrations of vestitol are accumulated by lucerne whatever may be the elicitor concentration, though the optimum concentration of elicitor, 0.1 mg/ml carbohydrate, is the same as that achieving maximum accumulation of medicarpin. In contrast, sativan accumulation continues to rise even at the highest concentration of elicitor used, the rise in its concentration corresponding with the decline in medicarpin accumulation. This shift towards sativan accumulation is interesting and must reflect differential effects of the elicitor preparation on the separate branches of the pathways leading to sativan and medicarpin (Fig. 1).

The effect on phytoalexin accumulation of elicitor from the V_1 (pathogenic) isolate was also determined in leaf tissue, and although synthesis of medicarpin was induced (Fig. 3), the concentration of elicitor required was significantly higher than was the case with the V_2 elicitor, and little accumulation occurred with elicitor concentrations below 1.0 mg/ml. In control tissue (distilled water only) none of the phytoalexins could be detected.

The crude elicitor preparations contained both carbohydrate and protein but the composition of the two elicitors is different, the ratio of carbohydrate to protein being 1 : 1 for the V_2 elicitor and 15 : 1 for the V_1. These ratios are consistent and have been observed over 15 different batches of elicitor prepared using several reisolates of each fungus. The difference in the abilities of V_1 and V_2 elicitors to induce phytoalexin accumulation in the host tissue must reflect differences either in the structure of the two elicitors or in the quantity of the structural determinant of elicitor activity each preparation contains. This difference in their composition led us to examine whether it is the carbohydrate or the protein moiety that is the determi-

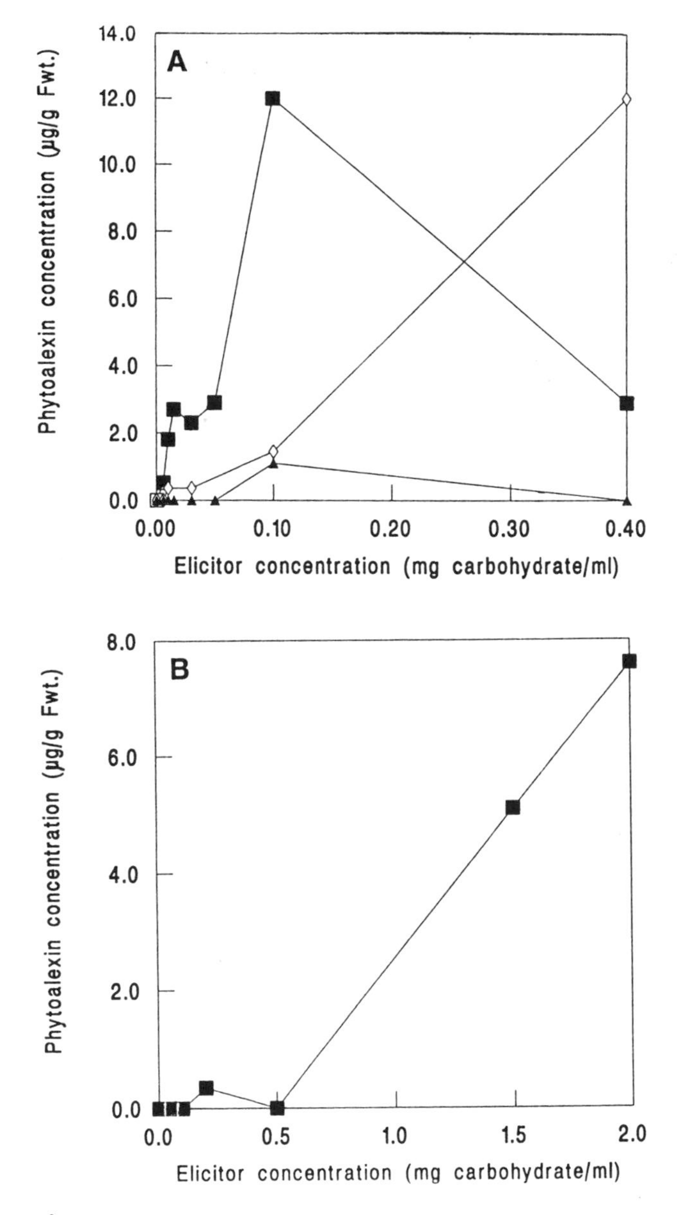

Figure 3 Phytoalexin accumulation in leaf tissue of lucerne (cv. Kabul) in response to treatment for 16 hr with various concentrations of (A) V_2 elicitor preparation, (B) V_1 elicitor. ▲-▲, vestitol; ■-■, medicarpin; ◇-◇, sativan.

nant of activity. We established that treatment of V_2 elicitor with periodate led to a loss of elicitor activity while trypsin treatment led to a 50% increase in activity compared to untreated elicitor (Smith and Milton, 1992). This indicated that it is the carbohydrate rather than the peptide moiety that determines activity, while the effect of trypsin in increasing elicitor activity suggests that a peptide inhibitor of phytoalexin synthesis is normally present in the preparation, and trypsin destroys it, or that the elicitor is a glycopeptide and trypsin cleaves the peptide backbone, making a carbohydrate determinant of activity more available to interact with the plant tissues.

The difference in composition of the two elicitors and their effects on phytoalexin accumulation is reflected in their effects on the activity of PAL, a key enzyme from the phenylpropanoid pathway leading to synthesis of medicarpin, sativan, and vestitol in lucerne (Fig. 1). In common with other tissues that have been characterized (see, for example, Daniel et al., 1988) induction of phytoalexin synthesis in lucerne is associated with increases in the activities of enzymes from the biosynthetic pathway (see later section for further details). In cell suspension cultures of lucerne, induction of phytoalexin synthesis by the V_2 elicitor (Fig. 4) is accompanied by and

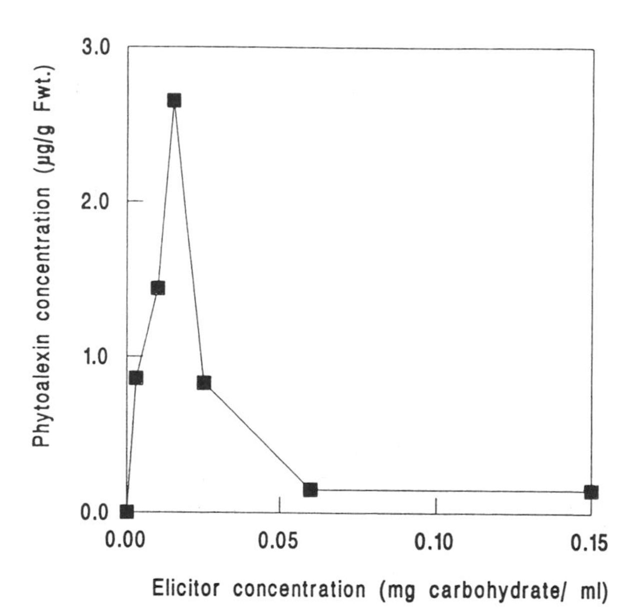

Figure 4 Sativan accumulation in cell suspension culture of lucerne (cv. Kabul) in response to 16-hr treatment with various concentrations of elicitor prepared from the V_2 isolate of *V. albo-atrum*.

dependent on an increase in activity of PAL, the enzyme that catalyzes the committed step on the phenylpropanoid pathway (Fig. 1). Treatment of cell cultures with cordycepin, an inhibitor of mRNA transcription, completely inhibits the increase in PAL activity normally associated with elicitor treatment (Fig. 6) (Little, 1989) and leads to failure of the fungal elicitor to induce phytoalexin synthesis. It was of interest, therefore, to determine whether the reduced rate of induced phytoalexin synthesis observed with V_1 elicitor resulted from a failure to increase the activity of this key enzyme.

Both leaf tissues and cell suspension cultures respond to treatment with V_2 elicitor with an increase in activity of PAL (Figs. 5 and 6). The concentration of elicitor required to give the maximum response in cell cultures (0.05 mg carbohydrate/ml) is, however, less than that for leaf tissue (0.1 mg/ml), presumably because access to the cells is easier in suspension culture than in leaf tissue. In leaf the concentration of elicitor that produces the maximum increase in PAL activity (0.1 mg/ml) is the same as that producing the maximum accumulation of medicarpin (Fig. 3), while in cell suspension a threefold higher concentration of elicitor is necessary to induce the maximum increase in PAL activity (Fig. 6) compared to that required to achieve maximum accumulation of sativan (0.15 mg/ml) (Fig. 4).

Surprisingly, treatment of both leaf and cell suspension cultures with V_1 elicitor caused a significant increase in PAL activity (Figs. 5 and 6) and for both tissues treatment with V_1 elicitor brought about a greater percentage of increase in activity than did the V_2 (Table 1). The data of Table 1 also indicate that whether V_1 or V_2 elicitor is used, suspension culture appears more responsive than leaf tissue.

Another feature of the response induced by both V_1 and V_2 elicitor is the presence in both dose–response curves of two peaks of activity, the one corresponding to the lower increase occurring at the lower concentration of elicitor in each case. Interestingly, when cordycepin at 100 μm is included in the cell culture to inhibit transcription of mRNA, the V_2 elicitor–induced increase in PAL activity is almost totally prevented at the higher elicitor concentration (Fig. 6), while there is only a small reduction in the induction achieved at the lower concentrations.

It is obvious from these results that the requirement for a greater concentration of V_1 elicitor to achieve induction of phytoalexin synthesis in lucerne compared to V_2 elicitor, and the lower quantities of phytoalexin accumulated in V_1 treated tissue, cannot be explained by a failure of the V_1 elicitor to induce the increase in PAL activity. In both leaf and cell suspensions, V_1 elicitor is actually more efficient in percent terms than the V_2 elicitor at increasing PAL activity (Table 1), and the absolute levels of activity induced by V_1 are higher than in tissues treated with V_2 elicitor (100

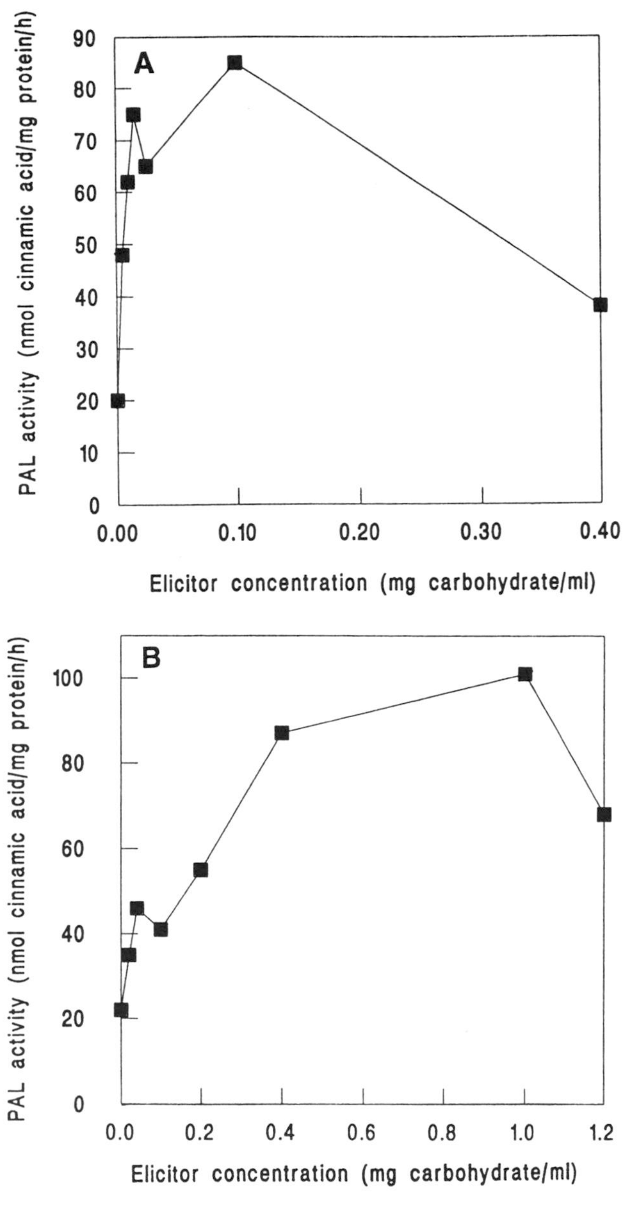

Figure 5 Activity of PAL in leaf tissue of lucerne (cv. Kabul) induced by 16-hr treatment with various concentrations of (A) V_2 elicitor preparation, (B) V_1 elicitor.

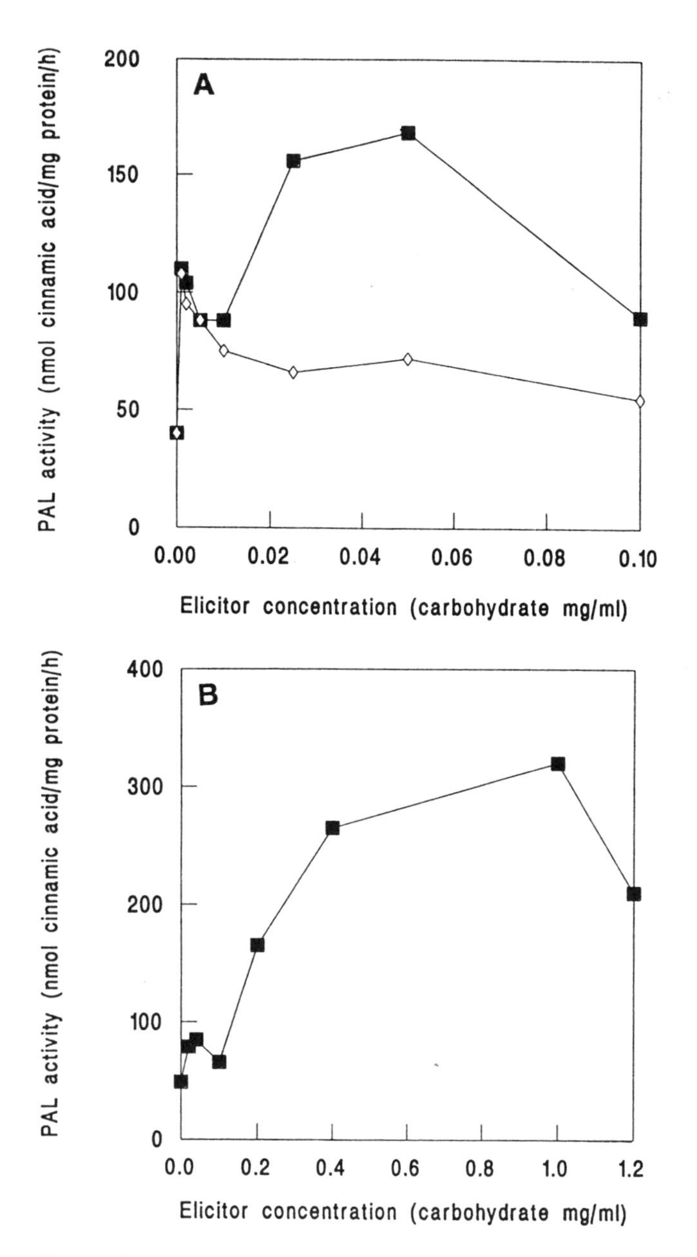

Figure 6 Activity of PAL in cell suspension cultures of lucerne induced by 16-hr treatment with various concentrations of (A) V_2 elicitor, (B) V_1 elicitor. In (A) the effects on the induction of activity of adding cordycepin at 100 μM are indicated, ◇-◇.

Table 1 PAL Activity in Leaf and Cell Suspension Cultures of Lucerne in Response to Treatment with V_1 and V_2 Elicitors

	Leaf		Cell suspension	
	PAL activity[a]	% Increase[b]	PAL activity[a]	% Increase[b]
Control	20		45	
Elicitor				
V1	100	400	320	611
V2	85	325	175	288

[a]Expressed in nmol cinnamic acid/mg protein/hr.
[b]Compared to control.

and 320 nmol cinammic acid/mg protein/hr for leaf and cell suspension cultures, respectively, treated with V_1 elicitor. Corresponding figures for V_2 elicitor are 85 and 175 nmol cinammic acid/mg protein/hr). Since the lower levels of activity induced by V_2 elicitor lead to synthesis of phytoalexins, then the greater levels of activity induced by the V_1 elicitor should certainly be adequate to support phytoalexin synthesis.

There are a number of possible explanations that could account for these results. In lucerne cell suspensions at least three different isoforms of PAL have been identified and the evidence indicates that PAL activity is encoded by a multigene family, as is the case in bean (Jorrin and Dixon, 1990). The three isoforms each have different K_m values and in response to treatment with an elicitor from *Colletotrichum lindemuthianum* a large relative increase in the form with the lower K_m value for phenylalanine was observed. In response to an elicitor from yeast, however, a different pattern of induction occurred and one of the other isoforms predominated. It is possible then that in the case of the elicitors from *Verticillium* the V_1 elicitor induces a different isoform of PAL, with a higher K_m than that induced by the V_2 elicitor, and that this isoform is less efficient at feeding phenylalanine substrate into the phytoalexin biosynthetic pathway. The result would be an increased PAL activity as determined under the assay conditions employed where the concentrations of phenylalanine is high but there is a failure to synthesize phytoalexins in vivo. We have no evidence at present that could evaluate this possibility.

It is also possible that both elicitor preparations from *V. albo-atrum* are a mixture of components and that different enzymes from the biosynthetic pathway are induced by different components of the mixture. This is certainly the case for a crude elicitor preparation from *Colletotrichum lindemuthianum* in which a number of polysaccharides, each containing galactose and mannose, were identified. When the mixture was fractionated each fraction was found to induce similar increases in activity of several of the enzymes from the phenylpropanoid pathway in cultures of *Phaseolus vulgaris*, but none of them by themselves were able to elicit synthesis of the phytoalexins phaseollin or kievitone, in contrast to the unfractionated mixture which did (Hamdan and Dixon, 1987). An integrated response in the host may depend on the presence in the elicitor preparation of several components from the microorganism, and if the V_1 elicitor preparation used in these present studies lacks one or more of the components that are present in the V_2 preparation, this may account for the differential phytoalexin synthesis observed.

Our studies relating to the purification of the elicitor determinants from the V_1 and V_2 elicitor preparations have so far indicated that each contains a number of different components, both peptide and carbohydrate (Evans, 1986; Smith et al., unpublished). Initial studies have failed to identify one component that could alone be responsible for the induction and phytoalexin elicitor activity has been located in several components, of various molecular weights (Evans, 1986). This is not surprising in view of the results with other elicitors, e.g., the glucan from *Phytophthora megasperma*, where the determinant of elicitor activity has been found to be distributed throughout the heterogeneous mixture of glucans present in the unfractionated mixture (Sharp et al., 1984). However, if any component of a mixture that is necessary for the induction of the complete response was missing from the V_1 preparation, then phytoalexin synthesis at the higher concentrations of elicitor employed (Fig. 4) would not have been achieved. This may mean that there is a lower quantity of a particular component present rather than a complete absence, but whatever the factor is, obviously it must affect a part of the pathway beyond the PAL-catalyzed step. Since that part of the pathway is unaffected, it will be interesting to assay these enzymes from the later part of the pathway though many of them are membrane-associated and are difficult to assay.

We are not in a position to evaluate the other possibility, i.e., that the V_1 preparation contains a suppressor molecule similar to that identified in chickpea (Kessmann and Barz, 1986) and which could inhibit a part of the pathway subsequent to the PAL-catalyzed step. If such a suppressor is present its inhibitory effects must reach a saturation point before the elicitor of PAL activity achieves its maximum effect and its effect must be at less than 100% inhibition of the pathway, because at the higher concentrations of V_1 used in our experiments medicarpin synthesis is eventually induced in leaf (Fig. 4). The possibility of a suppressor molecule being present is currently being examined by fractionation of the V_1 elicitor preparation as well as by assays in which the potential of the V_1 preparation to moderate the phytoalexin response induced by the V_2 elicitor is determined. If the V_1 preparation contains an inhibitor of the later stages of the biosynthetic pathway, then addition of the V_1 preparation would be expected to lead to inhibition of the elicitor activity of the V_2 preparation.

E. Future Prospects

There are a number of possible explanations for the difference in concentration of phytoalexins accumulated by lucerne in response to the nonpathogenic (V_2) and pathogenic (V_1) isolates of *V. albo-atrum*. In this respect, we have so far examined one of these possibilities, that the difference lies in differences in the structure of the component of the pathogen, the elicitor, that is perceived by the host and leads to phytoalexin production. Compared to the whole organism, working with an elicitor preparation, even an unfractionated one such as we have used, represents a simplification whereby the physiological and metabolic capabilities of the pathogen are absent. While this has been helpful in identifying a difference in one of the characteristics of the two isolates, i.e., the ability to induce PAL activity by both elicitors whereas only the V_2 elicitor is efficient at inducing phytoalexin synthesis, aspects such as the capacity of the fungus to metabolize the phytoalexins are not included in this approach.

In addition to studies aimed at further characterizing the two elicitor preparations in order to determine the structural determinant of elicitor activity and to examine whether a suppressor of phytoalexin synthesis is present in the V_1 preparation, experiments will have to be carried out with the fungus as the inducer. Although the complexity of the interaction will be increased greatly by this approach, it will take into account the responsiveness of the fungus to the defense mechanisms of the plant and therefore give a more detailed picture of the host–pathogen interaction.

IV. SIGNAL TRANSDUCTION

A. Background

Phytoalexins are not normally present in healthy plant tissues; they are synthesized following perception of the potential pathogen by the host in a process that requires increases in the activities of the enzymes responsible (Dixon and Harrison, 1990). Generally the induction process leading to phytoalexin synthesis is very rapid, and the increases in enzyme activity that are observed are known to result from increases in the rates of transcription of the specific genes encoding the enzymes of the particular biosynthetic pathway. Many legumes synthesize isoflavanoid phytoalexins and in these tissues increases occur in the activities of several enzymes of the phenylpropanoid pathway. Without the induced increases in the activities of these enzymes phytoalexin synthesis does not occur (Dixon and Harrison, 1990). In bean, for example, accumulation of mRNA for PAL, the first enzyme of the phenylpropanoid pathway, and for chalcone synthase (CHS), the first enzyme from the branch leading to isoflavanoid phytoalexins (Fig. 1), occurs within 10 min of treatment of cells with an elicitor from *Colletotrichum lindemuthianum* (Edwards et al., 1985). Such increases result from activation of the transcription rate of the genes which, in the case of the PAL and CHS genes, occurs within 5 min of the first exposure of cells to elicitor (Lawton and Lamb, 1987).

B. Receptor Sites

Such studies indicate that activation of gene transcription is one of the later events in a process that must begin with perception of the pathogen or elicitor by the cell. The perception process must involve interaction of a molecule that is characteristic of the pathogen, with a receptor in the host. Since these two events—binding of the elicitor and the subsequent activation of gene transcription—do not appear to occur on the same molecule, they must be linked by an intracellular signal transduction system that is capable of carrying specific information between the two leading to the biochemical response. The nature of the signal mechanism that mediates this induction of phytoalexin synthesis has not been characterized so far for any plant but the kinetics of the response in bean indicate that it is a very rapid process. Our own studies with lucerne have indicated that the mechanism is complex and that two signal systems may interact to control the induced synthesis of phytoalexins.

The location and nature of a receptor for the elicitor, whether located on the plasma membrane or at an intracellular site, will have consequences in any consideration of the mechanism of signal transduction involved. If the receptor is located on the plasma membrane, as is the case for the amine and peptide hormones of mammalian systems, then binding must lead to the generation of a second messenger that subsequently interacts with other elements of the cell. In contrast, if the elicitor is able to cross the plasma membrane and interact with a receptor located in the cytoplasm, in a manner similar to that established for the steroid hormones, then a more subtle form of transduction mechanism may operate, one in which the elicitor–receptor complex enters the nucleus and interacts with the DNA directly. In such a case as this, second-messenger molecules are not a feature of the signal transduction mechanism.

A comprehensive discussion of the experimental evidence relating to the location of receptor sites for elicitors of phytoalexin synthesis is beyond the scope of this chapter. It is relevant to note, however, that the most detailed studies to date, carried out in soybean using a specific heptaglucoside elicitor of glyceollin synthesis, have indicated the presence of a protein or glycoprotein receptor on the plasma membrane (Cosio et al., 1990) rather than at an internal receptor site. Hadwiger and his colleagues (1981), however, obtained some evidence that fungal elicitors can be transported into cells of pea, and Kendra et al. (1987) demonstrated a direct interaction of a chitosan elicitor with DNA.

When we first started our investigations, using cell suspension cultures of lucerne and elicitor from the V_2 isolate of *V. albo-atrum*, there had already been some studies in other systems aimed at establishing the role of second messengers in the induction process. The results, at least of some of them, appeared to indicate that the signal, represented by the interaction of elicitor and receptor, was transduced prior to activation of gene transcription rather than there being a direct interaction of elicitor with the genome. Although we do not have direct evidence of the location of the elicitor–receptor interaction in lucerne, our studies have shown that a second-messenger system, similar to that operating in the case of the mammalian amine and peptide hormones, must operate to mediate the phytoalexin response.

So while none of these early studies of other workers had been carried out with lucerne tissues, their findings appeared generally applicable and we considered the possibility that a second messenger could be involved in mediating induction of phytoalexin synthesis in lucerne.

C. Signal Systems

Among the second-messenger molecules that have been identified so far, three are encountered frequently: adenosine 3′,5′-cyclic monophosphate (cyclic AMP), Ca^{2+}, and diacylglycerol (DAG). Cyclic AMP is generated in signal systems by the activity of a receptor-regulated adenylyl cyclase, while Ca^{2+} and DAG are both products of the phosphatidylinositide signal pathway and are known to function in combination in specific agonist-induced responses (Berridge, 1987). All three of these second-messenger molecules have now been implicated as part of the mechanism mediating the induction of phytoalexin synthesis in lucerne.

D. Cyclic AMP in the Induction of Phytoalexin Synthesis

As long ago as 1976 cyclic AMP was implicated in the signal transduction mechanisms mediating phytoalexin synthesis when it was demonstrated that application of cyclic AMP to cells of sweet potato led to the synthesis of terpenoid phytoalexins (Oguni et al., 1976), and our own results are consistent with such a second-messenger role for cyclic AMP. For example, in lucerne cell suspension cultures in the absence of fungal elicitor, dibutyryl cyclic AMP, an analog of cyclic AMP that can cross membranes, was found to induce a fivefold increase in PAL activity that was also accompanied by accumulation of the phytoalexin sativan (Cooke et al., 1989). Normally neither medicarpin or sativan can be detected in lucerne cell suspension cultures unless they are challenged with the fungus or the fungal elicitor. The dependency of the elicitor-induced synthesis of phytoalexins upon an increase in activity of PAL was discussed earlier, and this demonstration of a direct effect of the cyclic AMP analog upon both the increase in PAL activity and phytoalexin accumulation indicates a potential for cyclic AMP to have a role as a second messenger in the response induced by the fungal elicitor.

Until recently the question of the existence of cyclic AMP in plants has been contentious, so that its role as a second messenger in agonist-controlled systems in plants hasn't always received appropriate consideration. However, there is now much evidence to indicate that cyclic AMP and its associated enzymes are present in plants (Brown, 1991), and in lucerne we have used a sensitive radioimmunoassay to detect cyclic AMP at a concentration of 2.5 pmol/g fresh wt cell suspension cultures (Cooke et al., 1989). That identification was subsequently confirmed by mass-analyzed ion kinetic energy mass spectroscopy. In response to treatment of lucerne cells with the V_2 elicitor the intracellular concentration of cyclic AMP rose to reach a maximum concentration of 29.5 pmol/g fresh wt within 4 min of the first challenge with elicitor, and the concentration returned to the basal figure within 7 min of the start of the treatment (Cooke et al., 1989) (Fig. 7).

The rapidity of the rise in cyclic AMP concentration that we observed is comparable to those that occur in responses in mammalian tissues that are controlled by the fast-acting hormones and in which cyclic AMP acts as a second messenger. Kurosaki et al. (1987a), working with suspension cultures of carrot, observed a similar rise in the intracellular concentration of cyclic AMP accompanying elicitor-induced phytoalexin synthesis, though in this case it only reached a maximum after 30 min of treatment of the cells with elicitor.

These same workers were able to demonstrate synthesis in carrot cells of the phytoalexin 6-methoxymellein in response to treatment with cholera toxin, an agonist of adenylyl cyclase that would be expected to increase the intracellular cyclic AMP concentration. The fact that cholera toxin stimulates phytoalexin synthesis implies both that adenylyl cyclase activity is present in carrot suspension cultures and that cyclic AMP is a second messenger in the induction of phytoalexin synthesis. In fact, adenylyl cyclase

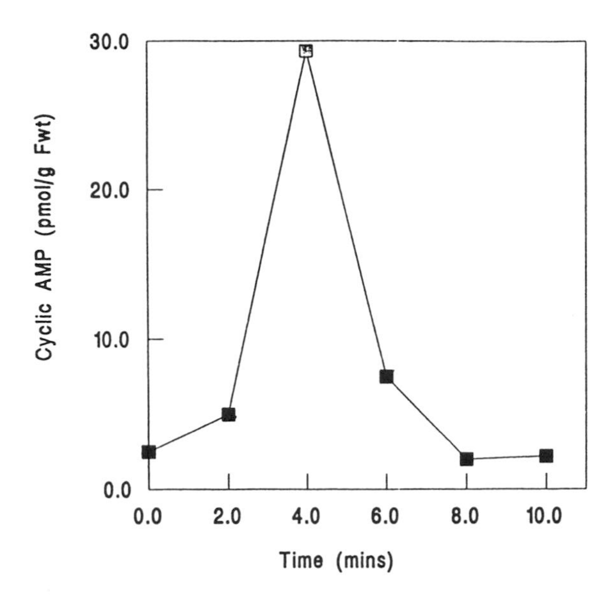

Figure 7 Change in intracellular cyclic AMP concentration in response to treatment of lucerne cell suspension cultures with V_2 elicitor (0.05 mg carbohydrate/ml).

has previously been partially purified from lucerne (Carricarte et al., 1988) and since that report we have also identified adenylyl cyclase activity in lucerne, in a plasma membrane–enriched fraction from cell suspension cultures. In vivo the activity of this enzyme shows a rapid transient rise shortly after treatment of cell suspensions with the V_2 elicitor (Cooke et al., 1994), a finding that, together with its location in a plasma membrane–enriched fraction, lends strong support to a model for signal transduction that features an elicitor-regulated adenylyl cyclase generating second-messenger cyclic AMP.

The activity of the degradative enzyme cyclic nucleotide phosphodiesterase was also observed in cell free homogenates of lucerne (Cooke et al., 1994; Robinson et al., 1992), so that both the means of synthesis and degradation of cyclic AMP have now been identified in lucerne. Both cyclic 2′,3′- and cyclic 3′,5′-AMP are hydrolyzed by the preparation, but we have shown that only the cyclic 3′,5′-AMP activity is regulated by calmodulin (CaM) since it is inhibited by CaM antagonists such as trifluoperazine, calmidazolium, and EGTA, and is stimulated by addition of Ca^{2+} and CaM (Robinson et al., 1992). We have isolated CaM from lucerne cell suspension cultures, using hydrophobic chromatography on phenylsepharose (Robinson et al., 1991), and have shown that it is capable of regulating the activity of the 3′,5′-cyclic nucleotide activity while the 2′,3′-clyclic nucleotide activity is unresponsive.

These studies have provided evidence that the enzymatic means for the synthesis and degradation of cyclic AMP are present in lucerne, and that in response to the fungal elicitor, the biological inducer of the phytoalexin response, a flux of cyclic AMP is generated within the cell. These elements, and their interaction with a cellular agonist, are essential characteristics, of course, of a second-messenger system based on cyclic AMP. The demonstration that a cyclic AMP analog (dibutyryl cyclic AMP) can induce a qualitatively similar response to that induced by the fungal elicitor adds to the evidence that cyclic AMP is involved in mediating the elicitor-induced phytoalexin response.

E. Ca^{2+} and Phytoalexin Synthesis

The results of our studies with lucerne have so far been entirely consistent with a second-messenger role for cyclic AMP in the induction of phytoalexin accumulation. However, some of our earlier work had indicated that induction of the response also depends on an elicitor-stimulated flux of Ca^{2+} across the plasma membrane (Little, 1989). Various related observations, including inhibition by Ca^{2+} antagonists such as La^{3+} and EGTA, of arachidonic acid–induced lubimin synthesis in potato (Zook et al., 1987) and inhibition of elicitor-induced 6-methoxymellein synthesis in carrot cells by the plasma membrane Ca^{2+} channel blocker verapamil (Kurosaki et al., 1987b) have implicated an involvement of intracellular Ca^{2+} fluxes in the induction process.

In many animal systems elevation of intracellular Ca^{2+} acts as a major second messenger in signal transduction systems and many effectors bring about their cellular responses though the generation of raised concentrations of cytosolic Ca^{2+} (Berridge, 1987). There are also some responses in which cyclic AMP and Ca^{2+} are known to operate in concert to control a single response (Rasmussen, 1981), and in this respect identification of a Ca^{2+}/CaM–responsive 3′,5′-cyclic AMP phosphodiesterase may prove to be a link between the signal molecules cyclic AMP and Ca^{2+} (Robinson et al., 1992).

In fact, Ca^{2+} has been shown to be an effective inducer of both an increase in PAL activity and phytoalexin accumulation in lucerne, with the response demonstrating a dose dependency on the ion with a 10-fold increase in PAL activity observed at a Ca^{2+} concentration of 3 mM (Little, 1989; Smith et al., 1989) (Fig. 8). Like the elicitor-induced increase in PAL activity, the effect of Ca^{2+} can be inhibited by the mRNA transcription inhibitor cordycepin indicating that Ca^{2+} causes an increase in gene tran-

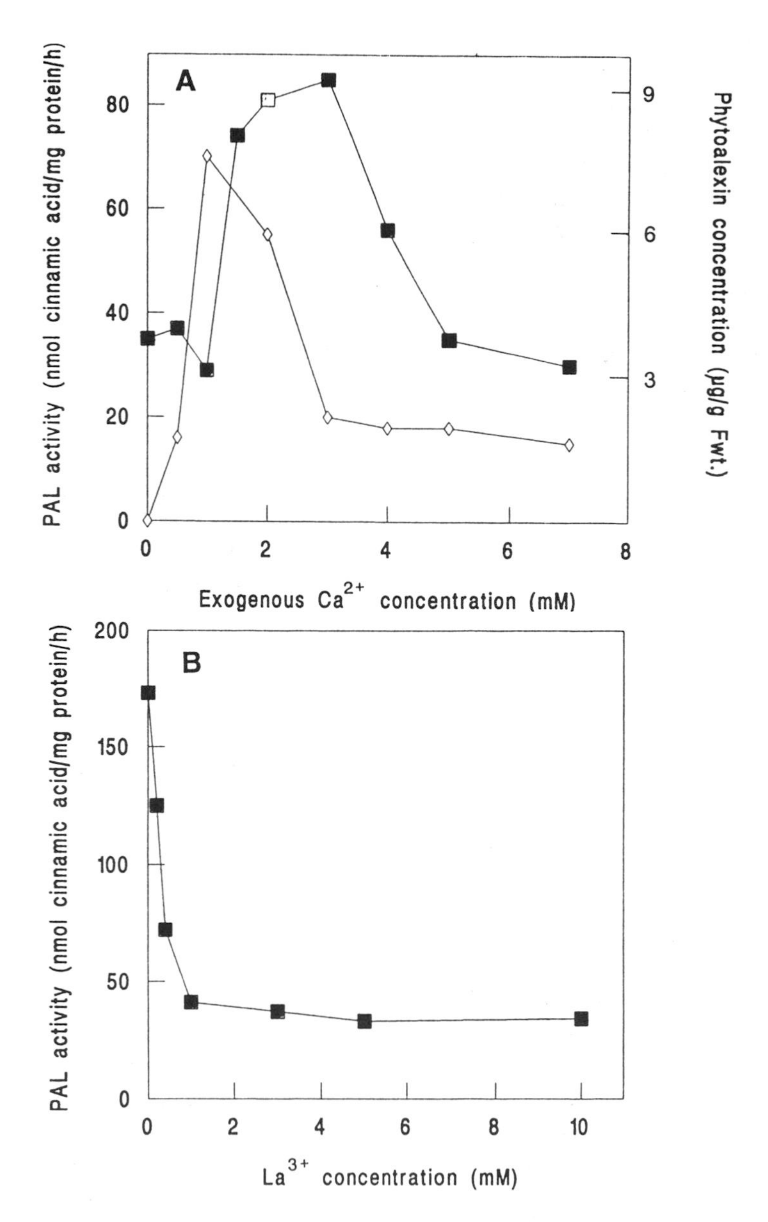

Figure 8 (A) Induction of PAL activity (■-■) and sativan production (◇ - ◇) in lucerne cell suspension cultures by treatment for 4 hr with various concentrations of Ca^{2+}. (B) Effect of the Ca^{2+} antagonist lanthanum on the induction of PAL activity by the elicitor from *V. albo-atrum*. The elicitor was present in each treatment at 0.05 mg of carbohydrate/ml, and PAL activity was measured after 4 hr.

scription and production of mRNA for PAL. By itself an effect of Ca^{2+} on the induction of phytoalexin accumulation does not provide direct evidence that Ca^{2+} has a second-messenger function in the fungal elicitor–induced response. However, compounds that are antagonist to Ca^{2+} and the phytoalexin response induced by Ca^{2+} were also found to be antagonistic to the effect of fungal elicitor in inducing the increase in PAL activity and phytoalexin accumulation. For example, La^{3+}, an ion which competitively inhibits Ca^{2+} movement through plasma membrane channels, effectively prevents the induction brought about by Ca^{2+} and the induction by fungal elicitor when it is added to cell suspension cultures (Fig. 8). Similarly, the metal ion chelator EGTA, or the Ca^{2+} channel blocker verapamil, will prevent the increase in PAL activity and the phytoalexin accumulation normally associated with elicitor treatment of cell suspension cultures (Little, 1989; Smith et al., 1989).

On the other hand, increasing the cytosolic Ca^{2+} concentration, by the addition of lucerne cell cultures of the calcium ionophore A23187 at 5 μM, enhances the ability of the fungal elicitor to increase PAL activity, the concentration of elicitor required to give the maximum increase in its presence reducing to only 10% of that required in the absence of the ionophore (Little, 1989). The concentration of exogenous Ca^{2+} required for maximum effect, in the absence of fungal elicitor, is likewise reduced in the presence of the ionophore, decreasing from 3 mM to less than 1 mM.

Such results provide indirect evidence that a flux of Ca^{2+}, generated across the plasma membrane in response to the elicitor, is a necessary part of the induction process, but they do not provide information regarding the duration or the size of the flow required. Such details will require the further development of intracellular Ca^{2+} imaging systems that are capable of measuring changes in the concentration of Ca^{2+}.

Neither do our present results establish what relationship exists between the requirement for an elicitor-generated flux of Ca^{2+} on the one hand and on the other the activation of gene transcription that is the result of elicitor treatment on the other. Nor is it possible to state whether the change in flux is the "primary" trigger in the sequence. We do have some evidence, however, to indicate that the flux of Ca^{2+} across the plasma membrane must be sustained for a period of about 50 min in order to achieve the maximum attainable elicitor-induced increase in PAL activity, but the way in which such a sustained Ca^{2+} flux can lead to activation of gene transcription will only become evident from further detailed biochemical studies. So, in our current studies, we are using a protoplast system to determine the flux of Ca^{2+} induced across the plasma membrane in response to the elicitor, and the effects of various agonists and antagonists of the signal system upon it, to determine the role of Ca^{2+} cycling across the membrane in the induction process.

F. The Phosphoinositide Signal System

The involvement of a Ca^{2+} flux in the induction process is clearly indicated by such results and, although there is more than one way by which a flux of Ca^{2+} across the plasma membrane may be generated, in many animal systems Ca^{2+}-mobilizing hormones generate second-messenger Ca^{2+} through the phosphoinositide signal transduction system outlined in Fig. 9. Two intracellular signal molecules are generated by this pathway: 1,2-diacylglycerol (DAG) and inositol-1,4,5-trisphosphate (IP_3) (Berridge, 1987). Both result from the activity of a receptor-mediated phosphoinositidase C that catalyzes the cleavage of phosphatidylinositol 4,5-bisphosphate in the plasma membrane. IP_3 is linked to Ca^{2+} as a second messenger because of its ability to release Ca^{2+} from internal sites, while DAG is capable of

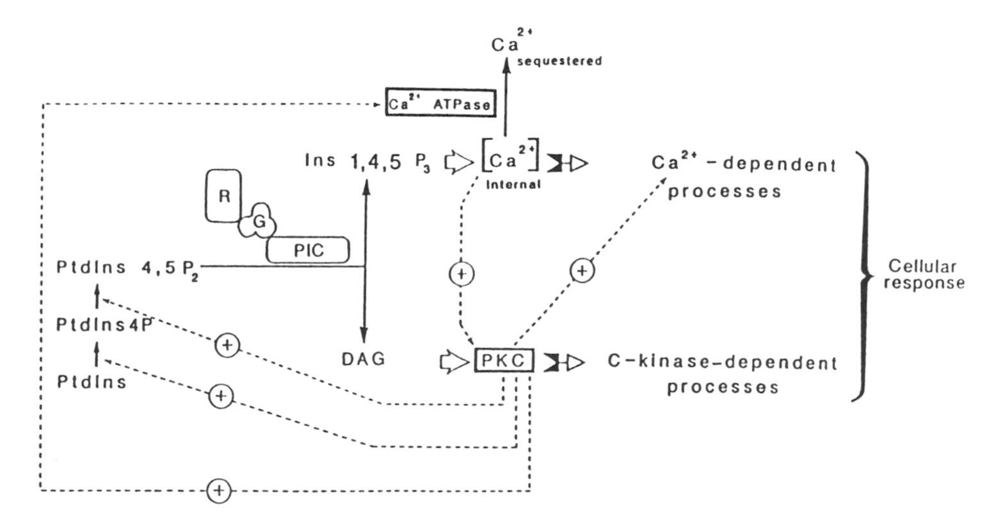

Figure 9 Phosphatidylinositide signal pathway. When an agonist binds to a receptor (R), it activates a G protein (G) which simulates phosphoinositidase C (PIC). Two intracellular signal molecules are generated from the hydrolysis of phosphatidylinositol 4,5-bisphosphate (PtdIns 4,5-P_2), diacylglycerol (DAG), and inositol 1,4,5-trisphosphate (Ins 1,4,5-P_3). Dashed lines represent the interaction of Ca^{2+} and protein kinase C (PKC) with various elements of the system, + indicates a promotion, "Ca^{2+} sequestered" refers to calcium ions which are removed to cellular compartments (vacuole, endoplasmic reticulum, mitochondria) and outside the cell via the activity of Ca^{2+} ATPases.

activating the membrane-associated protein kinase C (Nishizuka, 1986). Although the Ca^{2+} releasing activity of IP_3 appears to be restricted to internal sites (endoplasmic reticulum, tonoplast membrane), the product of its phosphorylation, inositol 1,3,4,5-tetrakisphosphate (IP_4), has been reported to stimulate entry of Ca^{2+} into the cell by way of its effect on Ca^{2+} channels in the plasma membrane (Irvine et al., 1988).

The number of studies in which the role of the phosphoinositide signal pathway in the activation of defense response genes has been examined is somewhat limited, though many of the elements of the system have now been identified in plants including phosphatidylinositol (PI), phosphatidylinositol 4-phosphate (PIP), and phosphatidylinositol 4,5-bisphosphate (PIP_2); Ca^{2+}-dependent phosphoinositidases (enzymes capable of hydrolyzing PIP_2); protein kinases similar to protein kinase C of mammalian systems; an IP_3-stimulated release of Ca^{2+} from intracellular sites; and the kinases and phosphates involved in phosphatidylinositide turnover (Drøbak, 1992).

Labeling experiments with lucerne in which cell suspension cultures have been incubated with [^{3}H]myoinositol and [^{32}P]orthophosphate have established the presence of PI, PIP, and PIP_2, and in response to challenge with V_2 elicitor we have been able to demonstrate a rapid turnover of ^{32}P-labeled PIP and PIP_2 (Walton et al., 1993). The response is rapid, with turnover occurring within 1 min of addition of elicitor. This is not part of a general turnover of membrane phospholipid since phosphatidylcholine and phosphatidylserine are unaffected by elicitor treatment. Such a turnover of PIP and PIP_2 in response to an agonist is typically observed in physiological responses mediated by the phosphoinositide signal transduction system and in these cases it represents an agonist-induced hydrolysis of PIP_2, catalyzed by the receptor-regulated phosphoinositidase C referred to earlier, and which results in the generation of the signal molecules DAG and IP_3. Kurosaki et al. (1987b) detected a breakdown of phosphatidylinositol lipids in carrot cells in response to a fungal elicitor, and the result with lucerne is significant since it implies that, in addition to the presence of the phosphoinositides in the membrane, there is also present a phosphoinositidase responsive to the fungal elicitor. Further, it points to the possibility that both DAG and IP_3 are generated by elicitor treatment.

In fact it has now been possible, using a specific binding protein assay, to identify IP_3 in extracts from lucerne suspension cultures (Cooke et al., 1991a; Walton et al., 1993), and the role of the phosphoinositide signal system in mediating elicitor-induced phytoalexin synthesis has been further indicated by measurement of the intracellular IP_3 concentration in response to the fungal elicitor.

When lucerne cells were treated with V_2 elicitor, the intracellular concentration of IP_3 rose within 1 min of treatment to reach a maximum of 60 pmol/g fresh weight (Cooke et al., 1991b). The occurrence of this transient increase (the concentration returned to basal level within 15 min) in IP_3 concentration together with the demonstrated turnover of PIP and PIP_2 is consistent with an agonist-induced phosphoinositide hydrolysis system operating during the early phase of the induction process. In this context compound 40/80, an inhibitor of phospholipase C, effectively inhibits the elicitor-induced increase in PAL activity and phytoalexin synthesis (Little, 1989), implying that induction of the response by the elicitor depends on the activity of the phosphoinositide signal transduction system.

A central transducer of the Ca^{2+} message, especially in the type of response described here in which sustained cycling of Ca^{2+} across the plasma membrane appears to be a feature, is protein kinase C, an enzyme that is activated by the increase in intracellular Ca^{2+} concentration and that subsequently affects the rate of cellular reactions by phosphorylating key enzymes (Nishizuka, 1986). 1-Oleoyl-2-acetylglycerol, a synthetic diacylglycerol that can activate protein kinase C in the same way that endogenous DAG produced by the phosphoinositide pathway does in vivo, at a concentration of 1 μM stimulated an increase in PAL activity and accumulation of phytoalexin to levels comparable with those achieved by the V_2 elicitor (Little, 1989). Similar increases were obtained when cells were treated with phorbol 12-myristate-13-acetate, a potent activator of protein kinase C. Induction of the phytoalexin response by activators of protein kinase C in a manner similar to that achieved by fungal elicitor implies that this type of kinase is involved in the induction process. In contrast, inhibition of the elicitor-induced response in lucerne by 1-(5-isoquinoline sulfonyl)-2-methylpiperazine and calphostin C, both of which are potent and selective inhibitors of protein kinase C, reinforces the results obtained with the protein kinase C activators (Smith et al., unpublished).

Collectively these results offer considerable evidence that an agonist-induced phosphoinositide hydrolysis operates during the elicitor-activated synthesis of phytoalexins in lucerne. We have been able to demonstrate the presence of the essential phosphoinositides and their turnover in response to fungal elicitor, while elicitor treatment of lucerne cells has been shown to cause a rapid transient increase in IP_3 concentration, characteristic of phosphoinositide signal transduction systems. The fact that compound 40/80 effectively inhibits elicitation of PAL activity and phytoalexin accumulation by the fungal elicitor, together with the fact that agonists and antagonists of protein kinase C are also agonists and antagonists of the elicitation by fungal elicitor, offers good support for a role for the phosphoinositide pathway in signal transduction mediating the phytoalexin response.

G. Future Development

The results from lucerne and the results of others have given us a good deal of encouragement in our efforts to characterize the signal system involved in the phytoalexin response. Evidence now exists that two signal systems, the cyclic AMP and the phosphoinositide pathways, are involved in mediating the phytoalexin response induced by the fungal elicitor. While at first glance there may appear to be a conflict in a model for signal transduction that features two signal systems, there are physiological responses in animal cells that are known to be regulated by the two systems operating together (Bolander, 1989). In these cases cross-regulation is known to occur and one of the areas which will continue to be of interest to us will be characterization of the way in which the two systems interact and examination of the system for evidence of regulatory elements between and within the pathways. This will require isolation of the relevant enzymes and determination of their kinetic parameters as well as identifying endogenous agonists and antagonists of their activities.

In this respect since protein kinases, those controlled by Ca^{2+}/CaM as well as the C- and cyclie AMP–dependent type, are central mediators of the second messengers Ca^{2+}, DAG, and cyclic AMP, we are currently examining lucerne tissue for the presence of these kinases. Our initial investigations have revealed the presence of both cyclic AMP and Ca^{2+}/CaM–dependent kinases, and it will be interesting to determine what their endogenous protein substrates are since these may be the ultimate targets of the transduction system.

In the future too it will be necessary to establish how the various elements of the transduction system are focused to affect the rate of gene transcription and to this end a number of studies of the molecular biology of the induction process are being carried out (Dixon and Harrison, 1990). The role of Ca^{2+} fluxes, an essential feature of the response, will continue to be an area of interest in which it will be important to determine the mechanism by which the fungal elicitor can induce such a change and how the flux affects the cell biochemistry.

V. EPILOG

Defense against disease in plants involves a number of mechanisms, some structural, some taking the form of preexisting chemical barriers, and some only being called into play once the attack by a pathogen has been detected by the plant. Of this latter category, the inducible defense responses, synthesis of phytoalexins has attracted a great deal of attention, while there are many cases of resistance in which phytoalexin synthesis is not the key event equally there are instances of resistance to a potential pathogen in which the ability of a host to rapidly synthesize phytoalexins is the major factor determining the outcome of the interaction. Various aspects of the response have been studied so far, including characterization of elicitor components; the kinetics of accumulation in the host; the biochemistry of the pathway; the ability of the pathogen to suppress synthesis or detoxify the phytoalexins; and, latterly, the molecular biology of the response. The complexity of host–pathogen interactions is such that no single approach can be expected to evaluate the role phytoalexin synthesis plays in the resistance of the host. In combination, however, particularly with the newer techniques of molecular biology available, that task is becoming easier.

REFERENCES

Aubé, C., and Sackston, W. E. (1964). Verticillium wilt of forage legumes in Canada. *Can. J. Plant Sci. 44*:427–432.

Berridge, M. J. (1987). Inositol trisphosphate and diacylglycerol: two interacting second messengers. *Ann. Rev. Biochem. 56*:159–193.

Bolander, F. M. (1989). *Molecular Endocrinology*. Academic Press, New York.

Bolwell, G. P., Bell, J., Cramer C., Schuch, W., Lamb, C. J., and Dixon, R. A. (1985). L-Phenylalanine ammonia-lyase from *Phaseolus vulgaris*. *Eur. J. Biochem. 149*:411–419.

Brown, E. G. (1991). Purines, pyrimidines, nucleosides and nucleotides. *Methods in Plant Biochemistry*, Vol. 5 (L. J. Rogers, ed.), Academic Press, London, pp. 53–90.

Carricarte, V. C., Bianchini, G. M., Muschietti, J. P., Tellez-Inon, M. T., Perticari, A., Torres, N., and Flawia, M. M. (1988). Adenylate cyclase activity in a higher plant, alfalfa (*Medicago sativa*). *Biochem. J. 249*:807–811.

Cocker, W., McMurry, T. B. H., and Staniland, P. A. (1965). A synthesis of demethylhomopterocarpin. *J. Chem. Soc.* 1034–1037.

Cooke, C. J., Newton, R. P., Smith, C. J., and Walton, T. J. (1989). Pathogenic elicitation of phytoalexin in lucerne tissue: involvement of cyclic AMP in the intracellular mechanism. *Biochem. Soc. Trans. 17*:919–920.

Cooke, C. J., Smith, C. J., Newton, R. P., and Walton, T. J. (1991a). Binding saturation analysis of inositol 1,4,5-trisphosphate in suspension cultures of lucerne cells. *Biochem. Soc. Trans. 19*:59S.

Cooke, C. J., Smith, C. J., Newton, R. P., and Walton, T. J. (1991b). Effects of phytoalexin elicitor on levels of inositol phosphates in lucerne cells in suspension culture. *Biochem. Soc. Trans. 19*:94S.

Cooke, C. J., Smith, C. J., Walton, T. J., and Newton, R. P. (1994). Evidence that cyclic AMP is involved in the hypersensitive response of *Medicago sativa* to a fungal elicitor. *Phytochemistry, 35*:889–895.

Cosio, E. G., Frey, T., Verduyn, R., van Boom, J., and Ebel, J. (1990). High affinity binding of a synthetic heptaglucoside and fungal phytoalexin elicitor to soyabean membranes. *FEBS Lett. 271*:223–226.

Cruickshank, I. A. M., and Perrin, D. R. (1960). Isolation of a phytoalexin from *Pisum sativum* L. *Nature 187*:779–800.

Daniel, S., Hinderer, W., and Barz, W. (1988). Elicitor-induced changes of enzyme activities related to isoflavone and pterocarpan accumulation in chickpea (*Cicer arietinum* L) cell suspension cultures. *Z. Naturforschung. C Bioscience 43c*: 536–544.

Degra, L., Salvi, G., Mariotti, D., De Lorenzo, D., and Cervone, F. (1988). A polygalacturonase-inhibiting protein in alfalfa callus cultures. *J. Plant Physiol. 133*:364–366.

Dewick, P. M. (1975). Pterocarpan biosynthesis: chalcone and isoflavone precursors of demethylhomopterocarpin and maackiain in *Trifolium pratense*. *Phytochemistry 14*:979–982.

Dewick, P. M. (1977). Biosynthesis of pterocarpan phytoalexins in *Trifolium pratense*. *Phytochemistry 16*:93–97.

Dewick, P. M., and Martin, M. (1976). Biosynthesis of isoflavanoid phytoalexins in *Medicago sativa*: the biosynthetic relationship between pterocarpans and 2′-hydroxyisoflavans. *J. Chem. Soc. Chem. Commun*. 637–638.

Dewick, P., and Martin, M. (1979). Biosynthesis of pterocarpan and isoflavanoid phytoalexins in *Medicago sativa*. The biochemical interconversion of pterocarpans and 2′-hydroxyisoflavans. *Phytochemistry 18*:591–596.

Dixon, R. A. (1986). The phytoalexin response: elicitation, signalling and control of host gene expression. *Biol. Rev. 61*:239–291.

Dixon, R. A., and Harrison, M. J. (1990). Activation, structure and organisation of genes involved in microbial defense in plants. *Adv. Genet. 28*:165–234.

Drøbak, B. K. (1992). Plant signal perception and transduction: the role of the phosphoinositide system. *Essays in Biochemistry*, Vol. 26 (K. F. Tipton, ed.), Portland Press, Colchester, pp. 27–37.

Edwards, C., and Strange, R. N. (1991). Separation and identification of phytoalexins from leaves of groundnut (*Arachis hypogaea*) and development of a method for their determination of reversed-phase high performance liquid chromatography. *J. Chromatogr. 547*:185–193.

Edwards, K., Cramer, C. L., Bolwell, G. P., Dixon, R. A., Schuch, W., and Lamb, C. J. (1985). Rapid transient induction of phenylalanine ammonia-lyase mRNA in elicitor-treated bean cells. *Proc. Nat. Acad. Sci. USA 82*:6731–6735.

Evans, D. J. (1986). PhD thesis, University of Wales.

Farkas, L., Gottsegen, A., Nogradi, M., and Antus, S. (1974). Synthesis of sophorol, violanone, lonchocarpan, claussequinone, philenopteran, leiocalycin and some other natural Isoflavanoids by the oxidative rearrangement of chalcones with thallium III nitrate. *J. Chem. Soc. Perkin I*:305–308.

Flood, J., Khan, F. Z., and Milton, J. M. (1978). The role of phytoalexins in *Verticillium* wilt of lucerne. *Ann. Appl. Biol. 89*:329–332.

Graham, J. H., Peaden, R. N., and Evans, D. W. (1977). *Verticillium* wilt of alfalfa found in the United States. *Plant Dis. Rep. 61*:337–340.

Hadwiger, L. A., Beckman, J. M., and Adams, M. J. (1981). Localisation of fungal components in the pea-*Fusarium* interaction detected immunochemically with antichitosan and antifungal cell wall antisera. *Plant Physiol. 67*:170–175.

Hahlbrock, K., and Scheel, D. (1989). Physiology and molecular biology of phenylpropanoid metabolism. *Ann. Rev. Plant Physiol. Plant Mol. Biol. 40*:347–369.

Hamdan, M. A. M. S., and Dixon, R. A. (1987). Differential patterns of protein synthesis in bean cells exposed to elicitor fractions from *Colletotrichum lindemuthianum. Physiol. Mol. Plant Pathol. 31*:105–121.

Hansen, H. R., and Weber, A. (1948). Plant diseases and pests in Denmark in 1945. *Tidsskrift for Planteavl 51*:434.

Hedlund, T. (1923). Om Nagra sjukdomar och skador pa vara lantbruksvaxter 1922. *Sveriges Allmanna Jordbrukstidskrift 5*:167–168 (cited by Lundin, P., and Jonsson, H., 1975, *Agri Hortique Genetica 33*:17–32).

Higgins, V. J. (1972). Role of the phytoalexin medicarpin in three leaf spot diseases of alfalfa. *Physiol. Plant Pathol. 2*:289–300.

Higgins, V., and Millar, R. (1968). Phytoalexin production by alfalfa in response to infection by *Colletotrichum phomoides, Helminthosporium turcicum, Stemphylium loti* and *S. botryosum. Phytopathology 58*:1377–1383.

Irvine, R. F., Moor, R. M., Pollock, W. K., Smith, P. M., and Wregett, K. A. (1988). Inositol phosphates: proliferation, metabolism and function. *Phil. Trans. R. Soc. London B320*:281–298.

Isaac, I. (1957a). Wilt of lucerne caused by species of *Verticillium. Ann. Appl. Biol. 45*:550–558.

Isaac, I. (1957b). *Verticillium* wilt of Brussel sprouts. *Ann. Appl. Biol. 45*:276–283.

Isaac, I., and Heale, J. B. (1961). Wilts of lucerne caused by species of *Verticillium.* III. Viability of *V. albo-atrum* carried with lucerne seed; effects of seed dressings and fumigants. *Ann. Appl. Biol. 49*:675–691.

Isaac, I., and Lloyd, A. T. E. (1959). Wilt of lucerne caused by species of *Verticillium.* II. Seasonal cycle of disease; range of pathogenecity; host-parasite relations; effects of seed dressings. *Ann. Appl. Biol. 47*:673–684.

Jorrin, J., and Dixon, R. A. (1990). Stress responses in Alfalfa (*Medicago sativa* L.) II. Purification characterisation and induction of phenylalanine ammonia-lyase isoforms from elicitor treated cell suspension cultures. *Plant Physiol. 92*:447–455.

Keen, N. (1978). Phytoalexins: efficient extraction from leaves by a facilitated diffusion technique. *Phytopathology 68*:1237–1239.

Keen, N. T. (1990). Phytoalexins and their elicitors. *American Chemical Society Symposium*, Vol. 439, pp. 114–129.

Kendra, D. F., Fritensky, B., Daniels, C. H., and Hadwiger, L. A. (1987). Disease resistance response genes in plants: expression and proposed mechanisms of induction. *Molecular Strategies for Crop Protection* (C. J. Arntzen and C. Ryan, eds.), Alan R. Liss, New York, pp. 13–24.

Kessmann, H., and Barz, W. (1986). Elicitation and suppression of phytoalexin and isoflavone accumulation in cotyledons of *Cicer aerietinum* L. as caused by wounding and by polymeric components from the fungus *Ascochyta rabiei. J. Phytopathol. 117*:321–335.

Khan, Z. F., and Milton, J. M. (1975). Phytoalexin production by lucerne (Medicago sativa L.) in response to infection by *Verticillium. Physiol. Plant Pathol. 7*:179–187.

Khan, Z. F., and Milton, J. M. (1978). Phytoalexin production and the resistance of lucerne (*Medicago sativa* L.) to *Verticillium albo-atrum*. *Physiol. Plant Pathol. 13*:215–221.

Khan, Z. F., and Milton, J. M. (1979). Some factors affecting the production of medicarpin and sativan by lucerne leaflets in response to *V. albo-atrum*. *Physiol. Plant Pathol. 14*:11–17.

Kreitlow, K. W. (1962). *Verticillium* wilt of alfalfa. A destructive disease in Britain and Europe not yet observed in the United States. United States Department of Agriculture, Agricultural Research Service, ARS 34-20:1–15.

Kurosaki, F., Tsurusawa, Y., and Nishi, A. (1987a). The elicitation of phytoalexins by Ca^{2+} and cyclic AMP in carrot cells. *Phytochemistry 26*:1919–1923.

Kurosaki, F., Tsurusawa, Y., and Nishi, A. (1987b). Breakdown of phosphatidylinositol during the elicitation of phytoalexin production in cultured carrot cells. *Plant Physiol. 85*:601–604.

Lawton, M. A., and Lamb, C. J. (1987). Transcriptional activation of plant defense genes by fungal elicitor, wounding and infection. *Mol. Cell. Biol. 7*:335–341.

Little, J. P. (1989). PhD thesis, University of Wales.

Mayama, S., Tani, T., Matsuura, Y., Veno, T., and Fukami, H. (1981). The production of phytoalexins by oat in response to crown rust. *Puccinia coronota* f. sp. *avenae*. *Physiol. Plant Pathol. 19*:217–226.

Murashige, T., and Skoog, F. (1962). A revised medium for rapid growth and bioassays with tobacco tissue cultures. *Physiologica Plantarum 15*:473–497.

Nelson, R. (1947). The specific pathogenesis of the *Verticillium* that causes wilt of peppermint. *Phytopathology 37*:17.

Newcombe, G., and Robb, J. (1989). The chronological development of a lipid-to-suberin response at *Verticillium* trapping sites in alfalfa. *Physiol. Mol. Plant Pathol. 34*:55–73.

Nishizuka, Y. (1986). Studies and perspectives of Protein kinase C. *Science 233*: 305–312.

Noble, M., Robertson, N. F., and Dowson, W. J. (1953). *Verticillium* wilt of lucerne in Britain. *Plant Pathol. 2*:31–33.

Oguni, I., Suzuki, K., and Uritani, I. (1976). Terpenoid induction in Sweet Potato roots by Adenosine 3′,5′-cyclic monophosphate. *Agric. Biol. Chem. 40*:1251–1252.

Onuorah, O. (1987). Isolation of pathogen-synthesised fraction that elicits phytoalexins in lucerne (*Medicago sativa* L.). *Acta Biologica Hungaria 38*:247–256.

Pierce, M., and Essenberg, M. (1987). Localization of phytoalexins in fluorescent mesophyll cells isolated from bacterial blight-infected cotton cotyledons and separated from other cells by fluorescence-activated cell sorting. *Physiol. Mol. Plant Pathol. 31*:273–290.

Rasmussen, H. (1981). *Calcium and cAMP as Synarchic Messengers*. Wiley, New York.

Richter, H., and Klinkowski, M. (1938). Wirtelpilz-Welkekrankheit an Luzerne und Exparsette (Erreger: *Verticillium albo-atrum* Rke. et Berth). *Nachrichtenblatt für den Deutschen Pflanzenschutzdienst Berlin 18*:57–58.

Robinson, P. S., Newton, R. P., Walton, T. J., and Smith, C. J. (1991). Calmodulin from lucerne: its potential role in signal transduction. *Biochem. Soc. Trans. 19*:190S.

Robinson, P. S., Newton, R. P., Walton, T. J., and Smith, C. J. (1992). Cyclic nucleotide phosphodiesterase activity in *Medicago sativa* L. *Biochem. Soc. Trans. 20*:3555.

Sharp, J. K., McNeil, M., and Albersheim, P. (1984). Host pathogen interactions 27. The primary structures of one elicitor active and seven elicitor inactive hexa (β-D-glucopyranosyl)-D-glucitols isolated from the mycelial walls of *Phytophthora* megasperma f. sp. *glycinea. J. Biol. Chem. 259*:11321–11336.

Smith, C. J. (1994). Phytoalexin accumulation: a defence response. *New Phytologist* (in press).

Smith, C. J., Newton, R. P., Mullins, C. J., and Walton, T. J. (1989). Plant host-pathogen interaction: elicitation of phenylalanine ammonia-lyase activity and its mediation by Ca^{2+}. *Biochem. Soc. Trans. 16*:1069–1070.

Smith, C. J., and Milton, J. M. (1992). Phytoalexin accumulation in the forage crop. *Medicago sativa*: a defence against the soil-borne wilt pathogen *Verticillium albo-atrum. Indian Phytopathol. 45*:1–6.

Van Etten, H. D., Mathews, D. E., and Mathews, P. S. (1989). Phytoalexin detoxification: importance for pathogenicity and practical application. *Ann. Rev. Phytopathol. 27*:143–164.

Waldmüller, T., and Grisebach, H. (1987). Effects of R-(1-amino-2-phenylethyl) phosphonic acid on glyceollin accumulation and expression of resistance to *Phytophthora megasperma* f. sp. *glycinea* in soyabean. *Planta 172*:424–430.

Williams, C. A., and Harborne, J. B. (1990). Isoflavanoids. *Methods in Plant Biochemistry Vol. 1, Plant Phenolics.* (P. M. Dey and J. B. Harborne, eds.), Academic Press, New York.

Zaki, A., Keen, N., and Erwin, D. (1972). Implication of vergosin and hemigossypol in the resistance of cotton to *Verticillium albo-atrum. Phytopathology 62*: 1402–1406.

Zook, M., Rush, J., and Kuc, J. (1987). A role for Ca^{2+} in the elicitation of Rishitin and Lubimin accumulation in potato tuber tissue. *Plant Physiol. 84*:520–525.

19

Stereoselective Synthesis of Spirovetivane-Type Phytoalexins

Chuzo Iwata and Yoshiji Takemoto
Osaka University, Osaka, Japan

I. INTRODUCTION

Phytoalexins are common in all the natural product groups, e.g., isoflavonoids, furanoterpenoids, sesquiterpenoids, polyacetylenes, and dihydrophenanthrenes (Tomiyama, 1971; Grisebach and Abel, 1978). They were defined in 1940 as defensive substances possessing antimicrobial properties which were produced by host plants after infection (Müller and Börger, 1940), but now they are recognized to be induced by products of microbial origin or stress treatment (injury, heat, UV light, etc.) as well as living microorganisms (Masamune et al., 1978). For example, representative sesquiterpene phytoalexins isolated from the Solanaceae family (potato, tobacco, etc.) are listed in Fig. 1 (**1–7**) (Stoessl et al., 1976). These compounds involve unique structures. In particular, phytoalexins (**4–7**) belong to spirovetivane sesquiterpenes bearing a spiro[4.5]decane skeleton.

Spirovetivanes are structurally characterized by the presence of a methyl group at C-10 and an isopropyl group at C-2 in the spiro[4.5]decane skeleton, and are furthermore divided into two groups depending on the relative configuration between the C-1–C-5 bond and methyl group at C-10. One group, represented by hinesol (**8**), and β-vetivone (**9**), has the cis configuration (C-10: β-methyl), which was isolated from vetiver oil as a fragrant principle. The other, as illustrated in Fig. 1 (**4–7**), has the trans configuration (C-10: α-methyl), which is known as spirovetivane-type phytoalexin. Since rishitin (**1**) was first isolated from tuber tissue of diseased white potatoes as phytoalexin in 1968 (Katsui et al., 1968), spirovetivane-type phytoalexins such as solavetivone (**5**), lubimin (**6**), and oxylubimin (**7**) have been successively discovered (Murai et al., 1982d).

As plants do not possess an immune system, these findings might be an answer to the question of how plants can protect themselves against pathogenic agents. In order to identify defense mechanisms of plants, much effort has been made for structural determination of these phytoalexins, rishitin (1) rishitinol (2) phytuberin (3)

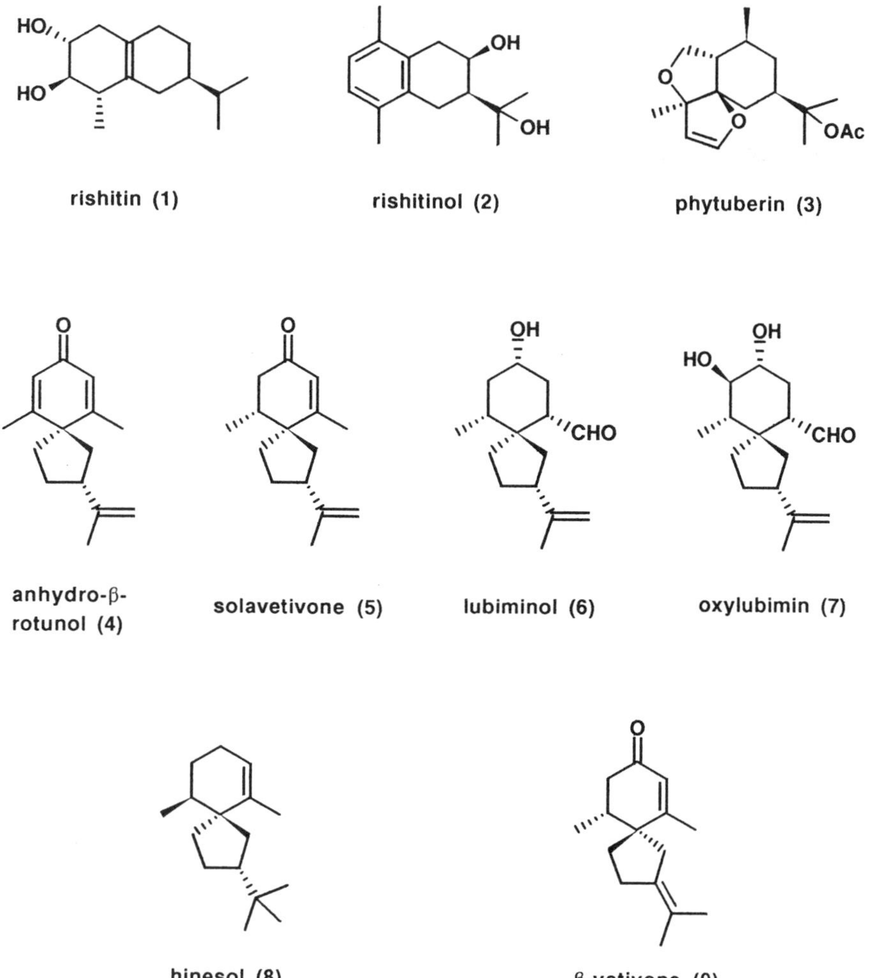

anhydro-β-rotunol (4) solavetivone (5) lubiminol (6) oxylubimin (7)

hinesol (8) β-vetivone (9)

Figure 1 Sesquiterpene phytoalexins isolated from Solanaceae.

total synthesis, and elucidation of biosynthetic pathway and their inductive mechanism. As a result, several groups succeeded in total synthesis of spirovetivane-type phytoalexins (**5–7**), which helps to reveal that these phytoalexins are biosynthetic intermediates of rishitin (**1**), exhibiting the strongest antifungal activity (Stoessl and Stothers, 1983; Coolbear and Threlfall, 1985). However, much remains unknown and further investigations should be able to reveal defense mechanisms of plants in detail. Major goals for phytoalexin research are the identification of new analogs, which are more effective and less toxic with a broader spectrum of antifungal or antibacterial activity. The aim is to use phytoalexins as agricultural chemicals and to elucidate the inductive or regulative mechanism of phytoalexins against pathogenic factors.

This chapter deals mainly with the organic synthesis of spirovetivane-type phytoalexins. For the purpose of preparing phytoalexin analogs and elucidating the biosynthetic pathway, it is imperative to synthesize all the highly oxygenated spirovetivane-type phytoalexins. Many synthetic strategies, especially spiroannulation techniques, have been developed in more than a decade by many synthetic chemists (Marshall et al., 1974; Krapcho, 1976, 1978; Hiroi, 1977; Murai, 1981). Herein we describe our racemic syntheses of solavetivone (**5**) and oxylubimin (**7**) and also some elaborations for asymmetrical synthesis of (−)-solavetivone in detail together with others' synthetic approaches. For a review of other phytoalexins and biochemical aspects, see Friend and Threlfall, 1976; Keen and Bruegger, 1977; Nakajima, 1978; and Tomiyama, 1981.

II. SYNTHETIC STUDIES ON SPIROVETIVANE PHYTOALEXINS

A. Synthesis of (±)-Solavetivone

Solavetivone (**5**) is a representative member of the spirovetivane-type phytoalexins, which was obtained from diseased potatoes in 1974 (Coxon et al., 1974) as well as from air-cured tobacco leaves in 1977 (Fujimori et al., 1977). Solavetivone has been proven to play an important role in the biosynthetic pathway leading to formation of rishitin (Murai et al., 1982a). Its structure was synthetically determined by Yamada et al. (1977). The crucial problem in the synthesis of (±)-solavetivone is the construction of the spiro[4.5]decane system, and thus far, employing different spirocyclization techniques, three groups have succeeded in its total synthesis.

The first synthesis was accomplished via utilization of intramolecular acid-catalyzed cyclization of acetal (**10**) (Fig. 2) (Yamada et al., 1977). The obtained compound (**11**) is a useful intermediate, which has also been converted to (±)-α-vetispirene and (±)-hinesol (Yamada et al., 1973b). Although incorporation of the three-carbon unit at C-2 of **12** proceeded nonstereoselectively to afford **13** and **14**, successive transformation of the minor adduct (**14**) via **15** produced (±)-solavetivone (**5**) as a sole product.

The second approach involves an acid-promoted π cyclization of bicyclic compounds (endo-**18** and exo-**18**), which were prepared from Diels–Alder adducts (endo-**16** and exo-**16**), respectively (Murai et al., 1981a,b, 1984a,b). Endo-**18** and exo-**18** showed similar reactivity under the acid

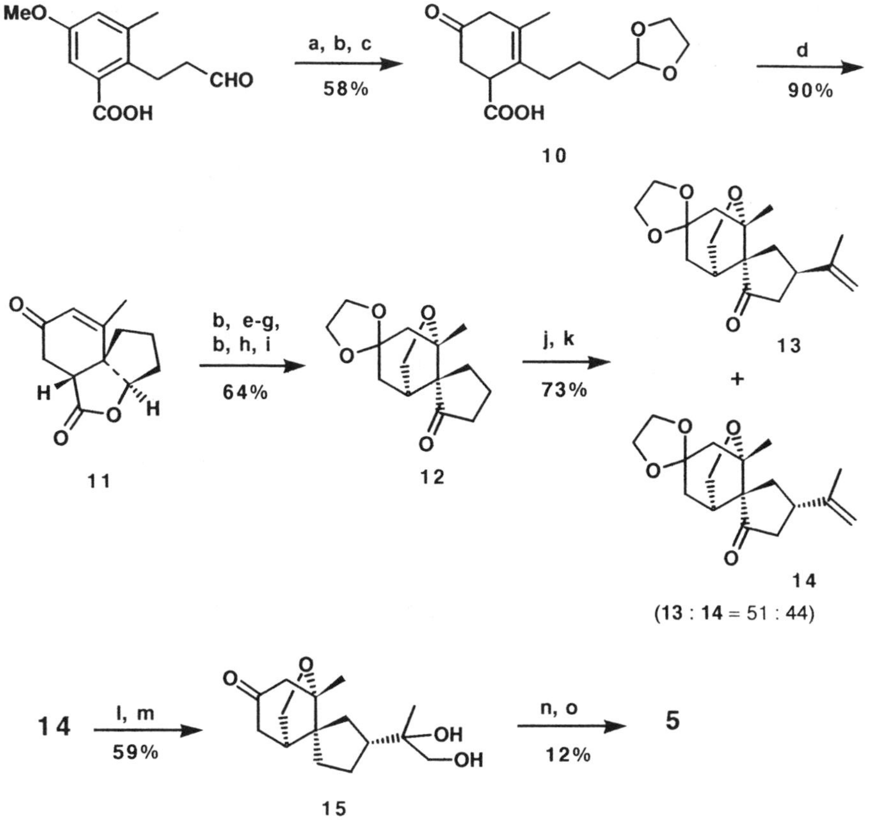

Figure 2 (a) $Ph_3P{=}CHOMe$; (b) $(CH_2OH)_2$, H^+; (c) Li, liq. NH_3, *tert*-BuOH, THF; aq. $(CO_2H)_2$; (d) 6 M HCl, DME, reflux; (e) $LiAlH_4$, THF; (f) 2 M HCl, DME, r.t.; (g) Ac_2O, pyridine; (h) KOH, MeOH; (i) CrO_3, pyridine; (j) LDA, THF; PhSSPh; mCPBA, CH_2Cl_2; $NaHCO_3$, toluene, reflux; (k) $CH_2{=}CMeMgBr$, CuI, THF; (l) $H_2NNH_2 \cdot H_2O$, KOH, DEG, reflux; (m) OsO_4, pyridine, THF, r.t.; $(CO_2H)_2$, aq. MeOH, r.t.; (n) $(PhO)_3P \cdot MeI$, $BF_3 \cdot Et_2O$, MeCN, r.t.; Zn, NH_4Cl, aq. EtOH, r.t.; (o) diimidazol-1-yl thioketone, MeCOEt; $(MeO)_3P$, reflux. r.t., room temperature.

conditions, yielding the aimed compound **(19)** and its dehydrated compound **(20)**, respectively. The major product **(19)** was heated with pyridine-modified alumina to furnish the desired compound **(5)** in 13.7% overall yield from the starting material (Fig. 3).

At the beginning of research, our first concern was to develop a general and more stereoselective methodology which could generate the spirocyclic system. Our first attempt, a photolysis of sodium salts of *p*-[2-(2-halogenomethyl-1,3-dioxolan-2-yl)ethyl]phenol, gave desired spirodienone, but only in moderate yields (Iwata et al., 1974). In the second stage, we tried a strategy utilizing a carbenoid intermediate (Iwata et al., 1981b, 1987). Starting from phenolic carboxylic acid **(21)**, the diazoketone **(22)** was prepared in 65% overall yield. When **22** was heated in the presence of cuprous chlo-

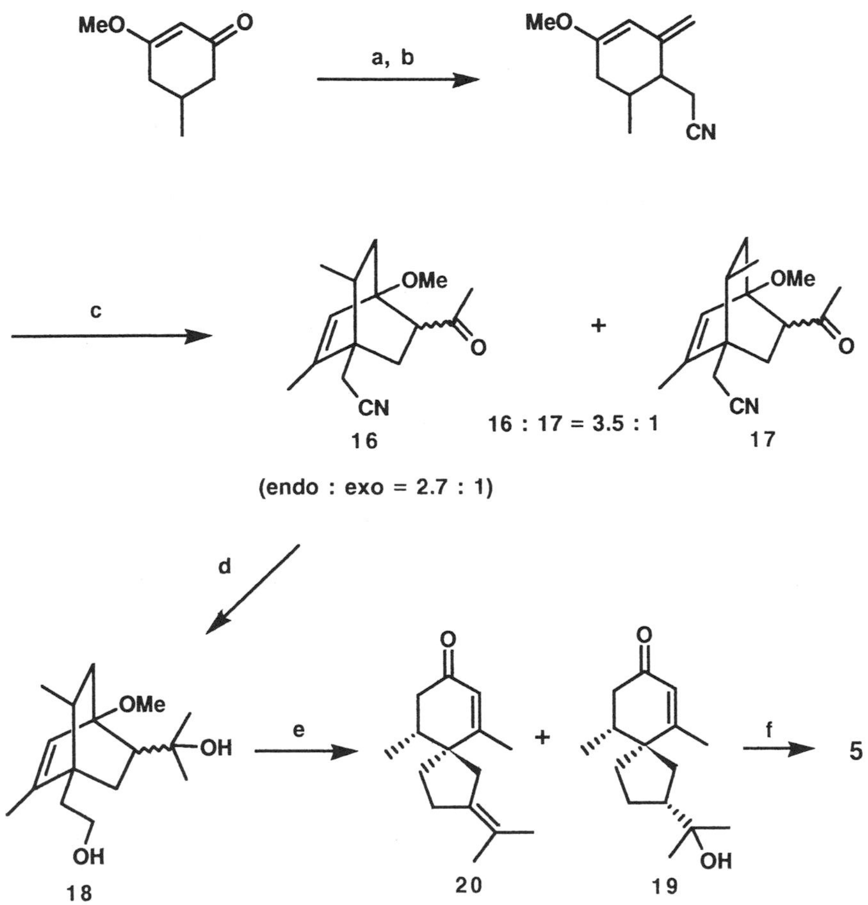

MeO O a, b MeO CN

c OMe O CN 16 + OMe O CN 17

16 : 17 = 3.5 : 1

(endo : exo = 2.7 : 1)

d

OMe OH OH 18 e O 20 + O OH 19 f 5

Figure 3 (a) LDA, HMPA, THF; $ClCH_2CN$ (94%); (b) $Ph_3P \cdot MeI$, NaH, DMSO (71%); (c) $MeCOCH{=}CH_2$, dichloromaleic anhydride, C_6H_6, reflux (**16** 59% and **17** 17%); (d) MeLi, Et_2O, $-78°C$; DIBAL, Et_2O, 0°C; $NaBH_4$, THF, H_2O, 0°C (endo 96%, exo 75%); (e) MsCl, pyridine; $(CO_2H)_2$, 33% aq. acetone [endo: **19**(63%), **20**(23%), exo: **19**(69%), **20**(26%)]; (f) pyridine-Al_2O_3, 220°C (60%).

ride in benzene, cyclization took place smoothly to give the aimed spirodienone **(23)** in 56% yield. This carbenoid-mediated cyclization proceeds under mild conditions and turns out to be applicable to many other phenolic diazoketones (Iwata et al., 1977, 1980). Inspection of minor product, dihydroazulene derivative **(24)**, indicates that the catalytic decomposition of phenolic α-diazoketones involves norcaradiene intermediate **(25)**. Reduction of the prochiral ketone **(23)** by lithium tri-*tert*-butoxyaluminum hydride gave monoalcohol **(26)**, which was subjected to Birch reduction providing possible four diastereoisomers **(27a, 27b, 27c, 27d)** in 78% yield in the ratio of 91 : 7 : 1 : 1 (Fig. 4). This stereoselectivity can be explained by the intramolecular proton migration from the hydroxyl group to formed dienolate dianion (Fig. 5). That is, two possible epimeric dianion (or radical anion) transition states, A (chair form) and B (boat or half-chair form), might be considered in the reduction process. There is a strong interaction between the alcohol oxygen and the methyl group at C-10 in the transition A but not in B. Therefore, the protonation would proceed predominantly via the transition state B, thus producing the isomer **27a** as a major product. In spite of numerous examples of the Birch reduction of α,β-unsaturated carbonyl compounds (Caine, 1976), there are few examples of neighboring group participation in the protonation at the β position of enones (McMurry et al., 1978), and so this is the first example of controlling the stereochemistry at the β carbon of enolate dianion by 1,4-asymmetrical induction of the hydroxyl group (Iwata et al., 1981a). Introduction of three carbon unit at C-2 of the major product **(27a)** was best achieved as follows: Thioacetalization of the ketone and mesylation of the alcohol were followed by the substituted reaction with sodium diethyl malonate anion to give **28** bearing desired stereochemistry in 55% overall yield. Conversion of malonyl group of **28** into allylic alcohol was carried out by a three-step sequence to afford allylic alcohol **(29)**. Removal of the hydroxyl group and sequential deprotection of the thioketal group of **29** provided (±)-solavetivone **(5)**.

B. Synthesis of (±)-Oxylubimin

Oxylubimin **(7)**, which was isolated from tuber tissue of white potatoes infected by *Phytophthora infestans*, is a most highly oxygenated spirovetivane-type phytoalexin bearing six asymmetrical carbon centers (Katsui et al., 1974; Stöessl et al., 1975). This natural product is not only a representative from the viewpoint of its strong biological activity and the complex stereochemistry, but it is also a promising biosynthetic intermediate in the major pathway to rishitin via solavetivone in vivo (Sato et al., 1978).

There have been two reports on the total synthesis of (±)-7. The first synthesis was achieved by Murai et al. in 1982 through a related approach

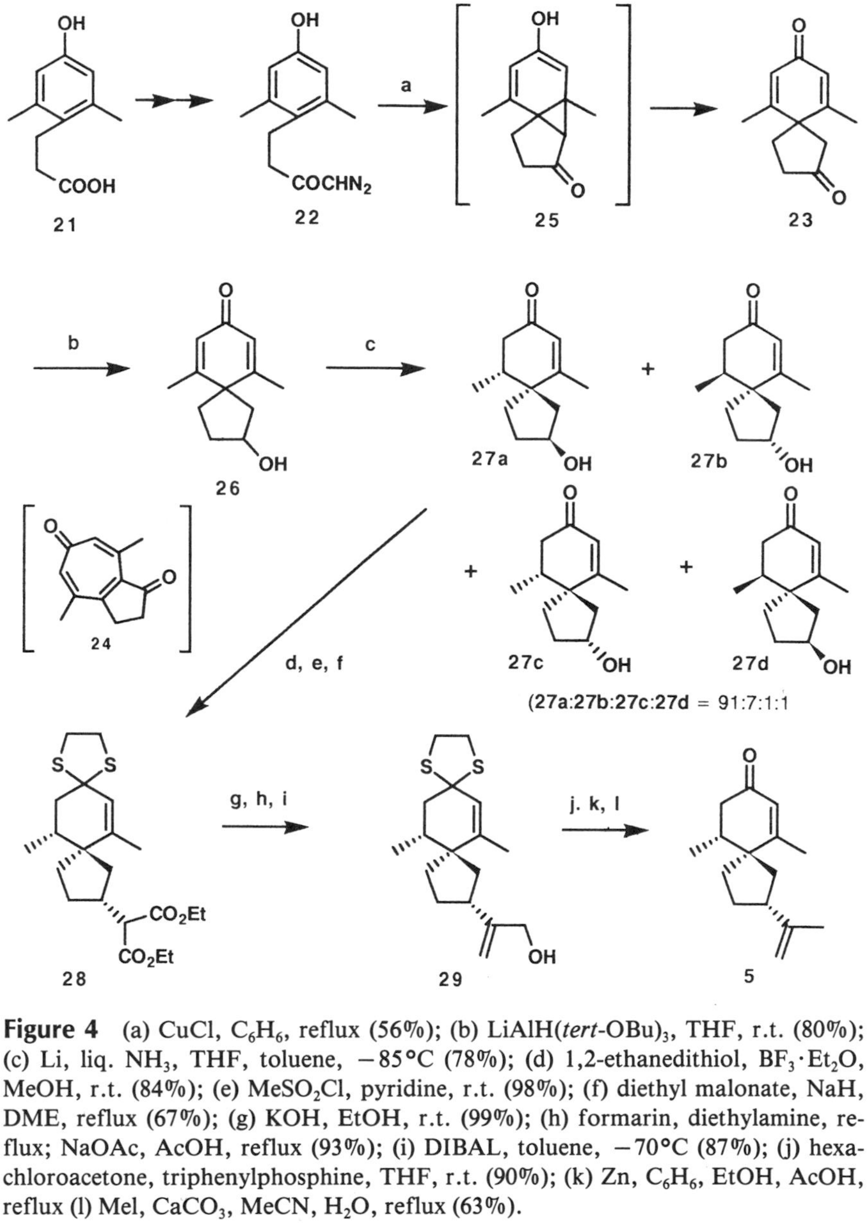

Figure 4 (a) CuCl, C_6H_6, reflux (56%); (b) LiAlH(*tert*-OBu)$_3$, THF, r.t. (80%); (c) Li, liq. NH_3, THF, toluene, −85°C (78%); (d) 1,2-ethanedithiol, $BF_3 \cdot Et_2O$, MeOH, r.t. (84%); (e) $MeSO_2Cl$, pyridine, r.t. (98%); (f) diethyl malonate, NaH, DME, reflux (67%); (g) KOH, EtOH, r.t. (99%); (h) formarin, diethylamine, reflux; NaOAc, AcOH, reflux (93%); (i) DIBAL, toluene, −70°C (87%); (j) hexachloroacetone, triphenylphosphine, THF, r.t. (90%); (k) Zn, C_6H_6, EtOH, AcOH, reflux (l) Mel, $CaCO_3$, MeCN, H_2O, reflux (63%).

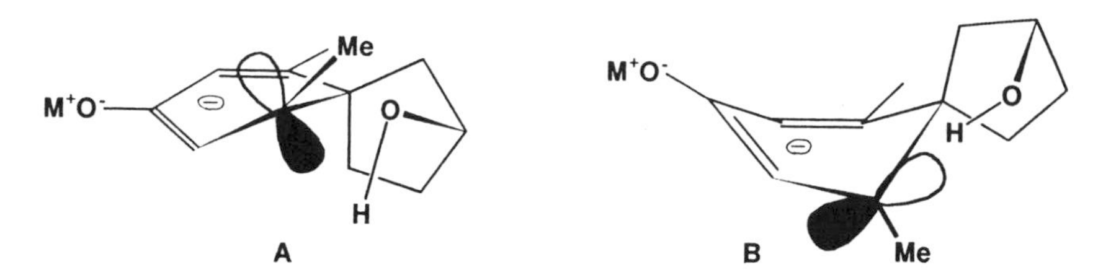

Figure 5 Intramolecular proton migration from the hydroxyl group to formed dienolate dianion. (A) Chair form, (B) Boat or half-chair form.

to solavetivone synthesis (Fig. 6) (Murai et al., 1982c, 1984c,d). The problem of stereoselectivity in a Diels–Alder reaction was nicely resolved by using bicyclic diene **(30)** to afford antiadduct **(31)** diastereoselectively as a mixture of endo-**31** and exo-**31**. However, the acid promoted π cyclization of **31** and α' hydroxylation of the major product **(32)** resulted in proceeding in low diastereoselectivity (**32**/**33** = 35 : 25 and **34**/**35** = 51 : 25). Subsequently, introducing two asymmetrical centers at C-6 and C-8 through hydrocyanation (**36**/**37** = 1 : 1) and borane reduction, enone **(34)** was converted to (±)-oxylubimin **(7)** and (±)-epioxylubimin **(38)**. Consequently, two natural products (**7** and **38**) which are epimerized by each other under basic conditions, were synthesized from **34** in the same way. They also succeeded in total synthesis of (±)-lubimin **(6)** and its diastereoisomers from **32**.

Our synthetic route for (±)-oxylubimin **(7)**, is shown in Fig. 7 (Iwata et al., 1985b, 1990). As the starting material, we employed the common intermediate **(27a)** in solavetivone synthesis. Our synthetic plan involved two key steps: 1.) the stereoselective α' hydroxylation of α,β-unsaturated ketone **(39)** into **40a**, and 2.) regio- and stereoselective hydride reduction of the α'-hydroxyenone (**40a** or **41**) into *trans*-diol **(42a)**. The first problem was overcome with a new hydroxylation method developed in our laboratory (Iwata et al., 1985a). By treatment of silyl enol ether derivative of enone **(39)** with triphenylphosphite ozonide (TPPO) and then triphenylphosphine, a diastereo mixture of **40a** and **40b** was obtained in 71% yield with high stereoselectivity (**40a**/**40b** = 8 : 1). On the other hand, oxidation of **39** with reagents such as *m*-chloroperbenzoic acid (MCPBA) (Rubottom and Gruber, 1978), MoOPH (Vedejs et al., 1978), and $Mn(OAc)_3$ (Dunlap et al., 1984) gave a mixture of **40a** and **40b** in moderate yields, but the stereoselectivity turned out to be unsatisfactory (Table 1). This high stereoselectivity using TPPO should be attributable to the reaction intermediate. Namely, in the TPPO oxidation the oxidant is considered to react simultaneously at two carbon centers, C-6 and C-9, in the silyl enol ether and

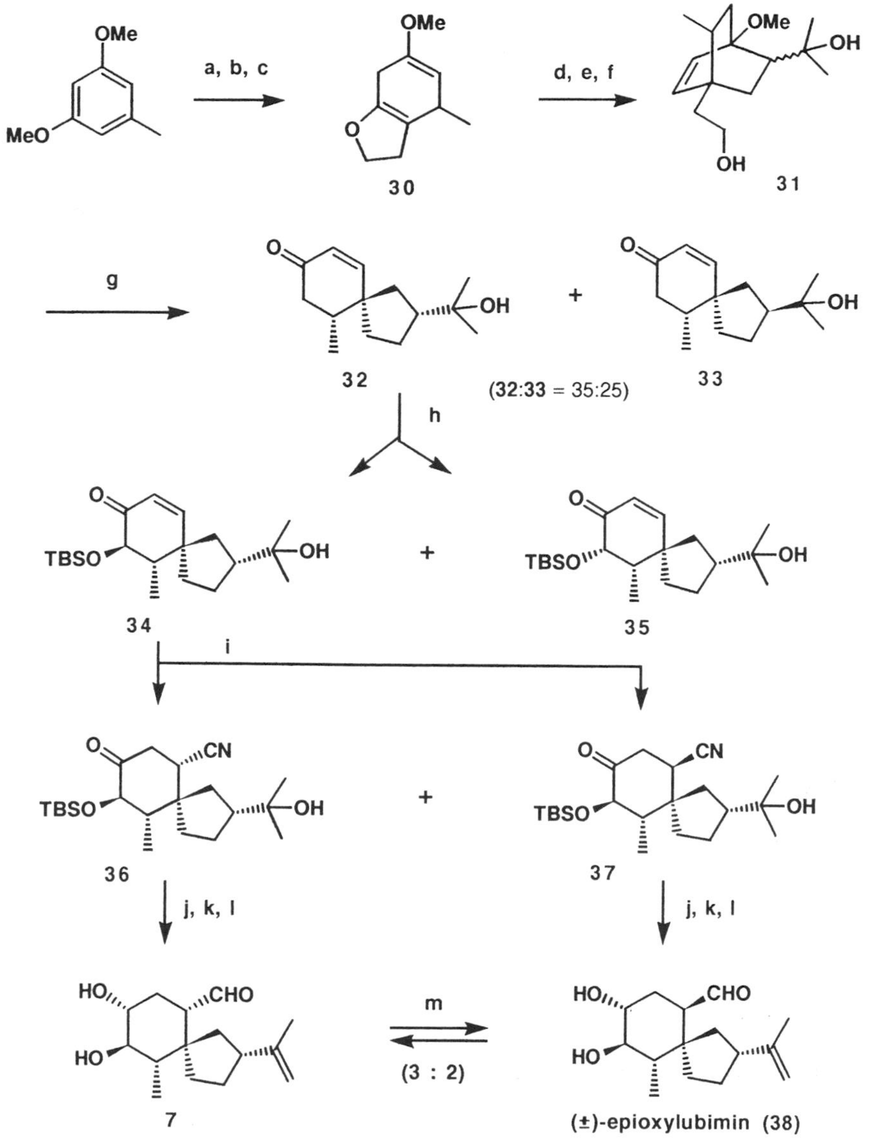

Figure 6 (a) $ClCH_2COCl$, $AlCl_3$, CS_2 (84%); (b) $NaBH_4$; H_2, Pd-C (100%); (c) Li, liq. NH_3, ether; EtOH (81%); (d) methyl acrylate, dichloromaleic anhydride, 150°C (70%); (e) $(CO_2H)_2$; Ac_2O, pyridine (70%); (f) $TsNHNH_2$; MeLi (71%); (g) MsCl; $(CO_2H)_2$, H_2O, methyl isobutyl ketone (60%); (h) LDA; TBSCl; $PhCO_3H$ (**34**: 51%, **35**: 25%); (i) HCN, Et_3Al (**36**: 40%, **37**: 39%); (j) $BH_3 \cdot NH_3$, aq. MeOH; TBSCl, imidazole, DMF (80% for **36**, 58% for **37**); (k) pyridine-Al_2O_3, 220°C; aq. HF, THF, MeCN (67%, 70%); (l) DIBAL (57%, 55%); (m) 5% KOH, MeOH.

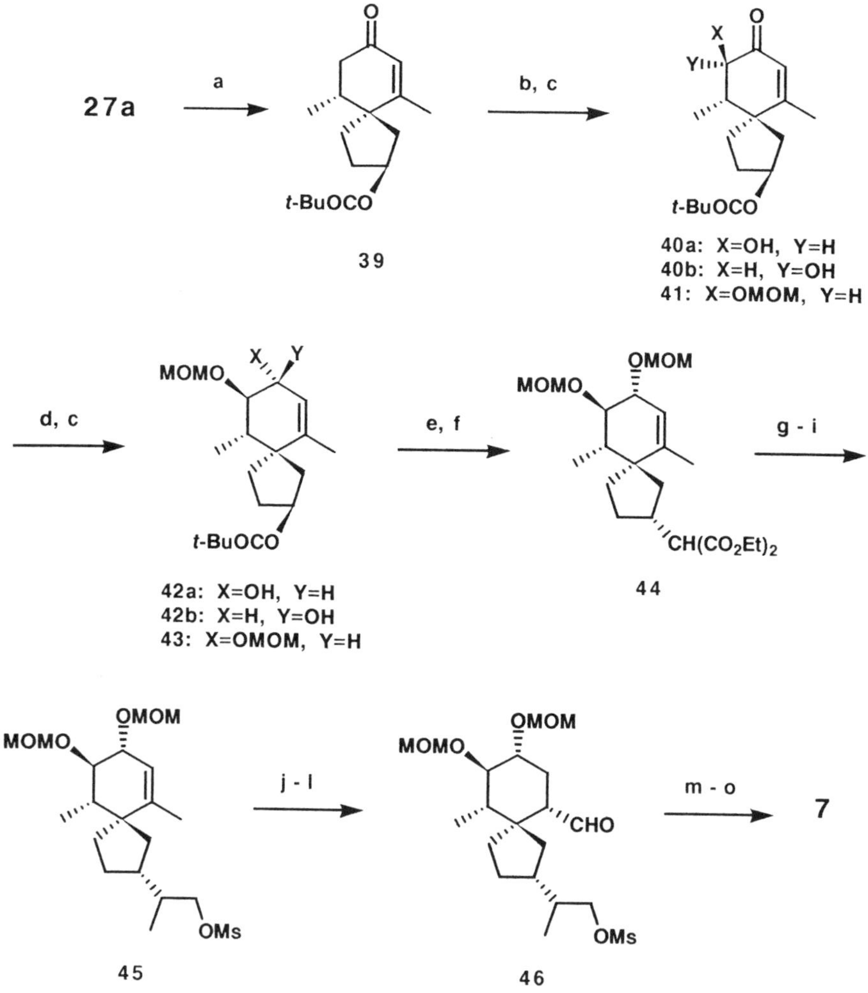

27a → (a) → 39 → (b, c) → 40a: X=OH, Y=H; 40b: X=H, Y=OH; 41: X=OMOM, Y=H

t-BuOCO

→ (d, c) → 42a: X=OH, Y=H; 42b: X=H, Y=OH; 43: X=OMOM, Y=H → (e, f) → 44 → (g - i)

MOMO, OMOM, $CH(CO_2Et)_2$

45 → (j - l) → 46 → (m - o) → 7

MOMO, OMOM, OMs, CHO

Figure 7 (a) *t*-BuCOCl, pyridine (92%); (b) LDA, DME, −18°C; TMSCl; TPPO, CH_2Cl_2, −50°C; Ph_3P, ether, r.t. (71%); (c) MOMCl, $(i\text{-}Pr)_2NEt$, CH_2Cl_2 (84% for **40a**, 91% for **42a**); (d) $NaBH_4$, $CeCl_3 \cdot 7H_2O$, MeOH, 0°C (90%); (e) MeLi, ether (95%); (f) MsCl, pyridine; NaH, $CH_2(CO_2Et)_2$, DME, reflux (97%); (g) NaH, DME; Red-Al, DME (74%); (h) $NaBH_4$, $CoCl_2 \cdot 6H_2O$, EtOH (78%); (i) MsCl, pyridine (79%); (j) SeO_2, xylene, reflux; $NaBH_4$, MeOH (70%); (k) H_2, Raney Ni (W2), EtOH (90%); (l) PCC, NaOAc, CH_2Cl_2 (82%); (m) $(CH_2SH)_2$, $BF_3 \cdot Et_2O$, CH_2Cl_2 (87%); (n) NaI, DBU, DME (66%); (o) MeI, $CaCO_3$, aq. MeCN, reflux (43%).

Table 1 α' Hydroxylation of the Enone (**39**) with Various Oxidants

Entry	Reaction conditions	X	Yield (%)	Ratio (**40a** : **40b**)
1	LDA, DME, −18°C; TMSCl; TPPO, CH_2Cl_2, −50°C, Ph_3P, Et_2O, room temp.	H	71	8 : 1
2	LDA, DME, −18°C; TMSCl; MCPBA, hexane, −15°C	TMS	63	1 : 1
3	LDA, THF, −78°C; MoOPH, −22°C; TBSCl, imidazole, DMF	TBS	52 (83)[a]	2 : 1
4	$Mn(OAc)_3$, C_6H_6, reflux	Ac	62	2 : 1

[a]Yield based on the consumed starting material.

form an endoperoxide (C) (Fig. 8). Because of severe 1,3-diaxial interaction between TPPO and the methylene unit at C-1 and C-4, the silyl enol ether was attacked from the top side of conformation A by the oxidant, leading to a predominant formation of **40a**. Conversion of α'-hydroxyenone (**40a**) in hand into *trans*-diol, the second crucial step, was achieved as follows. Since all attempts to reduce **40a** under various conditions resulted in the formation of a mixture of *trans*- and *cis*-diol without satisfactory stereoselectivity, methoxymethyl (MOM) ether (**41**) was reduced with $NaBH_4$ in methanol in the presence of cerium chloride at 0°C to give desired *trans*-diol (**42a**) as a major product (**42a**/**42b** = 6.4 : 1) (Table 2). Introduction of a three-carbon unit at the C-2 position in **43** was accomplished in a similar manner as described in solavetivone synthesis to afford **44** in 92% overall yield. Subsequently, by modified Marshall's method (Marshall et al., 1967), diester (**44**) was transformed into corresponding allylic alcohol in one step, followed by cobalt hydride reduction and mesylation to give mesylate (**45**). Successive reactions by allylic oxidation, catalytic hydrogenation of the C-6–C-7 double bond, and PCC oxidation provided the saturated aldehyde (**46**) as a single isomer concerning at C-6. Finally, derivation of **46** into (±)-oxylubimin (**7**) was straightforward, i.e., thioacetalization accompanied with deprotection of the bis-MOM ether, elimination of methanesulfonic acid with base, and deprotection of thioacetal group led to stereoselective formation of (±)-oxylubimin (**7**).

C. Other Phytoalexins

Based on our strategy described above, we attempted to synthesize a series of spirovetivane-type phytoalexins other than solavetivone (**5**), and oxylubi-

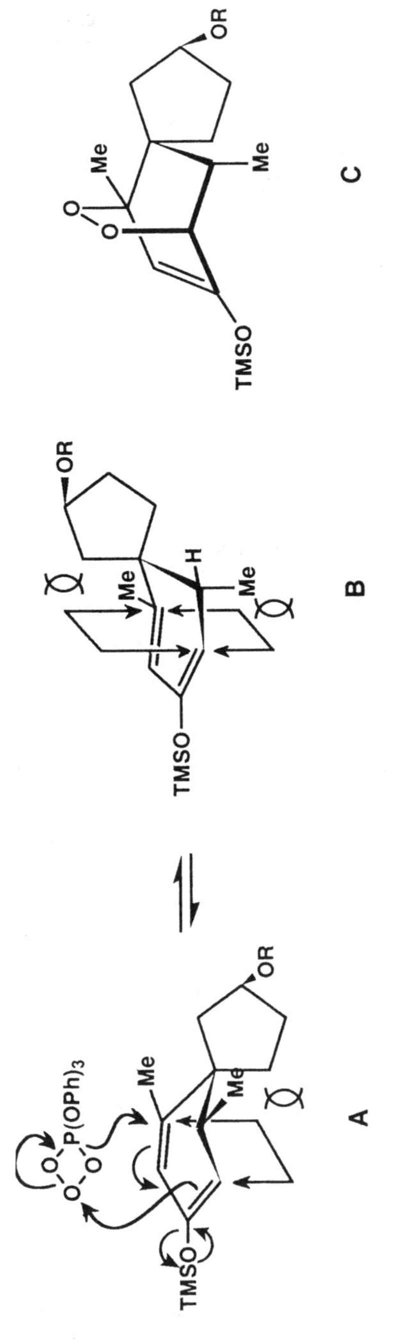

Figure 8 TPPO oxidation in which oxidant reacts simultaneously at C-6 and C-9 in the silyl enol ether, forming an endoperoxide.

Table 2 Reduction of α′-Hydroxyenone (**41**) Under Various Conditions

Entry	Reaction conditions	Yield (%)	Ratio[a] (**42a/42b**)
1	$Zn(BH_4)_2$, Et_2O, 0°C	79	1.9
2	$LiBH_3$(*n*-Bu), toluene, hexane, −78°C	84	1.2
3	$NaBH_3CH$, MeOH, 2 N HCl, room temp.	71	5.6
4	$NaBH_4$, $CeCl_3 \cdot 7H_2O$, MeOH, 0°C	90	6.4

[a]Ratio based on isolated yields.

min (**7**). Consequently, total syntheses of 3-hydroxysolavetivone (aglycone A_2) (**47**) (Iwata et al., 1986), lubiminol (**48**) (Iwata et al., 1984) and aglycone A_3 (**51**) (Iwata et al., 1981b) together with formal syntheses of 10-epioxylubimin (**38**) (Murai et al., 1982c), isolubimin (**49**) (Katsui et al., 1982), solanascone (**50**) (Fujimori et al., 1978), and aglycone A_4 (**52**) (Anderson et al., 1977) were accomplished. The results obtained in the synthetic studies of spirovetivane-type phytoalexins were briefly depicted in Fig. 9 (Iwata et al., 1988a,b; 1989).

D. Asymmetrical Synthesis of (−)-Solavetivone

There has been no report about asymmetrical synthesis of *trans*-spirovetivanes, although a few groups completed in total synthesis of *cis*-spirovetivanes, e.g., (+)-hinesol (**8**), (−)β-vetivone (**9**), starting from chiral natural products (Buddhasukh and Magnus, 1975; Deighton et al., 1975). Herein we would like to describe some synthetic studies for optically active *trans*-spirovetivanes.

Our synthetic strategy for asymmetrical synthesis of spirovetivane-type phytoalexins relies on vinylic sulfoxide (**53**) possessing chiral sulfinyl group (Fig. 10). The key reactions to construct chiral quaternary carbon centers in our route consist of the formation of chiral cyclopropane ring (**53** → **54**), the regioselective cleavage of the cyclopropane ring (**54** → **55**), and spiroannulation reaction of alkylmercury chloride (**55** → **56** and **57**). The obtained ketoester (**56**) can be transformed into solavetivone (**5**) and hinesol (**8**) via **58**. Another ketoester (**57**) is also a good intermediate for the synthesis of agarospirol. According to the strategy, we first examined a Michael reaction of Grignard reagents into chiral vinylic sulfoxides. From the preliminary studies, we know that a reaction of a vinylic sulfoxide with allyl magnesium bromide affords a mixture of monoallylated product and diallylated compound through a additive Pummerer-type reaction (Iwata et al.,

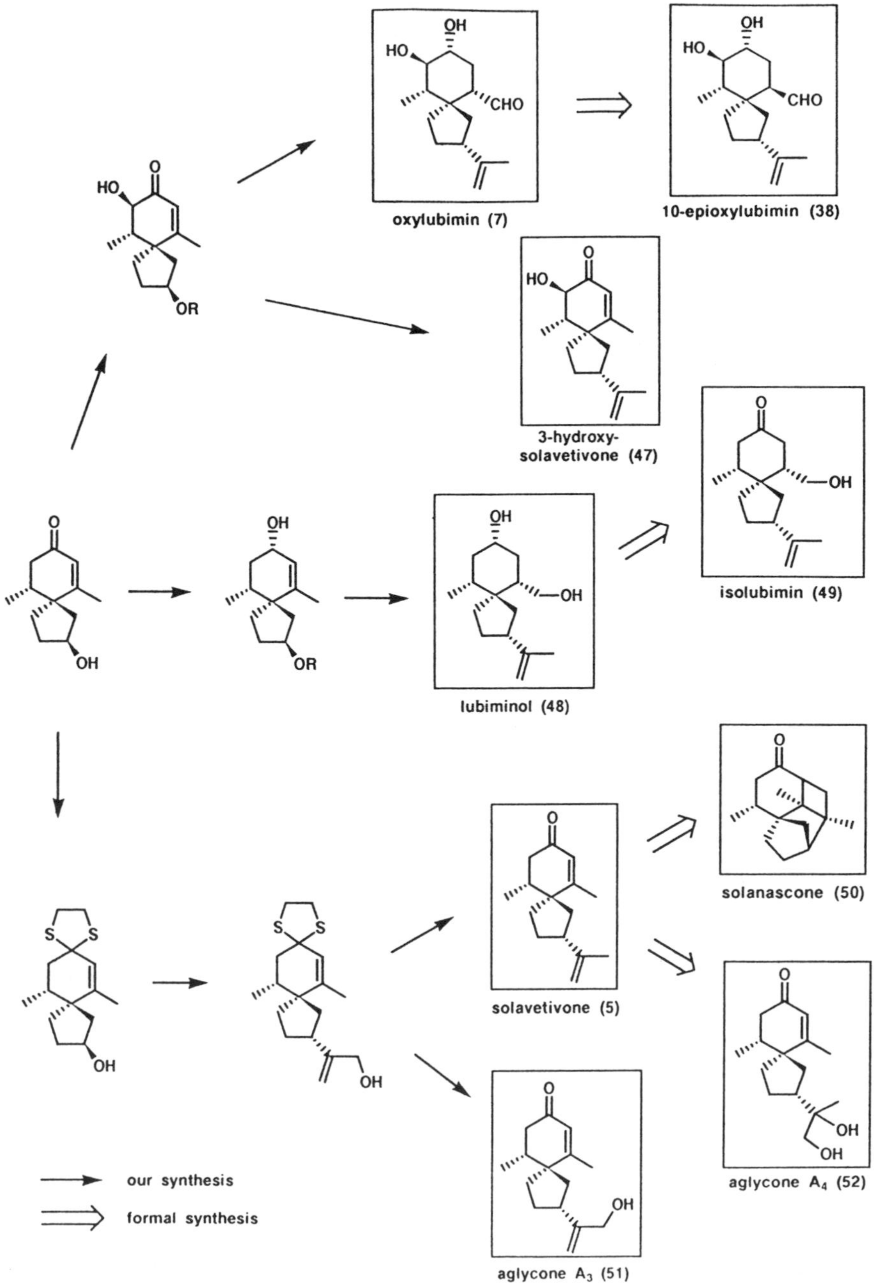

oxylubimin (7)

10-epioxylubimin (38)

3-hydroxy-solavetivone (47)

isolubimin (49)

lubiminol (48)

solanascone (50)

solavetivone (5)

aglycone A_4 (52)

aglycone A_3 (51)

our synthesis

formal synthesis

Figure 9 Results obtained in synthetic studies of spirovetivane-type phytoalexins.

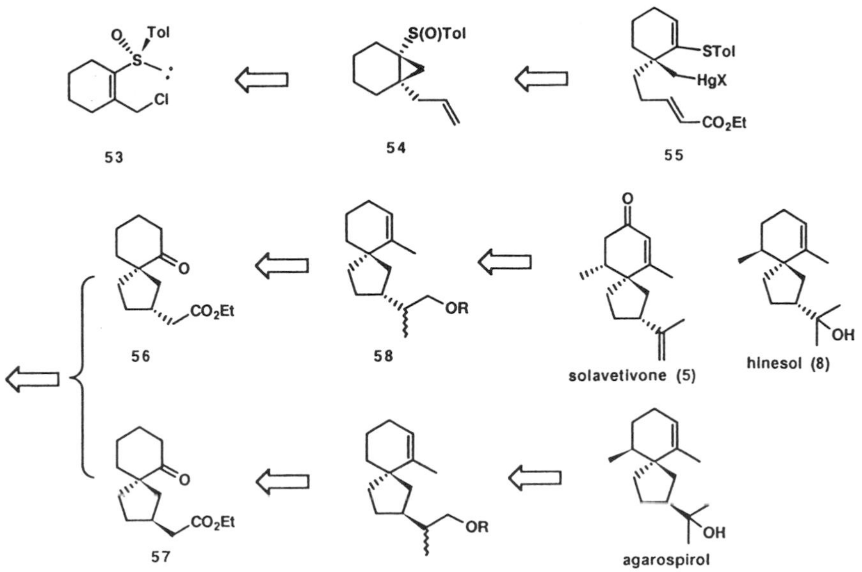

Figure 10 Synthetic strategy for optically active spirovetivane-type sesquiterpenes.

1991). Therefore, we undertook the above reaction with vinylic sulfoxide (**53**) bearing a chloromethyl group, hoping for a different mode of reaction (Fig. 11). The vinylic sulfoxide (**53**) was prepared from **59** by a four-step sequence, which was treated with allyl magnesium bromide in tetrahydrofuran at −78°C to provide cyclopropyl sulfoxide (**54**) in 66% yield as a single diastereoisomer along with coupling compound (**60**). The absolute configuration (Ss, R, S) of **54** was determined by an x-ray crystallographic study of its derivative. This highly diastereoselective cyclopropanation can be explained by the comparison of the energetic stability between two chelated transition models (A and B). Considering that the magnesium center of Grignard reagent can coordinate both to the oxygen atom of sulfinyl group and to the chloride atom at the γ position, conformer A in Fig. 12 is much more stable than conformer B because of the severe steric interaction between the tolyl and chloromethyl groups, and then the allyl group of the reagent is introduced exclusively from the α face. In order to resolve the second problem, i.e., regioselective cleavage of cyclopropane ring, sulfoxide (**54**) was transformed into cyclopropyl sulfide (**61**) in four steps. Among various electrophiles, reaction of **61** with mercury(II) trifluoroacetate in methylene chloride (CH_2Cl_2) in the presence of sodium acetate at room temperature gave rise to the desired α,β-unsaturated γ-sulfenylalkyl mer-

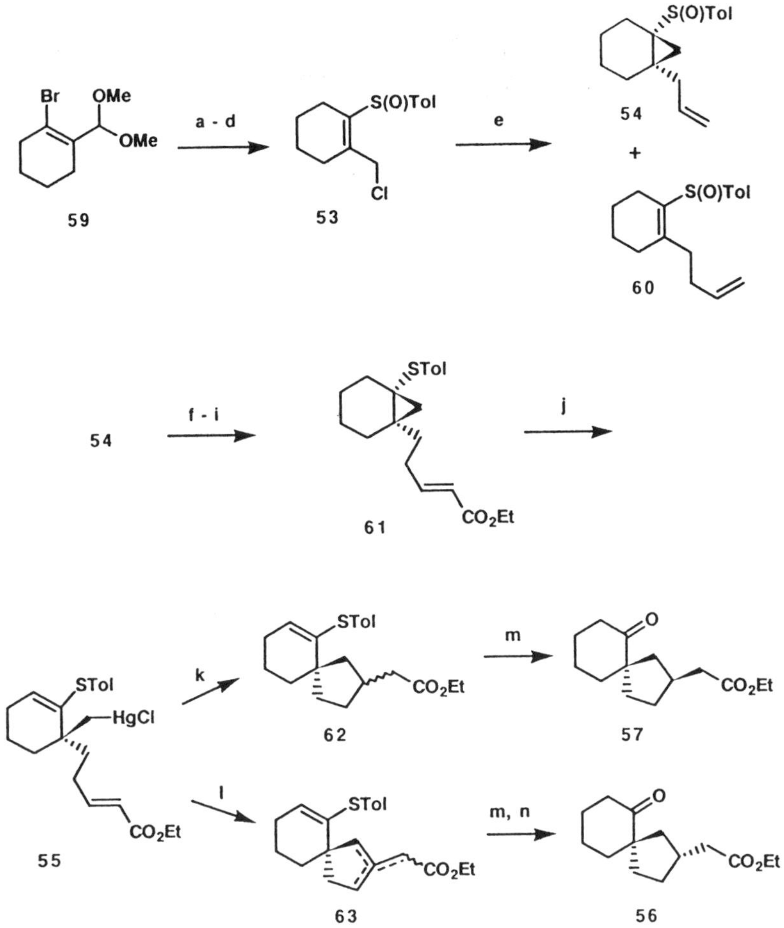

Figure 11 (a) *n*-BuLi; TolS(O)O-(*l*)-mentyl (91%); (b) *p*-TsOH, acetone, H_2O (97%); (c) $NaBH_4$, MeOH (100%); (d) MsCl, LiCl, 2,4,6-collidine, *n*-Bu_4NCl, DMF, 0°C (94%); (e) (Allyl)MgBr, THF, −78°C (82%); (f) $(CF_3CO)_2O$, NaI, acetone, 0°C (100%); (g) 9-BBN, THF, 0°C; 3M NaOH, 30% H_2O_2, r.t. (95%); (h) $(COCl)_2$, DMSO, CH_2Cl_2, −50°C; Et_3N, r.t. (87%); (i) $(EtO)_2POCH_2CO_2Et$, NaH, THF, r.t. (82%); (j) $Hg(OCOCF_3)_2$, NaOAc, CH_2Cl_2, r.t.; saturated aq. NaCl, r.t. (89%); (k) *n*-Bu_3SnH, CH_2Cl_2, −40°C → r.t. (86%); (l) Li_2PdCl_4, DMF, THF, reflux (91%); (m) 10% HCl, MeCN, 60°C (84% for **62** and **63**); (n) H_2, 10% Pd-C, EtOH, r.t. (100%).

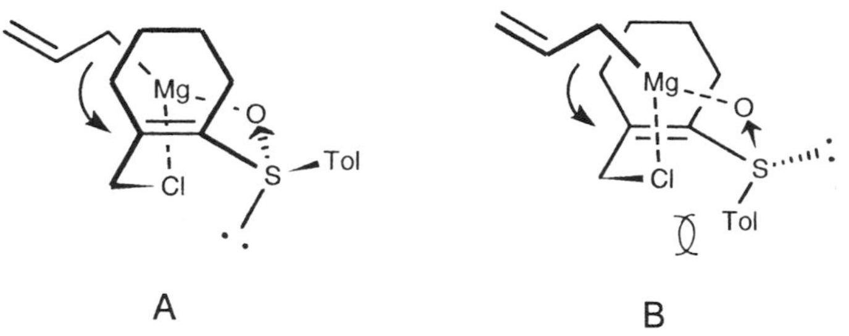

Figure 12 Plausible reaction mechanism of the asymmetric cyclopropanation.

cury chloride (**55**) in a highly regioselective manner without affecting α,β-unsaturated ester. Various reaction conditions were attempted to develop a new type of spiroannulation method. Sequential treatment of **55** by the radical cyclization (n-Bu_3SnH, AIBN) and acid-catalyzed hydrolysis gave ketoester (**56** and **57**) in 72% overall yield in the ratio of 11 : 89. On the other hand, palladium(II)-assisted cyclization followed by acid-catalyzed hydrolysis and catalytic hydrogenation also afforded ketoester (**56** and **57**) with the former as the major product in 76% overall yield (**56**/**57** = 83 : 17) (Imanishi et al., 1992). These obtained products (**56** and **57**) are useful intermediates for important natural products, e.g., natural (−)-solavetivone from **56** and unnatural (+)-agarospirol from **57**. We are now further elaborating our research with an aim for total synthesis of (−)-solavetivone (**5**) utilizing **56**.

III. EPILOG

As was detailed in the foregoing sections, considerable progress has been made in the total synthesis of naturally occurring spirovetivanes. Numerous strategies have been developed for construction of the spiro[4.5]decane ring system, and some routes have enabled the efficient regio- and/or stereospecific synthesis of *trans*-spirovetivane phytoalexins. However, structural modifications have not been reported in synthetic spirovetivanes except for a minor one, i.e., dideuterated (±)-solavetivone prepared by Murai et al., (1982a), who extended the research for establishing the biosynthetic production of rishitin from solavetivone via lubimin and oxylubimin on the basis of experiments with the (±)-[2,2-$^2H_2$]solavetivone. The synthetic methodology now available should allow the preparation of analogs having more profound structural modification, which provides us much opportunity to get more efficient drugs or prodrugs and to elucidate inductive and regulative mechanism of phytoalexin biosynthesis, including the isolation of enzymes responsible for the formation of spirovetivane phytoalexins.

An attempt has been made in this chapter to review the most significant synthetic advances reported over the last two decades. There can be little doubt that isolation of new types of phytoalexins from natural sources as well as development of new synthetic approaches will continue, and the next decade should see further advances in synthetic methodology and synthetic analogs, as well as the utilization of the enzymes involved in phytoalexin biosynthesis for the preparation of spirovetivane phytoalexins, resulting in improved plant disease therapy.

REFERENCES

Anderson, R. C., Gunn, D. M., Murray-Rust, J., Murray-Rust, P., and Roberts, J. S. (1977). Vetispirane sesquiterpene glucosides from flue-cured Virginia tobacco: structure, absolute stereochemistry, and synthesis. X-Ray structure of the p-bromobenzenesulphonate of one of the derived aglycones. *J. Chem. Soc., Chem. Commun.* (1): 27–28.

Buddhasukh, D., and Magnus, P. D. (1975). Synthesis of (+)-hinesol and 10-epi-(+)-hinesol. *J. Chem. Soc. Chem. Commun.* (23): 952–953.

Caine, D. (1976). Reduction and related reactions of α,β-unsaturated compounds with metals in liquid ammonia. *Org. React. 23*: 1–258.

Coolbear, T., and Threlfall, D. R. (1985). The biosynthesis of lubimin from [1-^{14}C]isopentenyl pyrophosphate by cell-free extracts of potato tuber tissue inoculated with an elicitor preparation from *Phytophthora infestans. Phytochemistry 24*(9): 1963–1971.

Coxon, D. T., Price, K. R., Howard, B., Osman, S. F., Kalan, E. B., and Zacharius, R. M. (1974). Two new vetispirane derivatives: stress metabolites from potato (*Solanum tuberosum*) tubers. *Tetrahedron Lett.* (34): 2921–2924.

Deighton, M., Hughes, C. R., and Ramage, R. (1975). Stereospecific synthesis of (−)-agarospirol and (−)-β-vetivone. *J. Chem. Soc., Chem. Commun.* (16): 662–663.

Dunlap, N. K., Sabol, M. R., and Watt, D. S. (1984). Oxidation of enones to α'-acetoxyenones using manganese triacetate. *Tetrahedron Lett. 25*(51): 5839–5842.

Friend, J., and Threlfall, D. R. (1976). *Biochemical Aspects of Plant–Parasite Relationships*. Academic Press, London.

Fujimori, T., Kasuga, R., Kaneko, H., and Noguchi, M. (1977). Isolation of solavetivone from *Nicotiana tabacum. Phytochemistry 16*(3): 392.

Fujimori, T., Kasuga, R., Kaneko, H., Sakamura, S., Noguchi, M., Furusaki, A., Hashiba, N., and Matsumoto, T. (1978). Solanascone: a novel sesquiterpene ketone from *Nicotiana tabacum*. X-Ray structure determination of the corresponding oxime. *J. Chem. Soc., Chem. Commun.* (13): 563–564.

Grisebach, H., and Ebel, J. (1978). Phytoalexins, chemical defense substances of higher plants? *Angewandte Chemie Int. Ed.* (*English*) *17*(9): 635–647.

Hiroi, K. (1977). Recent synthetic studies on carbocyclic spiro compounds. *Yuki Gousei Kagaku Kyokaishi 35*(12): 1029–1044.

Imanishi, T., Ohra, T., Sugiyama, K., Ueda, Y., Takemoto, Y., and Iwata, C. (1992). Novel asymmetric cyclopropanation utilizing sulfinyl chirality: application to construction of a spiro[4.5]decane system. *J. Chem. Soc., Chem. Commun.* (3): 269–270.

Iwata, C., Nakashita, Y., and Hirai, R. (1974). Studies on the syntheses of spirodienone compounds. IV. Photolysis of sodium salts of *p*-[2-(2-halogenomethyl-1,3-dioxolan-2-yl)-ethyl]phenol in N,N-dimethylformamide. *Chem. Pharm. Bull.* *22*(1): 239.

Iwata, C., Yamada, M., Shinoo, Y., Kobayashi, K., and Okada, H. (1977). Intramolecular cyclisation of phenolic α-diazoketones. Novel synthesis of the spiro-[4.5]decane carbon framework. *J. Chem. Soc., Chem. Commun.* (23): 888–889.

Iwata, C., Yamada, M., Shinoo, Y., Kobayashi, K., and Okada, H. (1980). Studies on the syntheses of spiro-dienone compounds. VII. Novel synthesis of the spiro[4.5]decane carbon framework. *Chem. Pharm. Bull.* *28*(6): 1932–1934.

Iwata, C., Miyashita, K., Ida, Y., and Yamada, M. (1981a). Effects of neighbouring hydroxy-groups in metal-ammonia reductions of α,β-unsaturated carbonyl compounds. *J. Chem. Soc., Chem. Commun.* (10): 461–463.

Iwata, C., Fusaka, T., Fujiwara, T., Tomita, K., and Yamada, M. (1981b). Total synthesis of (±)-solavetivone; X-ray crystal structure of 2-hydroxy-6,10-dimethylspiro[4.5]dec-6-en-8-one. *J. Chem. Soc., Chem. Commun.* (10): 463–465.

Iwata, C., Kubota, H., Yamada, M., Takemoto, Y., Uchida, S., Tanaka, T., and Imanishi, T. (1984). Stereocontrolled total synthesis of (±)-lubiminol, a spirovetivane phytoalexin. *Tetrahedron Lett.* *25* (31): 3339–3342.

Iwata, C., Takemoto, Y., Nakamura, A., and Imanishi, T. (1985a). Oxidation of 2-trimethylsilyloxy-1,3-diens with triphenyl phosphite ozonide. A regioselective α′-hydroxylation of α,β-unsaturated ketones. *Tetrahedron Lett.* *26*(27): 3227–3230.

Iwata, C., Takemoto, Y., Kubota, H., Kuroda, T., and Imanishi, T. (1985b). A stereoselective total synthesis of (±)-oxylubimin. *Tetrahedron Lett.* *26*(27): 3231–3234.

Iwata, C., Nakamura, A., Takemoto, Y., and Imanishi, T. (1986). Stereoselective total synthesis of (±)-3-hydroxysolavetivone. *Chem. Ind.* (20): 712–713.

Iwata, C., Yamada, M., Fusaka, T., Miyashita, K., Nakamura, A., Tanaka, T., Fujiwara, T., and Tomita, K. (1987). Total synthesis of (±)-solavetivone and aglycone A_3. Regio- and stereo-selective Birch reduction of 6,10-dimethyl-2-hydroxyspiro[4.5]deca-6,9-dien-8-one. *Chem. Pharm. Bull.* *35*(2): 544–552.

Iwata, C., Takemoto, Y., Kubota, H., Yamada, M., Uchida, S., Tanaka, T., and Imanishi, T. (1988a). Synthetic studies on spirovetivane phytoalexins. IV. A stereoselective synthesis of (±)-3-hydroxysolavetivone. *Chem. Pharm. Bull.* *36*(10): 2643–2646.

Iwata, C., Takemoto, Y., Kubota, H., Yamada, M., Uchida, S., Tanaka, T., and Imanishi, T. (1988b). Synthetic studies on spirovetivane phytoalexins. II. Stereoselective synthesis of (2RS,5RS,6SR,8RS,10SR)-6-hydroxymethyl-8-methoxymethoxy-10-methyl-2-pivaloyloxyspiro[4.5]decane. *Chem. Pharm. Bull.* *36*(11): 4581–4584.

Iwata, C., Takemoto, Y., Kubota, H., Yamada, M., Uchida, S., Tanaka, T., and Imanishi, T. (1989). Synthetic studies on spirovetivane phytoalexins. III. A total synthesis of (±)-lubiminol. *Chem. Pharm. Bull. 37*(4): 866–869.

Iwata, C., Takemoto, Y., Kubota, H., Kuroda, T., and Imanishi, T. (1990). Synthetic studies on spirovetivane phytoalexins. V. A stereoselective total synthesis of (±)-oxylubimin. *Chem. Pharm. Bull. 38*(2): 361–365.

Iwata, C., Maezaki, N., Kurumada, T., Fukuyama, H., Sugiyama, K., and Imanishi, T. (1991). New carbon-carbon bond formation by the Pummerer-type reaction of vinylic sulfoxides with allylmagnesium bromide. *J. Chem. Soc., Chem. Commun.* (19): 1408–1409.

Katsui, N., Murai, A., Takasugi, M., Imaizumi, K., Masamune, T., and Tomiyama, K. (1968). The structure of rishitin, a new antifungal compound from diseased potato tubers. *J. Chem. Soc., Chem. Commun.* (1): 43–44.

Katsui, N., Matsunaga, A., and Masamune, T. (1974). The structure of lubimin and oxylubimin, antifungal metabolites from diseased potato tubers. *Tetrahedron Lett.* (51/52): 4483–4486.

Katsui, N., Yagihashi, F., Murai, A., and Masamune, T. (1982). Structure of epilubimin, epioxylubimin, and isolubimin, spirovetivane stress metabolites in diseased potato. *Bull. Chem. Soc. Jpn. 55*(8): 2424–2427.

Keen, N. T., and Bruegger, B. (1977). *Host plant resistance to Pests.* ACS Symposium Series, No. 62 (P. A. Hedin, ed.), p. 1.

Krapcho, A. P. (1976). Synthesis of carbocyclic spiro compounds via rearrangement routes. *Synthesis* (7): 425–444.

Krapcho, A. P. (1978). Synthesis of carbocyclic spiro compounds via cycloaddition routes. *Synthesis* (2): 77–126.

Marshall, J. A., Anderson, N. H., and Hochstetler, A. R. (1967). The reduction of malonic enolates with lithium aluminum hydride. *J. Org. Chem. 32*(1): 113–118.

Marshall, J. A., Brady, S. F., and Anderson, N. H. (1974). The chemistry of spiro[4.5]decane sesquiterpenes. *Fortschritte Der Chemie Organischer Naturstoffe*, Vol. 31. Springer-Verlag. New York, pp. 283–376.

Masamune, T., Murai, A., and Katsui, N. (1978). Phytoalexin no taisha. *Kagaku to Seibutsu 16*(10): 648–660.

McMurry, J. E., Blaszczak, L. C., and Johnson, M. A. (1978). On the stereochemistry of reduction of 9-methyl-1-carboxy-Δ^4-3-octalone: a remarkable lithium-ammonia reduction. *Tetrahedron Lett.* (19): 1633–1634.

Müller, K. O., and Boger, H. (1940). *Arb. Biol. Reichsanstalt. Landw. Forstw. Berlin 23*: 189.

Murai, A. (1981). Recent advances in the synthesis of spirovetivane sesquiterpenes. *Yuki Gousei Kagaku Kyokaishi 39*(10): 893–908.

Murai, A., Sato, S., and Masamune, T. (1981a). Efficient synthesis of (±)-solavetivone. *J. Chem. Soc., Chem. Commun.* (17): 904–905.

Murai, A. Sato, S. and Masamune, T. (1981b). π-Cyclization: the synthesis of (±)-solavetivone and (±)-hinesol. *Tetrahedron Lett. 22*(11): 1033–1036.

Murai, A., Sato, S., Osada, A., Katsui, N., and Masamune, T. (1982a). Biosynthesis from solavetivone of the phytoalexin rishitin in potato. Implicit role of solavetivone as an activator. *J. Chem. Soc., Chem. Commun.* (1): 32–33.

Murai, A., Sato, S., and Masamune, T. (1982b). Total synthesis of (±)-15-norsolavetivone. *J. Chem. Soc., Chem. Commun.* (9): 511–512.

Murai, A., Sato, S., and Masamune, T. (1982c). Total synthesis of (±)-lubimin and (±)-oxylubimin. *J. Chem. Soc., Chem. Commun.* (9): 513–514.

Murai, A., Katsui, N., and Masamune, T. (1982d). Significant role of syntheses of phytoalexins. *Kagaku no Ryoiki 36*(10): 673–685.

Murai, A., Sato, S., and Masamune, T. (1984a). A general synthetic approach of spirovetivanes. The synthesis of (±)-solavetivone, (±)-hinesol, and related compounds. *Bull. Chem. Soc. Jpn. 57*(8): 2276–2281.

Murai, A., Sato, S., and Masamune, T. (1984b). Efficient synthesis of (±)-solavetivone and (±)-[8,8-$^2H_2$]solavetivone. *Bull. Chem. Soc. Jpn. 57*(8): 2282–2285.

Murai, A., Sato, S., and Masamune, T. (1984c). Total synthesis of (±)-lubimin and (±)-oxylubimin. I. Synthesis of (±)-15-norsolavetivone and related compounds. *Bull. Chem. Soc. Jpn. 57*(8): 2286–2290.

Murai, A., Sato, S., and Masamune, T. (1984d). Total synthesis of (±)-lubimin and (±)-oxylubimin. II. Transformation of (±)-15-norsolavetivone into (±)-lubimin, and (±)-oxylubimin, and related compounds. *Bull. Chem. Soc. Jpn. 57*(8): 2291–2294.

Nakajima, M. (1978). Phytoalexin no seisei to tenkan in seibutsu no seigyo kikou. *Kagaku Zokan 75*: 31–45.

Nystrom, J.-E., and Helquist, R. (1989). Short intramolecular Diels-Alder approach to spirovetivanes. Total synthesis of dl-hinesol. *J. Org. Chem. 54*(19): 4695–4698 and references cited therein.

Rubottom, G. M., and Gruber, J. M. (1978). m-Chloroperbenzoic acid oxidation of 2-trimethylsilyloxy-1,3-dienes. Synthesis of α-hydroxy and α-acetoxy enones. *J. Org. Chem. 43*(8): 1599–1602.

Sato, K., Ishiguri, Y., Doke, N., Tomiyama, K., Yagihashi, F., Murai, A., Katsui, N., and Masamune, T. (1978). Biosynthesis of the sesquiterpenoid phytoalexin rishitine from acetate via oxylubimin in potato. *Phytochemistry 17*(11): 1901–1902.

Stoessl, A., Stothers, J. B., and Ward, E. W. B. (1975). A 2,3-Dihydroxygermacrene and other stress metabolites of *Datura stramonium*. *J. Chem. Soc., Chem. Commun.* (11): 431–432.

Stoessl, A., Stothers, J. B., and Ward, E. W. B. (1976). Sesquiterpenoid stress compounds of the Solanaceae. *Phytochemistry 15*(6): 855–872.

Stoessl, A., and Stothers, J. B. (1983). Biosynthesis of antifungal stress metabolites from potato: observation of hydride shifts via the "β-hop" from incorporation of [2-$^2H_3$, 2-^{13}C]acetate and ^{13}Cmr. *Can. J. Chem. 61*(8): 1766–1769.

Tomiyama, K. (1971). Phytoalexin: chemistry and biology. *Kagaku no Ryoiki 31*(7): 658–668.

Tomiyama, K. (1981). Shokubutsukansen niokeru Kabinshosaiboushi no Seirigaku. *Seibutsukagaku 33*(1): 17–25.

Vedejs, E., Engler, D. A., and Telschow, J. E. (1978). Transition-metal peroxide reactions, synthesis of α-hydroxycarbonyl compounds from enolates. *J. Org. Chem.* *43*(2): 188–196.

Yamada, K., Nagase, H., Hayakawa, Y., Aoki, K., and Hirata, Y. (1973a). Synthetic studies on spirovetivanes. I. spirocondensation of a 4-(3′-formylpropyl)-3-cyclohexenone and stereospecific total synthesis of dl-β-vetivone. *Tetrahedron Lett.* (49): 4963–4966.

Yamada, K., Aoki, K., Nagase, H., Hayakawa, Y., and Hirata, Y. (1973b). Synthetic studies on spirovetivanes. II. Stereospecific total synthesis of dl-hinesol, dl-α-vetispirene, and dl-β-vetispirene (dl-β-isovetivenene). *Tetrahedron Lett.* (49): 4967–4970.

Yamada, K., Goto, S., Nagase, H., and Christensen, A. T. (1977). Total synthesis and stereostructure of (±)-solavetivone, a stress metabolite from infected potato tubers; X-ray crystal and molecular structure of an intermediate in the synthesis. *J. Chem. Soc., Chem. Commun.* (16): 554–555.

20

Enzymic Conversion of Furanosesquiterpene in *Ceratocystis fimbriata*-Infected Sweet Potato Root Tissue

Masayuki Fujita
Kagawa University, Kagawa, Japan

Hiromasa Inoue
Kyushu Dental College, Kitakyushu, Japan

K. Oba and Ikuzo Uritani*
Nagoya Women's University, Nagoya, Japan

I. INTRODUCTION

Sweet potato (*Ipomoea batatas* Lam.) root tissues accumulate bitter and oily substances in the mold-damaged regions when they are infected with pathogenic fungus such as *Ceratocystis fimbriata* Ell. and Halst. They are antifungal and Ehrlich's reagent-positive furanosesquiterpenes, which consist of several major components such as ipomeamarone [3 + 4′ + 2″] and ipomeamaronol [3 + 4′ + 3″] and many kinds of minor family compounds derived from them. Ipomeamarone was first isolated and determined in the structure among phytoalexins in the plant kingdom (Hiura, 1943; Kubota and Matsura, 1953). Its chemical structure is shown in Fig. 1. These furanosesquiterpenes are produced not only by infection of pathogenic fungi but also by treatment with poisonous heavy metal ions (Uritani et al., 1960) and mycotoxins (Fujita and Yoshizawa, 1987) in sweet potato root tissues. Therefore, various kinds of poisons, including mercuric chloride, have also been used as artificial phytoalexin inducers (elicitors) for the biochemical investigation on synthesis of furanosesquiterpenes in the root tissues (Oba and Uritani, 1981; Fujita and Yoshizawa, 1989; Fujita, 1985). Using various ^{14}C-labeled intermediates, it was elucidated that the furan-

*Current affiliation: Aichi Konan Gakuen School Corporation, Konan City, Japan

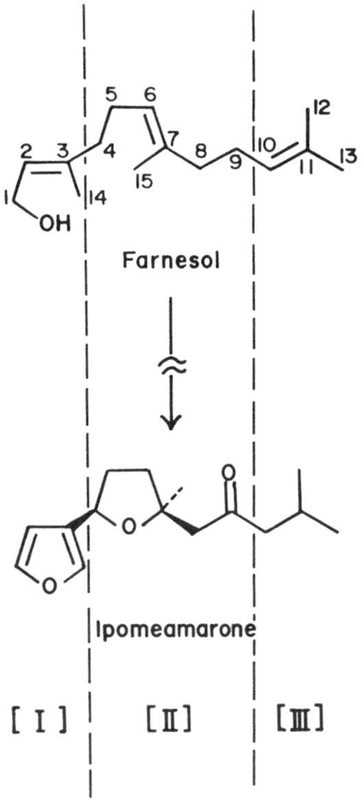

Figure 1 Chemical structures of farnesol and ipomeamarone. (I), Furan ring moiety; (II), tetrahydrofuran ring and the next ethylene moieties; (III), the rest (the isobutyl end of side chain) of the moiety.

osesquiterpenes were biosynthesized via acetyl CoA (Oba et al., 1970), 3-hydroxy-3-methylglutaryl-CoA (HMG-CoA), mevalonate (MVA) (Akazawa et al., 1962; Akazawa, 1964; Oshima and Uritani, 1968), MVA phosphate, MVA pyrophosphate, isopentenylpyrophosphate (Oshima and Uritani, 1969), geranyl pyrophosphate (Oguni and Uritani, 1969), farnesyl pyrophosphate and farnesol [1 + 1′ + 1″](Oguni and Uritani, 1970, 1971b), initiated from pyruvate, which was supplied by glycolysis. The metabolic pathway between acetyl-CoA and farnesylpyrophosphate is designated the first half-step and the pathway from farnesol to the phytoalexins the second half-step. They functionally bear their shares of responsibilities in the biosynthesis of the furanosesquiterpenes. The former plays a role in supplying farnesol as a material for carbon skeletons of ipomeamarone and ipomeamaronol. It has been clear that the amount of the supply is enzymatically controlled by HMG-CoA reductase as the rate-limiting enzyme in the pathway (Ito et al., 1979). Therefore, the amount of accumulation of furanosesquiterpenes is primarily controlled by that as quantitative pathway. In contrast to that, the latter contributes to the formation of the characteristic carbon skeleton of ipomeamarone and also confers the diversity of the minor family compounds which are derivatively produced by the bypath of synthesis of ipomeamarone (qualitative pathway).

Thirty-seven species of the sesquiterpenes and their decomposed products, which are considered to be metabolically derived from farnesol, have been isolated and identified as stress compounds in *C. fimbriata*-infected and mercuric chloride-treated sweet potato root tissues (Schneider et al., 1984). Among them, at least five kinds of sesquiterpenes, such as 9-hydroxyfarnesol [1 + 2′ + 1″], 9-oxofarnesol [1 + 3′ + 1″], 6-hydroxydendrolasin [3 + 2′ + 1″], 6-oxodendrolasin (component A_2) [3 + 3′ + 1″] (Burka et al., 1981; Ito et al., 1984), and dehydroipomeamarone [3 + 4′ + 1″], are regarded as intermediates in the most direct pathway between farnesol and ipomeamarone on the basis of the information from ^{14}C-tracer experiments. However, there may have been a few biochemical interpretations about the metabolic conversions among them, since few enzymological approaches to them have been carried out by other workers apart from Uritani's group. Therefore, it is not clear what kinds of enzymic systems regulatively catalyze the formation of characteristic carbon skeletons of ipomeamarone and ipomeamaronol and influence the composition of the minor family compounds produced by their bypath, and at what cell organelle their biosynthesis is done. In this chapter, we describe enzymologically the second half-step on the basis of some of the results obtained from our investigations and so forth.

As a matter of convenience, the structures of the furanosesquiterpenes described in this text were stated in combination of the numbers added to the partial ones in Fig. 2.

II. ENZYMIC FORMATION OF SKELETON OF IPOMEAMARONE IN THE SECOND HALF STEP

Ipomeamarone possesses one furan ring and one tetrahydrofuran ring in the skeleton. In order to interpret the metabolic formation of the characteristic carbon skeleton downstream from farnesol, it is significant to consider separately the expected enzymic conversions occurring in the three independent parts, namely, furan ring (I), tetrahydrofuran ring and the next ethylene (II), and the rest (the isobutyl end of side chain) (III) (see Fig. 1). We describe the possible enzymic pathways in the three parts successively. As described above, the chemical structures of sesquiterpenes are originally designated with the numbers shown in Fig. 2.

A. Furan Ring Moiety

The most plausible pathway of the carbon skeleton formation occurring in the furan ring moiety is shown in Fig. 2a. The authors considered that it was necessary that the hydroxylated C-l be enzymically oxidized to the aldehyde for the closing of the furan ring. Therefore, a farnesol dehydrogenase catalyzing the oxidation of farnesol to farnesal [2 + 1′ + 1″] was highly purified and characterized in detail from cut-injured sweet potato root tissues (Inoue et al., 1984). This enzyme is a homodimer consisting of subunits with mol wt 47,000 and needs $NADP^+$ as a cofactor for the activity. The activity was considerably decreased by some SH group inhibitors. The maximal activity was detected when *t,t*-farnesol was used as a substrate. The activity was reduced in the following order: *c,t*-farnesol (C_{15}) >

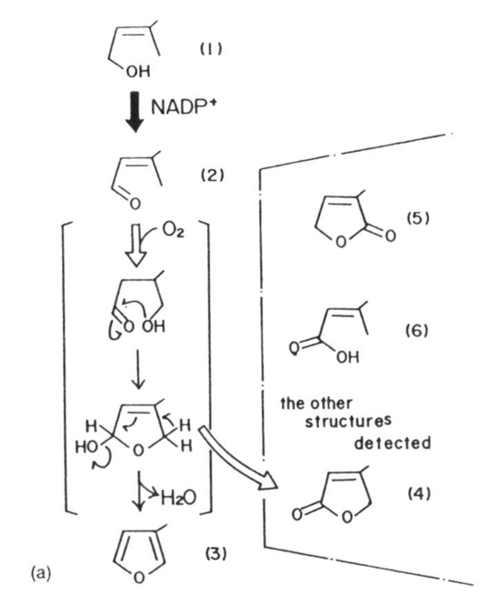

Figure 2 (a) Formation of furan ring of ipomeamarone. ⇒, Enzymic conversion proven; ⇒, enzymic conversion presumed; →, nonenzymic conversion presumed; [ ], undetected intermediates. (b) Ring closing and ring opening of tetrahydrofuran ring of ipomeamarone. ⇒, Enzymic conversion presumed; →, nonenzymic conversion presumed; [ ], undetected intermediates. (c) Reduction and oxygenation in the end of the side chain of ipomeamarone. ⇒, Enzymic conversion proven; ⇆, chemical equilibrium reaction.

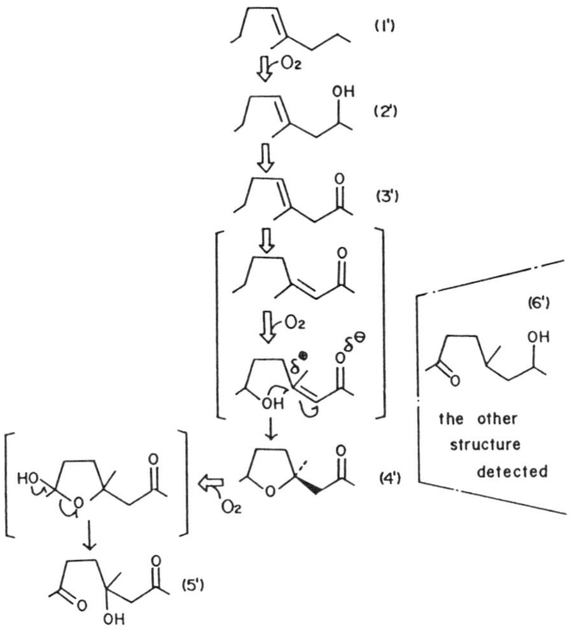

(1')

O_2

OH

(2')

O

(3')

O

O_2

$\delta^{\oplus}$

$\delta^{\ominus}$

O

O

OH

(6')

OH

O

the other
structure
detected

HO

O

O

O

O

(4')

O_2

O

(5')

O

OH

(b)

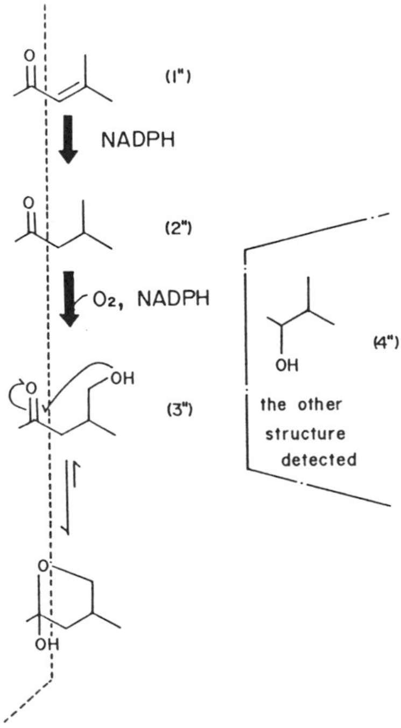

O

(1")

NADPH

O

(2")

O_2, NADPH

(4")

OH

O

OH

(3")

the other
structure
detected

O

OH

(c)

geraniol (C_{10}) > citronerol (C_{10}) > nerol (C_{10}) > decanol (C_{10}). Benzylalcohol, glycerol, and ethanol showed no activity. On the other hand, NADPH-dependent geraniol dehydrogenase was detected in orange juice vesicles and oxidized some monoterpene alcohols in the presence of $NADP^+$, but at a much slower rate than farnesol (Potty and Bruemmer, 1970). Time courses of the enzyme activity and amount of furanosesquiterpenes accumulated in sweet potato root tissues after cut injury and *C. fimbriata* infection are shown in Fig. 3a. Though a little enzyme activity was detected in the fresh tissue (without cut injury and *C. fimbriata* infection), it was dramatically increased with a little lag time (2–3 hr) after infection by *C. fimbriata* and about 20 hr before accumulation of furanosesquiterpenes, suggesting that this enzyme plays an important role in the biosynthesis of furanosesquiterpenes in diseased sweet potato root tissues. The above results suggest that oxidation at the C-l site of farnesol by the farnesol dehydrogenase participated in formation and accumulation of various furanosesquiterpenes. On the other hand, substitution of air phase with nitrogen gas suppressed the rate of the synthesis of ipomeamarone in vivo (Oguni and Uritani, 1971b). Thereafter, Burka and Thorsen (1982) proved that all three oxygen atoms of ipomeamarone were introduced from molecular oxygen but not from farnesol or water by tracer experiment using

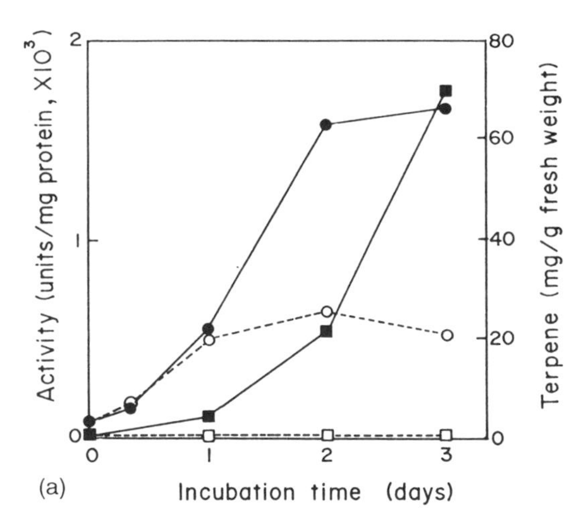

Figure 3 Changes in the activities of farnesol dehydrogenase (a), terpene reductase (b), ipomeamarone 12-hydroxylase (c), and the content of furanosesquiterpenes in response to fungal inoculation or cut injury. The enzymic activities in the cases of fungal inoculation (●) and cut injury (○), and the contents of furanosesquiterpenes in the cases of fungal inoculation (■) and cut injury (□).

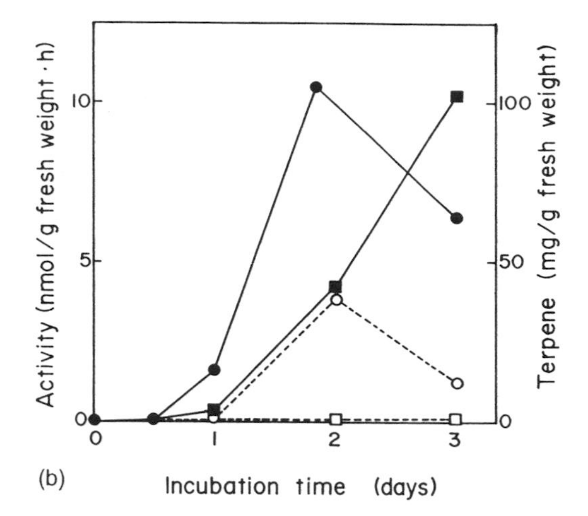

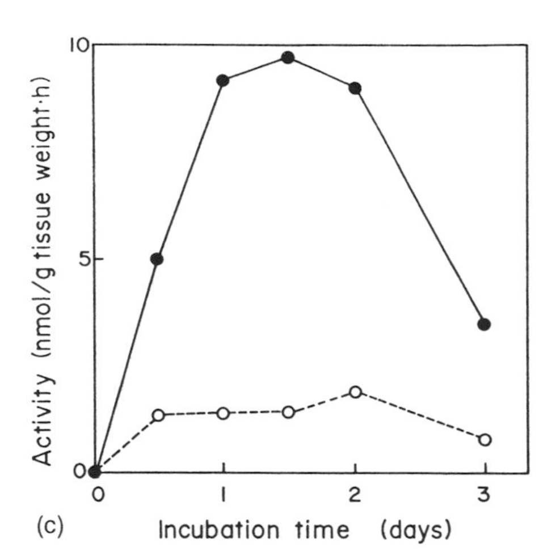

$^{18}O_2$. This evidence indicates that introduction of one oxygen atom from molecular oxygen by oxygenase occurs during the formation of the furan ring. The above evidence strongly suggests that C-1 oxidized to aldehyde with farnesol dehydrogenase nonenzymically binds to the oxygen atom of the hydroxyl group introduced to C-14 with oxygenase, so that hemiacetal is formed as a precursor of the furan ring. As soon as this hemiacetal is formed, one dehydration is expected to occur at C-1 and C-14 nonenzymically, so that furan ring is synthesized. These over all reactions show that an oxygen atom from farnesol is eliminated as a water molecule and another oxygen atom from molecular oxygen is utilized during cyclization of furan ring. Recently, two butenolides such as 6-oxodendrolasinoide [4 + 3′ + 1″] and ipomeamaronolide [5 + 3′ + 1″] have been isolated and identified in *C. fimbriata*-infected sweet potato root tissues (Schneider et al., 1984). The former compound might support the above hypothesis for the ring closing, since it appears to be a byproduct, to which the hemiacetal intermediate is immediately dehydrogenated but not dehydrated.

B. Tetrahydrofuran Ring and the Ethylene Moieties

There has been no reaction of terpene conversion elucidated enzymologically among ones speculated to occur in the tetrahydrofuran ring and the next ethylene moieties. However, the most reasonable metabolic pathway has been considered on the basis of some results from isotopic tracer experiments and organic chemical knowledges (Fig. 2b).

It has been proven with ^{14}C-tracer experiments that farnesol was converted to dehydroipomeamarone via 6-oxodendrolasin (Burka et al., 1981; Ito et al., 1984). As described above, since all three oxygen atoms of ipomeamarone originate from molecular oxygen (Burka and Thorsen, 1982), two oxygen atoms in the moieties are considered to be introduced by oxygenases. C-9 is first hydroxylated and then oxidated to ketone. The five kinds of sesquiterpenes having hydroxyl group at C-9—9-hydroxyfarnesol (4), 6-hydroxydendrolasin (Burka et al., 1981), 6-myoporol [3 + 6′ + 2″] (Burka and Iles, 1979), 6-dihydro-7-hydroxymyoporone [3 + 6′ + 4″] (Burka, 1978), and 9-hydroxyfarnesoic acid [6 + 2′ + 1″] (Schneider et al., 1984)—have been isolated and determined in their chemical structures. Burka et al. (1981) deduced that C-4 was hydroxylated and isomeric transition from the 6,7 double bond to the 7,8 one occurred, followed by Michael addition (1,4 addition) during the closing of the tetrahydrofuran ring. The most direct proof was recently given with experiments using deutriointermediates by Schneider et al., (1984). The introduction of oxygen atom by oxygenase seems to be a rate-limiting reaction for the closing of the ring, since no compound containing a hydroxyl group at C-4 has been detected yet.

Using ^{14}C-labeled precursors, Burka and Kuhnert (1977) demonstrated that the tetrahydrofuran ring of ipomeamarone was cleaved so that ipomeamarone was converted to 4-hydroxymyoporone [3 + 5′ + 2″]. Dehydroipomeamarone seems to be converted in a similar manner to 4-hydroxydehydromyoporone [3 + 5′ + 1″] too. Though the ring openings are expected to be triggered by the hydroxylation of C-4 by oxygenases, it has not been proven that an oxygen atom newly introduced originated from oxygen molecule and no corresponding intermediate in their reactions has also been detected yet. Therefore, the reaction of introduction of oxygen atom is also presumed to be rate-limiting step on these ring openings.

C. The Isobutyl End of Side Chain

The reactions for formation of the carbon skeleton in the rest moiety are shown in Fig. 2c. This pathway has been established enzymologically.

The 10,11 double bond originating from farnesol is saturated after the closing of the tetrahydrofuran ring, suggesting that it might play a role for the cyclization like C-9 ketone. The authors elucidated the properties of the enzyme catalyzing the saturation of dehydroipomeamarone using the cell-free system (Inoue and Uritani, 1980; Inoue et al., 1984). This enzyme was located in microsomal fraction and required NADPH for the activity. SH group–inhibiting reagents such as Cu^{2+}, monoiodoacetamide, and *N*-ethylmaleimide strongly reduced activity. It is likely that the SH group of the enzyme participates in the activity. On the other hand, addition of the same concentration of 4-hydroxydehydromyoporone as that of dehydroipomeamarone to the reaction mixture resulted in a 50% decrease in activity, suggesting that this enzyme could practically reduce both compounds as substrates in parallel. In contrast to the first half step forming a linear pathway, it has been elucidated through a lot of tracer experiments that the second half step was made up of complicated network systems for the biosynthesis of furanosesquiterpenes in sweet potato root tissues. The flexibility of the reductase in recognition of the substrate seems to support the complexity in the second half step. Time courses of the enzyme activity and accumulation of furanosesquiterpenes after *C. fimbriata* infection and cut injury are shown in Fig. 3b. The enzyme activity is not detected in the fresh root tissues but is found in the *C. fimbriata*-infected and cut-injured ones. The maximal activity of the former is two- to threefold higher than one of the latter. After infection with *C. fimbriata*, while furanosesquiterpenes accumulate with a lag time of 1 day, this enzyme activity appears at 0.5 day and then drastically increases in advance of the terpene accumulation. The above results suggest that the enzyme participates in the biosynthesis of the furanosesquiterpenes.

After the saturation of the double bond, C-12 of the side chain end is hydroxylated. The authors observed the hydroxylation at C-12 in vivo experiments using ^{14}C-labeled ipomeamarone (Oba et al., 1982). Furthermore, the biochemical properties in enzymological manner using the cell free system have been elucidated (Fujita et al., 1981, 1982). The enzyme catalyzing the reaction was located in the microsomal fraction, required NADPH as a cofactor for the activity, and was an SH enzyme. The inhibi-

Table 1 Inhibition of Ipomeamarone 12-Hydroxylase Activity by Carbon Monoxide and Its Reversal by Light

Gas mixture	Relative activity (%)[a]
Air	100
N_2 (100%)	25
O_2 (15%) + N_2 (85%)	90
CO (15%) + O_2 (15%) + N_2 (70%)	36
CO (15%) + O_2 (15%) + N_2 (70%) + light	70

[a]Relative activity was expressed as a percentage of the activity of air control.

tion by cytochrome c indicated the participation of a microsomal electron transport system in the activity. The enzyme activity required molecular oxygen and was severely inhibited by CO. The inhibition was efficiently suppressed by radiation of visible light during the reaction (Table 1). According to the above results, it was concluded that this enzyme reaction proceeded from a mixed function monooxygenase system containing cytochrome P-450 as a terminal constituent. Time courses of the enzyme activity after cut injury and *C. fimbriata* infection are shown in Fig. 3c. The enzyme activity was not detected in the fresh tissues but appeared and increased with little lag time after wounding and infection. However, the latter maximal activity was several fold higher than the former, suggesting that this enzyme also closely participated in the biosynthesis of furanosesquiterpenes. Tanaka et al. (1976) reported that cinnamic acid 4-hydroxylase was a cytochrome P-450–dependent enzyme system in sweet potato root tissues. However, no substrate competition between ipomeamarone 12-hydroxylase and cinnamic acid 4-hydroxylase strongly indicated that the two enzyme reactions were catalyzed by individual cytochromes P-450 (Fujita et al., 1982). The hydroxyl group introduced at C-12 is considered to contribute to the subsequent formation of a five-member ring (Schneider et al., 1984).

III. PARTICIPATION OF CYTOCHROME P-450 IN THE SECOND HALF STEP

As described above, at least three oxygenations are required in the formation of ipomeamarone (Burka and Thorsen, 1982) and then the introduction of one more oxygen atom by oxygenase is considered to facilitate the split in the tetrahydrofuran ring. However, there has been no enzymological evidence to this effect yet. In order to examine to what extent cytochrome(s)

P-450 participates in such oxygenations as ipomeamarone hydroxylation, furanosesquiterpene production and cytochrome(s) P-450 synthesis were determined in sweet potato root tissues treated with various chemicals (Fujita, 1985). Content of cytochrome P-450 was calculated from the CO difference spectrum of the microsomes under reduced condition (Fig. 4). The amount of production of furanosesquiterpenes and content of cytochrome P-450 are shown in Fig. 5 and in Table 2, respectively. These results indicate an intimate mutual relation between both factors. Therefore, there is no doubt that cytochrome(s) P-450 participates in multiple oxygenations during ipomeamarone biosynthesis and in the following oxygenative conversions including ipomeamarone 12-hydroxylation. As described above, the most plausible intermediates (hydroxylated forms) taking part in ring closing or ring opening of the furan and tetrahydrofuran rings of ipomeamarone have not yet been detected. This strongly indicates a possibility that the

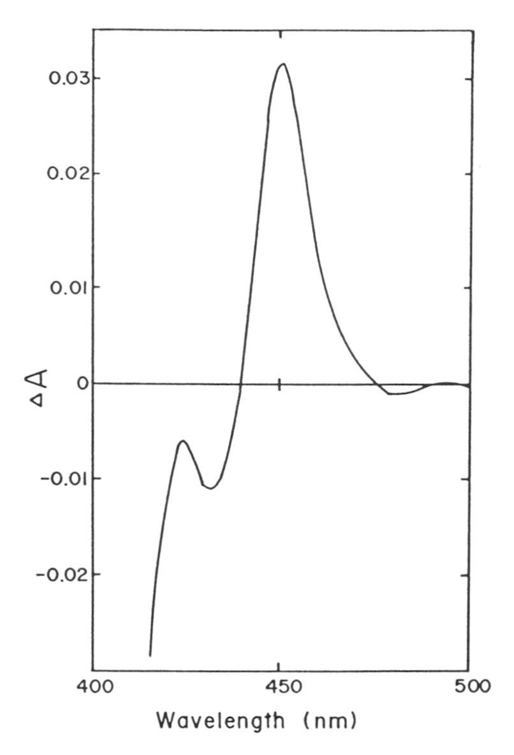

Figure 4 Difference spectrum of CO-bound − CO-unbound forms of dithionite-reduced microsomal fraction from diseased tissue.

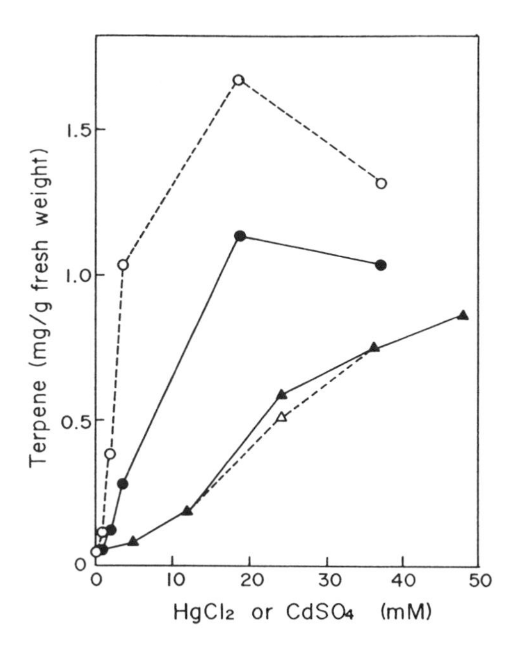

Figure 5 Terpene accumulation in discs of root tissue of sweet potato treated with $HgCl_2$ or $CdSO_4$, with or without phenyl isocyanide. ●, $HgCl_2$; ○, $HgCl_2$ + phenyl isocyanide; ▲, $CdSO_4$; △, $CdSO_4$ + phenyl isocyanide.

multiple monooxygenases minutely regulate in part the biosynthesis rate of ipomeamarone as the rate-limiting enzymes for the formation and degradation of the rings, in contrast to the HMG-CoA reductase controlling that in whole. Hence, each cytochrome P-450 and NADPH–cytochrome P-450 reductase should be purified as constituents for microsomal monooxygenase systems and the reconstitution system could allow examination of the individual furanosesquiterpene conversions in vitro enzyme systems. The authors have purified NADPH–cytochrome P-450 reductase to homogeneous on polyacrylamide gel electrophoresis and characterized (Fujita and Asahi, 1985b).

IV. INTRACELLULAR LOCALIZATION OF THE SECOND HALF STEP

Intracellular localization of ipomeamarone 12-hydroxylase has been examined with linear sucrose density gradient centrifugation and microscopic observation in detail, resulting in its being located in rough-surfaced endo-

Table 2 Effects of Treatment of Tissue Discs with Chemicals on Cytochrome P-450 and Microsomal Protein During Incubation of Discs of Root Tissue of Sweet Potato

Discs	Microsomal protein (mg/g fresh wt)	Cytochrome P-450 (pmol/g fresh wt)	Specific cytochrome P-450 amount (pmol/mg protein)[a]
Freshly prepared discs	0.213	2.74	12.9
Incubated discs[b]	0.500	37.0	74.1
Discs treated with[c]			
Phenyl isocyanide (PI)	0.541	42.9	79.2
$HgCl_2$[d]	0.538	88.6	165
$HgCl_2$[d] and PI	0.597	118.0	198
$CdSO_4$[d]	0.481	57.1	119
$CdSO_4$[d] and PI	0.565	60.4	107

[a]The specific amount in the microsomes.
[b]The discs were incubated for 2 days without treatment.
[c]The discs incubated for 1 day were treated with the chemicals incubated for 1 more day.
[d]The concentration was 3.68 mM.

plasmic reticulum (Fujita and Asahi, 1985a). The localization was apparently different from one of cinnamic acid 4-hydroxylase participating in the polyphenol metabolism in sweet potato root tissues. As described above, the terpene reductase was also located in postmitochondrial membraneous fraction. Multiple oxygenations in the furanosesquiterpene metabolism must also occur in microsomal fraction since they are considered to be dependent on participation of cytochrome P-450. This information strongly suggests that a lot of processes in the second half step are performed on microsomes. On the other hand, electron-dense vesicles containing furanosesquiterpenes were observed by electron microscopy and were recovered in 2% Ficoll fraction after discontinuous Ficoll density gradient centrifugation (Oba et al., 1984). These vesicles are enveloped in a single membrane in 7–10 nm thick and containing only a little antimycin A–insensitive NADPH–cytochrome c reductase activity, suggesting that they might originate from endoplasmic reticulum. The endoplasmic reticulum with heterogeneity in *C. fimbriata*-infected sweet potato root tissues was previously demonstrated by the distribution of RNA and various organelle marker enzymes after sucrose density gradient centrifugation (Fujita and Asahi, 1985a). Therefore, there might be highly differentiated endoplasmic reticulum participating in the biosynthesis, transport, and secretion of furanosesquiterpenes in *C. fimbriata*-infected sweet potato root tissues.

V. EPILOG

The biosynthesis of the furanosesquiterpenes in mold-damaged sweet potato root is divided into two parts; the first half step from acetyl-CoA to farnesyl pyrophosphate and the second half step after farnesol. The former is comparatively linear and the rate is strictly controlled by HMG–CoA reductase as a rate-limiting enzyme. The regulation in this step strongly influences the ebb and flow of metabolism in the latter step. Therefore, the former may be called the quantitative pathway. The latter forms network systems and produces ipomeamarone as well as various kinds of family compounds. The composition of the furanosesquiterpenes is variable under various conditions such as kinds of elicitors (Fujita, 1985; Fujita and Yoshizawa, 1987, 1989). Therefore, in contrast to the former, this step may be designated the qualitative pathway. In this chapter, the second half step was described enzymologically. In the step, it appears that cytochrome P-450–dependent mixed-function monooxygenation lurks as a rate-limiting reaction in ring closing and ring opening of the furan and tetrahydrofuran rings. This is supported by intimate mutual relation between the synthesis of cytochrome P-450 and the production of furanosesquiterpenes under diverse conditions. The hydroxylation at the end of side chain of ipomeamarone is the only reaction that has been enzymologically proved to depend on cytochrome P-450 in the furanosesquiterpene metabolism of sweet potato. Hence, it is necessary to isolate cytochromes P-450 and to examine the substrate specificities using the reconstitution systems. Since cytochromes P-450 participating in drug metabolism possess broad substrate specificities in animals, it is likely that the cytochrome P-450 catalyzing multiple substrates participates in the biosynthesis of furanosesquiterpenes in *C. fimbriata*-infected sweet potato root tissues. It was elucidated that the enzyme reducing dehydroipomeamarone to ipomeamarone also converted 4-hydroxydehydromyoporone to 4-hydroxymyoporone, suggesting that many kinds of enzymes participating in the second half step might have broad substrate specificities. For example, the farnesol dehydrogenase might actively oxidize 9-hydroxyfarnesol and 9-oxofarnesol to the corresponding aldehydes like farnesol. According to information on localization of the enzymes participating in the step, it seems to be located in endoplasmic reticulum. On the other hand, furanosesquiterpene-containing vesicles which were considered to be derived from endoplasmic reticulum were observed. The vesicles might participate in the transport of furanosesquiterpenes from the site for synthesis (noninfected region adjacent to infected region) to the site for accumulation (infected necrotic region).

REFERENCES

Akazawa, T. (1964). Biosynthesis of ipomeamarone. II. synthetic mechanism. *Arch. Biochem. Biophys. 105*(3): 512–516.

Akazawa, T., Uritani, I., and Akazawa, Y. (1962). Biosynthesis of ipomeamarone. I. the incorporation of acetate-2-C^{14} and mevalonate-2-C^{14} into ipomeamarone. *Arch. Biochem. Biophys. 99*(1): 52–59.

Burka, L. T. (1978). 1-(3′-Furyl)-6,7-dihydroxy-4,8-dimethylnonan-1-one, a stress metabolite from sweet potatoes (*Ipomoea batatas*). *Phytochemistry 17*(2): 317–318.

Burka, L. T., and Iles, J. (1979). Myoporone and related keto alcohols from stressed sweet potatoes. *Phytochemistry 18*(5): 873–874.

Burka, L. T., and Kuhnert, L. (1977). Biosynthesis of furanosesquiterpenoid stress metabolites in sweet potatoes (*Ipomoea batatas*). Oxidation of ipomeamarone to 4-hydroxymyoporone. *Phytochemistry 16*(12): 2022–2023.

Burka, L. T., and Thorsen, A. (1982). Biosynthesis of furanosesquiterpenoid stress metabolites in *Ipomoea batatas*: isotopic oxygen incorporation into ipomeamarone. *Phytochemistry 21*(4): 869–870.

Burka, L. T., Felice, L. J., and Jackson, S. W. (1981). 6-Oxodendrolasin, 6-hydroxydendrolasin, 9-oxofarnesol and 9-hydroxyfarnesol, stress metabolites of the sweet potato. *Phytochemistry 20*(4): 647–652.

Fujita, M. (1985). Stimulation of cytochrome P-450 synthesis in sliced sweet potato root tissue by chemicals applied to surface. *Agric. Biol. Chem. 49*(10): 3045–3047.

Fujita, M., and Asahi, T. (1985a). Different intracellular localization of two cytochrome P-450 systems, ipomeamarone 15-hydroxylase and cinnamic acid 4-hydroxylase, in sweet potato root tissue infected with *Ceratocystis fimbriata. Plant Cell Physiol. 26*(3): 389–395.

Fujita, M., and Asahi, T. (1985b). Purification and properties of sweet potato NADPH-cytochrome c (P-450) reductase. *Plant Cell Physiol. 26*(3): 397–405.

Fujita, M., and Yoshizawa, T. (1987). Induction of sweet potato phytoalexins by trichothecene mycotoxins. *Proc. Jpn. Assoc. Mycotoxicol. 25*: 29–30 (in Japanese).

Fujita, M., and Yoshizawa, T. (1989). Induction of phytoalexins by various mycotoxins and metabolism of mycotoxins in sweet potato root tissues. *J. Food Hyg. Soc. Jpn. 36*(6): 501–505 (in Japanese).

Fujita, M., Oba, K., and Uritani, I. (1981). Ipomeamarone 15-hydroxylase from cut-injured and *Ceratocystis fimbriata*-infected sweet potato root tissues. *Agric. Biol. Chem. 45*(8): 1911–1913.

Fujita, M., Oba, K., and Uritani, I. (1982). Properties of a mixed function oxygenase catalyzing ipomeamarone 15-hydroxylation in microsomes from cut-injured and *Ceratocystis fimbriata*-infected sweet potato root tissues. *Plant Physiol.* *70*(2): 573–578.

Hiura, M. (1943). Studies on storage and rot of sweet potato. 2, *Rep. Gifu Agric. Coll.* *50*(1): 1–5.

Inoue, H., and Uritani, I. (1979). Biosynthetic correlation of various phytoalexins in sweet potato root tissues infected by *Ceratocystis fimbriata*. *Plant Cell Physiol.* *20*(7): 1307–1314.

Inoue, H., and Uritani, I. (1980a). Conversion of 4-hydroxydehydromyoporone to other furanoterpenes in *Ceratocystis fimbriata*-infected sweet potato. *Agric. Biol. Chem.* *44*(8): 1935–1936.

Inoue, H., and Uritani, I. (1980b). Furano-sesquiterpene reductase from fungal-inoculated sweet potato root tissue. *Agric. Biol. Chem.* *44*(9): 2245–2247.

Inoue, H., Oba, K., Ando, M., and Uritani, I. (1984a). Enzymatic reduction of dehydroipomeamarone to ipomeamarone in sweet potato root tissue infected by *Ceratocystis fimbriata*. *Physiol. Plant Pathol.* *25*: 1–8.

Inoue, H., Tsuji, H., and Uritani, I. (1984b). Characterization and activity change of farnesol dehydrogenase in black rot fungus-infected sweet potato. *Agric. Biol. Chem.* *48*(3): 733–738.

Ito, R., Oba, K.,and Uritani, I. (1979). Mechanism for the induction of 3-hydroxy-3-methylglutaryl coenzyme A reductase in $HgCl_2$-treated sweet potato root tissue. *Plant Cell Physiol.* *20*(5): 867–874.

Ito, I., Kato, N., and Uritani, I. (1984). Biochemistry of two new sesquiterpenoid phytoalexins from sweet potato roots. *Agri. Biol. Chem.* *48*(1): 159–164.

Kubota, T., and Matsuura, T. (1953). Chemical studies on the black rot disease of sweet potato. *J. Chem. Soc. Jpn.* *74*: 101–109, 197–199, 248–251, 668–670.

Oba, K., and Uritani, I. (1981). Mechanism of furano-terpene production in sweet potato root tissue injured by chemical agents. *Agric. Biol. Chem.* *45*(7): 1635–1639.

Oba, K., Shibata, H., and Uritani, I. (1970). The mechanism supplying acetyl-CoA for terpene biosynthesis in sweet potato with black rot: incorporation of acetate-2-^{14}C, pyruvate-3-^{14}C, and citrate-2,4-^{14}C into ipomeamarone. *Plant Cell Physiol.* *11*(3): 507–510.

Oba, K., Oga, K., and Uritani, I. (1982). Metabolism of ipomeamarone in sweet potato root slices before and after treatment with mercuric chloride or infection with *Ceratocystis fimbriata*. *Phytochemistry* *21*(8): 1921–1925.

Oba, K., Nakamura, A., and Iwaikawa, Y. (1984). Isolation of furanoterpene-containing particles from *Ceratocystis fimbriata*-infected sweet potato root tissue. *J. Biochem.* *96*(6): 1951–1954.

Oguni, I., and Uritani, I. (1969). Biochemical studies on the terpene metabolism in sweet potato root tissue with black rot. Effect of C_{10}- and C_{15}-terpenols on acetate-2-^{14}C incorporation into ipomeamarone. *Agric. Biol. Chem.* *33*(1): 50–62.

Oguni, I., and Uritani, I. (1970). The incorporation of farnesol-2-^{14}C into ipomeamarone. *Agric. Biol. Chem.* *34*(1): 156–158.

Oguni, I., and Uritani, I. (1971a). Effect of monofluoroacetate on pyrubate-3-^{14}C

and glucose-U-^{14}C incorporation into ipomeamarone. *Agric. Biol. Chem.* *35*(12): 1980–1983.

Oguni, I., and Uritani, I. (1971b). Participation of farnesol in the biosynthesis of ipomeamarone. *Plant Cell Physiol.* *12*(4): 507–515.

Oguni, I., and Uritani, I. (1974). Dehydroipomeamarone as an intermediate in the biosynthesis of ipomeamarone, a phytoalexin from sweet potato root infected with *Ceratocystis fimbriata*. *Plant Physiol.* *53*(4): 649–652.

Oguni, I., and Uritani, I. (1974). Effect of (−)-hydroxycitrate on ipomeamarone biosynthesis from pyruvate in sweet potato with black rot. *Plant Cell Physiol.* *15*(1): 179–182.

Oshima, K., and Uritani, I. (1968). Participation of mevalonate in the biosynthetic pathway of ipomeamarone. *Agric. Biol. Chem.* *32*(9): 1146–1152.

Oshima, K., and Uritani, I. (1969). Enzymatic synthesis of isopentenyl pyrophosphate in sweet potato root tissue in response to infection by black rot fungus. *Plant Cell Physiol.* *10*(4): 827–843.

Potty, V. H., and Bruemmer, J. H. (1970). Oxidation of geraniol by an enzyme system from orange. *Phytochemistry* *9*(5): 1001–1007.

Schneider, J. A., Lee, J., Naya, Y., Nakanishi, K., Oba, K., and Uritani, I. (1984). The fate of the phytoalexin ipomeamarone: furanoterpenes and butenolides from *Ceratocystis fimbriata*-infected sweet potatoes. *Phytochemistry* *23*(4): 759–764.

Schneider, J. A., Lee, J., Yoshihara, K., Mizukawa, K., and Nakanishi, K. (1984). Biosynthetic studies of ipomeamarone. *J. Chem. Soc. Chem. Commun.* *6*: 372–374.

Tanaka, Y., Kojima, M., and Uritani, I. (1976). Properties, development and cellular-localization of cinnamic acid 4-hydroxylase in cut-injured sweet potato. *Plant Cell Physiol.* *15*(5): 843–854.

Uritani, I., Uritani, M., and Yamada, H. (1960). Similar metabolic alterations induced in sweet potato by poisonous chemicals and by *Ceratocystis fimbriata*. *Phytopathology* *50*(1): 30–34.

21

Defense Strategies of Tea (*Camellia sinensis*) Against Fungal Pathogens

B. N. Chakraborty, Usha Chakraborty, and A. Saha
University of North Bengal, Darjeeling, India

I. INTRODUCTION

The most widely consumed and the cheapest hot beverage in the world today is tea. Recent available statistics show that the total amount of world tea production reaches 2.5 million tons annually, and an average of 2 billion cups of tea are drunk every day (Yamanishi, 1991). Tea is made from the young leaves and unopened leaf buds of *Camellia sinensis* (L.) O, Kuntze, a species which includes some very distinct varieties. The cultivated varieties are classified into two main groups on the basis of foliar and growth characteristics: China tea, *Camellia sinensis* var. *sinensis*, and Assam tea or *C. sinensis* var. *assamica*. The former is a slow-growing dark tree with small, erect, comparatively narrow, dark green leaves which are resistant to cold; the latter is a rapidly growing, taller tree, with large drooping leaves and resistance to cold, but adapting well to tropical conditions.

The pests and blights which prey on the leaves are of vital importance, since any damage to the leaves defeats the whole purpose of its cultivation. The monocultural conditions under which tea is grown commercially makes the transmission of disease more difficult to control than where mixed or rotational crop cultivation is followed. The danger of disease is further aggravated by the fact that in many tea districts virtually no other form of cultivation is practiced and an epidemic outbreak of an easily transmitted disease can affect very large areas in a short period. The aerial surfaces of tea plants, like any other plant, are usually inhabited by a variety of microorganisms, many of which are capable of influencing the growth of foliar pathogens (Chakraborty and Chakraborty, 1988). The interaction between these microorganisms might result in the suppression of pathogen activity. Besides, it is likely that the tea plant, in the course of its adjustment to varying environments, has evolved a very effective defense mechanism which successfully wards off most of the fungal pathogens.

The common plant pathogens (fungi, bacteria, viruses) and pests (insects and other animals) induce some type of resistance in plants to subsequent challenges, both to the original as well as to other biotic agents (Chessin and Zipf, 1990). In general, the defenses of higher plants against any form of stress, whether biotic or abiotic, fall under two categories: preformed and induced. The inducible defense systems have been much more elaborately studied than the preformed ones. These have been termed as *alarms*. According to Chessin and Zipf (1990), alarm is a complex physiological phenomenon that comprises a series of steps, i.e., perturbation by a particular biotic or abiotic stress, followed by production of a systemically transmitted signal (electrical, chemical, or both) in the stressed tissue and finally the induction of a new morphological or physiological state which protects the target tissues from subsequent exposure to the same or other stresses. Among the inducible defenses photoalexin accumulation is believed to be an important early defense response in several plant–pathogen interactions (Ward, 1986; Van Etten et al., 1989). The preformed defense mechanism of plants, on the other hand, deals with internal chemical defenses which are already fully expressed in host tissues before infection and do not rise to higher levels in response to invading microorganisms (Schlösser, 1980). Various types of antifungal plant constituents such as saponins, unsaturated lactones, mustard oils, and cyanogenic and phenolic glycosides are widespread in the plant kingdom. Occurrence, distribution, and possible function of preformed antifungal compounds have been reviewed (Schonbeck and Schlösser, 1976; Schlösser, 1988). The role of preformed chemical barriers as an important part of the defense mechanisms of plants is now gaining wide acceptance and needs to be carefully looked into, before one can get a complete picture of the defense mechanisms of plants.

The study of defense strategies in tea was promoted by a few important considerations. First, tea, which forms the backbone of the economy of north eastern India, is subject to attacks by foliar pathogens like *Corticium invisum*, *C. theae*, *Exobasidum vexans*, *Colletotrichum camelliae*, *Pestalotiopsis theae*, and *Botrytis* sp. (Sarmah, 1960), resulting in a reduction of the quality and quantity of tea produced. Second, almost no work has been done on the defense mechanisms of tea plant. Our studies were based mainly on three foliar diseases: brown blight caused by *Colletotrichum camelliae*, grey blight caused by *Pestalotiopsis theae*, and leaf spot caused by *Bipolaris carbonum* (first reported by Chakraborty, 1987). The main objectives have been to determine the mechanism of defense of tea plant against these pathogens, with special emphasis on both preformed and induced chemical defenses. Since polyphenols are major constituents of tea leaves, their involvement in the defense mechanism either as preformed or induced chemicals seemed highly probable. All four forms of catechin (Fig.

1), i.e., epicatechin (EC) epicatechin gallate (ECG), epigallocatechin (EGC), and epigallocatechin gallate (EGCG), have already been reported from tea (Wang, 1991). It is also known that catechin is oxidatively cleaved to simpler phenols and phenolic acids like catechol, phloroglucinol, and protocatechuic acid (Fig. 2A–C). The enzyme catechin-2,3-dioxygenase was isolated from *Chaetomium cupreum*, which cleaved catechin into simpler

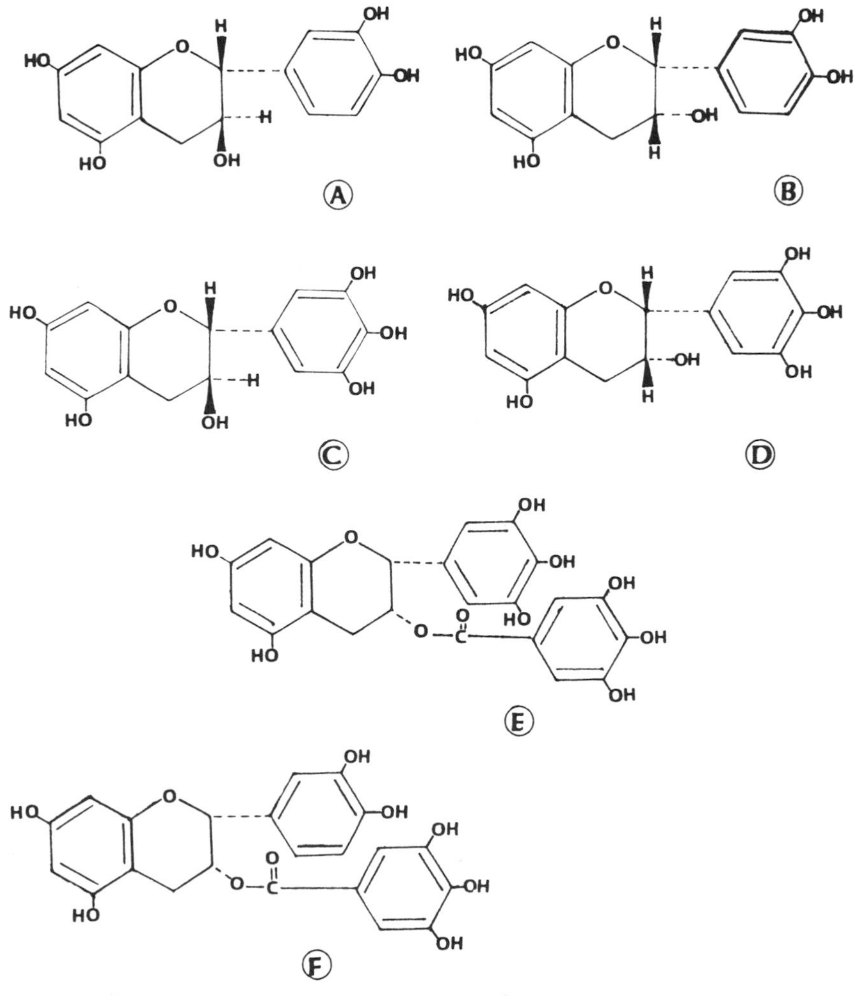

Figure 1 Chemical structures of catechin and its different forms. A, catechin; B, (-) epicatechin; C, gallocatechin; D, (-)epigallocatechin; E, (-)epigallocatechin gallate; F, (-)epicatechin gallate.

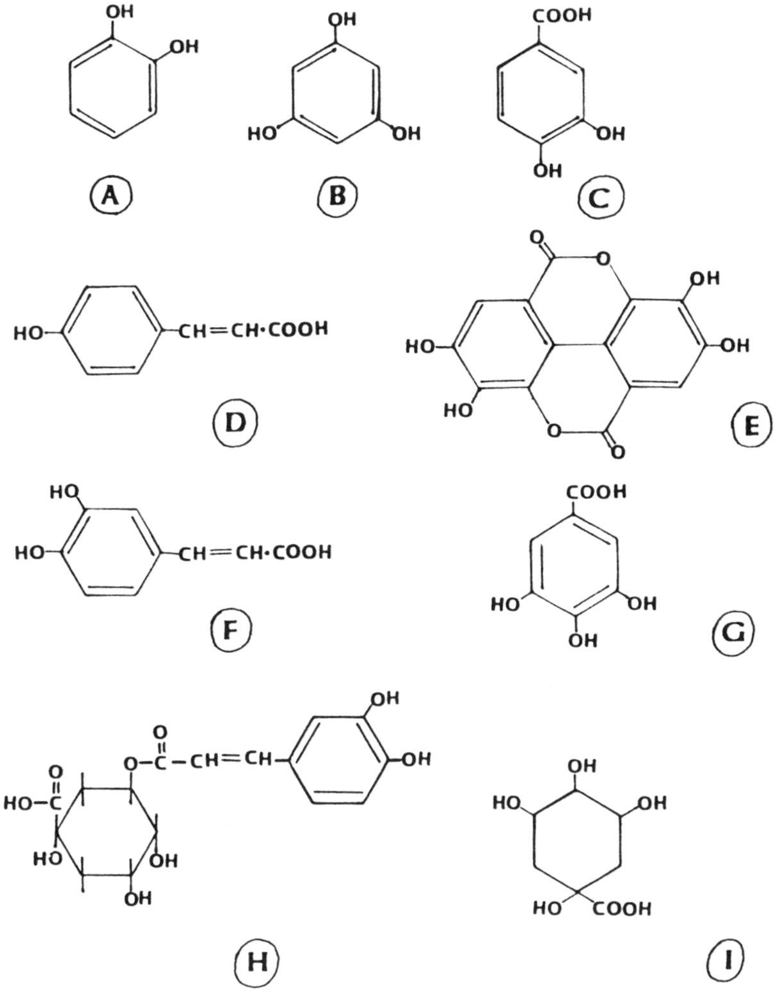

Figure 2 Chemical structures of some commonly occurring phenols and phenolic acids of tea plants. A, catechol; B, phloroglucinol; C, protocatechuic acid; D, *p*-coumaric acid; E, ellagic acid; F, caffeic acid; G, gallic acid; H, chlorogenic acid; I, quinic acid.

phenols as mentioned (Sambandam, 1982). A number of other phenols and phenolic acids have also been reported from tea, some of which are presented in Fig. 2D–I). Kawamura and Takeo (1989) demonstrated the antibacterial activity of tea catechin to *Streptococcus mutans*. Hamaya et al. (1984) established the presence of characteristic antifungal compounds in the leaf extracts of *Camellia japonica*. Conidial germination or growth of hyphae of many fungi, namely, *Pyricularia oryzae*, *Cochliobolus miyabeanus*, *Pestalotia longiseta*, *Gloeosporium theae-sinensis*, *Diaporthe citri*, etc. was inhibited. They isolated two triterpenoid saponins from the extract, which they named camellidin I and camellidin II, with molecular formulas $C_{55}H_{86}O_{25}$ (MW = 1146) and $C_{53}H_{84}O_{24}$ (MW = 1104), respectively. These were composed of 3β-hydroxy, 18β-acetoxy, 28-*nor*-olean-12-en-16-one, or 3β,18β-dihydroxy, 28-*nor*-olean-12-en-16-one as the sapogenin and D-glucuronic acid, D-glucose and 2 mol of D-galactose as the sugar moiety (Nagata et al, 1985). Nishino et al. (1986) further determined the sequence and configuration of the glycosides in the sugar moiety.

II. MATERIALS AND METHODS

A. Plant Material

Tea plants (18-month-old) of seven clonal varieties (TV-1, TV-9, TV-14, TV-16, TV-17, TV-18, and TV-26) were obtained from the clone house of Mohurgong and Gulma Tea Estate, Sukhna, Siliguri; and five clonal varieties (TV-19, TV-20, TV-23, TV-25, and Teen Ali-17/1/54) and two seed varieties (TS-449 and CP-1) were collected from the nursery of Dey's Tea Plants, Khaprail, Siliguri, West Bengal.

B. Fungal Culture

Colletotrichum camelliae Mass. and *Pestalotiopsis theae* Sawada were obtained from Tocklai Experimental Station, Jorhat, Assam. A virulent strain of *Bipolaris carbonum* (syn. *Helminthosporium carbonum*), anamorph of *Cochliobolus carbonum* Nelson, was obtained from the Departmental stock culture collection.

C. Inoculation Technique and Disease Assessment

A detached leaf inoculation technique as described by Dickens and Cook (1989) was used for artificial inoculation of tea leaves. Assessment of disease intensity was done on the basis of percent drops that resulted in lesion production after 24, 48, and 72 hr as described by Deverall and Wood (1961).

D. Collection and Bioassay of Leaf Diffusates

Leaf diffusates were obtained following the drop diffusate procedure of Müller (1958) with modifications. Drops of spore suspension were collected from leaf surfaces, combined, and centrifuged. Similarly in control, water drops were collected, combined, and centrifuged (exudate). Finally, the diffusates (spore-free supernatant) and exudates were passed through a sintered glass filter and bioassayed following the slide germination procedure as described by Trivedi and Sinha (1976). Usually 1.9 ml of diffusates was mixed with 0.1 ml of spore suspension of known concentration. Two drops (0.02 ml/drop) of suspension were placed separately at two ends of a grease-free glass slide. The slides were incubated in moist Petri dishes for 15 hr at 25 ± 1°C. Finally, germinated and ungerminated spores were stained with cotton blue in lactophenol and examined under the microscope.

E. Extraction of Antifungal Compound

Tea leaves were collected from the experimental garden and kept in plastic trays. Half of the total number of leaves were inoculated with conidial suspension of *B. carbonum* while the other half were uninoculated control. For the extraction of antifungal compound from infected tea leaves, the method of Keen (1978) was followed with modifications. Leaves were harvested after 6, 12, 24, 36, 48, and 72 hr of inoculation, weighed, and extracted with 40% aqueous ethanol in a rotary shaker for 24 hr. The filtrates were vacuum-concentrated to approximately one-half volume and extracted thrice with ethyl acetate and the organic layers were pooled and dehydrated with $MgSO_4$. Both the ethyl acetate and water fractions were taken to dryness separately and residues were dissolved in methanol.

F. Chromatographic Analysis

The water and ethyl acetate fractions of the extracts of both healthy and infected tea leaves were analyzed by thin layer chromatography (TLC) on silica gel G. The development of the chromatograms was carried out at room temperature and using a chloroform–methanol solvent system (9 : 1 v/v). Following evaporation of the solvent, the thin layer plates were observed under UV light and sprayed separately either with diazotized *p*-nitroaniline (Van Sumere et al., 1965), $FeCl_3$-$K_3Fe(CN)_6$, vanillin–H_2SO_4 (Stahl, 1967), or Folin–Ciocalteau phenol reagent (Harborne, 1973). Color reactions and R_f values were noted.

G. Fungitoxicity Assay

The extracts from leaf tissue of tea plants (TV-18 and TV-26) were bioassayed on TLC plates using *B. carbonum* as the test organism following the method of Hofmans and Fuchs (1970). Fungitoxicity was ascertained by the presence of inhibition zones, which appeared as white spots surrounded by a blackish background of mycelia.

The regions of thin layer chromatograms corresponding to these inhibitory zones were scraped and eluted in methanol. The eluants were rechromatographed on TLC plates to purify them and then eluted again. The eluants were tested for antifungal activities following spore germination testing as described by Werder and Kern (1985).

H. Preparation of Mycelial Wall Extract

Mycelial wall extract of *B. carbonum* was prepared following the method of Anderson-Prouty and Albersheim (1975). Mycelia (10-day-old culture) were collected on filter paper in a Buchner funnel and 20 g of fresh, packed cells were homogenized in an electric blender with 100 ml water. The resulting slurry was filtered and the residue was again homogenized in water. The mixture was centrifuged (1500*g*) for 5 min, the supernatant discarded, and the sedimented walls washed with 200 ml water and pelleted by centrifugation at least six times or until the supernatant was visually clear. Subsequently isolated cell walls were homogenized once in a 50-ml mixture of chloroform and methanol (1 : 1 v/v) and finally 50 ml acetone. This preparation when air-dried represented mycelial wall fraction. One gram of mycelial wall fraction was suspended in 100 ml of water and autoclaved for 30 min at 15 psi pressure. The autoclaved suspension was filtered, the filtrate centrifuged, and the supernatant concentrated to 10 ml under reduced pressure.

III. ANTIFUNGAL COMPOUNDS OF TEA

A. Varietal Resistance Test

Pathogenicity of the three selected foliar pathogens, i.e., *Colletotrichum camelliae*, *Pestalotiopsis theae*, and *Biopolaris carbonum*, were tested separately on different clonal and seed varieties of tea certified by Tocklai Experimental Station, Jorhat, Assam. Results (Table 1) revealed that, of the different TV varieties tested, TV-9 was moderately resistant while TV-18 was highly susceptible to both *C. camelliae* and *P. theae*. TV-18 and TV-26 were highly susceptible and resistant, respectively, to *B. carbonum*. With all three pathogens tested, it was seen that the younger leaves were

Table 1 Pathogenicity Test of Three Different Foliar Pathogens on Selective Tea Varieties Released from Tocklai Experimental Station

Varieties	Character[1]	% Lesion formed after inoculation with[2,3]		
		C. camelliae	*P. theae*	*B. carbonum*
Clones:				
TV-1	a,b	35.0 ± 2.6	44.6 ± 3.2	38.7 ± 2.4
TV-9	c	25.9 ± 1.8	18.3 ± 1.5	56.2 ± 2.7
TV-14	b	63.8 ± 2.2	27.8 ± 1.6	37.3 ± 1.6
TV-16	b	49.5 ± 1.9	16.4 ± 1.2	06.0 ± 0.8
TV-17	b	38.6 ± 2.8	27.1 ± 2.4	52.6 ± 3.5
TV-18	c	77.8 ± 2.6	70.5 ± 3.6	58.7 ± 2.8
TV-19	a	44.8 ± 1.6	52.3 ± 2.5	38.8 ± 1.6
TV-20	a,b	59.4 ± 3.4	21.9 ± 2.2	25.4 ± 2.5
TV-23	a,c	36.3 ± 1.8	63.6 ± 3.8	21.6 ± 1.5
TV-25	a,b,c	30.6 ± 1.6	19.7 ± 1.2	03.8 ± 0.2
TV-26	a,b	38.0 ± 2.4	22.3 ± 1.9	02.5 ± 0.3
Teen Ali 17/1/54	b,c	43.9 ± 2.5	28.7 ± 2.4	31.9 ± 1.2
Seed stocks:				
TS-449	a,b,c	51.6 ± 1.8	38.2 ± 3.2	35.6 ± 1.6
CP-1	a,c	46.7 ± 1.3	29.5 ± 2.4	25.9 ± 2.3

[1]Based on field observations: a, high yielding; b, submarginal lands and drought areas; c, infilling and interplanting (Bezbaruah and Singh, 1988).
[2]Average of three separate experiments (± = SE).
[3]72 hr after inoculation.

much more susceptible to diseases than the older leaves probably due to the variations in the texture and physical structure of the leaves. The older leaves have a very tough exterior, which probably makes penetration by the pathogen difficult. The trichomes which are abundant in older leaves also act as a barrier to certain fungi. This is found to be the case of the anthracnose fungus *Gloeosporium theae-sinensis*, which infects the plant through the trichome of young leaves. In many cases, the pathogen was inhibited from gaining entrance by a callosity which was produced by swelling of the trichome cell wall inward in such a way that it enveloped and preceded the invading hypha (Ando and Hamaya, 1986).

B. Biological Activities of Leaf Diffusates and Exudates

Tea varieties, resistant and susceptible to each of the three pathogens (*P. theae*, *C. camelliae*, and *B. carbonum*), were selected for further study. Leaf diffusates from the selected tea varieties were collected and assayed for their effect on spore germination and germ tube growth of the respective fungi. The results indicated that the diffusates from the resistant varieties were more fungitoxic than those from the susceptible varieties (Chakraborty and Saha, 1989). Leaf exudates of tea also contained some fungitoxic substances, which suggests the presence of antifungal compounds in the healthy leaves. Presumably, tea leaves contain some diffusible preformed antifungal compounds which play a role in their defense mechanism. Toda et al. (1989) also reported that extracts of Japanese green tea leaves inhibited the growth of various bacteria.

C. Detection and Identification of Antifungal Compounds in Tea Leaves

The results of drop diffusate tests strongly suggested that tea leaves contained both preformed and postinfectional antifungal compounds. To isolate these compounds the plants were inoculated with one of the pathogens, *B. carbonum*. After 72 hr of inoculation with the pathogen the leaves of resistant (TV-26) and susceptible (TV-18) varieties were extracted in aqueous methanol, concentrated in vacuum, fractionated with ethyl acetate, spotted on TLC plates, and developed in chloroform – methanol (9 : 1). On spraying the plates with Folin–Ciocalteau reagent, several spots appeared indicating the presence of phenolic compounds in the extracts of both healthy and infected leaves of the two varieties. In addition, the extract from healthy plants contained a brown-colored spot at R_f 0.8, developed when sprayed with vanillin-H_2SO_4.

These crude extracts were assayed for their antifungal activity by TLC plate bioassay method. Extract from healthy leaves of both varieties (TV-18 and TV-26) showed a prominent inhibition zone at R_f 0.8 in TLC plates (Fig. 3, zone A). These inhibition zones of TV-18 and TV-26 were of 20 and 25 mm diameter, respectively. There was no evidence of this inhibition zone after 72 hr of inoculation in TV-18, whereas in TV-26 the inhibition zone was faintly visible. In the extracts from inoculated leaves of both susceptible and resistant varieties, prominent inhibition zones appeared at R_f 0.6 (Fig. 3, zone B). This antifungal compound in resistant variety (TV-26) was approximately four times more than that of the susceptible variety (TV-18). In case of noninoculated control of both the varieties, much less antifungal activity was seen at the corresponding region.

The prominent inhibiting zone at R_f 0.8 was identified as catechin on the basis of its color reaction and R_f value (Kawamura and Takeo, 1989). Results of the present investigation also indicate that the catechin is presumably cleaved to some simpler phenols. The breakdown of this compound is almost complete in the susceptible variety but incomplete in the resistant variety as traces of catechin are located even after 72 hr of inoculation. The

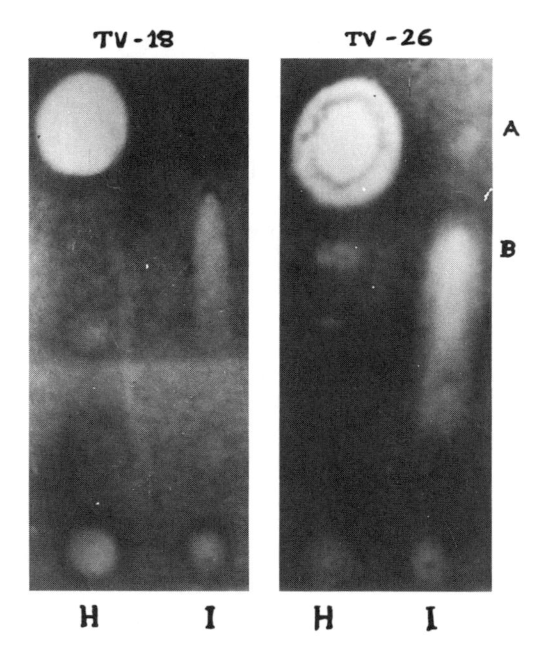

Figure 3 TLC bioassay of tea leaf extracts of susceptible (TV-18) and resistant (TV-26) varieties. H, extracts from healthy leaves; I, extracts from leaves inoculated with *B. carbonum* (48 hr after inoculation).

inhibitory zone B of the extracts from inoculated leaves at R_f 0.6 was identified to be catechol. Accumulation of catechol in resistant variety (TV-26) increased significantly (510 μg/g fresh wt tissue) in comparison to susceptible variety (187 μg/g fresh wt tissue) after 48 hr of inoculation with *B. carbonum*. In the susceptible variety, even though catechin is broken down completely, accumulation of catechol is not greater than the resistant variety. It seems probable that increased level of catechol is related to the resistance mechanism.

Since the results of the facilitated diffusion technique indicated the involvement of catechin as a preformed fungitoxic substance as well as accumulation of other antifungal phenols, mainly catechol, as a result of inoculation, the quantitative changes in phenol contents after 6, 12, 24, 36, and 48 hr of inoculation were determined in both varieties. At different times after inoculation, the pattern of changes in total phenolic content (ethyl acetate–soluble) of both the varieties differed significantly. The phenolic content of resistant cultivar (TV-26) showed a significant continuous increase from 24 hr of inoculation. In the uninoculated leaves of the same variety, the phenolics showed a gradual decrease with time. Both healthy and inoculated susceptible variety (TV-18) had less phenolic than the resistant variety, with the inoculated leaves having more than the healthy ones. The level of water-soluble phenolic compound was low and did not differ in resistant, susceptible, and noninoculated control plants (Fig. 4).

All these results therefore suggest that the healthy leaves of tea plants contain catechin, which is an antifungal polyphenol. Inoculation resulted in the breakdown of catechin with the simultaneous appearance of other antifungal phenols, mainly catechol. The concentration of this phenol kept increasing with time in the resistant variety up to 72 hr. Even though reports are available on the quantitative differences in phenol contents of resistant and susceptible varieties, most of them are conflicting with each other. Sridhar and Ou (1974) reported differences in total phenolic accumulation during the interaction of *Pyricularia oryzae* with rice. However, no differences were found in the phenolic content in the interaction of *Helminthosporium maydis* race T with N and T cytoplasms of a single maize genotype (Macri et al., 1974). On the other hand, Hammerschmidt and Nicholson (1977) demonstrated a clear difference between resistant and susceptible interactions of maize to *Colletotrichum graminicola* based on

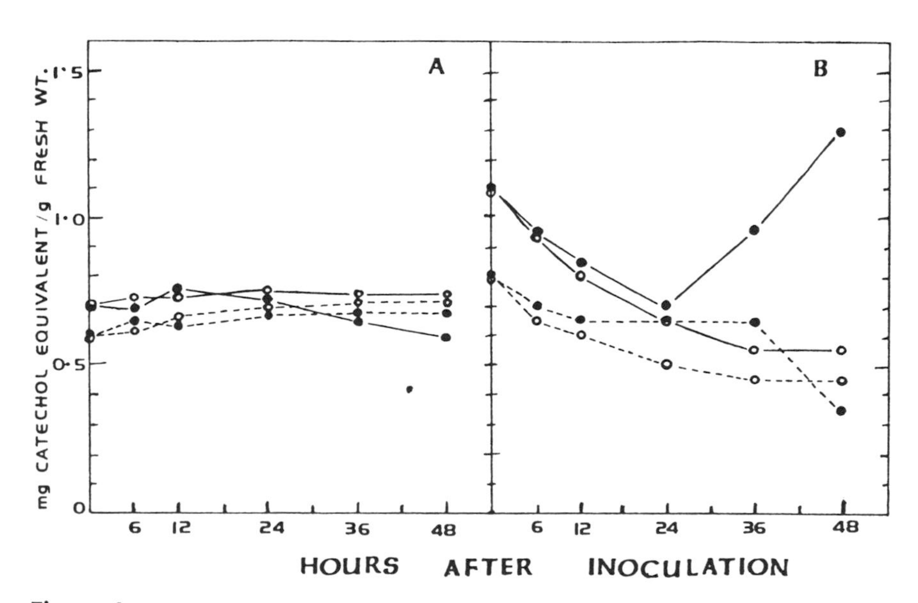

Figure 4 Changes in total water-soluble (A) and ethyl acetate-soluble (B) phenolics in extracts of tea leaf tissue of susceptible (TV-18) and resistant (TV-26) varieties. Healthy: TV-18 (○-----○); TV-26 (○ —— ○). Inoculated: TV-18 (●-----●); TV-26 (● —— ●).

accumulation of phenols. Apart from these conflicting results, further complications arise because phenols are not distributed uniformly in the infected tissues. Some phenols increased, while others decreased.

D. Elicitation of Antifungal Compound in Tea Leaves by Mycelial Wall Extract of *Bipolaris carbonum*

Having established the fact the pathogen *B. carbonum* induces the accumulation of phenols in the tea leaves and more rapidly in the resistant variety, the next question was whether elicitors could play any role in this induction. It was therefore decided to make mycelial cell wall extracts of the pathogen and to determine its effect on the induction of diffusible antifungal compounds in tea leaves. Cell wall extracts were prepared from the mycelia of *B. carbonum* as described under "Materials and Methods." To determine the effect of these wall extracts on the production of antifungal compounds, fresh spores of *B. carbonum* were suspended in its own mycelial wall extract and drops of suspension placed on leaf surfaces. After 48 hr, drops were collected, combined, and centrifuged. The supernatant was assayed for fungitoxicity by spore germination bioassay and also analyzed by TLC for detection of antifungal compound as described by Purkayastha and Ghosh (1983). It was found that fresh mycelial wall extract was highly stimulatory to growth of *B. carbonum* but inhibitory when collected from leaf surfaces after 48 hr incubation. Spores suspended in mycelial wall extract and incubated over leaf surfaces completely inhibited the germination of *B. carbonum*. The mycelial wall extract containing the elicitor very effectively induced the accumulation and subsequent diffusion of the antifungal compound in the resistant variety (TV-26).

E. Elicitation of Antifungal Compound by Chemical Treatment

Other than the biotic phytoalexin inducers (elicitors), chemicals of widely diverse natures are known to induce phytoalexin production in the plants. Under certain circumstances these chemicals might induce host resistance and could be used to control plant diseases (Purkayastha, 1986).

In the present investigation 17 compounds belonging to four groups (heavy metal salts, metabolic inhibitors, reducing agents, and amino acids) were selected. These compounds were bioassayed for their toxicity by spore germination tests. Results showed that the heavy metal salts were fungitoxic. At a concentration of 10^{-4} M, mercuric chloride, cupric chloride, nickel chloride, and cadmium chloride inhibited germination of *B. carbonum* completely while 75% inhibition in germination was observed with nickel nitrate. Cycloheximide and thioglycollic acid also inhibited germination of *B. carbonum* significantly. Other chemicals had no significant effect on the spore germination. These chemicals were further tested on susceptible variety (TV-18) for their effect on disease development, as described by Yanase and Takeda (1987). It was observed that nickel chloride and nickel nitrate were very successful in reducing disease development. There was no appearance of disease symptoms, even 2 weeks following inoculation. Significant increase of antifungal compounds in treated inoculated plant was observed which could be correlated with the reduction in disease symptoms. Keen et al. (1981) reported that sodium iodoacetate acts as an abiotic elicitor of glyceollin in primary leaves of cv. Harosoy soybeans and which is associated with the expression of resistance. Glyceollin production also increased significantly in soybean plants (cv. Soymax) after induction of resistance by sodium azide treatment (Chakraborty and Purkayastha, 1987). Sinha and associates in a series of experiments, successfully demonstrated chemical induction of disease resistance in rice infected with *Helminthosporium oryzae*, which they correlated with an increased accumulation of a phytoalexin-like substance in the treated plants (Sinha and Giri, 1979; Sinha and Hait, 1982; Giri and Sinha, 1983a,b). Ghosal and Purkayastha (1987) also observed a differential response of rice leaves to some abiotic elicitors of phytoalexin (momilactone A). Gibberellic acid, sodium azide, and penicillin induced production of momilactone A in rice. These compounds also reduced the sheath rot disease of rice. Among the six antibiotics used as foliar spray on a susceptible soybean cultivar ("Soymax"), cloxacillin induced maximum resistance against anthracnose (Purkayastha and Banerjee, 1990).

IV. EPILOG

The defense strategies of tea worked out in the course of the project pointed out the involvement of phenolics in disease resistance, some of which were preformed, and others were postinfectional. The latter could be defined as phytoalexins (Chakraborty et al., 1989). It is well known that plants defend themselves from microbial pathogens through different mechanisms including the synthesis of antimicrobial phytoalexins, the hypersensitive response, and cell wall modification (Halverson and Stacey, 1986). Such defense mechanisms may be induced by the expression of genes resulting from the recognition of a particular microbe by a host. The exchange of molecular signals between host and pathogen is considered to be one of the mechanisms of specificity. The signal molecules involved in defense responses need to be identified.

Preliminary work has been carried out on both biotic and abiotic elicitors in the course of our project. It is now proposed to purify and identify the fungal cell wall elicitors and establish their precise role in eliciting the defense response. Detailed work on cell wall modification relating to disease resistance in tea is another proposed area of research.

Compatibility in the host–parasite interaction in many instances is also dependent on the presence of cross-reactive antigens (CRA) between them. Mimicry of antigenic determinants in plants by various microbes is widespread and is important in the compatibility of plant–parasite interactions (Chakraborty, 1988). Except for a role in recognition phenomena (Dazzo and Brill, 1977, 1979), a direct and active role of CRA is still speculative but has been visualized in a number of host–parasite systems (Purkayastha, 1989). The role thus visualized for certain common antigens is one that provides both a stabilizing influence between interacting cells of different organisms and a basis for continued compatibility. It is believed that CRA forms a continuum between cells of host and parasite that favors the progressive growth and establishment of the parasite (DeVay et al., 1981). The role of CRA is probably subject to overriding effects of substances such as phytoalexins or other inhibitory substances already present in host tissues or induced by parasitic microorganisms. Alteration of specific host antigens by suitable treatment may also induce disease resistance in plants (Chakraborty and Purkayastha, 1987; Ghosal and Purkayastha, 1987; Purkayastha and Banerjee, 1990).

Keeping this in mind, it was decided to determine the presence of cross-reactive antigens between tea and its pathogens. Isolation and purification of CRA shared between tea plant and foliar pathogens as well as tissue and cellular locations of CRA is being carried out. In addition, further biochemical, physiological, and genetic research is in progress for a complete understanding of the defense mechanism of tea plant against a number of foliar pathogens.

ACKNOWLEDGMENTS

This work was supported in part by the Council of Scientific and Industrial Research, New Delhi. Financial assistance received from the University Grants Commission by A. Saha is also acknowledged.

REFERENCES

Albersheim, P., and Anderson-Prouty, A. J. (1975). Carbohydrates, proteins, cell surfaces and the biochemistry of pathogenesis. *Annu. Rev. Plant Physiol. 26*: 31–52.

Anderson-Prouty, A. J., and Albersheim, P. (1975). Host pathogen interactions. III. Isolation of a pathogen synthesized fraction rich in glucan that elicits a defense response in the pathogen's host. *Plant Physiol. 56*: 286–291.

Ando, Y., and Hamaya, E. (1986). Defense reaction of tea plant against infection of the tea anthracnose fungus. *Study Tea 69*: 35–43.

Bezbaruah, H. P., and Singh, I. D. (1988). Advances in the use of planting material. *Field Management in Tea* (S. K. Sarkar and H. P. Bezbaruah, eds.), Tocklai Experimental Station, Jorhat, India.

Chakraborty, B. N. (1987). A new record of leaf disease of tea (*Camellia sinensis* (L.) O. Ktze). *Curr. Sci. 56*(17): 900.

Chakraborty, B. N. (1988). Antigenic disparity. *Experimental and Conceptual Plant Pathology* (R. S. Singh, U. S. Singh, W. M. Hess, and D. J. Weber, eds.), Oxford and IBH Publishing Co. Pvt. Ltd., New Delhi, pp. 477–484.

Chakraborty, B. N., and Saha, A. (1989). Biological activity of leaf diffusates of tea in relation to resistance to *Bipolaris carbonum*. *Environ. Ecol. 7*(3): 717–720.

Chakraborty, B. N., and Purkayastha, R. P. (1987). Alteration of glyceollin synthesis and antigenic patterns after chemical induction of resistance in soybean to *Macrophomina phaseolina*. *Can. J. Microbiol. 33*: 647–651.

Chakraborty, B. N., Chakraborty, U., and Saha, A. (1989). Isolation and characterization of an antifungal compound from leaves challenged with *Pestalotiopsis theae*. *Indian Phytopathol. 42*(2): 302.

Chakraborty, B. N., and Chakraborty, U. (1988). Biological control of foliar diseases of tea. *Proceedings of 5th International Congress of Plant Pathology*, Kyoto, p. 344.

Chessin, M., and Zipf, A. E. (1990). Alarm system in higher plants. *Bot. Rev. 56*: 193–235.

Dazzo, F. B., and Brill, W. J. (1977). Receptor site on clover and alfalfa roots for *Rhizobium*. *Appl. Environ. Microbiol. 33*: 132–136.

Dazzo, F. B., and Brill, W. J. (1979). Bacterial polysaccharide which binds *Rhizobium trifolli* to clover root hairs. *J. Bacteriol. 137*: 1362–1373.

DeVay, J. E., Wakeman, R. J., Kavanagh, J. A., and Charudattan, R. (1981). The tissue and cellular location of a major cross-reactive antigen shared by cotton and soil-borne fungal parasites. *Physiol. Plant Pathol. 18*: 59–66.

Deverall, B. J., and Wood, R. K. S. (1961). Infection in bean plants (*Vicia faba* L.) with *Botrytis cinerea* and *B. fabae*. *Annals of Applied Biology. 49*: 461–472.

Dickens, J. S. W., and Cook, R. T. A. (1989). *Glomerella cingulata* on *Camellia*. *Plant Pathol. 39*: 75–85.

Ghosal, A., and Purkayastha, R. P. (1987). Biochemical response of rice (*Oryza sativa* L.) leaves to some abiotic elicitors of phytoalexin. *Indian J. Exp. Biol. 25*: 395–399.

Giri, D. N., and Sinha, A. K. (1983a). Effects of heavy metals on susceptibility of rice seedlings to brown spot disease. *Ann. Appl. Biol. 103*: 229–235.

Giri, D. N., and Sinha, A. K. (1983b). Control of brown spot disease of rice seedlings by treatment with a select group of chemicals. *Z. Pflanzenkrank. Pflanzenschutz 90*: 479–487.

Halverson, L. J., and Stacey, G. (1986). Signal exchange in plant-microbe interactions. *Microbiol. Rev. 50*(2): 193–225.

Hamaya, E., Tsushida, T., Nagata, T., Nishino, C., Enoki, N., and Manabe, S. (1984). Antifungal components of *Camellia* plants. *Ann. Phytopathol. Soc. Jpn. 50*(5): 628–636.

Hammerschmidt, R., and Nicholson, R. L. (1977). Resistance of maize to anthracnose, Changes in host phenols and pigments. *Phytopathology 67*: 251–258.

Harborne, J. B. (1973). *Phytochemical Methods*. Chapman and Hall, London.

Hofmans, A. I., and Fuchs, A. (1970). Direct bioautography on thin-layer chromatograms as a method for detecting fungitoxic substances. *J. Chromatogr. 51*: 327–329.

Kawamura, J., and Takeo, T. (1989). Antibacterial activity of tea catechin to *Streptococcus mutans*. *Nippon Shokuhin Kogyo Gakkaishi 36*(6): 463–467.

Keen, N. T. (1978). Phytoalexins: efficient extraction from leaves by a facilitated diffusion technique. *Phytopathology 68*: 1237–1239.

Keen, N. T., Ersek, T., Long, M., Bruegger, B., and Holliday, M. (1981). Inhibition of the hypersensitive reaction of soybean leaves to incompatible *Pseudomonas* spp. by blasticidin S, streptomycin or elevated temperature. *Physiol. Plant Pathol. 18*: 325–337.

Macri, F., Dilenna, P., and Vianelli, A. (1974). Preliminary research on peroxidase, polyphenol oxidase activity and phenol content in healthy and infected corn leaves, susceptible and resistant to *Helminthosporium maydis* race T. *Rev. Patologie Vegetale. 4*: 109–122.

Müller, K. O. (1958). Studies on phytoalexins. I. The formation and the immunological significance of phytoalexin produced by *Phaseolus vulgaris* in response to infections with *Sclerotima fructicola* and *Phytophthora infestans*. *Aust. J. Biol. Sci. 11*: 275–300.

Nagata, T., Tsushida, T., Hamaya, E., Enoki, N., Manabe, S., and Nishino, C. (1985). Camellidins, antifungal saponins isolated from *Camellia Japonica*. *Agric. J. Biol. Chem. 49*(4): 1181–1186.

Nishino, C., Manabe, S., Enoki, N., Nagata, T., Tsushida, T., and Hamaya, E. (1986). The structure of the tetrasaccharide unit of camellidinis, saponins possessing antifungal activity. *J. Chem. Soc. Chem. Commun.* 720–723.

Purkayastha, R. P. (1986). Elicitors and elicitation of phytoalexins. *Vistas in Plant Pathology* (A. Verma and J. P. Verma, eds.), Malhotra, New Delhi, pp. 25–44.

Purkayastha, R. P. (1989). Specificity and disease resistance in plants. Presidential address, Section of Botany, 76th Session of the Indian Science Congress, Madurai, pp. 1–32.

Purkayastha, R. P., and Banerjee, R. (1990). Immunoserological studies on cloxacillin-induced resistance of soybean against anthracnose. *Z. Pflanzenkrank. Pflanzenschutz 97*(4): 349–359.

Purkayastha, R. P., and Ghosh, S. (1983). Elicitation and inhibition of phytoalexin biosynthesis in *Myrothecium*-infected soybean leaves. *Indian J. Exp. Biol. 23*: 216–218.

Sambandam, T., Sivaswamy, N., and Mahadevan, A. (1982). Microbial degradation of phenolic substances. *Indian Rev. Life Sci. 2*: 1–18.

Sarmah, K. C. (1960). *Diseases of Tea and Associated Crops in North-East India*. Indian Tea Association, Memorandum No. 26, pp. 33–40.

Schlösser, E. (1988). Preformed chemical barriers in host-parasite incompatibility. *Experimental and Conceptual and Plant Pathology* (R. S. Singh, U. S. Singh, W. M. Hess, and D. J. Weber, eds.), Oxford and IBH Publishing Co. Pvt. Ltd., New Delhi, pp. 465–476.

Schlösser, E. W. (1980). Preformed internal chemical defenses. *Plant Disease and Advanced Treatise: How Plants Defend Themselves, Vol. 5* (J. G. Horsfall and E. B. Cowling, eds.), Academic Press, New York, pp. 161–177.

Schonbeck, F., and Schlösser, E. (1976). Preformed substances as potential protectants. *Encyclopedia of Plant Physiology, Vol. 4* (R. Heitfuss and P. H. Williams, eds.), Springer-Verlag, Berlin, pp. 651–678.

Sinha, A. K., and Giri, D. N. (1979). An approach to control brown spot of rice with chemicals known as phytoalexin inducers. *Curr. Sci. 48*: 782–784.

Sinha, A. K., and Hait, G. N. (1982). Host sensitization as a factor in the induction of resistance in rice plants against *Dreschlera* infection by seed treatment with phytoalexin inducers. *Trans. Br. Mycol. Soc. 79*(2): 213–219.

Sridhar, R., and Ou, S. H. (1974). Phenolic compounds detected in rice blast disease. *Biologia Plantarum 16*: 67–70.

Stahl, E. (1967). *Dünnschicht—Chromatographie*. Springer-Verlag, Berlin.

Toda, M., Okubo, S., Chrishi, R., and Shimamura, T. (1989). Antibacterial and bactericidal activities of Japanese green tea. *Jpn. J. Bacteriol. 44*(4): 669–672.

Trivedi, N., and Sinha, A. K. (1976). Resistance induced in rice plants against *Helminthosporium* infection by treatment with various fungal fluids. *Phytopathol. Z. 86*: 335–337.

Van Etten, H. D., Mathews, D. E., and Mathews, P. S. (1989). Phytoalexin detoxification: Importance for pathogenicity and practical implications. *Ann. Rev. Phytopathol. 27*: 143–164.

Van Sumere, C. F., Wolf, G., Teuchy, H., and Kint, J. (1965). A new thin-layer method for phenolic substances and coumarins. *J. Chromatogr. 22*: 48–60.

Wang, Z. N. (1991). Chinese famous teas and their characteristic constituents bioformation. *World Tea: International Symposium on Tea Science*, Shizuoka, Japan, pp. 23–33.

Ward, E. W. B. (1986). Biochemical mechanisms involved in resistance of plants to fungi. *Biology and Molecular Biology of Plant-Pathogen Interactions* (J. Bailey, ed.), Springer Verlag, Berlin. pp. 107–131.

Werder, J., and Kern, H. (1985). Resistance of maize to *Helminthosporium carbonum*: changes in host phenolics and their antifungal activity. *J. Plant Dis. Prot. 92*(5): 477–484.

Yamanishi, T. (1991). Flavor characteristics of various teas. *World Tea: International Symposium on Tea Science*, Shizuoka, Japan, pp. 1–11.

Yanase, Y., and Takeda, Y. (1987). Method for testing the resistance of tea gray blight caused by *Pestalotia longiseta* Spegazzini in tea breeding. *Bull. Natl. Res. Inst. Vegetable Ornamental Plants and Tea.* (Kanaya) *1*: 1–9.

22

Effects of Age-Related Resistance and Metalaxyl on Capsidiol Production in Pepper Plants Infected with *Phytophthora capsici*

Byung Kook Hwang
Korea University, Seoul, Republic of Korea

I. INTRODUCTION

A. Age-Related Resistance and Metalaxyl for Control of *Phytophthora* Blight

Phytophthora blight of pepper (*Capsicum annuum* L.) plants caused by *Phytophthora capsici* Leonian is a serious soil-borne disease in the major pepper-growing areas of the world (Leonian, 1922; Weber, 1932; Barksdale et al., 1984). It has been reported that *P. capsici* attacks plants such as pepper, eggplants, cucumber, honeydew melon, pumpkin, squash, tomato, and watermelon (Polach and Webster, 1972). *Phytophthora* rots of the lower stem of plants favored by high temperature, high soil moisture, and atmospheric humidity are common and most severe in low-lying, poorly drained areas. In pepper plants, a brown girdling rot of the stems becomes enlarged, the lower leaves drop, and eventually the whole plant wilts.

Phytophthora blight of pepper plants has been controlled by crop rotation, the use of resistant cultivars, and the application of systemic fungicides such as metalaxyl. Although this disease cannot be readily controlled by any one measure, growing resistant cultivars may be very effective in reducing damage from *P. capsici*. Recently, Kim et al. (1989) demonstrated an age-related resistance of pepper cultivars, which is distinctly expressed as pepper plants mature. Since the *Phytophthora* disease causes severe damage in pepper plants only at later growth stages in the field, the age-related resistance must be considered in breeding resistant pepper cultivars.

Our earlier studies demonstrated the expression of age-related resistance in pepper plants infected with *P. capsici* under controlled environmental conditions (Kim et al., 1989). We also suggested that the appearance of age-related resistance in the pepper cultivars may be due to the morphological and nutritional changes in tissues of pepper stems during aging (i.e., the drastic increase in amount of dry matter; the significant decrease in amounts of mineral nutrients such as nitrogen, phosphorus, potassium, calcium, and magnesium; and the lower contents of fructose, glucose, and sucrose in the stem tissues) (Jeun and Hwang, 1991). Such morphological and physiological changes in pepper stems of different ages associated with age-related resistance to *P. capsici* may be caused by the production of some proteins or enzymes considered to be the result of differential gene action (Hwang et al., 1991).

Metalaxyl [methyl *N*-(2-methyoxyacetyl)-*N*-(2,6-xylyl)-D,L-alaninate] is a systemic fungicide which is specifically active against members of the Oomycetes such as *Phytophthora* spp. and downy mildews. Recently, this fungicide has been intensively applied in pepper-growing countries for controlling *Phytophthora* blight caused by *P. capsici*. Metalaxyl has been shown to be highly effective against *Phytophthora* blight of pepper plants by inhibition of mycelial growth, germination of zoospores, zoospore release from sporangia, and sporangium formation, even at low concentrations (Sung and Hwang, 1988). When applied to pepper plants, either as a soil drench or as a foliar spray, metalaxyl exhibits both protective and curative effects against *Phytophthora* blight (Sung and Hwang, 1988).

B. Phytoalexin Production Associated with Age-Related Resistance and Metalaxyl Treatment

Such differences in the degree of resistance of pepper plants may also be due to preformed inhibitory substances present in plants and/or antimicrobial compounds, phytoalexins, accumulating after exposure to *P. capsici*. Recently, Hwang and Heitefuss (1982) demonstrated that preformed antifungal compounds in leaf extracts of barley, though present in trace quantities, may contribute to the expression of a gradual, decreased susceptibility or quantitative resistance to powdery mildew with increasing age of plants. In the soybean-*Phytophthora megasperma* f.sp. *glycinea* (Pmg) interaction, Paxton and Chamberlain (1969) reported that soybean plants became increasingly resistant to Pmg as they matured, despite greatly reduced phytoalexin (glyceollin) production. In contrast, Keen (1971) concluded that the highest glyceollin concentrations accumulated in infected older plants of soybeans. Lazarovits et al. (1981) also demonstrated that increased glyceollin production was associated with age-related resistance of soybean hypocotyls to the compatible race, whereas glyceollin production decreased with tissue age in the incompatible race interactions despite increased resistance.

Phytoalexins are produced by plants not only in response to interactions with fungi, bacteria, viruses, nematodes, and other living organisms, but also following treatment with many chemicals, irradiation by ultraviolet light, and exposure to the products of microbial metabolism (Bailey, 1982). In addition, phytoalexin production can also be induced by fungicides. Reilly and Klarman (1972) demonstrated that several fungicides induced the phytoalexin glyceollin in soybean [*Glycine* max (L.) Merr.] hypocotyl tissues. Cartwright et al. (1977) also reported that dichlorocyclopropanes may exert their systemic fungicidal activity against the rice blast disease by causing rice to synthesize the phytoalexins monilactones A and B.

Several years ago, it was demonstrated that control of *Phytophthora* rot of soybean hypocotyls by the systemic fungicide metalaxyl caused a hypersensitive response accompanied by glyceollin production (Ward et al., 1980; Börner et al., 1983). Subsequently, it was found that in plants only partially protected by metalaxyl glyceollin concentrations exceeded the EC_{90} in vitro in all lesions. However, spread of the pathogen was not restricted (Lazarovits and Ward, 1982), suggesting that glyceollin may not play a significant role in inhibiting spread of the fungus in metalaxyl-treated seedlings.

Stössl et al. (1972) first reported that pepper fruits produced the antifungal sesquiterpenoid phytoalexin capsidiol in response to infection by several pathogenic and nonpathogenic fungi. Other fungi and bacteria also elicit accumulation of capsidiol (Stössl et al., 1972; Ward et al., 1973). A series of extensive studies of the rate and magnitude of capsidiol accumulation coupled with ultrastructural studies to support a role for the compound in the restriction of fungal growth and development in infected pepper fruits and stems (Jones et al., 1974; Hwang et al., 1990). It also accumulated around sites of localized infection in leaves (Ward, 1976), but its accumulation or the sensitivity of fungi to capsidiol cannot explain the resistance or susceptibility of pepper to all fungi. The significance of capsidiol accumulation in compatible and incompatible interactions between ripening fruits and fungi has been assessed (Ward and Stössl, 1972; Ward et al., 1973; Stössl et al., 1973, 1977; Jones et al., 1975a, b). The relation between capsidiol concentration and speed of fungal invasion in stems of pepper cultivars susceptible or resistant to *P. capsici* has been assessed by Molot et al. (1981). However, they could not demonstrate a convincing connection between capsidiol induction at the infection front and inhibition of lesion development, or a clear difference in the reaction of the resistant cultivar.

The role of capsidiol in the age-related resistance of pepper plants to *P. capsici* has not been studied extensively. In the present studies, the involvement of postinfectionally formed capsidiol in the development of *P. capsici* in stems and roots of pepper plants at different growth stages was examined in relation to age-related resistance of pepper plants. It has also not been demonstrated that applications of the systemic fungicide metalaxyl for control of *Phytophthora* blight of pepper plants affect capsidiol production in infected pepper plants. In the experiments reported here, we examined capsidiol production in stems of pepper plants treated with metalaxyl and inoculated with *P. capsici*. We also compared capsidiol production in the susceptible and resistant cultivars treated with metalaxyl. The majority of results presented in this chapter have been reported elsewhere (Hwang and Sung, 1989; Hwang and Kim, 1990).

II. MATERIALS AND METHODS

A. Plant, Fungus, and Fungicide

The pepper (*Capsicum annuum* L.) cultivars used were Hanbyul and Kingkun, differing quantitatively in susceptibility to *Phytophthora* blight. Six or ten plants per pot were grown in plastic pots (5 × 15 × 10 cm) containing a mixture of steam-sterilized loam soil, sand, and peat (3 : 5 : 2 v/v). Fertilizer was applied at the rate of 0.27–0.13 g of actual N-P-K per pot at 3 weeks after planting. Plants were maintained at 25 ± 2°C during a 16-hr light period from 0600 to 2200 hr of about 5000 1x and at 16 ± 2°C during darkness.

An isolate of *P. capsici* was cultured on V-8 agar for 7 days at 25 ± 1°C. Zoospores were produced as described previously (Kim et al., 1989) and a zoospore suspension was used as inoculum. Zoospores in suspensions were counted with a haemocytometer, and the concentration was adjusted to 1 × 10^5/ml with sterile tap water.

The fungicide used in this study was metalaxyl, formulated as a 25% a.i. wettable powder. All concentrations are given as active ingredient (Ridomil, Seoul Agricultural Chemical Co.). Metalaxyl suspensions were prepared in sterile tap water to give appropriate concentrations. Metalaxyl treatments were performed by pouring the 20 ml of metalaxyl suspensions per pot uniformly over the surface a day before stem inoculation with *P. capsici*, except in the experiment on the effect of time of metalaxyl application.

B. Inoculation Procedure and Disease Assessment

A longitudinal wound, about 1 cm long and 1 mm deep, was made with a razor blade on each stem of pepper plants, approximately 1 cm (bottom) or 9 cm (middle) from the soil surface and at the top of stems. A small quantity of sterile cotton soaked in zoospore suspension (1 × 10^5/ml) for

30 min was placed on the wounded sites of stems. The inoculation sites were then covered with plastic tape to maintain the moist condition. Metalaxyl suspensions at various concentrations were poured into the soil before or after stem inoculation with *P. capsici.*

For root samples, plants at the six-leaf and first branched-flowering stages were inoculated by uniformly pouring 20 ml of a suspension of motile zoospores (1×10^5/ml) per pot over the surface of soil. Immediately after soil drenching, pots were placed in a large plastic tray filled with tap water.

Disease development, as measured by lesion length, in pepper plants was rated every day after inoculation. The data are the means of three replicates from the disease ratings of 15 infected plants.

C. Extraction and Estimation of Capsidiol

Fifteen stems of pepper plants per treatment replicate were harvested at various intervals after inoculation. Three replicates of each treatment were used. The infected stems were sliced transversely in 3-mm-thick sections for 4 cm upward (bottom and middle) or downward (top) from the inoculated sites. Comparable areas of healthy, metalaxyl-treated stems also were similarly sliced.

At different intervals following inoculation, soil was removed from the roots of 10 plants by using running tap water. The roots were carefully washed, blotted with filter paper, and sampled. The sliced stem sections or roots were placed in 250-ml Erlenmeyer flasks with 15 ml/g fresh wt of 40% aqueous ethanol and were then vacuum-infiltrated. The flasks were stoppered and placed on a reciprocal shaker (110 strokes/min) for 7 hr. After agitation by shaking, the stem or root tissues were removed by filtration and then dried overnight at 95°C. The filtrate was vacuum-concentrated at 40°C to approximately one-half volume. The concentrate was partitioned twice with ethyl acetate. The pooled ethyl acetate layer was dehydrated with $MgSO_4$ and evaporated to dryness at 40°C. Following transfer of the residue to vials with peroxide-free diethyl ether and evaporation of ether under nitrogen, the dry residue was dissolved in 0.5 ml ethyl acetate and then stored at −20°C until used for gas chromatographic analysis of capsidiol.

All capsidiol analyses were made by gaschromatography following Ward et al. (1973). The ethyl acetate sample was evaporated under nitrogen. The final residues were dissolved in a solution of methyl myristate (4 mM) in ethanol, with the ester serving as an internal standard. Aliquots (2–3 μl) from each ethanol solution were injected into a Packard model 419 gas–liquid chromatograph fitted with a glass column (182 cm long, 2 mm i.d.) containing Gas Chrom Q (80–120 mesh) coated with 3% SE30. The column was kept at 162°C, the injector at 192°C, and the flame ionization detector at 230°C. The carrier gas was nitrogen at a 40 ml/min flow rate with hydrogen and air at 30 and 300 ml/min, respectively. Retention times were 7.5 and 14.5 min for methyl myristate and capsidiol, respectively. Capsidiol in the samples was identified by cochromatography with authentic capsidiol, which was obtained from Dr. A. Stössl, Agriculture Canada, Research Center, London, Ontario, Canada.

III. RESULTS

A. Effect of Plant Age and Stem Part on Disease Development

About a day after inoculation with *P. capsici*, the inoculation sites of stems of pepper plants became greyish to brown, as previously described (Kim et al., 1989). Slight differences between susceptible cv. Hanbyul and resistant cv. Kingkun in lesion length were observed at the six-leaf stage (Fig. 1), and the differences were more apparent at the first branched-flowering stage (Fig. 2). When inoculated on three stem sites, lesions at the top of stems progressed more rapidly than those of the middle or bottom (Figs. 2 and 3).

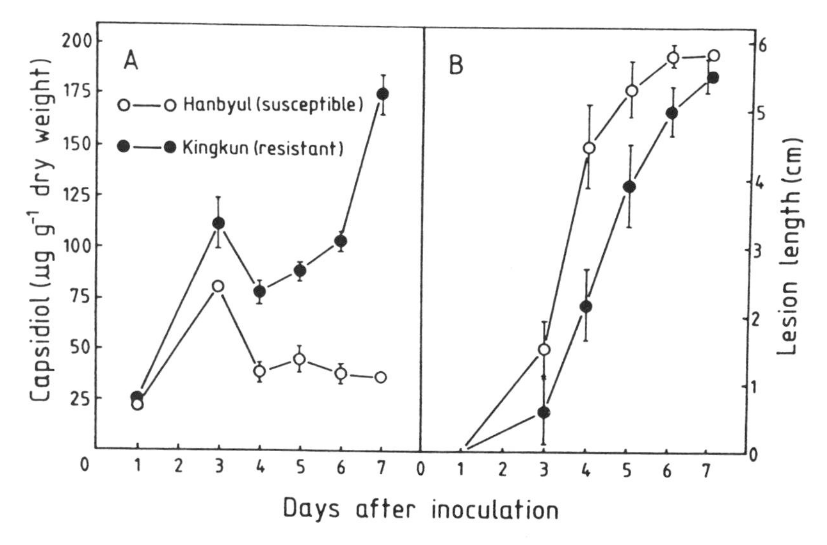

Figure 1 Time courses of (A) capsidiol accumulation and (B) disease development in the stems of pepper cultivars Hanbyul (susceptible) and Kingkun (resistant) at the six-leaf stage inoculated with zoospore suspensions of *P. capsici* at the bottom of stems. Vertical bars represent standard deviations.

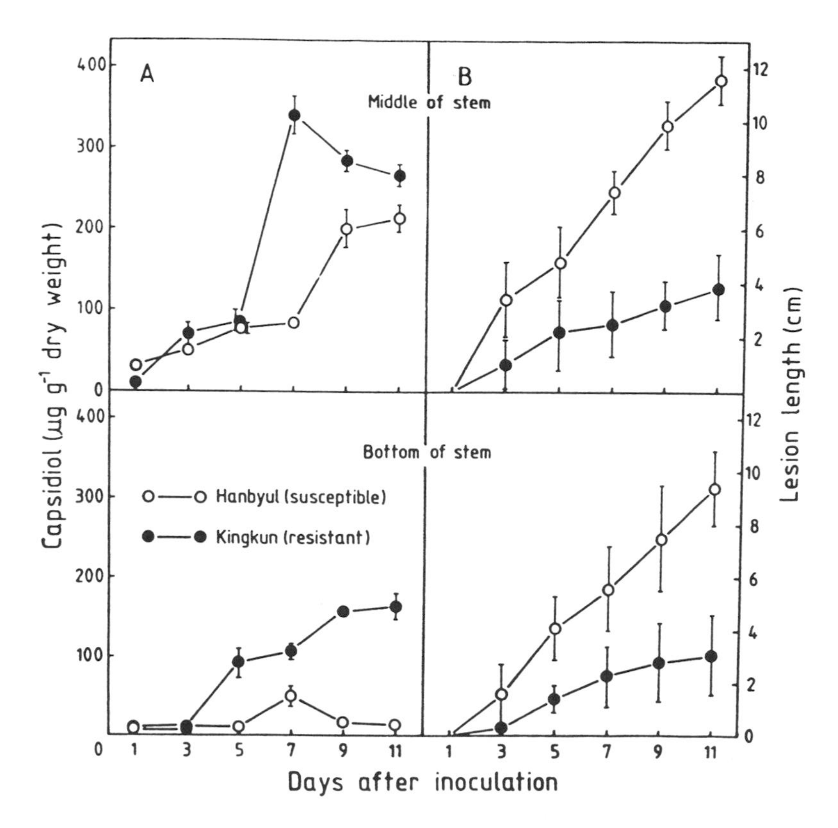

Figure 2 Time courses of (A) capsidiol accumulation and (B) disease development in the stems of pepper cultivars Hanbyul (susceptible) and Kingkun (resistant) at the first branched-flowering stage inoculated with zoospore suspensions of *P. capsici* at the middle and bottom of stems. Vertical bars represent standard deviations.

The ranks of lesion length were top > middle > bottom of stems at the first branched-flowering stage. The lesion development in Kingkun was much slower than that in Hanbyul, as the plants reached the first branched-flowering stage that is considered the approach of maturity.

B. Effect of Plant Age and Stem Part on Capsidiol Production in Stems

The concentrations of capsidiol were determined in the stems of the cv. Hanbyul and the cv. Kingkun infected by *P. capsici* at the six-leaf stage (Fig. 1). Capsidiol was not detected in stems from healthy plants. The amount of capsidiol in the stems could be measured only until 7 days after inoculation because all infected plants died. The lesion development in the

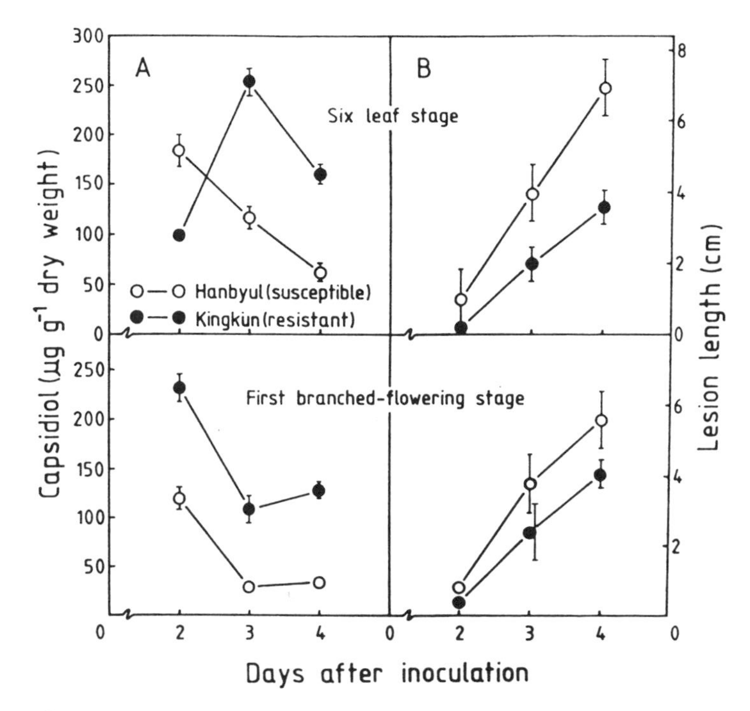

Figure 3 Time courses of (A) capsidiol accumulation and (B) disease development in the stems of pepper cultivars Hanbyul (susceptible) and Kingkun (resistant) at the six-leaf and first branched-flowering stages inoculated with zoospore suspensions of *P. capsici* at the top of stems. Vertical bars represent standard deviations.

stems of the cv. Hanbyul was more rapid than that of the cv. Kingkun. In contrast, the concentrations of capsidiol in the two cultivars were similar, with a maximum at the third day after inoculation. The resistant cv. Kingkun always contained more capsidiol than the susceptible cv. Hanbyul. In particular, a secondary accumulation of capsidiol at 7 days after inoculation occurred in the resistant cv. Kingkun. At the six-leaf stage, the level of capsidiol at the top of stems 2 days after inoculation was higher in Hanbyul than in Kingkun, but declined thereafter (Fig. 3).

The lesion development and capsidiol concentration at the top, middle, and bottom of stems of the susceptible cv. Hanbyul and the resistant cv. Kingkun at the first branched-flowering stage are compared in Figs. 2 and 3. The lesion length at the top, middle, and bottom of stems increased more rapidly in the susceptible cv. Hanbyul than in the resistant cv. Kingkun. The differences between the two cultivars became more conspicuous at the lower parts of stems. Capsidiol was first detected at all parts of stems a day after inoculation, thereafter accumulating more in Kingkun than in Hanbyul. At 7 days after inoculation, the levels of capsidiol at the middle and bottom of stems of the cv. Kingkun were about 2–3.5 times as high as those in the susceptible cv. Hanbyul. A high quantity of capsidiol was produced at the top of stems 2 days after inoculation, but thereafter it declined slowly. Significant differences between the two cultivars in capsidiol accumulation were observed at the bottom and middle of stems from 5 and 7 days after inoculation, respectively. More capsidiol was produced at the upper parts than at the bottom of stems.

C. Effect of Plant Age on Capsidiol Production in Roots

The concentrations of capsidiol which accumulated in roots of the susceptible cv. Hanbyul and the resistant cv. Kingkun at various intervals after inoculation with *P. capsici* are presented in Fig. 4. Capsidiol was not detectable in roots from healthy plants. At the six-leaf stage, the concentrations of capsidiol in roots of the two cultivars during the disease progress were similar to each other, except for a higher amount in the resistant cv. Kingkun at 6 days after inoculation. At the first branched-flowering stage, the accumulation of capsidiol in the roots gradually increased with the maxima

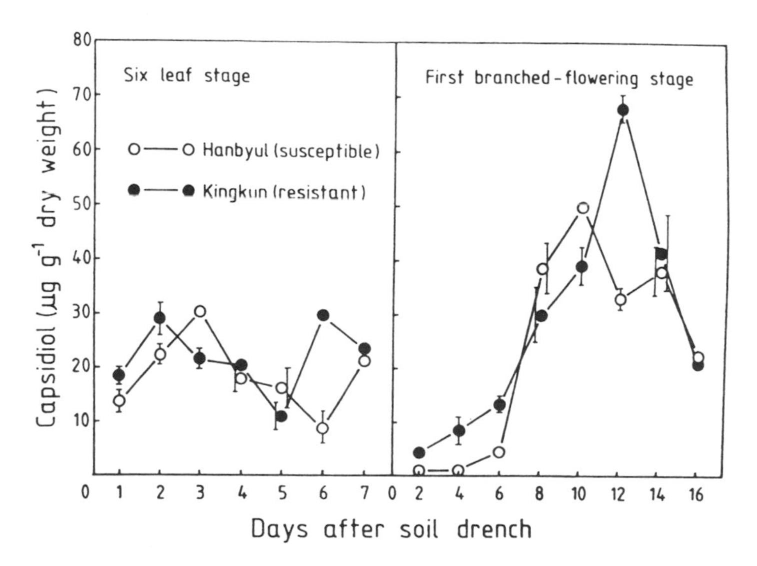

Figure 4 Time courses of capsidiol accumulation in roots of pepper cultivars Hanbyul (susceptible) and Kingkun (resistant) at the six-leaf and first branched-flowering stages drenched with zoospore suspensions of *P. capsici* in the soil. Vertical bars represent standard deviations.

at 10 and 12 days after inoculation in the cvs. Hanbyul and Kingkun, respectively, and declined afterward. In particular, higher quantities of capsidiol accumulated in Kingkun than those in Hanbyul until 6 days after inoculation. During *Phytophthora* infection in roots of pepper plants, capsidiol greatly accumulated at the first branched-flowering stage when compared to the six-leaf stage.

D. Effect of Concentration of Metalaxyl on Capsidiol Production

When metalaxyl was applied to the soil a day before stem inoculation of zoospore suspension of *P. capsici*, lesion development from the inoculation site of the stem decreased in the two cvs. Hanbyul and Kingkun, with increasing amounts of metalaxyl (Fig. 5A). Four days after stem inoculation at the eight-leaf stage, susceptible Hanbyul was more severely diseased than resistant Kingkun at 1 and 3 μg/ml. As the concentration of metalaxyl increased, disease progress gradually decreased, with the decline being more

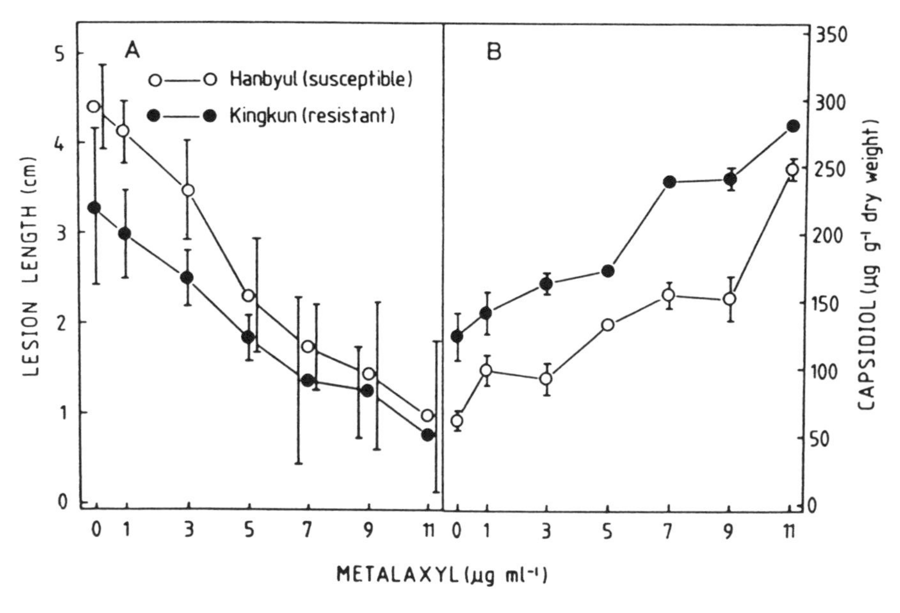

Figure 5 Influence of concentration of a metalaxyl drench in soil on (A) disease development and (B) capsidiol production in the stems of pepper cultivars Hanbyul (susceptible) and Kingkun (resistant). Plants were inoculated at the eight-leaf stage with zoospore suspensions of *P. capsici* at the bottom of the stem. Data were obtained on day 4 after stem inoculation. Vertical bars represent standard deviations.

pronounced in Hanbyul than in Kingkun. In particular, at the drench treatment of 11 μg/ml of metalaxyl, lesion development was strikingly inhibited to 1-cm lesion length in the two cultivars.

In contrast to the disease development in both cultivars, the production of capsidiol in pepper stems increased with increasing amounts of metalaxyl (Fig. 5B). In comparable inoculated, drenched plants, resistant Kingkun always contained more capsidiol in stems than susceptible Hanbyul, irrespective of metalaxyl concentration. However, capsidiol accumulation by metalaxyl was more noticeable in Hanbyul than in Kingkun. At 11 μg/ml of metalaxyl, the two cultivars had similar levels of capsidiol. Metalaxyl treatment did not induce capsidiol in healthy, uninoculated plants of both cultivars.

E. Effect of the Time of Metalaxyl Application on Capsidiol Production

When metalaxyl was applied to soil at 7 μg/ml 3 or 7 days before and at the time of stem inoculation on pepper plants at the 10-leaf stage, *Phytophthora* blight was inhibited in resistant Kingkun and greatly restricted in susceptible Hanbyul (Fig. 6). Small quantities of capsidiol accumulated similarly in the two pepper cultivars. When the soil drench of metalaxyl was delayed after inoculation, lesion development gradually progressed in both pepper cultivars and capsidiol accumulated in large amounts. In particular, the accumulation of capsidiol in metalaxyl-treated plants was much more marked in susceptible Hanbyul than in resistant Kingkun 1 or 2 days after inoculation.

F. Time Course of Capsidiol Production by Metalaxyl

As previously described, the retardation of lesion development was directly proportional to the amount of metalaxyl applied (Fig. 7). The accumulation of capsidiol in stems of the inoculated, undrenched controls increased to a maximum at 3 days after inoculation in the two cultivars and declined afterward. At 1 or 5 μg/ml of metalaxyl, accumulation of capsidiol increased more than in the inoculated, undrenched control, with a maximum at 2 days after inoculation in both cultivars. In these treatments, no significant differences between the two cultivars in capsidiol accumulation by metalaxyl were observed. Capsidiol production in both cultivars was about three times higher at 5 μg/ml of metalaxyl than at 1 μg/ml of metalaxyl at 2 days after inoculation. The concentration of capsidiol in stems decreased after maximum accumulation following 2 days of stem inoculation, regardless of metalaxyl treatment and pepper cultivar.

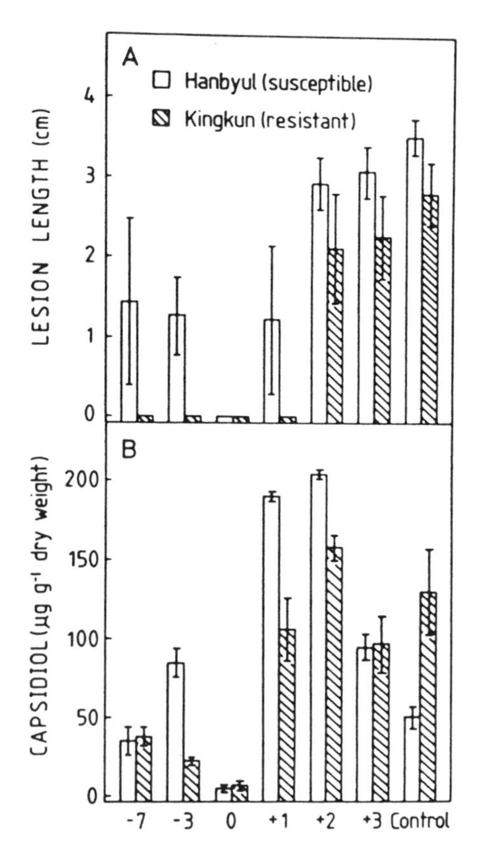

Figure 6 Influence of time of soil drench of metalaxyl relative to the time of inoculation on (A) disease development and (B) capsidiol production in the stems of pepper cultivars Hanbyul (susceptible) and Kingkun (resistant). Plants were inoculated at the 10-leaf stage with zoospore suspensions of *P. capsici* at the bottom of the stem. Data were obtained on day 4 after stem inoculation. Vertical bars represent standard deviations.

IV. DISCUSSION

A. Effect of Plant Age on Capsidiol Production

Infection systems used here allowed a precise determination of both disease development and capsidiol concentrations in *P. capsici*-infected tissues of pepper plants of the susceptible cv. Hanbyul and the resistant cv. Kingkun

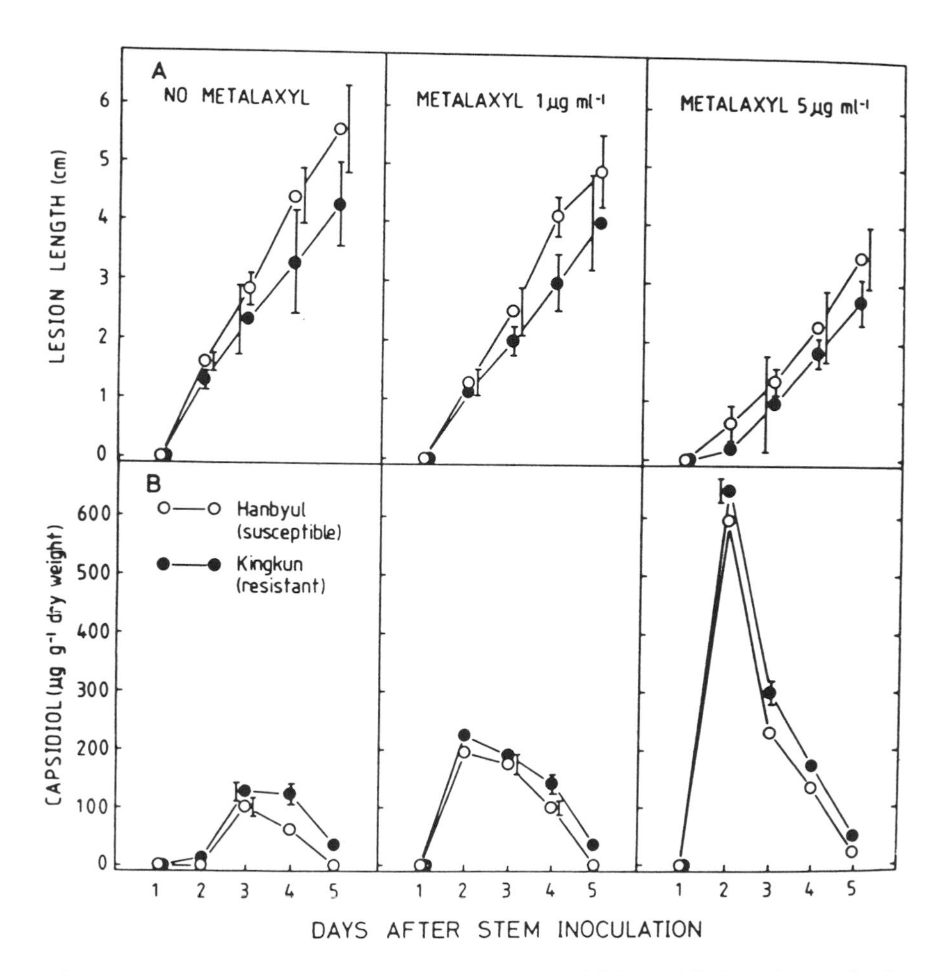

Figure 7 Time courses of (A) disease progress and (B) capsidiol production in the stems of pepper cultivars Hanbyul (susceptible) and Kingkun (resistant). Plants were inoculated at the eight-leaf stage with zoospore suspensions of *P. capsici* at the bottom of the stem after soil treatment with metalaxyl. Vertical bars represent standard deviations.

at different development stages. The resistance of pepper plants to *P. capsici* increased as they matured, irrespective of cultivar. The differences between susceptible and resistant responses to *P. capsici* were quantitative rather than qualitative because symptoms slowly developed even on resistant plants. Whatever the basis for age-related resistance in the different tissues of pepper plants, it appears that *P. capsici* has a preference for young tissues. In stem infections, for instance, the pathogen spread much more slowly at the bottom than in the middle of stems (Fig. 2). In contrast, more capsidiol was produced in the middle than at the bottom of the stem following wound inoculation. These findings indicating that capsidiol production decreased in older tissues of stems despite increased resistance do not support the concept of a general inverse relation between the amount of phytoalexin accumulating inside the host tissue and the rate of disease development. Accordingly, the increases in disease resistance in older tissues of stems may be derived from other mechanisms including possible morphological barriers and the production of inhibitory compounds other than capsidiol. Hahn et al. (1985) demonstrated that most of the phytoalexin accumulated in the epidermis in *P. megasperma* f. sp. *glycinea*-soybean system. Because there is more dry matter in the bottom tissue of pepper stems, the number of epidermal cells that can synthesize capsidiol would be less per unit dry weight at the bottom than in the middle of stems. On these assumptions, it should be considered that the higher amount of capsidiol may be produced per epidermal cell at the bottom than in the middle of stems.

Our data demonstrate that, in general, capsidiol concentrations in infected organs of pepper plants were correlated with the degree of resistance expressed to *P. capsici*. The resistant cv. Kingkun always contained more capsidiol in all infected organs such as stem and root than the susceptible cv. Hanbyul, suggesting that the resistant cultivar synthesizes more capsidiol in response to infection. Even the susceptible cultivar accumulated considerable capsidiol, especially when older plants were inoculated. As previously observed by Molot et al. (1981), however, capsidiol accumulation may not be the main factor restricting *P. capsici* growth in resistant pepper plants, but other antifungal compounds may be induced, together with capsidiol by the infection of *P. capsici*. It would appear that susceptible cultivars also have the ability to produce capsidiol, possibly being differently expressed as plants mature. Other mechanisms, such as the preferential degradation of capsidiol in susceptible plants, should not be excluded (Ward and Stössl, 1972; Stössl et al., 1977).

The basis for the increased resistance of mature pepper plants to *Phytophthora* blight seems to be distinguished from that of older tissues of the same stems. *Phytophthora* infection was reduced and capsidiol accumulated more in stems and roots at the first branched-flowering stage than at the six-leaf stage. These results suggest that capsidiol accumulation in infected stems and roots of mature pepper plants may contribute to the increase in resistance to *P. capsici*, being especially affected by host genotype and plant age. However, possible morphological barriers and other inhibitory compounds in mature pepper plants may also play important roles in retarding growth of *P. capsici*, as previously described in other host–pathogen combinations (Hwang and Heitefuss, 1982).

There were wide differences in the concentrations of capsidiol determined in different organs. Lower concentrations of capsidiol were found in roots than in stems. The production of low amounts of capsidiol in roots appear to be due to different biochemical interactions of host and pathogen in these organs. Tissues of the stems were uniformly necrotic and slightly sunken, but tissues of the roots were disintegrated in a watery rot and died. Small quantities of capsidiol (below 30 μg/g dry wt) in the young roots at the six-leaf stage may not contribute to the expression of resistance to *P. capsici*. In contrast, old roots of pepper plants at first branched-flowering stage responded to inoculation with zoospores of *P. capsici* by accumulating considerable quantities of capsidiol, indicating that capsidiol may contribute to the resistance expression of pepper plants at late growth stages.

Similarities in lesion development and time courses of capsidiol accumulation in root and stem of pepper plants at late growth stages would suggest that similar mechanisms govern the increased resistance of pepper plants to *P. capsici* with increasing age.

B. Effect of Metalaxyl on Capsidiol Production

In addition to the efficacy of metalaxyl for control of *P. capsici* on pepper plants (Sung and Hwang, 1988), metalaxyl treatment stimulated capsidiol production in pepper stems infected with the fungus. Whereas some fungicides are themselves capable of inducing phytoalexin production in plants (Reilly and Klarman, 1972), it seems likely that metalaxyl itself does not elicit capsidiol production in pepper stems but rather increases the capacity of pepper plants to synthesize capsidiol in response to infection by *P. capsici*. This observation is supported by the absence of capsidiol production in uninoculated, metalaxyl-treated plants in this study. Application of metalaxyl in pepper plants infected with *P. capsici* retarded disease development but did not change the rapidly growing stem lesions into brownish necrotic ones of the hypersensitive reaction. In contrast to the findings of Ward et al. (1980) that control by metalaxyl of *P. megasperma* f.sp. *glycinea* in soybean hypocotyls resulted in a hypersensitive resistant response accompanied by glyceollin production, our results suggest that metalaxyl can also stimulate phytoalexin production even in a susceptible response to *Phytophthora* infection in plants, even though there is no hypersensitive resistant response. Our previous electron microscopic observations that in metalaxyl-treated stems an electron-dense material was apposed in those sites of the host cell wall in most intimate contact with the fungal cell wall (Hwang et al., 1989, 1990) also provide the possibility that metalaxyl treatment may induce the plant defense reaction. In the control of rice blast by dichlorocyclopropanes, Cartwright et al. (1977) also suggested that the compound acts by sensitizing the plant such that subsequent inoculation with *P. oryzae* Br. and Cav. results in a resistant reaction.

With increasing metalaxyl concentrations, capsidiol production was stimulated in infected pepper stems. The increase followed similar patterns in the susceptible and resistant cultivars (Fig. 5). Inhibition of disease development as well as stimulation of capsidiol production by metalaxyl appear to be greatly influenced by the degree of resistance of pepper cultivars to *P. capsici*. Therefore, resistant pepper cultivars may possess a genetic basis for producing more capsidiol than susceptible ones, although the difference between the susceptible and resistant responses of pepper plants to *P. capsici* is quantitative rather than qualitative (Kim et al., 1989). Whereas large quantities of glyceollin produced in soybean by an incompatible host–pathogen interaction remained constant over a range of metalaxyl concentrations (Ward et al., 1980), it is of interest that capsidiol amounts in pepper increased significantly in infected stems of resistant cultivars with increasing doses of metalaxyl.

When compared with inoculated, metalaxyl-untreated controls, metalaxyl application before inoculation prevented disease development but did not stimulate capsidiol production. However, larger quantities of capsidiol, especially in susceptible Hanbyul, were produced in infected stems of pepper plants treated with metalaxyl at intervals during disease development (Fig. 6). These results suggest that metalaxyl may cause pepper stems to accelerate capsidiol production by affecting the host–pathogen interaction after the pathogen spreads into stem tissue. Our earlier finding, that metalaxyl is very effective for inhibition of mycelial growth of *P. capsici* (Sung and Hwang, 1988), raises the possibility, although unproven, that fungal cell wall polymers, which elicit capsidiol accumulation, may also be released in infected stem tissue from mycelium of *P. capsici* by metalaxyl treatment, as observed by Yoshikawa et al. (1981) in the *P.m.*f. sp. *glycinea*–soybean system.

The stimulation of capsidiol production by metalaxyl reached its maximum at 2 days after inoculation, when typical symptoms began to appear in wound-inoculated stems (Fig. 7). These results provide evidence that metalaxyl plays a critical role in inducing capsidiol production in the early stages of host–pathogen interaction. In contrast, capsidiol production in the inoculated, metalaxyl-untreated controls increased, with a maximum at 3 days after inoculation when considerable disease progress occurred, thus suggesting that *Phytophthora* infection may be important for stimulation of capsidiol production in the later stages of the host–pathogen interaction.

The amount of capsidiol decreased to base levels after the maximum accumulation. In particular, the effectiveness of metalaxyl treatment was conspicuous in reducing the amount of capsidiol in infected stems. The disappearance of capsidiol with time may not be due to degradation by *P. capsici* (Ward and Stössl, 1972; Stössl et al., 1973) but, rather, due to the metabolism of it by pepper tissue (Stössl et al., 1977) or possibly to a reaction with metalaxyl. Although unproven, it is also possible that some unknown factors at a late stage of the host–pathogen interaction may degrade or otherwise modify capsidiol in infected, metalaxyl-treated stems.

V. EPILOG

Yield losses of pepper plants caused by *Phytophthora* blight can be best minimized by a combination of practices including crop rotation, host resistance, and chemical application (Bruin and Edgington, 1981; Davidse, 1981; Shew, 1985; Sung and Hwang, 1988; Kim et al., 1989). However, due not only to economic burden but to the limit of cultivated land, the growers mainly rely on the use of resistant pepper cultivars and chemical control against the *Phytophthora* disease.

P. capsici severely infects pepper plants only at later growth stages in the fields. Therefore, a stable and durable type of resistance such as age-related resistance would be valuable for use in breeding programs of pepper cultivars. The expression of age-related resistance of pepper plants at later developmental stages may be due to the physiological and morphological changes in pepper stems during plant development (Jeun and Hwang, 1991). However, preformed antifungal substances, such as phenolics, may not contribute to the expression of age-related resistance of pepper plants, but rather phytoalexin "capsidiol," which accumulates after exposure to *P. capsici*, may play an important role in this resistance.

In the present studies, it has been demonstrated that capsidiol concentrations in infected stems and roots of pepper plants were correlated with the degree of resistance expressed to *P. capsici*. In comparable infected organs, the resistant cv. Kingkun always contained more capsidiol than the susceptible cv. Hanbyul. The stem and root of the two cultivars accumulated more capsidiol and became increasingly resistant as plants matured. These results suggest that capsidiol production has a role in increasing the resistance of pepper plants with aging.

A combination of direct fungitoxic and indirect effects involving activation of natural defense mechanisms has been demonstrated in pepper plants treated with the systemic fungicide metalaxyl (Börner et al., 1983; Hwang et al., 1989). Metalaxyl treatment not only produced direct effects in the fine structure of the fungal cell but also effected the defense reaction of pepper plants (Hwang et al., 1991). Due to the intimate nature of the interaction between *P. capsici* and pepper plants, fungicide interference with metabolism of the pathogen must inevitably lead to alteration of host physiology which results from disturbances at the pathogen–host interface. Enhanced capsidiol accumulation has also been reported in response to inoculation with *P. capsici* in pepper plants treated with metalaxyl.

In the present study, capsidiol production, associated with control of *Phytophthora* blight of pepper plants by metalaxyl, was determined in the infected stems of pepper cultivars resistant and susceptible to *P. capsici*. Increasing concentrations of metalaxyl in soil treatments a day before stem inoculation with *P. capsici* gradually retarded the lesion development on the stems of pepper plants but stimulated capsidiol production in the infected stems. Metalaxyl treatments did not change the rapidly growing stem lesions into the brownish necrotic ones of the hypersensitive reaction. In particular, the accumulation of capsidiol by metalaxyl treatment was more pronounced in the resistant cv. Kingkun than in the susceptible cv. Hanbyul. At metalaxyl concentrations of 1 and 5 μg/ml, lesions appeared on the stems 2 days after inoculation, with a maximum production of capsidiol. As the stem lesions developed and enlarged, the production of capsidiol in metalaxyl-treated, infected stems declined to a final base level similar to that in the infected control stems 5 days after inoculation. The metalaxyl treatments after 1 and 2 days of inoculation produced more capsidiol in the susceptible seedlings than before inoculation.

In conclusion, age-related resistance and metalaxyl treatment might make an important contribution to the effective prevention of *Phytophthora* blight in pepper plants by directly affecting the pathogen itself or indirectly enhancing capsidiol accumulation in infected tissue. However, there is a need for critical investigation of the natural sequence of events when age-related resistance is expressed in pepper plants in relation to capsidiol production. Further detailed research on the relative rates of fungal expansion and capsidiol production and degradation at the metalaxyl-treated infection site is also needed.

REFERENCES

Bailey, J. A. (1982). Mechanisms of phytoalexin accumulation. *Phytoalexins* (J. A. Bailey and J. W. Mansfield, eds.), Blackie, London, pp. 289–318.

Barksdale, T. H., Papavizas, G. S., and Johnston, S. A. (1984). Resistance to foliar blight and crown rot of pepper caused by *Phytophthora capsici*. *Plant Dis. 68*: 506–509.

Börner, H., Schatz, G., and Grisebach, H. (1983). Influence of the systemic fungicide metalaxyl on glyceollin accumulation in soybean infected with *Phytophthora megasperma* f.sp. glycinea. *Physiol. Plant Pathol. 23*: 145–152.

Bruin, G. C. A., and Edgington, L. V. (1981). Adaptive resistance to Peronosporales to metalaxyl. *Can. J. Plant Pathol. 3*: 201–206.

Cartwright, D., Langcake, P., Pryce, J., and Leworthy, D. P. (1977). Chemical activation of host defence mechanisms as a basis for crop protection. *Nature 267*: 511–513.

Davidse, L. C. (1981). Mechanism of action of metalaxyl in *Phytophthora megasperma* f.sp. *medicaginis*. *Netherlands J. Plant Pathol. 87*: 254–255.

Hahn, M. G., Bonhoff, A., and Grisebach, H. (1985). Quantitative location of the phytoalexin glyceollin I in relation to fungal hyphae in soybean roots infected with *Phytophthora megasperma* f.sp.*glycinea*. *Plant Physiol. 78*: 591–601.

Hwang, B. K., and Heitefuss, R. (1982). Antifungal compounds in leaves of spring barley in association with adult plant resistance to powdery mildew (*Erysiphe graminis* f.sp. *hordei*). *Phytopathol. Z. 104*: 272–278.

Hwang, B. K., and Kim, Y. J. (1990). Capsidiol production in pepper plants associated with age-related resistance to *Phytophthora capsici*. *Korean J. Plant Pathol. 6*: 193–200.

Hwang, B. K., and Sung, N. K. (1989). Effect of metalaxyl on capsidiol production in stems of pepper plants infected with *Phytophthora capsici*. *Plant Dis. 73*: 748–751.

Hwang, B. K., Kim, W. B., and Kim, W. K. (1989). Ultrastructure at the host-parasite interface of *Phytophthora capsici* in roots and stems of *Capsicum annuum*. *J. Phytopathol. 127*: 305–315.

Hwang, B. K., Ebrahim-Nesbat, F., Ibenthal, W. D., and Heitefuss, R. (1990). An ultrastructural study of the effect of metalaxyl on *Phytophthora capsici* infected stems of *Capsicum annuum*. *Pest. Sci. 29*: 151–162.

Hwang, B. K., Yoon, J. Y., Ibenthal, W. D., and Heitefuss, R. (1991). Soluble proteins, esterases and superoxide dismutase in stem tissue of pepper plants in relation to age-related resistance to *Phytophthora capsici*. *J. Phytopathol. 132*: 129–138.

Jeun, Y. C., and Hwang, B. K. (1991). Carbohydrate, amino acid, phenolic and mineral nutrient contents of pepper plants in relation to age-related resistance to *Phytophthora capsici*. *J. Phytopathol. 131*: 40–52.

Jones, D. R., Graham, W. G., and Ward, E. W. B. (1974). Ultrastructural changes in pepper cells in a compatible interaction with *Phytophthora capsici*. *Phytopathology 64*: 1084–1090.

Jones, D. R., Unwin, C. H., and Ward, E. W. B. (1975a). The significance of capsidiol induction in pepper fruit during an incompatible interaction with *Phytophthora infestans*. *Phytopathology 65*: 1286–1288.

Jones, D. R., Unwin, C. H., and Ward, E. W. B. (1975b). Capsidiol induction in pepper fruit during interactions with *Phytophthora capsici* and *Monilia fructicola*. *Phytopathology 65*: 1417–1421.

Keen, N. T. (1971). Hydroxyphaseollin production by soybeans resistant and susceptible to *Phytophthora megasperma* var. *sojae*. *Physiol. Plant Pathol. 1*: 265–275.

Kim, Y. J., Hwang, B. K., and Park, K. W. (1989). Expression of age-related resistance in pepper plants infected with *Phytophthora capsici*. *Plant Dis.* *73*: 745–747.

Lazarovits, G., and Ward, E. W. B. (1982). Relationship between localized glyceollin accumulation and metalaxyl treatment in the control of *Phytophthora* rot in soybean hypocotyls. *Phytopathology 72*: 1217–1221.

Lazarovits, G., Stössl, and Ward, E. W. B. (1981). Age-related changes in specificity and glyceollin production in hypocotyl reaction of soybeans to *Phytophthora megasperma* var. *sojae*. *Phytopathology 71*: 94–97.

Leonian, L. H. (1922). Stem and fruit blight of peppers caused by *Phytophthora capsici* sp. nov. *Phytopathology 12*: 401–408.

Molot, P. M., Mas, P., Conus, M., and Ferriere, H. (1981). Relations between capsidiol concentration, speed of fungal invasion and level of induced resistance in cultivars of pepper (*Capsicum annuum*) susceptible or resistant to *Phytophthora capsici*. *Physiol. Plant Pathol. 18*: 379–389.

Paxton, J. D., and Chamberlain, D. W. (1969). Phytoalexin production and disease resistance in soybean as affected by age. *Phytopathology 59*: 775–777.

Polach, F. J., and Webster, R. K. (1972). Identification of stems and inheritance of pathogenicity in *Phytophthora capsici*. *Phytopathology 62*: 20–26.

Reilly, J. J., and Klarman, W. L. (1972). The soybean phytoalexin, hydroxyphaseollin, induced by fungicides. *Phytopathology 62*: 1113–1115.

Shew, H. D. (1985). Response of *Phytophthora parasitica* var. *nicotiana* to metalaxyl exposure. *Plant Dis. 69*: 559–562.

Stössl, A., Unwin, C. H., and Ward, E. W. B. (1972). Postinfectional inhibitors from plants. I. Capsidiol, an antifungal compound from *Capsicum frutescens*. *Phytopathol. Z. 74*: 141–152.

Stössl, A., Unwin, C. H., and Ward, E. W. B. (1973). Postinfectional inhibitors from plants: fungal oxidation of capsidiol in pepper fruit. *Phytopathology 63*: 1225–1231.

Stössl, A., Robinson, J. R., Rock, G. L., and Ward, E. W. B. (1977). Metabolism of capsidiol by sweet pepper tissue: some possible implications for phytoalexin studies. *Phytopathology 67*: 64–66.

Sung, N. K., and Hwang, B. K. (1988). Comparative efficacy and in vitro activity of metalaxyl and metalaxyl-copper oxychloride mixture for control of *Phytophthora* blight of pepper plants. *Korean J. Plant Pathol. 4*: 185–196.

Ward, E. W. B. (1976). Capsidiol production in pepper leaves in incompatible interaction with fungi. *Phytopathology 66*: 175–176.

Ward, E. W. B., and Stössl, A. (1972). Postinfectional inhibitors from plants. III. Detoxification of capsidiol, an antifungal compound from peppers. *Phytopathology 62*: 1186–1187.

Ward, E. W. B., Unwin, C. H., and Stössl, A. (1973). Postinfectional inhibitors from plants. VII. Tolerance of capsidiol by fungal pathogens of pepper fruit. *Can. J. Bot. 51*: 2327–2332.

Ward, E. W. B., Lazarovits, G., Stössl, P., Barrie, S. D., and Unwin, C. H. (1980). Glyceollin production associated with control of *Phytophthora* rot of soybeans by the systemic fungicide, metalaxyl. *Phytopathology 70*: 738–740.

Weber, G. F. (1932). Blight of peppers in Florida caused by *Phytophthora capsici*. *Phytopathology 22*: 775–780.

Yoshikawa, M., Matama, M., and Masago, H. (1981). Release of a soluble phytoalexin elicitor from mycelial walls of *Phytophthora megasperma* var. *sojae* by soybean tissues. *Plant Physiol. 67*: 1032–1035.

23
Induced Chemical Resistance in *Sesamum indicum* Against *Alternaria sesami*

R. K. S. Chauhan
Jiwaji University, Gwalior, Madhya Pradesh, India

B. M. Kulshrestha
K. R. G. College, Gwalior, Madhya Pradesh, India

I. INTRODUCTION

This project was undertaken to investigate whether the fruits of *Sesamum indicum* L. could produce any phytoalexin in response to infection by *Alternaria sesami* (Kawamura) Mohanty and Beraha. *S. indicum* (Pedaliaceae), commonly known as til (sesame), is one of the important oil-yielding plants. The plant is susceptible to various diseases such as *Alternaria* blight caused by *A. sesami*, *Cercospora* leaf spot caused by *C. sesamicola*, *Phytophthora* blight caused by *P. parasitica* var. *sesami*, and root rot caused by *Macrophomina phaseolina*. However, no information is available so far regarding postinfectional antifungal compounds in sesame plant and their role in disease resistance.

II. METHODOLOGY

The seeds of *S. indicum* were procured from National Seed Corp. *Alternaria sesami* (Kawamura) Mohanty and Beraha was isolated from infected parts of *S. indicum*. Test organisms, *Aspergillus niger* (Van.) Tieghem, *Cladosporium cladosporioides* (Fres.) De Vries, *Colletotrichum capsici* (Syd.) Butler and Bisby, *Curvularia lunata* (Wakker) Boedijn, *Fusarium solani* (Mart.) Sacc. *Helminthosporium tetramera* McKinney, *Mycosphaerella rabiei* Kovachevsky, and *Rhizopus stolonifer* (Ehrenb. Ex Fr.) Lind were obtained from the Indian Type Culture Collection (IARI, Delhi).

Phytoalexin was detected from fruits following the drop diffusate technique. Thoroughly washed fruits were split open longitudinally and seeds were removed. These fruit halves were surface-sterilized with sodium hypochloride solution and washed twice with sterilized distilled water. Fruits were placed in a humid chamber, inoculated aseptically with the spore suspension of *A. sesami* (1×10^5 spores/ml), and incubated at 28 ± 2°C. The drops were collected centrifuged and the supernatant is collected.

The chemicals used for induction of phytoalexin, i.e., aureofungin, bavistine, 2,4-dinitrophenol, dithan M-45, mercuric chloride, sulfex, and thiobendezole are applied to fruit cavities as solutions of different ppm (200, 600, 1000). After incubation for 48 hr diffusates were collected, centrifuged, and the supernatant was tested for phytoalexins (Kulshrestha and Chauhan, 1988).

III. SESAME PHYTOALEXINS

A. Bioassay

Spore germination, germ tube growth, and radial mycelial growth bioassays were used in the present investigation (Kulshrestha, 1982; Chauhan and Kulshrestha, 1984). In different bioassays, *Alternaria sesami*, *Aspergillus niger*, *Cladosporium cladosporioides*, *Colletotrichum capsici*, *Curvularia lunata*, *Fusarium solani*, *Helminthosporium tetramera*, and *Mycosphaerella rabiei* were used as the test organisms. In spore germination bioassay, the effect of active principle from the host plant on spore germination of test organisms is observed and is done by estimating the inhibition of spore germination by slide germination method. To avoid influence by nutrients present in the spores during bioassay, germ tube growth bioassay is preferred (Müller, 1958). For this purpose, spores of different test organisms were kept in sterilized distilled water for the time period just enough to initiate germination. Water then removed by centrifugation and germinated spores were taken in the diffusate in a cavity slide and the effect on germ tube growth was observed. Radial mycelial growth bioassay consisted of measuring the diameters of colonies of test organisms grown on phytoalexin-amended media and their comparison with those of control ones (Langcake and Pryce, 1976). All the test organisms were found sensitive to varying degrees (Table 1).

It is clear that phytoalexins from sesame, besides reducing spore germination, frequently retard germ tube growth and further the growth of the mycelium, and thus make them less effective in causing infection. Therefore, phytoalexins appear to discourage infection by reducing germination of spores and subsequent growth of germ tubes.

B. Effect of Incubation Time on the Production of Sesame Phytoalexin

To determine the effect of incubation time on phytoalexin production in *S. indicum*, diffusates were collected after 4, 8, 16, 24, 32, 48, and 72 hr of

Table 1 Effect of Phytoalexin from Fruits of *S. indicum* on Spore Germination, Germ Tube Growth, and Radial Mycelial Growth of Fungi

	% Inhibition over control		
Fungus	Spore germination[a]	Germ tube growth[b]	Radial mycelial growth[c]
Alternaria sesami	63	37	48
Aspergillus niger	69	69	93
Cladosporium cladosporioides	41	50	70
Colletotrichum capsici	88	78	37
Curvularia lunata	69	77	70
Fusarium solani	65	66	56
Helminthosporium tetramera	100	88	88
Mycosphaerella rabiei	78	50	67

[a]Mean of 100 observations taken after 8 hr of incubation at 28 ± 2°C.
[b]Mean of 30 germ tubes observed after 24 hr.
[c]Mean of three replicates observed after 8 days of incubation.

inoculation. Phytoalexin activity was detected in the droplets only after 8 hr and concentration increased gradually up to 32 hr, reaching a maximum at 48 hr. There was no further increase or decrease in concentration of the phytoalexin up to 72 hr (Table 2).

C. Phytoalexin Production by Different Plant Parts

Fruit, leaf, and stem of sesame plant were taken to investigate the production of phytoalexin in these parts after inoculation with fungi. Phytoalexin

Table 2 Time Course of Phytoalexin Production

Incubation (hr)	% Inhibition in the spore germination of *A. sesami*
4	0
8	0
16	20
24	30
32	45
48	60
72	60

was detected in fruits, leaves, and stem of sesame plant although concentration was variable. Leaves and fruits produce more phytoalexin than stem (Table 3).

D. Studies on the Chemical Nature of Phytoalexin

Three solvent systems were employed for the separation of sesame phytoalexin on thin layer chromatograms, namely, chloroform–methanol (100 : 2), benzene–acetic acid–water (125 : 72 : 3), and butanol–acetic acid–water (4 : 1 : 5). Fluorescence under UV light (254 nm), R_f values, reaction with 1% ferric chloride and diazotized sulfanilic acid were the criteria used for chemical characterization of the phytoalexin (Table 4). Fluorescing spot was also tested for antifungal activity. The phytoalexin of sesame plant was found to be phenolic in nature. It completely inhibited spore germination of *A. sesami*.

E. Chemical Induction of Phytoalexin

Aureofungin, bavistin, 2,4-dinitrophenol, dithan M-45, mercuric chloride, sulfex, and thiobendazole were the chemicals tested for their ability to induce phytoalexin production in fruits of *S. indicum*. 2,4-Dinitrophenol and mercuric chloride induced phytoalexin production in higher concentrations (Table 5).

IV. EPILOG

The phytoalexin in *S. indicum* is found to be phenolic in nature. It is highly inhibitory to the spore germination, germ tube growth, and mycelial growth

Table 3 Phytoalexin production in fruits, leaves and stem of *S. indicum* in response to inoculation with *A. sesami*

Fungus	% Inhibition in the spore germination over control		
	Fruit	Leaf	Stem
Alternaria sesami	60	90	30
Colletotrichum capsici	88	80	20
Curvularia lunata	69	60	35
Fusarium solani	65	84	38
Helminthosporium tetramera	100	73	40
Mycosphaerella rabiei	78	61	30

Table 4 Chromatographic Characteristics of the Phytoalexin Isolated from Fruits of *S. indicum*

Solvent system	Spot no.	R_f values	Fluorescence under UV light (254 nm)	Reaction with 1% ferric chloride	Reaction with diazotized sulfanilic acid	% Inhibition of spore germination
Chloroform–methanol (100 : 2)	1	0.14	Bluish	Black	Pink	100
Benzene–Acetic acid–water	1	0.10	Bluish	Black	Pink	100
(125 : 72 : 3)	2	0.86	White	Brown	Yellow	30
Butanol–acetic acid–water	1	0.16	Bluish	Black	Pink	100
(4 : 1 : 5)	2	0.30	White	Brown	Yellow	58

Table 5 Chemical Induction of Phytoalexin in Fruits of *S. indicum*

Chemical	Concentration (ppm)	Phytoalexin production $R_f = 0.16$ butanol–acetic acid–water (4 : 1 : 5)
Aureofungin	200	–
	600	+
	1000	+ +
Bavistine	200	–
	600	–
	1000	–
2,4-Dinitrophenol	200	–
	600	+
	1000	+ + +
Dithan M-45	200	–
	600	–
	1000	+
Mercuric chloride	200	–
	600	+ +
	1000	+ + +
Sulfex	200	–
	600	–
	1000	+
Thiobendazole	200	–
	600	–
	1000	+
Sterilized distilled water (control)	–	–

–, Absence; +, presence in traces; + +, presence in moderate concentrations; + + +, presence in higher concentrations.

of some fungi tested. Phytoalexin was detected in the drops only after 8 hr of inoculation and the concentration increased up to 48 hr. *A. sesami* induced phytoalexin in fruits, leaves, and stem of sesame plant. Certain chemicals also could induce the phytoalexin. Further characterization of this compound, elucidation of its biosynthetic pathway, mechanism of action, and development of analogs for use in the crop protection are the future prospects.

ACKNOWLEDGMENTS

The authors thank Dr. S. Chauhan, Professor, School of Studies in Botany, Jiwaji University, Gwalior for her keen interest in the preparation of this manuscript.

REFERENCES

Chauhan, R. K. S., and Kulshrestha, B. M. (1984). Production of phytoalexin in *Sesamum indicum* against *Alternaria sesami. Indian Phytopathol. 37*: 482–485.

Kulshrestha, B. M. (1982). Studies on induced chemical resistance factors in certain plant diseases. PhD Thesis, Jiwaji University, Gwalior, India, pp. 255.

Kulshrestha, B. M., and Chauhan, R. K. S. (1988). Physicochemical nature of the phytoalexin produced by fruits of *Sesamum indicum* in response to inoculation with *Alternaria sesami* and its chemical induction. *Indian Phytopathol. 41*: 92–95.

Langcake, P., and Pryce, R. J. (1976). The production of resveraterol by *Vitis vinifera* and other members of the Vitaceae as a response to infection or injury. *Physiol. Plant Pathol. 9*: 77–86.

Müller, K. O. (1958). Studies on phytoalexins. III. The formation and the immunological significance of phytoalexin produced by *Phaseolus vulgaris* in response to infection with *Sclerotinia fructicola* and *Phytophthora infestans. Aust. J. Biol. Sci. 11*: 275–300.

24

Phytoalexins and Other Postinfectional Compounds of Some Economically Important Plants of India

M. Daniel
The Maharaja Sayajirao University of Baroda, Vadodara, Gujarat, India

I. INTRODUCTION

Studies on phytoalexins have taken rapid strides in the recent past, but still the available data are grossly inadequate to draw any valid conclusion on the distribution, metabolism, and role of these compounds in higher plants. Out of the 250,000 or more angiosperms, hardly a couple of hundred species have been analyzed for their phytoalexins. Except for the Fabaceae, no other family has been systematically screened for their defense reactions. Almost all the plants studied so far are the cultivated species. Nothing is known on the phytoalexins of wild plants, weeds, or tree species.

Study of phytoalexins in isolation, away from other disease resistance factors, is another reason for the ambiguity existing on the concept of phytoalexins. All the known phytoalexins and the preformed toxic substances (inhibitins) are secondary metabolites. Though the secondary metabolites in plants once were considered as waste products/excretory substances, the overwhelming evidence accumulated so far ascertains that the plants owe their survival to these compounds. It is these compounds which act as UV screens, deter herbivores, repel insects, and ward off pathogens. Some of them are responsible for effective pollination, dispersal of seeds, and dormancy. It is explicit that during evolution plants experimented with different defensive chemicals and therefore the various taxa elaborated distinct classes of compounds. Concurrent with the modifications to perfect the defense mechanisms were the attempts of the pathogens to detoxify these compounds. The struggle between plants and microbes/herbivores to outwit each other is the reason for the plethora of secondary metabolites found in nature. Most plants also have to face the threats of animals and therefore are compelled to produce the taste modifiers (bitter substances) along with antimicrobials. In this context a compound which possesses the ability to alter taste as well as inhibit fungal invasion would have been the compounds of choice for the plant. A number of secondary metabolites such as phenolics, tannins, flavonoids, sesquiterpenes, saponins, and iridoids fit well into this prescription.

Phytoalexins are produced as a result of triggering the biosynthetic pathways by the microbial invasion. It is therefore explicit that the cell should already possess the needed machinery for doing so. The switching on of a defense reaction means that the expression of the gene concerned is kept suppressed under normal conditions and this repression is removed once the signal from the pathogen is received. This gene would have formed de novo in evolution or would have been a modified form of a gene inherited from the ancestors. This also indicates that the precursors of phytoalexin would be present in the plant at any given instant and in this context the study of preformed substances assumes greater significance.

Our interest in phytoalexins was purely academic, i.e., to find out the defense reactions of plants. Therefore we based our studies on tree species. Trees are different from herbs in that they store more compounds for a comparatively longer time. The life span of leaves also is longer in trees than in herbs. The selection of plants was random. Later on some of the herbaceous perennials also were included. The physiological similarity of *Cuscuta* to fungal pathogens prompted us to include this angiosperm parasite in our study. The studies on different plants are at various stages of progress. All the results available with us are presented here, which collectively may shed some light on the defense reactions of plants.

II. TREES AND SHRUBS

A. *Tectona grandis* L.

T. grandis (teak; Verbenaceae) is one of the most important timber trees of India. The wood is used in high-class furniture, boat and ship building, houses, bridges, railway sleepers, etc. The bark of the tree is medicinally important as an astringent and the wood is used as an anthelmintic. The flowers and seeds are known diuretics.

Powdery mildew caused by *Uncinula tectonae* Salm. (Karadge et al., 1980), rust due to *Chaconia tectonae* Ramakr. K. and Ramakr. T.S. (Ramakrishnan and Ramakrishnan, 1949), and leaf spots due to *Cercospora tectonae* Stevens (Tirumalachar and Chupp, 1948) and *Sphaceloma tectonae* Wani and Tirum. (Wani and Tirumalachar, 1969) are the leaf diseases reported from *Tectona*. Wood rot caused by *Ganoderma applanatum* Pat. and *Daedalea flavida* Lev. are the diseases seen in the wood.

In Gujarat, leaf spot diseases appear during the months of July to

September. The brownish spots later enlarge to blackish brown patches. In extreme cases the whole leaf turns brownish black. *Curvularia clavata* Jain, the fungus isolated from the infected areas, was found to be the causative organism of this disease.

Chemical analysis of both healthy and infected leaves was carried out by standard procedures. The identities of flavonoids were confirmed by spectral analysis and cochromatography with authentic samples. Phenolic acids were identified on the basis of their spectral data, color reactions with diazotized *p*-nitraniline and sulfanilic acid, chromatographic properties, and cochromatography with standard compounds.

The healthy leaves of *Tectona* contained two flavones: 4′-OMe apigenin and luteolin. The phenolic acids present were syringic, sinapic, vanillic, melilotic, and gentisic acids. The other constituents of the leaves were quinones (lepachol and tectoquinone), proanthocyanidins, iridoids, alkaloids, and tannins. The infected leaves did not contain any flavone but a flavonol 3′,4′-dimethoxyquercetin instead and phenolic acids such as ferulic, vanillic, melilotic, and gentisic acids. They contained the same quinones as of the healthy leaves as well as proanthocyanidins, iridoids, alkaloids, and tannins.

There was no significant chemical difference between the diffusates of control and treated leaves when the healthy leaves were treated with the spore suspension of *C. clavata*. But when the leaves were treated with a nonpathogen, *Fusarium solani*, the diffusate contained *p*-hydroxybenzoic acid.

Mycelial growth assay of *F. solani* with *p*-hydroxybenzoic acid showed that this fungus is strongly inhibited at all concentrations. Maximum inhibition was noted at 1000 ppm. Spore germination and germ tube elongation of *F. solani* were also inhibited by *p*-hydroxybenzoic acid at 300, 500, and 1000 ppm concentrations. Maximums of 85% inhibition of spore germination and 64% inhibition of germ tube elongation were noted at 1000 ppm.

Similarly, the spore germination, germ tube elongation, and mycelial growth of *C. clavata* were inhibited by *p*-hydroxybenzoic acid at all concentrations. Maximum inhibition of mycelial growth was noted at 1000 ppm. At this concentration the inhibition of spore germination was 58% and inhibition of germ tube elongation 63% (Abraham, 1989).

The replacement of flavones by the flavonol 3′,4′-diOMe quercetin in diseased leaves appears to be highly significant. This involves a change in the biosynthetic pathway in which flavonols with more potent phenolic hydroxy groups are produced. Production of ferulic acid in place of syringic and sinapic acids also is noteworthy. Ferulic acid is known for its activity against *Poria weirii* (Li et al., 1972) and *Fusicoccum amygdali* (Borys and Childers, 1964). It is found to inactivate tungrovirus in vivo (Sridhar et al., 1979) and impart resistance in wheat varieties (Chigrin and Rozuru, 1969).

The production of *p*-hydroxybenzoic acid as a response to the non-pathogen is interesting. This supports the view expressed by Debnam and Smith (1976) that the non–pathogen-infested tissue produces more phytoalexins than the tissue infected by pathogen. The strong inhibition of mycelial growth, spore germination, and germ tube growth of *F. solani* and *C. clavata* proves the antifungal nature of *p*-hydroxybenzoic acid. The immediate production of this compound and its marked antifungal activity leads to the conclusion that *p*-hydroxybenzoic acid is indeed a phytoalexin. This phenolic acid had been reported as a phytoalexin in *Nectria*-infected apples (Swinburne, 1975).

Though *p*-hydroxybenzoic acid or its derivatives occur in plants as a component of lignin, its role in imparting resistance to the plant is proved by the studies on carrot slices using *Botrytis cinerea*. Here the infection leads to the production of inhibitors such as 6-methoxymellein, *p*-hydroxybenzoic acid, and falcarinol (Harding and Heale, 1981). The increase in *p*-hydroxybenzyl groups also leads to the increased production of lignin, which either directly or indirectly acts as a potential barrier to infection (Henderson and Friend, 1979).

B. *Cassia fistula* L.

C. fistula (Caesalpiniaceae) is a small to medium-sized deciduous tree commonly recommended in social forestry programs. The timber is used for making carts and agricultural implements. The leaves find application as an emollient for dropsy and rheumatic pains. The pulp from the pods is officially included in the British Pharmacopeia as an ingredient of senna and is used as a mild, pleasant, and safe purgative even for children and expectant mothers. The flowers of *C. fistula* are sometimes used as a food by natives. Its root is a tonic, febrifuge, and a strong purgative.

The common disease observed in *C. fistula* is powdery mildew caused by *Oidium* sp. (Salam and Rao, 1958) or *Phyllactinia corylea* (Pers.) Karst (Yadav, 1963). Other fungi reported from this plant are *Pestalotiopsis adusta* (Ell. Ev.) Steyart (Kanujia and Singh, 1978) and *Phoma comlanata* Tode ex. Fr. (Rajak and Rai, 1982).

The leaf spot disease of *C. fistula* is found in the months of January and February after which period the leaves are shed. The spots are small, irregular, and black. *Aspergillus niger* Van Tiegh. was the organism isolated from diseased areas. During pathogenicity tests the lesions begin to appear 6–8 days after inoculation with the pathogen.

The healthy leaves from the uninfected trees were found to contain three methoxyflavonols: 4′-OMe quercetin, 4′-OMe kaempferol, and 7,3′4′-triOMe quercetin; phenolic acids such as syringic and *p*-coumaric acids; proanthocyanidins; tannins and alkaloids. The diseased leaves contained the hydroxyflavonols quercetin and kaempferol, syringic and *o*-coumaric acids, alkaloids, proanthocyanidins, and tannins. The amount of flavonols in infected leaves were much higher than those of healthy leaves (Abraham and Daniel, 1988a).

The healthy leaves when exposed to the pathogen *A. niger* produced quercetin and kaempferol while the control contained the methylated derivatives of these compounds. The nonpathogen *F. solani* also evoked similar response when inoculated on healthy leaves.

Quercetin, one of the flavonols induced by the fungus, was assayed for its antifungal activity. This compound inhibited the spore germination of *A. niger* by 49% and germ tube elongation by 35% at 500 ppm. Mycelial growth also was inhibited at 100 and 500 ppm concentrations of quercetin (Abraham, 1989).

Evidently the process triggered by the infection of *C. fistula* leaves by *A. niger* is the demethylation of 4′-OMe kaempferol, 4′-OMe quercetin, and 7,3′,4′-triOMe quercetin to their hydroxylated parent compounds quercetin and kaempferol. The hydroxylated compounds are more toxic than their relatively unreactive methoxy derivatives.

The antifungal properties of quercetin and kaempferol were reported earlier by Sporoston (1957). Quercetin inhibited the growth of fungi such as *Daedalea quercina* and *Fomes annosus* (Alcubilla-Martin, 1970; Walchii and Scheck, 1976). Both quercetin and kaempferol were found to be absent from *Populus maximowiczii*, *P. laurifolia*, and other hybrids which were susceptible to *Dothichiza populea*, while they were located in the resistant varieties of *P. nigra* var. *italica* and its hybrids.

The induction of *o*-coumaric acid in place of *p*-coumaric acid by the pathogen and nonpathogen is interesting. Though *p*-coumaric acid is also known to possess antifungal activity (Trappe et al., 1973), the formation of *o*-coumaric acid had an added significance in that this phenolic acid produces coumarin by lactonization. Coumarin is a more potent antimicrobial agent (Berkenkemp, 1971). *o*-Coumaric acid has been reported to be a systemic fungicide (Gangulee and Kar, 1985).

C. *Morinda tomentosa* Heyne

M. tomentosa (Rubiaceae) is a small tree commonly found in the deciduous forests of Gujarat and other central and southern parts of India. The heartwood of this tree is durable, used for furnitures. A red dye "Suranj" is obtained from the roots of *M. tomentosa*. The leaves are used as a tonic and febrifuge. Its fruit is a deobstruent and is used against dysentery and asthma.

The diseases reported on *M. tomentosa* leaves are leaf spots caused by *Botryodiplodia theobromae* Pat. (Shreemali and Bilgrami, 1968), *Cercospora morindae* Syd, *Gleosporum morindae* Payak & Tirum. (Payak, 1953), and *Macrophoma morindae* Ramakr. & Sund. (Ramakrishnan and Sundaram, 1954).

The leaf spot disease of *Morinda* is noted in the months of October to December. The symptoms are blackish brown irregular patches on both the surfaces and at times the whole apical portion of the leaf turns dark. *Colletotrichium gleosporoides* Penzig and Sac. is the fungus isolated and found to be pathogenic.

The results obtained from *M. tomentosa* were similar to those of *C. fistula*. The healthy leaves contained the two methoxyflavonols 4′-OMe kaempferol and 3′,4′-diOMe quercetin, and the four phenolic acids vanillic, syringic, gentisic, and ferulic. The infected leaves contained the hydroxyflavonols kaempferol and quercetin along with four phenolic acids found in healthy leaves. The diffusates of both the pathogen- and non-pathogen-treated leaves contained quercetin and kaempferol (Abraham and Daniel, 1988a).

Quercetin at 500 ppm inhibited the spore germination of *C. gleosporoides* to 32% and germ tube elongation up to 53%. The mycelial growth of the fungus was inhibited at 100 and 500 ppm (Abraham, 1989).

D. *Eucalyptus globulus* Labill.

E. globulus (Myrtaceae) is a large tree extensively planted under aforestation programs throughout India. Its wood is useful in a variety of ways. The bark contains tannins and the leaves yield a volatile oil rich in 1,8-cineole, which is used as a mosquito repellant and local antiseptic.

A number of antifungal compounds are reported from various species of *Eucalyptus*. DaCosta and Rudman (1958) found that the outer heartwood of *E. microcorys* to be extremely resistant to decay-causing organisms such as *Coniophora carebella*, *Coriolus versicolor*, and *Fomes durus*. But methanol-extracted outer heartwood was promptly decayed by these fungi, and when the methanol extract was added to the heartwood of *E. regnans* susceptible to the fungus it conferred protection against decay. Inhibition of wood-rotting fungi by stilbenes and other polyphenols of *E. sideroxylon* has also been reported (Hart and Willis, 1974). Antimicrobial compounds are reported from *E. globulus* (Osborne and Thrower, 1964) and *E. triflora* (Egawa et al., 1977) also.

The common diseases found in *E. globulus* are leaf spot caused by

Pestalotiopsis funerea (Desm.) Stey. (Bilgrami, 1963) and root rot caused by *Ganoderma lucidum* Karst. (Bakshi, 1974).

E. globulus grown in Baroda develops leaf spot disease in the months of December and January. This disease is characterized by small, round, black spots scattered on the upper surface of the leaf. *Alternaria alternata* (Fr.) Keissler is isolated from the lesions and its pathogenicity is proved in the laboratory.

On chemical analysis, both the diseased and the noninfected leaves were found to contain the same compounds, i.e., 3′-OMe quercetin, 4′-OMe kaempferol, and vanillic, syringic, and *p*-hydroxybenzoic acids. The proanthocyanidins, iridoids, and alkaloids were also the same in both healthy and infected leaves.

When inoculated with the spores of the pathogen, the healthy leaves produced an additional compound fluorescing blue in UV light (long wavelength) and having an absorption maxima of 276, 285, 296, 330, and 341 nm in methanol. The chromatographic properties and the spectral data suggested this compound to be a coumarin. However, the nonpathogen used (*F. solani*) failed to induce any response in the healthy leaves. The diffusate from the treated leaves inhibited the mycelial growth of the pathogen (up to 54%) at 10 ml dilution. The inhibition of spore germination and germ tube elongation of *A. alternata* were 65% and 71%, respectively, at 10 ml dilution of the diffusate. This proves the blue fluorescing compound induced by *A. alternata* to be a phytoalexin (Abraham, 1989).

E. *Syzygium cumini* (L.) Skeels

S. cumini (Myrtaceae) is an evergreen tree found widely cultivated throughout India for its edible fruits. The fruit jambolan (Java plum) is used as fresh fruit as well as for jellies, wines, and cordials. The wood of *S. cumini* is hard and durable, used for agricultural implements and fuel. The bark of the tree is astringent due to tannins and is used against bronchitis, asthma, ulcers, and dysentery. The fruit juice and seed extract are hypoglycemic.

Tannins from *S. cumini* are found to inhibit the growth of fungi such as *Colletotrichium falcatum* and *Pyricularia oryzae* (Janardhanan et al., 1963).

The diseases reported in *S. cumini* are leaf spot caused by *Elsinoe kamatii* Tewari (Tewari, 1968), *Tripospermum juglandis* (Thumen) Hughes (Nath and Bhargawa, 1976), and *Phyllosticta eugeniae* Thirum. (Kamat and Kalani, 1964).

The leaf spot disease of *Syzygium* was found to occur in November and December. The diseased areas appear as small yellowish spots toward the base and tip of the leaf. The spots then grow in size and cover a large portion of the leaf blade, which becomes greyish yellow. The lesions are produced on both the surfaces of leaves but are more prominent on the upper surfaces. *Aspergillus niger* van Tieghum, the fungus isolated from the diseased portions of the leaf, was found to be the causative organism.

The flavonoid chemistry of the diseased leaves remained the same as that of the healthy leaves in containing 3′-OMe quercetin and myricetin. The healthy leaves contained vanillic, gallic, *p*-hydroxybenzoic, syringic, and *p*-coumaric acids, while in infected leaves, the *p*-coumaric acid is replaced by gentisic acid.

When phytoalexin was induced in the leaves of *S. cumini* with spores of the pathogen, the leachates contained a visibly colored pinkish brown compound (brown in UV, turning to pink with sodium carbonate spray and λ_{max}/MeOH 274, 318, 332, and 400 nm) was located in the diffusates. The color reactions and spectra denote this compound to be a benzoquinone. Studies on the characterization of this compound is in progress. There were no qualitative or quantitative differences in the diffusates when the leaves were treated with spores of the nonpathogen *F. solani*. Facilitated diffusion also showed the same result.

The diffusate from the pathogen-treated leaves containing the phytoalexin inhibited the mycelial growth of the pathogen at 1 and 5 ml dilution. A maximum of 35% inhibition of the mycelial growth was observed at 5 ml dilution of the diffusate. Spore germination and germ tube elongation of *A. niger* also was inhibited (87% and 48%, respectively) at 5 ml dilution (Abraham, 1989).

F. *Mangifera indica* L.

M. indica (Anacardiaceae) is the source of mango fruits. It is a tall evergreen tree cultivated not only for fruits but also for its timber and medicinal values. The wood is used for plywood and its bark for tanning. The ripe fruit is considered diuretic and laxative. The seed kernel is a medicine for asthma and diarrhea. Baked and sugared pulp of unripe fruits is recommended for cholera and plague patients.

Different parts of mango tree are known to suffer from a number of diseases. Apart from the fruit the leaves and inflorescences of the tree also are attacked by a number of pathogens. The important diseases are 1.) leaf spot caused by species of *Asterolibertina*, *Aureobasidium*, *Cercospora*, *Ciliochorella*, *Colletotrichium*, *Curvularia*, *Diplodia*, etc. (Mukherji and Bhasin, 1986), 2.) anthracnose by *Chaetomium mangiferae* Batista and Lima, and 3). powdery mildew by *Erysiphe cichoracearum* DC (Uppal and Patel, 1945).

The leaf spot disease found in Baroda is taken up for the study of phytoalexins. This disease occurs during January to March. The lesions are small in size, yellow–brown in color, and are seen scattered in the upper surface only.

Aspergillus niger van Tieghum was the fungus isolated from the diseased areas of leaves. The pathogenicity of this fungus is proved by tests.

The diseased leaves contained the same components, i.e., saponins, flavonols such as quercetin and quercetagetin, the xanthone mangiferin, and phenolic acids, vanillic, syringic, and *p*-hydroxybenzoic acids. There was no qualitative change between the diffusates of control and leaves treated with the pathogen. However, when the nonpathogen *F. solani* was used to examine the phytoalexin response, mangiferin was seen leaching out into the diffusate. This indicates the possible role of mangiferin as a phytoalexin. The antifungal activity of mangiferin against *Fusarium* has already been proved (Ghosal et al., 1977). It is found that the walls of *F. oxysporum* hyphae suspended in mangiferin were lysed within 72 hr. The mycelium turned black and the protoplasts contracted, got detached from the cell wall and were seen collected at one corner or the center of the cell. The seeds of safflower treated with mangiferin was resistant to *F. oxysporum*. In *Mangifera*, the content of mangiferin in flowers or shoots were very high when infected by *F. moniliformae* whereas the healthy flowers and shoots contained this compound in traces.

The increased production of mangiferin in response to *F. solani* may be the reason why this fungus is unable to infect the mango tree. It is also evident that mangiferin is less toxic to *A. niger* (Abraham, 1989).

G. *Anogeissus latifolia* Wall.

Anogeissus latifolia (Combretaceae) is a tall, magnificent tree found in the deciduous forests of western and southern India. This is one of the most important timber trees of Gujarat. It yields a gum, "gum ghatti," which is used as an adhesive, in calicoprinting, and in medicine. The leaves and twigs of *Anogeissus*, being tanniniferous, are used in the leather industry.

The fungus causing leaf spot in *A. latifolia* is reported to be *Pestalotiopsis versicolor* (speg.) Steyart (Agarwal and Ganguli, 1959). The disease, seen in Baroda, involves the appearance of small brownish spots in leaves, gradually spreading and ultimately drying up a large portion of the lamina. Sometimes the entire leaf is dried up. *Alternaria alternata* (Fr.) Kiessler was found to cause this disease.

The flavonols 3′4′-diOMe quercetin and 7,3′4′-triOMe quercetin were found in both infected and healthy leaves of *Anogeissus*. But the infected leaves also contained vitexin, a glycoflavone. The concentration of flavonoids was much higher in diseased leaves. The other phytochemicals, i.e., proanthocyanidins, tannins, and phenolic acids (vanillic and ferulic acids), also were common to both noninfected and infected leaves (Abraham and Daniel, 1990).

The diffusates from the leaves treated with spores of the pathogen and nonpathogen (*F. solani*) showed no qualitative differences from those of the control.

H. *Madhuca indica* Gmel.

M. indica (Sapotaceae), the source of mowra butter, is a large deciduous tree abundant in the forests of Gujarat and adjoining areas. Its wood is used for furnitures and construction work, and the fleshy sweet flowers are edible. A local liquor, "mowdi," is brewed from the flowers. The seed fat, mowra fat, is edible and used extensively for the making of margarines, chocolates, and soaps. This oil has medicinal importance due to its emollient properties and is used against skin diseases, rheumatism, piles, and hemorrhoids.

M. indica is infected by *Cylindrocladium scoparium* Morg (leaf spot; Nirwan and Singh, 1967), *Scopella echinulata* (Niessl.) Mains (leaf rust; Patil and Tirumalachar, 1971), and *Sphaceloma madhucae* Wani and Tirum. (leaf spot, Wani and Tirumalachar, 1971).

The spots observed here are small blackish brown patches growing to bigger patches covering a major portion of the leaf blade. The pathogenicity of *Colletotrichium gleosporoides* Penziz and Sacc., isolated from the lesions, was proved in the laboratory.

The healthy leaves of *Madhuca* contained a single flavonol (4′-OMe myricetin), phenolic acids such as *p*-hydroxybenzoic and vanillic acids, iridoids, and saponins. The infected leaves contained all these phytochemicals except for 4′-OMe myricetin which is replaced by myricetin. The diffusates from the leaves treated with spores of the pathogen and nonpathogen (*F. solani*) were alike in chemical constitution. However, when the pathogen-inoculated leaves were subjected to facilitated diffusion, the extract contained myricetin (Abraham, 1989).

I. *Heterophragma adenophyllum* Seem.

H. adenophyllum (Bignoniaceae) is a tall tree cultivated for its timber. The important diseases of this tree are 1.) powdery mildew caused by *Acrosporium* sp. (Patwardhan, 1966), 2.) leaf rust by *Phragmidiella heterophragmae* Tirum. and Mund. (Ling, 1951) and *Phyllospella stakmanii* Sathe (Sathe, 1965), and 3). leaf spot by *Sphaceloma heterophragmae* Wani and Tirum. (Wani and Tirumalachar, 1969).

The leaf spot disease prevalent in Baroda appears as brown spots which spread to the entire lamina of leaflet which in turn becomes cream-colored. The culture of diseased areas of leaf yielded *Botryodiploidia theobromae* Pat. The tests proved this organism to be the pathogen.

On analysis, the healthy leaves yielded 4′-OMe apigenin, benzoquinones, procyanidin, and syringic acid. The diseased leaves contained apigenin (in place of its methoxy derivative), the same quinones, procyanidin, and syringic acid. Drop diffusates with the pathogen contained apigenin (Chitra, unpublished). Further details are being investigated.

J. *Spathodea companulata* Beauv.

S. companulata (Bignoniaceae) is a tall, handsome tree cultivated for its beautiful red flowers. It yields a valuable red timber and the various parts of the plant are said to be useful in a number of ailments. *Glomerella cingulata* Spauld and Schrenk is reported to cause leaf spot disease in this plant (Ponnappa, 1967). But the black oval diseased lesions in leaf yielded *Curvularia prasadii* R.L. Mathur and B.L. Mathur, which is proved to be the pathogen.

The healthy leaves were found to contain 3′,4′-dimethoxyquercetin, vanillic acid, *p*-hydroxybenzoic acid, organic bases, and proanthocyanidins. The diseased leaves contained quercetin, vanillic acid, *p*-hydroxybenzoic acid, syringic and gentisic acids, bases, and proanthocyanidins. Quercetin is found to be produced when the healthy leaves were inoculated with the pathogen. Further studies are in progress (Chitra, unpublished).

K. *Zizyphus oenoplia* Mill.

Z. oenoplia (Rhamnaceae) is a straggling shrub, often semiscandent, producing edible drupes. The leaves are used against a wide variety of ailments such as tuberculosis, persistent fever, etc., and the seeds possess rejuvenating properties. Leaf rust due to *Catenulopsora zizyphi* Ramakr. & Subram. (Ramaskrishnan and Subramanian, 1946) is reported from this plant. The brownish diseased areas on the leaves of the plants in Baroda is found to be caused by *Alternaria alternata* (Fr.) Kiessler.

The healthy leaves of *Zizyphus* contained quercetin, 3′,4′-diOMe quercetin, vanillic acid, ferulic acid, *p*-coumaric acid, proanthocyanidins, alkaloids, and saponins. The diseased leaves contained quercetin in large amounts (but no methoxy derivatives) and the other phytochemicals of the healthy leaves (Abraham and Daniel, 1988b). The healthy leaves, when inoculated with the pathogen, produced quercetin in greater amounts.

I. *Carvia callosa* (Nees) Bremek.

C. callosa (*Strobilanthes callosus*, Acanthaceae) is a tall shrub growing profusely as an undergrowth in many deciduous moist forests in western India. The stem is used for the fibers it contains and the leaves exhibit expectorant properties. Leaf rust caused by *Aecidium carviae* Sathe (Sathe, 1966) is reported from Poona. The red-tinged diseased patches seen on the plants growing on Pavgadh hills near Baroda is found to be caused by *Aspergillus aculeatus* Lizuka.

The healthy leaves of *Carvia* yielded 7-OMe apigenin, 7-OMe luteolin, coumarins, vanillic acid, syringic acid, and proanthocyanidins. The diseased leaves contained 4′-OMe apigenin, 4′-OMe vitexin, 7-OMe luteolin, coumarins, vanillic acid, syringic acid, ferulic acid, and proanthocyanidins (Abraham and Daniel, 1988b). During pathogenicity trials it was also found that the diseased leaves produced 4′-OMe vitexin and ferulic acid.

4′-OMe vitexin is a glycoflavone in which 4′-OMe apigenin is linked to glucose by C-C bonds. This compound is water-soluble and is resistant to alkaline and acidic hydrolysis. Preliminary studies indicate that 4′-OMe vitexin is toxic to many fungi (Daniel, unpublished).

III. MEDICINAL PLANTS

A. *Adhatoda vasica* Nees

A. vasica (Acanthaceae) is a well-acclaimed medicinal plant of India. The leaves of this tall shrub yield two quinazoline alkaloids, vasicine and vasicinone, which are effective expectorants and bronchodilators. The whole plant is recommended for treatment of bronchitis, asthma, fever, and rheumatism.

A number of fungi are known to cause diseases in *Adhatoda*. Important pathogens reported are *Aecidium adhatodae* Syd. (Sydow et al., 1906) and *Chnoopsora butleri* Diet. and Syd. (Dietal, 1906) causing leaf rust; *Collelotrichium dematium* Grove (Roy, 1976) causing anthracnose; and *Cercospora adhatodae* Chowdhury (Chowdhury, 1948), *Colletotrichium capsici* Butler and Bisby, *Drechslera speciferum* Nicot. (Roy, 1976), *Phoma vasicae* Shreemali (Shreemali, 1972), and *Corynespora cassicola* Wei (Munjal and Gill, 1962) causing leaf spot.

The brown spot disease prevalent in plants in and around Baroda is caused by *Colletotrichium dematium* Groove. These spots enlarge as the infection proceeds and at times more than half of the leaf turns brownish black in color.

Both the healthy and infected leaves contained the flavonols kaempferol and quercetin, glycoflavones vitexin and isovitexin, and phenolic acids such as *p*-hydroxybenzoic, syringic, and *p*-coumaric acids. But the diseased leaves contain a flavone 7-OMe apigenin in addition. The amount of flavonoids and phenolic acids in the diseased leaves is much higher than in the noninfected ones. Between the two principal alkaloids vasicine and vasicinone, the latter was absent in the diseased leaves (Darshika and Daniel, 1992).

The production of 7-methoxyapigenin and the absence of vasicinone as a result of infection is proved when the healthy leaves were inoculated with the pathogen. But *Aspergillus niger*, the nonpathogen tested, failed to induce any chemical change in the leaves.

B. *Trianthema portulacastrum* L.

T. portulacastrum (Aizoaceae) is a succulent annual weed. The whole plant is used as a pot herb. The leaves are diuretic and used in edema and dropsy due to various causes. The roots possess cathartic and irritant properties.

Cercospora trianthemae Chiddarwar is reported to cause leaf spot in *Trianthema* (Chiddarwar, 1962). But the leaf spot disease occurring in Baroda is found to be due to *Fusarium* sp.

The healthy leaves of *Trianthema* contained 6,7-dimethoxy-3,5,4′-trihydroxyflavone, vanillic acid, *p*-hydroxybenzoic acid, and phytoecdysones. The diseased leaves, in addition to these compounds, contained quercetin and ferulic acid. By using drop diffusate technique it is found that the pathogen induces the formation of quercetin and ferulic acid. The nonpathogen *Aspergillus niger* failed to evoke any response in the healthy leaves (Darshika and Daniel, 1992).

Quercetin is found to inhibit spore germination of *Fusarium* at all concentrations (33% at 200 ppm, 40% at 400 ppm, and 70% at 1000 ppm). The germ tube growth also was found to decrease with increase in concentration of quercetin.

C. *Withania somnifera* Dunal

W. somnifera (Solanaceae) is a woody herbaceous perennial extensively used in different systems of medicine. It contains a number of alkaloids belonging to tropane, pyrrazole, or piperidine groups. Withaferins, a group of steroidal lactones, contribute to the bacteriostatic and antitumor properties of this plant. The whole plant is used as a rejuvenating tonic, sedative, and antiinflammatory drug. The roots are recommended for female disorders, rheumatism, dropsy, and in senile debility.

Leaf rust in *Withania* is caused by *Aecidium withaniae* Thuem. (Sydow, 1912) and leaf spot is by *Cercospora withaniae* H. & P. and *Colletotrichium capsici* Butler and Brisby (Pavgi and Singh, 1970). The black spot disease in the leaves observed in Baroda is found to be caused by *Chaetomium globosum* Kunze ex Fr.

Both the healthy and diseased leaves contained the same flavonoids (4′-OMe kaempferol and 3′4′-diOMe quercetin), phenolic acids, (vanillic, *p*-hydroxybenzoic, and ferulic acids), alkaloids, and withaferins. The concentration of flavonols and phenolic acids was very high in the diseased leaves. Both the pathogen and nonpathogen failed to yield any new compound when inoculated in healthy leaves (Darshika, unpublished).

D. *Tylophora asthmatica* W. & A.

T. asthmatica (*T. indica*, Asclepiadaceae) is a twiner exhibiting marked medicinal properties. The leaves of this plant contain a number of phenanthroindolizidine alkaloids, with tylophorine, tylocebrine, and tylophorinine being the principal ones among them. These alkaloids as well as the crude extract of the leaves possess powerful vesicant properties and therefore are used effectively in fighting asthma. Tylophorine has a paralyzing action on the heart muscle but a stimulating action on the muscles of blood vessels. All three principal alkaloids possess anticancer properties.

Cercospora tylophorina Rav. is reported from the diseased leaves of *Tylophora* (Rao, 1962). But the brownish black diseased spots on leaf are found to be caused by *Colletotrichium* sp.

The infected leaves contained quercetin in addition to kaempferol, gentisic, *p*-hydroxybenzoic, and caffeic acids present in the healthy leaves. The analysis of alkaloids in the diseased leaves indicated that they are devoid of the major alkaloid tylophorine. The leaves inoculated with pathogen also showed similar results. However, the nonpathogen *Aspergillus niger* failed to evoke any metabolic change in the leaves (Darshika, unpublished).

IV. *CUSCUTA* AND ITS HOSTS

Cuscuta is a leafless yellow twiner found parasitizing a wide variety of hosts. Out of the 150 species of this genus, only three are available in this part of the country, of which *C. reflexa* Roxb. and *C. chinensis* Lam. have been taken up for the present study. The hosts selected are *Cordia myxa* L. (Boraginaceae), *Streblus asper* Lour. (Urticaceae), *Clerodendron inerme* Gaertn. (Verbenaceae), and *Calotropis gigantea* Br. (Asclepiadaceae). The first three plants are hosts to *C. reflexa* and the fourth plant to *C. chinensis*. Both the leaves and bark of the hosts were analyzed for their phytochemicals and fungal pathogens. This study, though preliminary in nature, provides a number of interesting observations which would give insights to the defense reactions of various plants in question.

A. *Cordia myxa* L.

Two trees growing side by side, one infested with *Cuscuta* and the other noninfected, were taken up for chemical studies. The tree infected with *Cuscuta* possessed dark green leaves but seldom bore flowers or fruits, while the tree free of *Cuscuta* possessed pale green leaves and abundant flowers and fruits. Even the diseases of the two plants were different. The leaves of the *Cuscuta*-infested tree had small brown diseased spots caused by *Gleosporium* sp. while the other tree contained grey blisters on the leaf surface caused by *Botryodiplodia* sp.

On analysis, the leaves of noninfected *Cordia* yielded flavonols such as quercetin, 7,3′-diOMe quercetin, and 7,3′,4′-triOMe quercetin and phenolic acids like vanillic, *p*-hydroxybenzoic, melilotic, gentisic, and *p*-coumaric acids. They contained saponins and tannins also. The leaves of the tree with *Cuscuta* contained methoxyquercetins (7,3′-diOMe and 7,3′4′-triOMe) and syringic acid, ferulic acid, saponins, and tannins. The steroids of these two plants (which are being characterized) also are different. The study on the barks also yielded similar results (Julie, unpublished). The phytoalexin response of these plants is being investigated.

B. *Streblus asper* Lour.

In *Streblus* free of *Cuscuta*, the diseased areas appeared near the tip and margins of the leaves and were colored brown with a grey center. The culture of these spots yielded *Cladosporium* sp. The tree infected with *Cuscuta* had the diseased areas of the leaf colored brownish yellow on and around midrib, caused by *Chaetomium* sp. Flavonoids in both the leaves were in traces and could not be identified. The phenolic acids were also the same in both trees. There were differences in steroids between the two trees. The constituents of bark of the respective trees also were similar to those of leaves. The leaves of tree with *Cuscuta*, on inoculation with *Cladosporium*, produced quercetin and its derivatives in detectable amounts (Julie, unpublished).

C. *Clerodendron inerme* Gaertn

The plants with *Cuscuta* had white elliptic lesions in the lamina caused by *Alternaria* sp. whereas the *Clerodendron* resistant to this parasite had dark brown spots near the margin and near midrib caused by *Cladosporium* sp. The leaves of the former plant contained very low amounts of 4′-OMe apigenin and 7,4′-OMe apigenin while the leaves of latter plant produced large amounts of 4′-OMe apigenin and 7,4′-diOMe apigenin. The phenolic acids, steroids, and alkaloids were same in both the trees. The bark of

Cuscuta-infected plant produced scutellarein (a 6-hydroxyflavone), apigenin, and 4′-OMe apigenin, whereas the bark of the plant free of the angiosperm parasite possessed apigenin and 4′-OMe apigenin in traces. The steroids of both the plants were different (Julie, unpublished).

D. *Calotropis gigantea* Br.

The *Cuscuta*-infected plant possessed brown diseased spots on the leaves caused by *Gleosporium* sp. The plant free of *Cuscuta* was having brown patches with a pink center on the leaves caused by *Alternaria* sp. The leaves of the former plant produced methoxyquercetins and vanillic acid while the leaves of the latter plant contained quercetin, vanillic acid, ferulic acid, and *p*-hydroxybenzoic acid. The steroids, cardiac glycosides, and alkaloids were similar in both plants. The barks of the respective plants also were similar in chemical constitution (Julie, unpublished).

On chemical analysis, *C. reflexa* was found to possess chlorophylls (a and b) in traces, abundant carotenoids, 3–5% of waxes, saponins, alkaloids, and very little flavonoids. On the other hand, *C. chinensis* contained high amounts of flavonoids (quercetin and its methoxy derivatives), alkaloids, and negligible amounts of carotenoids, waxes, and saponins. Chlorophyll was absent from this species.

The results of studies conducted on all four hosts indicate that the plants resistant to *Cuscuta* are different chemically from the plants infested by the parasite. The major differences are in the flavonoids, phenolics, and triterpenoids. The resistant plants contained higher amounts of flavonoids always. Another interesting feature noted in the present study is the possible role of *Cuscuta* in deciding the resistance of the host part to certain fungal pathogens. The plants with *Cuscuta* are resistant to the fungal pathogens of the *Cuscuta*-resistant hosts. It is also observed that the leaves of *Cuscuta*-infected plant, when kept separate from *Cuscuta* and inoculated with the pathogen of the resistant plant, developed disease symptoms. *Cuscuta* plants, though very tender and possessing very little mechanical tissue, are resistant to fungal diseases. The disease-resistant factors may be higher wax content (in *C. reflexa*) or very high flavonoid content (in *C. chinensis*).

V. DISCUSSION

The reactions of the 16 plants to the fungal invasion can be grouped into three categories: 1.) demethylation of existing polyphenols, 2.) increase in concentration of phenolics, and 3). production of new compounds. In almost all cases the compounds converted/produced are phenolics, in which flavonoids form the major group followed by phenolic acids.

Demethylation of methoxyflavonoids is seen in seven plants: *Cassia*, *Morinda*, *Zizyphus*, *Madhuca*, *Trianthema*, *Spathodea*, and *Heterophragma*. Except for *Heterophragma* where flavones are located, all others produced flavonols, quercetin being the most common one. The preference to flavonoids may be explained by the fact that they are multipurpose compounds useful to the plant in a variety of ways such as UV screens, pigments, antimetabolites, and antimicrobials. It is known that the primitive angiosperms (Magnolidae) elaborated flavonols in leaves and during the course of evolution methoxylation is introduced, which is an advanced step seen in moderately advanced taxa (e.g., Rosidae). Reduction in flavonols and introduction of flavones, further advancements in flavonoid evolution, are seen in many of the Asteridae. Phenolics being antimetabolites, their increased concentration, especially flavonoids, in the cell sap may exert a harmful effect on the metabolism of plants. Therefore the mechanism in plants would have been to methylate these compounds to make them more lipophilic and store them in outer lipid layers of waxes and cutins. Such a placement of flavonoids does not interfere with their function as UV screens, antifeedants, or protective compounds. It has an added advantage that these compounds can be converted to more potent hydroxy derivatives and transport to the site of infection. This corroborates the observation that the leaves of *Citrus* cultivars resistant to Malesecco disease contain larger quantities of methylated flavonoids than the leaves of susceptible ones (Harborne, 1983).

The increase in the concentration of phenolics in diseased leaves is a well-known phenomenon and is seen in almost all the plants screened in the present project. In *Mangifera* and *Withania* the greater amounts of phenolics constitute the only detectable reaction the plant exhibits to invading fungi. These compounds are abundant on the tissues bordering diseased areas. Most of the diseased spots exhibit a red border indicating the accumulation of anthocyanins.

Production of new compounds is the mechanism of resistance expressed in *Tectona*, *Carvia*, *Anogeissus*, *Eucalyptus*, *Syzygium*, *Adhatoda*, and *Tylophora*. It is observed that each plant produces the typical compounds characteristic of the particular family to which it belongs. *Carvia* and *Anogeissus* produced glycoflavones which are otherwise seen in the primitive members of the Acanthaceae (Daniel and Sabnis, 1987) and Combretaceae. The Myrtaceae are known for their coumarins and quinones, and therefore *Eucalyptus* produced a coumarin and *Syzygium* a benzoquinone. Flavones are more common in the Acanthaceae and therefore *Adhatoda* could produce a flavone, while *Tylophora* belonging to the flavonol-rich Asclepiadaceae synthesized a flavonol. Though the Verbenaceae are known for their flavones and quinones, *Tectona* produced *p*-hydroxybenzoic acid, a phenolic acid widely distributed in the plant kingdom, but implicated in disease resistance. It is worth mentioning here that during drop diffusate studies, the leachates from the leaves kept as control contained simple phenol. This may indicate that during pathogenesis phenol would have been converted to *p*-hydroxybenzoic acid.

The studies on *Cuscuta* also emphasize the reliability of flavonoids as the resistant factors. The plants rich in flavones/flavonols are resistant to the attack of this parasite. The vulnerability of the hosts to the parasite may depend on the steroid constitution of the host also since these compounds are found to be different in the plants free of parasite and plants with parasite.

VI. EPILOG

The role of phenolics in the disease resistance of plants is proved beyond doubt. Some of these compounds act as phytoalexins and others work synergistically with the phytoalexins. The common flavonoids which are widely distributed in plant kingdom are definitely involved in disease resistance. Detailed studies on this aspect would be conducted in our laboratory.

More and more plants are to be screened before drawing conclusions on the distribution, properties, and metabolism of phytoalexins. The tendency to describe phytoalexins as the sole disease resistance factors is to be met with caution. It is definitely true that they form one of the very important resistant factors of plants. They have a major defensive role in a number of angiosperms but in many others they may have a supportive role. All these facts point to the highly complex nature of defense mechanism of plants.

Our future work on *Cuscuta* would be to find out whether this plant or any other angiosperm parasite induces phytoalexins on their hosts or not. The induction of resistant factors in the hosts of *Cuscuta* also is to be studied in detail. Another interesting feature to be probed is the nature of relationship between *Cuscuta* and its hosts. It may be a case of symbiosis also.

ACKNOWLEDGMENTS

I thank my research students (past and present) Joy Abraham, Darshika Parikh, Chitra Arya, and Julie Sebastian for permitting me to incorporate their unpublished results in this chapter.

REFERENCES

Abraham, K. J. (1989). Phytoalexins and other post-infectional compounds of some economically important trees of Gujarat. PhD thesis, M.S. University, Baroda.

Abraham, K. J., and Daniel, M. (1988a). Phytoalexins of *Cassia fistula* L. and *Morinda tomentosa* Heyne. *Natl. Acad. Sci. Lett. 11*(4): 101–102.

Abraham, K. J., and Daniel, M. (1988b). Phytoalexins and related post-infectional compounds of forest crops of Gujarat. *Adv. For. Res. India I*: 191–197.

Abraham, K. J., and Daniel, M. (1990). Chemical changes in the fungal infected leaves of *Zizyphus oenoplia* Mill and *Anogeissus latifolia* Wall. *Natl. Acad. Sci. Lett. 13*(7): 267–268.

Agarwal, G. P., and Ganguli, S. (1959). A leaf spot disease of *Anogeissus latifolia* Wall due to *Pestalotiopsis versicolor* (Speg) Steyaert. *Curr. Sci. 28*: 295–296.

Alcubilla-Martin, M. (1970). Fungus inhibitors in spruce bark. *Landwrit Forschung. 25*: 96–101.

Bakshi, B. K. (1974). Control of root disease in plantations in reforested stands. *Indian Forester 100*: 77–78.

Berkenkemp, B. (1971). Effect of coumarin on some pathogens of sweet clover. *Can. J. Plant Sci. 51*: 299–303.

Bilgrami, K. S. (1963). Leaf spot diseases of some common ornamental plants. *Proc. Natl. Acad. Sci. USA 33*(B): 429–452.

Borys, M. W., and Childers, N. F. (1964). The growth of *Fusicoccum amygdali* in vitro, as influenced by pH, phenolic compounds and unpurified extracts from peach twigs. *Acta Microbiologia Polond 13*: 211–220.

Chiddarwar, P. P. (1962). Contribution to our knowledge of the *Cercosporae* of Bombay state III. *Mycopathol. Mycol. Appl. 17*: 71–78.

Chigrin, V. V., and Rozuru, L. V. (1969). Changes in phenolic metabolism of spring wheat infected with stem rust. *Soc. Plant Physiol. 16*: 269–273.

Chowdhury, S. (1948). Some fungi from Assam III. *Indian J. Agric. Sci. 18*: 177–184.

DaCosta, E. W. B., and Rudman, F. (1958). The causes of natural durability in timber. I. The role of toxic extractives in the resistance of tallow wood (*Eucalyptus microcorys* F. Muell) to decay. *Aust. J. Biol. Sci. 11*: 45–57.

Daniel, M., and Sabnis, S. D. (1987). Chemosystematics of the Acanthaceae. *Proc. Indian Acad. Sci.* (*Plant Sci.*) *97*(4): 315–323.

Darshika, P., and Daniel, M. (1992). Changes in the chemical content of *Adhatoda* and *Trianthema* due to fungal diseases. *Indian J. Pharm. 54*(2): 73–75.

Debnam, J. R., and Smith, I. M. (1976). Changes in the isoflavones and pterocarpans of red clover on infection with *Sclerotinia trifoliarum* and *Botrytis cineria*. *Physiol. Plant Pathol. 9*: 9–23.

Dietal, P. (1906). Uber *Chnoopsora* eine neue Uridineen-Gattung *Ann. Mycol. 4*: 424.

Egawa, H., Tsutsui, D., Tatsuyama, K., and Hatta, T. (1977). Antifungal substances found in leaves of *Eucalyptus* species. *Experientia 33*: 889–890.

Gangulee, H. C., and Kar, A. K. (1985). *College Botany*. New Central Book Agency, Calcutta.

Ghosal, S., Biswas, K., Chakraborti, D. K., and Basuchaudhary, K. C. (1977). Control of *Fusarium* wilt of safflower by mangiferin. *Phytopathology 67*: 548–550.

Harborne, J. B. (1983). The flavonoids of the Rutales. *Chemistry and Chemical Taxonomy of the Rutales* (P. G. Waterman and M. F. Grundon, eds.), Academic Press, London, p. 149.

Harding, V. K., and Heale, J. B. (1981). The accumulation of inhibitor compounds in the induced resistance response of carrot root slices to *Botrytis cinerea. Physiol. Plant Pathol. 18*: 7–15.

Hart, J. H., and Willis, W. E. (1974). Inhibition of wood-rotting fungi by stilbenes and other polyphenols in *Eucalyptus sideroxylon. Phytopathology 64*: 939–948.

Henderson, S. J., and Friend, J. (1979). Increase in PAL and lignin like compounds as race specific responses of potato tubers to *Phytophthora infestans. Phytopathol. Z. 94*: 323–334.

Janardhanan, K. K., Ganguly, D., Baruah, J. B., and Rao, P. P. (1963). Fungitoxicity of extracts from tannin-bearing plants. *Curr. Sci. 32*: 226–227.

Kamat, M. N., and Kalani, I. K. (1964). Pathogenic fungi on *Syzygium cumini* Skeel. *J. Univ. Poona 28*: 122–124.

Kanujia, R. S., and Singh, C. S. (1978). Some new host records of parasitic fungi I. *Indian Phytopathol. 31*: 225–236.

Karadge, B. A., Chavan, P. D., and Thite, A. N. (1980). Changes in phenolic compounds of teak leaves induced by powdery mildew infection. *Indian Phytopathol. 33*: 114–116.

Li, C. Y., Lu, K. C., Trappe, J. M., and Bolle, W. B. (1972). *Poria weirii* inhibiting and other phenolic compounds in red alter and Doughlas fir. *Microbios 5*: 65–68.

Ling, L. (1951). Taxonomic notes on Asiatic smuts III. *Sydowia 5*: 40–48.

Mukherji, K. G., and Bhasin, J. (1986). *Plant Diseases of India: A Source Book.* Tata-McGraw-Hill, New Delhi.

Munjal, R. L., and Gill, H. S. (1962). Some dematiaceous Hyphomycetes from India. *Indian Phytopathol. 16*: 62–68.

Nath, V., and Bhargava, K. S. (1976). Some leaf spot fungi recorded in Goraphpur. *Indian Phytopathol. 29*: 461–462.

Nirwan, R. S., and Singh, B. B. (1967). Leaf spot of *Madhuca indica* caused by *Cylindrocladium scoparium. Indian Phytopathol. 20*: 169–170.

Osborne, L. D., and Thrower, L. B. (1964). Thiamine requirement of some wood-rotting fungi and its relation to natural durability of timber. *Trans. Br. Mycol. Soc. 47*: 601–611.

Patil, B. V., and Tirumalachar, M. J. (1971). Studies on *Elsinoe* and *Sphaceloma* diseases of plants in Maharashtra State (India). VII. *Sydowia 25*: 47–50.

Patwardhan, P. G. (1966). Some new records of powdery mildew fungi. *Plant Dis. Rep. 50*: 709–710.

Pavgi, M. S., and Singh, U. P. (1970). Parasitic fungi from North India. IX. *Sydowia 24*: 113–119.

Payak, M. M. (1953). Some new records of fungi from Bombay State. *Sci. Cult. 18*: 342–343.

Ponnappa, K. M. (1967). Some interesting fungi. I. Miscellaneous fungi. *Proc. Indian Acad. Sci. 66B*: 266–272.

Rajak, R. C., and Rai, M. K. (1982). Species of *Phoma* from legumes. *Indian Phytopathol. 35*: 609–612.

Ramakrishnan, T. S., and Ramakrishnan, K. (1949). *Chaconia tectonae* Ramakrishnan T.S. & K.Sp. Nov. on teak. *Indian Phytopathol. 2*(1): 17–19.

Ramakrishnan, T. S., and Sundaram, N. V. (1954). Additions to fungi of Madras. XVI. *Proc. Indian Acad. Sci. 40B*: 17–23.
Ramakrishnan, T. S., and Subramanian, C. L. (1946). On *Catenulopsora zizyphi* on *Zizyphys oenoplia* Mill. *Curr. Sci. 15*: 261–262.
Rao, V. G. (1962). Some new records of fungi imperfecti from India. *Sydowia 16*: 41–45.
Roy, A.K. (1976). Some new records of fungi on medicinal plants. *Curr. Sci. 45*: 464–465.
Salam, M. A., and Rao, P. N. (1958). Fungi from Hyderabad (India) III (*Oidium* sp.). *Indian Phytopathol. 11*: 126–129.
Sathe, A. V. (1965). Some new or revised species of *Physopella* (Uredinales) from India. *Sydowia 19*: 138–142.
Sathe, A. V. (1966). Some additions to rust fungi of India. *Mycopathol. Mycol. Appl. 29*: 333–338.
Shreemali, J. L. (1972). Two new pathogenic fungi causing diseases on Indian medicinal plants. *Indian J. Mycol. Plant Pathol. 3*: 84–85.
Shreemali, J. L., and Bilgrami, K. S. (1968). Range of variation in morphology of different isolates of *Botryodiplodia theobromae* Pat. *Indian Phytopathol. 21*: 351–360.
Sporoston, T., Jr. (1957). Studies on the disease resistance to *Impatiens Balsamina*. *Phytopathology 47*: 534–535.
Sridhar, R., Mohanty, S. K., and Anjaneyulu, A. (1979). In vivo inactivation of rice tungro virus by ferulic acid. *Phytopathol. Z. 94*: 279–281.
Swinburne, T. R. (1975). Microbial proteases as elicitors of benzoic acid accumulation in apples. *Phytopathol. Z. 82*: 152–162.
Sydow, H. (1912). *Fungi Indiae Orientalis* Pars. IV. *Ann. Mycol. 10*: 243–280.
Sydow, H., Sydow, P., and Butler, E. J. (1906). *Fungi Indiae Orientalis* Pars. I. *Ann. Mycol. 4*: 424–445.
Tewari, I. (1968). On the morphology and taxonomy of *Elsinoe kamati Sp. Nov.* on *Syzygium cumini* from India. *Nova Hedwigia 15*: 213–219.
Tirumalachar, M. J., and Chupp, C. (1948). Notes on some Cercosporae of India. *Mycologia 40*: 352–362.
Trappe, J. M., Li, C. Y., Lu, Y. C., and Bollen, W. B. (1973). Differential response of *Poria weirii* to phenolic acids from Douglas fir and red alder roots. *Forest Sci. 19*: 191–196.
Uppal, B. N., and Patel, M. K. (1945). Powdery mildew of Mango. *J. Univ. Bombay 9*: 12–16.
Walchii, O., and Scheck, E. (1976). Natural resistance of *Castanea sativa* wood to fungal attack and its cause. *Beih Mat. Organismen 3*: 77–89.
Wani, D. D., and Tirumalachar, M. J. (1969). Studies on *Elsinoe* and *Sphaceloma* diseases of plants in Maharashtra (India). VI. *Sydowia 23*: 257–260.
Wani, D. D., and Tirumalachar, M. J. (1971). Studies on *Elsinoe* and *Sphaceloma* diseases of plants in Maharashtra (India). VII. *Sydowia 25*: 47–50.
Yadav, A. S. (1963). Additions to the microfungi of Bihar I. Erysiphaceae. *Indian Phytopathol. 16*: 164–166.

25

Possible Role of Phytoalexin Inducer Chemicals in Plant Disease Control

Asoke Kumar Sinha
Bidhan Chandra Krishi Viswavidyalaya, Kalyani, West Bengal, India

I. INTRODUCTION

The main objective of plant pathology is to control an increasing number of diseases and to do so with maximal effectiveness. Since suitable resistant varieties are neither always available nor available against all kinds of pathogens, the grower often is forced to resort to chemical control based on the use of chemicals with direct toxic action on the pathogen. Such control is mainly available against fungi, which constitute the most important group among the plant pathogens. However, the lack of suitable fungicides against some of the major fungal pathogens and particularly the ecological hazards to which their regular and large-scale use may lead have worried concerned scientists, prompting them to look for some suitable and safer alternative approaches to plant disease control.

Most plants have a versatile, multicomponent defense, adequately equipped to provide them protection against most of their potential pathogens; only a few of them can overcome this defense and cause disease. The general assumption is that varieties within a host species are resistant when they possess one or more resistant genes and susceptible when they lack any such gene. Plant defense primarily depends on some need-based dynamic responses to attempted infection, mostly an inducible phenomenon, its qualitative and quantitative aspects being regulated by signals from the invading pathogen. Even a susceptible host variety is neither completely defenseless nor totally lacking in any genetic information for resistance. Even such a plant has its latent defense potential, which finds expression at certain growth stages and under certain stress conditions created during cropping. This naturally led to the feeling that by manipulation of cropping conditions or by creating certain stresses, it may be possible to activate this latent potential and put it into operation. In the background of great success achieved with active immunization, based on specific antigen–antibody reaction, in controlling some of the most serious human and animal diseases, this would not seem impossible, particularly because much similarity exists between the basic life processes in plants and animals. It is now well established that something analogous to active immunization of the vertebrates also occurs in plant systems, though its components and regulation may differ from those in animals (Kuć, 1982, 1987; Sequeira, 1983).

Evidence is now available to conclude that effective resistance can be developed in susceptible plants against their fungal, bacterial, and viral pathogens by the use of both biotic and abiotic agents, i.e., by 1.) prior inoculation with the same, related, or even unrelated pathogens of homologous or heterologous kind; 2.) prior treatment with cell-free microbial fluids containing their metabolites and/or cell constituents; or 3.) treatment with many synthetic compounds (Price, 1964; Matta, 1971, 1980; Wain and Carter, 1972; Goodman, 1980; Langcake, 1981; Lazarovits, 1988; Sinha, 1990, 1992). Resistance acquired by susceptible plant on treatment with various biotic agents was, in many instances, systemic in nature, lasting in effect and also broad spectrum in action, i.e., effective against more than one pathogen, may be of different kinds (Kuć, 1982, 1987; Mucharromah and Kuć, 1991). Particularly impressive had been the findings on tobacco (Matta, 1980; Kuć and Tuzun, 1983), cucumber (Jenns et al., 1979), and melons (Caruso and Kuć, 1977) where field level, long-term, systemic protection could be induced by and were effective against a broad range of fungi, some bacteria, and many viruses (Goodman, 1980; Hamilton, 1980). However, this approach had little acceptance from the growers probably because of heavy costs involved in the laboratory production of biotic inducer agents and also the logistic difficulties faced in delivering the agent to the crop, particularly when the cropped area is large.

The phytoalexin concept of Müller and Borger (1940), further developed with the incorporation of new findings over the next 50 years (Cruickshank, 1963, 1980; Dixon, 1986; Van Etten et al., 1989), probably provides the most rational explanation and also a basis for the dynamic defense of many plant species, primarily against their fungal and only occasionally against bacteria and nematode pathogens as well (Mansfield, 1982). Rapid postinfection production of phytoalexin-type antipathogenic principle in the affected tissue and its rapid accumulation to a toxic level results in resistant reaction. Such high-level accumulation of phytoalexin is linked with incompatible host–pathogen interaction. Though not universal in occurrence, phytoalexin accumulation is now accepted as one of the major dynamic defense mechanisms of the plant. Besides the pathogen itself and other infectious entities, 1.) its metabolites and cell constituents, 2.) some plant products, 3.) some synthetic compounds including even some fungicides and toxins, and also 4.) some stress factors can induce phytoalexin synthesis. Those in 1.) are termed as biotic "elicitor." While their presence and involvement in phytoalexin production have been reported in many cases, only a few could be isolated so far from the pathogen and characterized. More significant appears to be the fact that many synthetic compounds having no link either with the pathogen or host also share this ability. Many prefer the term *abiotic elicitor* for this group of compounds (Yoshikawa, 1978; Darvill and Albersheim, 1984), but for the purpose of this chapter the term *inducer* will be used to distinguish them from elicitors of biotic origin. Feelings have often been expressed that phytoalexins or their biotic elicitors can be directly used on host plants for disease control, particularly against those of fungal origin, but the exploratory studies yielded no significant information. However, during the last 50 years, many synthetic compounds of diverse nature had been tested on various crops, and with some of them considerable amelioration of disease symptoms was noticed, often linked with significant changes in host metabolism (Van-Andel, 1966; Langcake, 1981; Lazarovits, 1988; Sinha, 1992). At Bidhan Chandra Krishi Viswavidyalaya, Kalyani, we have continued exploratory work for more than a decade with a large group of synthetic compounds that are known phytoalexin inducers in the hope that these may trigger some active defense responses in the host plant including phytoalexin accumulation. Results in general have indicated great promise and may suggest an alternative approach to conventional chemical control. These findings are presented here in the background of what could so far be achieved in this direction with phytoalexins, their biotic elicitors, and synthetic chemicals with little involvement in the natural defense responses of plants.

II. PLANT DISEASE CONTROL BY NONCONVENTIONAL CHEMICAL AGENTS

Besides phytoalexins, the known resistance factor produced in plants in response to attempted infection by potential pathogens, mostly fungi, pathogen metabolites, or cell constituents that trigger host responses leading to phytoalexin production, i.e., elicitors, and miscellaneous synthetic compounds with or without a link to biosynthesis of phytoalexins in plants have been tested on many plant species with a view to exploring whether or not these biotic or abiotic agents may protect them from their pathogens. These findings are presented in the following pages and their implications, if any, discussed.

Phytoalexins

In spite of some anomalous results, there is now enough evidence to accept phytoalexins as a primary factor in plant defense, mostly against their fungal pathogens. They are neither universal in occurrence nor responsible for host defense in all cases. Rapid production of phytoalexin and its early accumulation to a toxic level at the site of infection is now known to be the basis of a major dynamic defense mechanism in many plants. About 300 phytoalexins have been characterized so far, mostly from dicotyledonous plants, and many more, for which good evidence is available, still remain to be characterized. Phytoalexins show great diversity in form, e.g., isoflavonoids, sesquiterpenoids, furanoterpenes, diterpenes, polyacetylenes, stilbene, etc. Some plant species produce more than one phytoalexin when infected. In a particular host–pathogen interaction, generally one phytoalexin plays a dominant role. If the phytoalexin is removed from the affected tissue, the normal resistance reaction is changed to a susceptible reaction (Klarman and Gerdemann, 1963). Similarly, if phytoalexin is added to the site of infection, there is a change from susceptible to resistant host reaction (Chamberlain and Paxton, 1968). Because of their broad-spectrum fungitoxic action and diverse forms, a logical consequence of these studies was to explore if phytoalexins could be directly utilized to protect plants from disease. However, few attempts were made in this direction and these did not indicate much promise in this respect. Weyrone could provide only some protection to French bean from rust (*Uromyces* sp.) and broad bean from chocolate spot (*Botrytis fabae*) (Fawcett et al., 1969) diseases. Ward et al. (1975) reported good control of late blight to tomato (*Phytophthora infestans*) with capsidiol but the protective effect lasted only up to 8 days. Viniferin also protected vines from downy mildew (*Plasmopara viticola*) for only 7 days (Langcake, 1981). A direct comparison between seven isoflavonoid phytoalexins and two widely used fungicides, benomyl and macozeb, led Rathmell and Smith (1980) to conclude that phytoalexins have little commercial potential for direct use in plant disease control like fungicides. This does not seem to be unlikely in view of their mild toxicity, unstable nature, and lack of movement in the plant tissue. Further, their complex ring structure does not make them suitable for easy isolation from the reacting host tissue and low-cost synthetic processes.

Phytoalexin Elicitors

Elicitors are molecules of biotic origin that signal plants to begin the process of phytoalexin biosynthesis. The fact that, besides the live inoculum of pathogen, its cell-free mycelium extract, spore germination fluid, culture filtrate, and so forth also induce phytoalexin synthesis makes this obvious. Since this is an area of much activity, many reviews are available (Keen, 1975; Yoshikawa, 1978; Bailey, 1982; Darvill and Albersheim, 1984; Dixon, 1986). Though the occurrence and functioning of elicitor-active compounds have been indicated for many plant–pathogen systems, only in a small number of cases could elicitors be isolated and characterized. It has been found that glucans, mostly those with branched structure, and glycoproteins constitute the most common types of elicitors. Besides these, polypeptides, lipids, lipopolysaccharides, lipoglycoprotein, fatty acids, and chitosan, the deacetylated derivative of chitin, are also elicitor-active. Some of these are extremely active at very low (10^{-10} M) doses. Though the occurrence of race-specific elicitors have been claimed for *Rhizopus stolonifer* (Stekoll and West, 1978) and *Phytophthora megasperma* f.sp. *glycinea* (Keen and Legrand, 1980), the evidence in support is not conclusive. Some enzymes of fungal origin, e.g., endopolygalacturonase from *R. stolonifer* and endopolygalacturonate lyase from *Erwinia carotovora*, and others of plant origin, e.g., chitinase and β-endoglucanase, also participate in the elicitation process, though not directly.

At the initial stage of elicitor-based research, it was felt that there might be some scope for their use in plant disease control. Though a hepta- and octasaccharide obtained from mycelial wall of *P. megasperma* f.sp. *glycinea* with elicitor activity could be synthesized (Sharp et al., 1984), because of the large molecules and often branched structures involved, elicitor synthesis cannot be easy and must be very expensive. Further, when foliar spray with some of the elicitors was tested as a possible method for plant disease control, no promise was indicated. This discouraged further attempts in this direction.

Miscellaneous Synthetic Compounds

Concurrent with studies to unravel the mysteries of host defense against plant pathogens have been attempts to study diverse, mostly nontoxic or mildly toxic synthetic compounds with a view to their potential applications in plant disease control. Although initial studies in this direction date back to the early part of this century, most serious studies were made during the last three decades (Dimond, 1963; VanAndel, 1966; Wain and Carter, 1972; Langcake, 1981; Lazarovits, 1988; Sinha, 1990, 1992). A variety of compounds with known or no effect on plant systems, particularly metal salts, amino acids, plant growth regulators, and miscellaneous compounds, had been tested for this purpose, mostly on major crop plants against their important fungal pathogens, and impressive results were recorded with some of them. The main idea was to look for nonhazardous but effective compounds that can possibly replace the conventional toxic agents in plant disease control. A general idea is given below, by group, of the protective effects achieved with such compounds.

Metal Salts

Among the metallic compounds, copper, zinc, and lithium salts and boron compounds have shown particularly good effect against many important crop diseases, e.g., lithium salts on cereals against powdery mildews (systemic) and rusts; copper salts against wheat rusts, brown spot of rice, and *Verticillium* wilt of cotton; zinc salts against leaf rusts of wheat and rust of linseed (Wain and Carter, 1972). Cupric chloride–induced resistance in cotton against *Verticillium* wilt was associated with increased phytoalexin production (Bell and Presley, 1969). Subramaniam (1963) achieved complete control of pigeon pea wilt (*Fusarium udum*) by seed treatment, foliage spray, or soil drench with 100 ppm manganese sulfate. Addition of micronutrients have often led to suppression of soil-borne pathogens, particularly those causing vascular wilts.

Amino Acids

Among the amino acids screened for their possible suppressive effect on plant diseases, some showed distinct promise, e.g., phenylalanine against apple scab (*Venturia inaequalis*); serine and threonine against cucumber scab (*Cladosporium cucumerinum*); serine against chocolate spot of bean (*Botrytis fabae*); and some sulfur-containing amino acids against pea root rot (*Aphanomyces euteiches*) (VanAndel, 1966; see references). Effective control of *Fusarium* wilt of tomato was achieved with root application of methionine (Jones and Woltz, 1969) and phenylalanine (Carrasco et al., 1978). In the later treatment, plants developed much increased phenol level. Promising results were recorded with some amino acid derivatives also. Presowing seed treatment with dodecyl-DL-alaninate provided rice plants long-lasting systemic protection from blast disease (Arimoto et al., 1976). Treatment with esters of *N*-allylglycine and *N*-allylsarcosine led tomato plants to acquire strong resistance to *Fusarium* wilt; they responded to infection with increased levels of phenolics and stimulated peroxidase activity (Kirino et al., 1980).

Plant Growth Regulators

Plant growth regulators have been particularly effective against vascular wilts, e.g., indoleacetic acid (IAA) 2,4-dichlorophenoxyacetic acid (2,4-D), naphthalene acetic acids (NAA), and 2,3,5-triiodobenzoic acid (TIBA) against *Fusarium* wilt of tomato; IAA and naphthalene acetamide (NAM) against *Verticillium* wilt of the same plant; 2,3,6-trichlorophenoxyacetic acid (2,3,6-T), and halogenated benzoic acids against Dutch elm disease (*Ceratostomella ulmi*); and benzoic acids, IAA, and TIBA against oak wilt (*Ceratocystis fagacearum*) (Wain and Carter, 1972). In all such effective treatments, plants showed inhibited growth. It has been suggested that these treatments led to a transformation of cell wall pectic compounds from their methylated state to calcium/magnesium pectates that are more rigid and less amenable to pectic enzyme action. Various growth retardants like cycocel (chlorocholine chloride), Phosphon-D, *N,N*-dimethylpyrrolidinium iodide, and *N,N*-didimethylpiperidinium iodide also inhibit the development of *Fusarium* or *Verticillium* wilt of tomato or both (Sinha and Wood, 1964; Buchnauer, 1971; Buchnauer and Erwin, 1973; Erwin, 1978). Resistance induced by the above iodides was associated with increased formation of a phytoalexin-type compound in *Verticillium* wilt–affected cotton plants. Some growth regulators also act against diseases other than wilt, such as chocolate spot of broad bean (*Botrytis fabae*), wheat rusts, and cucumber scab (Wain and Carter, 1972; see references).

Miscellaneous Organic Compounds

Some of the many synthetic compounds tested on various crops have shown promising results against their diseases. Primarily those cases where there is evidence or at least the suggestion that the induced protective effect in plants is mediated through changes in host metabolism merit consideration, and the details are presented here. Phenylthiourea effectively protects cucumber from scab (*C. cucumerinum*), and the protected plants respond to infection with increased PO activity and enhanced lignification (Sijpesteijn, 1969). With probenazole treatment, plants achieve substantial protection from both blast (Watanabe et al., 1979) and bacterial leaf blight (*Xanthomonas oryzae*) (Tomiku et al., 1979). When challenged with the pathogen, plants responded with higher accumulation of fungitoxic substances (Shimura et al., 1981) and increased PO activity and lignification (Iwata et al., 1980) as compared to the untreated plants. There is also the report that rice plants treated with WL28325, a prophylactic fungicide, get protected from blast, and this is associated with increased production of phytoalexins (momilactones) in response to infection (Cartwright et al., 1977, 1980). Ghosal and Purkayastha (1974) reported that treatment with gibberellic acid was effective in providing protection to rice plants which, in its protected state, developed much higher levels of PO as compared to the untreated plants.

Ethylene itself (Pegg, 1976) or its precursor, ethephon (Retig, 1974) effectively checks *Fusarium* wilt of tomato. Such protected plants also showed much increase in PPO and PO activities. Fosetyl-Al (TEPA = aluminum tris-*O*-ethyl phosphonate) has been found to be very effective against different diseases, particularly those caused by fungal pathogens, such as downy mildew of grapes (*Plasmopara viticola* (Lafon et al., 1977); *Phytophthora* infection of tomato, tobacco, pepper, and citrus (Vo-Thi-Hai et al., 1979; Bompeix et al., 1980; Gutter, 1983); and *Phomopsis* infection of grapes (Bertrand et al., 1977). No phytoalexin accumulated in the susceptible plants in the presence of Fosetyl-Al alone or when infected in its absence, but more than the normal levels of phytoalexin accumulate in the treated plants. Phosphorous acid, a breakdown product of fosetyl-Al, is more effective in this respect. Fenn and Coffey (1985) reported that fosetyl-induced protection coincided in time with the appearance of necrotic symptoms, accumulation of phosphorous acid, and increased PO activity. Application to the treated plants of α-aminooxyacetic acid, an inhibitor of aromatic biosynthesis and phenylpropanoid pathway, partially neutralizes the induced protective effect. Sodium azide treatment has been found to protect soybean plants from *Macrophomina phaseolina* infection (Chakraborty and Purkayastha, 1987). Similarly, cloxacillin and penicillin protect the same plant from *Colletotrichum dematium* infection (Purkayastha and Banerjee, 1990). In both sodium azide and cloxacillin treatment, but not with penicillin, the effect was associated with a change in the antigenic pattern of the host toward a reduced commonness between the host and the virulent pathogen isolate. In sodium azide treatment plants also responded to inoculation with *M. phaseolina* with increased phytoalexin production. A recent report states that a spray with oxalate or dibasic/tribasic potassium phosphate induces strong resistance in cucumber plants to four fungal, two bacterial, and two viral pathogens (Mucharromah and Kuć, 1991). However, the chemically induced protective effect was not as strong as that induced by prior inoculation with *Colletotrichum lagenarium*.

The above experimental findings show that many synthetic compounds without any evident toxicity to the pathogen can induce strong resistance in a host plant not only to one but sometimes to more than one pathogen, may be of different kinds, and may also be effective on more than one plant species. Such an induced protective effect is systemic in nature and has often been shown to be long-lasting in effect.

Synthetic Phytoalexin Inducer Compounds

Results presented so far, though often very significant and also suggestive in nature from the perspective of plant disease control, mostly involved one, rarely a few chemicals applied to a plant host. There was little justification in most cases for selecting and using a particular chemical on a particular host against a disease. In that context, the studies conducted in our laboratory are an exception. Studies conducted here for more than a decade (and still continuing) have involved a large group of synthetic compounds, tested on some important crop plants against some of their major fungal pathogens with different levels of parasitic specialization, are more broad-based in nature, and provide comprehensive information in the area of induced resistance in plants as a possible disease control measure. This is probably the first time that the test chemicals were used for some of their known properties, i.e., the ability to induce the production of phytoalexin-type resistant factor in plants (Uritani et al., 1960; Uehara, 1963; Condon et al., 1963; Perrin and Cruickshank, 1965; Schwochau and Hadwiger, 1968; Hadwiger, 1972). These include heavy metal salts, amino acids, plant growth regulators, metabolic inhibitors, reducing agents, antibiotics, drugs, and so forth. Since phytoalexins are now accepted as resistance factors having a major role in host defense in many plant–fungus interactions, it was felt that their induced production in (susceptible) plants by treatment with inducer chemicals might also trigger a series of metabolic responses that do not favor the pathogen. Such induced resistance may then become useful in plant disease control in the same way that immunization is effective against human and animal diseases.

The idea of exploring the possibility came from early studies in BCKV on biological induction of resistance in rich plants to brown spot disease incited by *Helminthosporium oryzae*. It was noted that optimum protection from *H. oryzae* was achieved when 3- to 4-week-old plants were inoculated with the mildly virulent isolate of this pathogen 2 days before challenge inoculation (Sinha and Trivedi, 1969; Sinha and Das, 1972). A mild isolate of *H. oryzae* had the strongest effect as inducer agent as compared to four other species of *Helminthosporium* or two minor pathogens of rice, namely, *Cercospora oryzae* and *Nigrospora oryzae* (Trivedi and Sinha, 1976a). Of the various kinds of fungal fluids, e.g., culture filtrate, mycelial extract, spore germination fluid, etc., that can also be used as a foliage spray to induce resistance in rice plants; maximum effect was achieved with spore germination fluid of the mild isolate, obtained from spore suspension with 3×10^6 conidia/ml, applied 2 days before challenge inoculation (Trivedi and Sinha, 1976b; Mukhopadhyay and Sinha, 1980). This was as effective as the optimum concentration of live inoculum ($\sim 5 \times 10^5$ conidia/ml). It was strikingly evident from these studies that, with minor exceptions, the magnitude of resistance induced by different treatments had good correlation with the level of fungitoxicity postinfectionally developed in the treated plants after challenge with the virulent isolate. The toxicity initially developed following the treatment was less significant in terms of both concentration and the final effect. Such observations suggested that the initial treatment with the live inoculum of a biotic agent or fluid containing its metabolites not only induced the accumulation of phytoalexin-type resistance factor(s) in the host tissue that might function as a passive defense causing some initial limitation to the challenger pathogen, but must have also activated the defense potential that usually remains latent in a susceptible host. In this background, testing of phytoalexin inducer compounds on

Table 1 Magnitude of Protective Effect (Resistance) Induced in Selected Crop Hosts Against Some of Their Major Pathogens by Treatment with Phytoalexin Inducers[a]

Treatment	Rice		Wheat	Chili	Chickpea		Groundnut	
	Brown spot (1)	Blast (2)	Wheat leaf spot (3)	Chili Damping-off (4)	Fusarium wilt (5)	Collar rot (6)	Early leaf spot (7)	Stem rot (8)
Ferric chloride	+++	++++	++	–	–	+	–	++
Cupric chloride	+++	++++	++++	–	++	++	+	++
Cupric sulfate			++++	–		++	+	++
Cadmium chloride	+++	+++	+++				–	–
Barium sulfate	+++	++	++++	–		++	+++	++++
Barium chloride	+++	++	++++	+++		+	–	++++
Zinc chloride	–	–	++++	–	++	++	++	++++
Zinc sulfate	–	–	++++			+	++	++++
Lithium sulfate	++++	+++	+++			+	+++	++++
Boric acid	+++			+++				+++
Mercuric sulfate	–	–	+++		++++	+++	–	–
Mercuric chloride			++++				+++	+
L-Cysteine	+++	+++	+++	++		+++	++	++++
Thioglycollic acid	++++	+++	++++	+++		+++		–
Sodium selenate	+++	+++	–				+	
Sodium selenite	++++	+++	++	+++			+	
Sodium sulfite	–	+++	++++	+++				

Cycloheximide	++++	++++	+++	+++	++++		+++	++++
Sodium malonate	+++	++++	+++	++	++			−
Sodium molybdate	+++	+++	++	+++				−
Sodium iodoacetate	+++	+++	++					
Nickel chloride						++		
Nickel nitrate	+++	++	++++		++			
Sodium fluoride	−	+++	++++	−	−			
p-Chloromercuricbenzoic acid	+++	++++	+++	+++			+++	−
Sodium azide	++		+++		−		++	−
Chitosan	++++	++++	−	−	++++		+++	++++
Indoleacetic acid	+++	+++	++++	++	++++		+++	++++
2,4-Dichlorophenoxyacetic acid	+++	+++	+++	+++	++++		+++	++++
2,4,5-Trichloro-phenoxyacetic acid	+++	−			++++		+++	++++
Cycocel	+++	−	+++		++++		+++	++++
DL-Phenylalanine		+++			++++		−	++++
DL-Methionine	+++	+++			++++			−

[a]Relative levels of protection achieved against crop diseases have been shown as follows: less than 30% reductions in symptoms/plant mortality = +, 30–49% reductions = ++, 50–69% reductions = +++, and 70% and above = ++++.

crop plants against their fungal pathogens seemed worth exploring. Detailed observations are given disease-wise. Table 1 gives an approximate idea of the magnitude of protective effect achieved against different crop diseases with different compounds and Table 2 summarizes the nature of changes in symptom expression achieved with strongly effective compounds and the associated changes in host metabolism.

Brown Spot of Rice (Helminthosporium oryzae)

More than 50 compounds, mostly known phytoalexin inducers and a few related compounds, had been tested on susceptible rice plants in different pot experiments for their possible effectiveness against artificial inoculation with the brown spot pathogen. In the initial experiments, plants were spray-treated with aqueous solutions of different chemicals, most at 10^{-4} M, and a few at lower doses (10^{-5} or 10^{-6} M). The concentrations initially tested were close to their optimum for phytoalexin induction in plants. Such treatment, administered 2 days before challenge inoculation at the age of 3 weeks by spraying conidial suspension to the leaves till dripping, resulted in appreciable reductions in leaf symptoms in most cases. Rarely, a compound totally failed. Only those compounds which could reduce symptoms moderately to considerably were further tested at a range of three concentrations, most at 10^{-3}, 10^{-4}, and 10^{-5} M, and cycloheximide and growth regulators at 10^{-4}, 10^{-5}, and 10^{-6} M. Of the three methods of application initially tested—i.e., presowing seed treatment by soaking seeds in chemical solution for 24 hr, 24-hr rootdip of seedlings in chemical solution before transplanting to field plots, and foliage spray 2 days prior to artificial inoculation—wet seed treatment appeared to provide a more significant as well as a more persistent protective effect. In all subsequent studies seed treatment was used as the preferred mode of application.

In seed treatment, compounds such as lithium sulfate, sodium selenite, thioglycollic acid, L-cysteine, *p*-chloromercuricbenzoic acid (*p*-cmb), and cycloheximide showed substantial protective action against brown spot. These reduced leaf symptoms by more than 70%. Compounds which showed somewhat less pronounced but still very significant effects and reduced symptoms by more than 50% and up to 70% included ferric chloride, nickel nitrate, cadmium chloride, barium chloride, sodium molybdate, sodium iodoacetate, IAA, 2,4-D, DL-methionine, and DL-*nor*-leucine (Giri and Sinha, 1983a, b). With many of the above compounds, their induced protective action appeared to be concentration-independent and showed little relation to their in vitro fungitoxic action, if any. These treatments adversely affected both the components of symptom expression with a substantially reduced lesion number and reduced lesion size. Treated plants developed comparatively fewer lesions of large size groups and many more

Table 2 Summary Observations on the Effects of a Diverse Group of Phytoalexin Inducer Chemicals, Used in Presowing Seed Treatment on Selected Crop Diseases

Compounds with very strong protective action	Host/pathogen (disease)	Postinfection effects of significance	Ref.
Li-sulfate, Na-selenite, L-cysteine, thioglycollic acid, cycloheximide, *p*-Cmb, Na-malonate, chitosan	Rice/*H. oryzae* (brown spot)	a) Many fewer lesions of reduced mean size	Sinha and Hait (1982)
		b) Significant increase in accumulation of phytoalexin-type compound	Giri and Sihna (1983a,b) Das (1992)
		c) Significant increases in total phenol, lignin, and protein contents and in PO activity as in resistant plants	Hait and Sihna (1987) Das (1992)
		d) Changes in antigenic pattern	Hait (1982)
Ferric chloride, cupric chloride, cycloheximide, Na-malonate, *p*-Cmb. 2,4-D, chitosan	Rice/*P. oryzae* (blast)	a) Many fewer lesions of reduced mean size, no coalescence of lesions	Sinha and Sengupta (1986) Das (1992)
		Significant increases in accumulation of fungitoxic substance(s), phenol and protein contents, and PO activity	Das (1992)
Cu-chloride, Cu-sulfate, Zn-chloride, Hg-chlroide, Hg-sulfate, Ba-chloride, Ba-sulfate, Ni-nitrate, Na-sulfite, Na-fluoride, and IAA	Wheat/*H. sativum* (leaf spot)	Many fewer lesions of reduced mean size	Hait and Sinha (1986)
		Moderate increase in the level of fungitoxic substances	Chakraborty and Sinha (1989)

Table 2 Continued

Compounds with very strong protective action	Host/pathogen (disease)	Postinfection effects of significance	Ref.
Ba-chloride, Boric acid, Na-sulfite, Na-molybdate, Na-selenite, Thioglycollic acid, 2,4-D,cycloheximide, Na-iodoacetate	Chili/*Pythium* sp. (damping-off)	Substantial reductions in seedling mortality	D. Sahana, unpublished information
Hg-sulfate, Cycloheximide, IAA, 2,4-D, 2,4,5-T, cycocel, chitosan, DL-phenylalanine, DL-methionine	Chickpea/*F. oxysporum* f.sp. *ciceri* (vascular wilt)	a) Marked reductions in disease symptoms and plant mortality b) Suppression of vascular colonization	Sahana (1991)
		Significant increases in total phenol, *O*-dihydroxyphenol, and protein contents, and in PPO and PO activities	Sahana, 1991 Chowdhury, 1992
Thioglycollic acid, Cu-chloride, Cu-sulfate, Zn-sulfate, Hg-nitrate	Chickpea/*S. rolfsii* (collar rot)	a) Marked reductions in plant mortality b) Significant reductions in pectolytic enzyme (PG and PMG) activity of the pathogen in the affected tissue	Sahana, 1991
2,4,5-T, Chitosan, Zn-chloride	Sugarbeet/*S. rolfsii* (root rot)	Marked reductions in plant mortality; and increase in root yield/ha and sugar yield/ha	S. Das unpublished information

Ba-chloride, Ba-nitrate, Zn-chloride, Zn-sulfate, Li-sulfate, IAA, 2,4-D, 2,4,5-T, cycocel, L-cysteine, DL-phenylalanine, cycloheximide, chitosan	Groundnut/*S. rolfsii* (stem rot)	a) Marked reductions in plant mortality	Acharya (1989)
		b) Reductions in pectolytic enzyme activity in affected tissue	Chowdhury (1992)
		c) Accumulation of calcium and magnesium and enhanced lignification	Chowdhury (1992)
		d) Increases in total phenol, *O*-dihydroxyphenol and total protein contents, and PPO and PO activities	
Ba-sulfate, Zn-sulfate, Zn-chloride, Li-sulfate and Hg-chloride, L-cysteine, chitosan, *P*-Cmb, cycloheximide, IAA, cycocel	Groundnut/*C. arachidicola* (early leaf spot)	a) Fewer lesions of reduced mean size coalescence between spots rare	S. Acharya (1989)
		b) Significant increases in pod yield	Chowdhury (1990)
		c) Significant reductions in inoculum buildup	
		d) Significant increases in total phenol and *O*-dihydroxyphenol contents and in PPO and PO activities	Chowdhury (1992)

of smaller size groups. Moderate to high levels of fungitoxicity that developed in the young rice seedlings in these treatments gradually declined and disappeared totally between 3 and 4 weeks. Challenge inoculation at this stage led the treated plants to freshly develop high to very high levels of fungitoxicity in vivo that showed good correlation to the level of protection achieved (Sinha and Hait, 1982). Such active response by the treated plants was believed to be due to a conditioning of their tissue, based on sensitization, under the influence of chemical treatment, so that they respond to challenge inoculation almost like resistant plants. Later, Hait and Sinha (1987) provided strong biochemical evidence to support this contention. In their experiment both untreated susceptible and resistant plants as well as susceptible plants receiving treatment with two highly effective compounds, namely, sodium selenite and L-cysteine, were inoculated at the age of 2, 4, 6, and 8 weeks. Leaf materials from both untreated and treated plants 3 days after each inoculation were used for analysis. It was observed that treated susceptible plants showed appreciably reduced total protein content and increased PPO activity as compared to the untreated plants. Such changes brought them closer to the conditions existing in resistant plants. Treated susceptible plants responded to challenge inoculation with significantly greater increase in both total phenol and protein contents as well as in PPO activities as compared to the control plants. Treated susceptible plants showed almost the same trends as those recorded for resistant plants. The induced effect was most pronounced in 4-week-old plants but declined thereafter. However, with sodium selenite treatment that yields a more persistent effect, induced effects remained appreciable up to 6 weeks in most respects, although not with L-cysteine. Further evidence of tissue conditioning came from gel and immune electrophoretic studies with plants in the above treatments (Hait, 1982). In both L-cysteine and sodium selenite treatments, plants appeared to have lost the three antigenic proteins that they shared with the virulent isolate of *H. oryzae*. This was noted only in selenite treatment but not with L-cysteine at 2 weeks; at 4 weeks the loss of proteins was evident in both treatments, but not thereafter. A point of further interest was that plants in both treatments showed two new protein bands corresponding to the two bands already present in resistant plants. These observations fit in well with the common antigen hypothesis (Doubly et al., 1960; DeVay and Adler, 1976).

Later studies by Das (1993) recorded strong protective action on rice plants with cupric chloride, sodium selenate, 2,4,5-T, and cycocel, and very strong action with chitosan, a β-1,4-linked glucosamine polymer derived from chitin by deacetylation. Kar and Sinha (1988) reported good protection from brown spot also with propantheline bromide, an anticholinergic drug, known as an effective inducer of phytoalexin (Hadwiger, 1972a).

Blast of Rice (Pyricularia oryzae)

Many of the compounds with strong protective action on rice plants against brown spot disease were also tested against blast disease in multilocation field trials at hot-spot locations in the states of Tripura, West Bengal, and Bihar. In most of these trials, infection was first detected after 5–7 weeks, but in one case infection came quite late, i.e., after 13 weeks. To prepare field plots for these experiments, nitrogen was added at a high dose, 100–120 kg/ha. Test compounds that provided very strong protection from blast in different trials included ferric chloride, cupric chloride, *p*-cmb, sodium malonate, cycloheximide, 2,4-D, and chitosan (Sinha and Sengupta, 1986; Das, 1993). Those with strong but less significant action were cadmium chloride, lithium sulfate, thioglycollic acid, L-cysteine, sodium selenite, sodium selenate, sodium molybdate, sodium iodoacetate, sodium fluoride, DL-methionine, DL-phenylalanine, and IAA. Plants in effective treatments showed many fewer blast lesions of reduced mean size, and lesions rarely coalesced. When infection occurred rather late, after 13 weeks in a trial, symptoms were assessed 2 weeks later, cupric chloride, *p*-cmb, and cycloheximide reduced leaf symptoms in the range of 85–89%. In treatments with sodium selenite, L-cysteine, and sodium malonate as well, the induced effect was quite strong and 60–71% reductions in leaf symptoms were noted. Seed-soaking treatment for 24 hr gave a better result than 48-hr treatment.

Biochemical studies with leaf samples from untreated and variously treated rice plants from the infected crop showed that plants treated with highly effective cupric chloride and sodium selenate and fairly effective IAA contained higher phenol and protein levels and more PO activity than the untreated plants. With minor exceptions, correlation existed between such increases and the magnitude of protection achieved with different chemicals.

Sheath Blight of Rice (Rhizoctonia solani)

In limited pot trials of exploratory nature, three compounds with strong effect against both brown spot and blast diseases of rice were also tested against sheath blight disease that has gained much importance in recent years. When 5-week-old pot-grown plants of susceptible nature were artificially inoculated, those in all three treatments developed significantly reduced symptoms as compared to fairly heavy symptoms developed in the untreated plants (Sarkar and Sinha, 1991). While cycloheximide showed fairly strong protective action against sheath blight, the other two had only moderate effects. Further studies are in progress.

Leaf Spot of Wheat (Helminthosporium sativum)

A large number of phytoalexin inducer chemicals were tested in pot trials against leaf spot disease of wheat incited by *Helminthosporium sativum*.

Out of about 40 compounds tested in wet seed treatment, the majority could provide 3- to 5-week-old wheat seedlings with strong to very strong systemic protection from artificial inoculation with *H. sativum* (Hait and Sinha, 1986; Chakraborty and Sinha, 1989). Those reducing leaf symptoms by 70% or more included the chlorides and sulfates of copper, zinc, mercury, and barium; nitrates of copper, zinc, nickel, and iron; and nickel chloride, sodium sulfite, sodium malonate, sodium fluoride, and IAA. Compounds such as lithium sulfate, L-cysteine, thioglycollic acid, sodium azide, sodium iodoacetate, sodium molybdate, *p*-cmb, cycloheximide, DL-phenylalanine, and 2,4-D also showed protective action that was strong but slightly less significant than that in the first group. These compounds primarily prevented the successful establishment of infection, as reflected in the very limited number of lesions developed, but many of them restricted lesion enlargement also. At the age of 2 weeks, diffusates from leaf, in most of the above treatments, showed high levels of fungitoxicity. This toxicity did not persist, declined gradually, and finally disappeared between 3 and 4 weeks. Inoculation at these stages led to the development of strong fungitoxicity in both treated and untreated plants, with the difference between their levels not sufficient to explain the difference in their leaf symptoms. The induced effect declined with time.

Damping Off of Chili (Pythium sp.)

Out of 15 compounds screened in small-scale nursery trials against damping off of chili, most showed significant effects in protecting chili seedlings from damage and mortality. The trials continued for 3 successive years. Compounds found to be highly effective in presowing wet seed treatment against the seedling disease include barium chloride, boric acid, sodium selenite, thioglycollic acid, sodium sulfite, sodium molybdate, cycloheximide, and 2,4-D (Sahana, unpublished information). In one or more experiments, these compounds reduced seedling mortality by 60% or more. Somewhat less significant effects were recorded with cadmium sulfate, L-cysteine, sodium malonate, *p*-cmb, and IAA.

Fusarium Wilt of Pigeon Pea (Fusarium udum)

Only a few of the 14 chemicals tested in wet seed treatment against *Fusarium* wilt of pigeon pea (i.e., chitosan, mercuric sulfate, cupric sulfate, and sodium sulfite) could significantly check symptom expression (37–48%) and reduce plant mortality (41–49%) (Maity, 1991). Root exudates and root extracts from susceptible plants that underwent such treatments showed appreciable fungitoxicity, measured against germination of both conidium and chlamydospore. Such toxicity in root extracts increased further in response to artificial inoculation with *F. udum*. Upon challenge, treated susceptible plants also recorded significant increases in flavonol content and peroxidase activity, and the increased levels came close to those of resistant plants. Susceptible plants, in treatments with chitosan and mercuric sulfate (the two most effective compounds), showed the loss of two of five antigenic proteins they share with the virulent isolate of *F. udum* but have acquired at the same time two new proteins corresponding to two already present in resistant plants. This is in conformity with the common antigen hypothesis.

Fusarium Wilt of Chickpea (Fusarium oxysporum *f. sp.* ciceri)

In pot experiments, successful control of vascular wilt of chickpea has been achieved by presowing wet seed treatment with some of the 40 chemicals tested. Highly effective among them are mercuric sulfate, cycloheximide, DL-phenylalanine, DL-methionine, IAA, 2,4-D, 2,4,5-T, cycocel, and chitosan (Sahana, 1991). When symptoms were finally assessed 5 weeks after inoculation, the above treatments reduced disease incidence by 55–68%, disease index by 79–89% against severe wilting in control plants, and plant mortality by 71–86%, and also limited both upward and lateral spread of the fungus in the host vascular system. Root exudates from such treated plants show considerable fungitoxicity and adversely affect both germination and germ tube growth of chlamydospores and conidia. The inhibitory effect was more pronounced on germ tube growth. The toxicity was also evident in the root extract and stem extract with further increases when plants become infected. Metal salts such as cupric chloride, zinc chloride, and nickel chloride also have good protective action against chickpea wilt, though less significant than the above compounds. Working with the highly effective compounds, Chowdhury (1992) reported that susceptible chickpea plants in these treatments respond to infection with significantly greater increases in total phenols, *O*-dihydroxyphenol, and total protein contents, as well as in PPO and PO activities. In these respects, responses of treated susceptible plants closely follow those of resistant plants.

Collar Rot of Chickpea (Sclerotium rolfsii)

The heavy mortality that characterizes *S. rolfsii* infection of chickpea plants in pot experiments is substantially checked by wet seed treatment with the sulfates and chlorides of copper and zinc, mercuric nitrate, L-cysteine, and thioglycollic acid (Sahana, 1991). These reduced plant mortality by 55–78%. In the treated plants, pectolytic enzyme (polygalacturonases) activity of the pathogen in the rot-affected tissue was found to be significantly lower than in the untreated plants. Compounds such as ferric chloride, cupric nitrate, zinc nitrate, barium sulfate, and barium nitrate also are fairly active on chickpea plants in limiting collar rot infection.

Root Rot of Sugarbeet (Sclerotium rolfsii)

Results from a replicated field trial with a mixed group of nine phytoalexin inducers used for seed treatment before sowing in *S. rolfsii*-infected soil showed that all except IAA could limit root rot of beet resulting from natural infection and contribute in some measure to improved plant health and better productivity (Das, personal communication). All-around strong effects were recorded with 2,4,5-T, chitosan, zinc chloride, and 2,4-D in that order. They reduced disease incidence by 61–72%, plant mortality by 25–72%, and increased root yield (ton/ha) by 32–61% and sugar yield (ton/ha) by 29–63%. Cycloheximide and cycocel effectively reduced both disease incidence and mortality by 55–65% and 46–50%, respectively, but caused only moderate increases in root yield and sugar yield.

Stem Rot of Groundnut (Sclerotium rolfsii)

In different series of experiments with pot-grown plants raised from both untreated and variously treated groundnut seeds, barium chloride, barium nitrate, zinc chloride, zinc sulfate, and lithium sulfate (Acharya, unpublished), IAA, 2,4-D, 2,4,5-T, cycocel, cycloheximide, chitosan, L-cysteine, and DL-phenylalanine substantially checked the progress of rotting resulting from artificial soil inoculation with *S. rolfsii*. Plant mortality was mostly markedly reduced (72–95%) in such effective treatments. Pectolytic enzyme activity on which the pathogenic activity of this pathogen is based was also significantly suppressed in these treatments. In treatments with organic compounds, affected tissue responds to challenge inoculation with accumulation of calcium and magnesium, and also enhanced lignification. It has further been shown that in these treatments plants respond to infection also with significantly greater increases in total phenol, *o*-dihydroxyphenol, and total protein contents, and with increased PPO and PO activity (Chowdhury, 1992).

Early Leaf Spot of Groundnut (Cercospora arachidicola)

In a series of field trials conducted at two locations, susceptible groundnut plants raised from seeds treated with different metal salts and boric acid exhibited moderate to strong resistance to natural infection with *C. arachidicola*. In comparison, fairly heavy infection and symptoms developed in plants raised from untreated seeds. Salts such as barium sulfate, barium nitrate, mercuric chloride, and lithium sulfate exercised strong protective effect and reduced leaf symptoms by 60% or more (Acharya, unpublished). In another series of field experiments, seed treatment with growth regulators, such as IAA, 2,4-D, 2,4,5-T, cycocel, as well as cycloheximide, led to substantial (62–67%) reductions in leaf symptoms (Chowdhury and Sinha, 1989). Later studies also achieved very strong reductions (73–87%) with chitosan, L-cysteine, and *p*-Cmb and a somewhat lesser effect with sodium azide (Chowdhury and Sinha, 1990). In all of the above cases, induced protective effect was systemic and long-lasting. In the treatments with growth regulators, pod yield was appreciably increased. Study of inoculum production (conidia) on a leaf at fixed position (sixth from the base) over a period of 10 days revealed that in groundnut plants receiving various treatments spore production appeared to be much limited as compared to that in the untreated plants, and the difference was quite significant. Generally, the drop in inoculum production has good correlation with the induced protection achieved with a treatment. Biochemical studies showed that groundnut plants in different effective treatments responded to natural infection with marked increases in total phenol and *o*-dihydroxyphenol contents, and less pronounced but still significant increases in total protein content and both PPO and PO activities (Acharya, unpublished; Chowdhury, 1992).

III. DISCUSSION

Results from both pot and field experiments using a large and diverse group of phytoalexin inducer chemicals, mostly in wet seed treatment, against some important crop diseases of fungal origin establish beyond doubt that such treatments can provide a rational and safe alternative to conventional chemical control using toxic compounds like fungicides that act directly on the pathogen. Based on these observations, the following generalizations seem possible:

1. While most of the phytoalexin inducer chemicals tested have the potential for providing some degree of protection to different plant species from their fungal pathogens, when used at rather low concentrations (10^{-6} to 10^{-4} M), many of them show substantial suppressive effects on symptom expression and plant morality.
2. Resistance induced in a particular plant species by such treatment appears to be broad spectrum in nature and may be active against a variety of fungal pathogens with different modes of pathogenesis as well as different levels of parasitic specialization. Many of the inducer compounds, different both in their chemical nature and their biological action, if any, can induce apparently similar resistance in a particular plant species against its one or more pathogens. The same compound may also be active on more than one plant species against their different pathogens.
3. For most of the effective compounds, the induced protective effect has little relation either to their in vitro toxicity (if any) or to their concentration gradient. Some show their maximum effect at the lowest concentration tested, often as low as 10^{-6} M.

4. The induced resistance developed in a plant species and the consequent protective effect against its pathogen(s) is systemic and mostly long-lasting in nature. The induced protection may remain active almost throughout crop life.

5. The effective compounds mostly appear to act on the pathogen not through any direct toxic action, if any, but through dynamic host-mediated responses that create in vivo an environment unfavorable for growth and/or activity of the fungus involved. Supporting evidence comes from biochemical studies.

6. In a highly effective treatment, susceptible host plant may lose all or some of its antigenic proteins that it has in common with the virulent isolate of the pathogen and/or acquire one or more new proteins corresponding to protein(s) the resistant plant already has. In terms of common antigen hypothesis, such changes in the host indicate its shift to reduced compatibility with the pathogen and/or increased resistance to it. With respect to the parameters commonly associated with a plant host's normal defense responses, such as phenolics, phytoalexin-type antifungal substance, lignin, and protein contents, as well as phenylalanine ammonia-lyase (PAL) PPO, and PO activities, susceptible plants in effective treatments initially recorded small to moderate increases, which declined with time. However, if and when challenged with pathogen, the same plants recorded significantly greater increases than the control plants. Their post-infection responses mostly show the same trends as characterize the resistant plant.

7. Effective compounds appear to condition susceptible host plant through activation of its latent defense potential and sensitization of its tissue, so that it becomes competent and also ready to interact with the challenger pathogen, if and when it would infect, with dynamic and vigorous responses like the resistant plant.

The above findings clearly establish the fact that even susceptible plants are not totally defenseless against a pathogen. Even in the absence of a specific gene for resistance, they also have some genetic information for defense reactions. This means that all plants have the genetic potential for disease resistance latent in them, and this can be activated and made operative by treatment with many different biotic or chemical agents. It is interesting to note that much similarity exists between the resistances induced by these agents. This has been experimentally demonstrated on wheat plants against *Drechslera sorokiniana* (Chakraborty and Sinha, 1984) and on cucumber against four fungal, two bacterial, and two viral pathogens (Mucharromah and Kuć, 1991). With both kinds of agents, induced resistance is time-dependent, systemic, and long-lasting. It is also nonspecific in respect to both the inducer agents and the pathogens against which it is active.

In both cases, the induced resistance appears to have developed in the host through a process of conditioning based on tissue sensitization. Such conditioned susceptible host tissue responds to challenge inoculation or infection with the pathogen with physiological changes that mostly indicate the same characteristic trends of the natural defense responses of the host species. These responses mostly involve stimulated production of phenolics, phytoalexin, lignin, protein, hydroxyproline-rich glycoprotein, and increased activity of PAL, chitinase, β-1,3-glucanase, PO, and PPO among others. Not all but some of them get activated and combine to provide the treated susceptible plant with a multicomponent and multilayered protective system in which the host tissue, now sensitized and made competent and ready to react dynamically, responds with changes in different components as may be required to contain the particular pathogen. The versatility of such induced resistance, i.e., systemic acquired resistance (SAR) in the treated plant, can only be explained by such coordinately regulated multicomponent systems. Results from our studies with phytoalexin inducer chemicals and also from other studies based on compounds with no known record of phytoalexin induction in plants (Langcake, 1981; Lazarovits, 1988) support this view. Various observations suggest that while many of the above components may be activated as a part of the tissue sensitization process following treatment and record small to moderate increases, these may constitute only a broad-spectrum protective effect of mild nature that is effective against more than one pathogen. When challenged with different pathogens, only a few of thee components will be moderately or vigorously stimulated and these will have a primary role in containing the pathogen. The magnitude of resistance ultimately expressed against different pathogens may then depend on the quality of the signal, i.e., nature of the elicitor, and its intensity, i.e., concentration of the elicitor, received from the pathogen. While the quality, i.e., the chemical nature of the elicitor from pathogen, will probably determine the defense components to be further activated, its strength will regulate the speed of response with respect to such components, and this will determine the level of protection achieved against a particular pathogen. The multicomponent nature of SAR in a plant may have another advantage, i.e., the different components may adversely affect the pathogen at different stages of its interaction with the host, as reported for chickpea wilt (Chowdhury, 1992).

Beside the large group of phytoalexin inducers tested by us, significant levels of protection of susceptible plants against their fungal pathogens have also been reported to be achieved with compounds like sodium azide, probenazole, WL28325, TEPA, gibberillic acid, cloxacillin, and penicillin among others (Langcake, 1981; Lazarovits, 1988). While sodium azide was the only known phytoalexin inducer among them, some others like piperidinium iodide, pyrrolidinium iodide, probenzaole, and trifluralin-like herbicides also appear to induce the production of a phytoalexin-type fungitoxic substance in healthy plants, though at low levels. However, plants in most of these treatments, except with cloxacillin and penicillin, responded to challenge with the pathogen with significantly high levels of such toxic compounds. It is felt, however, on the basis of above results that our search for sensitizer-type compounds need not be restricted exclusively to phytoalexin inducers and their related compounds. On the basis of his extensive studies on blast of rice with more than 150 chemicals, Rathmell (1982) concluded that any compound with the ability to cause tissue necrosis in plants would be able to protect rice plants from blast disease. Almost simultaneously, Kuć (1982) also concluded that both biotic and chemical agents that effectively induce resistance in plants have the common property of creating, through a mild metabolic perturbation, a common stress of persistent nature in the host. Only rarely has any adverse effect on seed germination or seedling health been recorded with a phytoalexin inducer chemical tested by us. Visible localized necrosis may occur when a live inoculum is used as the inducer agent but not generally with a microbial fluid or a dilute chemical solution used or seed treatment. In the case of chemical agents, the induced metabolic perturbation may not be such high magnitude as to cause cell necrosis but may be adequate to result in a certain byproduct(s) that functions as an endogenous elicitor and triggers a mild resistance response locally (i.e., in the neighboring cells). Either that byproduct or a secondary endogenous elicitor is translocated as a signal carrying information to distant parts of the host plant where it conditions the cells making them competent and also alert, through the process of sensitization. Such conditioned cells respond dynamically to infection like the cells of resistant plant. In the case of biotically induced resistance in plants, there is evidence of transmission of an immunity signal in cucumber (Jenns and Kuć, 1979) and tobacco (Kuć and Tuzun, 1983). In the latter, movement of such a signal through phloem tissue has also been reported (Jenns and Kuć, 1977). In no case of chemically induced resistance has such evidence been available until now. Compounds postulated to have a role as transmissible immunity signal include ethylene (Van Loon, 1982), salicylic acid (Ward et al., 1991), and oligogalacturoronide fractions of plant cell wall (Uknes et al., 1992). Both ethylene and salicylic acid can themselves induce many defense-related responses in plants and make them acquire SAR. However, no compelling evidence is available in support of such a role for them. Were such a role possible, these could probably be directly used for seed treatment or foliage spray for the purpose of disease control.

While screening a large number of synthetic organic compounds to identify phytoalexin inducers, Hadwiger and associates noted that the effective ones concomitantly induced new synthesis of RNA and protein and increased PAL activity, and these were correlated with phytoalexin synthesis (Hadwiger, 1971, 1972a, b; Hadwiger and Schwochau, 1970; Hess and Hadwiger, 1971). It was concluded that such a property of the effective compounds was related to their ability to bring about specific changes in the conformation of host cell DNA by intercalating into it or complexing with it. Heavy metals can also do the same job by complexing the host DNA. However, most of the phytoalexin inducers we have used in our attempts to induce resistance do not have any ability to complex with DNA (Hadwiger, personal communication), except for chitosan which showed very strong effect against all the diseases tested. It has a strong affinity with cell DNA. Chitosan protects pea pod tissue from its pathogen, *Fusarium solani* f.sp. *pisi*, as strongly as does prior inoculation with *F. solani* f.sp. *phaseoli*, a nonpathogen.

Within 15 min of such treatment or inoculation with the latter, chitosan can be detected within host cell cytoplasm and nucleus. Chitosan induces in pea tissue, besides phytoalexin production, increased protein synthesis and PAL activity (Hadwiger and Beckman, 1980) as well as enhanced lignification (Pearce and Ride, 1982)—all components of the host's natural defense response that are also induced by infection with *F. solani* f.sp. *phaseoli*. This shows that both biotically and chemically induced resistance function almost similarly. In our extensive studies with phytoalexin inducer chemicals for induction of host resistance in plants, it may seem quite natural to place the major emphasis on phytoalexin as the causal factor. However, most of them also coordinately induce the increased synthesis of phenolics, particularly *o*-dihydroxyphenols, proteins, and lignin, and enhanced activity of PAL and PO (all components of both induced and natural resistance in plants), thus broadening its scope of action. Some organic compounds like probenazole, TEPA, and WL28325, not known yet as phytoalexin inducers, induce strong resistance in some crop plants, and the protected plants respond to challenger pathogen with increased level and/or activity of some of the above defense components including phytoalexin synthesis (Vo-Thi-Hai et al., 1979; Iwata et al., 1980; Cartwright et al., 1980; Shimura et al., 1981). This does not happen in all cases. Phytoalexin inducer compounds may have certain advantages over others as the effective inducer of resistance because they stimulate early protein synthesis and PAL activity, a key enzyme in aromatic metabolism that regulates not only synthesis of some phytoalexins but also that of many toxic phenols and lignin.

Increased protein synthesis constitutes one of the responses of many plants with SAR to infection with their pathogen. This may involve increased synthesis of existing protein or de novo synthesis of some new protein including pathogenesis-related protein(s) (PR protein). Such proteins have been reported from tobacco and some other plants acquiring SAR (mostly against viral pathogens) by treatment with acetylsalicylic acid, aspirin, benzoic acid, 2,4-D, IAA, polyacrylic acid, etc. (Bozarth and Ford, 1988 and references therein). Some of them belong to know class of proteins, such as chitinase, β-1,3-glucanase, etc., with suggested involvement in induced resistance, but not with any clear evidence. Further evidence of possible involvement of proteins in induction of resistance in plants came from studies on the antigenic relationship between host cultivars and pathogen, commonness in antigenic protein standing for compatibility, and disparity for incompatibility between them (Doubly et al., 1960; DeVay and Adler, 1976). Investigations on chemically induced resistance in rice-*Helminthosporium oryzae* (Hait, 1982), pigeon pea-*Fusarium udum* (Maity, 1991), soybean-*Macrophomina phaseolina* (Chakraborty and Purkayastha, 1987)/*Colletotrichum dematium* (Purkayastha and Bannerjee, 1990) showed that a susceptible plant acquiring resistance lost all or some of the antigenic proteins it had in common with the pathogen, indicating a shift to reduced compatibility, but also developed (in rice and pigeon pea) one or more new proteins corresponding to some constitutively present in resistant plant, indicating a shift to greater resistance. Since alteration of specific antigen(s) in a susceptible host plant by chemical treatment appears to be linked with induced resistance, their immunochemical identification and regulation may provide a basis for their use in plant disease control. Purkayastha (1973) suggested a close relation between antigens, phytoalexins, and disease resistance.

Recognition of the potential pathogen by the host at the pathogen elicitor–host receptor level as "nonself" is deemed as the primary event that triggers a cascade of host responses leading to the expression of resistance (Callow, 1977; Yoshikawa, 1983). In the case of SAR in tomato to *Phytophthora* sp. (Vo-Thi-Hai et al., 1979) and in rice to *Pyricularia oryzae* (Sekizawa and Mase, 1980) developed, respectively, by treatment with TEPA and probenazole, it has been suggested that specific structural changes at the host–receptor surface following chemical treatment make such recognition of the pathogen by the susceptible plant possible. No clear evidence is available in support; however, the fact that many of the compounds can induce effective resistance in one or more plant species against one or more of their pathogens makes specific structural changes at the receptor level that would allow such recognition seem unlikely.

It is evident from the various observations that even susceptible plants having no gene for resistance specifically active against a particular pathogen have some genetic information for a generalized resistance against most of its pathogen. Such defense normally remains latent but can be activated by prior restricted inoculation with an infectious agent or treatment with a sensitizer-type chemical. According to Kuć (1987), resistance in plant is determined by the speed and magnitude with which genes for resistance mechanism are expressed and the activity of gene products rather than the presence or absence of such a gene. In spite of the undoubted promise of induced resistance, there has been no genetic analysis of the immunized plants. We know nothing about how the genes are activated. Recently, Ward et al. (1991) reported that the onset of SAR in tobacco to tobacco mosaic virus by restricted inoculation with the same virus or treatment with salicylic acid or 2,6-dichloroisonicotinic acid (INA), an immunizing agent, is correlated with coordinate gene expression and induction of nine classes of mRNA. Later, Uknes et al. (1992) made similar observations with respect to SAR in *Arabidopsis* sp. to *Pseudomonas syringae* pv. *tomato* and *Peronospora parasitica* by treatment with salicylic acid and INA or infection with the pathogen. In both types of treatment, onset of SAR was associated with high-level accumulation of three proteins and that genes corresponding to these proteins could also be induced by the same treatments. On pea pod tissue, chitosan, a natural elicitor of fungal origin, has been shown to induce many defense-related responses including the synthesis of 20 major proteins as can also be achieved by infection with *Fusarium solani* f.sp. *phaseoli*, a nonpathogen (Hadwiger and Waggoner, 1983). Disease resistance responses at the protein or mRNA level indicate that the number of genes associated with the disease resistance response is quite large. It has been shown that chitosan authentically mimics *F. solani* f.sp. *phaseoli* in enhancing the in vivo and in-vitro synthesis of all these proteins associated with host resistance in pea. So Hadwiger and Wagoner (1983) proposed that it would be useful to identify those genes whose activation most closely corresponds with resistance reaction and to investigate their regulatory properties. Only after it has been shown that the specific function of a gene contributes to or regulates host resistance or this has been genetically verified, can it be accepted as a disease resistance gene.

IV. EPILOG

Immunity provides the best form of defense for a plant. Substantial experimental findings leave little doubt about the fact that strong systemic resistance of a lasting nature can be induced in a susceptible plant cultivar against its pathogen by prior restricted inoculation with infectious agents of both homologous and heterologous nature, prior treatment with fluids containing their metabolites or cell constituents, or even treatment with synthetic chemicals. With its systemic and persistent nature, time dependence for optimization of the effect, and dynamic response to challenge with the pathogen, if and when that would occur, such biologically or chemically induced resistance outwardly resembles immunization of animals. However, induced resistance in plants with its nonspecific and broad-spectrum effect, differs from immunization of animals, which characteristically results in strong, specific action against a particular pathogen based on highly specific antigen–antibody reaction.

Great similarity exists between the biologically and chemically induced resistance in plants, at least in the expressive phase, though some differences may be there in the initial and/or determinative phase. Chemical agents may enjoy certain advantages over biotic agents in respect of their early and better absorption into the plant system, rapid distribution into its different parts, and also longer persistence in host cells. Extensive work by Kuć and associates (Kuć, 1987; Mucharromah and Kuć, 1991) on tobacco, cucumber, and melon illustrates both the characteristic features and effectiveness of biologically induced resistance. However, the use of biotic agents suffers from certain limitations. Costs involved in maintaining a laboratory with adequate technical hands for the production of large amounts of inoculum and logistical difficulties faced in properly delivering such inoculum to the corp, particularly under conditions of large-scale cultivation, make it not only cumbersome but uneconomic and noncompetitive. Also, such treatments do not fit in with the modern concept of agriculture. The above disadvantages may be overcome or bypassed by using synthetic, nontoxic chemicals which can mimic the signals from biotic agents in triggering a series of defense-related responses. Cultivators accustomed to the use of toxic chemicals for conventional chemical control would not find any difficulty in using the nonconventional chemicals. Their nonhazardous nature, simple use in seed treatment at micro doses, and broad-spectrum protective effect against a variety of plant pathogens both of different kinds and with different levels of parasitic specialization mark them as having great promise in plant disease control. These compounds act on the pathogen not directly but through host-mediated responses based on sensitization of susceptible host tissue and activation of its latent defense potential. Susceptible plants with SAR, when challenged with the pathogen, respond dynamically like the resistant plants, limiting symptom expression as a consequence. The conditioned and protected plants do not as such show any great change in any one of the defense-related responses but respond dynamically and vigorously to infection with significant increases in few or more components of defense responses, i.e., their responses occur at the right time and the right location where needed. Thus, any wasteful expenditure of energy is prevented, i.e., the induced resistance is also energy-efficient. This would help in the cultivation of commercially desirable but defense-susceptible crop varieties in preference to resistant ones, which are mostly not very good yielders.

It has been admitted that neither phytoalexins as such or their biotic elicitors can be of any use in plant disease control. Use of biotic agents also has many snags. Instead of wasting time on such biotic products, it would be sensible to look for synthetic chemicals which can mimic the natural signals from the biotic agents and trigger most, if not all, of the multicomponent defense responses of the host. Among the chemicals we tested, cupric chloride, lithium sulfate, L-cysteine, thioglycollic acid, barium sulfate, zinc chloride, sodium malonate, cycloheximide, *p*-Cmb, IAA, 2,4-D, 2,4,5-T, cycocel, and DL-phenylalanine showed strong inhibitory effect against more than one disease on different crops. Chitosan had strong effect against all the diseases tested. All of them are known phytoalexin inducers. However, search for sensitizer chemicals need not be kept restricted to phytoalexin inducers only; others may also have similar effect (Langcake, 1981; Lazarovits, 1988). However, phytoalexin inducer chemicals may enjoy some initial advantages over other chemicals in this respect.

Despite the many advantages, the use of sensitizer chemicals has yet to be accepted as a safe and effective alternative to conventional chemical control. This is a time for plant pathologists to realize their immense potential as a tool for plant disease control. With such synthetic chemicals, they can possibly plan a subtle, natural, and selective strategy for this purpose. Though mainly tested and found to be effective against fungal pathogens, indications are that they may also be effective against bacterial and viral pathogens for which no good control measure is yet known. Lazarovits's (1988) assertion that chemically induced resistance does not have the same broad base of protection as that induced by biotic agents does not appear to be valid in the light of more recent observations (Mucharromah and Kuć, 1991).

As the conventional chemical control constitutes an integral component of modern agriculture, its large-scale replacement with the use of nonconventional, sensitizer-type chemicals is out of the question, but their increasing use over the years may more and more reduce our dependence on toxic compounds. That would not be a small gain.

Immunization has opened an uncharted area of molecular biology related to the role of stress and infection on a plant's genome and its expression. Besides its practical utility, exploratory studies may also shed some light on its scientific significance. What are the basic events at a molecular level that trigger a series of responses that regulate and shape the induction of resistance in susceptible plants? Characterization of the immunity signal, factors that regulate its production, its cell-to-cell and long-range communication, and the hormonal mechanism that makes the plant respond to stress are points that need investigation and elucidation at the molecular level.

REFERENCES

Arimoto, Y., Homma, Y., Ohtsu, N., and Misato, T. (1976). Studies on chemically induced resistance of plants to disease. I. The effect of soaking rice seeds in dodecyl-DL-alanine hydrochloride on seedling infection by *Pyricularia oryzae*. *Ann. Phytopathol. Soc. Jpn. 42*: 397–400.

Bailey, J. (1982). Physiological and biochemical events associated with the expression of resistance to disease. *Active Defence Mechanisms in Plants* (Wood, R. K. S. ed.), Plenum Press, New York, pp. 39–65.

Bell, A. A., and Presley, J. T. (1969). Heat-inhibited or heat-killed conidia of *Verticillium albo-atrum* induce disease resistance and phytoalexin synthesis in cotton. *Phytopathology 59*: 1147–1151.

Bertrand, A., Ducret, J., Debourge, J. C., and Horriere, D. (1977). Etude des propriets d'une novelle famille de fongicides: less monoethyl phosphites metalliques. Characteristiques physiochimiques et proprites biologiques. *Phytiatr. Phytopharm. 26*: 3–18.

Bompeix, G., Ravise, A., Raynal, G., Fettouche, F., and Durand, M. C. (1980). Modalites de l'obtention des necroses-bioquantes sur feuillies detachees de tomate par l'action du tris-O-ethylphosphonate d'aluminium (phosethyl d'aluminum), hypotheses sur son mode d'action in vivo. *Ann. Phytopathol. 12*: 337–351.

Bozarth, R. F., and Ford, R. E. (1988). Viral interactions: induced resistance (cross protection) and viral interference among plant viruses. *Experimental and Conceptual Plant Pathology* (R. S. Singh, U. S. Singh, W. M. Hess, and D. J. Weber, eds.), Oxford–IBH, New Delhi, pp. 551–567.

Buchenauer, H., and Erwin, D. C. (1973). Reduction of the severity of *Verticillium* wilt of cotton by the growth retardants, N,N-dimethyl pyrrolidinium iodide associated with the production of antifungal compounds. *Z. Pflanzenkrankheit. Pflanzenschutz 80*: 576–586.

Buchenauer, H., and Erwin, D. C. (1976). Effect of plant growth regulator pydanon on *Verticillium* wilt of cotton and tomato. *Phytopathology 66*: 1140–1143.

Callow, J. A. (1977). Recognition, resistance and the role of plant lectins in host-parasite interactions. *Ad. Bot. Res. 4*: 1–49.

Carrasco, A., Boudet, A. M., and Marugo, G. (1978). Enhanced resistance of tomato plants to *Fusarium* by controlled stimulation of their natural phenolic production. *Physiol. Plant Pathol. 12*: 225–232.

Cartwright, D., Langcake, P., Pryce, R. J., and Ride, J. P. (1977). Chemical activation of host defence mechanism as a basis for crop protection. *Nature 267*: 511–513.

Cartwright, D., Langcake, P., and Ride, J. P. (1980). Phytoalexin production in rice and its enhancement by a dichloropropane fungicide. *Physiol. Plant Pathol. 47*: 259–267.

Caruso, F. L., and Kuć, J. (1977). Field protection of cucumber, watermelon, and muskmelon against *Colletotrichum lagenarium* by *Colletotrichum lagenarium* *Phytopathology 67*: 1290–1292.

Cassels, A. C., and Flynn, T. (1978). Studies on polyacrylic acid induced resistance to viral and non-viral plant pathogens. *Pest. Sci. 4*: 365–371.

Chakraborty, B. N., and Purkayastha, R. P. (1987). Alteration in glyceollin synthesis and antigenic pattern after chemical induction of resistance in soybean to *Macrophomina phaseollina. Can. J. Microbiol. 33*: 835–840.

Chakraborty, D., and Sinha, A. K. (1984). Similarity between the chemically and biologically induced resistance in wheat seedlings to *Drechalera sorokiniana. Z. Pflanzenkrankheit. Pflanzenschutz 91*: 59–64.

Chakraborty, D., and Sinha, A. K. (1989). Differential effects of anionic forms of selected heavy metal salts in seed treatment on *Helminthosporium* infection of wheat seedlings. *Indian Phytopathol. 42*: 157–160.

Chamberlain, J. W., and Paxton, J. D. (1968). Protection of soybean plants by phytoalexin. *Phytopathology 58*: 1349–1350.

Chowdhury, A. K. (1992). Management of selected fungal diseases of groundnut and chickpea by the use of unconventional chemicals. PhD thesis, Bidhan Chandra Krishi Viswavidyalaya, Kalyani.

Chowdhury, A. K., and Sinha, A. K. (1989). Control of early leaf spot of groundnut by treatment with a select group of chemicals. *Int. Arachis Newslett. 6*: 16–18.

Chowdhury, A. K., and Sinha, A. K. (1990). Control of foliar diseases of groundnut by seed treatment with non-conventional chemicals. *Int. Arachis Newslett. 8*: 16–17.

Cruickshank, I. A. M. (1963). Phytoalexins. *Ann. Rev. Phytopathol. 1*: 351–374.

Cruickshank, I. A. M. (1980). Defences triggered by the invader: chemical defence. *Plant Disease: An Advanced Treatise*, Vol. 5 (J. G. Horsfall and E. B. Cowling, eds.), Academic Press, New York, pp. 247–267.

Darvill, A. G., and Albersheim, P. (1984). Phytoalexins and their elicitors. *Ann. Rev. Plant Physiol. 35*: 243–275.

Das, A. R. (1993). Resistance induced in rice plant to brown spot and blast diseases by seed treatment with chemicals and the associated host responses. PhD thesis, Bidhan Chandra Krishi Viswavidyalaya, Kalyani.

Davis, D., and Dimond, A. E. (1953). Inducing disease resistance with plant growth regulators. *Phytopathology, 43*: 137–140.

DeVay, J. E., and Adler, H. E. (1976). Antigens common to host and parasite. *Ann. Rev. Microbiol. 30*: 147–163.

Dimond, A. E. (1963). The modes of action of chemotherapeutic agents in plants. *Perspectives of Biochemical Plant Pathology* (S. Rich, ed.), Connecticut Agricultural Experimental Station Bulletin 663, pp. 62–77.

Dixon, R. (1986). The phytoalexin responses: elicitation, signalling and control of host gene expression. *Biol. Rev. Cambridge Phil. Soc. 61*: 239–292.

Doubly, J. A., Flor, H. H., and Clagett, C. D. (1960). Relation of antigen of *Melampsora lini* and *Linum usitatissimum* to resistance and susceptibility. *Science 131*: 229.

Fawcett, C. H., Spencer, D. M., and Wain, R. L. (1969). The isolation and properties of a fungicidal component present in seedlings of *Vicia faba. Netherlands J. Plant Pathol. 75*: 72–81.

Fenn, M. E., and Coffey, M. D. (1985). Further evidence for the direct mode of action of fosetyl-Al and phosphorus acid. *Phytopathology 75*: 1064–1068.

Ghosal, A., and Purkayastha, R. P. (1984). Elicitation of momilactone by gibberillin in rice. *Curr. Sci. 53*: 506–507.

Giri, D. N., and Sinha, A. K. (1983a). Effects of heavy metal salts on susceptibility of rice seedlings to brown spot disease. *Ann. Appl. Biol. 103*: 229–235.

Giri, D. N., and Sinha, A. K. (1983b). Control of brown spot disease of rice seedlings by treatment with a group of chemicals. *Pflanzenkrankheit. Pflanzenschutz 90*: 479–487.

Goodman, R. N. (1980). Defences triggered by previous invaders: bacteria. *Plant Disease: An Advanced Treatise*, Vol. 5 (J. G. Horsfall and E. B. Cowling, eds.), Academic Press, New York, pp. 305–317.

Grinstein, A. N., Lisker, N., Katan, J., and Eshel Y. (1976). Effect of dinitroaniline herbicides on plant resistance to soil-borne pathogen. *Phytopathology 66*: 517–522.

Grinstein, A. N., Lisker, N., Katan, J., and Eshel, Y. (1981). "Trifluralin," a sensitizer for *Fusarium* resistance in tomatoes and cotton. *Phytoparasitica 9*: 235–236.

Guest, D. I. (1984). Modification of defence responses in tobacco and capsicum following treatment with Fosetyl-Al [aluminum tris (*O*-ethyl phosphonate)]. *Physiol. Plant Pathol. 25*: 125–134.

Gutter, Y. (1983). Supplementary anti-mold activity of phosethyl-Al, a new brown rot fungicide for citrus fruits. *Phytopathol. Z. 107*: 301–308.

Hadwiger, L. A., and Schwochau, M. E. (1971). Specificity of deoxyribonucleic acid intercalating compounds in the control of phenylalanine ammonia lyase and pisatin levels. *Plant Physiol. 47*: 346–351.

Hadwiger, L. A. (1972a). Increased levels of pisatin and phenylalanine ammonia lyase in *Pisum sativum* treated with antihistaminic, antiviral, antimalarial, tranquilizing or other drugs. *Biochem. Biophys. Res. Commun. 46*: 71–79.

Hadwiger, L. A. (1972b). Induction of phenylalanine ammonia lyase and pisatin by photosensitive psoralen compounds. *Plant Physiol. 49*: 779–782.

Hadwiger, L. A., and Beckman, J. A. (1980). Chitosan as a component of pea–*Fusarium solani* interactions. *Plant Physiol. 66*: 205–211.

Hadwiger, L. A., and Schwochau, M. E. (1970). Induction of phenylalanine ammonia lyase and pisatin in pea pod by poly-lysine, spermidine, or histone fractions. *Biochem. Biophys. Res. Commun. 38*: 683–691.

Hadwiger, L. A., and Wagoner, W. (1983). Electrophoretic patterns of pea and *Fusarium solani* protein synthesized in-vitro or in-vivo which characterize the compatible and incompatible interactions. *Physiol. Plant Pathol. 15*: 211–218.

Hait, G. N. (1982). Nature of seed-borne infection in selected crop diseases and control of seedling damage by seed treatment with non-toxic chemicals. PhD. Thesis, Bidhan Chandra Krishi Viswavidyalaya, Kalyani.

Hait, G. N., and Sinha, A. K. (1986). Protection of wheat seedlings from *Helminthosporium* infection by seed treatment with chemicals. *J. Phytopathol. 115*: 97–107.

Hait, G. N., and Sinha, A. K. (1987). Biochemical changes associated with induction of resistance in rice seedlings to *Helminthosporium oryzae* by seed treatment with chemicals. *Z. Pflanzenkrankheit. Pflanzenschutz 94*: 360–368.

Hamilton, R. I. (1980). Defence triggered by previous invaders: viruses. *Plant Diseases: An Advanced Treatise*, Vol. 5 (J. G. Horsfall and E. B. Cowling, eds.), Academic Press, New York, pp. 279–303.

Hess, S. L., and Hadwiger, L. A. (1971). The induction of phenylalanine ammonia lyase and phaseollin by 9-amino acridine and other deoxyribonucleic acid intercalating compounds. *Plant Physiol. 48*: 197–201.

Iwata, M., Suzuki, Y., Watanabe, T., Mase, S., and Sezikawa, Y. (1980). Effect of probenazole on the activities of enzymes related to the resistant reaction in rice plants. *Ann. Phytopathol. Soc. Jpn. 46*: 297–306.

Jenns, A. E. and Kuć, J. (1977). Localized infection with tobacco necrosis virus protects cucumber against *Colletotrichum lagenarium. Physiol. Plant Pathol. 11*: 207–212.

Jenns, A. E., and Kuć, J. (1979). Graft transmission of systemic resistance of cucumber to anthracnose induced by *Colletotrichum lagenarium* and tobacco necrosis virus. *Phytopathology, 69*: 753–756.

Jenns, A. E., Caruso, F. L., and Kuć, J. (1979). Non-specific resistance to pathogen induced systemically by local infection of cucumber with tobacco necrosis virus, *Colletotrichum lagenarium* or *Pseudomonas lachrymans. Phytopathologia Mediterranea 18*: 129–134.

Kar, S. C., and Sinha, A. K. (1988). Drugs suppress brown spot infection of rice. *Science and Culture, 54*: 276–277.

Keen, N. T. (1975). Specific elicitors of plant phytoalexin production: determinants of race specificity in the pathogen? *Science, 187*: 74–75.

Keen, N. T. (1986). Phytoalexins and their involvement in plant disease resistance. *Iowa State J. Res. 60*: 477–.

Keen, N. T., and Legrand, M. (1980). Surface glycoproteins: evidence that they may function as race-specific elicitors of *Phytophthora megasperma* f. sp. *glycinea. Physiol. Plant Pathol. 17*: 175–192.

Kirino, O., Oshita, H., Oishi, T., and Kato, T. (1980). Preventive activity of N-allylamino acids against *Fusarium* diseases and their mode of action. *Agric. Biol. Chem. 44*: 35–40.

Klarman, W. L., and Gerdemann, J. W. (1963). Induced susceptibility in soybean plants genetically resistant to *Phytophthora sojae. Phytopathology, 53*: 863–864.

Kuć, J. (1982). Plant immunization mechanisms and practical implications. *Active Defence Mechanisms in Plants* (R. K. S. Wood, ed.), Plenum Press, New York, pp. 275–298.

Kuć, J. (1987). Plant Immunization and its applicability to disease control. *Innovative Approaches to Plant Disease Control* (I. Chet, ed.), Wiley, New York, pp. 225–274.

Kuć, J. (1990). A case for self defence in plants against diseases. *Phytoparasitica 18*: 3–8.

Kuć, J., and Tuzun, S. (1983). Immunization for disease resistance in tobacco. *Rec. Adv. Tobacco Sci. 9*: 179–213.

Lafon, R., Bugaret, Y., and Bulit, J. (1977). Nouvelles perspectives de lutte contre le mildiou de la Vigne [*Plasmopara viticola* (B.C.) Berlese et de Toni] avec un fongicide systemique, l'ethylphosphite d'aluminium. *Phytiatr. Phytopharm. 26*: 19–40.

Langcake, P. (1981). Alternative chemical agents for controlling plant diseases. *Trans. R. Soc. London B 295*: 83–101.

Lazarovits, G. (1988). Induced resistance: xenobiotics. *Experimental and Conceptual Plant Pathology* (R. S. Singh, U. S. Singh, W. M. Hess, and D. J. Weber, eds.), Oxford–IBH, New Delhi, pp. 575–592.

Maity, B. R. (1991). Studies on *Fusarium* resistance in pigeon pea and its induction by chemical treatment. PhD thesis, Bidhan Chandra Krishi Viswavidyalaya, Kalyani.

Mansfield, J. W. (1982). The role of phytoalexin in disease resistance. *Phytoalexins* (J. W. Mansfield and J. A. Bailey, eds.), Blackie, London, pp. 319–322.

Matta, A. (1971). Microbial penetration and immunization of congenial host plants. *Ann. Rev. Phytopathol. 9*: 387–410.

Matta,, A. (1980). Defence triggered by previous diverse invaders. *Plant Disease: An Advanced Treatise*, Vol. 5 (J. G. Horsfall and E. B. Cowling, eds.), Academic Press, New York, pp. 345–361.

Mucharromah, E., and Kuć, J. (1991). Oxalate and phosphate induce systemic resistance against diseases caused by fungi, bacteria and viruses in cucumber. *Crop Prot. 10*: 265–270.

Mukhopadhyay, S., and Sinha, A. K. (1980). Spore germination fluid as inducer of resistance in rice plants against brown spot disease. *Trans. Br. Mycol. Soc. 74*: 69–72.

Müller, K. O., and Börger, H. (1940). Experimentelle Untersuchungen uber die *Phytophthora* resistenz der kartoffel. *Arbeiten. Biol. Reichsanst. Land. Forstwirtsch. 23*: 189–231.

Ossowski, O., Piloth, A., Garreg, P., and Lindberg, B. (1984). Synthesis of a glucoheptose that elicit phytoalexin accumulation in soybean. *J. Biol. Chem. 259*: 113–137.

Pearce, R. B., and Ride, J. P. (1982). Chitin and related compounds as elicitors of lignification response in wounded leaves. *Physiol. Plant Pathol. 20*: 119–123.

Pegg, G. F. (1976). The response of ethylene treated tomato plants to infection by *Verticillium albo-atrum*. *Physiol. Plant Pathol. 9*: 215–226.

Perrin, D., and Cruickshank, I. A. M. (1965). Studies on phytoalexins. VII. Chemical stimulation of pisatin formation in *Pisum sativum* 1. *Aust. J. Biol. Sci. 18*: 803–810.

Price, W. C. (1964). Strains, mutations, acquired immunity and interference. *Plant Virology* (M. K. Corbett and H. D. Sisler, eds.), University Press, Gainesville, FL, pp. 93–117.

Purkayastha, R. P. (1973). Phytoalexins: plant antigens and disease resistance. *Science and Culture, 39*: 528–535.

Purkayastha, R. P., and Bannerjee, R. (1990). Immunoserological studies on cloxa-

cillin-induced resistance to soybean against anthracnose. *J. Plant Dis. Prot. 97*: 349–359.
Rathmell, W. G. (1982). Active defence mechanisms of plants in relation to protection of crops from pathogens with chemicals. *Active Defence Mechanisms in Plants* (R. K. S. Wood, ed.), Plenum Press, New York, pp. 347–348.
Rathmell, W. G., and Smith, D. A. (1980). Lack of activity of selected isoflavonoid phytoalexins as protectant fungicides. *Pest. Sci. 11*: 568–572.
Retig, N. (1974). Changes in peroxidase and polyphenoloxidase associated with natural and induced resistance of tomato to *Fusarium* wilt. *Physiol. Plant Pathol. 4*: 145–150.
Sahana, D. (1991). *Control of chickpea diseases by seed treatment with non-toxic chemicals*. PhD thesis, Bidhan Chandra Krishi Viswavidyalaya, Kalyani.
Sarkar, M. L., and Sinha, A. K. (1991). Control of sheath blight of rice by unconventional chemicals. *Indian Phytopathol. 44*: 379–382.
Schwochau, M. E., and Hadwiger, L. A. (1968). Stimulation of pisatin production in *Pisum sativum* by Actinomycin D and other compounds. *Arch. Biochem. Biophys. 126*: 731–732.
Sekizawa, Y., and Mase, S. (1980). Recent progress in studies on nonfungicidal controlling agent probenazole with reference to the induced resistance mechanism of rice plant. *Rev. Plant Prot. Res. 13*: 114–121.
Sequeira, L. (1983). Mechanisms of induced resistance in plants. *Ann. Rev. Microbiol. 37*: 51–59.
Sharp, J., Albersheim, P., Ossowski, O., Pilotti, A., Garreg, P., and Lindberg, B. (1984). Comparison of the structure and elicitor activities of a synthetic and mycelial wall derived hexa-(β-D-glucopyranosyl)-D-glucitol. *J. Biol. Chem. 259*: 113–141.
Shimura, M., Iwata, M., Tashiro, N., Sekizawa, Y., Suzuki, Y., Mase, S., and Watanabe, T. (1981). Anti-conidial germination factor induced in the presence of probenazole in infected host leaves. I. Isolation and properties of four active substances. *Agric. Biol. Chem. 45*: 1431.
Sijpesteijn, A. K. (1969). Mode of action of phenylthiourea, a therapeutic agent for cucumber scab. *J. Sci. Food Agric. 20*: 403–405.
Sinha, A. K. (1976). Induced resistance as a factor in plant disease control. *Indian Biologist, 9*: 1–16.
Sinha, A. K. (1984). A new concept in plant disease control. *Science and Culture, 50*: 181–186.
Sinha, A. K. (1990). Basic research on induced resistance for crop disease management. *Basic Research for Crop Disease Management* (P. Vidyasekaran, ed.), Daya, New Delhi, pp. 87–100.
Sinha, A. K. (1992). An alternative approach to chemical control of plant diseases. *Plant Science in the Nineties* (R. D. Bannerjee, S. P. Sen, K. R. Samaddar, U. Sen, A. K. Sarkar, and A. K. Biswas, eds.), University of Kalyani, Kalyani, pp. 97–113.
Sinha, A. K., and Das, N. C. (1972). Induced resistance in rice plants against *Helminthosporium* infection. *Physiol. Plant Pathol. 8*: 401–410.
Sinha, A. K., and Hait, G. N. (1982). Host sensitization as a factor in the induction

of resistance in rice against *Drechslera* by seed treatment with phytoalexin inducers. *Trans. Br. Mycol. Soc. 79*: 213–219.

Sinha, A. K., and Sengupta, T. K. (1986). Use of unconventional chemicals in the control of blast of rice. *Proceedings of the 2nd International Conference on Plant Protection in the Tropics*, Genting Highlands, Malaysia, pp. 19–21.

Sinha, A. K., and Trivedi, N. (1969). Immunization of rice plants against *Helminthosporium* infection. *Nature, 223*: 963–964.

Sinha, A. K., and Wood, R. K. S. (1964a). Control of *Verticillium* wilt of tomato plants with "Cycocel" [(2-chloroethyl)trimethyl ammonium chloride]. *Nature, 202*: 824.

Sinha, A. K., and Wood, R. K. S. (1964b). The effect of growth substances on *Verticillium* wilt of tomato plants. *Ann. Appl. Biol. 60*: 117–128.

Stekoll, M. S., and West, C. A. (1978). Purification and properties of an elicitor of castor bean phytoalexin from culture filtrate of the fungus *Rhizopus stolonifer. Plant Physiol. 61*: 38–45.

Subramaniam, S. (1963). *Fusarium* wilt of pigeonpea. III. Manganese nutrition and disease resistance. *Proc. Indian Acad. Sci. (Plant Sci.), 57*: 259–274.

Tomiku, T., Fuzikawa, T., Sato, S., and Ando, S. (1979). Occurrence of *Xanthomonas oryzae* infection of rice and time and number of application of probenazole granules. *Nogyo Oyobi Engei 54*: 339–340.

Trivedi, N., and Sinha, A. K. (1976a). Factors affecting the induction of resistance in rice plants to *Helminthosporium oryzae. J. Soc. Exp. Agric. 1*: 20–24.

Trivedi, N., and Sinha, A. K. (1976b). Resistance induced in rice plants against *Helminthosporium* infection by treatment with various fungal fluids. *Phytopathol. Z. 86*: 335–344.

Trivedi, N., and Sinha, A. K. (1980). Effect of pre-inoculation treatments with some heavy metal salts and amino acids on brown spot disease in rice seedlings. *Proc. Indian Acad. Sci.* (*Plant Sci.*) *89*: 283–289.

Uehara, K. (1963). On the production of phytoalexin by metallic compounds. *Bull. Hiroshima Agric. Coll. 2*: 41–42.

Uknes, S., Mauch-Mani, Mayer, M., Potter, S., William, S., Drencher, S., Chandler, D., Susarenko, A., Ward, E., and Ryals, J. (1992). Acquired in *Arabidopsis. Plant Cell, 4*: 645–656.

Uritani, I., Uritani, M., and Yamada, H. (1960). Similar metabolic alterations induced in sweet potato by poisonous chemicals and by *Ceratostomella fimbriata. Phytopathology, 50*: 31–34.

VanAndel, O. M. (1966). Amino acids and plant diseases. *Ann. Rev. Phytopathol. 4*: 349–368.

Van Etten, H. D., Mathews, D. E., and Mathews, P. S. (1989). Phytoalexin detoxification: importance for pathogenicity and practical implications. *Ann. Rev. Phytopathol. 27*: 143–164.

Van Loon, L. C. (1983). The induction of pathogenesis-related proteins by pathogens and specific chemicals. *Netherlands J. Plant Pathol. 89*: 265–273.

Vo-Thi-Hai, G., Bompeix, G., and Ravise, A. (1979). Role du tris-*O*-ethyl phosphonate d'aluminium dan la stimulation des reactions du tomate contre le *Phytophthora capsici. C. R. Hebd. Seances Acad. Sci. C 288*: 1171–1174.

Wain, R. L., and Carter, G. L. (1972). Historical aspects. *Systemic Fungicides* (R. W. Marsh, ed.), Longmans, London, pp. 6–33.

Ward, E. W. B., Unwin, C. H., and Stoessl, A. (1975). Experimental control of late blight of tomatoes with capsidiol, the phytoalexin from pepper. *Phytopathology 65*: 168–169.

Ward, E. R., Uknes, S. J., Williams, S. C., Dincher, S. S., Wiederhold, D. L., Alexander, D. C., Ahl-Goy, P., Metraux, J. P., and Ryals, J. A. (1991). *Plant Cell, 3*: 1085–1094.

Watanabe, T., Sekizawa, Y., Shimura, M., Suzuki, Y., Matsumoto, K., Iwata, M., and Mase, S. (1979). Effects of probenazole (Oryzemate) on rice plants with reference to controlling rice blast. *J. Pest. Sci. 4*: 53–59.

Yoshikawa, M. (1978). Diverse modes of action of biotic and abiotic phytoalexin elicitors. *Nature, 275*: 546–547.

Yoshikawa, M. (1983). Macromolecules, recognition and the triggering of resistance. *Biochemical Plant Pathology* (J. Callow, ed.), Wiley, London, pp. 267–298.

Index

Abelmoschus esculentus, 25
Abscisic acid, 10, 16
Acanthaceae, 544, 549
Acer saccharum, 171
Acetonitrile, 202, 204, 240–242, 410, 411, 413
Acetophenone, 353
  4-alkoxy-, 417
  4-benzyloxy-2-hydroxy-, 415
  4-butoxy-, 417
  2-hydroxy-4,6-dimethoxy-, 263
Acetyl CoA, 216, 407, 468, 480
Acetyl salicyclic acid, 580
Achyla flagellata, 12
Acrocylindrium, 25
  oryzae, 21, 26
Acrosporium, 542
Actinomycin D, 8, 233, 245, 246
Adenosine monophosphate (AMP), 429
  cyclic-, 429, 430, 431, 438
  dibutyryl cyclic-, 430, 431
S-Adenosyl-L-methionine:bergaptol, 72
S-Adenosyl-L-methionine:xanthotoxol *O*-methyl transferase, 72
Adenyl cyclase, 429, 430, 431
Adhatoda, 544, 549
  vasica, 544
Aecidium
  adhatodae, 544
  carviae, 544
  withaniae, 545
Aflatoxin(s), 161, 218, 219
  B_1, 161, 218
  B_2, 218
  B_3, 218
  G_2, 218
Agarospirol, 461
Agglutination, conidia, 147, 148
Agrobacterium tumefaciens, 24, 64
Aizoaceae, 545
Ajmalicine, 12
Alarms, in defense systems, 486
Alanine, 144
Alfalfa, 48, 96, 222, 223, 391, 392, 405 (*see also Medicago sativa*)
Alkaloids, 535, 537, 539, 543, 545–548
Alkanals, 178
Alkanones, 178
Alkenals, 175, 178
  C_6–C_9–, 175
Alkyl *bis*-phenyl ether, 216, 217
Alkyl *bis*-phenyls, 209, 211, 214, 215, 220, 221
Allylic alcohol, 450
N-Allyl amino acids, 560
N-Allyl glycine, 560

N-Allyl sarcosine, 560
Alternaria, 547, 548
  alternata, 50, 539, 541, 543
  blight, 525
  brassicae, 238
  sesami, 525–531
Aluminum (Al), 144, 384
Aluminium *tris-O*-ethyl phosphonate (TEPA), 375, 561
  (*see also* Fosetyl-Al)
Amino acids in phytoalexin elicitation, 560, 563
γ-Aminobutyric acid, 144
Aminocyclopropane carboxylic acid (ACC), 106, 107
  synthase, 107
Aminoethoxy vinylglycine (AVG), 383
L-α-Aminohydrazino-β-phenyl propionic acid (AHPP), 383
α-Aminooxy acetic acid (AOA), 266, 275, 281, 379, 382, 383, 398, 399, 562
L-α-Aminooxy-β-phenyl propionic acid (AOPP), 380, 400
Anacardiaceae, 540
Anogeissus, 541, 549
  latifolia, 541
Anthocyanin(s), 124
  deoxy-, 193
Anthacnose, in
  Adhatoda, 544
  bean, 46
  Dioscorea, 336
  soybean, 27
Antibiotics, 2, 136, 288, 333, 497, 563
Antigens, 5, 20–27, 570
Aphanomyces euteiches, 12, 358, 560
Apigenin, 543, 548
  4′-OMe-, 535, 543, 544, 547, 548
  7-OMe-, 544, 545
  7,4′-diOMe-, 547
Apoplast, 187, 189, 401
Apple, 536
  scab, 560
Arabidopsis, 581
  thaliana, 237, 256
Arabinose, 147
Arachidins, 203, 208, 211, 212, 215, 217, 219
Arachidonic acid, 12, 13, 432
Arachis, 223, 361
  hypogea, 24, 50, 199, 287, 309
  (*see also* Groundnut)
Arginine, 144
Aromatic hydrocarbon fungicide (AHF), 328, 329
Aromatic substituent constant, 328
Arthrobactor luteus, 62
2-Arylbenzofuran, 377
Asclepiadaceae, 546, 549
Ascochyta
  imperfecta, 96
  pisi, 17
  rabiei, 15, 21, 397
Asparagine, 144
Aspartic acid, 144
Aspergillus, 218, 219, 222
  aculeatus, 544
  flavus, 161–181, 200, 217–220, 343, 361
  fumigatus, 361, 365
  japonicus, 102
  niger, 335, 343, 358, 359, 361, 525–528, 536, 537, 540, 541, 545, 546
  parasiticus, 161, 200, 218
Aspirin, 580
Asteridae, 549
Asterolibertina, 540
ATPases, 51–53, 75
Aureobasidium, 540
Aureofungin, 526, 528, 530
Aureol, 341, 355, 359, 360
Autofluorescence, 191–193
Auxins, 110
Avenalbumin, 14
Avena sativa, 14, 24

Bacillus cereus, 343
Bacteria, 117–119, 125, 153, 169, 183–198, 202, 242, 336, 344, 361, 505
  cultures, 118, 344
Bacterial pustule disease, soybean, 118
Barium chloride, 564, 567–569
Barium sulfate, 564, 568, 583
Barley, 42–43, 51, 52, 54
Batatasin(s), 336, 342–344, 348–351, 356, 361–363, 365, 366
  demethyl-, 342, 344, 348–350, 356, 361–363, 365
Bavistin, 526, 528, 530
B chromosomes, 201
Beloruskii rannii, 12
Benomyl, 558
Benzimidazole, 377, 382
Benzoic acid, 560, 580
  halogenated, 560
  p-hydroxy-, 535, 536, 539–545, 547–549
  2,3,5-triiodo-(TIBA), 560
Benzoquinone, 540, 543, 549
Benzyl amino purine, 10
Benzyl penicillin, 11
Bergaptol:*S*-adenosyl methionine, 72
Bibenzyl
  2′-hydroxy-3,5-dimethoxy-, 349
  3-hydroxy-5-methoxy-, 336, 348, 349
  2,4′-dihydroxy-3,5-dimethoxy-, 342, 349
Bignoniaceae, 543
Biochanin A, 15, 90, 120
Biosol-2, 141
Bipolaris carbonum, 486, 489–492, 494
Blackleg disease, crucifers, 230, 231, 254
Blight, bacterial, in cotton, 183–198
Boraginaceae, 546
Boric acid, 564, 568
Boron, 560
Botryodiplodia, 547
  natalensis, 268
  theobromae, 19, 343, 348, 350, 356, 538, 543
Botrytis, 191, 486
  cinerea, 3, 10, 14, 19, 20, 287, 288, 290, 291, 293, 294, 299, 300, 304, 308, 309, 318–326, 329, 357, 536
  fabae, 3, 19, 558, 560, 561
Brassenin, β-methoxy, 236
Brassica, 229–261
  adpressa, 247
  campestris, 229, 259
    var. *pekinensis*, 259
  carinata, 229–232, 247–250, 252
  insularis, 231
  juncea, 229–233, 235, 241, 243–246, 248–252
  naponigra, 260
  napus, 229–232, 235, 244, 246–253, 255
    var. *oleifera*, 229
  nigra, 229–232, 248–253
  oleracea, 229, 230, 232, 237, 247, 248, 250
  rapa, 229, 230, 237, 247, 248, 250, 251
Brassicanal
  A, 236
  B, 236
  C, 236
Brassilexin, 234–240, 244–255
Brassinin, 235–238, 240, 244, 245, 247–250, 254
  cyclo-, 234–240, 242, 244–250, 253, 254
  4-methoxy-, 235, 236, 238, 240, 242, 244–250
  spiro-, 235, 245
Brassitin, 236
  methoxy-, 236
Bremia lectucae, 384
Broadbean, 3, 191, 558, 561

Brussels sprouts, 405
Buckwheat, 402
2-Buten-1-ol, 176
2-Butoxy alcohol, 176

Cadalene, 168, 169, 178, 195
  biosynthesis, 193–194
  2,7-dihydroxy-(DHC), 164–166, 168–173, 178, 184–194
  2-hydroxy-7-methoxy-(HMC,DHMC), 164–166, 169–173, 178, 185–194
δ-Cadinene, 185, 194
Cadmium chloride, 12, 27, 496, 571–575
Cadmium sulfate, 478, 479
Caesalpiniaceae, 536
Caffeic acid, 487, 546
  O-methyl transferase, 124
Cajanol, 338, 352
Calcium (Ca,Ca^{++}), 13, 54, 74, 138, 144, 429, 431–433, 435, 438, 475
  chloride, 238
Callose, 72
Calmidazolium, 431
Calmodulin (CaM), 431, 432, 438
Calotropis gigantea, 546, 548
Calphostin C, 437
Camalexin(s), 236
  methoxy-1-, 236
Camelina sativa, 237, 251
Camellia
  japonica, 489
  sinensis, 485–501
    var. *assamica*, 485
    var. *sinensis*, 485
Camellidin I & II, 489
Camphene, 177
Canavalia ensiformis, 370, 388
Capsella bursa-pastoris, 252
Capsenone, 18
Capsicum
  annuum, 15, 503, 506
  frutescens, 522
Capsidiol, 9, 15, 18, 381, 384, 503–523, 558
  effect of metaloxyl, 512–514, 517–519
  plant age, 509–512, 514–517
  time course, 513–514
Carbon monoxide, 476, 477
Carotenoids, 548
Carrot, 430 (*see also Daucas carota*)
Carvia, 544, 549
  callosa, 544
Caryophyllene, 177
Casbene, 9, 14
  synthetase, 14
Cassia, 549
  fistula, 536
Catalase(s), 125, 169
Catechin, 486–489
  2,3-dioxygenase, 487
Catechol, 487, 488, 494
Catenulospora zizyphi, 543
Cauliflower mosaic virus, 184, 188
Cell aggregates in culture, 120, 121
Cellular responses, 85–115
  competency factors, 102–106
  conditioning, 102–106
  mediating, 107–110
  modulating, 107–110
Ceratocystis
  fagacearum, 560
  fimbriata, 4, 24, 467–483
Ceratostomella ulmi, 560
Cercospora, 210, 540
  adhatodae, 544
  arachidicola, 200, 208, 209, 215–217, 220–222
  beticola, 21
  morindae, 538
  oryzae, 563
  sesamicola, 525
  tectonae, 534
  trianthemae, 545
  tylophorina, 546
  withaniae, 545
Chaconia tectonae, 534

Chaetomium, 547
cupreum, 487
globosum, 14, 546
mangiferae, 540
Chalcone(s), 93, 354, 415
isomerase, (CHI), 5, 55, 86, 354, 407
synthase, (CHS), 5, 55, 86, 354, 357, 407
Charge transfer complexes (CTC), 318, 325, 328
Chickpea, 16, 397, 400, 564, 568, 573 (*see also Cicer arietinum*)
Chili, 572
Chitinase, 107, 401, 559, 577, 580
Chitosan, 10, 429, 558, 565, 567, 568, 570–575, 579, 581, 583
Chloramphenicol, 15
Chlorocholine chloride (Cycocel), 561, 565
Chlorogenic acid, 488
p-Chloromercuric benzoic acid (*p-Cmb*), 567, 569, 571–575, 583
Chnoopsora butleri, 544
Chocolate spot, broad bean, 558, 560, 561
Cholera toxin, 430
Chromosomes, B, 201
Chrysanthemum, 52
CHS gene, 54
Cicer arietinum, 15, 21 (*see also* Chickpea)
Ciliochorella, 540
Cinnamate-4-hydroxylase, 124, 216, 407, 476, 479
Cinnamic acid, 169, 216, 280, 281, 287, 356, 423–425
trans-, 407
Cinnamoyl-CoA ligase, 334, 354
Citronerol, 472
Citrullus vulgaris, 24
Citrus, 263–286, 384, 549, 561 (*see also* Orange)
aurantium, 263, 264
jambhiri, 264
limon, 263
macrophylla, 264
paradisi, 264
reticulata, 264
sinensis, 263, 264
Cladosporium, 207, 217, 218, 234, 235, 547
cladosporoides, 343, 525–528
cucumerinum, 203, 335, 343, 358–360, 560, 561
fulvum, 376, 401
herbarum, 203
Clerodendron inerme, 546, 547
Cloxacillin, 11, 27, 497, 562, 577, 578
Cochliobolus
carbonum, 401, 489
miyabeanus, 44, 489
Coffea arabica, 24
Collar rot, chickpea, 564, 573
Colletotrichum, 191, 546
camelliae, 486, 489, 491, 492
capsici, 525–528
corchori, 24
dematium, 544, 562, 580
var. *truncata*, 11, 24, 26, 27
falcatum, 539
gloeosporoides, 268, 538, 542
graminicola, 193, 495
lagenaricum, 562
lindemuthianum, 9, 10, 17, 55, 378, 400, 425
Combretaceae, 356, 541, 549
Competency factors, 102
Compositae, 4
Coniophora carebella, 538
Convolvulaceae, 4
Copper(Cu), 144, 560
chloride, 233, 241–243, 247, 248, 252 (*see also* Cupric chloride and Cuprous chloride)
sulfate, 15, 568
Corchorus capsularis, 24
Cordia, 547
myxa, 546, 547
Cordycepin, 422, 432
Coriolus versicolor, 538

Corn, 73 (*see also Zea mays*)
Corticum
  invisum, 486
  theae, 486
Corynebacterium insidiosum, 25
Corynespora cassicola, 544
Cotton, 129–198, 560 (*see also Gossypium* spp.)
  boll, 162, 163, 169, 170
  seed, 161–181
Cotyledon(s), 94–98, 189–191, 219
  bioassay, 102–106, 186, 201, 252, 253
4-Coumarate:CoA ligase, 5, 216, 407
o-Coumaric acid, 537
  4-methoxy-, 281
p-Courmaic acid, 281, 318, 356, 407, 488, 537, 540, 543–547
m-Coumaric acid, dihydro-, 356
p-Coumaroyl CoA, 216, 287, 334, 356, 406, 407
Coumarin(s), 275, 537, 539, 544
  6,7-dimethoxy- (*see* Scoparone)
  6,7-dimethyl pyrano- (*see* Xanthyletin)
  7,8-dimethyl pyrano- (*see* Seselin)
  7-hydroxy- (*see* Umbelliferone)
  7-methoxy-, 281
  6-methoxy-7-hydroxy- (*see* Scopoletin)
Coumestans, 337, 341, 351, 355, 359, 360
Coumestrol, 118–120, 341, 346
Cow pea, 51, 52, 123, 376–379, 385 (*see also Vigna unguiculata*)
Cristacarpin, 340, 348
Cross-protection phenomenon, 149–153
Cross-reactive antigens (CRA), 27, 497
Crown-6, 141
Crucifers, 187, 229–261
Cucumber, 222, 385, 503, 556, 562, 576
  scab, 560, 561
Cupric chloride, 9, 496, 560, 564, 567, 570–575, 583 (*see also* Copper chloride)
Cupric sulfate, 564, 567
Cuprous chloride, 449 (*see also* Copper chloride)
Culture darkening, 117–128
Curvularia, 540
  clavata, 535, 536
  lunata, 525–528
  prasadii, 543
  spicata, 210
Cuscuta, 534, 546–548, 550
  chinensis, 546, 548
  reflexa, 546, 548
Cyanogenic glycosides, 486
Cyclic AMP, 429, 430, 431, 432
  2′,3′-, 431
  3′,5′-, 431, 432
Cyclic nucleotide diphosphodiesterase, 431, 432
Cyclobrassinin, 234–240, 242, 245–250, 253, 254
  sulfoxide, 235, 236, 247, 249, 250, 254
Cyclodehydroisolubimin, 18
Cyclohexamone, 177
Cycloheximide, 233, 245, 246, 496, 565, 567–569, 571–575, 583
Cyclokievitone, 338, 339, 347, 355, 359, 360
  1′′,2′′-dehydro-, 359
  hydrate, 339, 347
  3′,4′,5′-trimethyl-, 359
Cyclopolyether derivatives, 141
Cyclopropane carboxylic acid,2,2-dichloro-3,3′-dimethyl-, 14, 375
Cycocel (*see* Chlorocholine chloride)
Cylindrocladium scoparium, 542
Cysteine, 144, 564, 567, 569–575, 583
Cystine, 144
Cytochrome *c*, 476
  reductase, 328, 329

Cytochrome *p*-450, 476, 479, 480
isoenzymes, 16
monooxygenase, 16
Cytokinin(s), 109, 110, 124

Daedalia
flavida, 534
quercina, 537
Dalbergioidin, 337, 347, 348, 352, 355, 359–361
5,2′-dimethoxy-, 337, 347, 352, 355
Daucas carota, 124 (*see also* Carrot)
Damping off, chili, 564
Datura stromonium, 465
1-Decanol, 176
N-Decyl aldehyde, 176
Defense genes, 107, 112, 291, 436
expression, 107
regulation, 107, 108
transcription, 291
Defense reaction, 41, 54, 64–65, 129, 136, 141
elicitation, 53
protein kinase in, 53–54
strategies of, 161–181, 485–501
suppression, 12, 41, 49–53, 142
Dendrolasin, 4
6-hydroxy-, 469, 474
6-oxo-, 469, 474
Dendrolasinoide, 6-oxo-, 474
Desoxy hemigossypol, 167
Detoxification, 5, 16–19, 155, 355, 356, 391–403
glycosides, 391–403
pterostilbene, 309
resveratol, 309
ϵ-viniferin, 309
Diacyl glycerol (DAG), 429, 435–438
Diadzein, 86, 90, 93, 94, 96–98, 118, 119, 121–123, 209, 211, 213, 215, 216, 337, 346, 352, 355, 359, 360, 400
biosynthesis, 93, 216, 355
conjugates, 90, 91, 95–97, 106, 109, 110, 400
8,2′-dihydroxy-, 337, 346, 352, 355
8,2′-dihydroxy dihydro-, 337, 346, 352, 355
2′-hydroxy-, 337, 346, 352, 354, 355, 359
2′-hydroxy dihydro-, 337, 346, 352, 355
Diaporthe citri, 264, 269, 489
Diazotised *p*-nitraniline (DPN), 203, 212–214, 490, 535
Diazotised sulphanilic acid (DSA), 203, 212–214, 535
Dibenzo-18, 141
Dibutylphthalate (DBP), 410, 411, 414, 416, 417
Dichlorocyclopropanes, 505, 518
Dichloromethoxyphenol (DCMP), 328
2,4-Dichlorophenoxy acetic acid (2,4-D), 123, 124, 560, 565, 567–569, 571–575, 580, 583
Dienols, 216–218, 220, 221
Diethyl amine, 451
Di-2-ethyl hexyl phthalate, 137
Diethylmalonate (DME), 451
Digitonin, 72
Dihomo-*y*-linolenic acid, 13
Dihydroazulene, 450
Dihydrostilbene(s), 334, 337, 342, 344, 348-351, 356, 357, 361–365
antimicrobial properties, 361
biosynthesis, 356, 357
chemotaxonomy, 358
growth inhibitory properties, 361–365
4-hydroxy-3,5-dimethoxy-, 342
3-hydroxy-5-methoxy-, 342
o-Dihydroxyphenol, 573–575, 579
2,4-Dinitrophenol, 526, 528, 530
Dioscorea, 333–335, 343, 348, 351, 356, 358
alata, 335, 336, 348, 356
botatas, 348, 351, 358, 361
bulbifera, 336, 349, 358, 366

[*Dioscorea*]
cayanensis, 334
chemotaxonomy, 358
decipiens, 336
dumentorum, 335, 336, 349, 358, 366
hispida, 336
mangenotiana, 336, 349, 358, 366
opposita, 348, 349
rotundata, 335, 336, 349, 365, 366
Dioscoreaceae, 348
Diplodia, 540
gossypina, 168
zeae, 70
Dipogon, *lignosus*, 348
Dithan M-45, 526, 528, 530
cDNA, 55, 195
chalcone isomerase, 55, 65
chalcone synthase, 55, 65
elicitor-releasing factor, 64
molecular cloning, 64
NADPH: isoflavone oxidoreductase, 403
phenyl alanine ammonia lyase, 55, 65
pine stilbene synthase, 357
soybean-β-1,3-endoglucanase, 64
DNAase I, 184
DNA plasmids, 184
Dodecyl aldehyde, 176
Dodecyl-DL-alaninate, 560
Dolichin (A&B), 340, 348
Dolichos biflorus, 348
Dothichiza populea, 537
Downy mildew, 504
tobacco, 41
vitis, 558, 561
Drechslera
sorokiniana, 576
speciferum, 544

EGTA, 54, 431, 432, 434
Eicosapentenoic acid, 12, 13
Electron transport inhibition, 75
Elicitors, 5, 6, 12, 15, 27, 53, 54, 61–67, 70–73, 77, 94, 117–122, 124, 148, 149, 155, 162, 166, 241–243, 409, 411, 416, 418–427, 436, 437, 467, 480, 556, 557, 558
abiotic, 6, 8, 233, 235, 241, 243, 244, 254, 497, 557
biotic, 86, 233, 242, 556
co-, 106
fatty acid, 167
primary, 99, 101
multiple, 100
pmg, 62
receptor complex, 53
releasing factor, 62
specific, 6
synergists, 99, 101, 102, 105, 167
V_1-, 418–427
V_2-, 418–427, 436, 437
volatile, 171–178
Ellagic acid, 488
Elsinoe kamatii, 539
β-1,3-Endoglucanase, 62, 63, 66, 72, 100, 108, 559
Endopolygalacturonate lyase, 559
Endopolygalacturonidase, 102
Epicatechin, 487
gallate, 487
Epigallocatechin, 487
gallate, 487
Epioxylubimin, 452
Erysiphe
graminis, 7, 42, 44, 47, 48
f.sp. *cichoracearum*, 540
f.sp. *hordei*, 42–45
f.sp. *tritici*, 45, 48
pisi, 44–47, 54
Erythrina, 357
Erwinia
caratovora, 102, 243, 559
ssp. *atroseptica*, 232
chrysanthemi, 243
pv. *dahliae*, 232
Escherichia coli, 343

Ethephon, 561
Ethylene, 9, 72, 106, 107, 246, 247, 383, 561, 578
N-Ethyl maleimide, 375
Eucalyptus, 538, 549
  globulus, 538, 539
  microcorys, 538
  regnans, 538
  sideroxylon, 538
  triflora, 538
Evans blue, 185
Exobasidium vexans, 486
Exopolysaccharides, 11

Fabaceae, 199, 253, 334, 345, 533 (*see also* Leguminoseae)
Falcarinol, 536
Farnesal, 470
Farnesoic acid, 9-hydroxy-, 474
Farnesol, 468–470, 472, 474, 475, 480
  dehydrogenase, 470, 472, 473, 480
  9-hydroxy-, 469, 474, 480
  9-oxo-, 469, 480
Farnesyl pyrophosphate, *trans*, *trans* (FPP), 185, 193–195, 468, 480
Fatty acids, 209, 210, 214–216, 558
Ferulic acid, 97, 168, 169, 535, 538, 542–548
Ferric chloride, 564, 567, 571–575
Flavone(s), 353, 535, 550
  6,7-dimethoxy-3,5,4′-trihydroxy-, 545
  6-hydroxy flavone, 548
Flavonoids, 208, 210, 213, 215, 216, 317, 324, 534, 541, 545, 546, 548, 549, 550
  demethylation, 17, 537, 549
  methoxylation, 549
Flavonol(s), 90, 91, 535, 537, 541, 542, 546, 549, 550, 573
Fluoresein, 184
Fomes
  annosus, 537
  durus, 538
Forage legumes, 391–403
Formarin, 451
Formononetin, 15, 90, 120, 209, 211, 213, 215, 216, 220, 407
  7-*O*-glycoside-6′′-*O*-malonate (FGM), 400
Fosetyl-Al, 264, 266, 277–280, 282, 290, 292, 293, 305–307, 375, 376, 383–384, 561, 562
Free radicals, 78, 187, 188, 328
French bean, 9, 52, 558 (*see also Phaseolus vulgaris*)
Furanocoumarins, 101
Furanosesquiterpenoids, 4, 467–483
  biosynthesis, 468, 475
  enzyme conversion, 467–483
Furanoterpenoids, 445, 558
Fusarium, 108, 541, 545, 560, 561
  moniliformae, 14, 541
  oxysporum, 189, 541
    f.sp. *ciceri*, 568, 573
    f.sp. *lycopersici*, 10
    f.sp. *pisi*, 17
    f.sp. *vasinfectum*, 19, 24, 25
  roseum, 393, 397
  sambucinum, 17
  semitectum, 24
  solani, 7, 18, 24, 525–528, 535–537, 539–542
    f.sp. *cucurbitae*, 358
    f.sp. *phaseoli*, 17, 356, 579, 581
    f.sp. *pisi*, 579
  sulphureum, 19
  udum, 25, 560, 572, 573, 580
Fusarium wilt, 561
  chickpea, 564, 573
  pigeonpea, 537, 572
  tomato, 560, 561
Fusicoccum amygdali, 535

Galactose, 138, 144, 146–148, 489
α-1,4-Galacturonides, 86
Gallic acid, 488, 540
Gallocatechin, 487
Gamma irradiation, 264, 265, 271–277, 282

Ganoderma
  applanatum, 534
  lucidum, 539
Gaseous phytoalexins, 171–178
Genes, 136
  antisense, 222
  avirulence, 20, 183
  bacterial blight resistance, cotton, 186
  CHS, 54–56, 428
  glucanase, 64
  glyceollin, 65
  PAL, 56, 428
  phosphorylated, 54, 57
  phytoalexin biosynthesis, 54
  Pda, 16, 201
  pisatin demethylase, 385
  resistance, 20
  Rps, 70, 85, 89, 92
  silencer regions, 54
  stilbene synthase, 309
  suppressor, 22
Genistein, 90–96, 119, 121–123, 337, 345, 347, 348, 352, 355, 359, 360
  biosynthesis, 355
  conjugates, 90, 96, 106
  8,2′-dihydroxy-, 337, 347, 352, 355
  8,2′-dihydroxy dihydro-, 337, 347, 352, 355
  2′-hydroxy-, 337, 347, 348, 352, 355, 359–361
  2′-methoxy-, 337, 347, 352, 355, 359, 361
Gentisic acid, 535, 538, 540, 543, 546, 547
Geraniol, 472
  dehydrogenase, 472
Geranyl pyrophosphate, 468
Gibberella pulicaris, 17, 18
Gibberellic acid, 11, 14, 26, 497, 561, 577
Gloeosporium, 547, 548
  morindae, 538
  theae-sinensis, 489, 492
Glomerella cingulata, 19, 543
Glucanase(s), 62, 107, 401, 577, 580
Glucans, 95
  wall-, 103
  yeast wall-, 103
  β(1 → 3)-, 63, 86, 100, 101
  β(1 → 6)-, 63, 86, 100, 101
Glucobrassicin, 236, 253
  4-methoxy-, 236
  neo-, 236
Glucosamine, 144
Glucose, 138, 144, 146–148, 489
  2-desoxy-, 293
Glucose-6-phosphate dehydrogenase, 187
β-Glucosidase, 393, 397
Glucosinolates, 245, 253, 254
Glucuronic acid, 489
Glutaryl CoA, 3-hydroxy-3-methyl (HMG CoA), 468
Glutathione, 106, 155
  S-transferase, 155, 156
Glycanase, 62, 77
  released elicitors, 62, 63
Glyceollin, 8, 26, 50, 61, 62, 71–73, 85, 92–94, 96, 98, 101–106, 117–119, 121–125, 193, 201, 354, 400, 497, 504, 505
  biosynthesis, 65, 86, 365
  elicitors, 71, 72, 86
  isomers, 73, 85, 120
Glycetine, 90
Glycine, 73, 334
  max, 24, 50, 61, 69, 117, 505 (*see also* soybean)
  sojae, 92
  wightii, 70
Glycinol, 340, 346, 348
  2,10-(dimethyl allyl-)-, 340
Glycolate oxidase, 329
Glycolic acid, 329
Glycopeptides, 49
Glycoprotein(s), 70, 71, 107, 138, 143, 144, 418
Glycosidases, 75
Glyoxylic acid, 329
Glyphosate, 397–400

Heavy metals, 6, 210
Helminthosporium, 563
  carbonum, 13, 208, 210, 217, 288, 401, 489
  maydis, 495
  oryzae, 14, 210, 497, 563, 567, 570, 580
  sativum, 406, 567, 571
  tetramera, 525–528
  turcicum, 70
  victoriae, 14
Helianthus annuus, 24
Heliocide H_2, 174
Hemigossypol, 133, 167
  6-dioxy-, 133
  6-methoxy-, 133
Hemigossypolone, 174
Hemileia vestatrix, 24, 29
Hendersonula toruloidea, 263, 268
Heptaglucosides, 15
Hepta-β-glucoside alditols, 15
Heptanal, 176
Heptanes, 146, 147
Hepta-β-D-glucopyranoside (G7), 62
1-Heptanol, 176
2-Heptanone, 177
3-Heptanone, 177
trans-2-Heptenal, 176
3-Hepten-1-ol, 176
Heterodera glycines, 201
Heterophragma, 549
  adenophyllum, 542
Hexachloroacetone, 451
2,4-Hexadienal, 176
n-Hexanal, 176
  cis-3-, 176
  trans-2-, 175, 176
Hex-2-enal, diethylacetal, 176
cis,2-Hexen-1-ol, 176
cis,3-Hexen-1-ol, 176
Hinesol, 445–447, 457
Histidine, 144
Homeostasis, 129, 155
Hordeum aestivum, 25
Host–parasite specificity, 41–60
HPLC profiling of phytoalexins, 97, 205
  crucifers, 234, 235, 240, 254
  groundnut, 205, 212–214
  Medicago, 410, 412–414, 416, 417
  soybean, 119, 120
  Vitis, 292
Hydrogen peroxide, 124, 125, 329
Hydroperoxide(s), 125
  organic, 125
Hydrophobicity, 325, 328
Hydroxycinnamic acids, 97–99
3-Hydroxy-3-methyl glutaryl CoA (HMG-CoA), 468
  reductase, 469, 478, 480
9-Hydroxy octadecadien-10,12-methyloate, 209, 211, 215, 220
13-Hydroxy octadecadien-9,10-methyloate, 209, 211, 215, 220
Hydroxystilbenes, 289, 317–331
  effect on cellular structure, 320–323
  effect on conidia, 320
  effect on fungal cells, 318–323
  effect on respiration, 318–320
Hypersensitive resistance, 249, 251, 252
Hypersensitive response (HR), 117, 125, 130, 183–189, 231, 251
Hyphantria cunea, 171

IAA oxidase, 169
Immunodepressors, 148
Immunosystem, 154, 155
Indole, 253
  acetic acid (IAA), 560, 565, 567–569, 571–575, 580
  3-aminoethyl-, 237
  phytoalexins, 233, 240, 244, 249, 253
3-Indole carbaldehyde, 237
Indole-3-carboxaldehyde, 238
Inducers, 70, 136, 137, 141, 293, 555–591
Infection inhibitor, 46
Inositol
  -1,4,5-triphosphate(IP_3), 435–437
  -1,3,4,5-tetrakisphosphate, (IP_4), 436

Invertase, 72, 73
Ipomoea batatas, 24, 467, 481
(*see also* sweet potato)
Ipomoeamarone, 4, 19, 467–470, 472, 474, 475, 477, 478, 480
dehydro-, 469, 475, 480
-12-hydroxylase, 472, 476, 478
Ipomeamaronol, 467–469
Ipomeamaronolide, 474
Iron(Fe), 145
Iridoids, 534, 535, 539, 542
Isobutyric acid, 185
Isoferrierin, 337, 347, 348, 352, 355, 359, 361
Isoflavan(s), 337, 351, 357–359, 392, 397
Isoflavanol, 407
2′,7-hydroxy,-4′-methoxy-, 407
Isoflavanone, 16, 337, 351, 352, 358, 360
7,4′-dimethoxy-2′-hydroxy-, 209, 211, 213, 215, 217, 221
7,2′-dihydroxy-4′-methoxy-, 209, 211, 213, 215
5-hydroxy-, 353
Isoflavone(s), 16, 90, 110, 118, 337, 351–353, 358, 359
biosynthesis, 216, 353, 354
conjugates, 90, 101, 110, 393
cycloderivatives, 338
5-deoxy-, 94
2′,7-dihydroxy,4′-methoxy-, 407
5-hydroxy-, 353
7-hydroxy,4′-methoxy-, 407
reductase, 222, 223
Isoflavonoids, 4, 183, 187, 317, 334, 344, 351, 353, 354, 358, 365, 445, 558
antimicrobial properties, 358–365
biosynthesis, 353–356
in chemotaxonomy, 357
detoxyfication, 356
UV spectra, 351–353
Isoflavonone, 337, 359
Isogenic isolines, soybean, 89
Isohemigossypol, 132–134, 139, 142, 146, 149–151, 154, 155
Isoleucine, 144
Isoliquiritigenin, 407
Isolubimin, 18, 457, 458
Isonicotinic acid (INA), 375
2,6-dichloro-, 375, 581
Isopentenyl pyrophosphate, 468
Isoprunetin, 337, 347, 352, 355, 359
2′-hydroxy-, 337, 347, 352, 355, 359
Isorhamnetin, 90–92
Isosojagol, 341, 346
Isosophoranone, 338, 352

Jasmonic acid, 105, 106
Juglone, 78

Kaempferol, 90–92, 537, 538, 546
4′-OMe-, 537–539, 546
Kievitol, 337, 338, 347, 355, 359, 360
2,3-dehydro-, 337, 347, 355, 359, 360
5-deoxy-, 338, 346, 355
Kievitone, 8, 17, 18, 337, 338, 344, 345–348, 352, 355, 358–360, 378, 380
cyclo-, 338, 339, 347, 355, 359, 360
2,3-dehydro-, 337, 347, 352, 355, 359
2′′,2′′-dehydrocyclo-, 339, 347
5-deoxy-, 337, 346, 352, 355, 359, 360
3′-(γ,γ-dimethylallyl)-, 338, 347, 352, 359, 360
hydrate, 17, 18, 337, 338, 347, 355, 359, 360
hydratase, 17, 356
4′-methoxy-, 338, 347, 352
2′,4′,5,7-tetramethyl-, 338, 359, 360
trimethoxy-, 338, 339, 359, 360
Kinetin, 109

La^{3+}, 432, 434
Lablab, 357
niger, 348
Lacinilene, 168, 178, 195
Lacinilene C (LAC,LC), 163–167, 169–173, 178, 184, 186–194
Lacinilene C7-methylether (LAC-ME,LCME), 163–167, 169–173, 178, 186, 187–194
Lactonisation, 169
Lactose, 144
Laminarin, 72
Laxifloran, 339, 348
Lectins, 144–148, 156
Leguminoseae, 4, 199, 334 (*see also* Fabaceae)
Lemna minor, 166, 167
Lepachol, 535
Leptosphaeria maculans, 230–235, 238, 239, 241–243, 249, 251, 252, 254–256
Lespedeza buergeri, 50
Lettuce, 384
Leucine, 144
Lignin, 72, 383, 384, 407, 536, 579
Limonone, 177
Linoleic acid, 178
Linolenic acid, 178
Linum usitatissimum, 22, 24
Lipid–suberin conjugates, 406
Lipoglycoproteins, 559
Lipopolysaccharide, 11, 559
Liquiritigenin, 407
Lithium salts, 560, 564, 567, 569, 571–575, 583
Lotus
bicolor, 50
corniculatus, 50
Lubimin, 9, 12, 13, 17, 18, 432, 445, 452, 461
15-dihydro-, 18
10-epioxy-, 457, 458
iso-, 18, 457, 458
oxy-, 445, 447, 450, 452, 455, 458, 461
Lubiminol, 446, 457, 458
Lucerne, 405–443 (*see also Medicago sativa*)
callus culture, 409
Lupin, 384
Luteolin, 535
7-OMe-, 544
Luteone, 337, 347, 352, 358
Lycopersicum esculentum, 25, 409, 418 (*see also* Tomato)
Lysine, 144
poly-L, 14

Maackiain, 8, 15, 16, 21, 317, 334, 341, 391–401
dihydro-(DHMaa), 393, 396
Macozeb, 558
Macrophoma morindae, 538
Macrophomina phaseolina, 21, 24, 26, 27, 525, 562, 580
Macroptilium
atropurpureum, 345, 348
bracteatum, 348
lathyroides, 348
marti, 345
Macrotyloma axillare, 348
Madhuca, 542, 549
indica, 542
Magnesium (Mg), 14, 145
Magnolidae, 549
Malacosoma californiapolviale, 171
Malate dehydrogenase, 184
Malonated conjugates, 91
Malonyl CoA, 216, 287, 334, 356, 407
Malonyl diadzein, 9, 193
Malonyl genistein, 91, 93
Manganese (Mn), 145
sulfate, 560
Mangifera, 541, 549
indica, 540
Mangiferin, 541
Mannitol, 190
Mannose, 138, 144, 146–148
Medicago sativa, 24, 50, 405–443 (*see also* Alfalfa and Lucerne)

Medicarpin, 8, 15, 16, 21, 48, 208, 211, 213, 215–217, 220–222, 334, 340, 346, 358, 359, 361, 391–399, 401, 406, 407, 411–413, 415–417, 419–422, 430
  biosynthesis, 216
  demethyl-, 209, 211, 213, 215, 217, 218, 220, 221
  demethylase, 222
  iso-, 340, 348, 358, 359, 361
Medicinal plants, 544–546
Melampsora lini, 22, 24
Melilotic acid, 535, 547
Mellein, 6-OMe, 430, 431, 536
Mercaptoethanol, 184
Mercuric acetate, 13
Mercuric chloride, 8, 10, 242, 342, 350, 362, 467, 478, 496, 526, 528, 530, 564, 567
Metalaxyl, 11, 503–523
  in phytoalexin production, 512–520
Metal salts, 560, 563
Methionine, 144, 560, 565, 568, 571–575
3-Methyl-1-butanol, 176
3-Methyl-2-butanol, 176
Methyl(3-indolylmethyl)-dithiocarbamate *S*-oxide, 238
Methyl linolenate, 209, 211, 214–217, 220, 221
Methyl myristate, 507, 508
O-Methyl transferases, 72
Mevalonate (MVA), 468
  phosphate, 468
  pyrophosphate, 468
Mevalonolactone, 193
  (5-^{3}H), 185, 194
Mevinolin, 382, 383
Microsomal fractions, 148
Milletia japonica, 50
Momilactone(s), 9, 14, 26, 497, 505, 561
Monilinia fructicola, 10, 19, 288
Monilicolin A, 10
Monoiodoacetamide, 475
Monooxygenases, 328, 378
Morinda, 538, 549
  tomentosa, 537, 538
Mucic acid, 293
Mucor, 328
Mustard oils, 486
Mycelial wall extract, 137, 162, 163, 233
Mycolaminarin, 64, 101, 103
Mycosphaerella
  ligulicola, 7, 50, 52, 53
  melonis, 7, 50
  pinoides, 7, 46, 47, 49–53, 55
  rabiei, 525–528
Mycotoxin, 467
Myo-inositol, 436
6-Myoporol, 474
Myoporone
  6-dihydro-7-hydroxy-, 474
  4-hydroxy-, 474, 480
  4-hydroxy dehydro-, 474, 475, 480
Myrcene, 174, 177
Myricetin, 540, 542
  4′-OMe-, 542
Myrosinase, 253
Myrothecium roridum, 10, 24, 26
Myrtaceae, 538, 539, 549

NADP, 470, 472
NADPH, 471, 475
  -cytochrome *c* reductase, 479
  -cytochrome *p*-450, 478
  -isoflavone oxidoreductase, 397
  -oxidoreductase, 397
  -reductase, 354
Naphthalene
  acetamide (NAM), 560
  acetic acid (NAA), 560
Nectria, 400, 536
  haematococca, 16, 17, 201, 216, 385
Necrosis, 189–193, 267
Neodunal, 341, 348
Neorautanea mitis, 348
Nerol, 472

Nickel
  chloride, 496, 497, 565
  nitrate, 496, 497, 565, 567
Nicotiana tabaccum, 376, 462
  (*see also* Tobacco)
Nigrospora oryzae, 563
Nitric oxide, 78
2-Nonanone, 177
1-Nonanol, 176
trans-2-Nonenal, 176
Nonyl aldehyde, 176
Nonyl phenol, 209, 211, 214, 215, 220
Norcaradiene derivatives, 450

Oat, 191 (*see also Avena sativa*)
Obligate parasitism, 41–44
Ocimene, 177
Octanol, 176
3-Octanone, 177
Octasaccharide, 559
trans-2-Octenal, 176
Oidium, 536
1-Oleoyl-2-acetylglycerol, 437
Oligogalacturonide elicitors, 11, 100, 578
O-Methyl transferase
  caffeic acid, 124
  S-adenosine-L-methionine: xanthotoxol, 72
  5-hydroxyferulic acid, 124
Ophiobolus, 24
Orange, 42 (*see also Citrus sinensis*)
Orchinol, 4
Orchis militaris, 4
Orthovanadate, 52
Oryza sativa, 24 (*see also* Rice)
Oryzalexin, 14
Oxadiazone, 10
Oxalic acid, 106
Oxidative phosphorylation, uncouplers, 317
Oxindoles, 237, 253
Oxygenase, 473, 474
Oxylubimin, 445–447, 450, 452, 455, 458, 461
Oxyrhynchus volubilis, 348

Parsley, 101, 107
Pda gene, 16, 201
Pea, 7, 16, 49, 51, 52, 56, 108, 187, 201, 581 (*see also Pisum sativum*)
  root rot, 560
  root rust, 560
Pectate lyase, 102, 105, 106
Pectinolytic enzymes, 14, 573
Pedaliaceae, 525
Penicillin, 497, 562, 577, 578
Penicillium
  crysogenum, 329
  digitatum, 264, 268, 271
  italicum, 268
4-Pentanoic acid, 177
2-Pentanone, 177
3-Pentanone, 177
1-Pentenol, 176
4-Penten-1-ol, 176
Pepper, 503–523, 561 (*see also Capsicum annuum*)
Peppermint, 405
Peronospora
  parasitica, 581
  tabacina, 57, 169
Peroxidase, 97, 120, 123–125, 169, 560, 573
  assay, 120
  isoenzymes, 97
Pestalotia longiseta, 489
Pestalotiopsis
  adusta, 536
  funerea, 539
  theae, 486, 489, 491, 492
  versicolor, 541
Phaeoisariopsis personata, 200, 220
Phaseol, 341, 346
Phaseollidin, 8, 17, 344, 346, 348, 355, 359, 360, 378, 386
  2(γ,γ-dimethyl allyl)-, 340, 346
  4(γ,γ-dimethyl allyl)-, 340, 346
  isoflavan, 339
  1-methoxyl-, 340, 348
Phaseolinae, 334, 345

Phaseollin, 3, 8–10, 17, 42, 48, 187, 317, 341, 344–346, 348, 359, 360, 392
  biosynthesis, 355
  hydroxy-, 17, 21, 42
  -isoflavan, 17, 334, 339, 346, 357, 359
Phaseolus, 333–335, 343, 346, 351, 354, 356, 357
  aureas, 335, 336, 344–346, 354, 357, 360, 365
  chemotaxonomy, 357
  coccineus, 335, 336, 345, 346, 348, 355, 357, 365
  lunatus, 335, 336, 345, 346, 354–356, 359, 365
  mungo, 335, 336, 344–346, 354–356, 365
  vigna complex, 334, 335, 357
  vulgaris, 9, 10, 17, 166, 335, 336, 344–346, 354, 357, 365 (*see also* French bean)
Phaseoluteone, 337, 347, 352, 359, 360
Phases of phytoalexin production, 44–49
Phenanthrene(s), 336, 348, 349, 356
  biosynthesis, 356
  dihydro-, 337, 343, 356
  2,7-dihydroxy-4,6-dimethoxy-9,10-dihydro-, 336, 348, 349
  6,7-dihydroxy-2,4-dimethoxy-9,10-dihydro-, 336, 348, 349
  2,7-dihydroxy-1,3,5-trimethoxy-, 9,10-dihydro-, 336
Phenanthro-indolizidines, 546
Phenol(s), 275, 488, 489, 550, 560, 570, 574, 575
  p-[2-(2-halogenomethyl-1,3-dioxalan-2-yl)ethyl], 449
Phenolic acids, 97, 487–489, 535–550
Phenolics, 72, 97, 98, 305, 376, 544, 549, 550, 577, 579
Phenoxy acetic acid
  2,4-dichloro-(2,4-D), 560
  2,3,6-trichloro-(2,3,6-T), 560
Phenyl alanine, 144, 216, 264, 266, 280, 281, 287, 351, 354, 356, 406, 407, 411, 425, 560, 565, 568, 569, 571–575, 583
Phenyl alanine ammonia lyase (PAL), 5, 10, 55, 73, 86, 107, 124, 216, 275, 281, 344, 348, 350, 354, 356, 379, 380, 383, 398, 401, 407, 409–411, 421–425, 430, 432–434, 437, 576, 577, 579
  assay, 410–412
  extraction, 411
  isoforms, 425
Phenyl isocyanide, 478, 479
Phenyloxy alkyl carbonic acid, 293
Phenylpropanoids, 85–115
  genetics, 92
  pathway, 85–115, 169, 406, 407, 421
  regulation, 90–92
Phenyl-*tert*-butyl nitrone, 188
Phenyl sepharose, 431
Phenyl thiourea, 561
Phloroglucinol, 487, 488
  reactive compounds, 174
Phoma
  arachidicola, 200, 208, 215, 220, 222
  comlanata, 536
  lingam, 230
  vasicae, 544
Phomopsis, 561
Phorbol-12-myristate-13-acetate, 437
Phosphatase, 54
Phosphatidyl choline, 436
Phosphatidyl inositol (PI), 436
Phosphatidyl inositol-4,5-bisphosphate (PIP_2), 435–437
Phosphatidyl inositol-4-phosphate (PIP), 436, 437
Phosphalidyl serine, 436
Phosphite, 305, 306

Phosphoinositidase C, 435, 436
Phosphoinositide signal system, 429, 435–437
Phospholipase C, 437
Phosphonate, 305, 375–390
Phosphon-D, 561
Phosphonic acid, 305, 379
  tritiated, 379
Phosphorous acid, 264, 266, 277, 280, 282, 305
Phototoxic phytoalexin, 188, 189
Phragmidiella heterophragmae, 542
Phyllactinia corylea, 536
Phyllospella stakmanii, 542
Phyllosticta eugeniae, 539
Phytoecdysones, 545
Phytoimmunity, 141
Phytophthora, 3, 75, 76, 101, 375–390, 580
  blight, 503, 525
  capsici, 19, 376, 379, 503–523
  cinnamomi, 384
  citrophthora, 263–286, 384
  cryptogea, 376–380
  fragarieae, 384
  infestans, 2, 12, 13, 23, 25, 119, 121, 193, 288, 450, 558
  megasperma, 360
    var. *sojae*, 10, 12, 21, 85–116, 193
    f.sp. *glycinea* (Pmg), 10, 13, 21, 50, 61, 63, 66, 201, 504, 516, 517, 559
    f.sp. *medicaginia*, 21
  nicotianae
    var. nicotianae, 381
  parasitica 12, 263, 384
    var. *sesami*, 525
  root rot, 69, 70, 74, 503, 505
  sojae, 69–72
  vignae, 21
Phytuberin, 381, 446
Phytuberol, 381
Pigeon pea, 572, 580
α-Pinene, 177, 178
β-Pinene, 177, 178
Pinosylvin, 342, 356, 357, 361
  dihydro-, 336, 342, 345, 348–351, 356, 358, 362, 365
  synthase, 356
Pinus, 356, 361
  sylvestris, 356, 357
Piperazine, 437
Piperidinium iodide, 577, 578
Pisatin, 3, 7, 8, 16, 44, 46, 49, 54, 108, 109, 201, 216, 317
  biosynthesis, 54
  demethylase, 16, 74, 201, 385
  synthesis, 49
Pisum sativum, 50 (*see also* Pea)
Plant moisture stress (PMS), 171
Plasmalemma fractions, 148
Plasmids
  DNA, 184
  R-1, 202
Plasmopara viticola, 288, 290, 291, 293, 300–309, 558, 561
Polyacetylenes, 4, 445, 558
Polyacrylamide gel electrophoresis (PAGE), 184
Polyacrylic acid, 580
Polyamines, 108, 109
Polygalacturonase, 406, 573
  endo-, 558
Polygenes, cotton, 186
Polyphenol oxidase (PPO), 124, 344, 351, 561, 570, 573–577
Polyphenols, 123, 124, 130, 486
Populus
  euroamericana, 171
  laurifolia, 537
  maximowiczii, 537
  nigra var. *italica*, 537
Poria weirii, 535
Post-infectional compounds, 533–553
Potassium phosphate, 562
Potato, 2, 3, 12, 23, 48, 55, 56, 164, 167, 169, 193, 431, 445, 447, 450 (*see also Solanum tuberosum*)

Powdery mildew, 542, 560
barley, 42–44
Cassia, 536
pea, 42, 44
Vitis, 308
Preformed glycosides, 393, 397–400
Primary recognition, microbes, 42
Proanthocyanidins, 535, 537, 539, 542–544
Probenazole, 375, 561, 577–580
Production potential, stilbenes, 294
Proline, 144
Propanoic acid,1,2,3-(2-acetyloxy) tricarboxylic-, 209, 211, 214, 215, 220
Propylene oxide (POx), 382, 383
Propantheline bromide, 570
Protease(s), 107
inhibitors, 107
Protein–phenol complexes, 318
Protein kinase(s), 53, 54, 435, 437, 438
Protein–lipopolysaccharide complex, 137, 418
Protocatechuic acid, 487, 488
Pseudomonas
aeruginosa, 343
cichorrii, 348
syringae
pv. *glycinea* (Psg), 117, 187
pv. *pisi*, 187
pv. *tomato*, 581
Psophocarpus tetragonolobus, 348
Psoralidin, 341, 346
Pterocarpan(s), 85, 337, 340, 351, 353, 354, 358–361, 406
biosynthesis, 354, 355
fuano-, 337, 341
6-hydroxy-, 16
3-hydroxy-9-methoxy- (*see* Medicarpin)
3-hydroxy-8,9-methylene dioxy- (*see* Maackiain)
3-hydroxy-22-thia(3-OHTP)-, 393, 394
3-methoxy-11-thia-(3-OMeTP), 393, 394
:NAPPH oxidoreductase, 397
oxidoreductase, 392
phytoalexins, 15, 401
11-thia-(TP), 393–396
Pterostilbene, 288, 289, 293, 309, 318, 320–323, 326, 328, 329
Puccinia, 191
arachidis, 200, 208, 209, 215, 217, 218, 220, 221
graminis, 25, 41
Putrescine, 108, 109
Pyricularia, 191
oryzae, 14, 489, 495, 518, 539, 567, 571, 580
Pyrrolidinium iodide, 578
N,*N*-dimethyl-, 561
Pythium, 568, 572
aphanidermatum, 12

Quercetagetin, 541
Quercetin, 90, 91, 537, 538, 541, 543, 545–548
3′,4′-dimethoxy-, 535, 538, 541, 543, 546
7,3′-dimethoxy-, 457
3′-methoxy-, 539, 540
4′-methoxy-, 537
7,3′,4′-trimethoxy-, 537, 541, 547
Quinazoline alkaloids, 544
Quinic acid, 488
Quinones, 155

Radioimmunoassay, 201, 202, 210, 430
Raffinose, 144
Rajania, 335
Raphanus sativus, 237, 247, 251
Receptor(s), 64, 66, 136, 142, 428, 429
sites, 63, 428, 429
substances, 144, 156, 429

Resistance
  age-related, 503–523
  insect, 75, 77
  reactions, 130, 144
Resveratrol, 5, 20, 210–212, 215, 217, 219, 287–289, 294–297, 300, 303–310, 318, 320, 322, 325, 329, 356
  biosynthesis, 356
  dihydro-, 342, 349, 356, 361
  trans-, 342, 361
Reverse-phase HPLC, 184, 210, 240, 241, 292
Rhamnaceae, 543
Rhamnose, 144
Rheum rhaponticum, 356
Rhizobium trifollii, 25
Rhizoctonia
  bataticola, 215
  repens, 4
  solani, 17, 70, 153, 154, 571
Rhizopus
  nigricans, 174
  stolonifer, 14, 525, 559
Rice, 14, 497, 560, 561, 563, 567, 570, 571, 580 (*see also Oryza sativa*)
  blast, 14, 505, 518, 560, 564, 567, 571, 578
  brown spot, 560, 564
  sheath blight, 571
Rishitin, 8, 12, 13, 21, 48, 193, 317, 381, 445, 446, 447, 450, 461
Rishitinol, 446
mRNA, 6, 27, 55, 107, 195, 354, 422, 432, 581
  chalcone isomerase, 55
  chalcone synthase, 428
  4-coumarate: CoA ligase, 72
  phenyl alanine ammonia lyase, 55, 56, 72, 428, 434
  pinosylvin synthase, 357
Root rot
  cotton, 153
  pea, 560
Rosidae, 549
Rotunol, anyhdro-β-, 446
Rps gene, 70, 85, 89, 92
Rubiaceae, 537
Rust, 560
  groundnut, 200
  linseed, 560
  wheat, 560, 561

Saccharomyces cervisiae, 11
Salicylic acid, 106, 222, 578, 581
Salix, 171
Saponins, 486, 489, 534, 541–543, 547, 548
Sapotoceae, 542
Saptoria nodorum, 17
Sarocladium oryzae, 25, 26
Sativan, 339, 358, 406, 407, 411, 412, 414–417, 419–422, 430, 433
Sclerotinia fructigena, 323
Sclerotium rolfsii, 568, 569, 573, 574
Scoparone, 263–286, 384
  biosynthesis, 280, 281
  production
    effect of fosetyl-Al, 277–280
    effect of phosphorous acid, 277–280
    radiation, 271–277
    temperature, 270, 271
Scopella echinulata, 542
Scopoletin, 163, 164, 168–173, 178
Scopolin, 169
Scutellarein, 548
Second messenger, 246, 429
Serine, 560
Sesamum indicum, 525–531
Seselin, 263
Sesquiterpene cyclase, 185, 195
Sesquiterpenoid phytoalexins, 12, 18, 132, 164, 183–198, 381, 383, 445, 558
  antibacterial activity, 187–189
Shikimate dehydrogenase, 124
Shikimate pathway, 398
Shikimic-polymalonic acid pathway, 216, 287

Signal(s), 98–110
  molecules, 55, 98–110
  perception, 99
  systems, 429, 432, 435–437, 438
  transduction, 53, 54, 99, 427–438
Silver nitrate, 233, 235, 241–243
Sinapic acid, 535
Sinapis alba, 247, 251
Sirodesmin PL, 233, 242, 245, 246
Sodium
  azide, 8, 14, 26, 497, 562, 565, 572, 577
  diethyl malonate anion, 450
  dodecyl sulfate, 11, 184
  fluoride, 565, 567, 571–575
  iodoacetate, 8, 497, 565, 568, 571–575
  malonate, 565, 567, 571–575, 583
  molybdate, 565, 568, 571–575
  nitrate, 15
  selenite, 8, 564, 567, 568, 570–575
  sulfite, 564, 567, 568
Solanaceae, 4, 334, 445, 446, 545
Solanescone, 457, 458
Solanum tuberosum, 25, 462 (*see also* Potato)
Solavetivone, 445–447, 450, 455, 457, 458, 461
  3-hydroxy-, 457, 458
  synthesis, 452
Sophoraisoflavone, 338, 352
Sorghum, 191
  bicolor, 336, 345, 361, 363, 364
Soybean, 26, 27, 42, 51, 52, 65, 66, 69–115, 117–128, 187, 193, 201, 385, 393, 400, 562, 580 (*see also Glycine max*)
Sphathodea, 549
  campanulata, 543
Spermidine, 108, 109
Spermine, 108, 109
Sphaceloma
  heterophragmae, 542
  madhucae, 542
  tectonae, 534
Spinacia oleracea, 318
Spiroannulation techniques, 447
Spirobrassinin, 236, 245
Spirodienone, 449, 450
Spirovetivane, 445
  phytoalexins, 445–466
  sesquiterpenes, 445
Staphylococcus aureus, 343
Stemphylium
  botryosum, 17, 392–397
  sarcinaeformae, 7, 47
Stenophylus
  angustifolia, 348
  stenocarpa, 348
Stereoselective synthesis, spirovetivanes, 445–466
Steroids, 547, 548
Stilbene(s), 208, 210, 212, 215, 219, 287–315, 317–331, 333–373, 558
  biosynthesis, 215, 216, 287, 290, 292–299, 303, 310, 356, 357
  3-chloro-4′-hydroxy-, 328
  detoxification, 309
  dihydro-, 334, 337, 342, 344, 348, 351, 356, 357, 361–365
  3,5-dimethoxy-4′-hydroxy-, 26, 318 (*see also* Pterostilbene)
  effect of fosetyl-Al, 305–307
  3-hydroxy-5-methoxy dihydro-, 342
  mode of action, 309, 317–331
  production potential, 294, 296, 300, 301, 309
  synthase, 5, 20, 287, 309, 356, 357
  4,3′,5′-trihydroxy-, 208, 212, 215, 256, 287, 318 (*see also* Resveratrol)
Stophostyles helvola, 348
Strawberry, 384
Streblus, 547
  asper, 546, 547
Strobilanthes callosus (*see Carvia callosa*)
Strontium, 13
Sugar beet, 21, 368
  root rot, 574
Sugar maple (*see Acer saccharum*)

Sulphex, 526, 528, 530
Superoxide ion, 78
Suppressor(s), 41, 50–52, 55, 56, 66, 142–144, 401
  gene, 22
Suspension cell cultures, 54, 55, 88, 118, 124, 411, 421, 422, 425, 430, 431
Sweet potato, 4, 429, 467–483 (*see also Ipomoea batatas*)
Synthesis inhibitors, 245, 246
Syringic acid, 535, 537–541, 544, 545
Systemic acquired resistance (SAR), 222, 310, 577, 582
Syzygium, 539, 549
  cumini, 539, 540

Tagetes, 189
Tamus, 335
Tannins, 534, 535, 537, 542, 547
Tea (*see Camellia sinensis*)
Tectona, 535, 536, 549
  grandis, 534
Tectoquinone, 535
Terpene reductase, 472, 479
Terpenoid phytoalexins, 429
Tetrachloroethane, 413
Thielaviopsis basicola, 25
Thiobendazole, 526, 528, 530
Thioglycollic acid, 496, 564, 567, 568, 571–575, 583
Thiophenes, 189
  dihydrobenzo- (DHBTP), 393, 394
Threonine, 144, 560
Tobacco, 2, 20, 23, 64, 169, 222, 309, 376, 381–384, 445, 447, 556, 561, 580, 581 (*see also Nicotiana tabaccum*)
  isogenic plants, 385
  mosaic virus, 42, 378
  transgenic plants, 5, 64, 309
α-Tocopherol acetate, 328
Tomato, 13, 21, 23, 191, 376, 401, 408, 418, 503, 558, 560, 580 (*see also Lycopersicum esculentum*)
Transaminase, 380
Transaminic acid, 137
Transgenic plants, 5, 20, 64, 222, 309
  tobacco, 5, 64, 309
Traumatic acid, 106
Trianthema, 545, 549
  portulacastrum, 545
2,3,5-Trichlorophenoxyacetic acid (2,3,5-T), 560, 565, 568, 569, 583
Tricophyton mentagyrophytes, 343
Tricyclazole, 375
Trifluoroperazine, 431
Trifluralin, 578
Trifolieae, 334
Trifolirhizin, 393, 397–399
Trifolium
  prattens, 50
  repens, 25, 50, 358
Trigonella, 334
2,3,5-Triiodobenzoic acid (TIBA), 560
Tripospermum juglandis, 539
Triticum aestivum, 25
Trypsin, 421
Tryptophan, 144
Tungrovirus, 535
Tylocebrine, 546
Tylophora, 546, 549
  asthmetica, 549
  indica, 549
Tylophorine, 546
Tylophorinine, 546
Tyloses, 135, 136, 140
Tyrosine, 144, 287, 328
Tyrosine ammonia lyase (TAL), 344, 348, 350

Ultraviolet light (UV), 6, 8, 210, 233, 241, 243, 245, 254, 282, 287, 288, 293, 295–297, 299, 301, 308, 445
  effect on scoparone production, 271–277
  effect on *Vitis* phytoalexin production, 293, 308

Umbelliferone, 264
Uncinula
  necator, 308
  tectonae, 534
Uromyces, 558
  appendiculatus, 42, 57
Urticaceae, 546
Ustilago maydis, 25

Valine, 144
Vasicine, 544, 545
Vasicinone, 544, 545
van der Waals volumes, 325, 328
Vanillic acid, 535–548
Venturia inaequalis, 70, 288, 560
Verapamil, 54, 432, 434
Verbenaceae, 534, 546
Verticillium, 405, 425
  albo-atrum, 13, 21, 174, 405–443
  dahliae, 131, 134, 135, 137–139, 142, 143, 145, 146, 148, 149, 152, 154, 156, 268, 405, 408
  wilt, 129–160, 560, 561
Vestitol, 334, 339, 358, 406, 407, 411, 412, 414–417, 419–421
  demethyl, 339, 346, 348, 359
  iso-, 339, 348
α-Vetispirene, 447
Vetiver oil, 445
β-Vetivone, 445, 446, 457
Vicia faba, 3, 50
Victorin, 14
Vigna, 333–335, 346, 357
  angularis, 335, 345, 346
  chemataxonomy, 357
  mungo, 336
  radiata, 335, 336
  sinensis, 50
  subterranea, 335, 345, 346, 348
  umbellata, 335, 345, 346
  unguiculata, 123, 335, 346, 354, 376 (*see also* Cowpea)
Vignafuran, 342, 347, 348, 355, 377, 378
Viniferins, 288, 293
  α-viniferin, 288, 289, 293, 309
  ε-viniferin, 288, 289, 293–297, 303, 304, 309, 310
Vitaceae, 317, 318, 320, 329, 356
Vitexin, 541
  4′-OMe-, 544
Vitis, 287–316
  acerifolia, 297, 301
  andersoni, 296, 297, 299, 301
  argentifolia, 296, 297, 299, 301
  champini, 296, 297, 299, 301
  cinerea, 296, 299, 301
  doaniana, 294, 295, 297, 301
  labrusca, 297
  longii, 296, 299
  riparia, 294–297, 299, 301
  rupestris, 294, 297, 299, 301, 302, 307
  treleasi, 296, 297, 299, 301
  vinifera, 287, 294–297, 299, 301, 302, 307 (*see also* Grapevine)
Voandozeia subterranea, 335
β-Vulgarin, 21

Water cress, 239
Wheat, 285, 560, 561, 564, 567, 571, 576
Wine analysis, 292
Withaferins, 545, 546
Withania, 545, 549
  somnifera, 545
Wound response, 167, 168
Wyerol, 19
Wyerone, 19
  acid, 19
  epoxide, 19

Xanthomonas
  campestris, 125, 126, 243
    pv. *campestris*, 187, 232
    pv. *glycines*, 118, 120–123
    pv. *malvacearum*, 24, 153, 154, 168, 183–198
  oryzae, 561
Xanthone, 541

Xanthotoxol, 72
Xanthoxylin, 263, 369
Xanthyletin, 263, 269
Xenobiotics, 154–155
Xylem, 132, 135, 138, 140–143, 151, 152
Xylose, 147

Zea mays, 25, 58
Zeatin, 109
Zinc, 56
  salts, 564, 567–569, 583
Zizyphus, 543, 549
  oenoplia, 543